ASTRONOMY AND
ASTROPHYSICS LIBRARY

Series Editors: M. Harwit, R. Kippenhahn, J.-P. Zahn

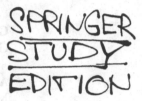

SPRINGER
STUDY
EDITION

ASTRONOMY AND
ASTROPHYSICS LIBRARY

Series Editors: M. Harwit, R. Kippenhahn, J.-P. Zahn

Tools of Radio Astronomy
K. Rohlfs

Physics of the Galaxy and Interstellar Matter
H. Scheffler and H. Elsässer

Galactic and Extragalactic Radio Astronomy, 2nd Edition
G.L. Verschuur and K.I. Kellermann, Editors

Astrophysical Concepts, 2nd Edition
M. Harwit

Observational Astrophysics
P. Léna

G.L. Verschuur K.I. Kellermann

Editors

With the Assistance of E. Bouton

Galactic and Extragalactic Radio Astronomy

Second Edition

With 207 Figures

Springer-Verlag

New York Berlin Heidelberg London Paris
Tokyo Hong Kong Barcelona Budapest

Gerrit L. Verschuur

Kenneth I. Kellermann

National Radio Astronomy Observatory, Charlottesville, VA 22901, USA

Series Editors

Martin Harwit
National Air and Space Museum
Smithsonian Institution
7th St. and Independence Ave. S.W.
Washington, D.C. 20560, USA

Rudolf Kippenhahn
Max-Planck-Institut für
Physik und Astrophysik
Institut für Astrophysik
Karl-Schwarzschild-Straße 1
D-8046 Garching,
Fed. Rep. of Germany

Jean-Paul Zahn
Université Paul Sabatier
Observatoires du Pic-du-Midi
et de Toulouse
14, Avenue Edouard-Belin
F-31400 Toulouse, France

Cover picture: The Elliptical Radio Galaxy M87 (Virgo A) jet and lobe structure. The image shows the complex filamentary structure as well as the radio jet (to the upper right of the image). (Observers: F.N. Owen, D.C. Hines. Courtesy NRAO/AUI.)

Library of Congress Cataloging-in-Publication Data
Galactic and extragalactic radio astronomy.
  (Astronomy and astrophysics library)
  Rev. ed. of: Galactic and extra-galactic radio
astronomy. 1974.
  Bibliography: p.
  Includes index.
  1. Radio astronomy.   I. Verschuur, Gerrit L.,
1937- . II. Kellermann, Kenneth I., 1937-
III. Galactic and extra-galactic radio astronomy.
IV. Series.
QB475.A1G35 1988   522'.682   87-24340
© 1974, 1988 by Springer-Verlag New York Inc.

Typeset by Asco Trade Typesetting Ltd., Hong Kong.
9 8 7 6 5 4 3 2 1

ISBN-13:978-0-387-97735-5   e-ISBN-13:978-1-4612-3936-9
DOI: 10.1007/978-1-4612-3936-9

Preface

Galactic and Extragalactic Radio Astronomy, Second Edition discusses radio observations of the universe beyond the solar system. It is intended for graduate students and practicing astronomers who wish to familiarize themselves with the wealth of astronomical phenomena that are "visible" at radio frequencies. The subject matter and general level essentially follow that of the first edition, but the content has been brought up to date to reflect the enormous growth in radio astronomy during the past 15 years. Indeed, the impact of radio observations of astronomical phenomena is now so great that the individual chapters could each be expanded into an entire book. Within the restraints of page limits, the authors have tried to emphasize fundamental ideas and to provide a guide to the literature for more detailed concepts. As in our previous book, the emphasis and completeness of the chapters reflects the interests and style of the individual authors.

The chapters have been reviewed by independent referees and we wish to thank E.M. Berkhuijsen, L. Blitz, J.M. Cordes, D.P. Cox, J. Crovisier, P.J. Diamond, J.R. Dickel, D.E. Hogg, W.M. Irvine, P.C. Myers, B. Partridge, R. Porcas, M.S. Roberts, R. Sancisi, and P.A. Shaver for their very substantial contributions which have improved the accuracy and clarity of the text. We also wish to thank P. Smiley, P. Weems, G. Kessler, R. Monk, and B. Cassell for their careful work in preparing the illustrative material, and the many secretaries involved who so carefully prepared the manuscript, and the National Radio Astronomy Observatory for its support throughout the preparation of this book. One of us (GLV) is particularly grateful to Drs. H. Hvatum and P. Vanden Bout. The National Radio Astronomy Observatory is operated by Associated Universities, Inc. under contract with the National Science Foundation.

GERRIT L. VERSCHUUR
KENNETH I. KELLERMANN

Contents

Contributors

Don C. Backer, Radio Astronomy Laboratory, University of California at Berkeley, Berkeley, California, USA

Robert L. Brown, National Radio Astronomy Observatory, Charlottesville, Virginia, USA

W. Butler Burton, University of Leiden, Sterrewacht Leiden, 2300 RA Leiden, The Netherlands

James J. Condon, National Radio Astronomy Observatory, Charlottesville, Virginia, USA

Riccardo Giovanelli, National Astronomy and Ionosphere Center, Arecibo Observatory, Arecibo, Puerto Rico

Mark A. Gordon, National Radio Astronomy Observatory, Tucson, Arizona, USA

Martha P. Haynes, National Astronomy and Ionosphere Center, Astronomy Department, Cornell University, Ithaca, New York, USA

Carl Heiles, Astronomy Department, University of California at Berkeley, Berkeley, California, USA

Robert M. Hjellming, National Radio Astronomy Observatory, Socorro, New Mexico, USA

Kenneth I. Kellermann, National Radio Astronomy Observatory, Charlottesville, Virginia, USA

Shrinivas R. Kulkarni, Henry M. Robinson Laboratory of Astrophysics, California Institute of Technology, Pasadena, California, USA

Harvey S. Liszt, National Radio Astronomy Observatory, Charlottesville, Virginia, USA

James M. Moran, Harvard-Smithsonian Center for Astrophysics, Cambridge, Massachusetts, USA

FRAZER N. OWEN, National Radio Astronomy Observatory, Soccoro, New Mexico, USA

MARK J. REID, Radio and Geoastronomy, Harvard-Smithsonian Center for Astrophysics, Cambridge, Massachusetts, USA

STEPHEN P. REYNOLDS, North Carolina State University, Department of Physics, Raleigh, North Carolina, USA

CHRIS J. SALTER, National Radio Astronomy Observatory, Charlottesville, Virginia, USA

BARRY E. TURNER, National Radio Astronomy Observatory, Charlottesville, Virginia, USA

JUAN M. USON, National Radio Astronomy Observatory, Socorro, New Mexico, USA

DAVID T. WILKINSON, Princeton University, Joseph Henry Laboratories, Department of Physics, Princeton, New Jersey, USA

LUCY M. ZIURYS, Five College Radio Astronomy Observatory, University of Massachusetts, Amherst, Massachusetts, USA

1. Galactic Nonthermal Continuum Emission

Chris J. Salter and Robert L. Brown

The third group [of radio static] is composed of a very steady hiss type static the origin of which is not yet known ... [however] the direction of arrival changes gradually throughout the day going almost completely around the compass in 24 hours.

Karl Jansky, 1932.

1.1 Introduction

1.1.1 Historical Preface

In the course of an investigation of shortwave inteference for the Bell Telephone Laboratories, Karl Jansky in 1930 built a rotating antenna array operable at 14.6-m wavelength (20.5 MHz) and used this instrument to study radio disturbances. He found that the static that was received could usually be categorized according to one of the following descriptions: intermittent and strong, intermittent and weak, or very steady and weak. The first two types of static resulted from thunderstorms which were either local or distant, respectively. However, the third type was conspicuous as it never became intense nor did it ever completely vanish. Moreover, Jansky (1932) noted that the apparent direction to this disturbance rotated through nearly 360° in 24 hours. Such periodic behavior indicated to him that the phenomenon was associated with the Sun, either directly or causally through some process occurring at the subsolar point. These inferences were sharply revised when, after an entire year of observation, it was found that the direction to the disturbance was not correlated with the position of the Sun, but rather referred to a fixed direction in space. Jansky (1933) established this direction to be near right ascension 18^h and declination $-10°$, i.e., in proximity to the galactic center. Before concluding that a cosmic source of radio radiation existed at the galactic center, other possibilities, such as the presence of a cosmic-ray source in that direction whose particles interacted with the Earth's atmosphere producing radio photons, had to be evaluated. Also the measured direction of origin was close to that of solar motion with respect to nearby stars. Even these hypotheses had to be dropped two years later when Jansky (1935) was able to establish that radiation could be received continuously as the antenna beam swept along the galactic equator. An intensity maximum appeared in the direction of the galactic center while a minimum was noted near the anticenter. A recent reanalysis of Jansky's original observations (Sullivan 1978) clearly demonstrates these points (Figure 1.1). The conclusion that the Galaxy itself was a source of intense radiation at frequencies of a few tens of megahertz was inescapable.

The origin of this radiation remained obscure. Jansky suggested that since the nearby stars seemed to be largely confined to the galactic plane—similar to the concentration of radio emission—perhaps merely long-wavelength stellar emission was being measured. In this model the radio intensity was simply proportional to the number of stars in the beam. The major problem with such an explanation was that at least one representative star, the Sun, appeared to be radio-quiet. As an

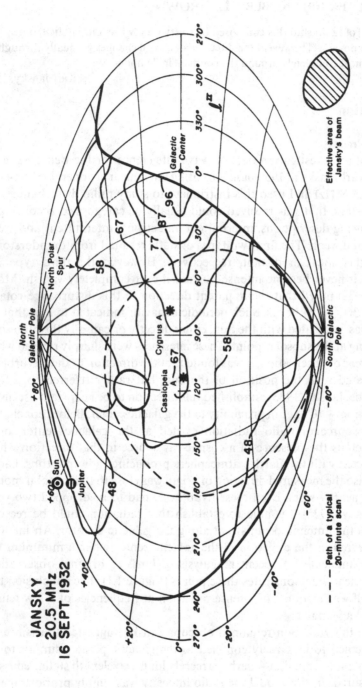

Fig. 1.1. A modern reduction of Jansky's data taken on 16 September 1932 (after Sullivan 1978). The contour map is in galactic coordinates in which 0° latitude corresponds to the plane of the Milky Way and 0° longitude corresponds to the galactic center. Contours are labeled in 1000 K.

alternative, Jansky proposed that the cosmic static could result from the thermal agitation of charged particles. He pointed out that such particles are found not only in stars but distributed throughout the Milky Way (see Chapter 2) Interstellar dust—which optical extinction measurements show is also concentrated toward the plane—served as a vehicle for two other early theories of the galactic radio emission. One involved radiation emitted by atoms recombining on the surface of grains, while the other considered, and rejected, radiation from the Rayleigh-Jeans tail of a blackbody distribution of 30 K dust (Whipple and Greenstein 1937).

In 1938 Grote Reber, operating a 9.4-m parabolic reflector, attempted to observe the "cosmic noise" at frequencies of 3300 and 910 MHz. If blackbody emission had been applicable, this should have been extremely intense at these frequencies, but he failed to detect any radiation to the limits of his sensitivity. At 162 MHz, however, emission was detected along the entire visible galactic equator. These new measurements confirmed Jansky's impression that the intensity was a function of galactic longitude and appeared most pronounced at the galactic center (Reber 1940a). Moreover, Reber (1940b) interpreted his fluxes as arising from thermal bremsstrahlung (free-free radiation) in a hot ($T_e = 10^4$ K), dense ($n_e = 1$ cm^{-3}) interstellar gas. Henyey and Keenan (1940) showed that while this interpretation was certainly consistent with the 162-MHz data, it did not agree with Jansky's lower-frequency measurements that would have required a thermal medium with $T_e = 1.5 \times 10^5$ K. Reber (1944) went on to interpret secondary maxima in the longitude distribution of the 162-MHz data, a result which confirmed that if stellar radiation produced the galactic radio background, then a population of "radio stars" quite distinct from the Sun was needed.

1.1.2 Early Surveys of Galactic Radio Emission

(a) Angular Distribution of the Cosmic Radio Waves

In the ten years following the classic work of Jansky and Reber, a number of radio surveys were made with increasing sensitivity at frequencies from 9 to 3000 MHz. From these surveys a moderately consistent picture emerged of the global properties of the cosmic radio emission. Representative of these early surveys is the first all-sky survey made by Droge and Priester (1956) shown in Figure 1.2. These authors combined their own northern hemisphere 200-MHz observations with those of the southern sky by Allen and Gum (1950). Regions of greatest intensity are confined to the galactic plane, with the maximum brightness at all frequencies occurring towards the galactic center. Further, several secondary maxima appear along the plane. Although the high-latitude emission is smoother, a "tongue" of enhanced brightness emerges to the north of the plane near $l = 30°$ and runs almost to the north galactic pole. This became known as the North Polar Spur. Figure 1.2 also illustrates that the positions of minimum emission are not at the poles but near $l = 230°, b = \pm 40°$.

As higher-resolution observations were made, the emission was seen to be even more concentrated to the plane and the effect of discrete sources became more pronounced. The existence of discrete sources had been recognized in the late 1940s and the question was asked as to what fraction of the disk radiation was due to the integrated effect of discrete sources. Perhaps all of it: the similarity between the

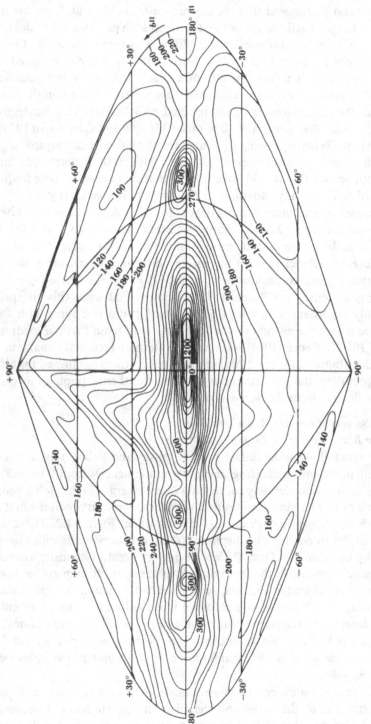

Fig. 1.2. A low-resolution (beam 17° × 17°) map of the distribution of brightness temperature at 200 MHz (Droge and Priester 1956).

concentration of radio emission to the galactic plane and the limited z-extent of normal-type stars suggested that the radio background might indeed originate in a distribution of "radio stars," i.e., objects distributed like faint G and M dwarfs but whose ratios of radio to optical luminosities were $\geqslant 10^7$ times that of the Sun. This explanation was supported by the fact that the density of early-type stars and novae was greatest in the Cygnus and Sagittarius regions, precisely the regions exhibiting radio maxima. However, to account for the phenomena that the intensity very near the galactic center greatly exceeds that elsewhere while the intensity contrast between the galactic poles and the galactic plane away from the Sgr-Cyg regions is not large, two distributions of radio stars were required. One of these had to be flattened and highly concentrated towards the galactic center, with the other being essentially isotropic. This second distribution could either be due to a superposition of extragalactic sources (Bolton and Westfold 1951) or represent an extended, quasi-spherical halo concentric with the galaxy (Shklovsky 1952). Shklovsky demonstrated that the occurrence of the emission minima at intermediate latitudes, rather than at the poles, favored the latter picture.

The brightness temperatures needed by the radio stars in such models were enormous, $T \sim 10^{16}$ K. Alternative models were considered in which "radio stars" contributed to galactic disk emission only at the lower frequencies, with the high-frequency radiation being almost entirely due to thermal free-free processes in the interstellar gas. Scheuer and Ryle (1953) tested this theory by looking for a narrow, low-latitude, thermal component using an interferometer operating at 81.5 and 210 MHz. They revealed that the emission is even more concentrated towards the plane ($<2°$) than previously thought. However, the electron temperature they derived for the hypothetical plasma was $\geqslant 18,000$ K, a value inconsistent with the optically measured temperature of $\sim 10,000$ K for HII regions.

(b) Spectral Distribution of the Cosmic Radio Waves
The wide range of frequencies at which the cosmic radiation could be detected quickly established that it represented continuum emission. It was recognized that measurement of the frequency spectrum of this continuum was vital to establishing the origin and nature of the radiation. From the earliest studies, attempts were made to fit the observed intensities to a power law spectrum, i.e., $I \propto \nu^\alpha$, where α is called the spectral index.

Piddington (1951) used measurements from 9.5 to 3000 MHz to examine the spectrum at three celestial positions: near the galactic center, at low latitude in the anticenter, and near the anticenter emission minimum at $b = -30°$. The spectrum near the center seemed complex. At the lowest frequencies he found a spectral index $\alpha \sim 2.0$, changing at intermediate frequencies to ~ -0.5 but increasing above 1000 MHz to 0.0. The spectra at the other two positions were identical, giving $\alpha \sim -0.7$ between 18 and 200 GHz. For comparison, a value of $\alpha = -0.1$ is expected for optically thin free-free emission from an ionized gas and $\alpha = +2.0$ for blackbody radiation. Even this early, it was clear that free-free emission could not explain the total spectrum of the cosmic emission over most of the sky and a nonthermal origin was indicated. However, Piddington and Minnett (1952) demonstrated that the "Cygnus X" complex, lying in the direction of a spiral arm, had a spectrum similar to that expected for free-free emission, suggesting that it represented the integrated radiation of all HII regions in the arm.

A later experiment made by Adgie and Smith (1956) derived a spectral index of $\alpha = -0.5 \pm 0.1$ away from the plane between 38 and 175 MHz. They found a flatter spectrum on the plane itself which they ascribed to a considerable proportion of the emission near the plane arising in HII regions.

(c) The Situation Circa-1956

By the mid-1950s, sufficient observations existed to pose a number of questions that were to direct subsequent investigations into the galactic radio continuum:

1. By what mechanism is the nonthermal component of the cosmic radio emission produced?
2. What contribution is made by discrete sources?
3. What fraction of the galactic continuum is thermal?
4. What is the relation of extended regions such as Cygnus X to the rest of the galactic radio background?
5. How is the radio background attentuated at low frequencies?
6. Does a quasi-sperical radio halo extend about the galaxy and what is the contribution of the integrated extragalactic component?
7. What is the North Polar Spur and is it unique?

It is the study of these and related questions that are the principal concerns of Sections 1.2 to 1.4. Interestingly (and disturbingly), a number of these questions have only partial answers today.

1.2 Physical Processes

1.2.1 Synchrotron Radiation

In the early days of the radio-star hypothesis, Alfvèn and Herlofson (1950) considered stellar models that might produce extremely high radio brightness temperatures. They concluded that the most likely process was by synchrotron radiation, i.e., radiation from electrons spiraling in a magnetic field. They suggested that many stars might be surrounded by an extensive region of intense magnetic field into which highly energetic electrons are continuously injected. In the same year, Kiepenheuer (1950) suggested a modified picture in which relativistic cosmic-ray electrons spiral around the interstellar magnetic field and their synchrotron radiation is responsible for the galactic radio background. Despite predicting approximately the observed intensity, this theory attracted little interest in the West, although it led to a number of important papers from scientists in the Soviet Union. In particular, Shklovsky (1953) noted that synchrotron radiation should be polarized; even when this prediction was verified for the radio and optical flux from the Crab Nebula, the mechanism still did not win general support. It was not until the high-resolution work of Mills (1959), which indicated that the radio continuum was correlated with the spiral arms, and the contemporary detection of linear polarization of the nonthermal background (Westerhout et al. 1962) that theories

Fig. 1.3. Synchrotron radiation as emitted in a narrow core by an electron spiraling around a magnetic field line.

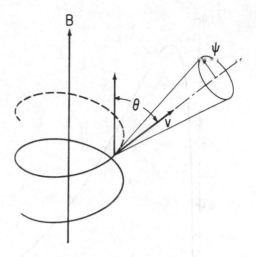

involving synchrotron emission from the ambient interstellar medium can be said to have won the day.

(a) Single Particle Emission

A number of basic reviews exist describing synchrotron radiation in the radio astronomical context (Ginzburg and Syrovatskii 1964, 1965, 1969, Pacholczyk 1970, Moffet 1975). Consequently, we present here just the basic results needed for the interpretation of observational data.

A relativistic electron passing through a region containing a magnetic field experiences the $\mathbf{v} \times \mathbf{B}$ Lorentz force causing it to move in a helical path around the field lines (Figure 1.3) and emit electromagnetic radiation. If the pitch angle of the electron in respect to the field is θ, and it has a total energy $E = mc^2/(1 - v^2/c^2)^{1/2} = \gamma mc^2$ (where γ is the Lorentz factor), it spirals around the field with a frequency

$$v_g = \frac{eB_\perp}{2\pi\gamma mc} = \frac{v_0}{\gamma} \tag{1.1}$$

where $B_\perp = B \sin \theta$ is the magnetic field component perpendicular to the electron's path and

$$v_0 = \frac{eB_\perp}{2\pi mc} = 2.80 B_\perp (\text{gauss}) \text{ MHz} \tag{1.2}$$

is the nonrelativistic electron gyrofrequency.

An observer will see the relativistic electron radiating into a narrow cone about its instantaneous velocity (Figure 1.3), where the half-angle of the cone is of the order

$$\Psi = 1/\gamma . \tag{1.3}$$

An observer intercepting this narrow beam will detect a radiation pulse for each revolution, while further away from the direction of electron velocity, the radiation is negligible. Our observer within the beam will, in fact, see a succession of pulses at the Doppler-shifted gyration frequency, v'_g, and all its harmonics, where

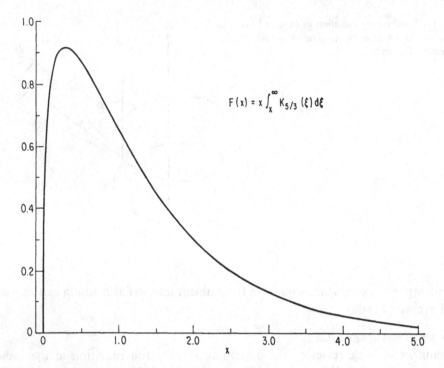

$$F(x) = x \int_x^\infty K_{5/3} (\xi) d\xi$$

Fig. 1.4. The synchrotron spectrum from a single electron as a function of $x = v/v_c$, where v_c is the synchrotron critical frequency defined in the text. (After Blumenthal and Gould 1970.)

$$v_g' \sim \frac{v_g}{\sin^2 \theta} = \frac{v_0}{\gamma \sin^2 \theta} \tag{1.4}$$

and the duration of each pulse is

$$\Delta t \sim \frac{1}{2\pi v_g \gamma^3} = \frac{1}{2\pi v_0 \gamma^2} . \tag{1.5}$$

Most energy will be radiated in harmonics whose frequencies are multiples of

$$(2\pi \Delta t)^{-1} \sim v_g \gamma^3 = v_0 \gamma^2 . \tag{1.6}$$

For an ultra-relativistic electron, the harmonics are so closely spaced that the spectrum is essentially a continuum. Detailed calculations show that the spectral power radiated in all directions and polarizations by the electron is

$$P(v) = \frac{\sqrt{3}\, e^3 B_\perp}{mc^2} \left[\left(\frac{v}{v_c} \right) \int_{v/v_c}^\infty K_{5/3}(\xi)\, d\xi \right] \tag{1.7}$$

where $K_{5/3}(\xi)$ is a modified Bessel function, and the shape of the quantity in square brackets is shown in Figure 1.4. The critical frequency, v_c, is seen to be near the emission maximum and is given by

$$v_c = \frac{3}{2}\gamma^2 v_0 = \frac{3e}{4\pi mc} B_\perp \gamma^2 = 4.21 B_\perp \gamma^2 \text{ MHz} = 16.08 \times 10^6 B_\perp E^2 \text{ MHz} \tag{1.8}$$

where E is the electron energy in GeV and $B_\perp$ is in gauss. For radio-frequency emission in typical astronomical magnetic fields of $B \sim 10^{-5}$ to 10^{-6} G, the electrons must have Lorentz factors of $\gamma \sim 10^3$ to 10^5, which are very energetic particles indeed! The rate of loss of energy by the electron to synchrotron emission is

$$\frac{dE}{dt} = -119.7 B_\perp^2 E \text{ GeV/yr} \tag{1.9}$$

and the time for an electron to lose half of its initial energy E_0 is

$$t_{1/2} = (119.7 B_\perp^2 E_0)^{-1} \text{ yr} . \tag{1.10}$$

The radiation from a single electron is elliptically polarized with the electric vector being a maximum in the direction perpendicular to the projection of the magnetic field on the plane of the sky. For an ensemble of monoenergetic particles, with randomly distributed pitch angles, the observed radiation is partially linearly polarized with degree of polarization

$$\Pi = \frac{K_{2/3}(v/v_c)}{\int_{v/v_c}^\infty K_{5/3}(\xi) \, d\xi} = \begin{cases} \frac{1}{2} & v \ll v_c \\ 1 - \frac{2}{3}(v_c/v) & v \gg v_c \end{cases} . \tag{1.11}$$

(b) Radiation from an Ensemble of Particles

The distant observer of synchrotron radiation from a region of uniform magnetic field B and spatial extent L will see emission from electrons having essentially the same pitch angle. Thus, if the ensemble of electrons is homogeneous and isotropic with an energy distribution per unit volume $N(E) \, dE$ in the interval E to $E + dE$, the specific intensity of the region is obtained from Equation (1.7):

$$I(v) = \frac{\sqrt{3} \, e^3 B_\perp L}{4\pi m c^2} \int_0^\infty N(E) \frac{v}{v_c} \int_{v/v_c}^\infty K_{5/3}(\xi) \, d\xi \, dE . \tag{1.12}$$

Let us assume (to be discussed below) that the distribution function $N(E)$ is a power law of index p between the energies E_1 and E_2, i.e.,

$$N(E) \, dE = N_0 E^{-p} \, dE \qquad (E_1 < E < E_2) . \tag{1.13}$$

Equation (1.12) then yields

$$I(v) = \frac{\sqrt{3} \, e^3 L}{8\pi m c^2} \left(\frac{3e}{4\pi m^3 c^5} \right)^{(p-1)/2} N_0 B_\perp^{(p+1)/2} v^{-(p-1)/2} a(p) \tag{1.14}$$

where $a(p)$ is a slowly varying function of p.

Note that as $B_\perp = B \sin \theta$, the emission is anisotropic for a uniform magnetic field. The most important result, however, is that a power law distribution of electrons radiates a power law emission spectrum $I(v) \propto v^\alpha$, where α and p are related by

$$p = 1 - 2\alpha . \tag{1.15}$$

If the specific intensity I is expressed as a brightness temperature T, these are related at radio frequencies by the Rayleigh-Jeans approximation,

$$I = \frac{2kTv^2}{c^2} \tag{1.16}$$

where k is Boltzmann's constant and the emission spectrum is given by $T(v) \propto v^\beta \propto v^{\alpha-2}$.

Table 1.1. Relative energies of the
synchrotron electrons.

Electron spectral index $\equiv p$	$(E_1/E_2)_{90\%}$
1.0	1620
1.5	117
2.0	56
2.5	22
3.0	15
4.0	8.9
5.0	6.1

The assumption of a power law energy spectrum for the relativistic electrons finds considerable application in radio astronomy as the radio spectra of many objects, both galactic and extragalactic, have a power law dependence. However, this is not the only justification for its use. From Figure 1.4, we see that the range of electron energies contributing significantly to the emission at a particular frequency depends strongly on the electron spectrum. It can be shown that 90% of the emission at a given frequency comes from a range of electron energies $(E_2/E_1)_{90\%}$ that narrows as the power law index p increases (Table 1.1). Normally, the power law approximation amounts to an assumption that the energy distribution of the electrons is a power law over an interval of ~ 5 to ~ 100 in energy.

The degree of polarization of the emission also depends on the electron power law index. For a uniform magnetic field, the degree of linear polarization is

$$\Pi = \frac{p+1}{p+\frac{7}{3}} \tag{1.17}$$

with the electric vector again being a maximum perpendicular to the projection of the magnetic field. For typical values of p ($p = 2$ to 4), $\Pi \simeq 0.7$.

(c) Influence of the Medium

The above results refer to processes occurring in a vacuum. If thermal plasma is present in the emitting region, or between the region and the observer, both the synchrotron radiation intensity and the degree of polarization (see Section 1.2.2) can be suppressed.

An ionized thermal gas can both emit and absorb electromagnetic radiation. A free electron passing through the Coulomb field of a proton may encounter a radio-frequency photon and absorb it, making the transition to another hypebolic orbit of slightly greater energy. If the thermal gas has path length L and is situated between the source and observer, the apparent synchrotron spectrum will vary as

$$I(\nu) \propto \nu^\alpha e^{-(\nu_A/\nu)^{2.1}} \tag{1.18}$$

where

$$\nu_A \sim \frac{n}{2} L^{1/2} \text{ MHz} \tag{1.19}$$

with n being the thermal electron density in cm^{-3} and L the path length through the absorbing plasma in parsecs. In the disk of the Milky Way, $n \sim 1$ cm^{-3} and $L = 10$ kpc so $v_A \sim 50$MHz.

If, however, the synchrotron emitting material and thermal gas are uniformly mixed, then the apparent synchrotron emission spectrum will vary as

$$I(v) \propto v^{\alpha + 2.1}[1 - e^{-(v_A/v)^{2.1}}] \; . \tag{1.20}$$

The presence of a thermal plasma in the emitting region also means that the refractive index of the medium is less than unity and the phase velocity of electromagnetic waves is greater than c. As a result, synchrotron radiation is suppressed at frequencies where the refractive index becomes significantly less than 1. This is often called the Tsytovich-Razin effect. The consequence is a low-frequency cutoff near

$$v_R = \frac{4cne}{3B_\perp} = \frac{2v_p^2}{3v_0}$$

$$\sim 2 \times 10^{-5} \frac{n}{B_\perp} \text{ MHz} \tag{1.21}$$

where $B_\perp$ is in gauss and n is in cm^{-3}, and $v_p = [(ne^2)/(\pi m)]^{1/2}$ is the plasma frequency. Again in the disk of the Milky Way $n \sim 1$ cm^{-3} and $B_\perp \sim 3 \times 10^{-6}$ G so $v_R \sim 10$ MHz.

It is of interest to compare this with the cutoff frequency for thermal absorption (Equation 1.19):

$$\frac{v_R}{v_A} = 4 \times 10^{-5} L^{-1/2} B_\perp^{-1} \; .$$

Thus the Tsytovich-Razin cutoff will only lie higher in frequency than the thermal cutoff when

$$B_\perp < 4 \times 10^{-5} L^{-1/2} \; . \tag{1.22}$$

1.2.2 The Faraday Effect
We saw in the previous section that synchrotron emission is expected to be partially linearly polarized. For an ordered field in a vacuum, a determination of the polarization position angle of the synchrotron emission from a celestial source immediately yields the orientation of its magnetic field component transverse to the line of sight, the field being orthogonal to the emitted position angle. In practice, things are not quite that simple, as the plane of polarization can be rotated during propagation by "the Faraday effect." A linearly polarized wave can be decomposed into two circularly polarized signals of opposite hand. If the wave encounters magneto-ionic material (plasma containing a magnetic field) while propagating through interstellar space, these two circularly polarized components have different phase velocities in the material. This effectively rotates the plane of polarization of the linearly polarized wave. In the wave's passage through the interstellar medium, its plane of polarization will be rotated by an amount Ψ given by

$$\Psi = 8.1 \times 10^5 \lambda^2 \int nB_\parallel dL \text{ radians} \tag{1.23}$$

where n is the electron density in cm^{-3}, λ is the wavelength in meters, $B_\parallel$ is the component of the magnetic field along the line of sight in gauss, and L is the path length in parsecs. Equation (1.23) is conveniently expressed as

$$\Psi = R\lambda^2 \tag{1.24}$$

where R is called the rotation measure and is given by

$$R = 8.1 \times 10^5 \int nB_\parallel dL \text{ rad } m^{-2} . \tag{1.25}$$

A positive value of R indicates that the component of the interstellar magnetic field along the line of sight is directed towards the observer. If the rotation measure has been determined for a source, the amount the plane of polarization has been rotated at any frequency can be predicted and the measured polarization position angles corrected to give the intrinsic position angle at emission and hence the orientation of the magnetic field component in the source transverse to the line of sight.

In practice, measured linear polarization percentages are often considerably lower than suggested by Equation (1.17). This can be due to a number of factors both instrumental and physical. Firstly, if Faraday rotation is large enough, the plane of polarization of the linearly polarized component can rotate significantly across the observing bandwidth, reducing the apparent degree of polarization. The finite beamwidth of a radio telescope can also have a similar effect if the orientation of the transverse magnetic field component in the source changes on a scale smaller than the telescope beam or a number of emission regions with different planes of polarization exist simultaneouslhy within the beam.

A case of depolarization that is particularly relevant to observations of the diffuse galactic background emission is when the synchrotron emitting region and the plasma causing Faraday rotation coexist. Even if the magnetic field in the region is uniform, the plane of polarization of radiation from the rear of the region can be rotated by a large angle before emerging from the front face of the region, resulting in large depolarization of the total emission from the region. This effect, often known as "front-back" depolarization, has been considered by Burn (1966). He showed that for a uniformly emitting slab of depth L containing a regular magnetic field with a component $B_\parallel$ along the line of sight and thermal plasma of electron density n, the observed degree of polarization $P(\lambda)$ is given by

$$P(\lambda) = \frac{P_0 \sin(R\lambda^2)}{R\lambda^2} \tag{1.26}$$

where P_0 is the intrinsic degree of polarization and R is the rotation measure of the slab,

$$R = 8.1 \times 10^5 nB_\parallel L \text{ rad } m^{-2} .$$

1.3 Total Intensity Observations of the Galactic Radio Continuum

The resolution of any radio telescope is $\sim (\lambda/D)$ radians, where D is the physical diameter of the telescope and λ is the observing wavelength. Thus, observations made with a given telescope will have coarser resolution at longer wavelengths.

This, coupled with the fact that the galactic nonthermal radiation is far stronger at long wavelengths, explains a bias that exists within presently available galactic surveys. Maps of the relatively weak, high-galactic-latitude emission have only been made at lower frequencies (< 1 GHz) and with relatively coarse resolution (> 0°.5). Conversely, at low latitudes where the emission is intense, surveys of relatively small areas exist with high resolution (< 0°.2) but almost all are at high frequency (> 1 GHz). The effects of this bias on our understanding would be greater were it not that at high latitudes most of the emission seems to posses very little small-scale structure. It does, however, have a considerable influence on our knowledge of the spectral properties of the galactic nonthermal radiation, especially at low latitudes where thermal emission makes a significant contribution to the total emission at high frequencies.

In the following two sections we will make the somewhat arbitrary distinction between surveys of resolution coarser than $\sim 0°.5$ and those of considerably higher resolution made on the galactic plane.

1.3.1 The Medium-Resolution Surveys

Over the past forty years many medium-resolution surveys of the general galactic background emission have been made. Most of these have mapped areas of particular interest or were limited to the observable sky at a given geographical location. Representative are the studies of the northern hemisphere at 10 MHz (Caswell 1976), 38 MHz (Williams et al. 1966), and 820 MHz (Berkhuijsen 1972) and surveys of the galactic loops at various frequencies (Section 1.3.3).

Rather surprisingly, only three independent attempts have been made to prepare all-sky maps of the galactic emission since the early work of Droge and Priester (1956) shown in Figure 1.2. Landecker and Wielebinski (1970) combined data taken between 85 and 178 MHz at various resolution to produce a 150-MHz picture of $\sim 3°$ resolution. At lower frequency, Cane (1978) used the northern hemisphere survey of Milogradov-Turin and Smith (1973) together with the southern sky observations of Mathewson et al. (1965) to obtain a 30-MHz map of about 9° resolution.

It was only in 1982 that an all-sky atlas was published containing data taken at a single frequency (408 MHz), standardized at a single resolution (0°.85). This survey (Haslam et al., 1982) was made using three of the world's largest parabolic reflectors and is shown in Figure 1.5. The global properties noted on the early low-resolution maps can be seen with greater detail and sensitivity. The galactic center serves as an approximate center of symmetry for the entire sky, with the low-latitude intensity dropping away steeply on each side up to $l \sim 60°$ and $l \sim 280°$. Substantial emission is present over the whole sky, although its distribution is far from uniform. Large-scale prominences, rising from low latitudes and extending nearly to the poles in both hemispheres, distort the uniformity of the high-latitude emission. These are known as the loops and spurs (Section 1.3.3).

The observed distribution of the cosmic radio radiation is often described in terms of a superposition of a number of components: the galactic center region (Chapter 7); the galactic disk centered on the galactic center; a population of discrete galactic sources (Chapters 2, 8, 9, and 10); relatively local features com-

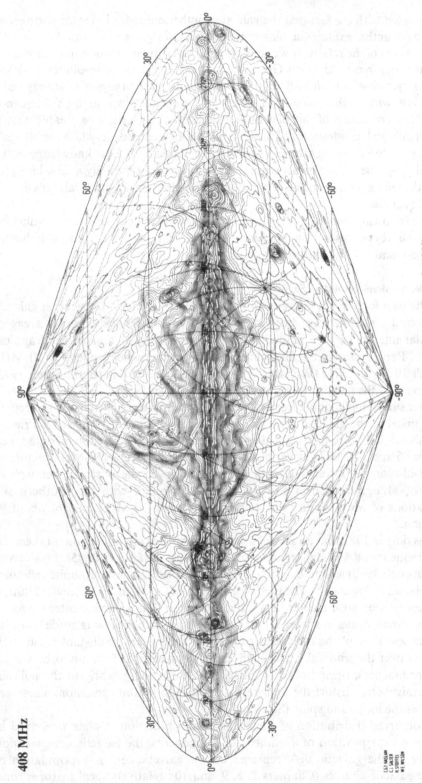

408 MHz

Fig. 15. The all-sky map at 408 MHz presented by Haslam et al. (1982).

prising the loops and spurs; a possible galactic halo symmetric about the galactic center; and an isotropic contribution due to extragalactic systems.

Before leaving medium-resolution studies, one further survey should be mentioned as it will extend the data base of all-sky surveys above 1 GHz. This is a series of observations at a frequency of 1.4 GHz and a resolution of $\sim 0°.6$ that is at present available for $\delta > -19°$ (Reich 1982, Reich and Reich 1985). This survey is being extended to cover southern declinations and will be complementary to the all-sky 408-MHz survey.

1.3.2 High-Resolution Surveys of the Galactic Plane

It is clear from Figure 1.5 that considerably higher resolution than $1°$ is necessary to fully resolve the emission at low galactic latitudes. A number of surveys of the galactic plane have been made with resolutions finer than $0°.2$, and these are suitable for separating the discrete source component from the smooth galactic disk emission. Such surveys have been performed mostly with large parabolic reflector telescopes.

At 2.7 GHz coverage of the galactic plane exists with resolutions between $4'$ and $8'$ for $l = 190°$ through $360°$ to $76°$, $|b| < 1°5$ (Day et al. 1972, Reich et al. 1985). The northern hemisphere observations are being extended to give complete longitude coverage at this frequency. At 5 GHz the region $l = 190°$ through $360°$ to $60°$ has been published with resolutions between $2'.5$ and $4'$ (Haynes et al. 1978, Altenhoff et al. 1978). The Cygnus X region ($l \sim 80°$) has also been mapped at 5 GHz with a similar resolution (Wendker 1984). Preliminary parts of a low-latitude 10-GHz survey of resolution $2'.7$ using the Nobeyama radio telescope have appeared recently (Sofue et al. 1984). A representative survey of a region of lower disk brightness is the 1.4-GHz survey of Kallas and Reich (1980), which covers $93° < l < 162°$; $|b| < 4°$ with a resolution of $9'$.

Below 1GHz, the sole high-resolution survey is that of Green (1974) made with the Molonglo Cross antenna. This has a resolution of $\sim 3'$ and covers $l = 195°$ through $360°$ to $55°$; $|b| < 3°$. Sadly, the survey was only published as a set of ruled-surface plots, although a useful longitude distribution of the disk component was included.

A typical example of galactic plane emission in the galactic center quadrant is shown in Figure 1.6. The relatively smooth disk component can be distinguished by its steep gradient away from the plane to the north and south. Small-diameter, discrete sources are superposed on the disk component, the majority in the galactic center quadrant being located at $|b| < 1°$. Altenhoff (1968) showed that the discrete galactic source component had a distribution in latitude with a half-width of $24'$, compared with the $2°$ half-width of the nonthermal background. Most of these bright sources represent HII regions or supernova remnants, described elsewhere in this volume. To data these high-resolution, low-latitude surveys have been mostly used as finder surveys for the study of these discrete sources, and their suitability for investigating the disk component has been largely neglected.

1.3.3 The Galactic Loops and Spurs

Figure 1.5 reveals several large-scale ridges of emission running away from the galactic plane to high latitudes. The most prominent one runs north from the plane

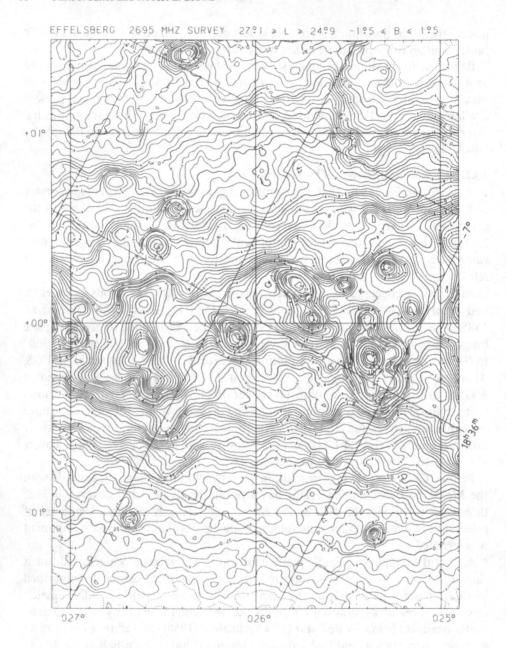

Fig. 1.6. A small section (2° × 2°) of the 2.7-GHz map of the galactic plane (Reich et al. 1985). The individual sources shown here are galactic HI regions and supernova remnants.

near $l = 30°$ and has been known since the earliest days of radio astronomy as the North Polar Spur. Quigley and Haslam (1965) demonstrated that some of the brightest of these spurs traced three small circles of large radius on the celestial sphere with remarkable accuracy (Table 1.2). These are now universally referred to as Loops I, II, and III. The North Polar Spur forms the brightest segment

Table 1.2. The most prominent galactic loops.[a]

	Diameter	Center *l*	*b*
Loop (North Polar Spur)	116° ± 4°	329° ± 1°5	+17°5 ± 3°
Loop II (Cetus Arc)	91° ± 4°	100° ± 2°	−32°5 ± 3°
Loop III	65° ± 3°	124° ± 2°	+15°5 ± 3°

[a] Berkhuijsen et al. (1971).

of Loop I. Additional loops have been suggested by a number of authors and morphologically similar structures have been identified in both HI and X-ray emission studies (e.g., Heiles 1979, Nousek et al. 1981; Chapter 3).

The continuum loops have a number of similarities to each other in addition to their small circle geometries. Berkhuijsen (1971) showed that between 240 and 820 MHz they possess step nonthermal continuum spectra having typical spectral indices of ~ −0.65. Apart from a main emission arc, each possesses an associated pattern of weaker filamentary-like ridges both inside the loop and extending for some distance outside the main ridge. The polarization properties of the loops will be dealt with in Section 1.5.4. It seems probable that the similar properties of the continuum loops imply a similar origin, and one can speculate whether the same origin might also apply to the coincident HI and X-ray loops. Loop I is not only the largest of these objects but also the brightest and best studied, and serves as the prototype of the class. The main ridge of Loop I shows a number of distinctive features. Its outer edge has a steeper gradient than the inner edge and brightness temperatures within the loop are higher than outside. Both of these results are consistent with the loop being a radiating shell, which together with its nonthermal spectrum has caused authors to suggest that it is the remnant of a nearby supernova remnant (Brown et al. 1960). The highest-resolution observations of Loop I (Sofue and Reich 1979) show that the loop follows its small circle faithfully down to $b \sim 3°$, but no continuum extension has been clearly identified at negative latitudes.

No optical filaments have been found on Loop I, although reports have appeared of diffuse optical emission. However, optical filaments have been suggested to be connected with Loops II and III (i.e., Elliott 1970). The three loops seem to have a profound influence on the optical polarization of nearby stars (Section 1.5.4), demonstrating both the local nature of the loops and the fact that they are foreground to these stars.

Perhaps the most significant observation for clarifying the nature of the galactic loops was the discovery of soft X-ray emission from Loop I. The X rays follow the radio continuum loop closely, being brightest in the North Polar Spur segment, but in contrast to the HI, they lie several degrees inside the brightest radio emission. The X-ray spectrum has a characteristic shape both on the intense X-ray spur and inside the loop, and this can be well fitted by a thermal spectrum with temperatures close to 3×10^6 K, similar to the spectrum of the Cygnus Loop supernova remnant.

This is, perhaps, the strongest evidence in support of an interpretation of the loops as being very evolved supernova remnants.

1.3.4 The Nonthermal Radiation and Galactic Structure

If the regions of high-radio-frequency emissivity are distributed spatially in similar fashion to the bright stars and gas in our galaxy—a reasonable assumption in the density wave model where enhancement of the magnetic field strength and gas and relativistic particle densities is to be expected behind the galactic spiral shock—then a substantial increase in the radio brightness distribution might be expected at longitudes where arms are seen tangentially. The first indications of such features were seen in the 50′-resolution, 85.5-MHz observations made with the Mills Cross (Mills 1959). However, the intense peaks predicted for isotropic emission were absent and only "steps" were seen in the brightness distribution. Mills' original longitude distribution and that at 408 MHz of Green (1974) are shown in Figure 1.7. On both, recognizable discrete sources have been subtracted and the resulting distribution is then averaged over regions $\Delta l = 0°.5$, $b = \pm 3°$. The largest of Mills' steps are also seen in the 408-MHz distribution, although agreement is less satisfactory for $l > 0°$. Brown and Hazard (1960) demonstrated that the steps could be modeled with the directional synchrotron process if the magnetic fields in the spiral arms have a moderate degree of irregularity.

Since Mills' discovery, a number of attempts have been made to interpret galactic nonthermal emission in terms of spiral structure. As the distribution of radiation contains no inherent distance information, considerable assumptions have to be made when attempting this operation. These assumptions have usually been of two types. Either the spiral structure based on other observations (i.e., HI, HII) is used with a model for the emissivity variations across and between arms to predict the expected synchrotron emission, which is then compared with continuum observations, or the observed profile is "unfolded" under assumed symmetries to give the distribution of emissivity directly. Typical of such studies is that of Phillipps et al. (1981a,b) and more recently Beuermann et al. (1985), which used the data of the Haslam et al. (1982) all-sky atlas.

In their very comprehensive work, Beuermann et al. (1985) used scans from the Haslam et al. (1982) survey made in latitude through the galactic plane at constant longitude to demonstrate that the galactic continuum emission has two distinct disk components: (1) a thin disk component of thickness roughly equivalent to the gaseous HI disk and (2) a much thicker disk component which encompasses the thin disk. The scale height of both disks perpendicular to the galactic plane increases with radial distance from the galactic center. In the case of the thin disk, the full equivalent width increases from a value of approximately 250 pc at galactocentric distances less than 8 kpc to 700 pc at the edge of the Galaxy 20 kpc from the center. The thick disk displays a similar widening with increasing galactocentric radius; the full equivalent width is nearly ten times greater than that of the thin disk at each radius. These properties are summarized in Table 1.3.

The physical origin and association of the two disk components may not be distinct. The thin radio disk (after subtraction of individual discrete sources) is composed of overlapping regions of both thermal and nonthermal emission

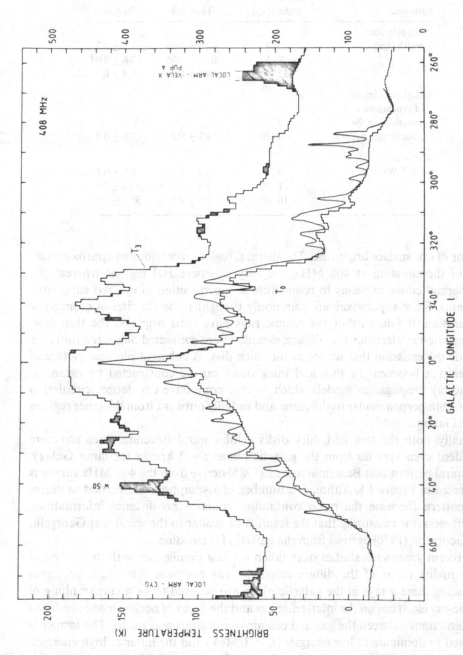

Fig. 1.7. The observed (solid curve) and corrected (histograms) 408-MHz brightness temperature along the galactic plane. T_3 is the mean of the two strips taken at $b = +3°$ and $b = -3°$; it refers to the left-hand temperature scale. T_0 is the mean background temperature for $|b| < 3°$; it refers to the right-hand temperature scale. The shaded histograms represent corrections made for known discrete sources. (After Beuermann et al. 1985).

Table 1.3. Physical parameters of the thin and thick radio disks at 408 MHz.

Parameter	Range of galactocentric radius (kpc)	Thin disk	Thick disk
Full equivalent width (kpc)	0–8	0.25 ± 0.03	2.3 ± 0.2
	8–12	0.37 ± 0.05	3.6 ± 0.04
	12–20	0.69 ± 0.09	6.3 ± 0.7
Radial scale length of synchrotron emissivity in the galactic plane (kpc)	4–10	3.3 ± 0.2	3.9 ± 0.3
Monochromatic power (10^{20} WHz^{-1})	0–1	0.6 ± 0.1	0.8 ± 0.1
	1–10	4.4 ± 0.6	43.3 ± 5.0
	10–20	2.6 ± 0.5	35.1 ± 7.0

regions of low surface brightness. The thermal regions, contributing approximately 40% of the emission at 408 MHz, are very extensive HII regions whereas the nonthermal emission seems to result from a superposition of evolved supernova remnants. Since supernovae are commonly thought to be the sites of cosmic-ray acceleration, it follows that the cosmic rays have their origin in the thin disk. The cosmic-ray electrons that diffuse, escape, or are convected out of the thin disk provide the emission that we see as the thick disk. A coherent physical picture of the relation between the thin and thick disks can be constructed by means of cosmic-ray propagation models which involve convective circulation, circulation that is both perpendicular to the plane and radially outward from the inner regions of the Galaxy.

Finally, both the thin and thick disks exhibit spiral structure. Such structure is evident even very far from the galactic plane, $z \sim 1$ kpc in the inner Galaxy. The spiral pattern that Beuermann et al. (1985) derive from the 408-MHz survey is illustrated in Figure 1.8. Although a number of assumptions are needed to derive this pattern (because the radio continuum contains no distance information), nevertheless it is reassuring that the result is so similar to the spiral that Georgelin and Georgelin (1976) derived from the galactic HI emission.

In recent years such studies have taken on new significance with the arrival of good-quality maps of the diffuse celestial γ-ray emission. The main processes producing these γ rays in the galactic plane are believed to be bremsstrahlung of cosmic-ray electrons on the interstellar gas and the decay of neutral pions produced in interactions between the gas and cosmic-ray protons and nuclei. The former is believed to dominate at low energies ($E < 1$ MeV) and the latter at high energies. As similar γ-ray distributions are found at low latitudes in both energy ranges, this suggests that the cosmic-ray heavy particle-to-electron ratio is constant over the galaxy. If so, the γ-ray emissivity is proportional to the product of the cosmic-ray intensity and the total gas density and is given by

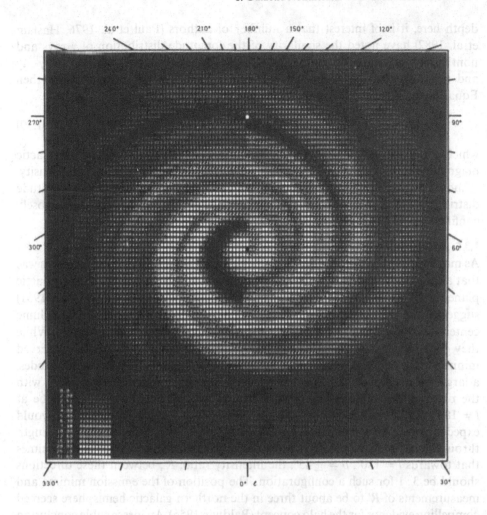

Fig. 1.8. A grey-scale representation of the face-on synchrotron brightness of the Galaxy. The position of the Sun is noted by a filled square. (After Beuermann et al. 1985.)

$$\varepsilon_\gamma \propto N_0 \rho \tag{1.27}$$

where the cosmic-ray energy distribution is

$$N(E)\,dE = N_0 E^{-p}\,dE$$

and ρ is the *total* gas density. From Equation (1.14) the synchrotron emissivity is given by

$$\varepsilon_R \propto N_0 B^{(p+1)/2} \ . \tag{1.28}$$

If the total gas distribution the galaxy is known, modeling the γ-ray emissivity distribution could in theory give the cosmic-ray intensity distribution. Then, for a reasonable choice of p, the magnetic field distribution could be recovered from models of the synchrotron emissivity. While we cannot consider this problem in depth here, it is of interest that a number of authors (Paul et al. 1976, Haslam et al. 1982) have noted the similarity of the longitude distribution of γ rays and nonthermal radio emission at low latitudes. If this is taken to signify that the γ-ray and radio nonthermal emissivities are in fixed ratio in the galactic plane, then Equations (1.27) and (1.28) imply that

$$\rho \propto B^{(p+1)/2} , \tag{1.29}$$

which for a value of $p \sim 3$ (Section 1.4.3) suggests that the square of galactic magnetic field in the plane is roughly proportional to the total gas density. It should be noted that the nonthermal radio emission has a much broader latitude distribution than the γ rays, implying that the interstellar gas is more closely confined to the galactic plane than the magnetic field and cosmic rays.

1.3.5 A Galactic Radio Halo?

As mentioned in the introduction, a remarkable feature of the early sky surveys was that although the distribution of radio brightness was concentrated to the galactic plane, it was still rather intense towards the poles. Westerhout and Oort (1951) suggested that this be interpreted either as emission from a large spherical volume centered on the galactic nucleus or as a superposition of extragalactic sources. While they preferred the latter conclusion, Shkovsky (1952) argued that, as the observed minima in the emission were not at the poles but at intermediate galactic latitudes, a large-scale spherical halo coextensive with the galaxy was more consistent with the observations. The minimum path length through such a halo would be at $l = 180°$, $b = 0°$, but due to the galactic disk emission at low latitudes one would expect an observed minimum closer to $l = 180°$, $b = \pm 45°$. Also, as the path length through a spherical halo of 16-kpc radius towards $l = 0$, $b = \pm 45°$ is three times that towards $l = 180°$, $b = \pm 45°$, the intensity ratio, R', between these directions should be $3 : 1$ for such a configuration. The position of the emission minima and measurements of R' to be about three in the northern galactic hemisphere seemed compelling evidence for the halo concept (Baldwin 1955). An inescapable conclusion associated with the halo concept is that such a large volume containing cosmic-ray particles would effectively isotropize these particles. The first detailed cosmic-ray studies conducted above the atmosphere indeed showed extreme isotropy for cosmic rays of energy greater than 10^{12} eV.

Although Mills (1959) found his data at 85 MHz consistent with a similar halo configuration to that proposed by Baldwin (1955), observational evidence against such a clear-cut picture soon appeared. Baldwin (1977) pointed out that the measurement of R' could be seriously affected in the north galactic hemisphere, as $l = 0°$, $b = +45°$ lies inside Loop I where an excess of emission might be expected. Also, Burke (1967) showed that R' in the southern hemisphere was not significantly different from unity, the result expected in the absence of a halo. A little later, Berkhuijsen (1971) used a spatial correlation between the total and polarized intensities from the North Polar Spur to derive the intensity of the emission underlying Loop I. She found by using this value for the underlying intensity that the temperatures at 810 MHz for $l = 0$, $b = +45$ and $l = 180°$, $b = +45$ were equal to within 10%, or 30% after allowance for the most probable extragalactic contribution. She concluded that a homogeneous spherical radio halo at 820 MHz either is not observed or has a radius of > 30 kpc. Further, she suggested from a similar consideration of the other loops and spurs that "nearly all the structure in the brightness distribution away from the galactic plane is caused by a few local features."

The radio halo problem was readdressed by Webster (1975) using low-resolution drift-scan observations at a number of frequencies between 17.5 and 408 MHz. He assumed that the halo is weakest in the anticenter and has a spectrum that is steeper than that of the disk emission. If a constant spectral index was assumed for the disk, Webster showed that a characteristic plot looking like a tuning fork should result when plotting the temperatures of a drift scan at one frequency against those at the same positions at another frequency. He found such a signature to be present in the data and he modeled it with a spherical halo of constant emissivity, concluding that the observations were consistent with a weak halo of diameter ~ 10–15 kpc and spectral index ~ -0.8. At meter wavelengths this halo had ~ 30 times smaller emissivity than the disk and similar total radiated power. He considered such a halo to be incapable of confining the cosmic rays. Webster (1978) extended his work by considering spheroidal halo models. He found a prolate or spherical halo would fit the data between 17.5 and 81.5 MHz, but needed a significantly oblate halo between 81.5 and 408 MHz. This would imply a decreasing scale height with increasing frequency, consistent with the expectations for aging effects caused by the energy losses of the cosmic-ray electrons.

The controversy over the existence of the halo is, however, clearly far from over. Bulanov et al. (1976) believe that if variations of the halo spectrum with position due to aging effects are taken into account, then Webster's method would give a halo of high luminosity. On the other hand, Milogradov-Turin (1985) feels that Webster's signature of the halo is produced by the presence of the galactic loops and spurs and the presence of a halo is not necessary. Baldwin (1977), using a very simple model for the disk emission, deduced a z-distribution of emissivity that had an extension out to at least 3 kpc above the plane at 408 MHz. He diplomatically chooses to call his extension to high z a "thick disk" rather than a halo, a description that would seem to describe the conclusions of Phillipps et al. (1981b) and Beuermann et al. (1985) rather well. Clearly, the status of the halo will continue to be a subject of speculation for some time yet.

1.4 The Spectrum of the Nonthermal Emission

As detailed above, the intensity observed by a radio telescope in any direction is a combination of radiation from various galactic components (i.e., disk, loop, and possibly halo emission) and an isotropic component due to the superposition of extragalactic radio sources (including the 2.7 K microwave background; Chapter 14). At meter, and longer, wavelengths this extragalactic component is expected to have a spectral index of about -0.75, which is steeper than that measured for the galactic disk. Using low-resolution observations at long wavelengths, and assuming the above spectral index, Bridle (1976) estimated a brightness temperature for the integrated extragalactic emission of 30 ± 7 K at 178 MHz, representing one-third to one-half of the minimum intensity observed at that frequency. At yet lower frequencies, the extragalactic component would be an even larger fraction of the total emission. Although this chapter is not the place to consider this topic further, the reader is referred to Simon (1977), who also predicted a turnover in the spectrum of the extragalactic component at 1 to 2 MHz due to synchrotron self-absorption in individual sources.

1.4.1 The Galactic Spectrum Above 10 MHz

Considering the strength of the nonthermal emission of the disk along the galactic plane, it might be expected that observations at low latitudes would be the principal tool used to study the continuum spectrum of this component. However, major complications are provided both by the contamination due to the population of galactic thermal and nonthermal sources and by distributed thermal emission from the interstellar plasma. It can be shown for a mixture of thermal and nonthermal radiation along the line of sight that the observed brightness temperature is given by,

$$T(v) = T_T(v) + A(v)T_N(v) , \qquad (1.30)$$

where T_T is the thermal contribution, T_N would be the nonthermal brightness temperature were the thermal component to be absent, and A is an absorption factor that accounts for absorption of the nonthermal radiation by the thermal gas. If the optical depth of the thermal gas is $\tau(v) \propto v^{2.1}$, its electron temperature is T_e, and the nonthermal emission has a power law spectrum, $T_N = T_0 v^\beta$, then

$$T(v) = (1 - e^{-\tau(v)})T_e + A(v)T_0 v^\beta . \qquad (1.31)$$

In practice, $A(v)$ depends on the relative disposition of thermal and nonthermal material along the line of sight. For example, if (1) the thermal material is behind the nonthermal, $A(v) = 1$; (2) the nonthermal material is behind the thermal, $A(v) = e^{-\tau(v)}$; (3) the thermal and nonthermal material are similarly distributed (uniform mixing), $A(v) = (1 - e^{-\tau(v)})/\tau(v)$. A number of studies (e.g., Westerhout 1958) have assumed the third case to apply and attempted to separate the thermal and nonthermal components at low latitude using observations at two or more separated frequencies. Note that if only two frequencies are used, values for the nonthermal spectral index and thermal electron temperature have to be assumed from other information. Conveniently, when $\tau \to 0$, $A \to 1$, and in these circumstances Equation (1.31) becomes

$$T(v) = \tau(v)T_e + T_0 v^\beta . \qquad (1.32)$$

Despite the problem of thermal emission, a number of attempts have been made to determine the nonthermal disk spectral index at high frequencies using low-latitude observations. Altenhoff (1968) believed that he had almost completely resolved the thermal contribution into discrete sources at 10' resolution. Through observations made betwen 234 MHz and 2.7 GHz, he found the nonthermal background spectral index to steepen from -0.6 at 300 MHz to -0.9 at 2.7 GHz. In view of this result, we refer the reader to Chapter 2 for a discussion of the diffuse thermal component and await with interest the interpretation of the new high-resolution, low-latitude surveys, described in Section 1.3.2. Penzias and Wilson (1966) used observations at $b = \pm 3°6$ between 408 MHz and 4.08 GHz to deduce a nonthermal spectral index of   0.9. Finally, Hirabayashi et al. (1972) combined observations made between 1.4 and 15 GHz. Although they found an average of 80% of the 15-GHz disk emission to be thermal for $24° < l < 48°$, $b = 0°$, they were able to derive a nonthermal spectral index of -1.0.

At low frequencies, spectral studies have had to be made at higher latitudes due both to the low resolution available and the absorption of the nonthermal emission by the thermal component at low latitudes [as frequency goes to zero, for uniform mixing $A(v) \rightarrow 1/\tau \propto v^{2.1}$]. The latter effect is graphically illustrated by the map of spectral indices between 38 and 404 MHz derived by Milogradov-Turin (1974). Thermal absorption at the lower frequency is seen as low spectral indices on the galactic plane, while at higher latitudes the values are very uniform, mostly lying between -0.48 and -0.58. A major series of observations made with scaled antennas at the University of Cambridge have defined the nonthermal spectrum from 13 MHz to 1.4 GHz (Bridle 1967, Sironi 1974, Webster 1974). The interpretation of these data relied on the fact that if the emission from an area of sky can be considered as consisting of an isotropic component and a spatially varying component of brightness temperature spectral index β, then a plot of brightness temperatures over the region at frequency v_1 against those at v_2 will yield a straight line of slope $(v_1/v_2)^{\beta}$. The Cambridge observers checked for variations of β with direction by considering separately data towards the local interarm region ($50° < l < 90°$) and towards the anticenter ($140° < l < 220°$). [Note, however, the conclusions of Webster (1975) on the "signature" of the halo; Section 1.3.4.] In good agreement with the higher-resolution observations described above, a spectral index of -0.8 was found above 400 MHz. However, below this frequency the spectral slope changes, giving a spectral index that is within 0.1 of -0.5 between 80 and 400 MHz, flattening somewhat further to nearer -0.4 between 13 and 80 MHz.

1.4.2 The Very-Low-Frequency Spectrum
Below 10 MHz a number of technical problems exist. Antennas have to be physically huge in order to achieve even medium resolution. Additionally, absorption in the ionosphere causes increasing difficulties as the frequency decreases and terrestrial interference becomes a major problem. While ground-based observations can be made down to about 2 MHz, even this is only possible from a few locations at solar minimum. All observations below this frequency have to be made from space vehicles.

Examples of maps made between 1 and 10 MHz can be found in Cane and Whitham (1977) and Novaco and Brown (1978). It seems certain that the effect of thermal absorption in the local interstellar medium becomes important at these frequencies, and even at high latitudes this shapes the appearance of the sky. Using observations down to 0.25 MHz, Novaco and Brown constructed spectra towards the galactic center and anticenter, as well as towards the two galactic poles. All spectra show a sharp turnover near 2–3 MHz, while below 1 MHz the radiation appears to be essentially isotropic, as expected for local thermal absorption. Between 1 and 4 MHz the north galactic pole appears brighter than the south, which

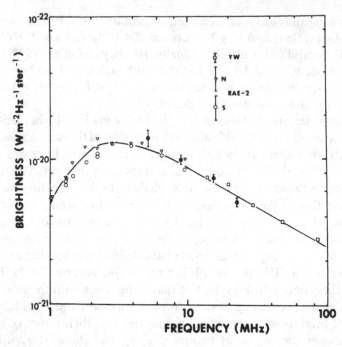

Fig. 1.9. Measurements of the low-frequency spectrum of the galactic background emission, showing the cutoff which results from absorption by thermal electrons in the interstellar medium. (After Cane 1979, with permission of the Royal Astronomical Society).

Novaco and Brown suggest may be due to the Sun being located somewhat to the north of the galactic plane.

Cane (1979) reported low-resolution observations of the north and south galactic poles, which remove earlier inconsistencies between the brightnesses in these two directions. She combines the best available data to produce the polar spectrum of the total emission between 1 and 100 MHz shown in Figure 1.9. Above 6 MHz the spectral index is -0.55 ± 0.03. She considers that this polar spectrum describes the average properties in the solar neighborhood adequately and fits the spectrum with the following model: (1) local nonthermal emission of spectral index α, (2) free-free absorption in thermal material uniformly mixed with the nonthermal emission, (3) an isotropic extragalactic component of spectral index -0.8. In her model, the extragalactic component is almost completely absorbed by the thermal material below 2 MHz and she does not believe it possible to test Simon's prediction of a synchrotron self-absorption cutoff in the spectrum of the extragalactic background. The fitted extragalactic component would represent a brightness temperature of 17.5 ± 4.5 K at 178 MHz, rather less than derived from the same spectral index by Bridle (1967). The local nonthermal emission has a spectral index of about -0.50.

1.4.3 The Energy Spectrum of Cosmic-Ray Electrons

Studies of the spatial and spectral distribution of cosmic-ray electrons in the galaxy are of great interest because they provide valuable information on the propagation, confinement, and production of very energetic particles. The relativistic cosmic electrons can be investigated both by direct measurement from outside the atmosphere and by inferences drawn from the nonthermal galactic radio continuum.

The cosmic-ray electrons, unlike the nucleonic component, reveal their presence in space via the synchrotron radiation they emit as they traverse the weak interstellar magnetic field. From this radiation it is possible to deduce directly the electron spectrum $[n(E) dE \propto E^{-p} dE]$ required to produce the observed background emission spectrum for a given value of the magnetic field. The radio intensity at a frequency v is related to the electron energy spectrum by Equation (1.12). Most studies of this kind have used the radio spectrum in the anticenter region, although Webber et al. (1980) preferred to use the spectrum towards the galactic poles. Their reasoning was that (1) at the poles thermal absorption sets in at a lower frequency, (2) the low resolution available at low frequencies is less important away from the galactic plane, (3) emission at the poles is most characteristic of local conditions. Reanalyzing the polar radio spectrum of Cane (1979), they found a spectral index for the disk emission of $\alpha = -0.57 \pm 0.03$ and, using Equation (1.15), believe that a value of $p = 2.14 \pm 0.06$ applies between 70 and 1200 MeV. This is considerably steeper than the electron spectrum deduced from the anticenter data and reflects the different radio spectra found in the two directions. The steep radio spectra found for the disk emission above 400 MHz suggest that above 1 GeV the electron spectrum should steepen, a value of $\alpha = -1$ implying $N(E) \propto E^{-3.0}$.

Experiments made from balloons and space vehicles have provided measures of the differential flux of cosmic-ray electrons from a few MeV up to several hundred GeV. The reduction of the cosmic-ray electron intensity at the earth by solar modulation is not believed to be significant above about 2 GeV, so at these energies the measured energy spectrum can be compared directly with that derived from radio observations. Until recently, considerable disagreement existed concerning both the measured spectral shape and the absolute electron flux above a few GeV. This seems to have been largely resolved by recent experiments (i.e., Golden et al. 1984, Tang 1984). In the energy range of 4.5 to 63.5 GeV, Golden et al. measured an electron intensity index $p = 3.15 \pm 0.2$, while Tang found an index of 2.7 around 10 GeV, increasing to 3.5 above 40 GeV.

Cosmic-ray nuclei have a different spectral slope to the electrons above 10 GeV of $p = 2.65$. This seems to be adequately explained by electrons and nuclei having a similar power law source spectrum, with the electron spectrum being steepened at these energies by synchrotron radiation and inverse Compton losses during propagation. Below 1 GeV the measured electron spectrum lies increasingly beneath that derived from radio observations, presumably due to solar modulation, and the ratio of expected to measured intensities as a function of energy gives valuable information on the modulation mechanisms. Badhwar et al. (1977) found that to reconcile the measured local cosmic-ray electron flux above a few GeV and the nonthermal radio emission, they had to invoke uncomfortably high interstellar magnetic fields compared with those derived from the rotation measures of pulsars. Rockstroh and Webber (1978) suggest that this problem can be resolved if account is taken of fluctuations in the interstellar magnetic field along the line of sight.

1.5 Linear Polarization of the Nonthermal Emission

We have seen that synchrotron radiation is emitted with a high degree of linear polarization. Equations (1.15) and (1.17) suggest an intrinsic polarization percentage of 72% for a spectral index $\alpha = -0.7$. At emission the electric vector of the polarized component will be perpendicular to the direction of the magnetic field as projected on the sky. This is opposite to the case of optical polarization of starlight where the Davis-Greenstein (1951) mechanism of orienting interstellar grains with respect to the field predicts the electric vector of the polarized light to be parallel to the field.

From the description of Faraday rotation in Section 1.2.2, it can be predicted that significant linear polarization will be observed for only rather local emission at meter and decimeter wavelengths. The interstellar medium contains not only magnetic fields and cosmic-ray electrons but also thermal electrons, and Faraday rotation within the emitting regions causes the radiation to suffer large "front-back" depolarization for long path lengths. For example, at a wavelength of $\lambda 20$ cm (1500 MHz), Equation (1.32) demonstrates that for a "typical" interstellar electron density of 0.03 cm^{-3}, and a magnetic field component along the line of sight of 3 μG, a polarized wave will have its plane of polarization rotated by 180° in a distance of 1000 pc. The equivalent distance at a wavelength of $\lambda 1$ m (300 MHz) is only about 40 pc! Thus, it is expected that emission from beyond a few hundred parsecs will reach the observer essentially unpolarized at these wavelengths.

1.5.1 Surveys of Linear Polarization of the Galactic Emission
Polarization of the galactic continuum emission was first clearly demonstrated by Westerhout et al. (1962). This detection was the final evidence needed to confirm the synchrotron radiation theory for the galactic nonthermal emission. In the succeeding years, polarization surveys were made between 240 and 1415 MHz, mainly by workers in Holland, England, and Australia. Lists of the available data can be found in Berkhuijen (1975) and Spoelstra (1984). Remarkably, no major surveys have been published since the compendium of Brouw and Spoelstra (1976). Further, only one major survey of the southern sky has been published (Mathewson and Milne 1965) and this was made over twenty years ago! Although the published data come nearest to complete sky coverage at 408 MHz, an all-sky linear polarization survey, available at a consistent resolution and frequency, is badly needed.

A major reason for the recent neglect of background polarization observations has to be that such measurements are among the most difficult ones made in radio astronomy. The linearly polarized component is invariably much weaker than the total intensity. Additionally, allowance has to be made for the Faraday rotation introduced as the radiation passes through the terrestrial ionosphere. Finally, all polarimeters have spurious responses in the presence of unpolarized signals, and radio telescopes and their polarimeters are no exception. Correction of the data for these instrumental effects is a difficult and painstaking operation.

A typical linear polarization survey of the northern sky is that at 1411 MHz shown in Figure 1.10 (Brouw and Spoelstra 1976). The resolution is 0°.6, although the data are somewhat undersampled. The polarized intensities are represented by vectors whose lengths are proportional to the intensities and whose orienta-

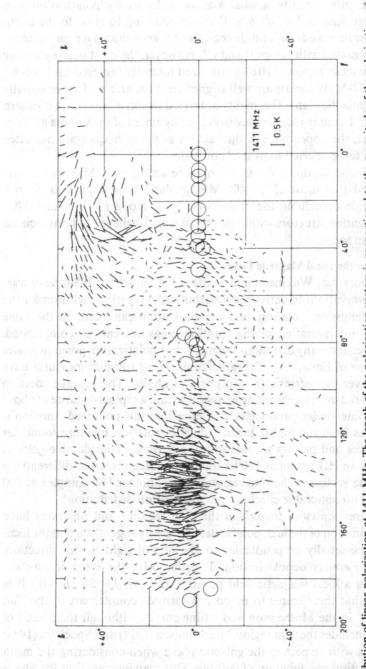

Fig. 1.10. Distribution of linear polarization at 1411 MHz. The length of the polarization vectors is proportional to the magnitude of the polarization while the orientation of these vectors is the polarization angle (i.e., the direction of the electric field). The circles indicate the positions of the brightest discrete sources. (After Brouw and Spoelstra 1976).

tions represent the polarization position angle of the electric field. The polarized intensities are distributed almost completely differently to the total intensity picture of Figure 1.5. The only instantly familiar feature is the strong polarization seen on the North Polar Spur at $l \sim 30°$, $b > 0°$ and continuing to close to the north galactic pole. Other intermediate-latitude regions of higher-than-average polarization seem to be connected with Loops II and III. However, the most striking feature of the map is the intense region of strong polarized intensity centered on $l \sim 140°$, $b \sim +10°$. The 1411-MHz vectors are well aligned over the area and are essentially orthogonal to the galactic plane. This highly polarized region is seen on all northern sky maps, although Faraday rotation destroys the alignment of the vectors at lower frequencies. In fact, the appearance of the vectors at lower frequencies has often caused this region to be referred to as the "fan region."

The situation in the southern sky is much more ambiguous. Mathewson and Milne (1965) found that most of the 408-MHz polarization seemed to lie in a band some 60° wide containing the great circle lying along $l = 160°$ and 340°. Some of the polarization structure within this band is now believed to be associated with the continuum loops.

1.5.2 Implications for the Local Magnetic Field

A number of authors (i.e., Wilkinson and Smith 1974, Spoelstra 1984) have used multifrequency observations to derive the distribution of rotation measures for the northern sky. As emission, rotation, and depolarization can occur in the same volumes of space, interpretation of the rotation measures can be complicated. In practice, however, for many directions the measured polarization position angles fit the simple λ^2 law of Equation (1.23) rather well. These rotation measures have been used to recover the intrinsic ($\lambda = 0$) position angles. The average rotation measure is about 8 rad m^{-2}, representing emission within a typical distance of about 200 pc for typical interstellar parameters. Evidence that this polarized emission is mostly local comes from the higher mean values of rotation measure found for extragalactic sources and pulsars whose radiation traverses considerable galactic path lengths. Also, an HII region at 140 to 300 pc has been shown to differentially rotate the 610-MHz vectors in the "fan region," while another HII complex at 700 to 1000 kpc shows no noticeable effect on the polarization distribution.

Rotation measures are close to zero in the "fan region," and observers have attempted an explanation of the high polarizations there by suggesting that the local magnetic field is essentially perpendicular to the line of sight in this direction, causing very low Faraday depolarization. Together with the intrinsic position angles, this implies a local magnetic field pointing towards $l \sim 50°$, $b \sim 0°$. It is rather disturbing that there seems to be no low-latitude counterpart of the "fan region" at $l \sim 320°$ in the Mathewson and Milne survey, although their band of high polarization includes the "fan region." Berkhuijsen (1971) and Spoelstra (1984) found asymmetries with respect to the galactic plane when considering the mean percentage polarization as a function of latitude. This may indicate that the Sun is situated somewhat above the plane.

The radio polarization studies provide considerable evidence that in addition to the uniform component of the local magnetic field, a random component also exists.

Spoelstra (1984) found that while the optical polarization directions for stars in the region $90° < l < 180°$, $b < 30°$ are generally orthogonal to the intrinsic radio polarization directions, the correlation breaks down for $0° < l < 90°$. He believes this to demonstrate that for $0° < l < 90°$ we are looking essentially along the uniform field, so the random component dominates the transverse field component. Further, the derived distribution of rotation measures displays much structure on the scale of a few degrees that can be explained by a random component to the field. If the average distance for the medium causing rotation is 200 pc, a scale size for fluctuations in the random field of about 20 pc is implied. Spoelstra as well as Wilkinson and Smith (1974) have modeled the depolarization of the galactic emission with frequency to derive fluctuation scale sizes in the range of 10 to 75 pc. They also found that the extrapolated intrinsic polarization percentages (after allowance for the unpolarized background) fell below the maximum given by Equation (1.17). This can be ascribed to the presence of the random component (Burn 1966) of the magnetic field of magnitude the same order as, or somewhat larger than, that of the uniform component.

1.5.3 Magnetic Fields in the Galactic Loops

Linear polarization of the radiation from the three main galactic loops has been found by several observers. As illustrated in Figure 1.10, the polarized intensities on the North Polar Spur segment of Loop I are by far the strongest, and much of this section will again focus on this feature. Spoelstra (1972) found that the polarized emission associated with Loops II and III appeared to be shifted to the inside of the continuum arcs, while for the North Polar Spur the two coincided. Wilkinson and Smith (1974) obtained the interesting result that the emission from Loop II and III depolarized less rapidly with wavelength for similar rotation measures than did the surrounding emission. This suggests that the loop radiation is undergoing "rotation without depolarization," indicative of the loops being discrete objects with most of the Faraday rotation occurring between them and the Sun. A similar conclusion holds for the North Polar Spur.

If a feature of a given constant degree of polarization is superposed on an isotropic unpolarized background, the plot of polarized intensity against total intensity across the region should give a straight line whose slope equals the degree of polarization. For the North Polar Spur, several authors have found this situation to hold above $b \sim 30°$. Berkhuijsen (1971) obtained a polarization percentage of about 70% at 1411 MHz above $b = +60°$. This is close to the maximum possible for synchrotron emission, indicating that the magnetic field is very regular within the loop. Certainly any random field component must be much smaller than in the general interstellar medium, as might be expected in the compressed shell of a supernova remnant. For the other loops, Berkhuijsen found somewhat lower polarization percentages between 15% and 40% at 1411 MHz. She also noted that the high polarized intensities extended outside the North Polar Spur to $l \sim 60°$, and from agreement in the polarization position angles of the vectors suggested that this was due to the influence of Loop I. Together with the evidence listed in Section 1.3.3, this indicates that the presence of the loop is felt considerably beyond its main arc. Spoelstra later demonstrated that several of the ridges inside Loop I are highly polarized. From the intrinsic polarization position angles he derived, Spoelstra concluded that for $b > 45°$ the projected magnetic field direction in the loop emission regions is parallel to the main arc, and that this order extends for at least $15°$ on either side of the ridge peak. However, for $b < 45°$ a projected magnetic field essentially perpendicular to the feature was derived, with the intrinsic radio vectors orthogonal to those at somewhat higher latitude.

Bingham (1967) pointed out that the influence of the North Polar Spur was also apparently present in the optical polarization of starlight. Above $b \sim 40°$, optical polarization vectors are essentially parallel to the Spur for at least $15°$ inside and outside of the feature. He believed that the effect set in for stars beyond 75 pc and concluded that the distance to the Spur is 100 ± 20 pc. Distances to the Spur from optical polarization are usually given the greatest credence, and most authors have suggested values between 50 and 120 pc by this method.

One piece of evidence, however, disturbs the tidy picture of spatial coincidence between the optically polarizing dust grains and the regions of nonthermal emission within the Spur. Spoelstra found that while both radio and optical data suggest a magnetic field parallel to the North Polar Spur above $b \sim 45°$, below this the optical polarization vectors continue parallel to the arc, where the radio polarization indicates an orthogonal magnetic field. He noted that the optical and radio polarization need not necessarily originate in the same volume of Loop I and conjectured that a low density of dust grains might exist within the synchrotron-emitting volume.

1.5.4 Rotation Measure Data and the Galactic Magnetic Field

Over the past twenty years the rotation measures of several hundred discrete radio sources have been determined from the large body of available polarization data (e.g., Tabara and Inoue 1980). Most of these sources are extragalactic objects or pulsars. Part of this Faraday rotation, often the majority, occurs within the interstellar medium of our galaxy, and the rotation measure values contain valuable information concerning the galactic magnetic field. While rotation measures of extragalactic systems include a contribution from the complete line of sight path through our galaxy, the pulsar data sample only that part of the interstellar medium between the pulsar and the observer. Although not strictly included within the main subject matter of this chapter, a brief summary will be given of the results of these studies with respect to the galactic magnetic field.

Analyses of rotation measure data for information on the galactic magnetic field have used a statistical approach to minimize the effect of the intrinsic rotation measures of the sources. Typical of such studies is that of Simard-Normandin and Kronberg (1980), who used the rotation measures of about 550 extragalactic sources. In accordance with other investigators, they find that a strong trend is apparent in the southern galactic hemisphere, with predominantly negative rotation measures being found in the region $0° < l < 180°$ and positive values mostly lying in $180° < l < 360°$. This strongly suggests an extensive, uniform component of the local magnetic field running toward $l \sim 90°$. In the northern hemisphere, however, the picture is more complicated. The sense of the rotation measures is positive for $0° < l < 90°$ and its sign alternates in successive quadrants. The probable solution to this asymmetry is that the northern hemisphere rotation measures in the first and fourth quadrants are strongly affected by the perturbation of the local magnetic field due to the North Polar Spur. It may be significant that a region of high rotation measure coincides in direction with Loop II, although Simard-Normandin and Kronberg prefer to interpret this as a feature of the large-scale dynamics of the galaxy rather than as being due to a local object. These authors consider the rotation measures at low latitude and fit simple models of the galactic field within the galaxy. They conclude that the data are consistent with either a circular or bisymmetric spiral model for the field, with a reversal of field sense occurring near the Sagittarius spiral arm. They find no evidence for any large-scale field reversal outside the solar radius.

It is not clear at present how the apparent local field direction of $l \sim 90°$ from rotation measure data is to be reconciled with the direction of $\sim 50°$ suggested by background polarization studies of the "fan region." It might be that the "fan region" is of very localized nature (there seems to be no corresponding feature at $l = 320°$).

The distribution of pulsar rotation measures agrees closely with that of extragalactic sources. A direction for the local uniform magnetic field towards $l = 94° \pm 11°$ was derived by Manchester (1974). The pulsar data show no evidence for any helical component to the local magnetic field. Simard-Normandin and Kronberg found that the pulsar rotation measures are consistently lower than those of angularly nearby extragalactic sources, suggesting that the regular magnetic field retains the same sense to very great distances from the Sun in most directions. (Note, however, their conclusion on a field reversal near the Sagittarius arm.) Pulsars are unique in allowing us to estimate, with accuracy, the strength of the component of the uniform magnetic field along the line of sight. Their rotation measures are proportional to $\int n B_\parallel \, dL$, while also measurable are their dispersion measures (Chapter 10) which are proportional to $\int n \, dL$. The ratio of these two quantities thus gives the mean magnetic field component along the line of sight, weighted by the electron density. Manchester (1974) fitted a simple model for the uniform field to measured values of the projected mean field component for pulsars with distances less than 2 kpc. He obtained a value of $2.2 \pm 0.4 \, \mu G$. From the residuals of the fit, he concluded that the field strength of the irregular component was comparable to that of the uniform field. (See also Chapter 3.)

1.6 The Galactic Nonthermal Radiation in Perspective

In terms of its primary physical parameters—mass, size, luminosity—the Milky Way is but a very typical spiral galaxy. The radio characteristics are remarkable: the total radio power of the Milky Way (at a fiducial frequency of 408 MHz, for example) is fifteen times larger than that of the nearby spiral M33 and is more than ten times greater than that of M31 (Berkhuijsen 1984). Among these three galaxies the Milky Way is the only galaxy which shows evidence for nonthermal emission from a thick disk, that is, disk emission with a z-height greatly in excess of that of the gas layer.

The presence of loops and filaments which tend to disrupt the order of the large-scale magnetic field also distinguishes the Galaxy from neighboring spirals. In the Galaxy the ratio of random to ordered field strengths is approximately unity whereas in M31 and M33 the ratio is more like 0.5. This disorder makes it difficult to identify the morphology of the large-scale galactic field: both a bisymmetric field which follows the spiral arms (Simard-Normandin and Kronberg 1980) and a closed structure (Vallee 1983) have been suggested. On the other hand, in M31 and M33 the field structure is more easily visible. In both galaxies a clear sinusoidal variation of rotation measure with azimuthal angle (Beck 1982) indicates that the field in these galaxies is bisymmetric along the spiral arms and connected to the intergalactic field. If this morphology also applies to the Galaxy, then the source of the disorder in the galactic field and its relation to the existence, maintenance, and dynamics of the unique thick galactic disk are central questions for investigation in the future.

Recommended Reading

Shkovsky, I.S. 1960. Cosmic Radio Waves. Cambridge: Harvard University Press. Introduction to early observations of the galactic continuum.

Pawsey, J.L. 1960. In A. Blaauw and M. Schmidt (eds.), Galactic Structure. Chicago: University of Chicago Press. 1960. Review of the observational situation at that date and the physical interpretation.

Heiles, C. 1976. Annu. Rev. Astron. Astrophys. 14:1. The interstellar magnetic field.

Cesarsky, C.J. 1980. Annu. Rev. Astron. Astrophys. 18:289. Cosmic-ray confinement in the galaxy.

van der Kruit, P.C. and R.J. Allen. 1976. Annu. Rev. Astron. Astrophys. 14:417. The radio continuum morphology of spiral galaxies.

Salter, C.J. 1983. Bull. Astron. Soc. India 11:1. The observations of Loop I (the North Polar Spur) and their interpretation.

Verschuur, G. 1979. Fund. Cosmic Phys. 5:113. A summary of radio techniques and measurements of the Galactic magnetic field.

References

Adgie, R., and F.G. Smith. 1956. Observatory 76:181.

Alfvèn, H., and N. Herlofson. 1950. Phys. Rev. 78:616.

Allen, C.W., and C.S. Gum. 1950. Aust. J. Sci. Res. 3A:224.

Altenhoff, W. 1968. In Y. Terzian (ed.), Interstellar Ionized Hydrogen. New York: Benjamin, p. 519.

Altenhoff, W., D. Downes, T. Pauls, and J. Schraml. 1978. Astron. Astrophys. Suppl. 35:23.

Badhwar, G.D., R.R. Daniel, and S.A. Stephens. 1977. Nature 365:424.

Baldwin, J.E. 1955. Mon. Not. Astron. Soc. 155:690.

Baldwin, J.E. 1977. In C.E. Fictel and F.W. Stecker (eds.), Structure and Content of the Galaxy and Galactic Gamma Rays. NASA Goddard Space Flight Center, Greenbelt, MD p. 206.

Beck, R. 1982. Astron. Astrophys. 106:121.

Berkhuijsen, E.M. 1971. Astron. Astrophys. 14:359.

Berkhuijsen, E.M. 1972. Astron. Astrophys. Suppl. 5:263.

Berkhuijsen, E.M. 1975. Astron. Astrophys. 40:311.

Berkhuijsen, E.M. 1984, Astron. Astrophys. 140:431.

Berkhuijsen, E.M., C.G.T. Haslam, and C.J. Salter. 1971 Astron. Astrophys. 14:252.

Beuermann, K., G. Kanback, and E.M. Berkhuijsen. 1985. Astron. Astrophys. 153:17.

Bingham, R.G. 1967. Mon. Not. R. Astron. Soc. 137:157.

Blumenthal, G.R., and R.J. Gould. 1970. Rev. Mod. Phys. 42:237.

Bolton, J.G., and K.C. Westfold. 1951. Aust. J. Sci. Res. 4:476.

Bridle, A.H. 1967. Mon. Not R. Astron. Soc. 136:219.

Brouw, W.N., and T.A.T. Spoelstra. 1976. Astron. Astrophys. Suppl. 26:129.

Brown, R.H., and C. Hazard. 1960. Observatory 80:137.

Brown, R.H., R.F.D. Davis, and C. Hazard. 1960. Observatory 80:191.

Bulanov, S.V., S.I. Syrovatskii, and V.A. Gogiel. 1976. Astrophys. Space Sci. 44:255.

Burke, B.F. 1967. Int. Astron. Union, Symp. 31:361.

Burn, B.J. 1966. Mon. Not. R. Astron. Soc. **133**:67.
Cane, H.V. 1978. Aust. J. Phys. **31**:561.
Cane, H.V. 1979. Mon. Not. R. Astron. Soc. **189**:465.
Cane, H.V., and P.S. Whitham. 1977. Mon. Not. R. Astron. Soc. **179**:21.
Caswell, J.L. 1976. Mon. Not. R. Astron. Soc. **177**:601.
Davis, L. Jr., and J.L. Greenstein. 1951. Astrophys. J. **114**:206.
Day, G.A., J.L. Caswell, and D.J. Cooke. 1972. Aust. J. Phys. Suppl. **25**:1.
Droge, F., and W. Priester. 1956. Z. Astrophys. **40**:236.
Elliott, K.H. 1970. Nature **226**:1236.
Georgelin, Y.M., and Georgelin, Y.P. 1976, Astron. Astrophys. **49**:57.
Ginzburg, V.L., and S.I. Syrovatskii. 1964. The Origin of Cosmic Rays. New York: Macmillan.
Ginzburg, V.L., and S.I. Syrovatskii. 1965. Annu. Rev. Astron. Astrophys. **3**:279.
Ginzburg, V.L., and S.I. Syrovatskii. 1969. Annu. Rev. Astron. Astrophys. **7**:375.
Golden, R.L., B.G. Mauger, G.D. Badhwar, R.R. Daniel, J.L. Lacy, S.A. Stephens, and J.E. Zipse. 1984. Astrophys. J. **287**:622.
Green, A.J. 1974. Astron. Astrophys. Suppl. **18**:267.
Haslam, C.G.T., C.J. Salter, H. Stoffel, and W.E. Wilson. 1982. Astron. Astrophys. Suppl. **47**:1.
Haynes, R.F., J.L. Caswell, and L.W.J. Simons. 1978. Aust. J. Phys. Suppl. **45**:1.
Heiles, C. 1979. Astrophys. J. **229**:533.
Henyey, L.G., and P.C. Keenan. 1940. Astrophys. J. **91**:625.
Hirabayashi, H., H. Yokoi, and M. Morimoto. 1972. Nature Phys. Sci. **237**:54.
Jansky, K.G. 1932. Proc. Inst. Rad. Eng. **20**:1920.
Jansky, K.G. 1933. Proc. Inst. Rad. Eng. **21**:1387.
Jansky, K.G. 1935. Proc. Inst. Rad. Eng. **23**:1158.
Kallas, E., and W. Reich. 1980. Astron. Astrophys. Suppl. **42**:227.
Kiepenheuer, K.O. 1950. Phys. Rev. **79**:738.
Landecker, T.L., and R. Wielebinski. 1970. Aust. J. Phys. Suppl. **16**:1.
Manchester, R.N. 1974. Astrophys. J. **188**:637.
Mathewson, D.S. and Milne, D.K. 1965, Australian J. Phys., **18**:635.
Mathewson, D.S., N.W. Broten, and D.J. Cole. 1965. Aust. J. Phys. **18**:635.
Mills, B.Y. 1959. Proc. Int. Astron. Union Symposium, No. 9. New York: Academic Press, p. 431.
Milogradov-Turin, J. 1974. Mem. Soc. Astron. Ital. **45**:85.
Milogradov-Turin, J. 1985. Int. Astron. Union, Symp. **106**:245.
Milogradov-Turin, J., and F.G. Smith. 1973. Mon. Not. R. Astron. Soc. **161**:269.
Moffet, A.T. 1975. Stars and Stellar Systems, Vol. 9. Chicago: University of Chicago Press, p. 211.
Nousek, J.A., L.L. Cowie, E. Hu, C.J. Lindblad, and G.P. Garmire. 1981. Astrophys. J. **24**:152.
Novaco, J.C., and L.W. Brown. 1978. Astrophys. J. **221**:114.
Pacholczyk, A.G. 1970. Radio Astrophysics. San Francisco: W.H. Freeman.
Paul, J., M. Casse, and C.J. Cesarsky. 1976. Astrophys. J. **207**:62.
Penzias, A.A., and R.W. Wilson. 1966. Astrophys. J. **146**:666.
Phillipps, S., S. Kearsey, J.L. Osborne, C.G.T. Haslam, and H. Stoffel. 1981a. Astron. Astrophys. **98**:286.
Phillipps, S., S. Kearsey, J.L. Osborne, C.G.T. Haslam, and H. Stoffel. 1981b. Astron. Astrophys. **103**:405.
Piddington, J.H. 1951. Mon. Not. R. Astron. Soc. **111**:45.
Piddington, J.H., and H.C. Minnett. 1952. Aust. J. Sci. Res. **5A**:17.
Quigley, M.J.S., and C.G.T. Haslam. 1965. Nature **208**:741.
Reber, G. 1940a. Astrophys. J. **91**:621.
Reber, G. 1940b. Proc. Inst. Rad. Eng. **28**:68.
Reber, G. 1944. Astrophys. J. **100**:279.
Reich, P., and W. Reich. 1986. Astron. Astrophys. Suppl. **63**:205.
Reich, W. 1982. Astron. Astrophys. Suppl. **48**:219.
Reich, W., E. Furst, P. Steffen, K. Reif, and C.G.T. Haslam. 1985. Astron. Astrophys. Suppl. **58**:197.
Rockstroh, J.M., and W.R. Webber. 1978. Astrophys. J. **224**:677.
Scheuer, P.A.G., and M. Ryle. 1953. Mon. Not. R. Astron. Soc. **113**:3.
Shklovsky, I.S. 1952. Astron. Zh. **29**:418.

Shklovsky, I.S. 1953. Dokl. Akad. Nauk SSSR **90**:983.

Simard-Normandin, M., and P.O. Kronberg. 1980. Astrophys. J. **242**:74.

Simon, A.J.B. 1977. Mon. Not. R. Astron. Soc. **180**:429.

Sironi, G. 1974. Mon. Not. R. Astron. Soc. **166**:345.

Sofue, Y., and W. Reich. 1979. Astron. Astrophys. Suppl. **38**:251.

Sofue, Y., H. Hirabayashi, K. Akabane, M. Inoue, T. Handa, and N. Nakai. 1984. Publ. Astron. Soc. Jpn. **36**:287.

Spoelstra, T.A.T. 1972. Astron. Astrophys. **21**:61.

Spoelstra, T.A.T. 1984. Astron. Astrophys. **135**:238.

Sullivan, W.T. 1978. Sky & Telescope **56**:101.

Tabara, H., and M. Inoue. 1980. Astron. Astrophys. Suppl. **39**:379.

Tang, K.-K. 1984. Astrophys. J. **278**:881.

Vallee, J.P. 1983. Astrophys. Lett. **23**:85.

Weaver, H. 1979. Inst. Astron. Union, Symp. **84**:295.

Webber, W.R., G.A. Simpson, and H.V. Cane. 1980. Astrophys. J. **236**:448.

Webster, A.S. 1974. Mon. Not. R. Astron. Soc. **166**:355.

Webster, A.S. 1975. Mon. Not. R. Astron. Soc. **171**:243.

Webster, A.S. 1978. Mon. Not. R. Astron. Soc. **185**:507.

Wendker, H.J. 1984. Astron. Astrophys. Suppl. **58**:291.

Westerhout, G. 1958. Bull. Astron. Inst. Neth. **14**:215.

Westerhout, G., and J.H. Oort. 1951. Bull. Astron. Inst. Neth. **11**:323.

Westerhout, G., C.L. Seeger, W.N. Brouw, and J. Tinbergen. 1962. Bull. Astron. Inst. Neth. **16**:187.

Whipple, F.L., and J.L. Greenstein. 1937. Proc. Natl. Acad. Sci. USA **23**:117.

Wilkinson, A., and F.G. Smith. 1974. Mon. Not. R. Astron. Soc. **167**:593.

Williams, P.J.S., S. Kenderdine, and J.E. Baldwin. 1966. Mem. R. Astron. Soc. **70**:53.

2. HII Regions and Radio Recombination Lines

MARK A. GORDON

2.1 Ionized Interstellar Hydrogen

2.1.1 Emission Nebulae

The hot clouds of ionized gas surrounding bright stars are known as emission nebulae, because their spectra exhibit emission lines. These are often classified further into bright diffuse nebulae, supernova remnants, and planetary nebulae. Diffuse nebulae are associated with the birth of stars whereas supernova remnants and planetary nebulae are associated with the death of stars. Diffuse nebulae have irregular shapes, often filaments, and lie in the plane of galaxies adjacent to the dark molecular clouds from which they formed. Supernova remnants are the remains of exploded stars, a catastrophic process associated with perhaps only about one percent of star deaths. Planetary nebulae are the extended atmospheres of dying, older stars and are found throughout the sphere in which a galaxy originally formed. All emit detectable radio emission.

The energy sources for emission nebulae are their central stars. Ultraviolet photons from these stars ionize the surrounding gas. For example, those with wavelengths shorter than 1102 Å can ionize carbon; with wavelengths shorter than 912 Å, hydrogen; with wavelengths shorter than 504 Å, helium. The newly created free electrons and nuclei then either recombine and emit new photons or they heat the gas by colliding with other atoms. By these processes, the radiant energy emitted by the star is transferred to the surrounding gas from which the stars were formed. These processes cause the surrounding gas to fluoresce.

The radiation field in a nebula weakens with increasing distance from the central stars. The spatial density of the ionizing photons decreases because of the spherical geometry of the nebula. Simply, there are fewer of them available to ionize the gas in each unit of volume. When the electrons and the ions recombine to emit new photons, the new photons can be emitted in any direction, this recombination process thereby *scattering* and diluting the original radiation field of the stars. Also, the energy radiated in the recombination process may be emitted in the form of two lower-energy photons rather than a single photon with the original ionizing energy. This process is called *incoherent* scattering. All of these processes weaken the ionizing ability of the radiation field and lead to an ionization boundary to the nebulae.

The shape of the diffuse nebula depends upon the distribution of the surrounding cold gas; the size, upon the amount of energy radiated by the central stars. If the radiation field is still strong at the edge of the cold cloud surrounding the newly formed stars, the photons escape into the the interstellar medium, and the nebulae are said to be density-bounded in these directions.

Nebulae emit both continuum and line emission in the radio region of the electromagnetic spectrum. The free electrons and ions emit and absorb continuum radiation detectable by radio telescopes. When they recombine, a small percentage of the resulting atoms will be in highly excited states. A few of the bound electrons will jump to adjacent lower states, emitting photons in the radio spectrum which are detectable as extremely weak spectral lines. Largely unaffected by interstellar dust, radio waves from optically obscured nebulae can be detected by astronomers.

This chapter will describe the characteristics of diffuse nebulae in the radio spectrum. In most diffuse nebulae, nearly all of the hydrogen is ionized. Such regions are also called HII regions. The relevant physics is the transfer of radio radiation through fully ionized plasmas, as well as through the bound levels of the atoms.

2.1.2 Diffuse Thermal Background

In addition to the discrete concentrations of ionized gas, there are spatially extended thermal sources of radio radiation. Chapter 3 describes the 21-cm emission of neutral hydrogen. There are also extended sources of other line emission and of free-free continuum emission, which can be detected by the dispersion of pulsar bursts, by the absorption of radiation from background nonthermal sources, and by line emission from highly excited states of hydrogen. The characteristics of this emission vary widely over our galaxy.

2.2 Radiation Transfer

To interpret radio observations quantitatively, it is necessary to understand how radiation travels through a medium. As a first step, we consider loss and gain mechanisms in the direction of the observer without regard to the details of atomic processes making up these mechanisms. We also assume all processes to be stationary, i.e., all parameters are time independent over the time scale of our observations.

Consider the situation sketched in Figure 2.1. The differential intensity dI contributed from an elemental volume of path length dx is

$$dI = \underbrace{-I\kappa\,dx}_{\text{absorption}} + \underbrace{j\,dx}_{\text{emission}} \qquad (2.1)$$

where κ is the linear absorption coefficient and j is the linear emission coefficient in the direction of observer. Integrating from the far side of the nebula toward the radio telescope, we find the observed specific intensity to be

$$I = \underbrace{I(0)e^{-\langle\kappa\rangle L}}_{\substack{\text{attenuated}\\\text{background}}} + \underbrace{\int_0^L \frac{j}{\kappa}e^{\kappa(x-K)}\kappa\,dx}_{\substack{\text{contribution}\\\text{from the nebula}}} + \underbrace{I(x > L)}_{\text{foreground}} \qquad (2.2)$$

Fig. 2.1. Background radiation $I(0)$ travels through the gas cloud, where it is either strengthened or weakened, toward the radio telescope.

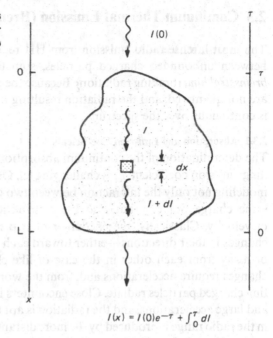

$$I(x) = I(0)e^{-\tau} + \int_0^\tau dI$$

where L is the near edge of the nebula. The parameter $\langle \kappa \rangle$ is the mean of the absorption coefficient over the physical length of the nebula. The radiation received by the radio telescope is the sum of (1) the attenuated background radiation, (2) the contribution from the nebula, and (3) the contribution of any foreground sources. Equation (2.2) is perhaps the most basic equation in astrophysics because it relates the physical processes in a medium, as described by κ and j, to the radiation intensity measured by the telescope. The remainder of this chapter is concerned with the evaluation of j and κ and their effects upon the observed radiation field.

In practice, it is convenient to lump κ and L together by defining a parameter, τ, called *optical depth*:

$$\tau \equiv \int_{x_1}^{x_2} \kappa \, dx \cong \langle \kappa \rangle (x_2 - x_1) . \tag{2.3}$$

In a sense the reciprocal of optical depth may be considered to be the probability of a photon at point x_1 reaching point x_2. It is useful because it reduces the number of unknowns in Equation (2.3) by combining $\langle \kappa \rangle$ and $(x_2 - x_1)$, and it is likely to be a direct observable. Note that, as Figure 2.1 shows, τ increases as the distance the photon travels, x, decreases. For our purpose, we will assume the near limit of the path to be the telescope itself, $x_2 = 0$, and designate optical path by the most distant point in the path, i.e., $\tau = \tau(x_1)$. In this notation the transfer equation given by Equation (2.2) becomes

$$I(x) = I(0)e^{-\tau(0)} + \int_0^{\tau(0)} \frac{j}{\kappa} e^{-t} dt + I(x > L) . \tag{2.4}$$

2.3 Continuum Thermal Emission (Bremsstrahlung)

The most intense radio emission from HII regions is that caused by interactions between unbounded charged particles, sometimes called free-free radiation or *bremsstrahlung* (braking radiation). Because the particles are free, their energy states are not quantized, and the radiation resulting from changes in their kinetic energy is continuous over the spectrum.

2.3.1 Absorption and Emission Coefficients

The determination of the continuum absorption coefficient, κ_c, and the corresponding emission coefficient, j_c, is challenging (cf. Oster 1961). The calculations require modeling not only the interaction between two charged particles, not always of the same charge polarity, but also the distribution of the particles as a function of velocity. Classically, the encounter of two moving charged particles involves changes in their directions—either toward each other for unlike-charge encounters or away from each other in the case of like-charge encounters. These direction changes require accelerations and, from the work of Hertz, we know that accelerating charged particles radiate. Close encounters involve substantial Coulomb forces and large accelerations, and the radiation is apt to be in the X-ray range. Radiation in the radio range is produced by the more distant encounters where Coulomb forces are small, and the particles can be considered to continue traveling in a straight line. In any case, the emission coefficient is determined by integrating the emission produced during each encounter, over a velocity distribution of particles (usually Maxwellian).

Some approximations are required to make this integration. The free electrons tend to shield the electric field of the ions, and thus the force field is effective only over some finite distance. Furthermore, it is usual to assume (1) that the energy radiated is small compared to the kinetic energy of the electron moving past the ion and (2) that the reciprocal of the radiated frequency is small compared to the time for the electron to undergo a 90° deflection. Physically, these assumptions imply that the electron-ion encounter is nearly adiabatic and that the period of the emitted wave train is short compared to the duration of the encounter.

With these assumptions, an expression for the free-free absorption coefficient in the radio domain is

$$\kappa_v = \left(\frac{N_e N_i}{v^2}\right) \cdot \left(\frac{8Z^2 e^6}{3\sqrt{3}m^3 c}\right) \cdot \left(\frac{\pi^{1/2}}{2}\right) \cdot \left(\frac{m^{3/2}}{kT}\right) \langle g \rangle \tag{2.5}$$

where N_e and N_i are the number densities of electrons and ions, v the wave frequency, Z the ion charge, e the electronic charge, m the electronic mass, c the speed of light, k Boltzmann's constant, T the kinetic temperature, and $\langle g \rangle$ the Gaunt factor averaged over a Maxwellian velocity distribution. All units are CGS. For temperatures less than 892,000 K,

$$\langle g \rangle \cong \frac{\sqrt{3}}{\pi} \ln\left[\left(\frac{2kT}{\delta m}\right)^{3/2} \cdot \frac{m}{\pi \delta Z e^2 v}\right] \tag{2.6}$$

and for temperatures greater than 892,000 K,

$$\langle g \rangle \cong \frac{\sqrt{3}}{\pi} \ln \left[\frac{2kT}{\pi v h \delta} \right] \qquad (2.7)$$

where δ is Euler's constant in the form $\exp(0.577)$ and the temperature being the division between classical and quantum-mechanical regimes.

Under some conditions of low frequencies and low temperatures, Equation (2.5) no longer holds, because the basic assumptions (1) and (2) above are violated. In this case, the Gaunt factors must be evaluated for each particular case. Oster (1970) has done this for a range of low temperatures and wave frequencies.

There is an approximation (Altenhoff et al. 1960) to Equation (2.5) which is often used in the analyses of radio observations of HII regions because of its simplicity:

$$\kappa_c \cong \frac{0.08235 N_e N_i}{v^{2.1} T_e^{1.35}} \qquad (2.8)$$

where v is in units of GHz, N is in units of cm^{-3}, and κ_c is in pc^{-1}. For densities and temperatures encountered in HII regions this approximation is accurate to within 5%. Mezger and Henderson (1967) tabulate a correction which improves the approximation.

Having the free-free absorption coefficient, we can calculate the emission coefficient by

$$j_v = \kappa_v B_v(T_e) \qquad (2.9)$$

where $B_v(T_e)$ is the Planck function which, if $hv < kT_e$, can be approximated by $2kT_e v^2/c^2$, the well-known Rayleigh-Jeans approximation.

2.3.2 Transfer Equation for Continuum Radiation

The substitution of the Rayleigh-Jeans approximation for the ratio j/κ and of the continuum optical depth τ_c are all that is needed to prepare Equation (2.4) to describe the observed free-free emission of HII regions in the radio range as a function of frequency

$$I_c(x) = I(0)e^{-\tau_c(0)} + \int_0^{\tau_c(0)} \frac{2kT_e v^2}{c^2} e^{-t} dt + I(x > L) \qquad (2.10)$$

where $\tau_c = \int \kappa_c dx$.

For simplicity, let us assume the nebula to be homogeneous in density and temperature and the foreground emission to be zero. Then Equation (2.10) can be evaluated by moving the integrand out from under the integral, and

$$I_c(x) = I(0)e^{-\tau_c(0)} + \frac{2kT_e v^2}{c^2} (1 - e^{-\tau_c(0)}) \qquad (2.11)$$

should describe the intensity of free-free emission of the nebulae observed at the radio telescope.

In practice, the radio astronomer uses units of temperature to measure the intensity of radiation. The actual *antenna temperature*, T_A, is less than the brightness temperature, T, by an efficiency factor, ε, and a beam dilution factor, W, defined as the ratio of the solid angle subtend by the ratio source to that of the radio beam.

$$I = \frac{2kT\nu^2}{c^2} = \frac{2kT_A\nu^2}{c^2}\frac{1}{\varepsilon W} \qquad (2.12)$$

Equation (2.11) becomes

$$T(x) = T(0)e^{-\tau_c(0)} + T_e(1 - e^{-\tau_c(0)}) , \qquad (2.13)$$

which can be compared directly with observations.

2.3.3 Low-Frequency Radiation from HII Regions
At low frequencies the frequency dependence of the free-free absorption coefficient given by Equation (2.8) causes τ_c to become much greater than unity for most HII regions. Here Equation (2.13) becomes

$$T(x) \cong T(0)e^{-\tau_c} + T_e . \qquad (2.14)$$

In the direction of the galactic plane, the background emission is mainly nonthermal and, as such, is intense at low frequencies. Moving the beam of the radio telescope across an HII region lying in this direction causes a decrease in the observed brightness temperature $T(x)$ because of absorption by the HII region. If the optical depth is sufficiently large, the background contribution to $T(x)$ is negligible, and one is able to measure the electron temperature of the HII region directly after correction for foreground emission, if any.

2.3.4 High-Frequency Radiation from HII Regions
At high frequencies the nonthermal radiation of the galactic background becomes negligible compared to the thermal radiation of the HII region. The ν^{-2} dependence makes $\tau_c < 1$ at high frequencies, and Equation (2.13) becomes

$$T(x) \cong T(0) + T_e\tau_c , \qquad (2.15)$$

which shows the antenna temperature to increase as the beam is scanned over the source.

2.3.5 Integrated Observations of HII Regions
It is often difficult to identify galactic HII regions lying at large distances from the Sun. Obscuration by interstellar gas and dust prevents the use of the tradiational optical emission lines of OII, OIII, and NII as diagnostics. What is needed is an identification technique which is insensitive to optical extinction. Fortunately, HII regions exhibit a unique variation of their radio emission as a function of frequency.

At any given frequency, the total spectral flux density, S, of an HII region is defined by the equation

$$S = \int_{source} I_\nu d\Omega \cong \frac{2k\nu^2}{c^2}\int T_{source} d\Omega \qquad (2.16)$$

where Ω is the solid angle subtended by the source. At frequencies where the beam is less than the source, the total spectral flux density is measured by mapping the source and performing the integration in Equation (2.16).

The basic transfer equation, Equation (2.11), can be rewritten in terms of S:

$$S(x) = S(0)e^{-\tau_c(0)} + S(1 - e^{-\tau_c(0)}) . \qquad (2.17)$$

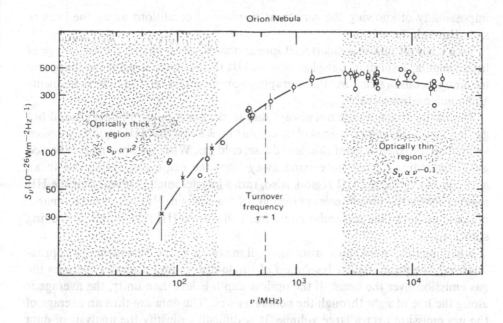

Fig. 2.2. Spectral flux density of the Orion Nebula plotted against frequency. The shaded regions mark the optically thick and thin regions of the spectrum. (Reprinted with permission by Gordon and Breach Science Publishers from: Terzian, Y. and Parrish A., *Astrophysical Letters*. Vol. 5(1970), pp. 261.

Neglecting the background $S(0)$, we can investigate the spectral flux density of an HII region as a function of frequency in the radio domain. At high frequencies where the gas is optically thin,

$$S(x) \cong S\tau_c \qquad \tau_c \ll 1 . \tag{2.18}$$

But $S \propto \nu^2$ and $\tau_c \propto \nu^{-2}$, and thus at high frequencies $S(x)$ is independent of frequency. [The Gaunt factor of Equation (2.6) makes $\tau_c \propto \nu^{-2.1}$, and thus $S(x)$ is really a weak function of frequency.]

Figure 2.2 shows a plot of S versus ν for the Orion Nebula. For frequencies less than 200 MHz, $S(x)$ varies as ν^2, as we should expect from Equations (2.16) and (2.17). For frequencies exceeding 2000 MHz, $S(x)$ varies as $\nu^{-0.1}$, as we would expect from Equations (2.8), (2.16), and (2.18).

The intermediate-frequency range is often known as the "turnover" range. In practice, it is usual to describe a thermal spectrum in terms of its *turnover frequency*, this being the frequency at which $\tau_c = 1$. (Some authors use $\tau_c = 1.5$.) Neglecting the background contribution, we use Equation (2.17) to calculate the flux decrease at the turnover frequency to be

$$S(x)_T = S(1 - e^{-1}) = 0.632S . \tag{2.19}$$

2.3.6 Reality and the "Homogeneous Nebula" Approximation

Traditionally, radio astronomers assumed that HII regions were homogeneous in both density and temperature. The justification for this "homogeneous nebula" approximation lay in the poor angular resolution of the telescopes and in the impossibility of knowing the variations in physical conditions along the lines of sight through the nebula.

Such a simple model is incorrect. Optical observations, having the advantage of better angular resolution, have long shown HII regions to be irregular in shape and inhomogeneous in structure. Photographs show most HII regions to have filaments and condensations of hot gas.

The irregular shapes are not always due to the obscuration of background hot gas by dust. Radio observations of molecular lines show that HII regions are often located on the edges of giant clouds of dense, cold gas. When a star forms, the heated gas surrounding it tends to expand away from the dense gas, giving rise to an asymmetrically shaped HII region. Also, radio interferometers reveal that the HII regions have structure on scales of seconds of arc, presumably the ionized remnants of dense eddies of the molecular cloud from which the HII region and its exciting stars formed.

Knowing this, should an astronomer still make use of the homogeneous approximation? The poor angular resolution of single-dish radio telescopes averages the gas emission over the beam. If the optical depth is less than unity, the average is along the line of sight through the nebula as well. The data are thus an average of the gas emission over a large volume. It is difficult to justify the analysis of data with a model which is substantially more complicated than can be tested by observation. Furthermore, distant HII regions are often obscured by the intervening dust and gas of the galactic plane. Optical observations cannot assist in the making of models for these nebulae, and their great distance from Earth gives them angular sizes which fall within the beam of the radio telescope. In these conditions, the homogeneous approximation is reasonable, providing that its limitations are kept in mind.

Subject to the uniformity assumptions, we can fit the theroretical spectrum given by Equation (2.17) to observations. The technique is to measure the flux density of the continuum emission at two different frequencies, preferably one in the optically thin region and one in the optically thick region. This procedure gives a system of two equations and two unknowns, T_e and N_e—providing that we know the physical size of the source (Wade 1958). Variations of this method involve procedures such as fitting the entire flux density spectrum by a least-squares fit (cf. Terzian et al. 1968) or the specification of the turnover frequency and the peak flux density.

In practice, the existence of temperature gradients within the HII region and its irregular shape will produce misleading results. The low-frequency flux comes from the anterior region of the HII region, whereas the high-frequency flux is an average along a line of sight through the HII region. The geometrical shape of the HII region will also influence the way radiation is transferred to the observer. Fitting the idealized Equation (2.17) to observed fluxes radiated from regions of different temperatures and path lengths will lead to values of mean temperature and density for the HII region which may not have much physical meaning.

2.3.7 Continuum Observations of HII Regions

Figure 2.3 shows a map of the radio emission from the Orion Nebula made at 23 GHz, a frequency at which the gas is optically thin. The angular resolution is 42 arcseconds, which corresponds to 0.10 pc at the distance of the nebula.

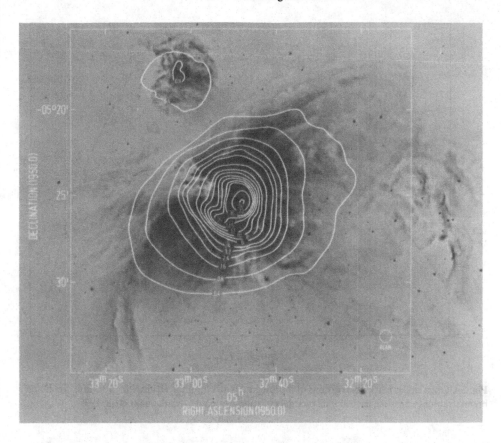

Fig. 2.3. The 23-GHz radio continuum contours, in units of main-beam brightness temperature, on an optical photo in Hα and [NII] of NGC 1976 (Orion A, M42), below, and NGC 1982 (M43), above. The angular resolution is 42″, which at the distance of Orion A, corresponds to a linear resolution of 0.10 pc. (Wilson and Pauls, 1984).

For comparison, the radio map is shown superimposed upon an optical photograph of the nebula. The two images correspond well, given differences in the angular resolution of the telescopes. One difference between the two images is the obscuration by the dust lane in the northeast corner of NGC 1976, which appears only faintly in the radio contours.

2.3.8 Thermal Emission from Dust

Embedded in the interstellar medium are dust grains, which therefore become part of the HII regions. At wavelengths longer than 3 mm, the continuum emission from the ionized gas dominates the radio emission. But, at shorter wavelengths, the emission from the dust begins to appear, rising well above the bremsstrahlung, reaching a maximum near 100 μm. Figure 2.4 shows this effect. The local radiation field of the HII region heats the dust grains to an equilibrium temperature depending upon their albedo, absorptivity, and emissivity. Usually this warm dust is associated with a dense region within the interstellar gas adjacent to the HII region, rather than mixed with the ionized gas of the nebula. Except for very short

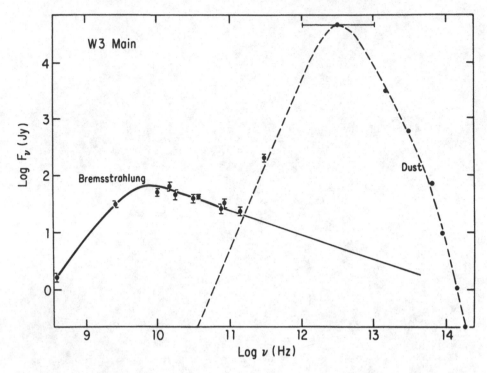

Fig. 2.4. Open circles mark observations of the integrated flux from the HII region W3 Main. Solid line: bremsstrahlung emission; dashed line: thermal emission from dust. (Malkamäki et al., 1979).

wavelengths, this emission is generally not observed for HII regions in the radio regime and will not be discussed further in this chapter.

2.4 Basic Theory of Radio Recombination Lines

In an ionized gas an electron and ion may pass close enough to each other that, with a suitable release of energy, they become bound, with the electron in an energy level having a large principal quantum number. Most of these newly bound electrons immediately jump to the ground state, releasing energy in emission lines known as *resonance lines.* In a few cases the electrons cascade downwards from level to level, releasing their energy in a series of lines known, perhaps misleadingly, as *recombination lines.* At any instant the number of electrons in these highly excited levels is approximately 10^{-5} of those in the ground state. In 1959, Kardashev suggested that these lines might be detectable at radio wavelengths.

Astrophysically, recombination lines are very important. The ratio of the energy radiated in the line to that in the underlying free-free continuum is a kind of partition function, giving the ratio of bound to free electrons and, hence, a measure of the gas temperature. Under some conditions the line can also be broadened by pressure effects, giving a measure of the gas density. In addition, Doppler effects in the form of the line shapes and center frequencies give information about the velocity fields of the gas.

2.4.1 Line Frequencies

Calculation of the line frequencies is simple. Levels having large principal quantum numbers differ little in energy. The line emission between adjacent levels appears in the radio region of the electromagnetic spectrum. While the spin and orbital angular momentum of the bound electrons do interact through Russell-Saunders coupling, the resulting fine structure has little effect on the overall line frequency in practice. The Doppler broadening is many hundreds of times larger than the fine-structure splitting throughout the radio range. We need only consider transitions between energy levels which are degenerate in their orbital angular momentum and are thus completely designated by their principal quantum numbers.

For large quantum states, the electric fields of hydrogenic atoms have only radial components described fully by Coulomb potentials. The well-known Rydberg formula adequately gives the line frequencies:

$$v = RcZ^2 \left[\frac{1}{n^2} - \frac{1}{(n + \Delta n)^2} \right] \tag{2.20}$$

where R is the Rydberg constant for the hydrogenic atom emitting the line, c the speed of light, Z the effective charge of the nucleus in units of the electronic charge e, n the lower principal quantum number, and Δn the change in n. For recombinations to singly ionized atoms, $Z = 1$; to doubly ionized atoms, $Z = 2$, etc. The Rydberg constant is a function of mass,

$$R = R_\infty \left(1 - \frac{m}{M} \right) \tag{2.21}$$

where R_∞ is the constant for infinite mass, m the electronic mass, and M the total mass (including the electrons) of the atomic species involved.

Numerically, combining Equations (2.20) and (2.21) and substituting the appropriate constants gives the rest frequency to five significant figures,

$$v(\text{GHz}) = 3.2898 \times 10^6 \left(1 - \frac{5.4876 \times 10^4}{\text{AMU}} \right) \times \left(\frac{1}{n^2} - \frac{1}{(n + \Delta n)^2} \right) Z^2 \tag{2.22}$$

where AMU is the atomic mass unit for the species.

As can be seen from Equation (2.20), recombination lines appear throughout the spectrum. In the radio domain where $\Delta n \ll n$,

$$v(n) \cong 2RcZ^2 \frac{\Delta n}{n^3} \tag{2.23}$$

and the approximate frequency separation between adjacent lines of the same frequency is

$$\Delta v = v(n) - v(n + 1)$$

$$\cong 6RcZ^2 \frac{\Delta n}{n^4}$$

$$\cong \frac{3v}{n} . \tag{2.24}$$

Table 2.1. Frequencies and frequency separations of α transitions.[a]

Principal quantum number of lower term, n	Line frequency (MHz)	Line wavelength	Frequency separation (MHz)
13	3×10^6	100 μm	690,000
28	3×10^5	1 mm	32,000
60	3×10^4	1 cm	1,500
130	3×10^3	10 cm	69
280	3×10^2	1 m	3.2
600	3×10	10 m	0.15

[a] After Kardashev (1959).

Each transition is identified by its lower principal quantum number n and its change Δn. For example, if $\Delta n = 1$, then the transition is called $n\alpha$; $\Delta n = 2$, $n\beta$; etc. The transition is further identified by its atomic species, viz., H157α, He109β. At a given frequency, the most intense transitions are those of the H$n\alpha$ type because of the great abundance of hydrogen and the large oscillator strengths of α transitions. Table 2.1 lists the frequencies and frequency separations for α transitions. As can be seen in Equation (2.24), the relative frequency separation decreases as the fourth power of quantum number.

How many quantized levels can an atom have in an HII region? There is some value of the principal quantum number, n_q, beyond which no bound levels exist. Clearly, no bound level can exist if the interval between depopulation processes is less than the orbital period of the bound electron (Brocklehurst and Seaton 1972). Equating the sum of all radiative and collisional rates out of a principal quantum level to the orbital frequency of a bound electron in that level, Shaver (1975) calculates the limiting level to be

$$n_q \cong \left[\frac{10^{20} T_e^{1.5}}{N_e} \exp\left(\frac{26}{T_e^{1/3}} \right) \right]^{1/8.2}. \qquad (2.25)$$

If we choose conditions appropriate for an HII region, $T_e = 10^4$ K and $N_e = 10^3$ cm^{-3}, Equation (2.25) gives $n_q \cong 740$. The rest frequency of the lowest-frequency recombination line from this source, the H739α line, is 16.3 MHz, according to Equation (2.22).

2.4.2 Shape Function

Observations of HII regions involve measurements of gas having a Maxwell-Boltzmann velocity distribution (we exclude large-scale turbulence for the moment). In the absence of magnetic fields, an HII region having an electron temperature of 10^4 K and an electron density of 10^2 cm^{-3} will thermalize in minutes following any perturbation in the velocity distribution of the charged particles. For larger electron densities, thermalization occurs in even shorter times.

Because the gas cloud can be considered to be composed of particles having a Maxwellian velocity distribution, the number of atoms with velocity components between v_x and $v_x + dv_x$ along the line of sight is

$$N(v_x) \, dv_x = N \sqrt{\frac{M}{2\pi kT}} e^{-Mv_x^2/2kT} \, dv_x \tag{2.26}$$

where N is the total number of particles and M in the mass of particles of that species. Using the classical Doppler formula* for the observed frequency

$$v = v_0 \left(1 - \frac{v_x}{c}\right) \tag{2.27}$$

and differentiating,

$$dv_x = -c \cdot \frac{dv}{v_0} \tag{2.28}$$

we convert Equation (2.26) into the specific intensity of the line, $I(v)$, on the assumption that the integrated intensity I is proportional to the total number of emitters N and that $dI(v)$ is proportional to $dN(v_x)$, that is, that the emitting gas is optically thin:

$$I(v) \, dv = I \sqrt{\frac{4\ln 2}{\pi}} \cdot \frac{1}{\Delta v} \exp\left[-4\ln 2 \left(\frac{v_0 - v}{\Delta v}\right)^2\right] dv \tag{2.29}$$

where v_0 is the line rest frequency and the Doppler width Δv is defined to be the full width of the line at half-intensity,

$$\Delta v \equiv \left(4\ln 2 \cdot \frac{2kT}{Mc^2}\right)^{1/2} v_0 \tag{2.30}$$

or, numerically,

$$\Delta v = 7.1634 \times 10^7 \left(\frac{T}{\text{AMU}}\right)^{1/2} v_0 \tag{2.31}$$

when T is in kelvins.

Equation (2.29) tells us that the emitted line will have a Gaussian shape when the gas is optically thin, as Figure 2.5 illustrates. It is convenient to define a line shape function $f(v)$ such that

$$I(v) \, dv \equiv I f(v) \, dv \; . \tag{2.32}$$

Because of its definition in Equation (2.32), $\int f(v) \, dv = 1$.

In this derivation, we have neglected three mechanisms which often distort the shapes of radio recombination lines emitted from HII regions. Large-scale velocities within the HII region, pressure braodening, and large optical depths can make the line shapes asymmetrical. Observed with the angular resolutions common to most filled-aperture telescopes, radio recombination lines remain remarkably Gaussian, however.

---

* Radio astronomical convention only in which the *frequency* shift is assumed to be proportional to the source velocity along the line of sight. The optical convention assumes the *wavelength* shift to be proportional to the source velocity.

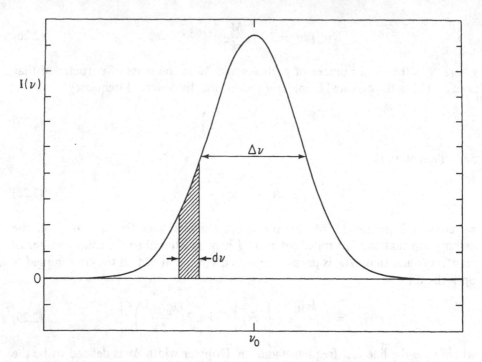

Fig. 2.5. A hypothetical Gaussian line shape described by Equation (2.29).

HII regions usually have large-scale velocity fields. The molecular clouds from which the young stars formed have density gradients, necessary to the process of star formation. The expansion of the newly ionized, hot gas against these gradients gives rise to asymmetrical flows of the hot gas. The mean velocities of small-scale eddies within the angular resolution of the beam are the same as the average velocity of the gas. Except for its independence of mass, this *microturbulence* acts similarly to the microvelocities of the Maxwellian distribution of the gas and broadens the Gaussian line. Large-scale flows or *macroturbulence*, on the other hand, themselves radiate Gaussian lines, which add together at the velocity offsets of the flows to produce an asymmetrical line profile for the HII region as a whole. Equation (2.29) can model microturbulence but not macroturbulence, which is discussed more fully in Section 2.5.4.

Pressure broadening will also produce non-Guassian line profiles. Such line profiles are a blend of Gaussian cores and Lorentzian wings. The total power emitted in a pressure-broadened line is the same as for a purely Gaussian profile, except at frequencies involving large opacities. Pressure broadening is discussed in more detail in Section 2.5.4.

In principle, the optical depth at the center of an emission line can be large enough to distort the line profile. Generally this effect is not important for radio recombination lines.

2.4.3 Absorption and Emission Coefficients
As with free-free emission, the calculation of absorption and emission coefficients is the major step necessary for the prediction of the line intensities as a function of

physical conditions in the gas. Absorption coefficients are additive. Since we have calculated κ_c in Equation (2.5), we need only calculate κ_L to determine the total absorption coefficient in the frequency range containing the recombination line.

We consider two energy levels designated by principal quantum numbers n' and n. Radiation interacts with such levels in two basic ways. First, an electron in the upper state n' may spontaneously radiate and jump to the lower state n. This type of process is called *spontaneous emission*, and its probability (s^{-1}) is designated $A_{n'n}$. Second, the electron may be induced to radiate by the ambient radiation field. This type of process is called *stimulated emission*, and its probability is designated $IB_{n'n}$. We will consider the stimulated emission term to be negative absorption, and thus the line absorption coefficient is

$$\kappa_L = \frac{h\nu}{4\pi} f(\nu)(N_n B_{nn'} - N_{n'} B_{n'n}) \ . \tag{2.33}$$

Here we assume the emission and absorption line profiles to be identical. The parameter N_i is the number of atoms having electrons in level i, which is given by the Boltzmann formula

$$\frac{N_{n'}}{N_n} = \frac{\omega_{n'}}{\omega_n} e^{(-h\nu/kT_{ex})} \tag{2.34}$$

where T_{ex} is the excitation temperature of the levels n' and n, or in terms of the electron temperature, T_e,

$$\frac{N_{n'}}{N_n} = \frac{b_{n'}}{b_n} \frac{\omega_{n'}}{\omega_n} e^{(-h\nu/kT_e)} . \tag{2.35}$$

Here, the departure coefficient b_i is the ratio of the actual number of atoms having electrons in level i to the number which would be there if the populations were in thermodynamic equilibrium at the electron temperature T_e. The ω_i is the statistical weight of level i. Using the known relationships between the transition probabilities and the above equations, we transform Equation (2.31) into

$$\kappa_L = \frac{h\nu}{4\pi} \cdot f(\nu) \cdot N_n B_{nn'} \left[1 - \frac{b_{n'}}{b_n} e^{(-h\nu/kT_e)} \right] . \tag{2.36}$$

For our applications, $h\nu < kT_e$, and we simplify Equation (2.36) to

$$\kappa_L \cong \kappa_L^* b_{n'} \left(1 - \frac{d \ln b_{n'}}{dn} \cdot \Delta n \frac{kT_e}{h\nu} \right) \tag{2.37}$$

where Δn is the change in principal quantum number. $d \ln b_{n'}/dn$ is the slope of the function b_n at the point $b_{n'}$, and κ_L^* is the linear absorption coefficient of the line under conditions of thermodynamic equilibrium,

$$\kappa_L^* = \frac{h^2 \nu^2}{4\pi kT_e} f(\nu) \cdot N_n^* B_{nn'} \qquad h\nu \ll kT_e \tag{2.38}$$

where N_n^* is the thermodynamic equilibrium (TE) population of level n. It is convenient to rewrite the Einstein coefficient $B_{nn'}$ in terms of the oscillator strength $f_{nn'}$:

$$\kappa_L^* = 1.070 \times 10^7 \Delta n \cdot \frac{f_{nn'}}{n} N_e N_i T_e^{-2.5} \exp\left(\frac{E}{kT_e}\right) \cdot f(v) \qquad hv < kT_e \qquad (2.39)$$

where we have also related the population N_n to those of the unbound states by means of the Saha-Boltzmann equation. Equation (2.39) has the population densities measured in cm^{-3}, T_e in K, and κ_L^* in pc^{-1}, and E is the energy of the upper quantum level. For hydrogen, $E/kT_e = (1.579 \times 10^5)/n^2/T_e$.

Oscillator strengths for quantum levels appropriate to radio recombination lines have been tabulated by Goldwire (1969) and by Menzel (1970). Menzel (1968) suggested the useful algorithm

$$f_{nn'} \cong nM\left(1 + 1.5\frac{\Delta n}{n}\right) \qquad (2.40)$$

where $M = 0.190775, 0.26332, 0.0081056$, and 0.0034918 for $\Delta n = 1, 2, 3$, and 4, respectively.

The line-emission coefficient can be related to κ_L simply by changing the definition

$$j_L \equiv N_{n'} A_{n'n} \frac{hv}{4\pi} f(v) \qquad (2.41)$$

to

$$j_L = \kappa_L^* b_{n'} B_v(T_e) \qquad (2.42)$$

by using the physical relationship between the Einstein transition probabilities, Equation (2.35), and the relationship between the Planck function, $B_v(T_e)$, and the linear emission and absorption coefficients as given by Equation (2.9).

2.4.4 Transfer Equation for Line Radiation

Following procedures used in Section 2.2, we can write that the intensity at some frequency within the recombination line is the sum of that in the underlying continuum (I_c) and that from the line (I_L):

$$I = I_L + I_c$$
$$= S[1 - e^{-(\tau_c + \tau_L)}] \qquad (2.43)$$

where the source function S is given by

$$S \equiv \frac{j_c + j_L}{\kappa_c + \kappa_L} . \qquad (2.44)$$

After substitution for the line-associated coefficients,

$$S = \frac{\kappa_c + \kappa_L^* b_{n'}}{\kappa_c + \kappa_L^* b_{n'} \gamma} B_v(T_e) \qquad (2.45)$$

where

$$\gamma = 1 - \frac{d \ln b_{n'}}{dn} \cdot \Delta n \cdot \frac{kT_e}{hv} . \qquad (2.46)$$

The line optical depth $\tau_L = \tau_L^* b_{n'} \gamma$, where the asterisk marks the value of the optical depth of the line under TE conditions.

Equation (2.43) provides a general description of the intensity of radiation emitted by a (homogeneous) thermal gas. At frequencies where no recombination lines can be detected, $\tau_L + \tau_c \cong \tau_c$, $S \cong B_v(T_e)$, and the equation describes the intensity of the free-free emission alone. If lines are emitted under TE conditions, γ and b_n are unity, $\tau_L = \tau_L^*$, and $S = B_v(T_e)$. On the other hand, if the gas is not in TE, then the b_n function must be determined from other calculations in order to predict the correct intensity of the radiation in the line. Such calculations are usually done assuming that any given level n is populated in statistical equilibrium, i.e., the sum of all processes populating a given level is equal to the sum of all processes depopulating that level. These calculations are a function of T_e and N_e, and have been performed for a wide range of astrophysical conditions by Brocklehurst (1970) and several others.

2.5 Refinements to the Transfer Equations

The general equations for the transfer of radiation, developed above, are applied to observations of radio recombination lines under either of two assumptions: local thermodynamic equilibrium (LTE) or statistical equilibrium (non-LTE). In either case, the most useful approach is to consider the ratio of the emission in the line to that in the underlying free-free continuum, thereby avoiding problems of absolute calibration of the intensity scale of the observations.

2.5.1 Local Thermodynamic Equilibrium

Physically, the term thermodynamic equilibrium (TE) describes a situation is which the energy exchange between the radiative and kinetic energy domains of a gas is so efficient that a single parameter, temperature, exactly describes the characteristics of both domains. While this situation cannot strictly occur in the open systems found in astronomy, there are localized situations which are so close that TE equations may be used without substantive quantitative errors. The term local thermodynamic equilibrium (LTE) describes these situations.

Spectroscopically, LTE refers to the circumstances in which the rates in and out of an energy level balance in detail, that is, the radiative rates into a level exactly equal the radiative rates out of that level, and similarly with the collisional rates. In LTE the departure coefficients b_n equal 1, by definition, and the line source function S equals the Planck function $B(T)$. Mathematically, LTE is the simplest spectroscopic case to deal with.

At frequencies in the line, the intensity is

$$I = I_L + I_c = B_v(T_e)[1 - e^{-(\tau_c + \tau_l^*)}] \tag{2.47}$$

and the intensity contributed by the line alone is

$$I_L = I - I_c = B_v(T_e)e^{-\tau_c(1 - e^{-\tau_l^*})} \tag{2.48}$$

$$\cong B(T_e)\tau_L^* \qquad \tau_L^*, \tau_c < 1 \tag{2.49}$$

where we have changed nomenclature by using I_L to indicate the specific intensity of the line component $I_L(v)$; I_c for $I_c(v)$, etc. Under the same conditions, $I_c \cong B_v(T_e)\tau_c$, and the ratio of the total energy emitted in the line to that emitted in the underlying continuum is

$$\int_{\text{line}} \frac{I_L\,dv}{I_c} = \int \frac{\tau_L^*\,dv}{\tau_c} \cong \int \frac{\kappa_L^*\,dv}{\kappa_c} . \qquad (2.50)$$

The observed integral on the left has the advantage that beam efficiencies and dilution factors apply more or less equally to numerator and denominator and hence more or less cancel. Thus the parameter is easy to measure.

Combining Equations (2.8) and (2.39) we find an expression relating the observations to the physical conditions in the gas:

$$\int \frac{I_L\,dv}{I_c} = 1.299 \times 10^5 \Delta n \cdot \frac{f_{nn'}}{n} v^{2.1} T_e^{-1.15} F \exp\left[\frac{1.579 \times 10^5}{n^2 T_e}\right] \qquad (2.51)$$

where dv must be measured in kHz. The factor F accounts for the fraction of the free-free emission due to interactions of He$^+$ with electrons:

$$F = \left(1 - \frac{N_{\text{He}}}{N_{\text{H}}}\right) \qquad (2.52)$$

where N_{He} and N_{H} are the relative number densities of helium and hydrogen ions, respectively. Observations of radio recombination lines in HII regions have established that $N_{\text{He}}/N_{\text{H}}$ is approximately 0.08, thereby giving a value for F of 0.92.

In Equation (2.51), note the variation of the line-to-continuum ratio with temperature and frequency. At any given frequency, additional atoms are ionized as the temperature increases, thereby increasing the free-free emission at the expense of the line emission. At any given temperature, the line-to-continuum ratio increases with frequency primarily because of the frequency dependence of the specific intensity of the line emission. The exponential term is nearly unity for lines in the radio range for all but very low temperatures.

In principle, Equation (2.51) provides an important tool in astronomy; the electron temperature of an optically obscured HII region can be calculated from simple observations—if the level populations are in thermodynamic equilibrium and if we ignore the effect of thermal and density gradients within the nebula.

2.5.2 Departures from LTE
Quite possibly, the levels involved in the radio recombination lines may not be in LTE. HII regions are optically thin at most radio wavelengths, and photons can easily escape from the gas. Also, collision cross sections have a steep dependence with principal quantum number, varying roughly as n^4, such that the influence of the kinetic energy field upon the level populations varies greatly over the radio range. In these circumstances, the coupling between the radiant and kinetic energy domains may not be adequate for the LTE equations to predict the intensities of the recombination lines accurately. Rather than balancing the rates in and out of

each level in detail, we must assume a statistical equilibrium to calculate the level populations.

For statistical equilibrium, we calculate the populations of the quantum levels from a system of equations in which all the ways out of a level n are equated with all the ways into that level:

$$N_n \sum_{n \neq m} P_{nm} = \sum_{n \neq m} N_m P_{mn} \qquad (2.53)$$

where P_{ij} is the transition rate from i to j and N_i is the population of level i. The system of equations described by Equation (2.53) is not closed; there is always one more unknown than equations. The necessary extra equation is provided by normalizing the populations N_i to the N_i^* expected in thermodynamic equilibrium; in other words, by creating the dimensionless variable $b_i \equiv N_i/N_i^*$. The solution of Equation (2.53) is then possible. For most situations encountered in radio recombination lines, the stimulated rate IB_{ij} is always less than the spontaneous rate A_{ij}. Also, collisions of the atoms with neutral particles are less important than those with electrons.

Figure 2.6 shows typical solutions to Equation (2.53). At large principal quantum numbers where, effectively, the atom's size is large, collisions dominate the transition rates and the departure coefficients are unity. At small quantum numbers, collisions have little effect upon the level populations, which are determined principally by the radiative rates. We visualize the electron orbiting the nucleus at a radius of $r = 0.529(n^2/Z)$ Å, an equation derived from Bohr's theory of atoms. The parameter Z is the atomic number. A hydrogen atom having its electron in a level of principal quantum number 200 has a radius 38,000 times greater than an atom with its electron in level 40. Its "target area" for collisions would be larger by a factor of approximately 10^9, from a classical point of view. The transition between the two asymptotes depends upon the gas density, that is, the extent to which the collisional rates are important.

With these population curves we can calculate the effect of non-LTE upon the intensities of radio recombination lines from HII regions. Lines formed by transitions between large quantum numbers will lie in the collision region, and their intensities will be the LTE values. On the other extreme, lines involving small quantum numbers will lie in the radiative region, and their intensities will be weaker than the LTE values, simply because there are fewer atoms in those levels, i.e., $b_n < 1$. In the transition region the situation is more complicated because of the substantial slope of the b_n curve as shown in the lower half of Figure 2.6. Here, although both upper and lower levels are underpopulated with respect to LTE, the upper level is slightly overpopulated with respect to the lower level. Thus we should expect an enhancement of the line intensity. And yet the b_n of the upper level is less than 1, which should weaken the line intensity. The point here is that in this region, these are two competing effects: line enhancement because of the slope of the b_n curve and line weakening because $b_n < 1$. Only quantitative analysis can predict whether a line formed in this region will be enhanced or weakened.

The correct form of the transfer equation for lines out of LTE is found by

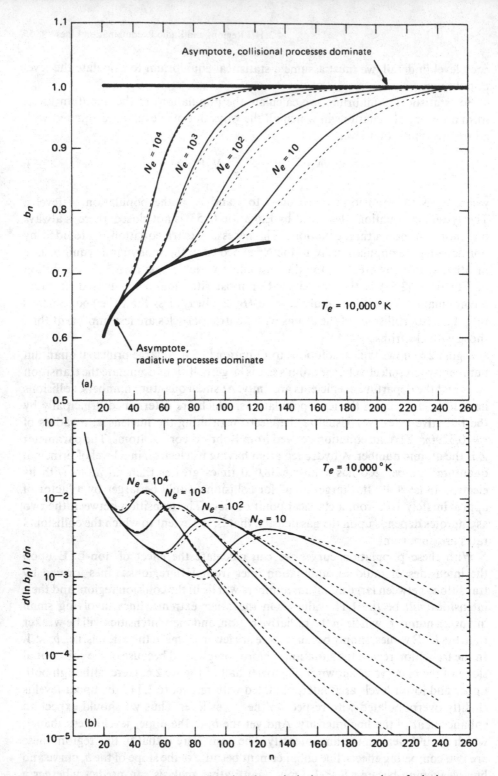

Fig. 2.6. *Top*: The population factor b_n is plotted against principal quantum number n for hydrogen gas having electron densities N_e. *Bottom*: The slope function is plotted against principal quantum number, for the above curves. Dashed lines mark the results obtained with other types of collision cross sections. (Sejnowski and Hjellming, 1969).

substituting the non-LTE source function (Equation 2.44) and line absorption coefficients (Equation 2.37) into the transfer equation (Equation 2.43). If τ_c and $\tau_L < 1$, as is the general case for HII regions emitting radio recombination lines in the centimeter range, the exponentials can be expanded to second order to give the line intensity as

$$I_L \cong I_L^* b_{n'} \left(1 - \frac{\tau_c}{2}\gamma\right) \quad \gamma \gg 1 \tag{2.54}$$

where γ is defined by Equation (2.46). Equation (2.54) shows clearly the effects of competing line-weakening and line-enhancing processes. If the term $\tau_c\gamma$ containing the slope of the b_n curve is very much less than 0, then $I_L > I_L^*$ and the line is enhanced. On the other hand, if $\tau_c\gamma$ is nearly 0, then $I_L < I_L^*$ and the line is weakened. In this case, the free-free optical depth τ_c can be considered to be an index to the number, N, of amplifiers along the line of sight, each having a gain of $-\gamma^{1/N}$. Substitution of number appropriate to HII regions into the correct transfer equation—the essence of which is given by Equation (2.54)—predicts that the lines will be enhanced over LTE values, and thus analysis by the LTE equation (Equation 2.51) will lead to an underestimate of the electron temperature.

The complete non-LTE form of Equation (2.51) is then

$$\int \frac{I_L dv}{I_c} = \int \frac{I_L^*}{I_c} \cdot b_{n'} \left(1 - \frac{\tau_c}{2}\gamma\right) dv \tag{2.55}$$

or in observational units

$$T_e^{1.15} = 1.229 \times 10^5 \frac{T_c}{P} \cdot \Delta n \cdot \frac{f_{nn'}}{n} v^{2.1} F b_{n'} \left(1 - \frac{\tau_c}{2}\gamma\right) \tag{2.56}$$

where the quantity P is the integrated antenna temperature over the line in units of $K \cdot kHz$. Equation (2.40) gives the oscillator strengths. Equation (2.56) gives a value for the electron temperature averaged over the thermal and density gradients along the line of sight through the nebula in some complex way—as is the case for all probes of distant nebulae.

2.5.3 Turbulence Broadening
Velocity fields within the radio beam broaden recombination lines. These fields, called microturbulence, involve packets of gas distributed in velocity space by collisions with each other. The number of them at each velocity interval is described by a Gaussian function. The overall Gaussian shape of a spectral line is then determined jointly by the microturbulence v_t and by the effective temperature of the gas, the overall width being

$$\Delta v = \left[\frac{2v}{c}\ln 2\left(\frac{2kT}{M} + \frac{2}{3}v_t^2\right)\right]^{1/2}. \tag{2.57}$$

Note that the line broadening due to microturbulence does not scale with either mass or temperature. Providing that the gas is optically thin at all velocities, turbulence broadening only widens the line profile; it does not change the integrated intensity of the spectral line.

2.5.4 Pressure Broadening

The presence of charged particles in HII regions in substantial densities means that recombination lines may be broadened by pressure effects, in this case the linear Stark effect. This interaction between the free, charged particles and the emitted wave train is usually calculated by means of one of two limiting approximations: *quasi-static* (when the wave distruption is so rapid that the particles can be assumed to be at rest during the disruption of the wave train) or *impact* (when the disruption occurs so slowly that the particle(s) moves substantial distances during the disruption). For recombination lines emitted from HII regions and from the interstellar medium, only impact broadening by free electrons is important.

For many years theoretical descriptions of Stark broadening always assumed adiabatic interactions between the perturbers and the emitted wave train. For recombination lines in the radio region of the spectrum this assumption is not valid, and the classical formulas cannot be used. Recombination lines involve transitions between levels degenerate in angular momentum. The particle-wave interaction can easily cause changes in the quantum numbers of the bound electron, thereby constituting nonadiabatic encounters. This kind of interaction results in cancellation of upper- and lower-state perturbations, causing the actual line broadening to be substantially less than would be predicted by the classical formulas.

Nonadiabatic encounters not only reduce Stark effects but also make them more difficult to calculate. Nonadibatic effects are particularly important for radio recombination lines, where the levels are separated by small amounts in energy, because the Stark effects arise in the impact regime. Here the perturber moves a long way before the emitted wave train is perturbed substantially, and the aspect angle between atom and perturber changes considerably during the phase change in the emitted wave train. Thus, there can be considerably coupling between the angular momentum of the atom and of the perturber which causes transitions between the degenerate energy levels.

The result of Stark effects is to redistribute the energy in the line over a larger frequency interval than would be expected from thermal and turbulent broadening alone. In particular, the energy is apt to be distributed into a line shape function known as a Voigt profile—a blend of a Gaussian core and nearly Lorentz wings. One method of characterizing the amount of Stark broadening in a spectral line is to find the ratio of the full widths at half-maximum of the Lorentz portion of the profile Δv^I to that of the Gaussian portion Δv^D for α-type lines (Brocklehurst and Seaton 1972):

$$\frac{\Delta v^I}{\Delta v} = 0.14 \left(\frac{n}{100}\right)^{7.4} \left(\frac{10^4}{T_e}\right)^{0.1} \left(\frac{N_e}{10^4}\right) \left(\frac{M}{M_H} \times \frac{2 \times 10^4}{T_D}\right)^{1/2} \quad (2.58)$$

where n is the principal quantum number, T_e is the electron temperature in K, T_D is the kinetic temperature in K required to account for the width of the Gaussian component (including the effects of electron temperature and microturbulence), M is the mass of the species, M_H is the mass of hydrogen, and N_e is the electron density in cm^{-3}. As an example, using reasonable values for an HII region of $T_e = 10^4$ K, $N_e = 10^4$, and $T_D = 2 \times 10^4$ K, we calculate that the ratio $\Delta v^I/\Delta v^D \cong 4$ for the H157α line (λ18 cm).

Table 2.2. Apparent electron temperatures for the Orion Nebula.[a]

Transition	Frequency (MHz)	$T_e(K)$
H110α	4874.157	6,990 ± 130
H138β	4897.779	9,150 ± 400
H158γ	4862.780	9,130 ± 900
H173δ	4909.394	10,260 ± 1000
H186ε	4910.858	10,430 ± 1600

[a] From Davies (1971).

2.6 Single-Dish Observations of Radio Recombination Lines from HII Regions

2.6.1 Early Observations of Integrated Line Intensities

Early observations of α-lines ($\Delta n = 1$) from HII regions tended to give unexpectedly low values for the electron temperature T_e, averaging 5800 K [e.g., the classic paper by Mezger and Höglund (1967)]. Many astronomers concluded that 10,000 K, which had been assumed to be the electron temperature of HII regions on the basis of optical observations, was too large. Others—Leo Goldberg (1966) in particular—felt that the 6000 K was an erroneous result caused by departures from LTE. Quantitatively, he believed the intensities of the lines observed at 6 cm were enhanced by an average of 80%, for which Equation (2.51) gave an underestimate of the true volume-averaged electron temperature.

Observations of higher-order transitions supported the non-LTE hypothesis. Table 2.2 lists one set of observations of the Orion Nebula near 6 cm. Because the beamwidth of the telescope was the same for each of these lines, these lines were produced by the same volume of gas—if the gas is optically thin. The apparent electron temperatures calculated from Equation (2.51) increase with principal quantum number, converging toward an asymptote of 10,000 K. This behavior is what we should expect if the lines are not formed in LTE. As Figure 2.7 illustrates, the highest-order transition involves the largest quantum number, the region of the b_n curve most dominated by collisions, the smallest enhancement effect, and the lowest apparent electron temperature. The α-transition, on the other hand, involves the smallest quantum number, the region of the b_n curve least influenced by collisions, the largest enhancement effect, and the highest electron temperature.

Other effects predicted by non-LTE seemed to be observed. For a given transition the line enhancement should vary as the beam is moved across the HII region, as shown by Equation (2.54). As the beam is moved away from the center of the HII region, the amount of gas along the line of sight decreases (parameterized by τ_c), and the maser effect should decrease. Figure 2.8 shows this effect through the parameter Q, defined to be the ratio of the apparent electron temperature calculated from the H109α line to that calculated from the H137β line. Q equals 1 if the gas is in LTE; values of Q less than 1 indicate departures from LTE. The map of Q for the HII region W49A shows Q to be smallest in the center, increasing toward the edges, as the non-LTE theory predicts.

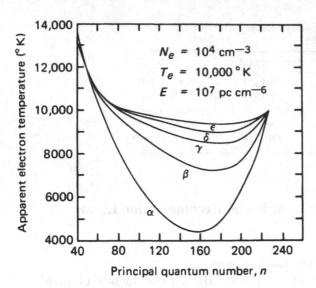

Fig. 2.7. Apparent electron temperature plotted against principal quantum number for $n\alpha$ through $n\varepsilon$ transitions. (Reprinted with permission by Gordon and Breach Science Publishers from: Goldberg, L. and Cesarsky, D.A., *Astrophysical Letters.* Vol. 6).

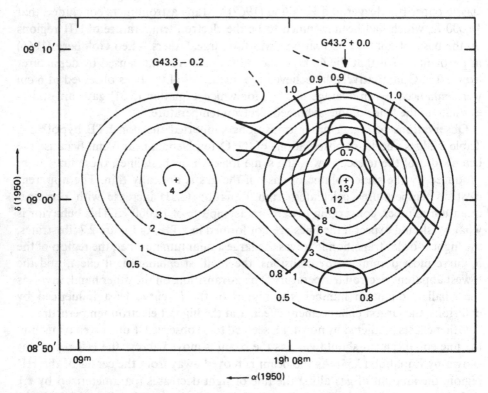

Fig. 2.8. Smoothed contours of the parameter Q over the HII region W49A. Q is the ratio of electron temperature calculated from the H109α line to that calculated from the H137β line, thereby constituting a measure of departures from LTE. (Gordon and Wallace, 1970).

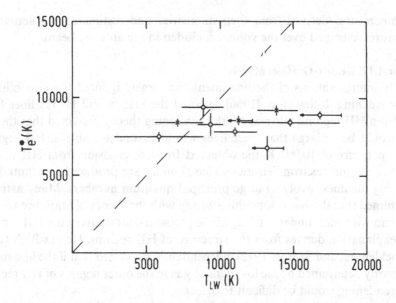

Fig. 2.9. A comparison of the apparent electron temperature derived from line intensities under the assumption of LTE with temperatures derived from the widths of H66α (circles) and H76α (crosses) hydrogen and helium recombination lines. (From Thum 1980. Copyright © by D. Reidel Publishing Company, Dordrecht, Holland.).

Another method of determining the electron temperature is independent of the line intensities. The method, originally devised by Baade et al. (1933), makes use of spectral lines emitted by atoms of two different masses. Since microturbulence also gives a Gaussian line shape, each line will have a width Δv due to microturbulence v_t which *does not* scale with mass and to thermal broadening which *does* scale with mass, as shown by Equation (2.57). An observational problem is that the line observations must be made with a high signal-to-noise ratio because the temperature depends upon the difference between the squares of the widths of the hydrogen and helium lines.

Comparing line widths of the 85α recombination lines of hydrogen and helium, Gordon Churchwell (1970) derived a temperature of 12,000 ± 500 K for the Orion Nebula and 12,480 ± 1000 K for IC 1795. These temperatures compare favorably with those calculated for the Orion Nebula from the higher-order lines listed in Table 2.2.

Similar results have been subsequently reported by others. Thum (1980) applied the line width technique to pairs of hydrogen and helium radio recombination lines observed near 1.4 and 2 cm. His results, shown in Figure 2.9, suggest that temperatures derived from the line widths exceed the apparent electron temperatures derived from the line intensities on the basis of LTE.

From the comparison of the intensity ratios of α with higher-order lines, from the comparison of apparent temperatures derived from radio recombination lines with those derived from optical lines, and from the comparison of temperatures derived from the line widths, one might conclude that Goldberg's suggestion was correct. It appeared that the radio recombination lines were not in LTE and that temperatures derived from their intensities underestimated the actual gas temperatures averaged over the volume included in the antenna beam.

2.6.2 Non-LTE Density-Gradient Models

The early interpretations of the line intensities largely ignored the possibility of Stark broadening. Lilley et al. (1966) detected the H156α and H158α lines (near 18 cm) from HII regions, whereas Stark broadening theory predicted that the line widths would be so large that these lines would be undetectable in HII regions. At a temperature of 10,000 K the observed free-free emission from HII regions required minimum electron densities which should have produced substantial line broadening for lines involving large principal quantum numbers. Many astronomers assumed that there was something wrong with the theory of Stark broadening rather than with their understanding of the physical conditions in the HII regions.

One explanation derives from the structure of HII regions. Hoang-Binh (1972) and Brocklehurst and Seaton (1972) independently suggested that if the line shapes are primarily determined by the low-density gas in the outer regions of HII regions, Stark broadening would be difficult to detect.

Brocklehurst and Seaton examined the available observations of radio recombination lines for the Orion Nebula and calculated the density gradient necessary to fit these data if the electron temperature was 10,000 K. This non-LTE model reconciled observations of the free-free emission and the absence of Stark broadening by noting that most of the line amplification occurs in the outer regions of the HII region. The susceptibility of the lines to the masing process generally increases as the density decreases, that is, in the outer regions closest to the observer. And at wavelengths where the gas is optically thin, the continuum emission which drives the masing process is also strongest in the outer regions. The greatest amount of the line amplification thus occurs in the low-density outer regions of the HII region where Stark broadening is least important, thereby resulting in predominantly Gaussian line profiles.

Brocklehurst and Seaton also suggested that observational effects associated with this process will make it difficult to detect the Lorentzian wings of the recombination lines. Because the regions where most of the line amplification occurs have low densities, the Stark-broadened wings will be weak in comparison with the strong Gaussian cores of the lines. This circumstance, they felt, makes it difficult for the observer to distinguish the weak wings from the background noise and the instrumentally caused undulations in the spectrometer baseline, thereby hiding the Stark broadening and resulting in an underestimate of the total power radiated in the recombination lines.

Improved observations required models with gradients in both density and temperature. Lockman and Brown (1975) suggested that the densest parts of the Orion Nebula must be somewhat cooler than the surrounding gas. Their model, summarized in Table 2.3, consists of only three regions, each with a unique temperature, density, and size. It also is hemispherical to account for the discovery that the HII region lies on the near side of a massive molecular cloud, being ionization-bounded on the far side and density-bounded on the near side. Figure 2.10 shows the model to fit the observations well.

Table 2.3. Lockman-Brown model of the Orion Nebula.[a]

Region	Temperature (K)	N_e(cm^{-3})	Angular size (arcmin)	Linear size (pc)
I	(7,500)	$10^{4.5}$	0.33	0.043
II	(10,000)	$10^{3.5}$	4.3	0.56
III	(12,500)	$(10^{2.3})$	19.1	2.5

[a] Quantities in parentheses are determined by the recombination-line observations.

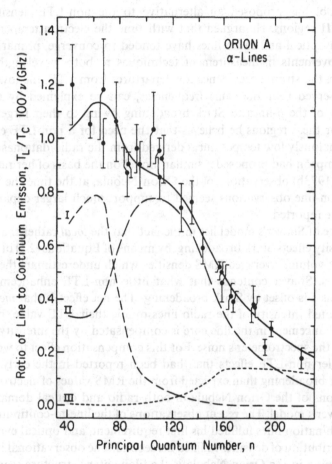

Fig. 2.10. The observed ratio of line to continuum intensities for α-type radio recombination lines of hydrogen from the Orion Nebula. Solid line: the predicted emission of the Lockman and Brown (1975) model; dashed lines: the predicted emission from each of the component regions listed in Table 2.3.

Recombination-line models for HII regions can be tested by comparing the number of ionizing photons necessary to sustain the models with those emitted from the observed exciting stars. The Lockman-Brown model for the Orion Nebula requires at least 4×10^{48} photons per second with energies exceeding the ionization potential of hydrogen, a flux expected from stars hotter than O7.5. The hottest star (θ^1 C) in the central Trapezium is O6; its ionizing flux exceeds this requirement by a factor of 4. In addition, slightly to the southeast of the center but still within the nebula is an 09 (θ^2 A) which, while radiating only half of the required flux, can help. The excess flux may be lost by leakage through the density-bounded near side of the nebula, and by absorption by embedded dust and helium. These considerations are equivalent to comparing the continuum flux of the model with the observed continuum emission from the nebula.

2.6.3 Clumped Models

Shaver (1980b) has proposed an alternative to the non-LTE density-gradient models of HII regions. He argued that with time the electron temperatures calculated from optical and radio lines have tended to converge, primarily because of the improvements in measurement techniques in both wavelength domains. Furthermore, the strongest evidence for departures from LTE, the low β/α ratios for lines observed near the same frequencies, can be explained by the greater susceptibility of the β-lines to Stark broadening owing to their larger quantum numbers. For these regions he believes that the need for a non-LTE explanation for the anomalously low temperatures derived from the radio data has weakened. Although Simpson had proposed a similar model on the basis of her radio (1973a) and optical (1973b) observations of the Orion Nebula, at the time the bulk of the recombination-line observations seemed to support much larger departures from LTE than she reported.

The essence of Shaver's model lies in the fact that the *local* rather than the RMS electron density effects Stark broadening; by means of Equations (2.5) or (2.8), radio data produce volume-averaged RMS densities which underestimate the maximum local densities. Shaver contends that what little non-LTE enhancement of line intensities exists is offset by Stark broadening. The net effect is that *measurements* of the integrated intensity of the radio lines mimic their LTE value, because the non-LTE enhancement in the line core is compensated by the intensity of the line wings lost in the spectrometer's noise. For this compensation effect to work, Shaver needed smaller non-LTE effects than had been reported in the early 1970s and greater Stark broadening than expected from the RMS values of electron densities.

Observations of the Orion Nebula in both radio and optical domains tend to support Shaver's model. The recent observations of the line-to-continuum ratios of radio recombination lines fulfilled his first requirement, and optical evidence for a clumped distribution of densities fulfilled his second. The observational justification for the clumping in the Orion Nebula is the filamentary structure seen in optical photographs and the wide range of electron densities deduced from optical spectra by Osterbrock and Flather (1959) shown in Figure 2.11.

The Shaver model for the Orion Nebula consists of five cylindrical regions having their lengths equal to their diameters and lying axially concentric with their axes normal to the line of sight. The size and density of each cylinder were chosen to fit the observed continuum spectrum and line widths. All of the regions were assumed to be isothermal with an electron temperature of 8200 K. Table 2.4 lists the model's parameters for each cylindrical zone. The radiation transfer for the lines was calculated for both LTE and non-LTE conditions, with the results "observed" with a spatial convolution function chosen to mimic the beamwidths of the telescopes

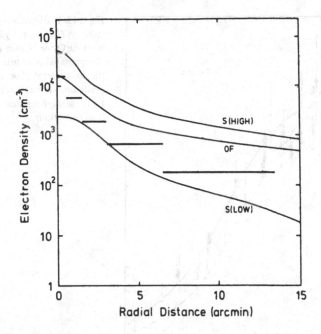

Fig. 2.11. Electron densities plotted against radial distance from the center of the Orion Nebula. S(HIGH) and S(LOW) refer to the high- and low-density components of the model by Simpson (1973b) and OF refers to the observations by Osterbrock and Flather (1959). The horizontal lines indicate the local densities used in the model by Shaver (1980b).

Table 2.4. Shaver model of the Orion Nebula.

Zone	I	II	III	IV	V
$\theta(')$	1	3	6	13	27
$2R(pc)$	0.13	0.39	0.79	1.70	3.54
$N_e(\text{rms})(10^4 \text{ cm}^{-3})$	0.7	0.3	0.08	0.02	0.006
$N_e(\text{local})(10^4 \text{ cm}^{-3})$	1.5	0.6	0.2	0.07	0.02
Filling factor	0.2	0.2	0.2	0.1	0.1

used for the observations. The non-LTE departure coefficients were those calculated by Brocklehurst (1970), and the Stark broadening employed formulas derived by Griem (1967).

The predictions of this model also agree well with observations. Figure 2.12 compares the predicted line-to-continuum ratios of α-lines with observations. While the non-LTE case fits the data well, even the LTE case differs by only a small amount. This difference between the two cases implies that the non-LTE amplification is weak, ranging from 15 to 25% for the observed α-lines.

2.6.4 Comparison of the Density-Gradient and Clumped Models
The definitive observations necessary to distinguish the two models are difficult to make with the accuracy necessary to distinguish them. The Lockman-Brown (LB) model for the Orion Nebula requires the electron density to decrease and the

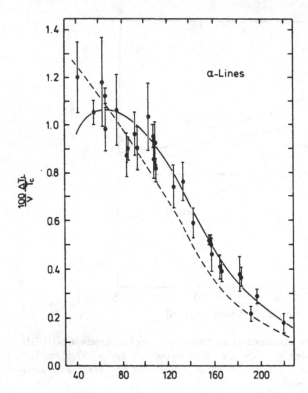

Fig. 2.12. The line-to-continuum ratio $(100/\nu)\Delta T_L/T_C$ for hydrogen α-lines from the Orion Nebula plotted against the principal quantum numbers n of the lower level. The curves are the predictions of Shaver (1980b) for non-LTE (solid line) and LTE (dashed line) conditions.

temperature to increase outward from the center of the HII region. The temperature gradient of this model predicts that the line-to-continuum ratio should somewhat decrease as the beam moves off the center of the HII region—an observation requiring a small beam and large signal-to-noise ratios. The isothermal Shaver model also has a density gradient but one with clumps; the ratio of local to RMS density is usually 2 for each of its five zones. The Shaver model predicts that the integrated intensity of α-lines should decrease at wavelengths shorter than 3 mm because of the increasing influence of the depletion factors regardless of the beam size. So also does the LB model, but because of the joint effects of the depletion factors *and* the smaller temperature of the nebula's core.

The high-resolution observations which are available appear to favor the Shaver model slightly. Wilson and Pauls (1984) have reported observations of H41α lines in the Orion Nebula with a beamwidth of 2 arcmin. These lines are probably free from Stark broadening and from maser effects. After correction for the level depletions, the derived electron temperatures fall in the range predicted by Lockman and Brown but, in conflict with their model, appear to decrease away from the center of the nebula.

That both models can fit the bulk of the radio observations is an assessment of the value of radio recombination lines as a probe of the *structure* of HII regions. The Orion Nebula, although small in comparison with many HII regions in our galaxy, is perhaps the easiest to observe because of its intensity due to its proximity to Earth. Even though the nebula can be spatially resolved by filled-aperture radio telescopes, the lines of sight through its gases probably traverse regions of varying physical conditions inaccessible to direct measurement. Worse are the possible consequences of comparing observations made with the varying, finite resolutions of different radio telescopes, as Hoang-Binh (1970) suggested and Lockman and Brown (1976) later demonstrated. It thus appears that, even for the unusually favorable conditions of the Orion Nebula, radio recombination lines are unable to determine a unique structure for the HII region. This shortcoming is also true for optical observations, unfortunately.

2.6.5 Empirical Determinations of Electron Temperature

On the positive side, modern observations suggest that radio recombination lines can predict the volume-averaged, density-weighted electron temperatures of HII regions within 15%; that is, the difference between the electron temperatures determined by LTE and non-LTE analyses is small in most cases.

Shaver (1980c) suggested that there is a unique frequency for which the LTE-derived electron temperature is approximately the volume-averaged temperature, independent of temperature, density, or structure of the HII region. This frequency f_c is

$$f_c \cong 0.081 E^{0.36} \tag{2.59}$$

where the emission measure E is defined as

$$E \equiv \int_0^L n_e^2 \, ds \tag{2.60}$$

where the integral is along the line of sight through the HII region and may be derived from the continuum observations by determining the optical depth of the free-free emission. At this frequency the non-LTE enhancement of the integrated intensity is offset by the depletion of the level populations, so that the observed line-to-continuum ratio can be interpreted by LTE analyses.

Odegard (1985) has suggested an alternative method to determine both the mean temperature and local electron density for an HII region from observations of two α-type recombination lines, widely separated in frequency. One line must be at a sufficiently high frequency that impact broadening and maser effects are negligible; the other, at a sufficiently low frequency that one or both of these effects are present but the continuum optical depth is still small. The same volume of the HII region should be observed for both frequencies; that is, the beamwidths should be the same. The iterative procedure involves assuming an initial value for the electron temperature, calculating the free-free optical depth from continuum observations, adjusting the electron density until the theory gives a line-to-continuum ratio equal to the value observed for the low-frequency line, using this value of n_e to obtain an estimate for the electron temperature consistent with that derived from the high-frequency observations, and re-iterating until the parameters converge.

Comparisons of the results of the iterative technique with computer models have been encouraging. Even with model HII regions having complicated structures in temperature and density, Odegard's technique produces accurate values of T_e and n_e averaged along the lines of sight through the model nebulae.

Table 2.5 lists the results of his analyses for a number of HII regions and compares

Table 2.5. Comparison of densities and temperatures for selected HII regions, (Odegard 1985).

Nebula	Derived T_e(K)	Derived $\log n_e$	$T_e(N^+)$	$T_e(O^{++})$	$\log n_e$	Zone(s)[a]
				Results from forbidden lines		
DR 21	$9,200 \pm 800$	4.50 ± 0.30	—	—	—	—
M17 S	$7,350 \pm 500$	3.58 ± 0.18	$7,000^b$	$8,800^b$	3.23 ± 0.06	O^{+2}
					$4.15 {+0.22 \atop -0.16}$	S^{+2}
S106	$10,900 \pm 600$	4.23 ± 0.09	$10,000$	—	$3.3\text{–}4.6^c$	S^{+1}
Orion A	$8,250 \pm 500$	3.38 ± 0.05	$8,330^d$	$8,800^d$	$2.9\text{–}3.8^c$	S^{+1}, O^{+1}, Cl^{+2}
			$8,900^d$	$8,850^d$	$2.65\text{–}3.9^c$	O^{+2}
Orion B	$7,750 \pm 500$	3.19 ± 0.08	—	—	~ 3.0	S^{+1}
Rosette	$8,900 \pm 1,600$	1.53 ± 0.31	$9,900$	—	<2.0	S^{+1}
NGC 7027	$13,700 \pm 2,500^e$	5.24 ± 0.16^e	$12,600$	$11,200$	5.37 ± 0.10	Ar^{+3}
	$14,000 \pm 2,500^f$	5.14 ± 0.11^f	—	$12,700$	$4.8 {+0.2 \atop -0.3}$	O^{+2}
NGC 6572	$<18,500$	>4.16	$12,900$	$10,600$	$4.32 {+0.17 \atop -0.28}$	S^{+1}, O^{+1}
IC 418	$8,750 \pm 1,250$	4.43 ± 0.72	$8,200$	$10,300$	$4.22 {+0.12 \atop -0.16}$	S^{+1}, O^{+1}
NGC 6543	$6,700 \pm 1,300$	3.93 ± 0.36	$8,500$	$8,000$	3.65 ± 0.08	O^{+2}

[a] The ionization zone or zones to which $\log n_e$ of the sixth column refers.
[b] From observations of M17 N.
[c] Range of $\log n_e$ from observations at several positions.
[d] Average of two or more observed positions.
[e] Obtained using H66α and H85α.
[f] Obtained using H66α and H90α.

them with optically determined values. Figures 2.13 and 2.14 show good agreement between electron temperatures and densities determined from the radio recombination lines and from optical nitrogen lines especially observations with small measurement errors. This agreement is a different situation than existed in the late 1960s and early 1970s.

Should these mean values be correct, one might ask why the results of Baade's technique described in Section 2.6.1 overestimate the electron temperatures. Such results would require the widths of the hydrogen lines to be too large with respect to the helium lines, for example, a situation which would occur if the hydrogen lines were to arise from regions of slightly greater turbulence or temperature than the helium lines.

2.7 Exploration of the Milky Way by Radio Recombination Lines

2.7.1 Galactic Distribution of HII Regions

Perhaps the most direct result of observations of radio recombination lines is the radial velocity indicated by the frequencies of the line centers. Figure 2.15 shows the positions of HII regions projected on the galactic plane by Georgelin and Georgelin (1976) on the basis of the radial velocities of HII regions. They used

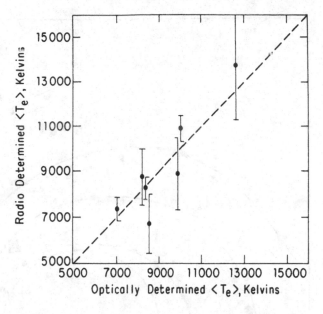

Fig. 2.13. The mean electron temperatures determined by Odegard (1985) from radio recombination lines plotted against temperatures determined optically from nitrogen emission.

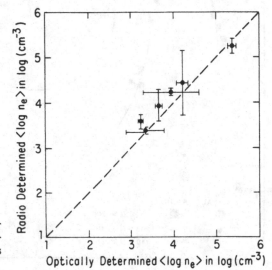

Fig. 2.14. The mean electron densities determined by Odegard (1985) from radio recombination lines plotted against densities obtained from forbidden optical lines.

optical Hα observations and radio recombination lines to establish unique distances for 100 HII regions on the basis of a galactic rotation curve, absorption of neutral hydrogen at 21 cm, and optical extinction. Radio recombination lines provided the bulk of the data for regions beyond a few kiloparsecs from the Sun, regions inaccessible to optical observations because of extinction by the interstellar gas lying in the galactic plane. These astronomers note that, at least in the outer regions, the giant HII regions appear to lie along spiral arms as is often seen in external galaxies.

Radio recombination lines have established the distribution of HII regions with distance from the galactic center. Figure 2.16 shows the data of Downes et al. (1980), reported by Wilson (1980). The HII regions predominantly lie within 4 and 9 kpc

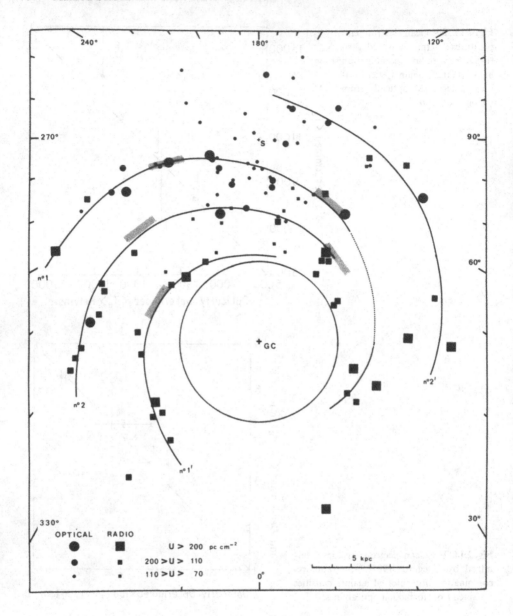

Fig. 2.15. The positions of large HII regions in the plane of the Milky Way determined from Hα and radio recombination lines. The plot symbols indicate the size of the nebulae in terms of the excitation parameter U, the cube of which is proportional to the flux of UV photons required to ionize the gas. Superimposed is a suggested spiral pattern. Hatched regions show intensity maxima in the radio continuum and in neutral hydrogen. (Georgelin and Georgelin, 1976).

Fig. 2.16. *Bottom*: The distribution of 114 HII regions per square kiloparsec (in percent of total) as a function of distance from the galactic center. *Middle*: The same distribution weighted by the cube of the excitation parameter. *Top*: The distribution of CO emission per sequare kiloparsec (in percent of total). Northern Hemisphere data. HII region data from Wilson (1980); CO data from Sanders et al. (1984).

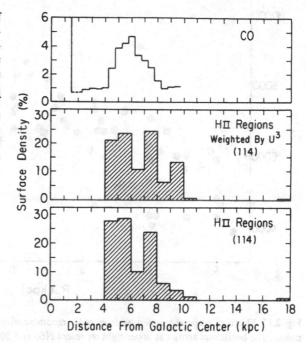

from the galactic center. When weighted by the cube of the excitation parameter, the distribution shows the surface density of ionization to be more uniformly distributed over this range. Observations of carbon monoxide show this range to contain molecular clouds, thereby statistically linking the cold clouds with HII regions and with star formation.

2.7.2 Galactocentric Variations of Electron Temperatures

The line-to-continuum ratios of radio recombination lines have been used to investigate the galactic variation of the temperatures of HII regions. Shaver et al. (1983) derived electron temperatures for 67 HII regions seen from the southern hemisphere. Figure 2.17 shows their electron temperatures to increase with distance from the galactic center, an effect first noted by Churchwell and Walmsley (1975).

Having determined the temperatures of HII regions, Shaver et al. were able to derive abundances from the optical spectra of thirty-three of the same sources. They argue that the radio data provide temperatures accurate to 5% RMS, which then permit the derivation of accurate abundances from the highly-temperature-sensitive optical spectra. Figure 2.18 shows the metallicity to decrease with increasing distance from the galactic center, an effect which could be responsible for the galactocentric gradient in the temperatures of the HII regions (Panagia 1979).

2.7.3 Abundance Ratio of Helium to Hydrogen

Almost from the first detection of radio recombination lines, their importance for investigating the large-scale distribution of helium was recognized. Many papers have appeared (see the review by Mezger 1980). Early observations showed a large scatter in the ratios of He/H from one HII region to another. The observations suggested that the line ratios tended to increase with distance from the galactic

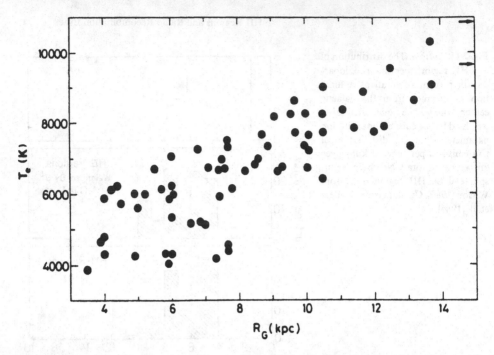

Fig. 2.17. Electron temperatures derived from radio recombination lines plotted against galactocentric radius. The horizontal arrows at upper right represent N66 and 30 Doradus in the Magellanic Clouds. (Shaver et al., 1983, with permission of the Royal Astronomical Society).

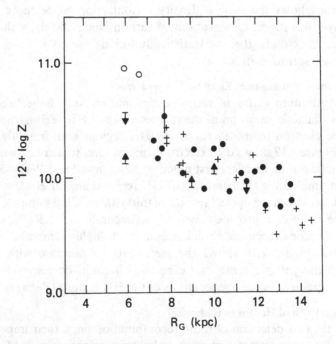

Fig. 2.18. The variation of metallicity with galactocentric radius. The parameter Z equals 25[0]/[H]. Open circles denote the nebulae S38 and S48; vertical arrows, N66 and 30 Doradus; crosses, earlier optical values derived by Peimbert (1979). (Shaver et al., 1983, with permission of the Royal Astronomical Society).

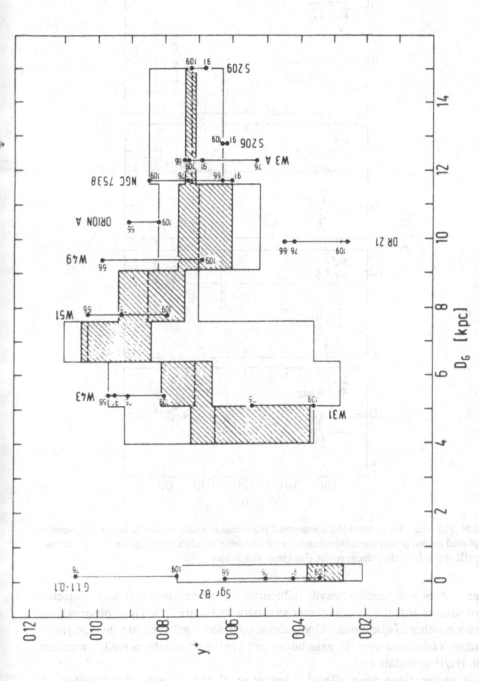

Fig. 2.19. The variation of the helium abundance, y^+, with galactocentric distance as determined from radio recombination lines. Boxes show the range of 109α (5 GHz) observations. Shaded boxes exclude the extreme 109α points. Dashed horizontal lines mark median values. The dots on the vertical lines mark the high-frequency observations, identified by their principal quantum numbers. (Thum et al. 1980).

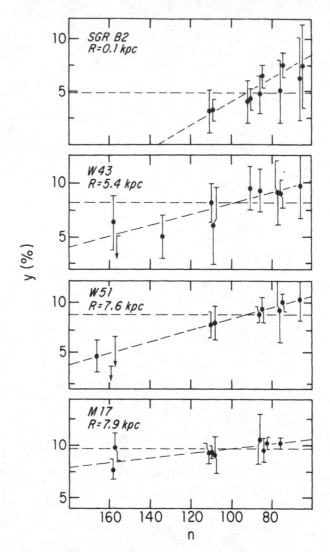

Fig. 2.20. The ratio y (in percent) of the integrated intensities of hydrogen and helium recombination lines plotted against principal quantum number n for a number of galactic HII regions. R is the distance of the HII region from the galactic center. (Lockman and Brown, 1982).

center—albeit a suggestion heavily influenced by anomalously low ratios detected in two sources near the galactic center as shown by Figure 2.19. These observations raised a number of questions. Why is the scatter so large? Why are the line ratios a function of distance from the galactic center? Are the observations really a measure of the He/H abundances?

Two explanations were offered. Mezger et al. (1974) and Churchwell et al. (1978) suggested a geometrical effect in which the selective absorption of Lyman continuum photons was making the He$^+$ and H$^+$ ionization zones spatially non-coincident, causing the line ratios to vary from one source to another. They believed that the galactic gradient in the line ratios was in fact a real abundance effect. Alternatively, Brown and Gomez-Gonzales (1975) suggested a radiation transfer

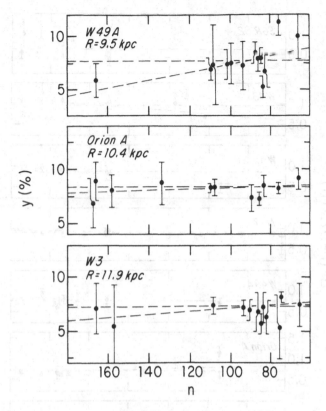

Fig. 2.20 continued

effect, such that recombination lines produced by high-emission-measure HII regions and passing through cold, partially ionized regions lying in the peripheries of these HII regions were subject to unpredictable amounts of masing.

If the masing hypothesis were true, it would mean that radio recombination lines would be useless as probes of He/H abundances. The extent of the masing would be dependent upon the conditions and geometry peculiar to individual sources, giving varying ratios of He/H from source to source. Furthermore, the observed ratios of He^+/H^+ would vary with principal quantum numbers, that is, with the particular transition observed. The anomalously low line ratios observed in sources near the galactic center could well be due to the large amount of partially ionized cold gas lying between them and observers.

Distinguishing between these two hypotheses is not easy. Smith and Mezger (1976) noted that the observations available to them did not show the variation of the line ratios with frequency or quantum number predicted by the non-LTE model. Also, the values of He/H determined for the Orion Nebula from radio recombination lines agree with optical determinations, showing that the radio values are not substantially in error for that nebula, at least. (See the review by Mezger 1980.)

More recent observations have complicated the situation. Lockman and Brown (1982) have shown that Y^+, the ratio of the integrated intensities of helium and hydrogen recombination lines (Figure 2.20), and the ratio of their widths vary inversely with principal quantum number (Figure 2.21)—independently of the

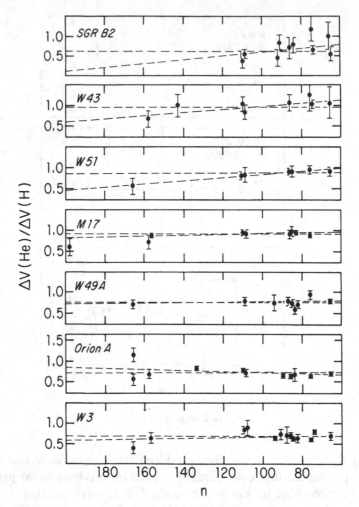

Fig. 2.21. The ratio of the widths of helium and hydrogen lines plotted against principal quantum number *n*. (Lockman and Brown, 1982).

beamwidth of the radio telescope. The measured value of Y^+ would seem to depend upon the particular recombination lines used for the measurements.

The explanation for this effect is less clear. This phenomenon evidently appears in all HII regions, its size varying inversely with distance from the galactic center. Lockman and Brown suggest the cause to be observations, arising from the difficulties in compensating for Stark broadening of the hydrogen lines and from confusion of the spectra from the HII regions with those from the diffuse interstellar medium (ISM). At low frequencies and large principal quantum numbers, the Stark-broadened line wings could blend with instrumental irregularities in the baseline of the spectrometer, resulting in an underestimate of integrated fluxes and widths of the broader hydrogen lines relative to the narrower helium lines and an overestimate of *y*. In the same sense, the spectra from the faint, diffuse ionized

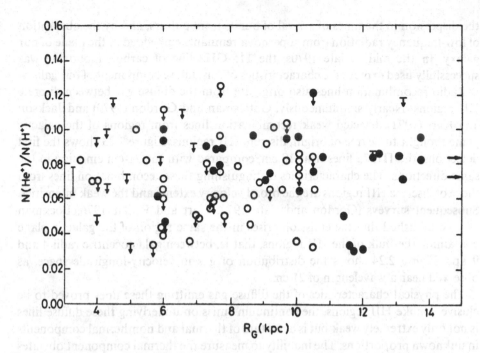

Fig. 2.22. The variation of the ratio [He⁺]/[H⁺] with galactocentric distance. Open and filled circles denote radio and optical measurements, respectively. (Shaver et al., 1983, with permission of the Royal Astronomical Society).

regions in the galactic plane could blend with those of HII regions preferentially at low frequencies and as an inverse function of distance from the galactic center. This spectral confusion would appear as a distorted baseline which, if mistakenly removed by the usual polynomial fit, would alter the integrated intensities of the "corrected" hydrogen and helium lines.

Finally, an optical survey of helium in our galaxy, based only upon temperatures derived from radio recombination lines, shows no helium gradient when added to the radio data. Shaver et al. (1983) show the ratio of helium to hydrogen to be 0.074 ± 0.003 with no apparent gradient in galactocentric distance, as indicated by Figure 2.22. Presumably the scatter is due to varying geometries within the HII regions, causing varying degrees of ionization of helium with respect to hydrogen. These data do not include the galactic center sources, which have anomalously weak helium recombination lines with respect to hydrogen and strongly influence the gradient found in the radio data of Figure 2.19.

2.7.4 The Diffuse Component

Our galaxy contains a great deal of gas of not in the form of stars or other discrete objects. The presence of this diffuse "interstellar medium" was first suggested by the optical absorption lines of calcium and sodium detected in stellar spectra in the early part of this century. Later, the ISM was seen by the emission of the 21-cm line of atomic hydrogen, which permitted measurements of its distribution along the line of sight by the Doppler effect. Still later, the ISM made known its presence by the dispersion in the times of arrival of bursts from pulsars, and by the absorption of low-frequency radiation from supernova remnants embedded in the plane of our galaxy. In the mid to late 1970s, the 115-GHz line of carbon monoxide was successfully used explore the characteristics of this diffuse component of our galaxy.

Radio recombination lines also originate from the diffuse gas between discrete HII regions. Nearly simultaneously, Gottesman and Gordon (1970) and Jackson and Kerr (1971) detected weak recombination lines from regions of the galactic plane thought to be free of bright, discrete HII regions. Figure 2.23 shows the first detections, the H157α lines near 18 cm, compared with the 21-cm emission in the same directions. The characteristics distinguishing these recombination lines from those of discrete HII regions are the broad velocity extents and the weak intensities. Subsequent surveys (Gordon and Cato 1972, Hart and Pedlar 1976, Lockman 1976) established that the emission arises in the same region of the galactic plane containing the bulk of the HII regions, that is, between galactocentric radii 4 and 9 kpc. Figure 2.24 shows the distribution of gas in velocity-longitude space, as observed near a wavelength of 21 cm.

The physical characteristics of the diffuse gas emitting these lines proved to be elusive. Unlike HII regions, the continuum emission underlying these diffuse lines is not only extremely weak but is a mixture of thermal and nonthermal components in unknown proportions. The inability to measure the thermal component obviates the technique of deriving an effective electron temperature by comparing the ratio of bound and unbound atoms along the line of sight, the principle underlying Equation (2.53). Consequently, one can interpret the observations only by

$$\int T_L \, dl = 1.070 \times 10^4 \cdot \Delta n \frac{f_{nn'}}{n}$$

$$\int N_{\mathrm{HII}} N_e \cdot b_n \cdot T_e^{-1.5} \exp\left(\frac{1.579 \times 10^5}{n^2 T_e}\right) dl \qquad \tau_L, \tau_c < 1 \qquad (2.61)$$

where the brightness temperature T_L of the line excess to the continuum background is in K, v is in kHz, the densities N_i are in cm^{-3}, and the path length l is in pc.

Equation (2.61) shows that the basic information contained in the diffuse recombination lines lies within the integral. For most circumstances within the ISM, the exponential term is close to unity for values of n appropriate for the centimeter wavelength range. The velocity extent of the line profile determines the path length l from the rotation curve of our galaxy. The observations then determine a value of the quantity $\langle N_{\mathrm{HII}} N_e \cdot b_n \cdot T_e^{-1.5} \rangle$ averaged over the path l, in which the ionization cannot be separated from the electron temperature.

One possible technique for separating the physical parameters is to use the free-free absorption, the absorption coefficient for which is given in an approximate form by Equation (2.8). Measurement of the turnover frequency for a supernova remnant lying in the plane of the galaxy, on the far side of the ionization zone, determines the integral

$$\int N_e N_i T_e^{-1.35} \, dl \qquad (2.62)$$

along the line of sight. Because the electron temperature has a different exponent

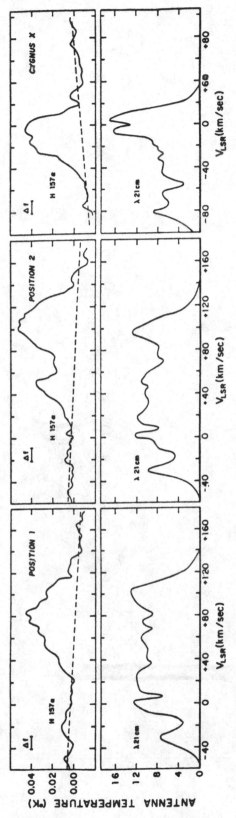

Fig. 2.23. Juxtaposition of the first detection of radio recombination lines from the diffuse interstellar medium against the 21-cm emission lines of atomic hydrogen. (Gottesman and Gordon 1970).

79

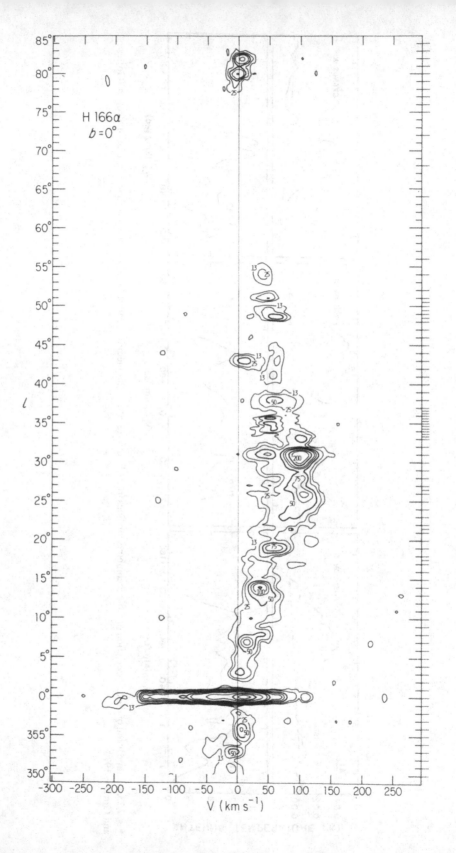

H 166α
b = 0°

l

V (km s⁻¹)

than for the line radiation, it should be possible to separate average density from average temperature. Cesarsky and Cesarsky (1973) performed this experiment by observing recombination lines in the direction of the supernova remnant 3C391 and comparing their emission with the observed turnover frequency for the source. They concluded that the average electron temperature must be less than 400 K.

The experimental uncertainties associated with the measurements of the weak lines, with estimates of the low-frequency spectrum emitted by 3C391, with the probable fluctuations in physical conditions along the line of sight to the source, and with the appropriate value for the departure coefficients make this result a tentative one in spite of the care with which this experiment was performed. In addition, if the clouds in the interstellar medium are approximately in pressure equilibrium and have the same fractional ionization, the cold clouds will contribute much more to the line emission than the hot clouds along the line of sight (Cesarsky and Cesarsky 1971). Without knowing the detailed distribution along the line of sight, we cannot say what this temperature refers to.

The fact that cold clouds can emit more intense recombination lines than hot clouds suggests another approach to determining the physical conditions along the line of sight. Shaver (1975) predicted that if the dominant component were cold clouds, meter-wavelength recombination lines should be unusually strong as a result of maser amplification of the strong background continuum of the galactic plane. The important factor here is the frequency dependence of the stimulated emission factor of Equation (2.46). Considering the observed recombination lines and the limits for the meter-wavelength lines, Shaver (1976b) concludes that the most likely explanations for the line emission from the diffuse component are low-density ($\cong 5$ to 10 cm^{-3}), warm ($\cong 5000$ K), and large (20 to 150 pc) HII regions along the lines of sight. The presence of spatially extended, partially ionized cold regions appears to be ruled out.

2.8 Radio Observations of HII Regions with High Angular Resolution

2.8.1 Continuum Observations

In the late 1960s radio astronomers identified a new kind of HII region, called *compact* HII regions (Mezger et al. 1967). Observationally they are characterized by thermal continuum spectra with turnovers at higher frequencies than usual. These regions have average electron densities equal to or greater than 10^4 cm^{-4} and emission measures equal to or greater than 10^7 pc cm^{-6} (Mezger 1968). Earlier identifications included DR21 and dense sources embedded in IC1795 (W3) and W49A. In most cases, compact HII regions are associated with OH and with H_2O masers. The high emission measures and small sizes suggest that compact HII

◁ ——————————————————————————————

Fig. 2.24. H166α emission from lines of sight along the galactic plane, observed with the 140-ft telescope. Contour units are millikelvins of antenna temperature. Tick marks along the right ordinate mark longitudes of the observations. (Copyright © by D. Reidel Publishing Company. Dordrecht, Holland. From Lockman 1980).

regions are involved with the early stages of star formation. In a survey of ninety-one candidates, Wink et al. (1982) found twenty-one sources containing one or more compact components, of which four fit the above definition of compact HII regions.

Synthesis telescopes have revealed new characteristics to some of these sources. Dreher and Welch (1981) found that the ultracompact HII region W3(OH) has an ionized shell with an electron density of approximately 10^5 cm^{-3} and a radius of about 10^{16} cm, which probably surrounds a newly formed bright star and may be as young as 300 years. Similarly, Turner and Matthews (1984) observed six ultracompact HII regions and found that four had shell structures. The mechanism for the shells is unknown. Possibilities include a spherical ionization front radiatively driven or maintained by a newly formed central star (Davidson and Harwit 1967) and an ionization front resulting from the impact of a hot stellar wind upon the surrounding gas (e.g., Shull 1980).

Not all structures within compact HII regions are spherically symmetric. Rodriguez et al. (1982) observed a part of the HII region NGC 6334 which appears split in two, perhaps by the breaking out of a previously confined ionized plasma. Reid and Ho (1985) observed an HII region complex and found a cometlike component, suggesting the rupture of a protostellar ionization shell perhaps by a process known as the "champagne" model (Tenorio-Tagle 1979).

The advent of high-resolution observations in molecular lines has shown us that it may no longer be possible, and certainly may not be desirable, to interpret the morphologies of compact and ultracompact HII regions alone. These phenomena involve many factors, including the characteristics of the dense gas from which the HII regions form. The evolution of these sources is dynamic and complex (Lada 1985), and our knowledge of this evolution at many wavelengths will be fundamental to understanding just how stars form.

2.8.2 Recombination-Line Observations

High-resolution studies of radio recombination lines with synthesis instruments show that the line-to-continuum ratio can vary greatly from one component of an HII region to another. van Gorkom (1980) reports that, at 5 GHz (H109α), this ratio varies from 1 to 9% when observed on an angular scale of 6 arcseconds. For single-dish observations this ratio is usually about 5%. Furthermore, van Gorkom reports that the smaller ratios appear to be associated with the most compact components. Here calculations show that the variations in this ratio could be wholly due to variations in the continuum optical depth of the compact components, without need of a variation in the electron temperature. No strong maser effects would be expected for the H109α lines at the high densities ($\leq 10^4$ cm^{-3}) characteristic of these components; Figure 2.6 shows that the departure coefficients are probably close to 1 and their gradients are small.

Comparison of the synthesis with single-dish observations tell us where the line emission arises. Most HII regions consist of dense clumps embedded in a lower-density gas, very probably in a hierarchical arrangement. The synthesis telescopes are insensitive to extended components and thereby exclude radiation from them. Churchwell et al. (1978) noted that the integrated continuum flux measured by the Westerbork array represented 35% of the single-dish value, whereas the line flux was only 25% of the single-dish value. The conclusion is that much of the line emission must come from the extended components in the HII regions.

While synthesis telescopes often reveal a chaotic structure to radio sources inaccessible to the lower angular resolution of single dishes, there are exceptions. Lacy et al. (1980) reported observations of the [NeII] fine-structure line (12.8 μm) from the galactic center with an angular resolution of about 4 arcseconds. The infrared (IR) data suggested the region to consist of a number of individual clouds. Subsequent observations (van Gorkom et al. 1985) of the H76α (15 GHz) line with the Very Large Array telescope, with a resolution of about 5 arcseconds, showed that this thermal source is actually well ordered in velocity and smooth in intensity. The radial velocities are consistent with those measured from the [NeII] lines, suggesting that the IR data come from a smooth, well-ordered source rather than from individual clouds.

2.9 Other Aspects of Radio Recombination Lines

2.9.1 Low Frequencies

At frequencies below 200 MHz, the level populations become sensitive to the radiation of background sources. Shaver (1975) calculated that under certain conditions the stimulated emission can alter the population of high-n levels in such a way as to make conditions favorable for strong non-LTE masing of low-frequency lines. Competing effects which weaken these lines are pressure broadening, optical depth effects, and the underpopulation of these levels with reference to LTE. For atoms with multiple electrons, dielectronic recombination (see Section 2.9.3) could affect the level populations to make low-frequency recombination lines detectable in absorption against strong radio sources (Shaver 1976a). On balance, the most favorable conditions exist for cold clouds lying in front of strong continuum sources, such as the cold component of the ISM where the electron temperature is low and the emission measure high. The most favorable frequency range would be 50 to 200 MHz, depending greatly upon the physical conditions along the line of sight to the background source.

Such strong lines in this frequency range have not been observed toward strong continuum sources (Shaver 1976b), leading Shaver to conclude that the sources of the observed recombination lines from the diffuse ISM are warm, extended regions of electron densities exceeding 5 cm^{-3} rather than old, partially ionized clouds.

New observations have reopened the subject of low-frequency recombination lines. Konovalenko and Sodin (1980) reported the detection of a weak absorption line at 26.13 MHz in the nonthermal emission from the Cassiopeia supernova remnant, which they suggested might be due to a hyperfine transition of ^{14}N. Later, Blake et al. (1980) suggested that the line could be due to a 631α recombination line of a heavy element, such as carbon, on the basis of the known relative abundance of nitrogen in the ISM. Konovalenko and Sodin (1981) confirmed this suggestion by detecting the adjacent 630α line of carbon at 26.254 MHz and the C640α line at 25.04 MHz, and Konovalenko (1984) later observed a transition at 16.7 MHz.

Combining these observations with previous theoretical work by Shaver,

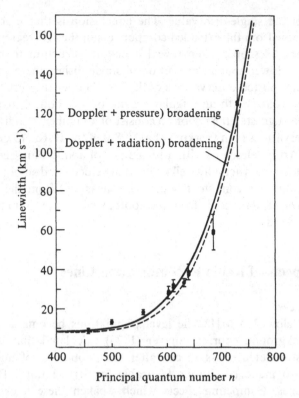

Fig. 2.25. Comparison of the calculated and observed line widths for α-type recombination lines observed at low frequencies. The curve for pressure broadening assumed an electron temperature of 50 K and an electron density of 0.18 cm^{-3}; for radiation broadening, a brightness temperature of 2000 K at 100 MHz was assumed. The Doppler component is 8 km s^{-1}. (Anantharamaiah et al., 1986. Reprinted by permission from Nature, Vol. 315, pp. 647. Copyright © 1986 Macmillan Magazines Limited).

Konovalenko and Sodin predicted that the recombination lines of carbon should appear in emission in the frequency range 100–200 MHz when observed toward Cas A. Perhaps equally interesting is their suggestion that the recombination lines are consistent with the existence of a cold (50 K) cloud of ionized carbon ($n_e \cong 0.1$ cm^{-3}) along the line of sight to Cas A. Its free-free absorption could explain the observed decrease in the low-frequency continuum emission of Cas A just as well as the hypothesized traditional HII region.

In 1986 Anantharamaiah et al. reported detections toward Cas A at four more frequencies: 26, 38, 52, and 68 MHz. These spectra result from averaging together the α-type transitions contained in each band. They found that the velocity profile of these absorption lines matches the HI absorption profile associated with the Perseus arm of our galaxy. The line width implied a lower limit of 150 pc for the distance of the gas from Cas A, based upon radiation damping. Also interesting was the discovery that the width of the 68-MHz line was four times smaller than that of the 26-MHz line, suggesting either pressure or radiation broadening for the low-frequency lines as shown in Figure 2.25. Figure 2.26 compares the integrated intensities of these lines with theoretical predictions of Walmsley and Watson (1982)

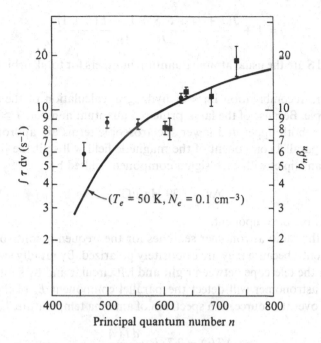

Fig. 2.26. Comparison of the integrated intensities of the absorbed radio recombination lines with a model of Walmsley and Watson (1982). The ordinate is in terms of net optical depth, which is proportional to the line intensities. The curve $b_n\beta_n$ is normalized to a value of 10 at a quantum number of 600. (Anantharamaiah et al., 1986. Reprinted by permission from Nature, Vol. 315, pp. 647. Copyright © 1986 Macmillan Magazines Limited).

based upon a dielectronic recombination process for a cloud with a temperature of 50 K and an electron density of 0.1 cm^{-3}. Agreement is good but additional observations are needed for lines of quantum numbers below 500.

2.9.2 Zeeman Effect

Radio recombination lines are subject to the same Zeeman effect as the 21-cm hydrogen line. For principal quantum numbers $30 < n < 300$, the formulas for the 21-cm line are appropriate for the $n\alpha$ and $n\beta$ transitions of hydrogen, helium, and carbon (Greve and Pauls 1980).

The magnetic fields along the line of sight will split the line radiation into three components. Viewed perpendicular to the magnetic field, the spectral line will be linearly polarized (*pi* component) with its electric vector parallel to the field. Viewed along the direction of the magnetic field, the spectral line will be split into two components (*sigma* components) of opposite circular polarization, one on either side of the unshifted pi component, v_0, by an amount

$$v - v_0 = \frac{1}{4\pi}\frac{eB}{m}[(g_{n'} - g_n)M_{Jn'} \pm g_n] \quad (2.63)$$

where the frequencies v and v_0 are in Hz when the electronic charge e, the magnetic field strength B, and the electronic mass e are in CGS units. $M_{Jn'}$ is the magnetic quantum number for the component of Jn' along the field. The Landé g-factor is

$$g = 1 + \frac{J(J + 1) + S(S + 1) - L(L + 1)}{2J(J + 1)} \tag{2.64}$$

where J, L, and S are the usual atomic quantum numbers for total, orbital, and spin momenta.

For the radio recombination lines of hydrogen, calculation of the amount of splitting is simple. Because of the large principal quantum number, $J \cong L > 1$ and the g-factors for both upper and lower spectroscopic terms are approximately 1. Therefore, the parallel component of the magnetic field will split the line into the classical Zeeman triplet with each sigma component shifted by

$$\Delta v = 1.39 \text{ Hz}/\mu\text{G} \tag{2.65}$$

to either side of the pi component.

In practice, the radio astronomer searches for the frequency shifts between the sigma components, because they are oppositely polarized. By rapidly switching the polarization of the telescope between right and left circular and by subtracting the difference, the astronomer will detect the parallel component $B_{\parallel}$ of the magnetic field, averaged over the source, as a spectrum of antenna temperature $T(v)$ given by

$$\Delta T(v) = 2.78 B_{\parallel} \cdot \frac{dT(v)}{dv} \tag{2.66}$$

where $T(v)$ is the profile of the recombination line. The slope of the line profile, $dT(v)/dv$, is an important factor because the temperature difference will be greatest at those frequencies where the line profile is steepest.

Attempts to detect the Zeeman effect in hydrogen and carbon recombination lines, while unsuccessful, have suceeded in placing limits on the magnetic fields in HII regions and dark clouds. The weakness of the recombination lines makes these observations difficult. Troland and Heiles (1977) were able to set a limit of 210 μG for the parallel field in the Orion Nebula, a region characterized by a mean density of 10^4 cm^{-3}. Although it was hoped that a search for the Zeeman effect in carbon recombination lines might be more successful because their line widths are smaller and the densities of the dark clouds greater by two orders of magnitude, Silverglate (1984b) was only able to establish an upper limit of about 1 mG.

Magnetic fields are expected to play an important role in the dynamics of the ISM. If the fields are frozen into the gas, the magnetic field strength should vary as a function of density. Zeeman observations of the 21-cm line of hydrogen show the field in the interstellar medium to be about 2 μG at a density of about 1 cm^{-3}. When combined with the limits from the recombination lines, the Zeeman observations cover a density range of six orders of magnitude and are consistent with a power law of $B \propto n^{\alpha}$, where the exponent $\alpha < \frac{2}{3}$. These limits are consistent with the theoretical predictions of $\frac{1}{3} \leqslant \alpha \leqslant \frac{1}{2}$ if the gas contracts to high densities along magnetic field lines (Mouschovias 1976).

2.9.3 The Sun
Recombination lines in principle can also be detected in the Sun. Dupree (1968) calculated that the highly stripped ions in the solar atmosphere may exhibit

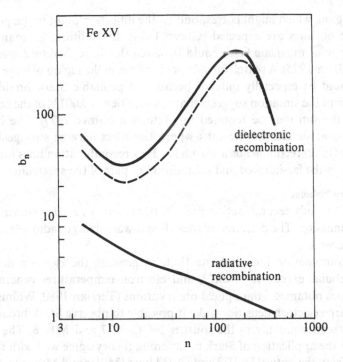

Fig. 2.27. Calculated population curves for FeXV with dielectronic recombination (upper curves) and without (lower curve). The dashed curve shows the population without cascades from the upper levels. (Dupree, 1968).

unexpectedly large populations of their high quantum states. These overpopulations can give rise to detectable recombination lines both in emission and absorption.

The overpopulation phenomenon depends upon dielectronic recombination, a process in which an ion with one or more electrons recombines with a free electron, resulting in a doubly excited atom. The doubly excited state then decays via a radiative transition, leaving the atom in a singly excited state and the gas with an overpopulation of its high quantum states. While not applicable to hydrogen, this process can be effective for the highly stripped ions like CIII, OIV, OV, NeVII, and SiXI in the solar atmosphere (Berger and Simon 1972) and possibly for CII in interstellar dark clouds.

A population curve illustrates the situation. Figure 2.27 shows calculations made by Dupree for FeXV. She noted that the region $20 < n < 200$ could lead to greatly enhanced emission lines because of the overpopulation of the upper with respect to the lower quantum terms and that the region $n > 200$ could lead to enhanced absorption lines. No line would appear at the inflection point near $n = 200$. Subsequent theoretical studies (Shore 1969) discovered that FeXV required a different kind of stabilizing transition than Dupree used for her calculations, so that her results overestimated this effect for FeXV. Even so, Dupree's calculations are qualitatively correct and illustrate the situation for other ions.

By 1985 no radio recombination lines have been observed in the Sun, with limits of about 0.1% of the solar continuum. All of the observations have been made in the 3-mm region, which might correspond to the inflection point in the population curve where no lines are expected (Greve 1974). In addition, Zeeman splitting from intense solar magnetic fields could broaden the lines, thereby reducing their intensities (Greve 1975). A search for asborption lines in the region of large quantum numbers would be especially difficult because of probable Stark broadening. A re-evaluation of the situation suggests that sensitivities of 0.001% of the continuum emission of the sun may be required for detection (Greve 1977). The best hope appears to be at shorter wavelengths where the effect may be strongest. In fact, Hoang-Binh (1982) recommends a search for Hnα recombination lines for $5 \leqslant n \leqslant 20$, which lie in the far-infrared and submillimeter part of the spectrum.

2.9.4 Planetary Nebulae

It is difficult to study recombination lines from planetary nebulae because of their weak radio emission. The detection of these lines awaited large radio telescopes and sensitive receivers.

Radio recombination lines indicate that, in general, the electron densities of planetary nebulae exceed 10^{-4} cm^{-3}, and electron temperatures generally agree with the values obtained from optical observations (Terzian 1980, Walmsley et al. 1981). The large electron densities make it possible to observe Stark broadening of high-n lines, seen especially in the sources NGC 7027 and M1-78. The densities derived from the application of Stark broadening theory agree well with the values determined from the optical [OIII] and [NII] lines (Miller and Mathews 1972) and with the radio continuum (Walmsley et al. 1981).

Radio recombination lines are also effective for determining Y, the ratio of helium to hydrogen abundance in planetary nebulae. Walmsley et al. find Y to be 0.14, when the fractional ionization of helium is taken into account. This value agrees with the optical values determined by Miller and Mathews (1972) and by Shields (1978).

2.9.5 Dark Clouds

The discovery of radio recombination lines from carbon, in conjunction with studies of molecular clouds, has led to a new understanding of the star formation process. The new concept was that many HII regions, as sites of newly formed stars, were small parts of large, cold molecular clouds, parts made visible by the fluorescence caused by the newly formed stars.

Palmer et al. (1967) reported the detection of an anomalous recombination line from the sources NGC 2024 and IC 1795. The weak, unusually narrow line appeared to the high-frequency side of the helium recombination line, thereby suggesting a recombination line from a heavier element. Complicating the observation was that while the line was seen in the spectra of other HII regions, its frequency varied with respect to the adjacent helium line. On the basis of relative cosmic abundances, it was likely that the line was from carbon but the variation in its apparent radial velocity with respect to helium and hydrogen was difficult to understand.

Subsequent observations eventually solved most of the mystery. The intensity of the anomalous lines was observed to vary with wavelength in a different way than the hydrogen and helium recombination lines, prompting Goldberg and Dupree (1967) to explore dielectronic recombination as a possible mechanism to populate

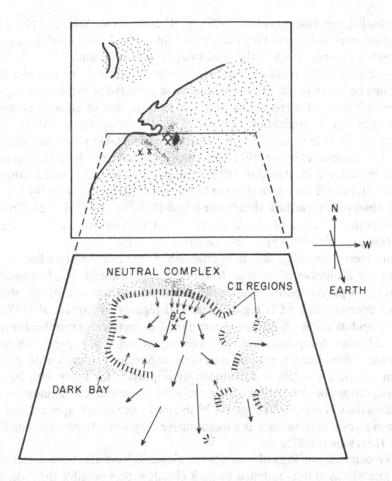

Fig. 2.28. A schematic diagram of the Orion Nebula, illustrating the locations and dynamics of the CII and HII regions with respect to the star cluster θ^1 Ori. (Balick et al., 1974).

the quantum levels of carbon in HII regions, but without success. However, by 1974, an analysis (Dupree 1974) of all available observations confirmed the suggestion (Palmer and Zuckerman 1968) that the anomalous lines were simple recombination lines of carbon emitted by dense ($n_H \cong 10^4$ cm^{-3}), cool ($T \cong 50$ K) regions associated with the HII regions, often as shells surrounding them. CII regions adjacent to HII regions are ionized by photons with energies between 11.256 and 13.595 eV—the ionization potentials of CI and HI—leaking from the HII regions. The radial velocities of CI recombination lines differ from those of the HI and HeI lines simply because they come from different regions. Figure 2.28 schematically shows the situation for the Orion Nebula (Balick et al. 1974).

CI lines have also been detected from dark nebulae radiating no HI or HeI lines. Brown and Knapp (1974) detected carbon recombination lines from the dark nebula near the star ρ Oph. Subsequently, they and coworkers reported similar detections in NGC 2023, M78, S140, and Mon R2. Pankonin and Walmsley (1978) added NGC

7023 to this list; van Gorkom et al. (1979), NGC 2068, NGC 2071, and NGC 2170. van Gorkom et al. noted that they detected CI lines in only six out of twenty-seven dark clouds searched; evidently the phenomenon is uncommon.

CII ionization zones require special excitation conditions. The ionizing flux of energy must be less than the 13.595-eV ionization potential of hydrogen but greater than the 11.256 eV of carbon. The putative uniform flux of galactic cosmic-ray photons provides an ionization rate of 10^{-15} s^{-1}, adequate to ionize enough hydrogen and carbon in dark clouds to produce detectable radio recombination lines for *both* elements (Brown 1973), but these lines have not both been generally observed in dark clouds (Gordon 1973). One possible selective mechanism is the UV radiation from B stars embedded in these clouds (Brown et al. 1974). Indeed, a B star is observed in the dark cloud near ρ Oph (Vrba et al. 1975). A confirmation of this hypothesis would be the detection of a CII zone around the B star, an observation possible with a synthesis telescope.

Recent observations of CI lines have established the variation of the line flux with frequency for a number of sources. It appears that neither a purely stimulated-emission nor a purely spontaneous-emission model can account for the observed frequency dependence for CII regions near HII regions. Watson et al. (1980) and Walmsley and Watson (1982) have reexamined dielectronic recombination as a viable mechanism for populating the levels of carbon. These papers show that dielectronic recombination may be extremely important for levels with principal quantum numbers $n > 250$, and qualitatively important for levels with $20 > n \geqslant 250$. Observations of the carbon $n'\beta/n\alpha$ ratio may qualitatively illuminate conditions within the CII regions, as Figure 2.29 shows. In any case, it appears that using the carbon recombination lines as a quantitative diagnostic to physical conditions within CII regions is difficult.

Radio recombination lines of even heavier elements have also been detected. The narrow line widths of lines radiated by dark clouds makes possible their identification. Chaisson (1975) and Pankonin and Walmsley (1978) reported observations of a recombination line from sulfur in ρ Oph and Lynds 1630. Later, Pankonin et al. (1977) identified a line in NGC 2024 as also due to sulfur, and Silverglate (1984a) reported the detection of S166α lines in W48, S87, and S88B. On the basis of relative cosmic abundances, these Z^+ regions probably also contain recombination lines from magnesium, silicon, and iron, which would be difficult to identify because of the small separation in line frequencies (see Equation 2.24) and because their abundance in the gas phase of the ISM may have been depleted by condensation into dust grains.

The ionization regions of carbon and heavier elements have been distinguished from HII regions by the name partially ionized medium, or PIM. This phase of the ISM is difficult to study because the weak lines are difficult to observe and the excitation conditions are usually too poorly known to model with confidence.

2.9.6 Galaxies and Quasars

Following the detection of radio recombination lines from HII regions in the Milky Way, a natural course was to search for the lines in external galaxies and quasars under the assumption that stimulated emission would make the lines detectable. The first detection was the H166α line in the peculiar galaxy M82, made with

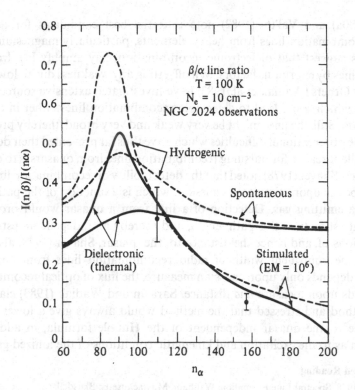

Fig. 2.29. The predicted ratio of the intensities of a C$n'\beta$ to C$n\alpha$ line of carbon plotted against n, where $n' = 1.26n$, for conditions expected for the CII zone in NGC 2024. Dashed curves include dielectronic recombination in calculations of level populations; solid curves do not. Lower curves (stimulated) include a background continuum produced by an intense HII region; upper curves (spontaneous) do not. (Walmsley and Watson, 1982).

the Westerbork synthesis telescope (Shaver et al. 1977), later confirmed at the Algonquin Radio Observatory with the H102α line (Bell and Seaquist 1977), at Haystack Observatory with the H92α line (Chaisson and Rodriguez 1977), and at the Max Planck Institute with the H110α line (Shaver et al. 1978). Analysis showed that the lines arise in ionized gas between the nucleus and the observer, stimulated by the nonthermal background emission. Bell and Seaquist (1978) suggested that the average characteristics of the source region are: temperature, 5000 K; electron density, 150 cm^{-3}; and emission measure, 2×10^6 cm^{-6} pc.

Subsequently, Bell and Seaquist (1978) detected the H102α line from another peculiar galaxy, NGC 253, later confirmed with the H112α line by Mebold et al. (1980). Here also the emission appeared to be entirely due to stimulated emission from the ionized gas lying in front of the nonthermal nucleus. Model fitting suggests average parameters of temperature, 5000 K; electron density, 2×10^3 cm^{-3}; and emission measure, 10^6 cm^{-6} pc.

Recombination lines from only one other galaxy have been detected at this writing, the Markarian-Seyfert galaxy Mrk 668-OQ 208 (Bell and Seaquist 1980) with the H83α and H99α lines. These observations are also consistent with lines stimulated by the nonthermal emission from the galactic nucleus.

Recombination lines from quasars have proved to be elusive, unfortunately.

Shaver (1980a) and Val'tts (1982) note that the best candidates for detection may be recombination lines from heavy elements, particularly magnesium. Their calculations suggest that dielectronic recombination may amplify low-frequency (< 1 GHz) lines by enormous factors, well offsetting any weakness due to low cosmic abundance. Others (Wadiak et al. 1983) have investigated extensive source models for quasars which could lead to detectable recombination lines either in emission or absorption. Still, the lines might be very weak and very broad, thereby presenting considerable observational difficulties which have thus far prevented their detection.

One excellent reason for pursuing recombination lines from quasars is to estimate their distance. Shaver (1978) noted that the detectability of recombination lines from quasars depends upon stimulated emission, which is a function of the path length through the emitting gas. Detection of a line from a quasar would provide an independent estimate of this path length, and thereby an independent estimate of the dimensions of, and hence the distance to, the quasar. Shaver (1978) also noted that while the equivalent width of radio recombination lines from quasars in absorption depends only upon emission measure, the flux of optical recombination lines depends upon measure *and* distance. Sarazin and Wadiak (1983) elaborated on this method and stressed that the method would always give a lower limit to the distance to the quasar independent of the Hubble formula, in addition to information as to the excitation and kinematic conditions of the ionized gas.

Recommended Reading

Jefferies, J.T. 1968. Spectral Line Formation. Waltham, Massachusetts: Blaisdell.
Osterbrock, D.E. 1974. Astrophysics of Gaseous Nebula. San Francisco: Freeman.
Shaver, P.A. (ed.) 1980. Radio Recombination Lines. Dordrecht: Reidel.
Spitzer, Lyman, Jr. 1978. Physical Processes in the Interstellar Medium. New York: Wiley.
Vallee, J.P. 1985. "Les Gax Ionisés dans Notre Galaxie (Revue des Régions HII, HeII, CII, SII Interstellaires)," J. Roy. Astron. Soc. Can. **80**:16.

References

Altenhoff, W., P.G. Mezger, H. Wendker, and G. Westerhout. 1960. Veröff. Sternwarte, Bonn, No. 59, p. 48.
Anantharamaiah, K.R., W.C. Erickson, and V. Radhakrishnan. 1986. Nature **315**:647.
Baade, W.F. Goos, P.P. Koch, and R. Minkowski. 1933. Z. Astrophys. **6**:355.
Balick, B., R.H. Gammon, and R.M. Hjellming. 1974. Publ. Astron. Soc. Pacific **86**:616.
Bell, M.B., and E.R. Seaquist. 1977. Astron. Astrophys. **56**:461.
Bell, M.B., and E.R. Seaquist. 1978. Astrophys. J. **223**:378.
Bell, M.B., and E.R. Seaquist. 1980. Astrophys. J. **238**:818.
Berger, P.S., and M. Simon. 1972. Astrophys. J. **171**:191.
Blake, D.H., R.M. Crutcher, and W.D. Watson. 1980. Nature **287**:707.
Brocklehurst, M. 1870. Mon. Not. R. Astron. Soc. **148**:417.
Brocklehurst, M., and M.J. Seaton. 1972. Mon. Not. R. Astron. Soc. **157**:179.
Brown, R.L. 1973. Astrophys. J. **194**:692.
Brown, R.L., and G.R. Knapp. 1974. Astrophys. J. **189**:253.
Brown, R.L., R.H. Gammon, G.R. Knapp, and B. Balick. 1974. Astrophys. J. **189**:253.
Brown, R.L., and J. Gomez-Gonzales 1975. Astrophys. J. **200**:598.
Cesarsky, C.J., and D.A. Cesarsky. 1971. Astrophys. J. **169**:293.
Cesarsky, D.A., and C.J. Cesarsky. 1973. Astrophys. J. **184**:83.
Chaisson, E.J. 1975. Astrophys. J. **197**:L65.
Chaisson, E.J., and L.F. Rodriguez. 1977. Astrophys. J. **214**:L111.
Churchwell, E.B., and C.M. Walmsley. 1975. Astron. Astrophys. **38**:451.
Churchwell, E.B., L.F. Smith, J. Mathis, P.G. Mezger, and W. Huchtmeier. 1978. Astron. Astrophys. **70**:719.

Davidson, K., and M. Harwit. 1967. Astrophys. J. **148**:443.
Davies, R.D. 1971. Astrophys. J. **163**:479.
Downes, D., T.L. Wilson, J. Bieging, and J. Wink. 1980. Astron. Astrophys. Suppl. Ser. **40**:379.
Dreher, J.W., and W.J. Welch. 1981. Astrophys. J. **245**:857.
Dupree, A.K. 1968. Astrophys. J. **152**:L125.
Dupree, A.K. 1971. Astrophys. J. **173**:293.
Dupree, A.K. 1974. Astrophys. J. **187**:25.
Georgelin, Y.M., and Y.P. Georgelin. 1976. Astron. Astrophys. **49**:57.
Goldberg, L. 1966. Astrophys. J. **144**:1225.
Goldberg, L., and D.A. Cesarsky. 1970. Astrophys. Lett. **6**:93.
Goldberg, L., and A.K. Dupree. 1967. Nature **215**:41.
Goldwire, H.C., Jr. 1969. Astrophys. J. Suppl., No. 152, **17**:445.
Goldon, M.A. 1973. Astrophys. J. **184**:77.
Goldon, M.A., and T. Cato. 1972. Astrophys. J. **176**:587.
Gordon, M.A., and E. Churchwell. 1970. Astron. Astrophys. **9**:307.
Goldon, M.A., and D.C. Wallace. 1970. Astrophys. J. **167**:235.
Gottesman, S.T., and M.A. Gordon. 1970. Astrophys. J. **162**:L93.
Greve, A. 1974. Solar Phys. **36**:85.
Greve, A. 1975. Solar. Phys. **40**:329.
Greve, A. 1977. Solar Phys. **52**:423.
Greve, A., and T. Pauls. 1980. Astron. Astrophys. **82**:388.
Griem, H. 1967. Astrophys. J. **148**:547.
Hart, L., and A. Pedlar. 1976. Mon Not. R. Astron. Soc. **176**:547.
Hoang-Binh, D. 1970. Astrophys. Lett. **6**:151.
Hoang-Binh, D. 1972. Mem. Soc. Roy. Sci. Liège **3**:367.
Hoang-Binh, D. 1982. Astron. Astrophys. **112**:L3.
Jackson, P.D., and F.J. Kerr. 1971. Astrophys. J. **168**:29.
Jackson, P.D., and F.J. Kerr. 1975. Astrophys. J. **196**:723.
Kardashev, N.S. 1959. Sov. Astron.—AJ **3**:813.
Konovalenko, A.A. 1984. Sov. Astron. Lett. **10**:846.
Konovalenko, A.A., and L.G. Sodin. 1980. Nature **283**:360.
Konovalenko, A.A., and L.G. Sodin. 1981. Nature **294**:135.
Lacy, J.H., C.H. Townes, T.R. Geballe, and D.J. Hollenbach. 1980. Astrophys. J. **241**:132.
Lada, C.J. 1985. Annu. Rev. Astron. Astrophys. **23**:267.
Lilley, A.E., D.H. Menzel, H. Penfield, and B. Zuckerman. 1966. Nature **209**:468.
Lockman, F.J. 1976. Astrophys. J. **209**:429.
Lockman, F.J. 1980. In P.A. Shaver (ed.), Radio Recombination Lines. Dordrecht: Reidel.
Lockman, F.J., and R.L. Brown. 1975. Astrophys. J. **201**:134.
Lockman, F.J., and R.L. Brown. 1976. Astrophys. J. **207**:436.
Lockman. F.J., and R.L. Brown. 1982. Astrophys. J. **259**:595.
Maddalena, R.J. 1985. Unpublished Ph.D. dissertation, Columbia University.
Malkamaki, L., G. Sandell, K. Mattila, and K.-H. Gebler. 1979. Astron. Astrophys. **71**:198.
Mebold, U., P.A. Shaver, M.B. Bell, and E.R. Seaquist. 1980. Astron. Astrophys. **82**:272.
Menzel, D.H. 1968. Nature **218**:756.
Menzel, D.H. 1970. Astrophys. J. Suppl., No. 161, **18**:221.
Mezger, P.G. 1968. In Y. Terzian (ed.), Interstellar Ionized Hydrogen. New York: Benjamin.
Mezger, P.G. 1980. In P.A. Shaver (ed.), Radio Recombination Lines. Dordrecht: Reidel.
Mezger, P.G., and A.P. Henderson. 1967. Astrophys. J. **147**:471.
Mezger, P.G., and B. Höglund. 1967. Astrophys. J. **147**:490.
Mezger, P.G., W. Altenhoff, J. Schraml, B.R. Burke, E.C. Reifenstein, and T.L. Wilson. 1967. Astrophys. J. **150**:L157.
Mezger, P.G., L.F. Smith, and E.B. Churchwell. 1974. Astron. Astrophys. **47**:34.
Miller, J.S., and W.G. Mathews. 1972. Astrophys. J. **172**:593.
Mouschovias, T.C. 1976. Astrophys. J. **207**:141.
Odegard, N. 1985. Astrophys. J. Suppl. **57**:571.
Oster, L. 1961. Rev. Mod. Phys. **33**:525.

Oster, L. 1970. Astron. Astrophys. **9**:318.

Osterbrock, D.E., and E. Flather. 1959. Astrophys. J. **129**:26.

Palmer, P., B.M. Zukerman, H. Penfield, A.E. Lilley, and P.G. Mezger. 1967. Nature **215**:40.

Palmer, P., and B.M. Zukerman. 1968. Astron. J. **73**:S196.

Panagia, N. 1979. Mem. Soc. Astron. Ital. **50**:79.

Pankonin, V., and C.M. Walmsley. 1978. Astron. Astrophys. **64**:333.

Pankonin, V., C.M. Walmsley, T.L. Wilson, P. Thomasson. 1977. Astron. Astrophys. **57**:341.

Peimbert, M. 1979. *In* W.B. Burton (ed.), The Large-Scale Characteristics of the Galaxy. Dordrecht: Reidel.

Reid, M.J., and P.T.P. Ho. 1985. Astrophys. J. **288**:L17.

Rodriguez, L.F., J. Canto, and J.M. Moran. 1982. Astrophys. J. **255**:103.

Sanders, D.B., P.M. Solomon, and N.Z. Scoville. 1984. Astrophys. J. **276**:182.

Sarazin, C.L., and E.J. Wadiak. 1983. Astron. Astrophys. **123**:L1.

Sejnowski, T.J., and R.M. Hjellming. 1969. Astrophys. J. **156**:915.

Shaver, P.A. 1975. Pramana **5**:1.

Shaver, P.A. 1976a. Astron. Astrophys. **46**:127.

Shaver, P.A. 1976b. Astron. Astrophys. **49**:1.

Shaver, P.A. 1978. Astron. Astrophys. **68**:97.

Shaver, P.A. 1980a. In P.A. Shaver (ed.), Radio Recombination Lines. Dordrecht: Reidel, p. 247f.

Shaver, P.A. 1980b. Astron. Astrophys. **90**:34.

Shaver, P.A. 1980c. Astron. Astrophys. **91**:279.

Shaver, P.A., E. Churchwell, and A.H. Rots. 1977. Astron. Astrophys. **55**:435.

Shaver, P.A., E. Churchwell, and C.M. Walmsley. 1978. Astron. Astrophys. **64**:1.

Shaver, P.A., R.X. McGee, L.M. Newton, A.C. Danks, and S.R. Pottasch. 1983. Mon. Not. R. Astron. Soc. **204**:53.

Shields, G.A. 1978. Astrophys. J. **219**:565.

Shore, B.W. 1969. Astrophys. J. **158**:1205.

Shull, J.M. 1980. Astrophys. J. **238**:860.

Silverglate, P.R. 1984a. Astrophys. J. **278**:604.

Silverglate, P.R. 1984b. Astrophys. J. **279**:694.

Simpson, J.P. 1973a. Astrophys. Space Sci. **20**:187.

Simpson, J.P. 1973b. Publ. Astron. Soc. Pacific **85**:479.

Smith, L.F., and P.G. Mezger. 1976. Astron. Astrophys. **53**:165.

Tenorio-Tagle, G. 1979. Astron. Astrophys. **71**:59.

Terzian, Y. 1980. In P.A. Shaver (ed.), Radio Recombination Lines. Dordrecht: Reidel, p. 75f.

Terzian, Y., P.G. Mezger, and J. Schraml. 1968. Astrophys. Lett. **1**:153.

Terzian, Y., and A. Parrish. 1970. Astrophys. Lett. **5**:261.

Thum, C. 1980. In P.A. Shaver (ed.), Radio Recombination Lines. Dordrecht: Reidel, Fig. 9, p. 50.

Thum, C., P.G. Mezger, and V. Pankonin. 1980. Astron. Astrophys. **87**:269.

Troland, T.H., and C. Heiles. 1977. Astrophys. J. **214**:703.

Turner, B.E., and H.E. Matthews. 1984. Astrophys. J. **277**:164.

Val'tts, I.E. 1982. Soviet Astron.—AJ. **27**:18.

van Gorkom, J. H. 1980. *In* P.A. Shaver (ed.), Radio Recombination Lines. Dordrecht: Reidel.

van Gorkom, J.H., U.J. Schwarz, and J.D. Bregman. 1985. *In* H. van Woerden, R.J. Allen, W.B. Burton, (eds.), The Milky Way Galaxy. Dordrecht: Reidel.

van Gorkom, J.H., P.A. Shaver, and W.M. Goss. 1979. Astron. Astrophys. **76**:1.

Vrba, F.J., K.M. Strom, S.E. Strom, and G.L. Grasdalen. 1975. Astrophys. J. **197**:77.

Wade, C.M. 1958. Aust. J. Phys. **11**:388.

Wadiak, E.J., C.L. Sarazin, and R.L. Brown. 1983. Astrophys. J. Suppl. **53**:351.

Walmsley, C.M., and W.D. Watson. 1982. Astrophys. J. **260**:317.

Walmsley, C.M., E. Churchwell, and Y. Terzian. 1981. Astron. Astrophys. **96**:278.

Watson, W.D., L.R. Western, and R.B. Christensen, 1980. Astrophys. J. **240**:956.

Wilson, T.L. 1980. In P.A. Shaver (ed.), Radio Recombination Lines. Dordrecht: Reidel, p. 205.

Wilson, T.L., and T. Pauls. 1984. Astron. Astrophys. **138**:225.

Wink, J.E., W.J. Altenhoff, and P.G. Mezger. 1982. Astron. Astrophys. **108**:227.

Zuckerman, B.M., and P. Palmer. 1968. Astrophys. J. **153**:L145.

3. Neutral Hydrogen and the Diffuse Interstellar Medium

SHRINIVAS R. KULKARNI and CARL HEILES

3.1 Introduction: Fundamentals

Neutral atomic hydrogen (HI) is an important component of the interstellar medium (ISM). The Galaxy has been estimated to contain about 4.8×10^9 $M_\odot$ of HI (Henderson et al. 1982)*. Estimates for the total amount of H_2 range from 3.5×10^9 $M_\odot$ (Sanders et al. 1984), nearly equal to the HI estimate, to a value that is only 25% of the HI estimate (Bloemen et al. 1986). Galactic HI constitutes about 4.4% of the mass of the visible matter (Bahcall et al. 1983). The mean surface density distribution of HI is roughly constant from about 4 kpc to 20 kpc; however, HI dominates H_2 in mass beyond Galactocentric radius 8 kpc (Blitz et al. 1983). Unlike H_2, HI is not concentrated in a small number of giant clouds. Estimates of the "filling factor", the fraction of the Galactic interstellar space occupied by hydrogen, range from 20% to 90%. These factors establish the preeminence of atomic hydrogen in the dynamics and evolution of the ISM.

Our knowledge of the properties of Galactic HI have mainly resulted from the study of the 21-cm line of HI. In the HI atom, the magnetic moment of the proton interacts with the combined magnetic field generated by the orbiting electron and the magnetic moment of the electron. This interaction results in the hyperfine splitting of all the energy levels of the hydrogen atom. The 21-cm line is the result of the hyperfine splitting of the ground state ($1^2S_{1/2}$). This particular hyperfine transition is sometimes referred to as the "spin-flip" transition since it is dominated by the interaction of the two magnetic moments. Hyperfine lines are magnetic dipole transitions. This, together with the low frequency of the 21-cm line, makes the Einstein A coefficient for the 21-cm line exceedingly small: 2.85×10^{-15} s^{-1}. Thus an HI atom, after being excited, typically waits 12 million years before spontaneously decaying.

The 21-cm line of hydrogen was the first spectral line to be discovered at radio wavelengths. Towards the end of the Second World War, the Dutch astronomer van de Hulst predicted the detectability of the interstellar 21-cm line of HI. Seven years later, Ewen and Purcell of Harvard University made the first detection of 21-cm line radiation from interstellar HI. The Dutch effort, though initiated earlier than that of the Harvard group, suffered a serious setback from fire in their laboratory. The very next day after the discovery, Purcell cabled fellow colleagues in Holland and Australia. The discovery papers from the Harvard group (H.I. Ewen

---

* Throughout this article R_o, the radius of the solar circle, is assumed to be 10 kpc.

and E.M. Purcell) and the Leiden group (C.A. Muller and J.H. Oort) were published together with a short cable confirming the detection from the Australian group (W.N. Christansen and J.V. Hindman). An account of this discovery and other such fine examples of cooperation in the early days of radio astronomy can be found in a book celebrating the golden anniversary of Jansky's discovery of cosmic radio waves (Sullivan 1984).

The frequency of this important line has been accurately measured using hydrogen masers and is 1,420,405,751.786 ± 0.01 Hz. Deuterium, like hydrogen, has a hyperfine transition, but at about 327 MHz. This important line, despite many painstaking efforts, has so far eluded detection at radio wavelengths.

3.1.1 Radiative Transfer and Excitation of the HI Line

The usual form of the equation of radiative transfer is (Spitzer 1978)

$$\frac{dI(v)}{ds} = j(v) - \kappa(v)I(v) \tag{3.1}$$

where $I(v)$ is the specific intensity at frequency v, s is the distance from the observer along the line of sight, and $\kappa(v)$ and $j(v)$ are the volume absorption coefficient and emissivity, respectively. The physical interpretation of Equation (3.1) is quite straightforward: over a small distance along the line of sight, $I(v)$ is increased by a source contribution, $j(v)$, and decreased by local absorption of the incoming beam, $\kappa(v)I(v)$.

At centimeter wavelengths the Rayleigh-Jeans limit applies, and it is more convenient to use T_B, the brightness temperature, which is linearly related to the specific intensity: $I(v) = \frac{(2v^2 k T_B)}{c^2}$. The equation of radiative transfer in the HI line can then be written as

$$\frac{dT_B(v)}{d\tau(v)} = T_s - T_B(v) \tag{3.2}$$

where T_s is the "excitation temperature" (often called the "spin temperature" for the 21-cm line) and is a measure of the emissivity of the material (defined below) and τ is the opacity along the line of sight. This equation immediately shows the advantage of using T_B: in the limit of large τ, $T_B = T_s$. Thus the brightness temperature of a blackbody is equal to its physical temperature.

Consider an HI cloud along a line of sight. Let the number density of atoms in the upper and the lower level be n_2 and n_1, respectively, and let g_2 and g_1 be the corresponding statistical weights. T_s is defined by

$$\frac{n_2/g_2}{n_1/g_1} = e^{-E/kT_s} . \tag{3.3}$$

For the HI atom, $g_2 = 3$ and $g_1 = 1$; E/k, the level separation, is a mere 0.07 K. The relations between T_s and other "temperatures" are discussed below. For the special situation of HI gas in thermodynamic equilibrium at temperature T, $T_s = T$.

In the most general case, both T_s and τ can vary from position to position along the line of sight and then the solution to Equation (3.2) is nontrivial. However, for

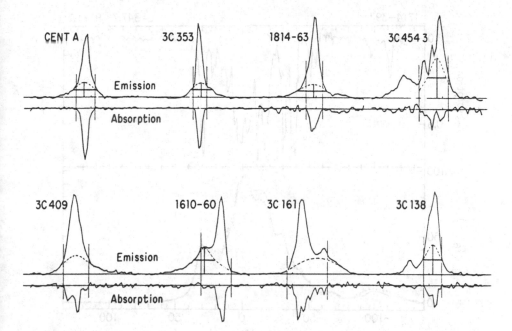

Fig. 3.1. Comparison of eight emission and absorption spectra obtained at intermediate latitdues (Radhakrishnan et al. 1972). The observational equivalent of $\Delta T_B(v)$, as defined in Equation (3.5), is the quantity plotted. The velocity limits of the absorption spectra are marked by thin vertical lines. The dashed lines indicate a fit to an optically thin component. The typical velocity dispersion is about 9 km s^{-1}.

the simple case of an isolated, single, homogeneous HI cloud, Equation (3.2) can be solved to yield

$$T_B(v) = T_{bg}(v)e^{-\tau(v)} + T_s(1 - e^{-\tau(v)}) \tag{3.4}$$

where $T_{bg}(v)$ is the brightness temperature of radiation incident of the far side of the cloud. In an actual measurement, we plot $\Delta T_B(v) = T_B(v) - T_{bg}$:

$$\Delta T_B(v) = T_B(v) - T_{bg} = (T_s - T_{bg})(1 - e^{-\tau(v)}) . \tag{3.5}$$

Examples of $\Delta T_B(v)$ appear in Figures 3.1 and 3.2. Here $\tau(v)$ is the optical depth at velocity v and is related to $N(v)$, the column density (number of atoms with a velocity v in a cylinder of base 1 cm^{-2}) by

$$\tau(v) = \frac{N(v)}{C \times T_s} . \tag{3.6}$$

Here v is assumed to be in km s^{-1} and the constant C is 1.83×10^{18} cm^{-2} K^{-1} (km/s)$^{-1}$.

In order to gain a physical feeling for these equations, we consider two extreme limits of the single cloud with no background radiation (i.e., $T_{bg} \ll T_s$):

(1) Optically thin case [$\tau(v) \ll 1$]. In this case, $T_B(v) = T_s\tau(v) = N(v)/C$ and the measured brightness temperature is proportional to the column density per unit

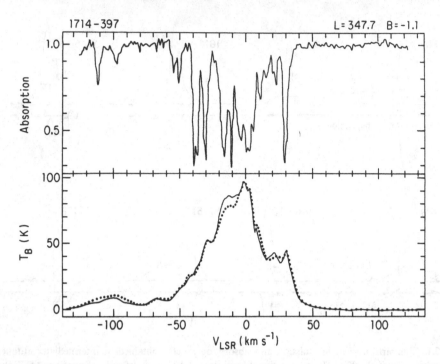

Fig. 3.2. The HI emission (top) and absorption (bottom) spectra towards 1714-397 (Dickey et al., 1983). Note the sharp absorption features compared to the emission spectrum.

velocity. Physically what this means is that almost all of the 21-cm photons emitted by spontaneous emission escape the cloud without being absorbed. The emission rate is practically independent of T_s since E/kT_s is exceedingly small for any reasonable T_s. Thus the the number of photons leaving the cloud is a direct measure of the HI column density.

(2) Optically thick case $[\tau(v) \gg 1]$. In this case $T_B = T_s$, i.e., the brightness temperature is simply the spin temperature. Owing to the high optical depth, 21-cm photons emitted within the interior of the cloud get absorbed by foreground HI atoms. Only the photons emitted within $\tau \lesssim 1$ of the front surface manage to leave the cloud. Thus, the observed brightness temperature is independent of the column density of the cloud and depends only on the temperature of the cloud.

Equation 3.5 makes it clear that we can see HI either in "emission" or in "absorption" depending on whether T_s is larger or smaller than T_{bg}, respectively. In the former case, the emission from the foreground HI more than makes up for the attenuation of the background radiation, and vice versa for the latter case.

For the special situation of a cloud having $T_s = T_{bg}$, the emission from the foreground is exactly compensated by the absorption of the background and thus there is no net emission. For example, an HI cloud at the same temperature as the 3 K cosmic background radiation is not detectable. This effect can be crucial in some cases involving low-density extragalactic clouds, as discussed below. However, the 3 K radiation can be neglected for galactic clouds, because $T_s \gg 3$ K.

What determines T_s? A particularly lucid discussion of this point is given by Field (1958). First, HI atoms reside in the cosmic blackbody, which by itself would make $T_s = 3$ K. If this were the only excitation mechanism, the 21 cm line would remain unobservable except towards continuum radio sources, where T_{bg} would depart from 3 K. Fortunately, however, there are two other excitation mechanisms at work: collisions and Lyα radiation.

If collisions were the only influence, then it is clear from thermodynamic considerations that the excitation temperature would equal the gas kinetic temperature. Collisions compete with the 3 K radiation. Because there is no unique temperature, we have a non-LTE situation and must use statistical equilibrium to calculate the excitation temperature. Let P_{12} and Q_{12} be the probability that an atom in state 1 jumps to state 2 by collisional and radiative excitation, respectively; and the same letters with reversed subscripts denote probabilities in the opposite direction. Then time-independent statistical equilibrium requires equality of the rates in each unit volume:

$$n_1(P_{12} + Q_{12}) = n_2(P_{21} + Q_{21}) . \tag{3.7}$$

The collisional rates P can be expressed in terms of a cross section σ; at $T_k = 1000$ K, $\sigma_{21} \approx 6$ Bohr-orbit areas, decreasing slowly with increasing temperature and decreasing rapidly for $T \lesssim 300$ K (Allison and Dalgarno 1969).

The radiative rates are expressed with the usual Einstein coefficients A and B, together with the mean intensity J averaged over the spectral line. In the Rayleigh-Jeans approximation, J is more conveniently expressed by the brightness temperature averaged over all directions and over the spectral line, $\langle \bar{T}_B \rangle$. We have, for the downward rate,

$$Q_{21} = A_{21} + JB_{21} = \left(1 + \frac{k\langle \bar{T}_B \rangle}{h\nu}\right) A_{21} \tag{3.8}$$

with a similar equation, but without the A term, for Q_{12}.

Putting all this together requires some care in the algebra, because no first-order terms in $h\nu/kT$ can be neglected. The excitation temperature is a weighted mean of the kinetic and radiation temperatures:

$$T_s = \frac{T_k + y\langle \bar{T}_B \rangle}{1 + y} \tag{3.9}$$

where

$$y = \frac{kT_k}{h\nu} \frac{A_{21}}{P_{21}} . \tag{3.10}$$

Neglecting electron collisions, which is justified if the ionization fraction [HII/HI] $\lesssim 0.05$,

$$y \sim \frac{(T/1000)}{(n_{HI}/0.2)} \quad (T \gtrsim 300 \text{ K}) . \tag{3.11}$$

Collisions dominate if $y \gg 1$, in which case $T_s = T_k$. This is certainly the case for HI in cold clouds (the CNM). However, it is not the case for the warm HI (the WNM). For typical WNM parameters ($n \approx 0.4$ cm^{-3}, $T \approx 8000$ K), $y \approx 4$ and $T_s \approx T_k/5$. It is not generally realized that a major component of Galactic HI is so far out of collisional equilibrium in the 21-cm line.

There also occur extragalactic situations where collisional equilibrium is not obtained. The currently most spectacular example is the large intergalactic HI cloud in the M96 group (Schneider et al. 1983). The volume density in this cloud is of order 5×10^{-3} cm^{-3}, which at *any* T_k is small enough so that collisional excitation is ineffective.

Nevertheless, in both the galactic WNM and the M96 cloud, the excitation temperature still remains close to the kinetic temperature because of Lyα excitation. Lyα radiation connects the ground electronic state (principal quantum number $n = 1$) with the next higher electronic state ($n = 2$). The Lyα line lies in the UV at a wavelength of 1216 Å. Of course, its "ground state" really consists of two levels, i.e., the ones responsible for the 21-cm line. Thus, Lyα transitions from the lower level of the 21-cm line have a slightly shorter wavelength than those from the upper level. If the intensity of the Lyα line at these two wavelengths differs, then the upward transition rates out of the two 21-cm levels also differ. Thus, Lyα radiation can effect the excitation temperature of the 21-cm line.

Lyα radiation holds a place of unique importance in interstellar physics. In HI gas, essentially all of the atoms lie in the $n = 1$ state because the Lyα Einstein A is 6×10^8 s^{-1}. This means that an average atom can stay in the $n = 2$ electronic state for only $A^{-1} = 2 \times 10^{-9}$ s before it drops to the $n = 1$ electronic state. This is enormously shorter than the time a typical atom spends in the $n = 1$ electronic state before being reexcited to the $n = 2$ state. Thus, essentially all of the atoms are in the $n = 1$ state.

The net result is that the optical depth to Lyα photons is huge. The ratio of Lyα to 21-cm-line optical depth is 2.4×10^9 ($T_s/1000$ K). For example, the M96 cloud has $N_{HI} = 6 \times 10^{19}$ cm^{-2} and $\tau_{Ly\alpha} \approx 2 \times 10^6$. A typical Ly$\alpha$ photon is scattered thousands of times before it can escape from such a cloud or be destroyed. This large number of scatterings couples the radiation field in the Lyα line to the matter, because upon each scattering the atom and photon recoil so as to conserve momentum.

Field (1959) has given an elegant analytical solution to this complicated problem in radiative transfer, together with a lucid physical interpretation of the results. After many scatterings, the Lyα line flattens in the middle, but has a slight slope such that the mean intensity $J \propto \exp -(h\nu_{Ly\alpha}/kT_k)$. In other words, its color temperature is T_k. *This is just the exact dependence required to drive the 21-cm-line excitation temperature towards T_k!* However, we note the following qualification: if the Lyα optical depth is small over the turbulence length scale, then the Lyα color temperature will tend toward the temperature indicated by the total line width, which always exceeds T_k.

Whether or not Lyα radiation can override the 3 K blackbody radiation in the excitation of the 21-cm line depends on the intensity of the Lyα radiation. Lyα photons cannot work their way into the cloud from the outside because the Lyα optical depth is so large. Thus, the Lyα photons must be generated inside the cloud itself. A Lyα photon is created whenever a proton and electron recombine to form an atom. In statistical equilibrium, the rate of recombination is equal to the rate of ionization of HI atoms. Such ionization occurs by cosmic-ray particles and extreme UV (EUV) and soft X-ray photons.

Deguchi and Watson (1985) have treated this problem in detail. The ionization rate is large enough in the Galactic environment to make $T_s \approx T_k$ for almost any HI cloud observable in the 21-cm line. It is probably intense enough inside the M96 cloud to force T_s close to T_k, although there is some uncertainty because the intergalactic flux of energetic photons and cosmic-ray particles is not well known. In fact, Deguchi and Watson point out that an independent measurement of T_s from HI in absorption would provide valuable information on the density of these ionizing particles in intergalactic space. In addition, they point out the interesting fact that the column density in the M96 cloud may be near the minimum for which Lyα excitation can be effective. If so, similar clouds with lower column densities have $T_s \approx 3$ K and are therefore undetectable with the 21-cm line in emission. This is unfortunate: the existence of individual, isolated clouds of primordial HI that have not yet formed galaxies would be important for theories of galaxy formation.

3.1.2 Observations of the Diffuse Medium

(a) Brief Review of the Four Phases: CNM, WNM, WIM, HIM

Until recently, the diffuse ISM was thought to consist of just three major components: a cold ($T \sim 80$ K), dense, and largely neutral HI (cold medium or CNM); a warm ($T \lesssim 8000$ K) HI, either surrounding the cold clouds in an envelope or pervading much of the space as an "intercloud medium" (warm medium or WNM); and a hot ($T \sim 10^6$ K), highly ionized medium (the hot ionized medium or HIM). Recently, diffuse Hα emission line data (Reynolds 1984 and references therein) indicate that a nontrivial fraction of the interstellar volume is filled by a highly ionized, warm ($T \sim 8000$ K) hydrogen (warm ionized medium or WIM). These four phases are discussed in detail in later sections.

The basic data that differentiate CNM and WNM are beautifully illustrated in Figure 3.1. This figure exhibits both emission spectra, i.e., $T_{bg} \ll T_s$ in Equation (3.5), and absorption spectra with $T_{bg} \gg T_s$. The narrow emission components have associated absorption features. However, the broad components, shown in dotted lines, do not exhibit corresponding absorption features. Since the opacity is inversely proportional to T_s, the narrow components are cold (the CNM) and the wide components are warm (the WNM).

(i) The WNM (see Section 3.3.2). Roughly half of all HI is WNM. It has a low volume density and fills a substantial fraction of the interstellar volume. Cooling rates are small at low densities, so the total power required to heat the gas is small compared to that required for other phases; however, the heating agent has yet to be identified.

The distribution of the WNM can be determined from HI emission data. Mebold (1972) decomposed about 1200 emission spectra at $b \sim 30°$ and $0° < l < 360°$ into narrow ($\sigma < 5$ km s^{-1}) and wide ($5 < \sigma < 17$ km s^{-1}) Gaussian components and found:

1. In almost all directions the wide-σ component HI is present. In no direction can one obtain an emission profile containing only narrow-σ components. About 40%

of the total HI lies in the wide component. Thus warm HI is undoubtedly widespread and has a non-negligible filling factor.

2. The wide-σ HI has a z-scale height twice as large as the narrow-σ HI. The variation of the mean velocity with l was used to obtain the z-distribution of the wide component: $n_w(z) \approx 0.20e^{-(|z|/220)}$ cm^{-3}. The uncertainty in the scale height is large—about 70 pc. More recently, Lockman (1984) found a long exponential tail (scale height 480 pc) for part of the WNM.

3. An upper limit of $T_k = 9600$ K, corresponding to a thermal velocity disperison of 8.8 km s^{-1}, was derived for the wide component from the increase in the observed velocity dispersion from $b = 90°$ to $b = 30°$. A more detailed analysis shows that the velocity distribution of the WNM has a long tail (Kulkarni and Fich 1985). It is probably this tail which leads to Lockman's large scale height component.

Heiles (1980) has extended this analysis for $|b| > 10°$ and finds that the wide-σ component is indeed present along almost all lines of sight outside of 8 kpc Galactic radius and in quantities predicted by (2) above. He also reports finding large holes, occupying 10% to 20% of the volume, in which there is a deficiency of wide-σ HI. The wide-σ HI is also absent on the very local scale of about 100 pc. This local "hole" in the HI is also found from UV absorption line data and from reddening studies.

(ii) The CNM (see Section 3.3.3). The CNM is distributed in relatively dense clouds that occupy an insignificant fraction of the interstellar volume. Owing to the high volume density, the cooling rate per atom is high and the power required to keep the clouds at their observed temperature is large, nearly 10^{42} erg s^{-1} over the whole Galaxy.

Narrow features in the emission spectrum are almost always easily detected in the absorption spectrum. As an example, we show in Figure 3.2 a low-latitude absorption spectrum from our VLA survey. Notice the complexity and the distinctness of the absorption dips. Each sharp absorption feature is supposed to arise in the same cold gas that was responsible for optical interstellar lines and hence the nomenclature, "HI clouds".

Both HI emission and absorption data are needed in order to understand HI clouds, and we discuss these matters extensively in Section 3.3. The discussion in Section 3.2 shows that clouds are filaments and/or sheets, rather than spherical. Here we note three general properties:

1. The narrow features, unlike the WNM, are not seen along every line of sight. At intermediate latitudes, a strong absorption feature can be seen typically in only one out of three directions. Thus the strongly absorbing clouds have a much smaller filling factor than the WNM.

2. The velocity widths of the absorption features are typically narrower than those of the corresponding emission features.

3. The cloud-cloud velocity dispersion is approximately 6.9 km s^{-1}. This is somewhat smaller than velocity dispersion of the WNM.

Historically, the first interpretation was that the clouds represented isothermal fluctuations of the interstellar HI gas. However, it was soon realized that emission and absorption spectra were not similar. Shuter and Verschuur (1964) suggested that temperatures differences were the prime cause of this dissimilarity. Clark (1965) formalized this interpretation in his now famous "raisin-pudding" model of the ISM. In this model, clouds are discrete concentrations of cold HI confined by a much warmer, pervasive intercloud medium HI; pressure equilibrium between these two components was invoked for stability reasons.

(iii) The WIM (see Section 3.3.5). Observations of Hα emission force us to include yet another medium—a warm ionized medium (WIM). The Hα emission appears to be widely spread, constitutes $\sim 30\%$ of the mass of diffuse gas, and is energetically important: to maintain its ionization over the whole Galaxy requires $\gtrsim 10^{42}$ erg s^{-1}.

(iv) The HIM. Our qualitative understanding of the ISM changed dramatically with the early-1970s detections of diffuse soft X-ray emission and of OVI UV absorption lines. The interpretation of these observations is that a large fraction of the interstellar space is occupied by the HIM within which the WNM and CNM exist in pressure equilibrium [the McKee and Ostriker (1977) "three-phase" model; Section 3.1.6]. However, there is no consensus on the filling factor of HIM on a Galactic scale. The simple reasons are that the X-ray data penetrate no further than ~ 100 pc and that the OVI lines do not appear to come from a widely distributed component [see review by Cowie and Songaila (1986)].

(v) Global Characteristics of the CNM, WNM, WIM, and HIM. There is much disagreement concerning the fundamental parameters of the four components. First, while there is overall agreement that the Sun itself is imbedded in a local "bubble" of HIM, there is no agreement as to whether the local ISM is highly typical, with the HIM filling the lion's share of the interstellar volume. The topology of HIM is unclear: does the HIM reside in connected "tunnels", as expected in a supernova-dominated ISM, or does it exist in many independent bubbles? Second, the spatial relation of WNM to CNM is controversial: in some models, WNM forms a "sheath" around the CNM clouds, and in other models, the CNM clouds are immersed in an all pervasive WNM. Third, there is no consensus about the topology, shapes, and sizes of the CNM "clouds," nor is it clear whether the clouds are transient structures. Fourth, there is a variety of observational evidence for electrons in the ISM, but we do not know the degree of ionization of the CNM and WNM. Finally, the morphology of the WIM is unknown because only a tiny fraction of the sky has been mapped in Hα. Future progress of ISM theories depends crucially on answering these questions.

(b) Galactic Distribution of HI

To zeroth order, the run of $N_H(b)$, the integrated column density of HI at galactic latitude b, shows that the HI layer can be approximated by a stratified horizontal atmosphere model, i.e., $N_H(b) = N_{H,\perp}/\sin(|b|)$, where $N_{H,\perp} \equiv \langle N_H(b)\sin(|b|)\rangle$ is the mean column density reduced to the pole; the total vertical column density is $2N_{H,\perp}$. We will henceforth use the subscript "$\perp$" to indicate any mean column density reduced to the pole. In more detail, $N_{H,\perp}$ systematically decreases with increasing $|b|$. This is shown by two contrasting determinations of $N_{H,\perp}$: $\sim 3 \times 10^{20}$ cm^{-2} from

104   Shrinivas R. Kulkarni and Carl Heiles

data around $|b| \lesssim 45°$ (Heiles 1976a); and a smaller value, $\sim 2 \times 10^{20}$ cm^{-2}, directly *at* the galactic poles. This is one of several data indicating that the Sun is located in a local hole or a region of HI deficiency.

The detailed vertical distribution of the HI disk is best studied by looking at distant HI. Lockman (1984) has studied HI associated with the "tangent points" and finds that 8% of the HI emission comes from gas with a $|z|$-height > 480 pc. (Note: Lockman's value for $N_{H,\perp}$ is too low because of 21-cm-line opacity effects.) In detail, the HI layer may be approximated by the sum of two Gaussians with small scale heights and an exponential with a large scale height of about 480 pc. The vertical distribution of *local* HI may also be obtained by comparing the amount of HI inferred from Lyα absorption towards high-latitude OB stars with 21-cm observations (Lockman et al. 1986). Such observations, though meager when compared to the 21-cm data, confirm Lockman's (1984) picture.

On a larger scale, one must rely on low-latitude HI emission data (e.g., Biltz et al. 1983), which do not discriminate between CNM and WNM. The HI has a constant surface density from Galactocentric radius $R \sim 4$ kpc to ~ 20 kpc, beyond which there is a decrease. Galactic HI is still detectable at distances as large as $R \sim 30$ kpc! The scale height is roughly constant inside the solar circle. Beyond the solar circle the HI layer flares, reaching scale heights of 1 to 2 kpc. The HI disk is reasonably flat out to about 18 kpc, beyond which it is badly warped. In the outer Galaxy a clear spiral pattern in HI can be discerned. Efforts to map the detailed distribution of HI in the inner Galaxy have not been equally successful, primarily due to the ambiguity between "near" and "far" points (Chapter 7).

3.1.3 Interstellar Pressure

Pressure *equality* is the cornerstone of all models of ISM: the pressure in each phase is assumed to be equal to the mean interstellar pressure. The theoretical basis for this assumption is that the time to equalize any pressure imbalance, the sound crossing time, is smaller than all other time scales such as the mean time between supernovae shocks, the recombination time scale, and the cooling time scale.

The *value* of the mean interstellar pressure is a crucial parameter. In particular, in thermal equilibrium the WNM and the CNM can coexist only over a limited range of pressure (Figure 3.3). Since both the WNM and the CNM are observed to exist, the interstellar pressure must lie within this small range of pressure, at least near the solar circle—if thermal equilibrium actually is obtained.

Before we plunge ahead into a discussion of the observations we would like to caution the reader that almost everyone, observer and theorist alike, assumes that interstellar pressure is simply the standard gas kinetic pressure, $P = nkT$. However, in the interstellar medium, the pressures due to magnetic fields $(B^2/8\pi)$ and to cosmic rays are comparable to nkT (Spitzer 1978, p. 234). Furthermore, the observed line width is larger than the thermal value, so the "turbulent pressure" exceeds the thermal value. Unfortunately, the interaction of these pressure components with each other is not understood theoretically. Anyway, in order to verify the assumption of pressure equality one must measure the *total* pressure in each phase, and in addition develop the necessary theory. Unfortunately, most of our pressure probes (see below) measure only the *particle* pressure.

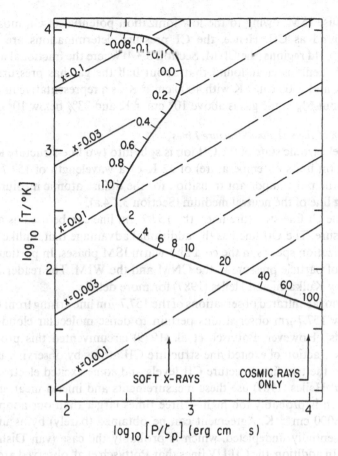

Fig. 3.3. Temperature in thermal equilibrium vs. (P/ζ_p) for "average depletion" and EIP heating by cosmic rays (Draine 1978). The dashed lines are lines of constant ionization fraction n_e/n_H. The numbers at selected points on the curve indicate the ratio of grain heating to cosmic-ray heating. Two values of P/ζ_p are indicated. Both assume $nT = 3000$ cm^{-3}. "Cosmic rays only" assumes cosmic-ray ionization with $\zeta_t = 0.7 \times 10^{-16}$ s^{-1}. "Soft X-rays" assumes soft X-ray ionization with $\zeta_t = 13 \times 10^{-16}$ s^{-1} (Section 3.1.4.b).

A small note on nomenclature: unlike standard physics textbooks, here it is usually more convenient to drop Boltzmann's constant k and redefine pressure as simply nT. Thus the units of pressure are cm^{-3} K. Note that n is the total particle density, and hence for a completely ionized medium such as the WIM, $P = 2n_e T$.

We now review the meager measurements of interstellar pressure.

(a) Excitation of CI Fine Structure

The ground electronic state of CI is split into three fine-structure levels. Jenkins et al. (1983) measured the fractional population of these levels by observing the ultraviolet transitions between the ground and some excited electronic states. The population ratios are determined by a balance between the collisional excitation rate, which depends both on n and T, and the radiative decay. Modeling this microphysics, Jenkins et al. obtained an estimate of the interstellar pressure.

In the diffuse ISM, owing to the low ionization potential of CI, most gaseous carbon is found as CII. Hence, the CI pressure determinations are biased to high-density, cold regions (the CNM; Section 3.3.4) where the fractional abundance of CI is high. Jenkins et al. found that about half the gas has pressure between 2500 cm^{-3} K and 6500 cm^{-3} K with 4000 cm^{-3} K as a representative intermediate pressure. About 6% of the gas is above 10^4 cm^{-3} K and 33% below 10^3 cm^{-3} K.

(b) Excitation of the CII Fine-Structure Lines

The ground electronic state of the CII ion is split into two fine-structure levels with a separation (in units of temperature) of 92 K or a wavelength of 157.7 μm. This transition is the most important transition for the diffuse atomic medium: it is the main cooling line of the neutral medium (Section 3.1.4.a).

Just like the CI fine-structure lines, the 157.7-μm line can be used as a probe of particle pressure. The CII line has the additional advantage that, unlike CI, it is a majority ionization species in the cold and warm ISM phases. In particular, it is a good probe of particle pressure in the CNM and the WIM. The reader is referred to a review by Kulkarni and Heiles (1987) for more details.

There are no far-infrared observations of the 157.7-μm line arising from the diffuse ISM (the few 157.7-μm observations pertain to dense molecular clouds and not diffuse clouds). However, Pottasch et al. (1979) circumvented this problem and measured the fraction of excited fine-structure CII atoms by observing ultraviolet lines between the two fine-structure CII levels and some excited electronic states. Kulkarni and Heiles (1987) use these measurements and infer a mean interstellar pressure which is probably too high—three times larger than our adopted mean pressure of 3000 cm^{-3} K. Agreement can be obtained (barely) by assuming that carbon is essentially undepleted, which is probably the case (van Dishoeck and Black 1986). In addition, the CII UV lines that Pottasch et al. observed are optically thick and there is some reasonable doubt about the accuracy with which the column density of the excited CII ions can be inferred. The best way to probe the pressure in the ISM would be by observations of the 157.7-μm line directly in the far infrared.

(c) Pressure of the WIM

We postpone a detailed discussion of this method to Section 3.3.6. In that section, we use DM and Hα observations to infer the true electron density of the WIM to be about 0.25 cm^{-3}. The temperature of the WIM has been measured to be about 8000 K (Section 3.3.5.c). Thus the mean pressure in the WIM is $\sim 4000 \text{ cm}^{-3}$ K— close to our assumed mean pressure.

To summarize, we have estimates of the CNM and WIM *particle* pressures: there are no estimates for the WNM. In the absence of disturbances (e.g., shocks, H-ionization by starlight), pressure equality is expected from theory. The CI data confirm this, although there are also wide pressure variations. A large-scale, long-term global average would almost certainly yield pressure equality among the three phases CNM, WNM, and WIM.

In this article, we generally frame our discussion in terms of a uniform ISM pressure and adopt a value of 3000 cm^{-3} K. However, it is important to remember that there do exist wide fluctuations.

3.1.4 Theory of the Kinetic Temperature of Diffuse Interstellar Gas

The temperature of any object settles at the point where the rate of energy loss from cooling equals the rate of gain from heating. Thus, the theoretical discussion of this topic requires consideration of the relevant heating and cooling processes; see the review by Dalgarno and McCray (1972).

(a) Cooling Processes

We first consider cooling processes, which are well understood because they rely on purely atomic processes. The most important process is collisional excitation of fine-structure transitions of heavy-element atoms or ions by H atoms and by electrons. The excitation energy is lost from the gas when the heavy element spontaneously emits a photon in the transition. At low densities, every collisional excitation results in the emission of a photon. Thus, the volumetric cooling rate is proportional to the collisional rate, i.e., to the *square* of the volume density. Electrons as colliding particles have much higher cross sections than HI atoms, and both are important in the WNM and the CNM.

For the transition of a particular atom or ion to be an effective coolant it must satisfy three criteria:

1. The heavy element must be sufficiently abundant, and must exist in the gas phase instead of being "depleted" onto solid grains.
2. The transition must be in the favored ionization state of the element (in diffuse interstellar gas, elements are generally ionized if their ionization potentials are smaller than that of HI, 13.6 eV).
3. The transition must have a small enough energy so that, at typical temperatures of interstellar gas, the exponential energy factor $e^{-(E/kT)}$ is not unreasonably small (this is why we consider only fine-structure transitions for interstellar cooling processes).

There is one element that satisfies these criteria better than any other: carbon. In the absence of depletion, the 157.7-μm fine-structure line of ionized carbon (Section 3.1.4.d) dominates all other cooling processes as long as $T_k < 600$ K; above 600 K, FeII takes over. However, with depletion (van Dishoeck and Black 1986), gaseous Fe is often so underabundant that CII dominates the cooling for $T_k \lesssim 10^4$ K. Above 10^4 K, collisional excitation of O ions and, in addition, H atoms (even with the energy exponential factor $\approx 10^{-6}$!) causes a dramatic increase in cooling rate.

In the above paragraphs we have implicitly assumed that the cooling photon is not hindered from escaping by the opacity of the cooling line itself. The opacity of the CII line, τ_{CII}, is not always negligible. If all C is in the gas phase, i.e., no depletion onto dust grains so that $[C/H] = 4 \times 10^{-4}$, then

$$\tau_{CII} = \frac{T_k \tau_{HI}}{181 \text{ K}} \frac{\delta v_{HI}}{\delta v_{CII}}. \tag{3.12}$$

Here τ_{HI} is the optical depth in the 21-cm line and δv is the velocity width of a line. Note that $T_k \tau_{HI} > T_B$, and in the Galactic plane we typically observe $T_B > 100$ K; thus, near $b = 0°$, the optical depth of the CII line cannot be neglected and may be high. For individual diffuse clouds, τ_{CII} is nearly always small.

So much for theory—is there *observational* confirmation of this cooling mechanism? Pottasch et al. (1979) used interstellar CII UV absorption lines in stellar spectra to measure the interstellar column density of fine-structure excited CII in nine clouds towards eight stars. They also used UV absorption data to measure the column density of HI nuclei. The result was (Energy loss per H nucleus) $\approx 1.0 \times 10^{-25}$ erg cm^{-3} s^{-1}. In principle, this is an observational determination of the required heating rate per H atom, because in statistical equilibrium heating must equal cooling. However, the UV line is optically thick, which leads to large uncertainties. The best way to determine this quantity would be directly from the 157.7-μm cooling transition itself.

(b) Heating Processes

Heating processes are more difficult than cooling processes because they depend on energy input from external sources. Energy can come in three forms: starlight, energetic ionizing particles (EIPs: cosmic rays and photons), and mechanical energy (shocks and waves). The physical principles involved with the former two are better understood than those with the latter because they involve straightforward atomic processes. Mechanical heating by sound waves (Spitzer 1982) and by hydromagnetic waves (Zweibel and Josafatsson 1983) might well be important. However, these mechanisms are complicated and hard to understand, and we shall not discuss them further.

(i) Starlight. The classical, but ineffective, heating mechanism involves ionization by starlight of heavy-element atoms and ions having ionization potentials less than 13.6 eV. UV starlight, typically with photons distributed roughly uniformly with energy below 13.6 eV, ionizes a CI atom (ionization potential 11.3 eV). The typical energy of the liberated electron is 1.2 eV, corresponding to a temperature of 14,000 K. The electron then merrily collides with a large number of CII ions, exciting the cooling transition discussed above, and finally (after about 10^5 CII excitations) recombines with a CII ion. The resulting CI atom is quickly ionized. The rate of CI ionization, and thus the heating rate, is limited not by the intensity of UV starlight but instead by the rate at which CI atoms are formed by recombination. This makes the volumetric heating rate *independent* of the starlight intensity and proportional to the *square* of the volume density, just like the cooling rate. Thus the equilibrium temperature is independent of physical conditions; it is about 16 K. Only Verschuur's cold clouds (see Crovisier and Kazès 1980) have been found to be this cold. Most clouds are much warmer, and this classical mechanism fails by at least two orders of magnitude.

Heating by photoelectric emission from dust grains dominates all other starlight-related mechanisms (Draine 1978). Grains are photoionized by UV starlight. If the UV intensity is large, then photoejection continues until the grain is left with such a large positive charge that no further ejections can occur. In this case the photoejection rate, and thus the heating rate, is limited by the rate at which the free electrons can recombine with the grains; the volumetric heating rate varies as the approximate *square* of the ISM density. In the opposite case of low UV intensity, the heating rate is limited by the UV intensity itself; the volumetric heating rate varies *linearly* with the density. Grains also cool the gas, because the solid grains are themselves much colder than the gas. This means that at sufficiently high gas temperatures, the grains are more efficient at cooling than at heating. Thus, if grains were the only heating *and* cooling process, the temperature would equilibrate, typically between 5000 and 10,000 K.

The effectiveness of photoelectric emission from grains depends on the photoelectric threshold and yield, and their wavelength dependence. These parameters depend not only on the material, but also on a grain's size, shape, and electron charge. In particular, the parameters' variation with grain size is unknown. For both the cold gas in diffuse clouds (the CNM) and the low-density WNM, only the most optimistic values for grain parameters will suffice to produce the observed temperatures (de Jong 1980; Section 3.1.4.c), and the heating mechanisms remain elusively uncertain. It is conceivable that the population of small grains discovered by the Infrared Astronomical Satellite (IRAS) could ease this disturbing situation.

Starlight is attenuated by dust. However, it is not seriously attenuated in the interiors of diffuse clouds. The most effective photons are in the UV where an extinction of one magnitude corresponds to an HI column density of only about 3×10^{20} cm^{-2}. However, this underestimates the penetrating power of UV photons because most of the extinction results from scattering, not absorption; in other words, grains have high albedos in the UV. The true penetration depth depends on the angular distribution of scattering, which is not accurately known, but the depth is likely to be higher than the above value by a factor of about four. Since the interstellar UV radiation field is roughly isotropic and can enter a cloud from all directions, a typical cloud should be heated effectively throughout its volume.

(ii) Energetic Ionizing Particles (EIPs). EIP heating may be important at low volume densities, because the volumetric heating rates vary only *linearly* with density. The effectiveness of EIPs depends mainly on a single parameter, ζ, the ionization rate per H atom. Ionization can be produced by either cosmic rays or soft X rays.

There are two important values for ζ. One is the *primary* ionization rate ζ_p. The secondary electrons thereby produced cause further ionization; thus the *total* ionization rate ζ_t exceeds the primary rate. The ratio $\zeta_t/\zeta_p \approx 1.6$ for cosmic rays and ~ 2 to 6 for X rays, depending on the energy (Dalgarno and McCray 1972).

There is a big difference between the two ionizing agents: cosmic rays can penetrate clouds, while soft X rays cannot. The HI column density penetration depth for soft X-ray photons of energy E_x is roughly $1.4 \times 10^{17}(E_x/E_0)^{2.5}$ cm^{-2}, where E_0 is the ionization potential of HI. This includes absorption from a cosmic mix of heavy elements having no depletion (Cruddace et al. 1974), so it is an underestimate. According to Draine (1978), observational data imply an average $E_x \sim 74$ eV; this implies a penetration depth of approximately 10^{19} cm^{-2}, about 100 times smaller than the corresponding depth for UV photons and about three times smaller than the edge-to-center HI column density in the median HI cloud.

The situation for cosmic rays is not absolutely clear-cut. Many authors have questioned whether cosmic rays are subject to various instabilities that prevent their penetration into clouds. However, Cesarsky and Völk (1978) show that the instabilities are probably *not* effective, a result which is confirmed by the studies of cosmic ray-induced thermal and ionization equilibrium in *dense molecular* clouds

(Glassgold and Langer 1973). Thus, only simple particle-particle interactions should be considered, for which a 10-MeV cosmic ray can penetrate $\sim 10^{23}$ cm^{-2}, much larger than the HI column density of a typical HI cloud.

The best value of ζ for cosmic rays probably results from the comprehensive fits of theoretical chemical models to observed molecular abundances in clouds by van Dishoeck and Black (1986). The abundance of oxygen-containing molecules is directly proportional to ζ_t, because these molecules form from ion-molecule reactions of which the first is charge exchange of O with H$^+$; H$^+$ is, in turn, produced by cosmic-ray ionization of H. They find $\zeta_t \approx 7 \times 10^{-17}$ s^{-1}.

In the WNM, soft X rays exist to an extent depending on the proximity of sources (hot stars and the HIM) and the amount of intervening gas. Thus, ζ from soft X rays is likely to vary considerably from place to place. Draine (1978) adopts $\zeta_p = 5 \times 10^{-16}$ s^{-1} ($\zeta_t \approx 13 \times 10^{-16}$ s^{-1}); clearly, this value can be no more than a rough average.

(c) The Equilibrium Temperature

Draine (1978) incorporates reasonable estimates for the heating and cooling processes discussed above to calculate temperatures in both equilibrium and time-dependent models. It is theoretically most straightforward to express the equilibrium temperature as a function of the ratio (P/ζ_p), shown for his "standard" grain parameters and "average" depletion of heavy elements onto grains by the solid curve in Figure 3.3. The numbers on the curve are ratios of grain to EIP heating rate. EIPs dominate at high temperatures, i.e., low densities, and above ~ 4000 K the grains cool instead of heat. The dashed lines are lines of constant ionization fraction $x_e = n_e/n_H$.

The "Z" portion of the curve occurs near the density at which linearly density-dependent heating processes become important. If a large fraction of the heavy elements is depleted onto grains, the curve moves to the right at low temperatures (smaller cooling rate by heavy elements in the gas phase, thus higher equilibrium temperature for the same pressure) and the "Z" character of the curve changes: the curve becomes, in essence, a step function with infinite slope at the critical density. See Draine (1978) for details.

Near the critical density, the "Z" shape leads to *three* equilibrium temperatures at the same pressure—and thus three equilibrium densities. Only *two* of these equilibria are *thermally stable* (Field 1965). The middle one, where the curve has slope greater than 1, is unstable because ($d \log p/d \log n$) < 0. Thus, if a parcel of gas undergoes a small density *decrease* and remains on the equilibrium curve, its pressure *increases* relative to its surroundings, so that it will tend to expand and its density will decrease even more.

In the case of high depletion, where the curve is essentially a step function, T is single-valued and, formally, two phases cannot coexist in pressure equilibrium. However, the dividing line is so narrow that, for all practical purposes, the pressures are identical on each side of the step. This case is analogous to the equilibrium between a saturated vapor and its liquid or solid phase.

Where does the ISM fit on Figure 3.3 with $P = 3000$ cm^{-3} K (4.1×10^{-13} erg

cm^{-3})? In clouds (the CNM), ζ_t should equal its cosmic-ray value, 0.7×10^{-16} s^{-1} ($\zeta_p = 0.44 \times 10^{-16}$ s^{-1}), so $\log(P/\zeta_p) = 4.0$. This is far to the right of the "Z", and only one phase—the cold gas—can exist in thermal equilibrium. The equilibrium temperature is ~ 50 K which, as we shall see (Section 3.3.3), is perhaps a bit on the low side—illustrating the fact that photoelectric grain heating is not over-whelmingly powerful. In the warm diffuse medium, soft X rays might make ζ_t as high as 1.3×10^{-15} s^{-1}. Figure 3.3 is not strictly applicable to soft X rays, but can be used with the appropriate value of $\log(P/\zeta_p) \approx 2.8$. This is just barely low enough to allow the existence of the warm phase (WNM); as mentioned earlier, the heating agent for the WNM is on shaky ground.

(d) Is Statistical Equilibrium Always Obtained?

Above, we have implicitly assumed that the kinetic temperature is in statistical equilibrium—that the heating rates and cooling rates are equal. This is not necessarily the case, because at least some heating processes (e.g., shocks) are impulsive, and cooling time scales are not necessarily short compared to the time interval between impulsive events.

In calculating the cooling rate, we include cooling from CII alone. Most cooling calculations have included silicon and iron (e.g., Dalgarno and McCray 1972). However, we now know that silicon and iron are depleted onto grains to such an extent (van Dishoeck and Black 1986) that they can be neglected to first order relative to carbon. Carbon is barely depleted in most of the ISM. Electron and H-atom cross sections for CII are from Hayes and Nussbaumer (1984) and Launay and Roueff (1977). Below 10^4 K and in the absence of heating, the cooling time τ_c [$= (d \ln T/dt)^{-1}$] is approximately

$$\tau_c \approx 4.2 \times 10^5 \left(\frac{T}{1000}\right)^{2.5} \left[\left(\frac{T}{1000}\right)^{0.58} + \frac{x_e}{0.0079}\right]^{-1} \exp \frac{92}{T} \text{ years .} \quad (3.13)$$

Here we have assumed pressure equilibrium with $nT = 3000$ cm^{-3} K. τ_c is inversely proportional to this assumed pressure, which may be much higher behind shocks and lower in regions of high magnetic field.

The equilibrium value of x_e can be easily calculated; for $\zeta_t = 10^{-16}$ s^{-1} and $T = 1000$ K, $x_e \approx 0.005$. However, as the gas cools, x_e will not retain its ionization-equilibrium value (which decreases with T), because the recombination rate is too slow. We will see below that a significant fraction of HI is in the WNM, probably at $T \approx 8000$ K. As an illustration, if x_e is its equilibrium value (≈ 0.03), $\tau_c \approx 10^7$ yr. This is long on interstellar time scales—long, for example, compared to the expected interval between supernova shocks. Thus the WNM is *not* likely to be in thermal equilibrium.

3.1.5 Elements of Interstellar Gas Dynamics

The motions of interstellar gas are governed by pressure and gravity. The diffuse gas has a small residual ionization from the heavy elements, which is enough to make the gas a perfect conductor in its interaction with the magnetic field. This means that the field lines are "frozen into" the gas. In addition, cosmic-ray particles are tied to the magnetic field. Thus, all are locked together and the gas is affected not only by its own pressure, but also by magnetic field pressure and cosmic ray pressure. Here we briefly discuss two straightforward gas-dynamical situations, static equilibrium and shocks; see Spitzer (1978) for details.

(a) Static Equilibrium: The z-Structure in the Galaxy

The plane-parallel nature of the Galactic disk makes a one-dimensional formulation appropriate. The situation is analogous to a standard planetary atmosphere, except for the pressure contributions from the magnetic field and cosmic rays and the variation of gravitational field with z. We assume that the magnetic field lines run perpendicular to the z-direction so that only magnetic pressure $B^2/8\pi$, and not magnetic tension, is important. In hydrostatic equilibrium, the pressure gradient must balance the gravitational force:

$$\frac{d}{dz}(P_{gas} + P_{mag} + P_{cr}) = \rho g_z . \tag{3.14}$$

Here P_{mag} and P_{cr} are the magnetic and cosmic-ray pressures, respectively, and g_z is the local z-component of the Galactic gravitational field. P_{gas} is the gas pressure, including not only the thermal component but also that part arising from macroscopic motions: $P_{gas} = \frac{1}{3}\rho\langle v^2 \rangle$, where ρ is the mass density of the gas and $\langle v^2 \rangle$ is the full (three-dimensional) mean square random velocity of the gas. If the pressure ratios are independent of z, if $\langle v^2 \rangle$ is independent of z (i.e., the atmosphere is "isothermal"), and if $g_z \propto z$ (valid for $|z| \lesssim 500$ pc), then the solution is $\rho \propto e^{-(z/h)^2}$, where the scale height h is

$$h = \left[\frac{2(P_{tot}/P_{gas})\langle v^2 \rangle}{-3dg_z/dz}\right]^{1/2} \approx 100\left(\frac{P_{tot}}{P_{gas}}\right)^{1/2}\left(\frac{\langle v^2 \rangle^{1/2}}{10 \text{ km s}^{-1}}\right) \text{pc} . \tag{3.15}$$

Here P_{tot} is the total pressure, $P_{gas} + P_{mag} + P_{cr}$.

The validity of the constant-pressure-ratio assumption is questionable (Section 3.4.3.a). In addition, Badhwar and Stephens (1977) show that this assumption violates the observed intensity of radio synchrotron radiation towards the Galactic poles. Nevertheless, this equation is still useful in giving an idea of how difficult it is for gas to get very high above the Galactic plane. For a significant amount of gas to achieve a height of 1 kpc, which as we shall see is observed, either $\langle v^2 \rangle^{1/2}$ must be about 100 km s^{-1} or the ratio (P_{tot}/P_{gas}) must be about 100.

This arrangement, with a horizontal magnetic field, is subject to Parker's (1986) instability. In this instability, a gas cloud at finite z is pulled toward the Galactic plane by the gravitational force of the stars. The magnetic field is tied to the gas, so the field lines are pulled down by the cloud and become tilted. Other gas slides down the tilted field lines, increasing the mass of the cloud and providing positive feedback. This instability has been invoked to explain the "beads on a string" nature arrangement of giant HII regions in spiral galaxies (Mouschovias et al. 1974).

(b) Static Equilibrium: A Uniform Spherical Magnetized Cloud

Here we consider the idealized case of a spherical cloud of radius R having uniform gas density and magnetic field. The magnetic field outside the cloud is assumed to decay as a dipole field; the value of the coefficient of the magnetic term depends sensitively on the assumed field geometry. We write the equation of equilibrium, derived from the virial theorem, as a balance between the forces tending to make the cloud smaller (gravity and external pressure P_0) vs the forces tending to make it larger (magnetic and internal pressure):

$$\frac{4\pi}{15} GR^2 \rho^2 + P_0 = p + \frac{B^2}{12\pi} \qquad (3.16)$$

where G is the gravitational constant and B the magnetic field strength. P and P_0 must include not only thermal motions, but also macroscopic motions and cosmic-ray pressure. Macroscopic motions can be represented by a fictitious temperature, which is chosen to reproduce the measured line width of the gas.

For typical *diffuse* clouds, the gravitational term is negligible. If there is no magnetic field, then equilibrium demands pressure equality across the boundary of the cloud, i.e., $P_0 = P$. If there is a magnetic field, then equilibrium still demands pressure equality across the cloud boundary, with the magnetic pressure term included (its exact value depends on the field geometry both inside the cloud and at the cloud boundary). These are most reasonable solutions and, together with the theorist's usual neglect of the magnetic field, lead to the oft-used assumption that the gas pressure in the diffuse ISM is everywhere the same.

If gravity is not negligible, and gas pressures are small compared to magnetic pressure, then equilibrium requires a balance between magnetic and gravitational forces. Such equilibrium leads to the requirement

$$B = 3.4 \frac{N_H}{10^{21}\ \text{cm}^2} \mu G\ . \qquad (3.17)$$

This is interesting because it shows that the maximum permitted field strength is simply related to the column density of a cloud. Some clouds are observed with comparable field strengths, which shows that magnetic forces cannot be neglected.

(c) Transient Features: Shocks
A sufficiently large pressure difference drives a shock into the ambient gas with a supersonic velocity specified by the Mach number M, equal to the ratio of the shock velocity to the adiabatic sound velocity $\sqrt{(5kT_1/3\mu_1)}$ in the ambient gas (μ is the mean weight per gas particle). The density, pressure, and temperature jump discontinuously across a shock.

If the gas has no time to radiate away any of the thermal energy gained by its compression by the shock, the shock is called *adiabatic* and, if the shock is "strong" ($M \gg 1$), the following equations apply:

$$\frac{P_2}{P_1} = \frac{5}{4} M^2$$

$$\frac{\rho_2}{\rho_1} = 4 \qquad (3.18)$$

$$\frac{T_2}{T_1} = \frac{5}{16} M^2$$

Here, the subscript 1 refers to the ambient gas and 2 refers to the gas that has passed through the shock.

Let us consider a typical numerical example which might occur, for example, for a shock driven by the pressure difference across an HII/HI region boundary. For $T_1 = 100$ K, we expect $M \approx 10$, for which $T_2 = 3100$ K and $P_2/P_1 = 125$. If P_1 were equal to the "standard" interstellar pressure used in Equation (3.16), then the cooling time $\tau_c \approx 10^4$ yr. This is very short; the HI column density that remains hot is so small as to be detectable only with great difficulty by 21-cm-line observers.

In a short time the gas comes back into thermal equilibrium and cools back down to its original temperature, 100 K. This cooling occurs at roughly constant pressure. The column density of the heated gas is so small that, for observational purposes, the shock appears *isothermal*—the gas on each side of the shock has the same temperature. The relations in Equation (3.18) do not apply in this case; instead we have

$$\frac{P_2}{P_1} = \frac{5}{3}M^2$$

$$\frac{\rho_2}{\rho_1} = \frac{5}{3}M^2 \qquad (3.19)$$

$$\frac{T_2}{T_1} = 1 \quad \text{(of course!)}$$

In these equations, M should be calculated as defined above, using the adiabatic sound speed in the ambient material.

Now, in our numerical example above with $M = 10$, the density jump $\rho_2/\rho_1 = 167$, not 4! The cooling behind the shock makes a huge difference in the character of the shock. Such large density increases may lead to shock-generated star formation, if they really occur.

In fact, however, they often do *not* occur because of magnetic fields. Suppose that a magnetic field B_1 resides in the ambient gas, pointing perpendicular to the shock velocity. In the one-dimensional shock compression, $B \propto \rho$ so that in the postshock gas, $B_2 = 167B_1$. While the magnetic field increases by a factor of 167, the magnetic *pressure* increases by a factor of 167^2! In other words, the ratio of magnetic to gas pressure is 167 times larger in the postshock gas than in the ambient gas. This large increase in the *relative* importance of magnetic pressure means that magnetic forces almost certainly dominate the gas dynamics of the isothermal shock, thus totally invalidating the results expressed in equation (3.19).

In this case, it is the *total* (gas plus magnetic) pressure that remains roughly constant during the cooling behind the shock. If initially the magnetic pressure behind the shock is negligible, the gas becomes denser as it cools to keep the total pressure roughly constant. As the gas cools and becomes denser, the point is eventually reached where the magnetic pressure dominates. At that point, further decreases in gas temperature are not accompanied by increases in gas density because the required constant pressure is maintained by the magnetic field. In the limiting case the density jump across the shock is given by

$$\frac{\rho_2}{\rho_1} = 2^{1/2} \frac{v_{shock}}{v_{Al}} \tag{3.20}$$

where v_{Al} is the Alfvén velocity in the ambient gas $(B_1^2/4\pi\rho_1)$.

3.1.6 Two Famous Models of the Interstellar Medium

Historically, our knowledge of the ISM ballooned during the 1960s because of high-quality optical, 21-cm line, pulsar dispersion, and radio continuum data. This set the stage for the development of comprehensive models of the ISM. The first model (Field, Goldsmith, and Habing 1969, hereafter FGH) invoked static equilibrium in the z direction and used the thermodynamic properties of the interstellar gas to derive the properties of a *two-phase* quiescent ISM undisturbed by supernovae. Later, a seminal paper by Cox and Smith (1974) showed that supernova explosions can greatly modify the general qualitative aspect of the ISM and add a *third phase*. Additional observational data in the form of diffuse soft X-ray emission and UV absorption lines from interstellar OVI showed that, at least near the Sun, nontrivial fractions of the interstellar volume are filled with the third phase—a hot, rarefied gas having $T \approx 5 \times 10^5$ K. McKee and Ostriker (1977; hereafter MO) developed a detailed theory of a three-phase supernova-dominated ISM. The latest development (Ikeuchi, Habe, and Tanaka 1984, hereafter IHT) includes the possibility of time variability and encompasses the full range of supernova rates; for a considerable range in supernova rate, the ISM is found to cycle between two-phase and three-phase states.

The FGH model features statistical equilibrium on all scales of length and time. A primary requirement of FGH was keeping clouds from expanding rapidly in the absence of gravitational forces. Equation (3.16) shows that this requires an external pressure. Since clouds are locations where density is high, pressure equality requires that the less dense confining medium have a higher temperature than the cloud gas. In other words, the existence of stable clouds requires the simultaneous existence of at least one additional gas phase at the same pressure. Figure 3.3 shows that this situation—the cold cloud (CNM) and the warmer "intercloud" (WNM) phases existing at the same pressure—does in fact obtain in the "Z" portion of the curve, which occurs over a small range in gas pressure. These are the two phases of FGH. FGH postulated that about half the mass of the ISM was heated to the WNM phase. Nearly all of the interstellar volume was filled with WNM; a small fraction was occupied by the CNM clouds. Unfortunately, as shown in Figure 3.3, reasonable soft X-ray intensities are just barely sufficient to push the ISM into the "Z" region. More efficient photoelectric grain heating, or another heating mechanism, would ease the situation.

Formally, in the FGH model the ISM pressure decreases monotonically with z. Above the z-height that corresponds to the lower left corner of the "Z" on Figure 3.3 ($\log P/\zeta_p \lesssim 2.5$), WNM exists; below this height, CNM exists. Strictly speaking, then, the FGH model predicts vertical stratification of the two phases. Going from this strict stratification to the observed situation, in which the CNM and the WNM are mixed over a large range of z-heights, requires stirring by macroscopic velocities. FGH invoke this stirring with a wave of the hand. This is a major weakness in the model, because stirring in the required amount can be produced only by supernovae, which in turn leads to production of the HIM. However, the HIM has a very long cooling time. Thus, the production of the HIM alters the basic structure of the ISM.

In its classic, pure form the FGH model had electrons for pulsar dispersion residing in the WNM. It used EIPs to both heat and ionize the WNM. To produce the required temperature and the then accepted value of pulsar dispersion measure toward the Galactic pole (DM$_\perp$; Section 3.3.5.a), a very high value of ζ_t, $\sim 8 \times 10^{-16}$, was needed. With today's value for DM$_\perp$, we would need $\zeta_t \sim 10^{-14}$ s^{-1}, which is unacceptably high.

Observational data now argue that the electrons reside instead in the WIM (Section 3.3.5). The WIM is probably produced by stellar photoionization (Section 3.3.5.d), which was neglected by FGH. Accordingly, we must consider *a modified FGH model* in which there is a nonionizing heating mechanism for the WNM— either mechanical energy or photoelectric heating of grains. Furthermore, as emphasized in Section 3.1.4.d, the WNM need not be in thermal equilibrium because of the long cooling time. This modified FGH model does not require a high value of ζ; neither ionization nor thermal equilibrium of the WNM is needed.

The MO model features a long-term global equilibrium among the production and destruction rates of both energy and cold clouds, with large spatial and temporal variations resulting from the impulsive nature of the main source of energy input, supernovae. A third gas phase appears: the HIM, with $T \approx 5 \times 10^5$ K. This phase is heated by the mechanical energy released by supernova explosions, and it fills the inside of the expanding shells caused by the explosions. A shell expands to a very large radius in the ISM. If the supernova rate is high enough, shells collide and form an interconnecting network of tunnels of hot gas. Cold clouds are produced at the periphery of an expanding shell, where the swept-up gas cools very rapidly after it enters the associated shock (Section 3.1.5.c). The formation of cold clouds in the shock is balanced by the destruction of cold clouds that were overrun by the shock and remain inside of the shell, evaporated away by the HIM. In MO's theory, the energy input to the ISM from the supernova explosion is balanced primarily by radiation from the HIM; in Cox's (1981) variant, the energy goes primarily into maintaining a WNM. The observed macroscopic random velocities of cold clouds come from the leftover expansion velocities of old shells.

MO predict many observable properties of the ISM given: (a) the assumption of balance between production and destruction of cold clouds by supernovae; (b) the assumption of balance between supernova energy production and radiation by the HIM; and (c) the supernova rate. The theory predicts the observed interstellar pressure and is reasonably successful in accounting for the observed hot diffuse gas in space. Soft X-ray emission comes from the HIM. OVI absorption comes from the somewhat denser hot gas located at the conductive interface between clouds and HIM; however, the fact that only $\sim 10\%$ of cloud interfaces show OVI (Cowie and Songaila 1986) violates the model. Electrons for pulsar dispersion and Hα emission come from the peripheries of the clouds.

MO clouds are in pressure equilibrium and have an onionskin structure. The outer skin is a warm ionized medium (WIM), with the same temperature as the

WNM but more highly ionized. It is produced by UV photons originating in the HIM and in OB stars shining on the outer surface of the cloud. Soft X rays, produced by emission from the HIM, can work their way further into the cloud because the photoionization cross section decreases with increasing energy; they produce the WNM. These two onionskins are very thin and contain about 4% of the mass of the cloud. Only half of the gas in these onionskins is neutral.

Thus, MO predict that only 2% of the HI is warm, very much smaller than observed. This is the biggest observational problem with the model. A related observational problem involves the very clumpy HI distribution predicted by the MO model: the observed distribution seems much smoother, especially for the WNM (Jahoda et al. 1985). These observational contraints are very strong. They might be accommodated by increasing the ratio of surface area (where the WNM resides, at the periphery of clouds) to volume. This could be accomplished with large numbers of small spherical clouds, or by the more realistic (Section 3.2) alternative of sheets or filaments instead of spherical clouds. However, an increased area/volume ratio might make it more difficult for clouds to survive when they are exposed to supernova explosions, evaporated away, or ionized by UV and soft X-ray photons.

There is an additional problem: the HIM should occupy most of the volume of interstellar space. The filling factor of HIM is not well known, but many observers believe it is no larger than $\sim 50\%$ (Section 3.3.6).

Neither the FGH nor the MO model satisfies the totality of observational data. A real appreciation of this point can be obtained from the comprehensive review by Cox and Reynolds (1987). Perhaps this is a result of the solar neighborhood being in IHT's cyclically varying region of parameter space. However, the supernova rate increases substantially toward the Galactic interior, which might place the Galactic interior in their "runaway" region of parameter space; in this regime, only the HIM can exist. Alternatively, the solar neighborhood might have a supernova rate just short of that required to produce a three-phase ISM so that we have a mixture of the FGH and MO models. If so, the MO model would apply in the inner Galaxy and the modified FGH model in the outer Galaxy, with a broad transition zone which includes the solar circle.

Heiles (1987a) has applied these ideas to spiral galaxies, including the fact that there are two types of supernovae, Type I and Type II (associated with Population II and I stars, respectively). The old Population II stars are distributed in a disk whose density decreases exponentially with Galactic radius; the young Population I stars are formed in clusters, which are concentrated near Galactocentric radius 6 kpc with the molecular clouds. Heiles finds that the model predicts that Type II supernovae dominate the ISM inside a few kpc galactic radius, and may be responsible for the HI "holes" found in many galaxies (Brinks and Bajaja 1986). Theoretically, Type I supernovae dominate the ISM by large factors inside galactic radius ~ 12 kpc; he finds this at variance with observations. There seems to be some fundamental problem with either the theoretical models or the observational input parameters. Nevertheless, the idea that supernovae dominate galactic interiors, but not exteriors, remains appealing and eminently reasonable.

3.2 Structure of the Diffuse Interstellar Medium: HI Emission

Examination of the structure of HI requires isolating individual features in both angle and velocity. HI is so ubiquitous that this is most easily accomplished by observing away from the galactic plane, at $|b| \gtrsim 10°$. This biases our studies of individual features to those that lie relatively near the Sun; with z-heights of 100 pc, individual features typically lie no further distant than 600 pc.

Our discussion of the effects of supernovae in Section 3.1.6 shows that the Sun may lie in a broad transition region between supernova-dominated and quiescent ISM models. However, our understanding of ISM processes is not yet sufficient to indicate how individual features are affected by supernova rates, or how the general structure of the ISM might change with the degree of domination by supernovae. Clearly, our observational restriction to study regions near the Sun is very unfortunate.

The sky is very large—it contains more than 40,000 deg^2. Fully sampling it with a 25-m telescope having a beam of half-power beam width (HPBW) of 36 arcmin requires several hundred thousand individual observations. These data have been obtained, in the north at the Hat Creek Observatory (Heiles and Habing 1974) and in the south at the Instituto Argentino de Radioastronomía (Colomb et al. 1980) and at Parkes, Australia (Cleary et al. 1979). Display of such an enormous data set is best done photographically.

3.2.1 Low-Velocity Gas

Figure 3.4, taken from Colomb et al. (1980), shows the HI structure on the sky in Galactic coordinates for three LSR velocity ranges: two small ranges centered at -8 and $+8$ km s^{-1}, which exhibit some prominent individual features, and a large range around 0 km s^{-1} to exhibit nearly the full amount of gas, irrespective of velocity. In addition, Figure 3.4(d) contains sketches of prominent OB associations and dark clouds above declination $-30°$; regions of enhanced diffuse X-ray emission and radio continuum loops; and the Orion Nebula.

Figure 3.4 constitutes a capsule summary of our knowledge about the structure of HI and its relation to other interstellar phenomena. In it we see the following.

(a) Shells

In Figure 3.4(a), at -8 km s^{-1}, there are two prominent shells: one centered near $(l, b) = (320°, 15°)$, i.e., near the Sco-Oph OB association, located just outside of Radio Loop I, otherwise known as the North Polar Spur; and another at high negative latitudes running from $l \approx 90°$ to $180°$, which runs fairly close to Radio Loop II. In Figure 3.4(b), at $+8$ km s^{-1}, there are also two prominent shells: one running from $(l, b) \approx (170°, -40°)$ to $(220°, 40°)$, known as the Eridanus loop, and one encircling $(l, b) \approx (122°, 28°)$ (the North Celestial Pole).

The North Polar Spur shell is the most spectacular in the sky. It also illustrates an important kinematic property of many shells: expansion. Expansion of a shell exhibits itself as a change in apparent shell diameter with velocity: the measured velocity is only the line-of-sight component, so that the "polar cap" approaching the observer is a small disk observed at the full expansion velocity, while the "equator" moves perpendicular to the observer so is seen as a large circle with no Doppler shift.

This example illustrates an important property of expanding shells and super-shells: in nearly all cases only one hemisphere is observable. At the velocity of -8 km s^{-1} in Figure 3.4, we see a circle slightly smaller than the full "equator." The approaching "polar cap" of this shell is observed at velocities as high as -30 km s^{-1}. However, the *receding* "polar cap" is not observed at all. Statistically, there is no tendency favoring approaching over receding hemispheres. Presumably, most shells are produced by energetic stellar winds and supernovae; many are associated with stellar associations [see summary by Heiles (1984)]. There are two explanations for seeing only one hemisphere: there might have been too little gas on one side of the stars to form a cool shell; or there might have been too much to allow the shell to have reached a large enough radius to be observable. Choosing between these should be possible if we knew the detailed distribution of dense molecular clouds in the vicinity of the explosion center, which is in principle derivable from CO observations.

The diameter of the North Polar Spur shell is about 120°, which is huge! Because of this large angular diameter, it is clear that approaching face of this shell is nearby—and, indeed, some parts of it are seen as nearby interstellar gas observed in optical absorption lines against nearby stars (Crutcher 1982), and some as interstellar gas flowing through the solar system observed in Lyα emission from spacecraft (Bertaux et al. 1985). This implies that some parts of the shell have already passed by us.

Figure 3.4(d) shows that enhanced diffuse X-ray emission is observed from inside this shell. The X rays come from hot gas which has probably been repeatedly heated by supernova explosions and energetic stellar winds from massive stars in the Sco-Oph association. The reader is referred to Heiles et al. (1980) for observational details and Borken and Iwan (1977) for theoretical aspects.

The other prominent shell in Figure 3.4(a) also exhibits some change in size with velocity. It lies close to Radio Loop II and may be associated with it. However, there is no independent evidence pointing towards an association, and the accuracy of the superposition is not so striking as it is for the North Polar Spur shell.

The Eridanus shell, in Figure 3.4(b), is another one exhibiting a clear change in diameter with velocity (Heiles 1976b). It also encircles hot, X-ray-emitting gas, and like the North Polar Spur shell, it is undoubtedly the result of energy released by massive stars. Many radio astronomers would argue that the absence of continuum radio emission shows that the energy came from stellar winds instead of a supernova, because many supernova remnants emit diffuse synchrotron radiation. However, one must view such an argument with suspicion because there is no proof whatsoever that *all* supernova remnants emit radio continuum.

A number of HI synthesis studies have been done towards specific HII regions or supernova remnants. Much of the recent work has been done at the Dominion Radio Astronomy Observatory in Penticton, Canada (e.g., Landecker et al. 1982), and some at Cambridge (e.g., Read 1980). G78.2 + 2.1 appears to have exploded within a slab of HI and generated a hole in the slab with a ring of HI at the edge.

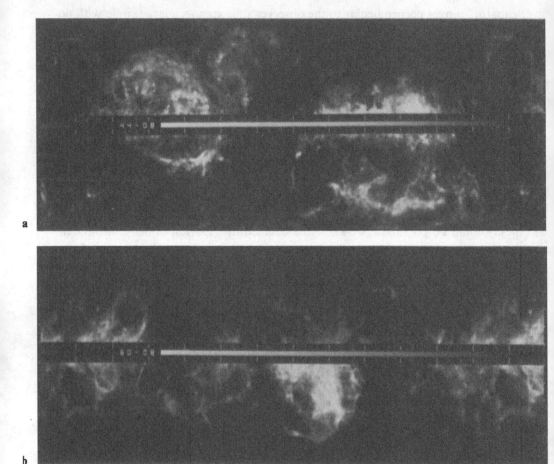

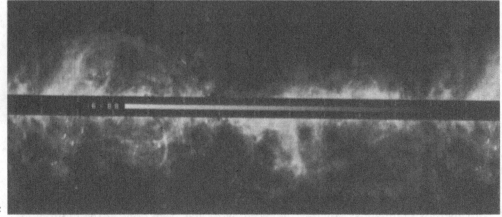

Fig. 3.4. (a) HI column density in a small velocity range centered at -8 km s^{-1}. (b) HI column density in a small velocity range centered at $+8$ km s^{-1}. (c) HI column density in the range -38 to $+38$ km s^{-1}. (d) Crude sketches of prominent OB associations (Blaauw 1964), dark clouds (Lynds, 1962), regions of enhanced diffuse X-ray emission (McCammon et al., 1983), and radio continuum loops (Berkhuijsen et al., 1971). Coordinates for panels a, b, and c are the same as shown in panel d.

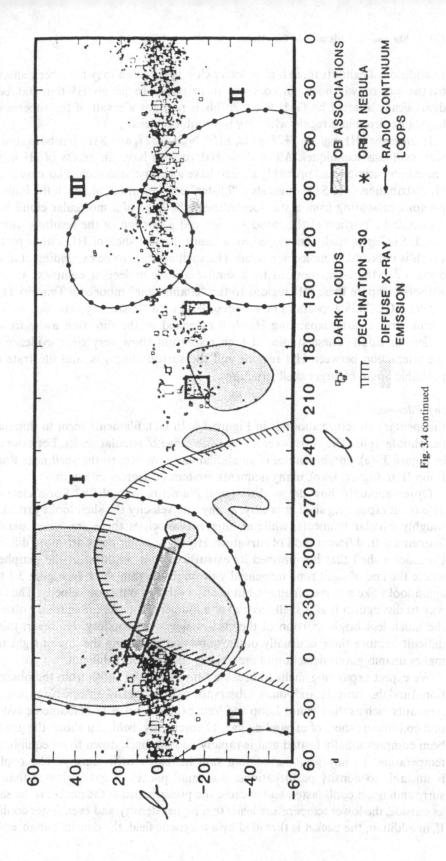

Fig. 3.4 continued

LEGEND:
- DARK CLOUDS
- DECLINATION −30
- DIFFUSE X-RAY EMISSION
- OB ASSOCIATIONS
- ORION NEBULA
- RADIO CONTINUUM LOOPS

In addition, it exhibits three high-velocity cloudlets, which may have been affected by the supernova shock. In contrast, there seems to be no HI that has been dynamically affected by G116.9 + 0.2; this is perhaps a result of the supernovae having occurred in a region with very low ambient density.

HI near the HII regions S125, S142, S159/NGC 7538, and S184 has been studied with synthesis techniques. All of these HII regions have an excess of HI in the immediate vicinity, and probably all also have an associated molecular cloud. The HI distribution near S125 suggests a "blister" configuration, in which the ionizing photons emanating from a star located near the edge of a molecular cloud have dissociated a portion of the cloud and ionized a portion of the resulting atomic cloud. S142 is partially enveloped by a hemispherical shell of HI, within part of which is imbedded a molecluar cloud. The shell appears to be expanding at about 6 km s^{-1}. S184 is surrounded by a similar atomic/molecular complex, and the authors compare these HII regions to the "champagne" models of Tenorio-Tagle (1979). NGC 7538 appears to be well described by a "blister"-type model, with an asymmetric, clearly expanding HI shell extended in the direction away from a molecular cloud. These studies at high resolution show very clear evidence for the interaction between HII regions and the surrounding gas, and illustrate the probable origin of larger shell structures.

(b) Filaments

Filamentary structure abounds in Figure 3.4. In fact, filaments seem to dominate the photographs. They exist over a very wide range of angular scales. For example, in Figure 3.4(a), an abundance of small filaments connect to the shell near Radio Loop II; in Figure 3.4(b), many filaments are tens of degrees in length.

Observationally, how do we distinguish filaments from shells? For a clear-cut case of an expanding shell, it is easy: at any one velocity the shell looks just like a roughly circular filaments, while at adjacent velocities there are other, parallel filaments with different radii of curvature. However, other cases are more difficult. Consider a shell that has finished its expansion. It is brightest at the periphery, where the line of sight runs tangentially through the thin shell. In Figure 3.4 this again looks like a curved filament but is observable at only one velocity. The only way to distinguish it as a shell, instead of a filament, is to unambiguously observe the much less bright portion of the shell inside the periphery. However, this is difficult because there is usually other, unrelated gas along the line of sight that makes unambiguous detection of the weaker emission very difficult.

We expect expanding shells to break up into filaments, both from the observational and theoretical standpoints. Observationally, optically observable supernova remnants such as the Cygnus Loop and Vela exhibit a network of filaments instead of a continuous sheet of expanding gas. Theoretically, behind a shock the gas has been compressionally heated and is rapidly cooling back down to its equilibrium temperature. In any gas, the cooling rate increases with density. The cooling is unstable to density perturbations: if a small packet of gas is denser than its surroundings, it cools faster, and because the pressure inside the packet is the same as outside, the lower temperature leads to a higher density and even faster cooling. If, in addition, the packet is threaded by a magnetic field, the density enhancement will be filamentary along the magnetic field lines, because the magnetic field eliminates thermal conduction across the field lines. Many of the CO clouds discovered by Magnani et al. (1985) are associated with HI filaments rather than large accumulations of HI, and we suspect that the molecules are formed in such shock-produced filaments.

As a result, we expect expanding shells to form filaments. These filaments should persist after the shell stops expanding. They are not necessarily straight—at least, optically observable filaments in supernova remnants are not straight, but seem to wander randomly. Thus, a curved HI filament might *not* be the periphery of a shell, but instead might simply be a condensation formed during the thermally unstable cooling.

Are HI filaments really filaments, or are they curved sheets viewed tangentially? This remains an observational question without an answer.

(c) Major HI Concentrations Around Dark Clouds and Newly Formed Stars
For $|b| > 10°$, Figure 3.4(c) shows the total HI column density and Figure 3.4(d) the distribution of related interstellar objects. There are three major regions containing young stellar objects (OB associations and HII regions) and molecular clouds: Ophiuchus, Perseus/Taurus, and Orion [centered near $(l, b) = (0°, 15°)$, $(170°, -20°)$, and $(210°, -15°)$, respectively]. Every one of these regions is enveloped by a large HI concentration with high column density. Approximate typical physical properties of the HI concentrations can be easily derived from HI column density maps and are as follows: $N_H \sim 1.0 \times 10^{21}$ cm^{-2}; $\langle n_H \rangle \sim 2.5$ cm^{-3} (volume average; actual value probably higher because of clumping); linear diameter ~ 120 pc; mass $\sim 1.0 \times 10^5$ M$_\odot$.

These parameters vary widely from one region to another. Orion, for example, is enveloped by about 7×10^4 M$_\odot$ of HI with a linear diameter of about 125 pc (Gordon 1971), while the associations I Mon and II Mon (not shown on our figures because $|b| < 10°$) are enveloped by HI masses of about 1.5×10^5 and 2×10^4 M$_\odot$, respectively (Raimond 1966). At least in some cases an equivalent mass is contained in molecular clouds; in Orion, for example, Thaddeus (1982) reported a total of about 3.3×10^5 M$_\odot$ in H$_2$ (a somewhat uncertain figure, because it is derived from CO observations using a controversial CO/H$_2$ conversion factor).

The virial equilibrium of these clouds is interesting because no term in Equation (3.16) can be neglected. Consider Orion as an example. The gravitational term, including only the self-gravity of the HI and not the molecular clouds that lie within, is equivalent to $nT \approx 600$ cm^{-3} K. With volume-average gas densities of 2.5 cm^{-3} and macroscopic motions equivalent to a Doppler temperature of about 1600 K, internal motions dominate gravity by a factor of about 6. Finally, when we include a point mass of H$_2$ clouds at the center with mass equal to the total HI mass, the factor 6 reduces to 2.5. Even though this neglects the magnetic field, which is strong, the balance is probably close and it is likely that these clouds are in virial equilibrium with *all* terms in Equation (3.16) contributing.

(d) Shapes of HI Clouds
Most astronomers think of interstellar clouds as being similar to those big, puffy things in the sky—good approximations to spheres. Spheres are usually treated by

theorists for the sake of convenience, but—like Madison Avenue advertisements—this repetitive literature tends to make us believe in them. However, spherical blobs are rare in the HI maps. Instead, long filaments abound and spherical shells, sometimes expanding, are also seen. The shells are sheets. The filaments may simply be sheets seen edge-on. The basic point is that HI clouds are more likely to be two-dimensional (i.e., sheets) or even one-dimensional (i.e., cylinders).

Given the complexity of the 21-cm sky, a *quantitative* study of HI cloud shapes is not easy. As the discussion below shows, most of HI clouds are filamentary with sharp edges, and hence high resolution is crucial for cloud studies. There have been only two statistically adequate high-resolution studies of the 21-cm sky (Heiles 1967, Verschuur 1974). Both these studies mapped well over one hundred square degrees of the sky at 10-arcmin resolution. The demarcation of clouds is clearly a subjective process—a statement best appreciated by looking at the contour diagrams presented in these two papers. However, the conclusions of both these studies are similar, viz., most of HI is organized in either large sheets or filaments with large axial ratios.

Despite such studies, the common notion of spherical clouds with a standard radius (the so-called "standard cloud") still persists. While we do not offer any panacea by the way of specific numerical data, we do wish to emphasize that the detailed studies of specific regions of the sky have not yielded the postulated spherical clouds. Instead, they have consistently found filaments and sheets.

(e) Structure of HI Clouds

The clouds discussed above were mapped with poor angular resolution, and we might well expect them to have internal structure. Evidence that HI clouds do indeed have structure first came from the fact that the emission line widths are systematically larger than absorption line widths (Section 3.1.2.a). A simple interpretation of this fact, which appeared as early as 1972, is that the HI clouds are not isothermal; the entire cloud appears in emission whereas only the colder portions of the cloud appear in absorption (Radhakrishnan et al. 1972). It is an interesting historical point that this interpretation was conveniently forgotten until very recently, when high-angular-resolution observations forced its renaissance.

The internal structure of HI clouds can be probed by high-resolution mapping of the 21-cm sky. It is easiest to use the small-scale structure in background continuum sources as high-resolution probes. A number of observers have used double-lobed radio sources as background sources for such studies and find very little HI structure on scale lengths below 1.5 pc (see Crovisier et al. 1985).

Internal structure can also be studied by high-resolution mapping of HI in *emission*. This is best accomplished by aperture synthesis techniques, although they are laborious. Studies of "random samples" of the ISM have been done by Read (1980), by Joncas et al. (1985), and by Kalberla et al. (1985). Even at small scales, extended filaments and/or sheets dominate the distribution of HI. Occasionally there are sharp edges, probably indicative of shocks in the gas. Also seen are small clumps within the filaments and/or sheets. In agreement with earlier absorption studies, these observers find little structure below scale lengths of 1 pc.

The field studied by Kalberla et al. also contains 3C147, a bright point radio source. This is important, because for the first time temperatures could be derived independently for the various components. We postpone the discussion of these matters to Section 3.3.4. The 3C147 field is nonideal in some respects. It is at low latitude ($b \sim 10°$) near the galactic anticenter. Thus the distance is uncertain and probably fairly large; they assume 500 pc. The large distance makes aperture synthesis techniques mandatory, which makes the mapping very difficult. Studies at higher latitudes could use single-dish techniques to attain the same linear resolution. Furthermore, the distance uncertainty makes the derived parameters uncertain. The dearth of information on cloud structure needs to be filled by additional studies such as this.

3.2.2 Intermediate-Velocity Gas

The lion's share of HI lies at LSR velocities below 40 km s^{-1}, which is roughly the range pictured in Figure 3.4(c). However, some of the gas lies at higher negative velocities: "intermediate velocities," extending to about -90 km s^{-1}, and "high velocities," extending to more negative velocities, currently up to about -500 km s^{-1} (Giovanelli 1980). At high positive Galactic latitudes, about 50% of the HI is in this disturbed state. High-velocity gas (HVG) is briefly discussed in Chapter 7 of this volume. Here we concentrate on intermediate-velocity gas (IVG), because it is clearly related by physical interaction to the local low-velocity gas.

Negative-velocity gas was originally discovered by the Dutch radio astronomers. The Dutch work on IVG (Wesselius and Fejes 1973) remains the best and most comprehensive. The principal result is that IVG is located primarily at high positive Galactic latitudes in the general direction of the Galactic anticenter, extending from $l \approx 120°$ to $300°$. This coincides with a "hole" in the low-velocity gas (LVG) distribution. The anticorrelation between LVG and IVG is very striking, even on small angular scales in some regions, where the LVG and IVG fit together like pieces of a jigsaw puzzle. The IVG column density is close to what the LVG column density would be if the LVG "hole" didn't exist. Thus, Wesselius and Fejes argue that the LVG has been "displaced" by IVG—that some agent affected the LVG and changed its velocity without changing the total HI column density. Heiles (1984) has presented photographs of the IVG distribution which make it look like a portion of an expanding circular shell.

The only way to determine distances to anomalous-velocity HI is through associated optical absorption lines. Unfortunately, the technique is not absolutely reliable because optical lines are not necessarily strongest at the locations of highest HI column density (Heiles 1974). Limits on distance to some portions of the IVG have been obtained with a few stars (Wesselius and Fejes 1973) and indicate a distance of a few hundred pc. The general scale of the IVG features, some 60°, then implies a linear diameter of a few hundred pc.

In some cases there are striking associations between LVG and gas at abnormal velocities. The most striking example involves the HI associated with Radio Loop II, discussed above in connection with Figure 3.4(a). Here, HVG gas lies precisely on top of LVG (Cohen 1981); the association is particularly striking and detailed, and can hardly be due to chance. A less convincing example involves gas on small angular scales inside the "anticenter shell," a 30° diameter shell covering a very large range of velocities (Heiles 1984). Mirabel (1982) shows that an excess of high-negative-velocity gas, ranging to -200 km s^{-1}, coincides with a disturbance of lower-velocity gas in the vicinity of $(l, b) = (185°, -11°)$. This argues for direct collision, and it is conceivable that the energy thereby released produced the much larger anticenter shell. The probable source of the negative-velocity gas is fragmentation from the Magellanic stream. Finally, Heiles (1984) discusses a rather loose association between HVG, IVG, and possibly even LVG at high positive latitudes; this association remains to be proved convincingly.

3.2.3 Supershells and "Worms"

Figure 3.4 shows a number of shells that are tens of degree in diameter. The fact that they are so large in angular size implies that they could easily be resolved and detected at much larger distances. Heiles (1984) searched for such objects in the Galactic plane and discovered quite a large number of them. Most surprising was the discovery of a population of extremely large shells, "supershells," ranging up to 2 kpc in diameter. The Sun may be located just inside the boundary of a supershell (Lindblad et al. 1973). These objects are truly remarkable. If produced by the intantaneous release of explosive energy, the required energies range up to 4×10^{53} erg— equivalent to 400 supernovae. Energy released from a large number of stars in the form of stellar winds and supernovae explosions may, in fact, be the energy source for most or all supershells. However, the large energies and the fact that usually only one hemisphere of an expanding supershell is visible may imply another mechanism in some cases—specifically, the collision of infalling gas with the gas in the galactic plane.

Large shells or supershells are also seen in other galaxies. The spectacular optical photographs of the Large Magellanic Cloud in Hα (Davies et al. 1976) exhibit many impressive structures which have been analyzed by Goudis and Meaburn (1978) and Meaburn (1980). The detailed Westerbork synthesis HI maps of M31 and statistical analysis of shells by Brinks and Bajaja (1986) are particularly impressive. Other extragalactic shells are summarized by Heiles (1984). Lists of shells in the Galaxy are unsuitable for statistical purposes because of observational selection effects.

Heiles (1984) has observed vertical pillars of gas in the inner Galaxy, which to a cursory glance look like worms crawling vertically out of the Galactic plane. These are probably the manifestations of what would otherwise be large shells. Large shells should not exist in the inner Galaxy, because their sizes would be larger than the thickness of the HI disk. An expanding shell would be expected to "break through" the disk, punching a hole entirely through the HI layer (Heiles 1987a). These holes would look like shells without tops, and this is the most straightforward explanation of the "worms."

3.3 Temperature and Ionization of the Diffuse Interstellar Medium

3.3.1 HI Absorption: Techniques, Data, and Analysis

HI absorption data are a convenient probe of gas temperature, and have been crucial to our understanding of the thermodynamics and structure of diffuse clouds as well as the warm medium. They are also fundamental to the determination of pulsar distances and other Galactic radio objects such as HII regions, supernova remnants, and exotic objects without normal optical counterparts.

(a) Techniques and Data

Both single dishes and interferometers can be employed to obtain HI absorption spectra. We discuss both these techniques and show that, in general, interferometric measurements are more reliable than single-dish techniques.

Consider an isothermal HI cloud and a radio source of flux density S behind it. In the single-dish technique, we assume that the beamwidth of the telescope is much smaller than the angular size of the cloud but larger than that of the radio source. Then the antenna temperature measured by the telescope in the direction towards the source (the "on-source" spectrum; e.g., Figures 3.1 and 3.2) is given by the equivalent of Equation (3.5),

$$\Delta T_{A,\,\mathrm{on}}(v) = (T_s - T_{\mathrm{src}})(1 - e^{-\tau(v)}) \ . \tag{3.21}$$

Here, we implicitly assume that the beam efficiency of the telescope is unity so that, for an extended source such as the HI cloud, the antenna temperature is equal to the brightness temperature. T_B and T_{bg} in Equation (3.5)—which are brightness temperatures—are replaced by their antenna temperature equivalents T_A and T_{src}. T_{src} is the antenna temperature of the compact source and is given by $S \times G$ where G is the gain of the telescope. G for a single 25-m VLA antenna is ~ 0.1 K Jy^{-1} and for the giant Arecibo reflector, ~ 8 K Jy^{-1}. We have neglected background contributions such as the 3 K cosmic background radiation since such contributions are present in both the "on" and the "off" spectra and do not affect the determination of the optical depth spectrum.

The single-dish technique consists of obtaining an "off-source" emission spectrum, $\Delta T_{A,\,\mathrm{off}}(v)$, in a direction displaced by approximately one beamwidth so as to obtain an independent measure of T_s. Thus, an important assumption is being made: the cloud has little structure over one beamwidth. Under this assumption, the absorption spectrum is readily obtained as:

$$(1 - e^{-\tau(v)}) = \frac{(\Delta T_{A,\,\mathrm{off}}(v) - \Delta T_{A,\,\mathrm{on}}(v))}{T_{\mathrm{src}}} \ . \tag{3.22}$$

In practice, the off-source spectrum is formed from several off-spectra taken around the source in a pattern such as a cross or hexagon. Linear gradients in $T_B(v)$ are cancelled out by the use of such patterns. However, structure on the scale of the beamwidth is not, which is a fundamental limitation on single-dish work. For example, even for the Arecibo dish, a fluctuation of 10% in adjacent emission spectra is typical for $|b| > 10°$. This variation increases at low latitudes, towards intermediate-and high-velocity clouds, and of course is worse for larger beamwidths and weaker background sources. For this reason, essentially all single-dish HI absorption data towards weak or low-latitude sources are unreliable.

Interferometers surmount these difficulties because of their ability to act as *spatial filters*. An interferometer with a baseline B responds only to structures in the sky with spatial frequency of $\sim B/\lambda$, where λ is the wavelength. Most of the HI features in the sky do not have much structure on angular scales below about an arcmin (see Crovisier et al. 1985), whereas most background radio sources are generally smaller than 1 arcmin. Hence an interferometer with a fringe spacing smaller than 1 arcmin but bigger than the size of the background source resolves the foreground HI emission and responds only to the background source. Thus interferometers measure the absorption spectrum $(1 - e^{-\tau(v)})$ directly. The ability of interferometers to form images (aperture synthesis) is not critical for HI absorption measurements. Thus the sensitive Arecibo interferometer is as useful as the multielement Very Large Array (VLA). In fact, a large amount of HI data have been obtained from the VLA in the so-called "phased mode" in which the VLA is essentially reduced to a three-element interferometer (Dickey et al. 1983).

Interferometric surveys are always more reliable than single-dish surveys. However, the Arecibo single-dish surveys provide reliable absorption data because of the very narrow Arecibo beam and the associated high antenna gain. Kulkarni et al. (1985) compare Arecibo single-dish data with Arecibo interferometric data and conclude that, at least for $|b| > 10°$ and strong sources ($S > 3$ Jy), the Arecibo single-dish data are reliable.

Pulsars are ideal background sources for HI absorption measurements. The "on-source" and the "off-source" HI emission spectra can be obtained using a single dish without moving the telescope beam by simply recording the HI spectra separately when the pulsar is on and off. Unfortunately, most pulsars are weak objects at 21 cm (< 1 mJy), and hence HI absorption spectra for only about 40 of the 500 known pulsars have been measured.

(b) Derived Spin Temperature: Definition and Limitations

Our earlier discussion of the excitation temperature of the 21-cm line was pursued in the context of a single, isothermal HI cloud. In practice this simple situation is rarely encountered. When more than one HI cloud is present along a given line of sight, the measured spin temperature is related in a complicated way to the true spin temperatures.

Consider an isothermal HI cloud with a spin temperature T_s and a column density $N_H(v)$. The velocity dependence of N_H is a consequence of both thermal and macroscopic motions. Observationally we can measure $\tau(v)$ and $T_B(v)$. In the spirit of Equation (3.5), and assuming $T_{bg} \ll T_s$, we define $T_n(v)$, the *naively-derived spin temperature*, in terms of measured quantities as:

$$T_n(v) = T_B(v)/(1 - e^{-\tau(v)}) . \tag{3.23}$$

For the simple case of a single, isothermal cloud that we considered in Section 3.1.1, $T_n(v) = T_s$.

Complications arise when there is more than one parcel of HI with overlapping velocities along the same line of sight. For example, consider the case of two isothermal clouds with the same velocity distribution, but with two different spin temperatures. The parameters of the cloud nearer to the observer will be denoted with the subscript "1" and those of the background cloud by "2." Application of Equation (3.23) yields

$$T_n(v) = \frac{T_{s,1}(1 - e^{-\tau_1}) + T_{s,2}(1 - e^{-\tau_2})e^{-\tau_1}}{(1 - e^{-\tau_1 - \tau_2})} \tag{3.24}$$

where, for clarity, the velocity dependence of τ has been dropped out.

It should be clear from Equation (3.24) that the deduced spin temperature is now a function of v and ranges *from $T_{s,1}$ through $T_{s,2}$*. Thus, literally interpreting $T_n(v)$ as spin temperature would lead us to the false conclusion that HI at temperatures intermediate to the two temperatures exist. For this reason we wish to strongly caution the reader against accepting results based on blind analyses of $T_n(v)$.

We now consider three illustrative cases to specifically understand the pitfalls and limitations of $T_n(v)$:

Case (a): Only the foreground cloud is optically thick, i.e., $\tau_1 \gg 1$ and $\tau_2 \ll 1$. Due to the velocity dependence of N_H, these inequalities are satisfied over only a limited range of velocity. Over this limited range of velocities, the optically thick foreground cloud absorbs most of the emission from the background cloud and, not surprisingly, $T_n \approx T_{s,1}$.

Case (b): Only the background cloud is optically thick, i.e., $\tau_1 \ll 1$ and $\tau_2 \gg 1$. In this case, $T_n \approx T_{s,1}\tau_1 + T_{s,2}$. Thus, the naively-derived spin temperature of the cloud is increased by the contribution of the much warmer foreground HI. This case probably occurs often in practice, with a cold cloud lying behind much warmer foreground material, which could be either the cloud's warmer envelope or the WNM HI.

Case (c): Both clouds are optically thin. Then it easy to show that T_n is equal to the column-density-weighted *harmonic mean* temperature. In order to appreciate the importance of this unusual weighting, consider the specific example in which equal column densities of cold ($T_{s,1} = 80$ K) and warm ($T_{s,2} = 8000$ K) gas exist along a line of sight. In this case, $T_n \approx 160$ K. This is much higher than the temperature of the cold cloud and ridiculously lower than the temperature of the warm HI. Admittedly, this is somewhat of an extreme case since the column density per unit velocity of WNM is typically less than that of CNM. However, the nature of this result is general in the sense that T_n always *overestimates* the temperatures of CNM but *underestimates* the temperature of WNM. As observational examples, Kalberla et al. (1985) carefully separated the two contributions for five absorption components and found T_n/T_s to range from 2.1 to 4.9.

3.3.2 Temperature of Warm HI

Warm HI (the WNM) is probably the least well understood component of the ISM. Nearly everything we know about the WNM has come from 21-cm-line data. However, these data provide little in the way of direct measurements of temperatures and volume densities. The best diagnostics for physical conditions are provided by high-spectral-resolution UV absorption line data, which are sparse; and by high-spectral-resolution mapping data of the far-IR cooling lines of carbon, silicon, and iron, which are nonexistent and likely to remain so for the foreseeable future. Thus, while we have measurements of the column density of the WNM, we have only educated guesses about its volume density, ionization fraction, and to some extent even temperature.

At present, the temperature of WNM can be estimated using two different techniques: 21-cm emission/absorption and UV absorption line studies. The first technique has contributed most of the information. However, it is difficult because the 21-cm-line optical depth is inversely proportional to temperature, so optical depths are small. The measurements involve weak absorption with relatively strong emission, and systematic effects are important. The most important instrumental effect is stray radiation (see Kalberla et al. 1980), which adds extraneous emission and makes T_n (Equation 3.23) an upper limit. The most important interpretive bias results from temperature variations along the line of sight; these make T_n a lower limit to the temperature of WNM. These effects push T_n in opposite directions, which prevents even the determination of reliable limits for T_n of the WNM.

Many attempts have been made to measure the optical depth of WNM. However, in most cases no absorption has been detected so that only upper limits to the optical depth, and thus lower limits to T_n, can be derived. Many of these lower limits lie in the neighborhood of 3000 K or below (e.g., Mebold et al. 1982). The highest *measured* temperature of WNM is ~ 6000 K towards the strong source Cyg A (Kalberla et al. 1980). The highest 3σ lower limit is 10^4 K towards 3C123 (Kulkarni et al. 1985); however, this needs to be confirmed by independent observations before it can be accepted.

Payne et al. (1983) have statistically analyzed lower limits on temperature for the sensitive Arecibo HI absorption survey. Along every line of sight they divide HI into gas with and without detectable absorption. The latter gas, which has high temperature, has been named "Not Strong Absorbing" (NSA) HI. Nearly 50% of the NSA gas detected by them has velocity that is distinctly different from that of the clouds. Thus it has been named "Independent" NSA HI. This gas constitutes about 37% of the total HI in their sample. Their statistical analysis shows that most of the independent NSA gas has spin temperature ≈ 5300 K, with no more than 30% of the gas lying below 5000 K.

The study of UV absorption lines of various ions of the WNM can, in principle yield the temperature. Several papers presented at the IAU Colloquium on the Local Interstellar Medium (e.g., York and Frisch 1984) find $T \gtrsim 6000$ K for a handful of nearby (distance $\lesssim 200$ pc) stars. This technique is reliable and needs to be applied in a large number of directions.

The most straightforward conclusion from these data is the following: *the temperature of most of the WNM lies in the range 5000 to 8000 K.* We emphasize this conclusions in italics, because many past data provided only lower limits on temperature, which led many people to erroneously think that the WNM actually had temperatures just above the lower limits, i.e., of order 1000 K. Nevertheless, our conclusion that most of the WNM lies above 5000 K is *not absolutely firm.* It is based on only *one* 21-cm-line measurement, *one* statistical analysis of 21-cm-line measurements, and a few UV results on nearby stars. Further observational confirmation of this conclusion is *crucial.*

Theoretically, both CNM and WNM can coexist in static thermal equilibrium only over a limited range of pressures (Figure 3.3). Above a critical pressure, the warm phase must undergo a phase transition into the cold phase. Then a valid question is whether very much WNM gas exists at $z = 0$ pc, where the pressure is the highest. Unfortunately—and rather surprisingly—there are few definitive data on this point because of the confusion induced by the plethora of absorption components near $b = 0°$. This question is crucial for the theoretical models of the ISM, and observational data are urgently required.

3.3.3 Temperature of the Cold HI (CNM)

Cold HI (the CNM) is easily observed in absorption spectra. For any given emission-absorption pair, we find that the naively-derived spin temperature $T_n(v)$ is not constant across the absorption feature. This is quite apparent in Figure 3.1: clearly, $T_n(v)$ increases on either side of the velocity at which maximum absorption occurs. This is a simple consequence of the fact that absorption lines are narrower than their emission counterparts. The simplest and in fact the correct interpretation of the velocity dependence of $T_n(v)$ is that the *HI clouds are not isothermal blobs*. Traditionally, the smallest $T_n(v)$, i.e., T_n at the velocity of maximum absorption, is called the spin temperature of the cloud. We will denote this temperature by $T_{n,\,min}$ and the peak absorption by τ_{max}.

The absorption features are found to be well represented by Gaussian functions in optical depth τ_v. The distribution function of $\sigma_v(\tau)$, the velocity dispersion of the absorption features, *peaks* at about 0.75 km s^{-1}; the *mean* $\sigma_v(\tau)$ is ~ 1.7 km s^{-1} and is significantly larger than the peak value because the distribution function has a long tail. In contrast, the emission features are broader, peaking at ~ 2.2 km s^{-1} (Crovisier 1981). No cloud has been observed with $\sigma_v(\tau) < 0.4$ km s^{-1} and very few with $\sigma_v(\tau) > 4$ km s^{-1}. The lower limit is probably unaffected by observational biases.

Both turbulence and thermal motions contribute to the observed width of the absorption spectra. The velocity dispersion in the HI line from thermal motions alone is $\sigma_{th}(T) = \sqrt{(T/121)}$ km s^{-1}. The ratio $[\sigma_v(\tau)/\sigma_{th}(T_{n,\,min})]$ indicates the degree of turbulent motion. It has a mean value of 1.3 and ranges up to about 2.2; there appears to be no dependence on τ_{max} (Payne et al. 1983). These derived ratios are always underestimates, because $T_{n,\,min}$ is an overestimate of the coldest temperature (Section 3.3.1.b); this is underscored by the fact that many ratios have values smaller than unity!

The interpretation of these ratios is fraught with complications. This ratio is equal to unity for an isothermal blob of gas with no macroscopic motions and no magnetic field. However, clouds are not isothermal blobs (Section 3.3.4) and, in that case, the ratio depends on the detailed temperature structure. In addition, clouds apparently have macroscopic velocities, the origin of which is not understood. The presence of magnetic fields introduces yet another free parameter—the Alfvén speed. For these reasons, we are skeptical of simple analyses of line widths.

(a) Classical Absorption Studies: the T-τ Relation

Lazareff (1975) first noted that $T_{n,\,min}$'s are inversely related to τ_{max}. This is the famous "T-τ" relation. A fit to the sensitive Arecibo data yields:

$$T_{n,\,min} = T_0(1 - e^{-\tau_{max}})^{-\alpha} \tag{3.25}$$

with $T_0 = 55 \pm 7$ K and $\alpha = 0.34 \pm 0.05$ (Payne et al. 1983).

The scatter in this relationship is large. In addition, the fit is systematically high at large τ_{max}'s because it is primarily determined for τ_{max} between 0.03 and 0.3. Furthermore, it is important to realize that the least-squares fit for the T-τ relation is done to the *logarithm* of the measured quantities. Thus, the points scatter above and below the fit with equal *multiplicative factors* instead of with equal *arithmetic differences.* For example, for $\tau_{max} = 0.1$, the T-τ relation predicts $T_{n,min} = 120$ K; the observed scatter is a factor of two, so observed values range from about 60 K to 240 K. This logarithmic fit gives more weight to data points with lower $T_{n,min}$'s and, in effect, is a fit to the *lower envelope* of $T_{n,min}$'s. The large scatter and this nonlinear weighting should be borne in mind when using the T-τ relationship for statistical purposes.

Values of $T_{n,min}$ range from 20 K to about 250 K and those of τ_{max} from 0.01 to ~ 2. The lowest observed τ_{max} is limited by sensitivity. Nevertheless, theoretically we expect a *real* decrease in the number of clouds with smaller τ_{max}'s (i.e., $T_{n,min} \gtrsim 250$ K) if clouds are in stable thermal equilibrium (Figure 3.3).

Radial velocity data, together with differential Galactic rotation, can be used to extract $\langle |z| \rangle$ of the CNM layer, although the *shape* of the vertical layer cannot be convincingly extracted. Belfort and Crovisier (1984) find $\langle |z| \rangle \sim 100$ pc. They also find that the derived scale height is a function of the optical depth of the clouds: the scale height for $\tau_{max} > 0.1$ is 88 ± 13 pc, while that for $\tau_{max} < 0.1$ is larger— 229 ± 48 pc. The T-τ relation enables us to conclude that *the clouds get warmer at higher* $|z|$. (But see Section 3.3.4 for an alternative interpretation.) This trend, of larger scale heights for warmer CNM gas, is to some small extent consistent with the fact that the scale height of WNM HI is large and has a long exponential tail (Section 3.1.2.b). Apart from this exponential tail, the scale height of WNM, with $T_k \gtrsim 5000$ K, is comparable to that of clouds with $\tau_{max} \lesssim 0.1$ K.

(b) Statistics of Clouds

The procedure to derive the statistics of absorption features has been nicely presented by Crovisier (1981) and applied to Arecibo data by Payne et al. (1983). The number of clouds with $\tau_{max} > \tau$ along a line of sight, reduced to $|b| = 90°$, is well represented by $P(\tau > \tau_{max}) = 0.3\tau_{max}^{-0.4}$. At $b = 0°$, this translates to

$$\textit{Number of clouds per kpc with } \tau > \tau_{max} = 3.0\tau_{max}^{-0.4} \ . \tag{3.26}$$

Using the T-τ relation and a velocity width 1.3 times the thermal width, the corresponding probabilities for $T_{n,min}$ and N_H can be obtained: they are $P(T < T_{n,min}) \propto T_{n,min}^{1.2}$ and $P(N > N_{cloud}) \propto N_{cloud}^{-0.8}$, respectively. Both of these derived probabilities are consistent with the ones derived directly from the data by Payne et al. (1983). These probability relations are valid over the range $\tau_{max} \approx 0.02$ to 1.0; this corresponds to ranges of temperature and column density of 208 to 55 K and 0.32×10^{20} to 2.2×10^{20} cm^{-2}, respectively. The fact that very few of the clouds have $T \gtrsim 210$ K underscores the fact that the CNM and WNM are distinctly different components: the WNM has $T \gtrsim 5000$ K (Section 3.3.2).

At $b = 0°$, we have *Number of clouds per kpc with* $N_H > N_{cloud} = 5.7 \ N_{cloud}^{-0.8}$. Here N_{cloud} is expressed in units of 10^{20} cm^{-2}. The median cloud has $\tau_{max} \sim 0.07$, $T_{n,min} \sim 135$ K, and $N_H \sim 0.6 \times 10^{20}$ cm^{-2}. If it is in pressure equilibrium with the rest of the ISM at $nT = 3000$ cm^{-3} K, it has $n_{HI} \sim 22$ cm^{-3} and a diameter of ~ 0.9 pc.

The distribution of column densities of diffuse clouds has been obtained using two other methods: interstellar reddening and optical absorption lines. These methods are ill suited for the cloud statistics. The observed reddening is simply the *integrated* column density of dust. For this reason, the analysis is necessarily crude: the spectrum of diffuse clouds is approximated by "standard" and "large" clouds (Spitzer 1978). However, in light of the large range of N_{HI} that is actually observed, we seriously question the value of this crude approximation. For example, the *medlun* HI cloud has $N_{HI} = 0.6 \times 10^{70}$ cm^{-7}, nearly four times smaller than a "standard" cloud. *Thus the "standard" cloud is not even representative of the cloud population!* The reddening of a median HI cloud is so small ($E_{B-V} = 0.01^m$) that reddening data cannot be effectively used to study diffuse clouds.

Hobbs (1974) has attempted to use KI absorption lines to measure the column density distribution function. There are two problems in using KI lines for this purpose. (1) KI lines are weak. Hobbs finds an average of 4.6 clouds per kpc. The above relations then imply that he observes only those clouds having $N_{HI} > 10^{20}$ cm^{-2}—nearly twice N_{HI} for the median cloud. Thus, Hobbs observes fewer than half the interstellar clouds. (2) While KI absorption lines are good *tracers* of clouds with large column densities, they cannot directly *measure* column densities. Hobbs uses an empirical quadratic relation between N_{HI} (obtained from Lyα measurements) and N_{KI} to determine column densities. We question this scheme since the Lyα-derived N_{HI} refers to the *total* column density of HI, which may include more than one diffuse cloud and definitely includes a substantial contribution from WNM. In order to obtain N_{HI} from the KI data properly, a knowledge of depletion of metals and electron density in clouds is needed (Section 3.3.5.b).

It has been popular among theorists and observers alike to derive the cloud "size-spectrum" by using the observed column density distribution function together with the assumptions of constant volume density and spherical clouds. These assumptions are blatantly incorrect. Firstly, clouds are not spherical. Instead, most clouds are either filaments or sheets. Secondly, low-N_{HI} clouds have higher temperatures, and thus lower volume densities if the pressures are constant; constant gas pressure is a fundamental tenet of all models of the ISM. In our considered opinion, calculations and models based on these false assumptions should not be taken seriously at all.

3.3.4 Temperature Structure of HI Clouds

In Section 3.2.1.e, we described the high-resolution study of the 3C147 field by Kalberla et al. (1985). The presence of 3C147 allowed these authors to derive temperatures independently for the various HI components. They conclude:

(a) The clumps, which are imbedded in the filaments and/or sheets, are primarily responsible for HI absorption. They have densities of ~ 20–50 cm^{-3} and spin temperatures (T_{cl}) between 30 and 80 K, averaging 40 K.

(b) The filaments and/or sheets are associated with lukewarm (say, $\gtrsim 500$ K) HI envelopes; the envelopes account for 80% of the HI emission.

It is interesting to compare these temperatures derived for HI with those derived for $H_2 (T_{H_2})$ from UV studies. T_{H_2} averages about 80 K (Spitzer 1978), considerably higher than the 40 K which typifies T_{cl}. This implies that substantial amounts of H_2 have temperatures higher than the clumps—i.e., that much of the H_2 exists outside the clumps.

(a) Temperature Structure and the T-τ Relation

Given the fact that clouds have temperature structure, how should the T-τ relation be interpreted? Its original *data* involved $T_{n,\min}$, the single-dish naively-derived temperature of an HI cloud (Section 3.3.3). However, $T_{n,\min}$ is larger than the real temperature of clumps T_{cl}. Its original *interpretation* involved the assumption of homogeneous clouds, but the picture of HI clouds is now complex: clumps are immersed in filaments and/or sheets, which are themselves immersed in lukewarm envelopes; possibly the whole complex is surrounded by WNM.

The clumps are primarily responsible for HI absorption. Thus $\tau_{cl} \approx \tau_{\max}$. In contrast, all three components are responsible for the HI emission; thus $T_{n,\min}$ depends not only on T_{cl} but also on the relative amounts of clump, filament/sheet, envelope, and WNM HI. Kalberla et al. (1985), after accounting for the envelope and WNM emission, find that the colder clumps appear to have larger τ_{cl}. Unfortunately, their sample numbers only five. Clearly, we need more data before we can observationally study the relation between the *intrinsic* parameters T_{cl} and τ_{cl}.

Thus our discussion must be based on previous single-dish measurements and model fitting. Payne et al. (1983) fitted the observed $T_{n,\min}$ and $\tau_{\max}$ to a model in which the clumps are assumed to be at a fixed *spin* temperature $(T_{s,cl})$ and surrounded by a lukewarm envelope of constant column density, contributing a fixed *brightness* temperature $(T_{B,env})$. We imagine $T_{s,cl}$ to be the same as T_{cl}, and $T_{B,env}$ to arise from emission from the envelopes and the WNM. In their model, the clump column density N_{cl} is allowed to vary and this in turn leads to variation in $\tau_{\max}$. This model leads to a fit as good as that obtained from the T-τ relation; the best fit parameters are $T_{s,cl} = 55$ K and $T_{B,env} = 4$ K.

Liszt (1983), in his numerical simulations of observables for various cloud models, rejects the above clump-envelope model because it disagrees in detail with the T-τ relation. However, we believe his rejection is inappropriate: it is based on disagreement with the T-τ relation, not on disagreement with the *data*. (Recall our words of caution about the T-τ relation in Section 3.3.3.a.) A plot of the *observed data* versus the clump-envelope model simulation shows satisfactory agreement (Payne et al. 1983). In the clump-envelope model, $T_{s,cl}$ is the quantity that determines the exponent α (Equation 3.25). With the lukewarm and warm HI contributing most of the emission in the typical case, the T-τ relation is really a fit to clouds having the *lowest values* of $T_{s,cl}$.

All of this discussion emphasizes that we need an unambiguous verification of the clump-envelope model. This requires a reliable separation of clumps and envelopes in many areas of sky. Mebold et al. (1982) have attempted to do this decomposition but their adopted method is controversial. One way to separate clumps and envelopes is through high-resolution HI maps. The best way, which must await proper spaceborne instrumentation, would employ mapping of the far-IR cooling lines. The clumps are cold, and cool primarily by emission of the 157-μm line of CII. The warmer envelopes and the WNM are warm; above 600 K, FeII (26 and 35 μm) and SiII (35 μm) also become important coolants.

(b) Physical Implications

The mean velocity dispersion of HI in emission is ~ 3.5 km s^{-1}, which implies that a large fraction of the envelopes have temperatures below about 1300 K. The envelopes exhibit no absorption, and are warmer than ~ 300 K. Thus the envelopes are lukewarm—say, 500 K. If we seriously believe in thermal equilibrium (Figure 3.3), then the envelopes cannot exist in this temperature range. *Thus either the envelopes or their heating agent is transient.* It is plausible that the envelopes are the conductive interfaces between the clumps and either the WNM or the HIM. Penston and Brown (1970) have treated this problem theoretically and find that the column density in the interface is much lower than observed for the envelopes; however, magnetic fields, which were not included in their calculations, may change their results. In our opinion this challenging set of problems needs immediate theoretical and observational attention.

3.3.5 Ionization of the Diffuse Interstellar Medium

Ionized gas is a major constituent of the diffuse interstellar medium, and the power required to keep it ionized is comparable to the power injected by supernovae. Until recently, not much was known about this medium, primarily due to lack of probes. However, with the advent of sensitive Fabry-Perot spectrometer observations of optical emission lines, we have come to recognize this medium as a phase in its own right. Not much is known about the origin of this gas, nor is its relation to the neutral gas well understood.

(a) Observations

Ionized gas in the ISM is revealed by a variety of observations, each sensitive to a combination of the electron density and temperature. These will be reviewed briefly here.

(i) Dispersion of Pulsar Signals. The group velocity of an electromagnetic signal of frequency v traveling through ionized gas is $V_g = c(1 - v_p^2/v^2)^{1/2}$, where $v_p = 8.97 n_e^{1/2}$ kHz is the plasma frequency. Owing to this dispersion in the ISM, a pulse at higher radio frequency arrives earlier than at lower frequencies. The observed rate of change of delay with frequency provides a measurement of $DM = \int n_e \, dl$, the integrated column density of electrons to the pulsar.

DM is the simplest observable quantity since it directly measures the column density of electrons. HI absorption data have been used to obtain kinematic distances for two dozen pulsars. These data, in conjunction with DMs, have been used to establish the Galactic mean electron density distribution (Lyne et al. 1985):

$$\langle n_e \rangle = [0.025 + 0.015e^{-|z|/70}]\left[\frac{2}{1 + R/R_0}\right] \text{cm}^{-3} . \tag{3.27}$$

Here and below, the average $\langle \ \rangle$ is meant to denote the *volume-averaged* quantity, i.e., in this case the electron density that would be obtained by distributing the electrons uniformly in space. The *true* electron density is larger. Here R is the Galactocentric distance in kpc and $R_0 = 10$ kpc. The first term scales inversely with the assumed R_0 whereas the second term is independent of R_0.

The first term describes a widely distributed component with an exponential scale height $\gtrsim 1000$ pc (Vivekanand and Narayan 1982). The value of this scale height is a lower limit because the scale height of electrons' probes, i.e., the pulsars, is much smaller, only ~ 400 pc. Similarly, only a lower limit can be placed on $DM_\perp$, the mean vertical column density of electrons. Manchester and Taylor (1977) make allowance for this and obtain $DM_\perp \gtrsim 30$ cm^{-3} pc. The second term in the above equation describes, in a statistical way, the contribution by discrete, bright HII regions and hence is of not great interest in this chapter.

(ii) Optical Recombination Emission. In the ISM, a free electron eventually recombines with a proton and in the process emits a host of recombination-line photons. Velocity resolution is required to separate solar system gas from interstellar gas. Nearly all useful work has employed a wide-beam (5′ to 50′) Fabry-Perot spectrometer to observe Hα λ6563 Å (see Reynolds 1984). The observed Hα intensity is $I = 0.36 \int n_e^2 T_4^{-0.9} \, dl$ Rayleighs (R), where T_4 is the temperature in units of 10^4 K and dl is path length in parsecs. I is directly proportional to the total recombination rate which, in steady state, is equal to the total ionization rate of H atoms. Thus, for an assumed constant temperature, the emission measure EM, i.e., $\int n_e^2 \, dl$, can be obtained.

The Hα emission appears to be widely spread with a disk distribution. From the observed intensity at various latitudes, the intensity reduced to the pole, i.e., $I_\perp \equiv \langle I(b)\sin(b)\rangle$, is found to be between 0.5 and 1.7 R. The lower limit is mainly weighted by measurements at high latitudes; $I_\perp = 1$ R is a good mean fit. This translates to $EM_\perp = 2.8$ cm^{-6} pc for $T_4 = 1$. From latitude scans in the Perseus arm, Reynolds (1986) finds that the $|z|$-distribution of EM is well represented by an exponential with scale height 300 pc. If we assume that the local scale height of the WIM is also 300 pc, then the observed $EM_\perp$ requires $\langle n_e^2 \rangle \approx 0.009 e^{-|z|/300}$ cm^{-6}.

(iii) Low-frequency Radio Absorption. Low-frequency radio waves, emanating either from the Galactic synchrotron background or a strong radio source, are absorbed by free electrons in encounters with positive ions. This process, commonly referred to as "free-free" absorption, is exactly the inverse of the bremsstrahlung emission process. The optical depth $\tau_{ff} \propto g_{ff} \int n_e^2 T^{-3/2} v^{-2} \, dl$. After allowing for variation in g_{ff}, the Gaunt factor, $\tau_{ff} \propto v^{-2.1} T^{-1.35}$ EM; this approximation is valid for radio frequencies. From observations at different frequencies, the emission measure can be estimated for an assumed temperature.

In order to see significant absorption from the diffuse ISM, one has to go down to frequencies as low as a few MHz! Only on some rare occasions does the Earth's ionosphere allow astronomers to view the low-frequency heavens, and then too only from a few special places such as Tasmania and Penticton, Canada. Almost all the ground-based work in this difficult field has been done in Tasmania and mainly by the pioneering radio astronomer Grote Reber and his colleagues. Even in Tasmania, useful observations are limited to frequencies greater than 2 MHz. Observations at lower frequencies have been obtained from spacecraft, which unfortunately are confused by the lack of any reasonable angular resolution.

The observational situation is uncertain. From observations of extragalactic sources at 10 MHz with an array at Cambridge, Bridle and Venugopal (1969) found that the optical depth reduced to the north Galactic pole was $\tau_{ff,\perp}(10 \text{ MHz}) = 0.1 \pm 0.02$. From observations of the diffuse background of the sky visible from Tasmania, Cane (1979) derived a value about 2.5 times smaller whereas Ellis (1982), using essentially the same data, derived a value 5 times smaller than the Cambridge value. The interpretation of diffuse background measurements is complicated because the source (i.e., the relativistic electrons) and the absorber are spatially mixed. The interpretation of the Cambridge observations is unambiguous. However, unlike the observations at Tasmania, the Cambridge observations are at the relatively higher frequency of 10 MHz where the free-free optical depths are much smaller and hence systematic errors could play a larger role.

(b) Ionization Within Diffuse Clouds

So far we have not discussed the ionization within diffuse clouds. The reason is that there are simply no reliable measurements of x_e, the ionization fraction! We caution the reader against trusting values of x_e that are quoted in papers dealing with analysis of optical and UV absorption lines. In all cases of which we are aware, these values are *assumed* to be equal to the fractional ratio of [C/H] because starlight ionizes any heavy element with ionization potential less than that of HI, and carbon is the most abundant such element. Furthermore, some authors assume that much of the carbon is depleted on grains, but use the full cosmic abundance of carbon for their assumed value of x_e!

The above assumption is incorrect. There are two sources of electrons within diffuse clouds: ionization of carbon by starlight and ionization of hydrogen and helium by EIPs (Section 3.1.4.b). For typical diffuse cloud conditions ($n_H \sim 40 \text{ cm}^{-3}$, $T \sim 100$ K), cosmic-ray ionization dominates carbon by a factor of ~ 2; its contribution is $x_e \sim 6 \times 10^{-4} \zeta_{t,-16}^{0.5} (n_H/40)^{-0.5} (T/100)^{0.35}$, where $\zeta_{t,-16}$ is the total cosmic-ray ionization rate per H atom in units of 10^{-16} s^{-1}.

The predicted electron density in the CNM is too low to account for the observed Hα emission or the pulsar dispersion measures or the low-frequency free-free absorption.

(c) Where Do the Electrons Reside?

(i) In the WNM? In the classic FGH model of the ISM, pulsar dispersion arises from the weakly ionized WNM. We first consider whether this hypothesis is consistent with observations. Doing so is not straightforward, because there are no direct observations of the ionization fraction within the WNM. The traditional probes of EM and free-free absorption are insensitive to a partially ionized, warm, tenuous medium. Nevertheless, we can use a combination of theory and DM data to show that the WNM is not the dominant source of interstellar electrons.

HI in the WNM can be ionized by soft X rays or cosmic rays. Whether soft X rays can actually be effective depends on the relative distribution of the WNM and the sources of soft X rays. Soft X rays with a mean energy of 74 eV cannot penetrate a column density of more than $\sim 10^{19} \text{ cm}^{-2}$, and will be effective only if the mean distance between sources $\lesssim 20$ pc.

In ionization equilibrium, $x_e \approx 0.03 \zeta_{t,-16}^{0.5}$; here, the WNM is assumed to be at a mean pressure of 3000 cm^{-3} K and a temperature of 8000 K. From the vertical column density of HI (Sections 3.1.2.a and 3.3.2), the corresponding vertical column density of electrons is $N_{e\perp} \sim 0.5 \times 10^{19} \zeta_{t,-16}^{0.5} \text{ cm}^{-2}$. This value for $N_{e\perp}$ is nearly 20 times smaller than the observed vertical density of electrons, DM$_\perp$. In addition, the implied EM$_\perp$ and τ_{ff}(10 MHz) are orders of magnitude below the observed values (Section 3.3.5.a). A more refined calculation, taking into account the actual z-

distribution of the WNM, results in a similar discrepancy; nor is the conclusion affected when the larger soft X-ray ionization rate of $\zeta_{p,-16} = 5\,\mathrm{s}^{-1}$ ($\zeta_{t,-16} = 13\,\mathrm{s}^{-1}$) is used. Only by assuming a mean pressure as low as $1000\ \mathrm{cm}^{-3}\,\mathrm{K}$ and $\zeta_{t,-16}$ as high as ~ 40 can the observed $\mathrm{DM}_\perp$ be reproduced; even with these unacceptable values the observed $\mathrm{EM}_\perp$ cannot be accounted for. Thus the simple conclusion is that the WNM is not a major supplier of interstellar electrons.

The conclusion that the WNM is not the *main* source of electrons in the ISM is very important. In particular, it implies that the chief ionizing agent is not EIPs. Hence, the WNM heating and ionization agents need not be one and the same, as in the FGH model. However, this conclusion by itself does not invalidate the FGH model: the electrons must arise from another, poorly specified agent—the same agent that produces electrons in the MO model. With this agent, we have the modified FGH model introduced in Section 3.1.6.

(ii) In the WIM? Since neither the WNM nor the CNM contribute a dominant fraction of interstellar electrons, we are forced to the conclusion that almost all the electrons arise in another phase. Following Reynolds, we identify this phase to be the WIM.

Observationally, the WIM is identified as the phase that produces the diffuse Hα emission. The WIM clearly has a non-negligible filling factor, because diffuse Hα is seen along most lines of sight. Many collisionally excited, metastable optical emission lines of ions such as NII and SII arise from this phase (Reynolds 1985 and references therein). The WIM must be warm: the line widths of the optical lines suggest $T \sim 8000$ K and the detection of the NII metastable line requires $T \gtrsim 3000$ K. The ratios [NI/NII] and [OI/OII] are tied to the [HI/HII] ratio; the lack of detection of NI suggests that the medium is substantially ionized: $x_e > 0.75$.

Any model for interstellar electrons must explain the observed $\mathrm{EM}_\perp$, $\mathrm{DM}_\perp$, and $\tau_{ff,\perp}$. Theoretically, the ratio $I_\perp(\mathrm{H}\alpha)/\tau_{ff,\perp} \sim T^{0.45}$. Adopting Ellis's value for $\tau_{ff,\perp}$ (Section 3.3.5.a), we derive a temperature of 4400 K; adopting Cane's reduces this to 3000 K. In our opinion, these temperature estimates are consistent with the 8000 K derived for the WIM from optical observations, especially given the observational uncertainties in the radio-frequency optical depth. This derived temperature is considerably higher than the temperature of the CNM. Hence we can safely conclude that both the Hα and the free-free absorption arise from one and the same medium—the WIM. Nevertheless, the observational data, especially the free-free absorption data, need to be improved so that we can state this conclusion with absolute certainty.

In Section 3.3.6, we show that DM and EM data are consistent if they both arise in the WIM. We defer this discussion until then, because it is inextricably tied to our estimate of the filling factors of WNM and WIM—a topic so important that it requires its own separate section.

(d) The Energy Source for the WIM
The energy requirement to maintain the WIM is high. Locally, the number of recombinations in a column perpendicular to the galactic plane is about $4 \times 10^6\ \mathrm{cm}^2\,\mathrm{s}^{-1}$. To maintain this over the whole Galaxy requires $> 10^{42}\ \mathrm{erg\,s}^{-1}$, comparable to *all* the power injected by Galactic supernovae. Clearly, then,

supernova-related processes are totally inadequate. Possible sources include pho-toionization by hot stars and ionization by shocks. Shocks are unlikely. First, strong shocks are ruled out by the temperatures inferred from the observed widths of the optical emission lines, and the upper limits of NI and OI and the strength of NII relative to Hα do not call for substantial amounts of the ISM to be weakly shocked. Secondly, the various line ratios are extremely sensitive to shock speeds and the observed ratios do not vary substantially from one region to another. Finally, a large shock-ionization rate simply transfers the energetics question to the agent that drives the shocks.

All the principal optical data have been nicely explained in a model of the WIM with steady-state equilibrium between photoionization by diffuse Lyman continua of hot stars and recombination (Mathis 1986). In this model, the ratio [NII]/Hα is extremely sensitive to the assumed temperature and the observations constrain the temperature of the WIM to a narrow range around 7500 K. Observations indicate that the excitation of these lines is quite unlike that of the bright HII regions, in that one observers a stronger [SII]/Hα, a much weaker [OIII]/Hα, but a normal [NII]/Hα. This trend is, in fact, predicted by Mathis's model and results from the diffuse nature of the ionizing source(s). The close agreement between Mathis's steady-state model and observations suggests that the WIM is in thermal and ionization equilibrium.

What are the sources of the diffuse Lyman continua? Potential sources include O stars, B stars, hot white dwarfs, nuclei of planetary nebulae, and QSOs. Of these, only O stars have more than enough energy to account for the observed recombination rate: more precisely, O stars have more than five times the ionizing flux demanded by the Hα observations.

However, O stars are not distributed uniformly. Most are clustered in time and space, with all of their ionizing photons taken up by HII regions in the immediate vicinity. Observationally, it appears that 50% of the O stars are in this condition, being immersed in bright, ionization-bounded HII regions that are readily detect-able on the Palomer Sky Survey photographs (Torres-Peimbert et al. 1974). This implies that the remaining 50% have escaped from their dense, star-forming clouds. But even for these the situation is uncertain. An O star surrounded by enough neutral matter forms an "ionization-bounded" HII region whose extent is limited by the number of ionizing photons emitted by the star. If all O stars are surrounded by ionization-bounded HII regions, then there are no photons left to produce the WIM. Elmegreen (1976) has studied the effect of single O stars on "normal" interstellar gas and found that the HI clouds are rapidly ionized, leaving behind debris that often results in ionization-bounded HII regions.

Clearly, the crucial point is whether too many isolated O stars are surrounded by ionization-bounded HII regions. The observational data are sparse: HII regions around three stars (ζ Oph, λ Ori, and α Vir) appear to be ionization bounded, while those around δ Ori and ζ Pup are density bounded if viewed in their *immediate vicinity* (Reynolds and Ogden 1979). These last two are special cases—they lie within the Orion and Gum shells, respectively. Such shells are thought to be a result of supernovae formed by the death of massive stars in the clusters. The young stars that are now present keep the inside edges of the shell ionized. Such combination "supernova shell–HII region" structures appear in the spectacular Hα photographs of Sivan (1974). They are very different from the standard HII regions because the interior has been swept clean and consists of low-density, hot gas; most of the emission measure arises at the shell edges where the Lyman continuum photons are completely absorbed. Thus the HII regions surrounding these stars are also ionization bounded.

We conclude that the fraction of O stars that are incapable of ionizing the general ISM is higher than found by Torres-Peimbert et al. (1974). However, we cannot conclude that O stars are *completely* ineffective in ionizing large volumes within the HI cloud layer, because O stars are so powerful that only a small fraction of the total population would produce significant ionization. The question of the ionizing agent *in the cloud layer* is important and needs to be definitively resolved. Possible photoionization agents other than O stars include the more widely distributed B stars and other hot stars such as hot white dwarfs and nuclei of planetary nebula.

We believe that O stars are effective outside the cloud layer. A significant fraction of the O stars belong to the so-called "runaway" subgroup, with high peculiar velocities. Such stars escape the cloud layer in about one-third of their main-sequence lifetimes. Once outside the layer, the cloud debris mechanism of Elmegreen is no longer operative and the stars should be able to ionize a large volume of the ISM. This is borne out observationally: there are indications that the high-$|z|$ O stars are not ionization bounded (Reynolds 1982). In addition to the O stars, nuclei of planetary nebula, hot white dwarfs, and QSO UV light also contribute to the ionization flux at high $|z|$'s. Collectively, there should be no problem in maintaining the WIM outside the cloud layer.

3.3.6 Filling Factors of the WIM, WNM, and CNM

The filling factors of the various phases of the ISM are *very* important numbers. If HI dominates the interstellar space, then the ISM is best described by the modified FGH model; if the HIM dominates, then the ISM is best described by the MO model. For this reason their values are controversial and are hotly debated by astronomers at various meetings. In this section we derive the ISM pressure and the filling factors of CNM, WNM, and WIM under various assumptions. We shall see that our approach yields reasonable results and helps to round out the overall picture of the ISM. Nevertheless, the uncertainties involved in our assumptions and parameter values prevent us from making definitive statements about whether the modified FGH model or the MO model applies.

The most basic assumption is the conclusion stated in Section 3.3.5.c, namely, that the pulsar dispersions and Hα emission are produced by a single electron component that resides in the WIM. In that section, we promised a proof that the DM and EM data are consistent with a single WIM electron component. Here we assume such a component and use the DM and EM data to estimate physical parameters in the ISM. These physical parameters are in agreement with our expectations, based on independent data and on theoretical arguments. This consistency constitutes the promised proof.

This most basic assumption is all we require to derive the WIM filling factor and its z-dependence, $\phi(z)$, from the scale heights of $\langle n_e \rangle$ (from DM observations) and

$\langle n_e^2 \rangle$ (from Hα observations; see Section 3.3.5.a). The exponential scale heights for these components are 1000 and 300 pc, respectively. The *true* electron density, n_e, is related to the *volume-averaged* electron density, $\langle n_e \rangle$, by $\langle n_e \rangle(z) = \phi(z)n_e(z)$; similarly, $\langle n_e^2 \rangle(z) = \phi(z)n_e(z)^2$. A little algebra yields for the WIM:

$$n_e \sim 0.27 e^{-|z|/428} \text{ cm}^{-3} \, , \tag{3.28}$$

and

$$\phi \sim 0.11 e^{|z|/748} \, . \tag{3.29}$$

In Section 3.3.5.d, we concluded that the WIM is largely ionized and its temperature is ~ 7500 K. Its gas pressure is $2n_e T$. Thus we obtain the run of WIM gas pressure with z:

$$P \sim 4000 \, e^{-|z|/428} \text{ cm}^{-3} \text{ K} \, . \tag{3.30}$$

It is important to note that the above equations are only statistical in nature and should not be interpreted too literally. At any given $|z|$, the pressure in the ISM probably varies by at least factors of 2.

We now assume that *gas* pressure equilibrium exists accross all the phases of the ISM. Since Equation (3.30) provides the WIM pressure as a function of $|z|$, our assumption tells us the pressure of *all* gas components as a function of $|z|$: CNM, WNM, and WIM. However, we stress that *strict* gas pressure equilibrium is unlikely to be achieved (Section 3.1.3). Elmegreen (1976) has argued that the WIM may have an overpressure by a factor of 4.

Given our assumption of pressure balance among all gas components in the ISM, Equation (3.30) should also describe the run of ambient pressure of the ISM. It is, then, a very important equation—if it is correct. In fact, it is remarkably consistent with our meager, independent knowledge about the interstellar pressure. It predicts the mean pressure within the cloud layer ($|z| \sim 100$ pc) to be ~ 3200 cm^{-3} K—in excellent agreement with the observational determination (Section 3.1.3). This agreement is remarkable because Equation (3.30) was derived with very little or no input physics but only some basic algebra. Note that the pressure scale height is significantly larger than the classical WNM scale height of 200 pc. It permits observed HI clouds to be in pressure equilibrium even at high z's (Lockman 1984).

We now use the assumption of gas pressure equality among the various phases, together with Equation (3.30), to derive the filling factors of the CNM and WNM. This is straightforward: knowing the pressures and temperatures of these components, we derive the true densities and then, from the volume-averaged densities, the filling factors.

For the CNM, the filling factor is complicated by the inhomogeneous structure of clouds (Sections 3.2.1.e and 3.3.4.a). For example, if we assume that the CNM contains equal amounts of HI at 40 K and 400 K, then the filling factor of the former is a negligible $\sim 5 \times 10^{-3}$ and of the latter a non-negligible 0.05. Both these estimates apply only to the cloud layer ($|z| < 100$ pc) and are essentially independent of z.

For the WNM, we assume that the vertical distribution is best described by the sum of two components: a Gaussian of scale height 250 pc and an exponential of scale height 480 pc (Lockman 1984; also Section 3.1.2.b). Since the temperatures of the WNM and WIM are very nearly equal, the mass density of the WNM is twice that of the WIM under our assumption of gas pressure equality. The sum of the filling factors of the WNM and WIM at $z = 0$ is ~ 0.4 and gradually rises to about 0.6 at $z = 1$ kpc.

Under the assumption of gas pressure equality, we find $\sim 50\%$ of the interstellar volume to be occupied by the WNM and WIM. If these numbers were reliable, we could conclude that the HIM occupied the other 50% so that the ISM would be borderline between the modified FGH and MO models. However, because of magnetic and cosmic-ray pressures, the *gas* pressures in the WNM and WIM are likely to be different. In particular, the gas pressure in the WNM is likely to be smaller than that in the WIM. In this case, the filling factor of the WNM increases, and the WNM and WIM would occupy substantially more than 50% of the interstellar volume.

3.4 Interstellar Magnetic Fields

3.4.1 Methods of Measurement
Methods fit into two categories: methods to measure the *direction* of the component lying *perpendicular* to the line of sight ($B_\perp$, the projected direction on the plane of the sky), and methods to measure the *strength* of the component of magnetic field lying *parallel* to the line of sight, $B_\parallel$. For details, see Verschuur (1979).

(a) $B_\perp$: Linear Polarization of Starlight
Historically, the very existence of the interstellar magnetic field was deduced from the discovery of the polarization of starlight. The polarization is produced by the alignment of spinning dust grains in the interstellar magnetic field. The original "Davis-Greenstein" theory of grain alignment involves dissipation of the *thermal* spin energy of grains by the imaginary component of magnetic susceptibility in the grain (see Jones and Spitzer 1967). However, this theory requires magnetic field strengths that are much higher than those observed. This important point is apparently not generally appreciated, because some authors still derive field strengths using this theory. Weak fields can align grains because the spin energy of grains is much larger than thermal—it is *suprathermal* because of the unbalanced *time average* torque on a grain produced by a small number of sites where H_2 molecules are produced and ejected with large kinetic energy, spinning the grain up like a steam turbine. Suprathermal grains are generally aligned such that the direction of polarization of starlight lies parallel to $B_\perp$. The theory, which is presented in an elegant paper by Purcell (1979), involves a remarkably wide range of physical principles.

(b) $B_\perp$: Linear Polarization of Synchrotron Radiation
Synchrotron radiation is produced by the acceleration of relativistic electrons gyrating in a magnetic field. The direction of linear polarization is parallel to the direction of acceleration. Thus, the polarization is perpendicular to the magnetic field, and the observed polarization is perpendicular to $B_\perp$. If the field is perfectly

uniform, the linear polarization is large, typically $\sim 70\%$ (Ginzburg and Syrovatskii 1969). In "real astronomy," the magnetic field is unlikely to be perfectly uniform. In this case, as the uniform field direction changes from being perpendicular to parallel to the observer's line of sight, the direction of linear polarization remains perpendicular to $B_\perp$ but the intensity of the polarization goes to zero.

Derivation of $B_\perp$ from synchrotron radiation is subject to two caveats. One is Faraday rotation, which changes the angle of polarization at long wavelengths (see below). The other is that the region in which the relativistic electrons reside may not be the same as that in which the diffuse interstellar matter resides. This point is well illustrated by the North Polar Spur, where the directions of $B_\perp$ derived from the polarization of starlight and from the polarization of synchrotron radiation are perpendicular for $b \lesssim 40°$ (Spoelstra 1971).

(c) $B_\perp$: Linear Polarization of Radio-Wavelength Spectral Lines
It has been predicted (Kylafis 1983, Deguchi and Watson 1984), but not yet observed, that molecular spectral lines at radio wavelengths should be linearly polarized parallel to $B_\perp$. The polarization depends on a number of complicating factors, including the characteristics of the local velocity and radiation fields; it is insensitive to the field strength. This may be the only way to probe the field structure in dense molecular clouds and protostars.

(d) $B_\parallel$: Farady Rotation
The rotation measure RM $\propto \int n_e B_\parallel \, dl$ is derived from the Faraday rotation of the plane of linear polarization of a background radio continuum source. Along the line of sight all of the electrons rotate either clockwise or counterclockwise, depending on the direction of the magnetic field. This causes the index of refraction, and thus the phase velocity, to be different for the two senses of circular polarization. Linear polarization with a specific position angle is produced by combining the two circular polarizations with a specific phase difference. Because the phase difference changes with wavelength, the position angle also changes with wavelength, λ; it is proportional to RMλ^2 (see Spitzer 1978). RMs can be measured for ordinary radio sources that exhibit linear polarization if a single emitting region dominates over the wavelength range analyzed and if the region is "Faraday thin"; see the very important cautions emphasized by Vallée (1980). Many pulsars are also linearly polarized. Pulsars provide the additional important piece of information, dispersion measure (DM $\propto \int n_e \, dl$). The ratio RM/DM provides the magnetic field strength directly, with no ambiguity about the electron density.

(e) $B_\parallel$: Zeeman Splitting
The other way to measure $B_\parallel$ is with Zeeman splitting. An external magnetic field splits the upper level of the 21-cm line into three levels. The splitting between the highest and the lowest levels is 2.8 B_{tot} Hz, where B_{tot} is the *total* field strength in μG. In all diffuse clouds, this splitting is much smaller than the typical line width. In this limiting case, the *observed* splitting is $2.8 B_\parallel$ Hz—proportional only to the *parallel* component of magnetic field. To detect the splitting, one observes the difference between the two circular polarizations. This difference is the frequency derivative of the line profile with an amplitude proportional to $(B_\parallel / \delta v)$, where δv is

the line width. Thus high frequency resolution is not required. Zeeman splitting was first detected in the 21-cm line seen in absorption against the classical strong radio sources by Verschuur; the early history is an interesting story of the frustrations involved with inadequate instrumentation (Verschuur 1979). Zeeman splitting is now being observed in the 21-cm line seen in emission, a very difficult undertaking because of the instrumental problems involved (see Troland and Heiles 1982a).

Zeeman splitting has been searched for in H recombination lines (Troland and Heiles 1977) and C recombination lines (Silverglate 1984). It has been observed in several centimeter-wavelength OH lines. Any molecule having an electron with an unpaired electron spin yields Zeeman splitting comparable to the H atom. In practice, the only such molecules having centimeter-wavelength lines are those exhibiting Λ-doubling (the hydrides). Several such molecules have millimeter-wavelength lines, but Doppler widths are so large that the ratio of splitting to line width is too small to be detectable with system temperatures currently available.

(f) Faraday Rotation Versus Zeeman Splitting

These two techniques for measuring $B_{\parallel}$ sample different kinds of region. Zeeman splitting favors high HI column density and narrow line width, so it samples the cold HI clouds—the CNM. Faraday rotation samples ionized regions, the same as sampled by pulsar DMs. Section 3.3.5 shows that pulsar DMs are produced mainly by the WIM. Measuring the magnetic field strength in the WNM is virtually impossible: Zeeman splitting is too difficult to measure with the WNM's wide, weak HI lines, and Faraday rotation is small because of the low ionization (Section 3.3.5.c).

The fact that Zeeman splitting and Faraday rotation sample different regions is elegantly illustrated by the observational results towards the Crab Nebula, against which Zeeman splitting has been detected in HI absorption, and the associated pulsar, for which the Faraday rotation has been measured. Zeeman splitting shows a field directed towards the Earth, while Faraday rotation shows the opposite.

There is confusion with regard to signs. We adopt the IEEE convention for circular polarization, not the definition of classical optics. In the IEEE convention, a right-hand circularly (RHC) polarized wave rotates clockwise as it propagates away from the observer. In Zeeman splitting, if the RHC component is observed at a higher frequency, then the magnetic field points towards the observer. A field pointing *towards* the observer is a *negative* magnetic field. In Faraday rotation, if the position angle of the plane of polarization increases with increasing wavelength (corresponding to counterclockwise rotation against the plane of the sky: the position angle is measured eastward from north), then the RM is positive. This corresponds to a field pointed towards the observer. Thus, *positive* RMs correspond to *negative* magnetic fields (!).

3.4.2 Fields in External Galaxies

Sofue et al. (1986) have recently reviewed the field, so we omit the details and stress one fundamental point: the conclusions, at the moment, are premature. Attempts to determine magnetic field geometry and strength in external galaxies have been based on measurements of linear polarization at optical and radio wavelengths.

Optical measurements are, in nearly all cases, plagued by contamination from scattered light.

The radio emission is polarized because it is nonthermal. In the absence of Faraday rotation, the observed polarization is perpendicular to $B_\perp$. At 6-cm wavelength, where a number of galaxies have been measured with reasonable accuracy, Faraday rotation is expected to be small so that, to zeroth order, the observed polarization vectors trace out the direction of $B_\perp$. The results show that, again to zeroth order, the field lines are circular or, perhaps, aligned along the spiral arms. We believe these zeroth-order results. Furthermore, they are in accord with results on our Galaxy (Section 3.4.3).

The problems arise with attempts to go beyond this zeroth-order picture. Typically, the assumption is made that the magnetic field is either circular or aligned with the arms, and that departures of the observed polarization angles from those expected under this assumption arise from Faraday rotation. This allows the determination of Faraday rotation from measurements of polarization at a single wavelength! Such determinations are used to infer the direction of $B_\parallel$ and, with suitable assumptions about electron density, the strength of $B_\parallel$. In this way, it has been concluded that many galaxies show a bisymmetric field structure, which can be produced if a primordial field is "wound up" by differential rotation; the field reverses several times with galactocentric radius and alternates in sign from one spiral arm to the next.

In our opinion, these interpretations are premature for two reasons. The first is that in many cases the position angles do not show the behavior expected under the model in a clear and unambiguous fashion. The second is that M81 is the only galaxy where good polarization measurements at more than one wavelength exist, and the angles of departure do not vary as λ^2 (Beck et al. 1985). In principle, accurate RM measurements would allow much to be learned about magnetic fields in external galaxies, and with today's availability of sensitive aperture synthesis instruments the field is ripe for investigation.

3.4.3 The Galactic Field on Scales Above 100 pc
(a) The Field Strength
Rough information on the large-scale behavior of the magnetic field strength can be obtained from models fitting the all-sky brightness distribution of the diffuse Galactic synchrotron emission (Phillips et al. 1981, hereafter PKOHS; Beuermann et al. 1985, hereafter BKB). These models provide estimates of the volume emissivity as a function of Galactocentric radius R and height above the plane z. The emissivity decreases outwards with R. If one fits this with an exponential, the radial exponential scale length is about 3.9 kpc or somewhat less.

The emissivity decreases with $|z|$. PKOHS model this in a manner that can be represented roughly with a thick disk plus a long, linearly decreasing $|z|$ tail. In contrast, BKB model it with two disks, a thin and a thick disk with roughly equal volume emissivities at $z = 0$, but no long tail. In both models, the $|z|$-scale heights of the disks *increase* with R, roughly as $\exp(R/10 \text{ kpc})$. At the solar circle, the disk scale height of PKOHS is roughly 0.6 kpc and the long tail extends to ~ 8 kpc; the disk scale heights of BKB are ~ 0.2 and 1.8 kpc.

The strength of the local magnetic field can be obtained from the derived emissivity near the Sun, together with the observed energy spectrum of the electron component of cosmic rays. The local emissivity is not well determined, because the models must deconvolve intensities built up along long path lengths. PKOHS and BKB derive values for the local emissivity that differ by a factor of ~ 3.5. The derived values for the total local field strength are about 4 and 9 μG for PKOHS and BKB, respectively. If we assume an equal split between uniform and random component of magnetic field, each component would be $2^{1/2}$ times smaller than the total. The resulting value for PKOHS is about 3 μG. This is larger than the strength of the uniform component derived from pulsar RMs and DMs, which is typically ~ 1.6 μG (e.g. Manchester and Taylor 1977).

The synchrotron emissivity depends not only on the magnetic field strength, but also on the energy density of the electron component of the cosmic rays. In the direction of R, Bloemen et al. (1986) have derived the distribution of the electron cosmic rays from observations of diffuse Galactic gamma rays. Folding their result together with the R-dependence of synchrotron emissivity, we have roughly $B \propto \exp(-R/20 \text{ kpc})$—roughly independent of R.

In the $|z|$ direction, we have no information on the variation of relativistic electrons. Perhaps the most reasonable assumption is that the energy density of cosmic rays is equal to that of the magnetic field. This is close to the "minimum energy" configuration required to generate the observed radiation. If, in addition, we assume that the energy of the *electron* component of the cosmic rays (which is a negligible fraction of the *total*) is proportional to the magnetic energy density, then the emissivity varies as $B^{7/2}$ (Ginzburg and Syrovatskii 1969). For PKOHS's single disk, the local B scale height would be about 2.4 kpc. For BKB's disks, the thin disk would raise the field strength by only $2^{2/7} \approx 1.2$ at $z = 0$, where the two disks have equal synchrotron emissivities, so that the B scale height would be determined by the thick disk and would be ~ 6 kpc.

While these numbers are subject to considerable uncertainty because of the assumptions involved, the B scale height must be considerably larger than the scale height of synchrotron emissivity because the synchrotron emissivity depends on a high power of B. Thus, the scale height of the magnetic field must be larger than that of the ISM. This has ramifications for the z-component of gravitational equilibrium (Section 3.1.5).

(b) The Field Direction

Optical polarization showed long ago that the field lines lie parallel to the Galactic plane; the comprehensive map by Mathewson and Ford (1970) exhibits more details. Towards $l_0 \approx 80°$ the polarization vectors tend to "focus" as we look down the field and see the field lines receding into the distance. Some analyses of Faraday rotation data are in agreement with the result, but others are in disagreement, yielding $l_0 \approx 100°$ (Simard-Normandin and Kronberg 1979, Thomson and Nelson 1980, Inoue and Tabara 1981). To zeroth order, then, the field is circular—just as it is in external galaxies. However, the exact value of l_0 is more than just a detail: the local spiral arm has $l_0 < 90°$, so if in fact $l_0 \approx 100°$ then the field is significantly tilted with respect to the spiral arms.

The above optical and RM data sets refer, in fact, to regions within a few kpc of the Sun. One would like to extrapolate and conclude that the magnetic field is roughly circular everywhere in the Galaxy. However, there is an absence of large RMs toward the Galactic interior. This is important, for it indicates that the Galaxy cannot have a roughly circular uniform field that points in the same direction everywhere. In the Galactic plane near, say, $l = 30°$, such a field would make every volume element along the line of sight contribute in the same sense to the RM of an extragalactic source, so that with the long path lengths of tens of kiloparsecs we would expect RMs to be systematically larger than ~ 500 rad m^{-2}. Such large RMs are not observed. In addition, Manchester and Taylor (1977) point out that closely spaced pulsars, located at different distances, sometimes have different algebraic signs of RM.

These facts indicate the existence of either one or more field reversals or a significant radial field component inside the solar circle (Heiles 1976c). Alternatively, the uniform component of the field may vanish, leaving only a random component. Analyses of RMs of extragalactic sources probably favor field reversal(s) inside the solar circle, but the data are not yet good enough to determine this unambiguously, much less the number and location(s) of the field reversals. There is no evidence for field reversals *outside* the solar circle (Vallée 1983).

Clearly, there are no widely accepted conclusions in this field. This is partly because of inadequate statistics and differences in techniques of analysis, but probably equally as important is real structure in the magnetic field on various length scales. Some analyses are weakened by the inclusion of incorrect data and by excess zeal in matching observations to models; no paper in this field can be read without critical care. Further progress requires a quantum jump in the number of RMs, particularly for extragalactic sources in the Galactic plane and for pulsars everywhere.

3.4.4 The Galactic Field on Scales of 100 pc and Below
On small scales we must have confidence that the detected field is associated with one particular interstellar object along the line of sight. Faraday rotation, which integrates over the whole line of sight, is of limited utility but has nevertheless delineated several "magnetic bubbles" (Vallée 1984) where RMs are enhanced in a limited area. Detection of these bubbles is difficult if the RM enhancement is small. In our opinion, all of the bubbles except the Monogem Ring $[(l, b) \approx (203°, 11°)]$ are statistically sound.

However, there is a serious question about the identification of the Gum Nebula bubble with the Vela supernova remnant. Vallée and Bignell (1983) ascribe an RM of about 130 rad m^{-2} to the Vela remnant itself, while the Vela pulsar has RM $= 34$ rad m^{-2}, four times smaller. If the pulsar lies at the center of the remnant, its RM should equal the ordinary interstellar contribution plus half the contribution of the whole remnant. In fact, at the distance of 400 pc, the ordinary interstellar contribution alone amounts to nearly 34 rad m^{-2}. This argues that the Vela pulsar lies on the very near side of the Gum Nebula.

Optical polarization reveals a tendency for magnetic field lines to lie either parallel or perpendicular to filaments or elongated clouds. On large scales this is spectacularly the case for the North Polar Spur HI shell, and may also be the case for the HI shell located in the vicinity of Radio Loop II (see Heiles and Jenkins 1976). Vrba et al. (1976) find parallel alignment in dust filaments in Ophiuchus and R Corona Australis. The opposite tendency, with perpendicular alignment, is observed in Taurus (Moneti et al. 1984) and probably near Orion (Heiles 1987b). L204, a long dust filament, presents the appearance of perpendicular alignment on the large scale, but shows parallel alignment in some regions (McCutcheon et al. 1986).

A remarkable result has been found by Cohen et al. (1984), who compiled optical polarization data obtained from stars behind bipolar flows of protostars. There is a very definite tendency for the bipolar flows—having *arcminute* angular scales—to be aligned with the optical polarization vectors—having *tens of arcminute* or *degree* angular scales. Thus, the *large-scale, diffuse* interstellar field lies parallel to the bipolar flow and is the primary factor that determines its direction! Theorists have put forward two processes that can explain this result. In one, Alfvén waves transfer angular momentum from the collapsing protostellar cloud to the surrounding medium. This process is more efficient for the component of angular momentum that is perpendicular to the field lines in the surrounding medium (Mouschovias and Paleologou 1980), which leads to the observed alignment. In the other, a spherically symmetric outflow from the protostar blows a cavity whose shape is influenced by the pressure in the external medium (Königl 1982); the pressure of a magnetic field is anisotropic, in the sense that also leads to the observed alignment.

HI Zeeman splitting has been used to measure the field strength in interstellar shocks (Troland and Heiles 1982b). One of these is the Eridanus shell, expanding at about 23 km s^{-1} (Section 3.2.1.a). If there were no magnetic field, the HI should rise in density behind the shock by a factor of several hundred, as outlined in our discussion of isothermal shocks in Section 3.1.5.c. However, the magnetic field (measured value $\approx 7 \ \mu G$) has prevented this large increase. The actual factor is only 3 or so. In addition, the magnetic field is being stretched on very large scales by the expansion of the shell as a whole. This distortion is now becoming sufficiently large to allow the magnetic field to play a role deaccelerating the overall shell expansion.

On small scales, the field structure has been mapped for HI in absorption against strong continuum sources using aperture synthesis techniques. Against Cas A, Schwarz et al. (1986) used the WSRT to map the Perseus arm field. The field is enhanced in HI clumps that are associated with H_2CO clumps; measured values range from 20 to 40 μG, but actual $B_\parallel$'s are 1.5 to 2 times larger because the clumps constitute only a fraction of the total HI. It is not known whether the HI surrounds the molecular clumps or simply lies in proximity. In a surprising result, Heiles and Stevens (1986) used OH Zeeman splitting to find that the observed field strengths in the associated *molecular* clumps, which have higher volume densities, are about five times smaller than in the atomic clumps. This goes against the expected tendency of field strength to increase with gas density. Neither the atomic nor the molecular clumps is in virial equilibrium because self-gravity is negligible, the magnetic field is high, and there is apparently insufficient external pressure. The clumps may be transient condensations located behind a passing shock wave.

Heiles (1987b) has recently reviewed the observational situation in detail. Because of flux freezing, we generally expect an increase in magnetic field strength with volume density. Magnetic field strengths show *no* evidence of increase over the density range 0.1 to 100 cm^{-3}. The lower densities are sampled by Faraday rotation, and the higher densities by Zeeman splitting. Thus, there is no significant increase in magnetic field strength between the WIM and CNM. This constancy can be rationalized with theory, in that magnetic field enhancement is not expected if density enhancement occurs by relatively quiescent streaming of low-density gas along the field lines. However, such quiescent streaming is probably rare: we have observational evidence for shocks, and for field enhancement behind shocks. Thus we do expect enhancement. The magnetic field dominates the gas dynamics in at least some of cold HI clouds, so we cannot expect larger field strengths in the CNM. Thus the real question is why the field is relatively strong in the WIM/WNM where gas density is low.

3.5 Summary

Our goal has been to describe the present state of observational knowledge in light of the present state of theoretical development and to point out areas that are ripe for future work. The above discussion shows that *every theory violates some observations.* However, this does not mean that every theory is totally invalid! Instead, it means that no theory is complete and no theory applies to all situations. In this brief summary, we cannot reiterate all of the inconsistencies, and therefore we shall restrict the summary to a list of questions that we feel are most important and need immediate attention.

In the supernova-dominated McKee-Ostriker (1977) (MO) model, the HI lies in clouds and the hot intercloud medium (HIM) pervades most of space; the ISM is like Swiss cheese, with the HI filling the holes. In the quiescent Field, Goldsmith, and Habing (1969) (FGH) model the ISM is again like Swiss cheese, but instead the warm HI (the WNM) is the cheese and the cold HI clouds (the CNM) the holes. Is the real topology of the ISM consistent with either picture? The applicable model almost certainly changes with Galactocentric radius R and it is not clear where in the Galaxy the transition occurs. Clearly, the time is ripe for a systematic study of the variation of the different phases of the ISM as a function of R—including the study of external galaxies.

Locally, the filling factor of WNM is estimated to be $\gtrsim 0.4$ and that of the warm ionized medium (WIM) about 0.1—with large uncertainties. However, there is no observational proof that the WNM exists at low $|z|$'s. If the ISM is HIM-dominated at low $|z|$'s, then the pressure scale height should be large ($\gtrsim 1$ kpc) and there should be no dramatic changes in the vertical structure of HI at smaller $|z|$'s. Thus observations that can determine the $|z|$-distribution of WNM, and CNM properties as well, are sorely needed. In particular, the determination of the amount of WNM at $z = 0$ is crucial. In contrast to the CNM, the origin of the velocity dispersion of WNM and its long tail is not well understood. Is some WNM closely associated with CNM, as in some models? If so, what is the fraction? This fraction determines the scale height of that portion of the WNM not associated with clouds.

It is not clear whether the WIM and WNM are spatially related. The energy input needed to maintain the WIM is enormous and the sources are unknown. High-resolution maps in the Hα line should shed light on the sources of the ionization. Popular possibilities include isolated OB stars, or hot white dwarfs and nuclei of planetary nebulae. Observations are needed to determine whether the ionizing photons from isolated OB stars are used up by surrounding low-density, ionization-bounded HII regions.

Despite many advances in the observations of the ISM, there has been little advance on the heating and ionizing agents of HI. Our conclusion that the temperature of the WNM lies between 5000 and 8000 K desperately needs reliable observational confirmation. In the "modified FGH" model introduced in Section 3.1.6, the heating mechanism for the WNM is not necessarily also an ionizing agent. Thus, an agent other than energetic ionizing particles may dominate—but we have no idea what this agent might be. The lukewarm ($T \gtrsim 250$ K) HI, which probably exists at cloud peripheries, is thermally unstable. This implies that the cloud envelopes or the heating agents are transient events, or that we do not understand cloud boundaries. The measurement of HI ionization within clouds is important since only energetic X rays or low-energy cosmic rays can penetrate clouds; this means that reliable low-frequency absorption and UV measurements are needed towards HI clouds. It is equally important to determine the ionization fraction of the WNM, which occupies a large fraction of the interstellar volume and may contribute significantly to the average interstellar electron density; potential ionizing agents have to be distributed evenly.

Moving to smaller scales, we find that HI clouds are not spherical. In an HIM-dominated ISM, the surface area and shape of CNM governs the total evaporation rate. For this reason, it is important to determine whether they are filaments or sheets. In addition, clouds are not homogeneous isothermal structures. The only detailed study shows that they consist of small-scale clumps immersed in luke-warm envelopes. A realistic theoretical model for HI clouds is desperately needed. Observationally we need to study the properties and distribution of the envelope and the clumps *separately*. Detailed mapping in the various infrared cooling lines offers great promise, but the technology does not yet exist.

The magnetic field is neglected in most discussions of the ISM. However, it is strong enough to be dynamically important, and in some cases dominant. The overall properties of the large-scale Galactic field remain uncertain: the number of field reversals inside the solar circle and the variation of field strength with R and z are unknown. On smaller scales the field is relatively uniform. It seems to be fairly strong in the WIM, where Faraday rotation occurs—but why? It is sometimes parallel, sometimes perpendicular to filamentary clouds; what causes this, and what are the dynamic ramifications? The field strength is observed to drastically limit the density increase behind at least some isothermal shocks. Does this always occur, even if the field lies parallel to the shock velocity? What are the ramifications for formation of dense molecular clouds and for star formation?

Acknowledgment. This work was supported in part by National Science Foundation grant no. AST 85-13422, awarded to C. Heiles, and a Robert A. Millikan Fellowship to S.R. Kulkarni.

Recommended Reading

Spitzer, L. 1978. Physical Processes in the Interstellar Medium. New York: Wiley Interscience.

Hollenbach, D., and H. Thronson. 1987. Interstellar Processes. Dordrecht: Reidel.

Beck, R., and R. Gräve. 1987. Interstellar Magnetic Fields. Berlin: Springer-Verlag.

Cowie, L.L., and A. Songaila. 1986, Annu. Rev. Astron. Astrophys. **24**:499.

Cox, D., and R.J. Reynolds. 1987, Annu. Rev. Astron. Astrophys. **25**:303.

Lockman, F.J. 1988. Annu. Rev. Astron. Astrophys. In preparation.

References

Allison, A.C., and A. Dalgarno. 1969. Astrophys. J. **158**:423.
Badhwar, G.D., and S.A. Stephens. 1977, Astrophys. J. **212**:494.
Bahcall, J.N., M. Schmidt, and R.M. Soneira. 1983. Astrophys. J. **265**:730.
Beck, R., U. Klein, and M. Krause. 1985. Astron. Astrophys. **152**:237.
Belfort, P., and J. Crovisier. 1984. Astron. Astrophys. **136**:368.
Berkhuijsen, E.M., C.G.T. Haslam, and C.J. Salter. 1971. Astron. Astrophys. **14**:252.
Bertaux, J.L., R. Lallement, V.G. Kurt, and E.N. Mironova. 1985. Astron. Astrophys. **150**:1.
Beuermann, K., G. Kanbach, and E.M. Berkhuijsen. 1985. Astron. Astrophys. **153**:17.
Blaauw, A. 1964. Annu. Rev. Astron. Astrophys. **2**:213.
Blitz, L., M. Fich, and S. Kulkarni. 1983. Science **220**:1233.
Bloemen, J.B.G.M., A.W. Strong, L. Blitz, R.S. Cohen, T.M. Dame, D.A. Grabelsky, W. Hermsen, F. Lebrun, H.A. Mayer-Hasselwander, and P. Thaddeus. 1986. Astron. Astrophys. **154**:25.
Borken, R.J., and D.C. Iwan. 1977. Astrophys. J. **218**:511.
Bridle, A.H., and V.R. Venugopal. 1969. Nature **224**:545.
Brinks, E., and E. Bajaja. 1986. Astron. Astrophys. **169**:14.
Cane, H.V. 1979. Mon. Not. R. Astron. Soc. **189**:465.
Cesarsky, C.J., and H.J. Völk. 1978. Astron. Astrophys. **70**:367.
Clark, B.G. 1965. Astrophys. J., **142**:1398.
Cleary, M.N., C. Heiles, and C.G.T. Haslam. 1979. Astron. Astrophys. Suppl., **36**:95.
Cohen, R.J. 1981, Mon. Not. R. Astron. Soc. **196**:835.
Cohen, R.J., P.R. Rowland, and M.M. Blair. 1984. Mon. Not. R. Astron. Soc. **210**:425.
Colomb, F.R., W.G.L. Poppel, and C. Heiles. 1980. Astron. Astrophys. Suppl. **40**:47.
Cowie, L.L., and A. Songaila. 1986. Annu. Rev. Astron. Astrophys. **24**:499.
Cox, D.P. 1981. Astrophys. J. **245**:534.
Cox, D.P., and R.J. Reynolds. 1987. Annu. Rev. Astron. Astrophys. **25**:303.
Cox, D.P., and B.W. Smith. 1974, Astrophys. J. **189**:L105.
Crovisier, J. 1981. Astron. Astrophys. **94**:162.
Crovisier, J. and I. Kazès. 1980. Astron. Astrophys. **88**:329.
Crovisier, J., J.M. Dickey, and I. Kazes. 1985. Astron. Astrophys. **146**:223.
Cruddace, R., F. Paresce, S. Bowyer, and M. Lampton. 1974. Astrophys. J. **187**:497.
Crutcher, R.M. 1982. Astrophys. J. **254**:82.
Dalgarno, A., and R.A. McCray. 1972. Annu. Rev. Astron. Astrophys. **10**:375.
Davies, R.D., K.H. Elliot, and J. Meaburn. 1976. Mem. R. Astron. Soc. **81**:89.
Deguchi, S., and W.D. Watson. 1984. Astrophys. J. **285**:126.
Deguchi, S., and W.D. Watson. 1985. Astrophys. J. **290**:578.
de Jong, T. 1980. Highlights of Astronomy **5**:301.
Dickey, J.M., S.R. Kulkarni, J.H. van Gorkom, and C.E. Heiles. 1983. Astrophys. J. Suppl. **53**:591.
Draine, B.T. 1978. Astrophys. J. Suppl., **36**:595.
Ellis, G.R.A. 1982. Aust. J. Phys. **35**:91.
Elmegreen, B.G. 1976. Astrophys. J. **205**:405.
Field, G.B. 1958. Proc. Inst. Rad. Eng. **46**:240.
Field, G.B. 1959. Astrophys. J. **129**:551.
Field, G.B. 1965. Astrophys. J. **142**:531.
Field, G.B., D.W. Goldsmith, and H.J. Habing. 1969. Astrophys. J. **155**:L149.

Ginzburg, V.L., and S.I. Syrovatskii. 1969. Annu. Rev. Astron. Astrophys. 7:375; see also Annu. Rev. Astron. Astrophys. 3:297.

Giovanelli, R. 1980. Astron. J. 85:1155.

Glassgold, A.E., and W.D. Langer. 1973. Astrophys. J. 186:859.

Gordon, C.P. 1971. Astron. J. 75:914.

Goudis, C., and J. Meaburn. 1978. Astron. Astrophys. 68:189.

Hayes, M.A., and H. Nussbaumer. 1984. Astron. Astrophys. 134:193.

Heiles, C. 1967. Astrophys. J. Suppl. 15:97.

Heiles, C. 1974. Astrophys. J. 193:L31.

Heiles, C. 1976a. Astrophys. J. 204:379.

Heiles, C. 1976b. Astrophys. J. 208:L137.

Heiles, C. 1976c. Annu. Rev. Astron. Astrophys. 14:1.

Heiles, C. 1980. Astrophys. J. 235:833.

Heiles, C. 1984. Astrophys. J. 55:585.

Heiles, C. 1987a. Astrophys. J., in press.

Heiles, C. 1987b. Interstellar Processes. D. Hollenbach, H. Thronson (eds). Dordrecht: Reidel, p. 171.

Heiles, C., and H.J. Habing. 1974. Astron. Astrophys. Suppl. 14:1.

Heiles, C., and E.B. Jenkins. 1976. Astron. Astrophys. 46:333.

Heiles, C., and M. Stevens. 1986. Astrophys. J. 301:331.

Heiles, C., Y-H. Chu, R.J. Reynolds, I. Yegingil, and T.H. Troland. 1980, Astrophys. J. 242:533.

Henderson, A.P., P.D. Jackson, and F.J. Kerr. 1982. Astrophys. J. 263:116.

Hobbs, L. 1974. Astrophys. J. 191:395.

Ikeuchi, S., A. Habe, and Y.D. Tanaka. 1984. Mon. Not. R. Astron. Soc. 207:909.

Inoue, M., and H. Tabara. 1981. Publ. Astron. Soc. Jpn. 33:603.

Jahoda, K., D. McCammon, J.M. Dickey, and J.F. Lockman. 1985. Astrophys. J. 290:229.

Jenkins, E.B., M. Jura, and M. Loewenstein. 1983. Astrophys. J. 270:88.

Joncas, G., P.E. Dewdney, L.A. Higgs, and J.R. Roy. 1985. Astrophys. J. 298:596.

Jones, R.V., and L. Spitzer. 1967. Astrophys. J. 147:943.

Kalberla, P.M.W., U. Mebold, and W. Reich. 1980. Astron. Astrophys. 82:275.

Kalberla, P.M.W., U.J. Schwarz, and W.M. Goss. 1985. Astron. Astrophys. 144:27.

Königl, A. 1982. Astrophys. J. 261:115.

Kulkarni, S.R., J.M. Dickey, C. Heiles, 1985. Astrophys. J. 291:716.

Kulkarni, S.R., and M. Fich. 1985. Astrophys. J. 289:792.

Kulkarni, S.R., and C. Heiles, 1987. Interstellar Processes. D. Hollenbach, H. Thronson (eds). Dordrecht: D. Reidel, p. 87.

Kulkarni, S.R., K.C. Turner, C. Heiles, and J.M. Dickey. 1985. Astrophys. J. Suppl. 57:631.

Kylafis, N.D. 1983. Astrophys. J. 275:135.

Landecker, T.L., R.S. Roger, and P.E. Dewdney. 1982. Astron. J. 87:1379.

Launay, J.M., and E. Roueff. 1977. J. Phys. B 10:879.

Lazareff, B. 1975. Astron. Astrophys. 42:225.

Lindblad, P.O., K. Grape, A. Sandquist, and J. Scheler. 1973. Astron. Astrophys. 24:309.

Liszt, H. 1983. Astrophys. J. 275:163.

Lockman, F.J. 1984. Astrophys. J. 283:90.

Lockman, F.J., 1988. Annu. Rev. Astron. Astrophys., in preparation.

Lockman, F.J., L.M. Hobbs, and M.J. Shull. 1986. Astrophys. J. 307:380.

Lynds, B.T. 1962. Astrophys. J. Suppl. 7:1.

Lyne, A.G., R.N. Manchester, and J.H. Taylor. 1985. Mon. Not. R. Astron. Soc. 213:613.

Magnani, L., L. Blitz, and L. Mundy. 1985. Astrophys. J. 295:402.

Manchester, R.N., and J.H. Taylor. 1977. Pulsars. San Francisco: W.H. Freeman and Co.

Mathewson, D.S., and V.L. Ford. 1970. Mem. R. Astron. Soc. 74:139.

Mathis, J.S. 1986. Astrophys. J. 301:423.

McCammon, D., D.N. Burrows, W.T. Sanders, and W.L. Kraushaar. 1983. Astrophys. J. 269:107.

McCutcheon, W.H., F.J. Vrba, R.L. Dickman, and D.P. Clemens. 1986. Astrophys. J. 309:619.

McKee, C.F., and J.P. Ostriker. 1977. Astrophys. J. 218:148.

Meaburn, J. 1980. Mon. Not. R. Astron. Soc. 192:365.

Mebold, U. 1972. Astron. Astrophys. **19**:13.

Mebold, U., A. Winnberg, P.M.W. Kalberla, and W.M. Goss. 1982. Astron. Astrophys. **115**:223.

Mirabel, I.F. 1982. Astrophys. J. **256**:112.

Moneti, A., J.L. Pipher, H.L. Helfer, R.S. McMillan, and M.L. Perry. 1984. Astrophys. J. **282**:508.

Mouschovias, T.C., and E.V. Paleologou. 1980. Astrophys. J. **237**:877.

Mouschovias, T.C., F.H. Shu, and P.R. Woodward. 1974. Astron. Astrophys. **33**:73.

Parker, E.N. 1966. Astrophys. J. **145**:811.

Payne, H.E., E.E. Salpeter, and Y. Terzian. 1983. Astrophys. J. **272**:540.

Penston, M.V., and F.E. Brown. 1970. Mon. Not. R. Astron. Soc. **150**:373.

Phillips, S., S. Kearsey, J.L. Osborne, C.G.T. Haslam, and H. Stoffel. 1981. Astron Astrophys. **103**:405.

Pottasch, S.R., P.R. Wesselius, and R.J. van Duinen. 1979. Astron. Astrophys. **74**:L15.

Purcell, E.M. 1979. Astrophys. J. **231**:404.

Radhakrishnan, V., J.D. Murray, P. Lockhart, and R.P.J. Whittle. 1972. Astrophys. J. Suppl. Ser. **24**:15.

Raimond, E. 1966. Bull. Astron. Inst. Neth. **18**:191.

Read, P.L. 1980. Mon. Not. R. Astron. Soc. **193**:487.

Reynolds, R.J. 1982. Astron. J. **87**:306.

Reynolds, R.J. 1984. Astrophys. J. **282**:191.

Reynolds, R.J. 1985. Astrophys. J. **294**:256.

Reynolds, R.J. 1986. *Gaseous Halos of Galaxies*. Bregman, J.N., F.J. Lockman (eds). Green Bank: NRAO.

Reynolds, R.J., and P.M. Ogden. 1979. Astrophys. J. **229**:942.

Sanders, D.B., P.M. Solomon, and N.Z. Scoville. 1984. Astrophys. J. **276**:182.

Schneider, S.E., G. Helou, E.E. Salpeter, and Y. Terzian. 1983. Astrophys. J. **273**:L1.

Schwarz, U.J., and H. van Woerden. 1974. In F.J. Kerr and S.C. Simonson (eds.), IAU Symposium on Galactic Radio Astronomy, Reidel: Dordrecht.

Schwarz, U.J., T.H. Troland, J.S. Albinson, J.D. Bregman, W.M. Gross, and C. Heiles. 1986. Astrophys. J. **301**:320.

Shuter, W.L.H., and G.L. Verschuur. 1964. Mon. Not. R. Astron. Soc. **127**:387.

Silverglate, P.R. 1984. Astrophys. J. **279**:694.

Simard-Normandin, M., and P.P. Kronberg. 1979. Astrophys. J. **242**:74.

Sivan, J.P. 1974. Astron. Astrophys. Suppl. **16**:163.

Sofue, Y., M. Fujimoto, and R. Wielebinski. 1986. Annu. Rev. Astron. Astrophys. **24**:459.

Spitzer, L., Jr. 1978. Physical Processes in the Interstellar Medium. New York: John Wiley and Sons.

Spitzer, L., Jr. 1982. Astrophys. J. **262**:315.

Spoelstra, T.A.T. 1971. Astron. Astrophys. **13**:237.

Sullivan, W.T. 1984. The Early Years of Radio Astronomy. Cambridge: Cambridge University Press.

Tenorio-Tagle, G. 1979. Astron. Astrophys. **71**:59.

Thaddeus, P. 1982. Ann. N.Y. Acad. Sci. **395**:9.

Thomson, R.C., and A.H. Nelson. 1980. Mon. Not. R. Astron. Soc. **191**:863.

Torres-Peimbert, S., A. Lazcano-Aranjo, and M. Peimbert. 1974. Astrophys. J. **191**:401.

Troland, T.H., and C. Heiles. 1977. Astrophys. J. **214**:703.

Troland, T.H., and C. Heiles. 1982a. Astrophys. J. **252**:179.

Troland, T.H., and C. Heiles. 1982b. Astrophys. J. **260**:L19.

Vallée, J.P. 1980. Astron. Astrophys. **86**:251.

Vallée, J.P. 1983. Astron. Astrophys. **124**:147.

Vallée, J.P. 1984. Astron. Astrophys. **136**:373.

Vallée, J.P., and R.C. Bignell. 1983. Astrophys. J. **272**:131.

van Dishoeck, E., and J.H. Black. 1986. Astrophys. J. Suppl. **62**:109.

Verschuur, G.L. 1974. Astrophys. J. Suppl. Ser. **27**:283.

Verschuur, G.L. 1979. Fund. Cosmic Phys. **5**:113.

Vivekanand, M., and R. Narayan. 1982. J. Astrophys. Astron. **3**:399.

Vrba, F.J., S.E. Strom, and K.M. Strom. 1976. Astron. J. **81**:958.

Wesselius, P.R., and I. Fejes. 1973. Astron. Astrophys. **24**:15.

York, D.G., and P. Frisch. 1984. Int. Astron. Union Colloquium No. **81**, p. 3 (NASA Conf. Publ. 2345).

Zweibel, E.G., and K. Josafatsson. 1983. Astrophys. J. **270**:511.

4. Molecules as Probes of the Interstellar Medium and of Star Formation

BARRY E. TURNER

4.1 Introduction

4.1.1 Historical Perspective

Until the discovery of interstellar molecules at radio wavelengths in the late 1960s, the study of the interstellar medium (ISM) in its global aspect was restricted to use of the 21-cm line of HI and to optical absorption lines of several atoms (Na, Ca, Fe, K, Ti) toward nearby hot stars. To avoid excessive extinction, the optical techniques were limited both to distances of <1 kpc from the Sun and to diffuse clouds (visual extinction <2 magn). They provided no indication of temperatures. The 21-cm line, although visible throughout the Milky Way, also sampled gas only in diffuse regions, although over a wide range of temperatures. Indeed, prior to the use of the 21-cm line, it had already been deduced theoretically from a consideration of heating and cooling mechanisms in the ISM that the pressure versus density relation would have a minimum in pressure at a density of ~ 1 cm^{-3}, so that, in principle, the ISM could exist as two phases in pressure equilibrium, one being a hot (intercloud) medium, the other a cold (cloud) phase, each in thermal equilibrium at the same pressure. Nevertheless, for quite some time the ISM continued to be modeled as a relatively smoothly distributed disk of gas with a canonical temperature of 100 K and densities ranging from 0.1 cm^{-3} to as high as 100 cm^{-3} in extreme cases where the HI line was severely self-absorbed. Although discrete HI clouds were recognized, optical observations such as the study of the patchiness in extinction near the sun (Fitzgerald 1968) established only that the dust in clouds must be at least forty times as dense as in the intercloud regions. By 1971, 21-cm studies indicated a range of temperatures from 30 to 200 K for the cloud component, with a median near 80 K, and a temperature of ~ 8000 K for the hot intercloud component. The cloud temperatures were higher than predicted and suggested depletion of certain cooling elements. Thus by 1971, theory and observation recognized the existence of an interstellar cloud component, but densities in excess of ~ 25 cm^{-3} were not observable, and the ISM was viewed as an idealized two-phase (later, three-phase) system in which the extreme conditions of density and temperature necessarily attending star formation were neither observed nor predicted.

Although the formation of stars obviously implies that the ISM evolves through very dense phases, so little was perceived about these regions that it was not expected that complex molecules could even exist in the ISM, because densities were regarded as too low to form them efficiently, and protection against photodissociation by the general interstellar UV field was regarded as inadequate.

In the eighteen years that complex interstellar molecules have been extensively studied, our view of the ISM has been radically changed. We now recognize that by mass about one-half of the entire Milky Way ISM is molecular in form, and in large regions (the "molecular annulus," between 3 and 7 kpc from the galactic center) about 90% of the ISM by mass is molecular. Using molecules as probes, we are now able to trace densities as high as 10^{10} cm^{-3} (limited only by spatial resolution) and temperatures ranging from < 10 K to several 1000 K. In this review, we emphasize what has been learned about the ISM and star formation using molecules as diagnostic probes. The astrophysics and chemistry of the molecules themselves is the subject of Chapter 5.

4.1.2 The Central Role of Molecular Spectroscopy

Unlike atoms, molecules can probe a wide range of physical conditions (temperature and density) by observations of their emission or absorption over the many transitions accessible at radio and submillimeter wavelengths. Ground-state rotational energy levels for typical small molecules are characterized by energies $E/k \sim \hbar^2/2Ik \approx 0.1$ to 1000 K. Thus, the large number of energy levels that typify most interstellar molecules, all connected by radio-wavelength transitions, have sufficiently small energies to be appreciably excited under all ranges of interstellar conditions, either by collisions or by radiation. Some molecules, particularly diatomic species with simple single-ladder energy level schemes and many allowed transitions, serve as good densitometers—the observed populations of the levels depend strongly on the density. Frequently used are CS, HCN, HC$_3$N, and HNCO. Other molecules, particularly symmetric tops with several energy ladders that have forbidden transitions between them, act as sensitive thermometers. Examples are NH$_3$, CH$_3$CN, and CH$_3$C$_2$H. CO, because of its ubiquity and low dipole moment, has been the chosen species to trace the general distribution of molecular gas, although its lines are usually saturated, and ^{13}CO is needed to obtain better information on cloud sizes and distributions. By contrast, only H among the abundant atoms has magnetic hyperfine splitting in the radio domain, and only C has (two) fine-structure transitions in the far-IR region, at which wavelengths it can be observed throughout the Milky Way. Thus, this review will focus almost entirely on what has been learned from molecular observations.

A detailed analysis of how temperature, density, mass, and internal motions of interstellar clouds are derived from molecular data is not possible here. The reader is referred to Chapter 9 of *Galactic and Extragalactic Radio Astronomy* (Turner 1974). Here, we summarize only a few essential points.

The measured intensity I_v of an emission line between two energy levels u and l is given by

$$I_v = B_v(T_{ul})(1 - e^{-\tau_v}) \tag{4.1}$$

where B_v is the Planck function and $\tau_v = N_l \alpha_v$ is the opacity, with α_v the absorption coefficient and N_l the column density in level l. For $hv \ll kT$, Equation (4.1) can be written in terms of the excess brightness temperature

$$\Delta T_B = (T_{ul} - T_{bg})(1 - e^{-\tau_v}) \ . \tag{4.2}$$

For $\tau_v \gg 1$, $\Delta T_B = T_{ul} - T_{bg}$ so that the excitation temperature T_{ul} is directly obtainable assuming that T_{bg} (usually the 3 K background) is known. For $\tau_v \ll 1$, $\Delta T_B = N_l \alpha_v T_{ul}$. The frequency dependence of α_v depends upon the radial velocity distribution of molecules along the line of sight: for a Gaussian profile, the integrated coefficient is

$$\alpha = \int \alpha_v dv = \frac{2\sqrt{\ln 2} \lambda^2 A}{8\pi^{3/2}} \frac{g_u}{g_l}[1 - \exp(-hv/kT_{ul})]$$

where A is the Einstein coefficient of spontaneous emission. For $\tau \ll 1$ and $hv \ll kT_{ul}$, the abundance can therefore by determined as

$$N_l = \frac{8\pi^{3/2}}{2\sqrt{\ln 2}} \frac{g_l}{g_u} \frac{k}{hcA} \int \Delta T_B dv \qquad (4.3)$$

where the excess brightness temperature is integrated over velocity (km s^{-1}).

To understand the physics of molecular clouds and their relation to star formation, we must determine total density ($\approx$density of H_2 molecules) and kinetic temperature T_k. If collisions dominate the exchange of populations between states, then all T_{ul} would equal T_k (and all lines with $\tau \gg 1$ would have the same intensity). In practice, the single molecular line believed to be universally optically thick is the $J = 1{-}0$ transition of ^{12}CO, and its small A coefficient ensures that collisions dominate its excitation. Thus T_k is usually equated to the brightness of the ^{12}CO line. Corroborating evidence is obtained from symmetric-top species such as CH_3CN (Hollis 1982) and CH_3C_2H (Kuiper et al. 1984).

Densities are obtained from molecules which radiate weaker lines than those of CO yet which may be ascertained (e.g., by isotope ratio brightnesses) to be optically thick. Many molecules fall into this category, and the reason is that $T_{ul} < T_k$ because collisional excitation rates are smaller than the spontaneous emission rate (A) of deexcitation. The higher the dipole moment, the higher must be the density, n, to elevate T_{ul} toward T_k. Thus different molecules sample different density regimes. The value of n that makes T_{ul} the average of T_{bg} (the cosmic background) and T_k is

$$n = \frac{A}{\langle \sigma v \rangle [1 - \exp(-hv/kT_{bg})]}$$

where $\langle \sigma v \rangle$ is the velocity-averaged collision rate. Typically, $\langle \sigma v \rangle \approx 10^{-15} \bar{v}$, and $\bar{v} = (8kT_k/\pi m)^{1/2}$ is the mean relative speed of the two colliding particles. Since T_{ul} is usually much closer to T_k than to T_{bg}, this value of n is a lower limit. Nevertheless, it has values of 10^5 to 10^6 cm^{-3} for many typical molecules with transitions in the 1-to 3-mm region. The required density for CS ($J = 3{-}2$) to heat up to 50 K is $n > 10^7$ cm^{-3}. This simple two-level analysis ignores the complications of multiple-quantum collisions ($\Delta J > 1$) and the interaction of the various levels through the effects of radiative trapping. These are usually included in detailed analyses which solve the rate equations simultaneously for many levels of a molecule and which find the combination of (T_k, n) which matches the observed brightnesses of many different transitions which are typically observed for a given molecule.

A second important way of deducing densities and masses of clouds utilizes only

^{12}CO and ^{13}CO. This method is important in estimating masses of large cloud regions whose densities may be too low to excite detectable emission from other molecules (all of which have larger A coefficients than CO) or when the abundances of other molecules may be too low (CO is the most ubiquitous and abundant species other than H_2). While ^{12}CO is optically thick, ^{13}CO is typically only marginally so. Assuming ^{13}CO is in fact optically thin and using Equation (4.3) leads to

$$N(^{13}\text{CO}) = 5 \times 10^{15} \, T_{01} \int T_B \, dv \, \text{cm}^{-2} \qquad (4.4)$$

where T_{01} and T_B are in K and dv is in km s^{-1}. T_{01} comes directly from the ^{12}CO observations. The conversion from N_l to the total ^{13}CO abundance $N(^{13}\text{CO})$ uses the rotational partition function and assumes a Boltzmann distribution with temperature T_{01}. The conversion from $N(^{13}\text{CO})$ to the desired quantity $N(\text{H}_2)$ is controversial and has been extensively discussed for many years. In the modern approach, $N(^{13}\text{CO})$ is found to correlate with optical extinction A_v in the visual, a linear relation obtaining up to $A_v \approx 6\text{–}8^m$ (e.g., Dickman 1978, Encrenaz et al. 1975). At higher A_v, the relationship appears to fail (Frerking et al. 1982), although recent results of Herbst and Dickman (1983) suggest that the linearity may extend to $A_v = 18^m$ in ρ Oph. A_v is then related to $N(\text{H}_2)$ by $N(\text{H} + \text{H}_2)(\text{cm}^{-2}) = 0.94 \times 10^{21} A_v(\text{magn})$ (cf. Frerking et al. 1982), which is equivalent to a gas-to-dust ratio of roughly 100 by mass. The resulting "best" value (for the Taurus cloud) is currently given as $N(\text{H}_2)/10^{21} = [N(^{13}\text{CO})/1.4 \times 10^{15}] + 1$ (Frerking et al. 1982). From the resulting values of $N(\text{H}_2)$, the mass of the molecular cloud can be estimated because its scale size is determined from mapping and a knowledge of the distance, which is at least approximately known for most sources.

A recent variant of the above method is the "standard cloud" approach, which makes use of the fact that P_{12}/P_{13} appears uniquely related to A_v in the "cold" cloud TMC-1 ($T_k \simeq 10$ K) for $A_v \lesssim 3^m$, and similarly (with a different relationship) for the "warm" cloud ρ Oph ($T_k \approx 25$ K), for $A_v \lesssim 4^m$. Here, $P_{12} = \int T_B(^{12}\text{CO}) \, dv$. Thus, any measured value of P_{12}/P_{13} that lies above the "saturation" value determines A_v and hence $N(\text{H}_2)$. The advantage of this method is that it requires no assumptions about excitation or optical thinness of ^{13}CO. The disadvantage is that the single "standard" cold and warm clouds upon which it is based may not be standard at all. Indeed, P_{12}/P_{13} is $\sim 7\text{–}10$ in the galactic center, ~ 5.5 in the molecular annulus ($5 \leqslant R \leqslant 6$ kpc), and ~ 10 in the vicinity of the Sun. These differences may, however, reflect true differences in the intrinsic $^{12}\text{CO}/^{13}\text{CO}$ abundance ratio, which would have to be corrected for in applying this method. So far, this method has given quite similar results to that based on Equation (4.4) (Rickard et al. 1985).

For condensed regions within clouds where ^{13}CO is optically thick, C^{18}O has been used along with the empirically determined relationship $N(\text{H}_2)/10^{21} = [N(C^{18}\text{O})/1.7 \times 10^{14}] + a$ where $a = 1.3$ and 3.9 for the Taurus and ρ Oph clouds, respectively (Frerking et al. 1982). In such condensed cores, an extrapolation of the $N(C^{18}\text{O})$ vs. A_v relation, or even the relation between $N(\text{H}_2)$ and A_v, may be unwarranted. Thus, analyses of the excitation of many other species, the most popular being CS, H_2CO, and NH_3, are used. This involves a solution of the coupled statistical equilibrium and radiative transfer equations.

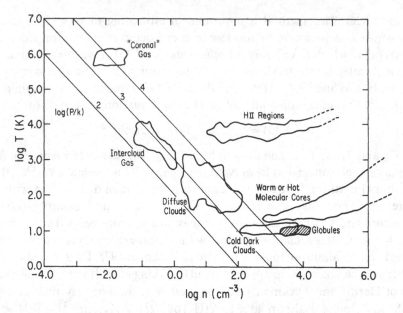

Fig. 4.1. The physical regimes of the interstellar medium, as currently observed.

4.2 The Large-Scale Morphology of the ISM

4.2.1 The Physical Regimes of the ISM

In the past decade, surveys at optical and radio wavelengths have yielded hundreds of determinations of (T_k, n) and have identified many distinct regimes of the ISM. Myers (1978) first summarized these, and we extend his summary here. Although the range of physical parameters characterizing the ISM may well form a continuum of values, there is some indication (based on identifiable heating and cooling mechanisms) that distinct regimes or phases do occur. We discuss seven such regimes here, some in pressure equilibrium and some not. They are shown in Figure 4.1.

(a) "Coronal" Gas

Coronal gas refers to a hot, highly diffuse component revealed by absorption lines of atoms in high stages of ionization seen in the far-UV spectra of OB stars; the most conspicuous of these lines belong to the OVI ion, whose abundance peaks in the range $5.3 \leqslant \log T \leqslant 5.9$, which, along with $-2.3 \leqslant \log n \leqslant -1.5$, characterizes these regions (Jenkins 1978a, b). The filling factor for coronal gas is uncertain, but likely is 0.2 to 0.5 (Myers 1978). The presence of very hot components with $5 \leqslant \log T \leqslant 7$ has been confirmed by other UV line observations (cf. Hartquist and Snijders 1982) and in the soft X-ray continuum and lines. A large fraction of the hot medium is contained inside bubbles, shells, and supershells. The pressure p in these regions is given by $p/k \lesssim 10^4$ cm^{-3} K.

(b) Intercloud Gas

Intercloud gas is observed in the form of broad, low-intensity emission features in the 21-cm HI line toward extragalactic sources (Davies and Cummings 1975,

Lazareff 1975, Kulkarni and Heiles, this volume, Chapter 3). This gas is described by $2.9 \leqslant \log T \leqslant 4$, $-1.0 \leqslant \log n \leqslant 0$, although the current observations are insensitive to gas hotter than $\sim 10^4$ K. This component has a pressure $p/k = nT \simeq 10^3$ cm^{-3} K.

(c) Diffuse Interstellar Clouds

Diffuse interstellar clouds are defined here as those with column densities $2 \times 10^{19} \leqslant N \leqslant 2 \times 10^{21}$ cm^{-2}, where $N = N(\text{H}) + 2N(\text{H}_2)$ is inversely correlated with T: $\log N = a - b \log T$. The lower limit on N is the currently lowest detectable value (by 21-cm absorption techniques: Davies and Cummings 1975, Lazareff 1975), while the upper limit represents the value above which T becomes roughly independent of density (A_v becomes $> 1^m$ and the UV field becomes unimportant). At the low-N range, these clouds have no detected molecular content, while the higher-N clouds have appreciable H$_2$, observed optically (Savage et al. 1977), and CO, observed both optically (Wannier et al. 1982) and at 2.6 mm (Knapp and Jura 1976, Blitz et al. 1984). The threshold for onset of detectable CO is $A_v \simeq 0.3^m$ (Blitz et al. 1984). Other molecules such as CH$^+$, CH, C$_2$, and CN are also seen optically, but more complex molecules are not. The size and number distribution of these clouds is poorly known; Spitzer (1978) argues a "standard" size of ~ 5 pc, although some exist as large as 35 pc. The CO clouds detected by Blitz et al. (1984) are typically 2 pc in diameter.

A controversial question is how diffuse clouds may relate to dense molecular clouds, for example, whether they are "halos" surrounding the denser clouds. In favor of this suggestion is the fact that HI and CO emission lines correlate in velocity but not in line intensity (Kazes and Crovisier 1981), consistent with certain CH observations that many diffuse clouds may actually be the outer portions of dense clouds (Federman and Willson 1982). High-resolution strip maps across projected edges of dense CO clouds (Wannier et al. 1983) indicate the presence of warm (> 100 K) HI halos 0.5 to several pc thick. This is suggestive of diffuse gas at the boundaries of dense clouds. Absorption lines of HCO$^+$, CS, C$_2$, and HCN seen at millimeter wavelengths toward HII regions indicate the presence of these molecules in quite low-density material, possibly surrounding dense clouds (Nyman 1983, 1984). Absorption lines of H$_2$CO and NH$_3$ toward Cas A indicate similar diffuse cloudlets, some also present in HI data. Against the hypothesis is that even the larger diffuse clouds appear to be much too numerous to be halos around dense clouds. Further, the fine-structure transitions of CI and CII, which are well matched to the physical conditions of diffuse clouds, in fact reveal that the CI and CII gas is mixed throughout the denser molecular clouds much like CO (Phillips and Huggins 1981, Crawford et al. 1985). These species, at least, do not trace a diffuse gas component and are not distributed like halos.

Diffuse clouds are cooled by collisional excitation of the fine-structure levels of C$^+$ and, if $T > 100$ K, also by the rotational levels of H$_2$. Heating by cosmic rays is inadequate, even if the ionization rate is as high as 10^{-16} s^{-1}, which is unlikely. Starlight is the apparent heat source. Energy released by the photodissociation of H$_2$ (and its reformation) is probably insufficient, but continuum absorption of starlight by grains, with the attendant photoejection of electrons, probably suffices. A combination of cooling by C$^+$ and heating by photoejection on grains leads to an expected $T \propto n^{-1}$ dependence, as observed, and thus to an expected constant pressure for all diffuse clouds. A representative pressure for diffuse clouds is $p/k \simeq 3700$ cm^{-3} K.

(d) Cold Dark Clouds

Cold dark clouds refer to well-defined regions of extinction, $\sim 2 \leqslant A_v \leqslant 25^m$, which are not associated with emission or reflection nebulosity (i.e., regions containing early-type stars). These clouds are largely molecular [$H_2/(H + H_2) \simeq 0.9$] except for possibly a halo of HI. The temperature is remarkably constant, 10 ± 3 K over the range 10^2 to 10^4 cm^{-3} in density, and the line widths are smaller and much more constant over these clouds than over other types. Dense cold cores are typical within these clouds, perhaps 10 to 100 times as dense as the overall cloud. These cold cores are not associated with the presence of embedded stars.

The clouds are cooled by collisional excitation of CO rotational transitions (cooling by grains is unimportant). Heating by cosmic rays, by gravitational collapse, or by ambipolar diffusion are all possible, but there is no observational evidence for gravitational collapse in the observed line shapes, and fields as large as 100 μG (probably ruled out by observation) are needed for adequate heating by ambipolar diffusion. A temperature of ~ 10 K is predicted if there is equilibrium between cosmic-ray heating and CO cooling.

(e) Globules

Globules as defined by Bok et al. (1971) and Bok (1977) have been found by molecular observations to have (T, n) similar to the denser cold dark clouds. The mass and size distribution seem to fit onto the lower end of the dark cloud distribution (Figure 4.1). Bok (1977) estimates there are 25,000 large globules in the galaxy. Heating and cooling are dominated by the same two processes as for cold dark clouds. The physical conditions in the dense cores of cold dark clouds are indistinguishable from those of many globule cores (Figure 4.1), although globule envelopes are much less extensive than dark cloud envelopes.

(f) Giant Molecular Clouds

Giant Molecular Clouds (GMCs) or Cloud Complexes were recognized ten years ago as the most massive entities in the galaxy, when they were discovered via CO emission as a result of global mapping projects undertaken with the Columbia and Texas telescopes. Prominent examples are M17, Cep OB3, Sgr B2, Orion, W49, Ser OB1, Cyg OB1, Cyg OB2, Per OB2, Mon OB1, Mon OB2, and CMa OB1. The OB1 designation reveals that these objects are nearly always associated with open clusters of OB stars. A clear-cut definition of a GMC is not possible, as their boundaries are defined by observational sensitivity (many giant clouds are known to merge, e.g., Orion-Monoceros, Taurus-Perseus clouds). Their presently deduced sizes range from ~ 20 to ~ 200 pc with a median size of ~ 45 pc. Figure 4.2, taken from Maddelena et al. (1986), shows the outlines of GMCs observed in the Orion-Monoceros region via the $J = 1-0$ transition of CO. This region illustrates why the definition of a GMC is arbitrary, because it is possible that all the clouds actually form part of a single cloud complex. The Orion A/Orion B clouds are connected with no abrupt velocity change. λ Ori and Orion B are at roughly the same distance

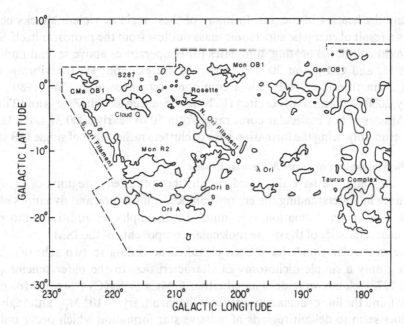

Fig. 4.2. The Orion system of GMCs (Maddelena et al., 1986).

from the Sun. The Northern Filament and Orion B are close to each other and have similar velocities. The Southern Filament and Mon R2 also have the same velocities and distances. It has been suggested that the large loop formed by the two filaments, Mon R2, and Orion A and B represents a magnetic bubble expelled from the galactic plane, which lies ~ 100 pc to the north of Orion A. Finally, the fragmented clouds around the star λ Ori appear to comprise an expanding shell of molecular gas.

Estimates of mean densities in GMCs range from < 100 cm^{-3} (see Dame et al. 1986) to 10^3 cm^{-3} (Blitz and Shu 1980). Masses range from 10^3 to 10^6 or 10^7 $M_\odot$. Indeed, there seems to be a mass threshold of $\sim 10^3$ $M_\odot$ above which massive (OB) stars form in association with the molecular cloud and below which they do not. Later-type (low mass) stars do occur embedded in all types of clouds (GMCs, cold dark clouds, and even in one or two globules), but their heating effect is negligible. Over most of their volume (i.e., in their outer regions), GMCs have the same physical properties as cold dark clouds or possibly the denser portions of diffuse clouds. There is one central difference, however: GMCs nearly always have clusters of OB stars either adjacent to them or embedded within them, often in multiple hot cores [but see Myers et al. (1986) for exceptions].

(g) Hot Molecular Cores in GMCs

Hot molecular cores are regions of high (T, n) embedded in GMCs, which are the sites of ongoing or incipient massive star formation. Because of limited spatial resolution, the full range of (T, n) has not yet been determined, but the highest-resolution observations using interferometers now reveal T as high as 2000 K and n as high as 10^8 cm^{-3} in these cores. Higher values clearly occur. For $n \gtrsim 10^4$ cm^{-3}, gas collisions with warm grains heated by the embedded protostars or OB stars dominate the heating of the gas. In many of these regions, violent shocks occur, often as a result of energetic supersonic mass outflow from the protostar itself. Such mechanical collisional heating dominates for temperatures above several hundred K. Hot H_2 and CO gas at 2000 K (observed, for example, via the vibrationally excited 2-μm transition of H_2) is a common occurrence (e.g., Elias 1980), and recently, 2000 K vibrationally excited HCN has been seen as well (Ziurys and Turner 1986). Masses of hot molecular cores range from 50 to nearly 1000 $M_\odot$, the larger ones presumably being the formation sites of clusters rather than of single OB stars.

4.2.2 The Morphology of Dense Molecular Clouds

In the foregoing, we have discussed the distinct interstellar regions of (T, n), of importance in understanding the energy balance, equilibrium, and dynamics of the ISM. As sites of star formation, it is important to emphasize additional morphological characteristics of the dense molecular components of the ISM.

Dense molecular clouds have been classified according to two schemes, both involving only a simple dichotomy of characteristics. In the older scheme (e.g., Turner 1978), clouds were designated either GMCs or SMCs ("Small Molecular Clouds"), and the line of demarcation was rather arbitrary, $\sim 10^3\ M_\odot$, although this mass does seem to delimit regions of massive star formation which occur only in association with clouds exceeding this mass. In a slightly newer classification scheme (Evans 1978), clouds are described as Group A if kinetic temperatures within them nowhere exceed ~ 20 K and as Group B if they contain warm or hot cores exceeding 20 K. These latter cores are readily identified as massive star formation regions. It happens that Group A clouds bear an almost exact 1:1 correspondence with SMCs, and Group B clouds a similar (but slightly lower) correspondence with GMCs. In Table 4.1 we summarize the physical characteristics in terms of Group A and B

Table 4.1. Properties of molecular clouds.

Property	Group A (cold)		Group B (warm)	
	Envelope	Core(s)	Envelope	Core(s)
T_k (K)	10	10	10–20	20–100
n (cm^{-3})	1(2)–1(3)	1(4)–1(5)	1(2)–1(3)	3(4)–1(6)
Diameter (pc)	1–10	0.1–1	10–200	1–3
M ($M_\odot$)	10–1000	1–100	1(3)–1(5)	1(2)–5(3)
M_j ($M_\odot$)	8–24	~2	?	2–4
Strong IR?	No	No	No	Yes
Morphological types	Globules Cold dark clouds (SMC)		"Giant Molecular Clouds" (GMC)	
Star formation?	Low-mass stars only ~None earlier than A0		Copious low-mass stars Usually OB stars and clusters	
No. in galaxy	Globules: ~25,000 Dark clouds: ?		4000	
Total mass	?		4(9) $M_\odot$	
Motions (Δv)	0.5–3.0 km s^{-1} (supersonic)	0.2–0.4 km s^{-1} (subsonic)	3–15 km s^{-1} (supersonic)	

terminology, but the correspondence with SMCs and GMCs should be borne in mind.

Several points should be emphasized.

1. The delimiting temperature of 20 K is physically meaningful. Molecular gas containing no internal heating sources, at densities $n > 10^2\ \mathrm{cm^{-3}}$, and opaque to UV will equilibrate at $T_k \simeq 10$ K under cosmic-ray heating and CO rotational cooling. Typical protostellar heating sources representing eventual or actual OB stars will elevate the adjacent gas to $\gtrsim 20$ K over a region of a few parsecs, typical of the size of warm cores. Outside the warm cores in GMCs, the gas has properties similar to those of SMCs.
2. SMCs appear never to have associated stars of spectral type earlier than $\sim$A0 or perhaps late B, and may have no associated star formation at all. About half of all SMCs are observed to have associated low-mass stars, usually T Tauri type. GMCs are presumed to have low-mass star formation (although this is proven only for the Orion GMC), and usually have OB star formation as well.
3. The Jeans mass (M_j) is always exceeded in both cloud groups. Thus, they are gravitationally unstable and either should be collapsing at roughly free-fall rate or are supported by some internal agent.
4. Internal motions, measured by line widths, are always supersonic throughout GMCs, and nearly always so in SMCs. In SMCs, any subsonic motions are restricted to small, localized clumps or cores.

4.2.3 GMCs and Spiral Arms

In other galaxies, OB stars trace spiral arms. In our galaxy, OB stars are associated with GMCs. Thus, GMCs should be associated with spiral arms in galaxies, although the question remains whether GMCs, which do not have associated OB stars, might be found between spiral arms as well.

There is a growing amount of direct observational evidence for GMCs in spiral arms. In several galaxies, CO emission appears enhanced in spiral arms (M31: Boulanger et al. 1981; IC 342: Rickard and Palmer, manuscript in preparation; M51: Rickard and Turner, manuscript in preparation; Lo et al. 1985). In the Milky Way, the survey of Cohen et al. (1980) shows decided contrasts in temperature rather than abundance. Also, boundaries of individual CO clouds are not well distinguished, so whether they are in fact GMCs or SMCs is not clear. Surveys of OH in the plane of the Milky Way (Turner 1979, 1982, 1983) have utilized the peculiar property of the 1720-MHz satellite line which is seen in emission only at threshold gas temperatures $\gtrsim 15$ K. This emission emulates the "spiral arm" pattern of the warmer CO gas and cuts off just at temperatures below those expected both in GMCs and in spiral arms, as a result of global heating by the spiral density wave (sdw) shock, or perhaps by the OB star distribution. The sizes of the OH clouds seen in this emission are those of GMCs, not SMCs. Thus we conclude that molecular GMCs indeed occur in spiral arms.

GMCs may also occur between spiral arms. If so, they are colder than those within the arms. Certainly SMCs exist between (and probably within) spiral arms, as shown by their local distribution near the Sun.

The lifetime of GMCs is important in deciding if they exist in interarm regions. Several arguments suggest that GMC lifetimes are short compared with 10^8 yr, the time needed to form a GMC by *random* collisions between smaller clouds (Kwan 1979):

1. OB stars that form in GMCs probably destroy the GMCs in a few tens of millions of years, a time scale suggested both theoretically and observationally. Bash et al. (1977) find that 90% of open clusters containing O stars have associated molecular clouds, while < 10% of clusters whose youngest stars are B0 to B4 have associated clouds. Taking into account the main-sequence lifetimes of these stars, one deduces that OB stars disrupt the parent GMC in $\sim 1.5 \times 10^7$ yr.
2. If GMCs are confined to spiral arms, their lifetimes must be shorter than 4×10^7 yr. Otherwise they would diffuse into the interarm regions.
3. GMCs are observed to contain clumps. Random collisions between these clumps, at rates determined by their observed random motions, ensure that these clumps must coalesce into a centrally condensed cloud structure in $\sim 1 \times 10^7$ yr (Blitz and Shu 1980). Such structures are not seen, with one exception.
4. GMCs appear windswept and irregular. Supersonic dissipation should smooth the edges in $< 1 \times 10^8$ yr.
5. The observed terminal velocity profile for GMCs on an (l, v) diagram is consistent with lifetimes shorter than 4×10^7 yr if GMCs are formed in spiral arms and launched from them with ballistic orbits (Bash et al. 1977).

These arguments are suggestive of a "short" GMC lifetime, but are not conclusive. Point (1) is invalid if OB star formation is slow getting started after a GMC is formed, or shuts off for a period of time. Point (3) assumes no fresh turbulent input into the cloud, which is probably invalid (see Section 4.4.2).

If GMC lifetimes are indeed "short" as suggested, then they cannot be built up by *random* collisions of smaller clouds (Kwan 1979) because growth times greater than 2×10^8 yr are needed to produce a mass spectrum with $\geqslant 0.1$ of the interstellar mass in clouds whose mass exceeds $10^5\ M_{\odot}$, as is observed. Furthermore, an SMC collision time of $> 2 \times 10^8$ yr is about the time for one galactic rotation, so that all SMCs would have to pass through the sdw shock at least once before agglomerating into a GMC. Thus, we conclude that GMCs are *formed* by the sdw shock. They could form directly from diffuse HI gas as it passes through the sdw shock (Elmegreen 1979) or from the *enhanced* collision rate between SMCs as they pass through the sdw shock (Norman and Silk 1980) or even by magnetic instabilities which can speed up the rate of coalescence within sdw shocks (Blitz and Shu 1980). In any case, the formed GMCs are strongly momentum-coupled to the swept-up shocked material, and thus they are launched from the spiral arm with postshock velocities (Bash et al. 1977).

In summary, either atomic gas or SMCs which presumably result from the disruption of previous GMCs are compressed and form GMCs in sdw shocks. The GMCs are launched with postshock velocities. There may be a delay time t_1 before onset of OB star formation, followed by a period of time t_2 before the OB stars disrupt the cloud. The lifetime $t = t_1 + t_2$ is probably of order a few $\times 10^7$ yr. Since $t_2 \approx 1.5 \times 10^7$ yr, then $t_1 \lesssim 2.5 \times 10^7$ yr. Such a delay time seems consistent with the observation that three out of sixteen "local" GMCs have no OB stars (Blitz 1980). If some SMC clouds are actually parts of GMCs (inadequately mapped), the quantitative argument improves.

4.3 The Nature of Star-Forming Regions

We have seen how the global distribution of GMCs and SMCs suggests why OB star formation may trace spiral arms, while low-mass star formation ("field" stars) may be more uniformly distributed throughout the galactic disk. Before investigating the central question of star formation—how molecular clouds actually collapse to form stars—it is necessary to examine in detail what is known of the morphology, dynamics, and physical conditions of the star-forming cores themselves. These cores, or "clumps," are obviously the link between the smooth, low-density GMCs with scale sizes of ~ 50 pc and stars, or the still elusive protostars. Most of the more complex molecules have been detected only in these clumps. Typically many different molecules must be studied to obtain reliable estimates of core parameters. NH_3 is one of the most useful, providing estimates of T via its metastable lines and n via its nonmetastable lines.

4.3.1 Small Star-Forming Cores in SMCs and GMCs

The definition of a small or low-mass core (one not associated with appreciable heat sources) is somewhat arbitrary, because it is not easily identified by CO maps, being no hotter than the surrounding more diffuse cloud gas. In the nearer clouds, cores are recognized as regions of enhanced visual extinction and are usually seen in molecular lines sensitive to $n \geqslant 10^4$ cm^{-3} (CS, HC_3N, NH_3, and H_2CO). This density and the fact that these cores are 10 to 100 times smaller than CO clouds serve as defining properties. The size of the core depends entirely on the defining density, since cores have no sharply delineated boundaries; $n \sim r^{-p}$ with $1 \leqslant p \leqslant 3$ is representative for core regions over ~ 0.1 to 0.5 pc in size. By comparison, an isothermal gas sphere in hydrostatic equilibrium has $p = 2$, while $p = \frac{3}{2}$ corresponds to free-fall collapse. Observations are not presently accurate enough to distinguish these cases.

The internal structure of low-mass cores is poorly known, but a few have been detected in the $\lambda 1.3$ cm line of NH_3 at the VLA (B335, B5, and L1172: cf. Myers 1986). At $\sim 6''$ resolution, these cores seem to consist not of several independent clumps, but instead of a rather smooth density and temperature structure, of sizes ranging from 2×10^{16} to 3×10^{17} cm. Rotation in all such cores is too small to have dynamical consequences.

A survey of ~ 100 dense low-mass cores in nearby SMCs has been made in lines of CO and NH_3 (Myers et al. 1983, Myers and Benson 1983, Myers 1983). About 30 of these have been studied in NH_3. They appear to form a relatively homogeneous group with typical properties summarized by: size ≈ 0.05 to 0.2 pc, density $= 10^4$ to 10^5 cm^{-3}, $T_k = 9$ to 12 K, $\Delta v = 0.2$ to 0.4 km s^{-1}, $M = 0.3$ to $10\ M_\odot$ (Myers 1985a, b). Thus the internal motions are subsonic. Low-mass cores are often found near T Tauri stars and low-luminosity ($\lesssim 10\ L_\odot$) IR stars as seen by IRAS. About

30% of such cores have embedded (low-mass) stars, 20% have such stars nearby, and 50% have no such stars very near. The stars embedded in low-mass cores have remarkably cold IR spectra. The cores with no such stars have lower turbulence, the values of Δv being almost entirely thermal. Stellar detection statistics suggest that these cores will very soon (within $\sim 10^5$ yr) form low-mass stars (Myers 1985b); the "waiting" time before a core reaches $n \sim 10^4$ cm^{-3} and star-forming collapse begins is comparable to or less than the free-fall time (10^5 yr for the above parameters).

Low-mass cores that do contain low-mass stars show two types of interaction. (1) *CO outflows* are seen in seven of eleven such cores. Since such flows may often be difficult to detect because of the aspect angle, it is possible that all such cores have outflows. Thus, stars in low-mass cores are probably younger than T Tauri stars, which do not exhibit such behavior. It has been suggested that these young stars disrupt their parent cores, perhaps via the outflows, before becoming visible as T Tauri stars. (2) *NH$_3$ line broadening* occurs for low-mass cores with embedded stars. The nonthermal component of Δv is ~ 0.4 km s^{-1} for cores with stars, 0.2 km s^{-1} for cores without stars. The mechanical energy of a core having 0.4 km s^{-1} turbulence exceeds that having 0.2 km s^{-1} turbulence by $\sim 10^{42}$ erg, or $<1\%$ of the typical energy of a molecular outflow (cf. Section 4.3.4). Only a small coupling efficiency is implied.

Apart from the velocity structure that accompanies outflow, and the basic turbulent component, little dynamical information exists for low-mass cores that bears on the question of rotation or of infall. Indeed, most low-mass cores seem to be in virial equilibrium, self-gravity being balanced by turbulence.

Low-mass cores have been studied almost exclusively in nearby SMCs. Very little is known about the prevalence of low-mass cores in GMCs.

4.3.2 Massive Star-Forming Cores in GMCs

Massive cores have greater size, density, temperature, velocity dispersion, and mass than the low-mass cores, and are found near younger stars of earlier (i.e., OB) type. They appear to have a much greater range in their properties than the low-mass cores, varying by factors of 10 or more in (n, T) and in size. Limited spatial resolution may account for some of the apparent diversity: because nearly all massive cores are 1 kpc or more in distance, most current single-dish millimeter-wave studies cannot resolve structure on scales less than ~ 0.3 pc. Thus, massive cores have more ambiguous estimates of (n, T) and mass than do low-mass cores. A survey of ten massive cores near ultracompact HII regions and/or luminous ($\geqslant 10^4$ $L_\odot$) IR sources (Ho et al. 1981a) gave median values of $n = 3 \times 10^3$ cm^{-3}, $M = 300$ $M_\odot$, but n is surely underestimated by limited resolution. The well-studied and nearby massive cores in S140, M17, and Ori B have $T = 30$ to 50 K and n at least 10^6 cm^{-3} in clumps whose sizes are 5×10^3 to 10×10^3 AU and whose masses are in the range of 0.5 to 5 $M_\odot$, within the range expected for low-mass protostars. The overall core masses are much higher. The detailed studies of the Ori(KL) core with recent interferometry (Section 4.3.6) no doubt present a more realistic picture, although the presence of massive stars and protostars in the vicinity of the hot core may have distorted its properties. Emission from shocked H$_2$ gas is regularly observed as outflows from massive cores and indicates $T_k = 2000$ K in the outflow itself. Non-outflowing gas near the protostellar sources is heated to several hundred K in many regions studied via the excitation of several molecules. A summary of properties for massive cores is: size ≈ 0.1 to 3 pc, $n = 10^4$ to 10^6 cm^{-3}, $T_k = 30$ to 300 K, $\Delta v \approx 1$ to 4 km s^{-1}, $M = 10$ to 10^5 $M_\odot$, although high-resolution studies already indicate higher (n, T). Independent of limited resolution, these cores are 100 to 1000 times more massive than low-mass cores. They are also at least 10 to 1000 times denser, 3 to 10 times hotter, and 3 to 10 times more turbulent than low-mass cores. They have highly supersonic turbulence [possibly because they are usually found near compact HII regions and luminous ($\geqslant 10^4$ $L_\odot$) IR sources]. Under limited ($\sim 1'$) resolution, massive cores do not exhibit a simple $n \propto r^{-p}$ behavior; in M17, S140, and Ori B, Mundy (1984) found that core sizes do not decrease significantly with increasing n, which suggests significant clumpy structure unresolved at $1'$ (0.1 to 0.3 pc size scales).

Dynamically, many massive cores have velocity gradients of ~ 0.1 to 1 km s^{-1} pc^{-1}, indicative of rotation, but not large enough to account for the nonthermal part of the line broadening. The rotational contribution rarely equals the turbulent component. At very high resolution, rotation may become more important, as is observed in two clumps north of Ori(KL), one in ON-1 and one in NGC 2071. However, at the scale size of $1'$ (0.3 pc at 1 kpc), massive cores, like small cores, give the appearance of virial equilibrium. This does not exclude the possibility that some regions are collapsing, especially on small scales, since collapse speeds are only slightly different from virial speeds. However, as seen in Section 4.4, cores appear to be supported globally on the ~ 1-pc scale and are not collapsing at anything like the free-fall rate.

Just as for low-mass cores, high-velocity outflows are quite common in high-mass cores. Often at the centers of such outflows are small, dense elongated structures which are widely interpreted as disks or toroids, generally $\lesssim 0.1$ pc in diameter. In many cases the "disks" are perpendicular to the outflow direction, and may collimate the flows.

The temperature structure of massive cores is poorly known, with the exception of the Ori(BN) object (Scoville et al. 1983), which has an ionized wind at $T = 10^4$ K out to a radius of 20 AU from the central star, a possible shock front with $T = 3000$ K at $\geqslant 3$ AU, a 900 K dust envelope around the ionized region at 20 to 25 AU, and an ambient molecular core at 600 K at $\geqslant 25$ AU. Thermal structure is not known in such detail for any core for scales of 10^2 to 10^4 AU. Massive cores containing OB stars certainly have higher CO temperatures than their surroundings, typically ~ 40 K at 1 pc from the peak, declining to ~ 20 K at 5 pc (Evans et al. 1981, 1982). Detailed studies of the energetics of several massive cores (e.g., S255, S235, and S140) have shown that the heating can be accounted for solely by input from nearby stars and embedded IR sources. The energy flow is through the dust, which is cooled by far-IR radiation. The gas cooling rate (by CO lines) is much less than the dust cooling rate. Other potential heating mechanisms (ambipolar diffusion, gravitational collapse) appear generally to be unimportant, as there is no dynamical evidence for collapse except possibly in a few cases (Walker et al. 1986).

4.3.3 Large-Scale Relationships of Star-Forming Regions
Star formation and interaction with associated molecular cloud material may take place on scales ranging from a few hundred parsecs typical of OB associations/GMC

complexes to the few hundredths of a parsec observed in dense cores. Major clues about star-forming processes arise from the study of systematic effects operating on the cores which actually form the stars.

(a) Low-Mass-Star-Forming Regions

The powerful forces arising from HII regions, supernova remnants, and radiative effects intrinsic to regions of OB stars do not occur in low-mass-star-forming regions. Larson (1981) pictures small cores as arising from turbulent motions and surviving because they happen to be self-gravitating. Shu (1985) proposes that cores condense as interior cloud regions lose magnetic support due to ambipolar diffusion, on a time scale of $\sim 10^7$ yr. Either massive or small cores appear (e.g., Blitz and Thaddeus 1980) to move at virial speeds with respect to each other, massive cores with relatively little "background" effect because the density contrast may be smaller. Collisions between cores are in any case frequent and can inhibit or promote collapse depending on details.

Perhaps the most relevant question we can ask about low-mass cores is what their distribution is in nearby SMCs or GMCs. Is it random or clumped? Is it correlated with ages of existing nearby stars? A detailed study of the $15° \times 15°$ Taurus-Auriga region (Myers and Benson 1983) reveals that the ~ 130 young T Tauri type stars cluster in seven distinct groups. Dense cores, as identified by NH_3 emission, show a close association with the star groups, more specific than the association of general extinction and star groups. As we shall see, these dense cores have mass sufficient to form one or two low-mass stars and the stability character-istics of a region which has recently begun, or will soon begin, to collapse. An important conclusion is that the number of known cores is consistent with the number of stars (~ 30 to 50) expected to form in the next free-fall time. The expected number comes from the distribution of stellar ages in Taurus-Auriga, which implies the onset of star formation about 6×10^6 yr ago, followed by continuous exponen-tial growth in numbers of stars with a doubling time of 6×10^5 yr. The free-fall time of a core is $\sim 2 \times 10^5$ yr. Since the number of young stars in this region is presently 100 to 200, we expect 30 to 50 new stars should form in the next 2×10^5 yr, agreeing with the number of dense cores. By contrast (see Section 4.4.1), dividing the total mass of molecular clouds in the Galaxy by the typical free-fall time gives an expected star formation rate of 30 M_0/yr, in poor agreement with the one to four stars observed to form per year.

The Taurus-Auriga region is the most studied low-mass-star-forming region and shows reasonable agreement between number of dense cores and number of stars expected to form in the next free-fall time. One other important point about low-mass-star-forming regions stems from the study by Larson (1982) of the masses and ages of T Tauri stars in three clusters. He finds a remarkable trend in the spatial concentration of these stars as a function of age of the cluster. In Taurus the mean stellar mass is $< 1\ M_\odot$ and the mean stellar age is 6×10^5 yr; the stars are dispersed over 40 pc and occur in clumps. In Orion, which is more centrally condensed than Taurus, the mean mass and age are $> 1\ M_\odot$ and 1×10^6 yr. Finally, the still more condensed cluster NGC 2264 has even more massive stars and a mean age of 2×10^6 yr. [An exception to this trend may be the ρ Oph cluster (Wilking and Lada

1983), a highly compact (1 pc × 2 pc) and young system which, however, is forming only low mass ($\sim 1 \, M_\odot$) stars.]

(b) Massive-Star-Forming Regions
(i) The Question of Sequential Star Formation. Perhaps the most important, and controversial, aspect of massive-star formation within GMCs is whether it is sequential. It is expected that massive stars should, in some way, affect nearby star formation, since they (1) ionize the nearby gas, thus driving shocks into the molecular gas; (2) accelerate (and shock) the nearby gas by stellar winds; and (3) heat and accelerate the nearby gas by supernova explosions. A qualitative prediction is that of "sequential" massive-star formation (Elmegreen and Lada 1977) passing in bursts through a GMC until it is entirely dispersed. Properties of OB associations near the Sun are summarized by Blaauw (1964) and those of the related GMCs by Blitz (1980). Often, the association members divide into subgroups, 10 to 20 pc apart, each $\sim 10^3 \, M_\odot$. Age differences are a few million years. Younger stars appear more closely associated with molecular cloud material. The youngest members often produce HII regions on the edges of the parent GMCs and adjacent to sites of subsequent active star formation. The GMCs are themselves often elongated, with lengths of 80 to 100 pc, and the subgroups are frequently strung out along the continuations of the long axes of these clouds. Thus, the OB associations in these cases conform to a pattern in which the birth of successive generations of massive stars progressively erodes away the parent GMC. A good example is that of W3; progressing from west to east, there are first stellar subgroups of decreasing ages, next in line an HII region, and then a GMC with several massive cores lying along a ridge perpendicular to the line joining the subgroups and the GMC. Evidence against the sequential picture is, however, accumulating from several regions which have been examined in detail, among them S128 (Ho et al. 1981b) and Cep OB3 (Sargent et al. 1983); both morphologies and age sequences of the protostars seem inconsistent with the sequential picture. Embedded cores in the Per OB2 and Ser OB1/M17 associations are so far apart that the effects of an earlier generation of stars could not have caused their formation. In the NGC 6334 complex, stars appear to be turning on over such a wide area that external effects alone cannot have been responsible. In addition, evidence (Section 4.3.3.b [iii]) that massive-star-forming cores occur in the more central regions of clouds as well as at the edges (e.g., Orion, W3, and S255) suggests that mechanisms other than sequential ones are important. Finally, another problem is that the trigger which initiates massive-star formation is as yet unknown (but see Section 4.5.2) although it is often surmised to be the galactic sdw (Section 4.2.3). If so, star formation should continue preferentially in one particular direction, whereas Blaauw (1984a, b) finds that star formation in the Orion arm of the galaxy is progressing in many different directions at once. Star formation seems to be initiated in a fairly random manner. Growing evidence (cf. Sargent 1985) also suggests that stochastic processes are responsible for formation of massive stars within each subgroup. Even the accepted ages of OB associations have recently been questioned, and it appears in some cases that within each consecutive subgroup progressively more massive stars are formed with time, as we discussed above for more quiescent regions.

(ii) Other Global Effects on Massive-Star Formation. The Orion system of GMCs (Figure 4.2) illustrates several systematic, large-scale phenomena relevant to massive-star formation (Maddelena et al. 1986, Maddelena and Thaddeus 1985). On the western edges of both Ori A and Ori B GMCs lies a sharply defined ridge in which both density and temperature are approximately two times higher than eastward of the ridge. Star formation is greatly enhanced along the ridge. The probable origin is the Orion OB association which lies to the west and south of these GMCs. The western and southern cloud edges thus encounter a higher intensity of starlight, of stellar winds, and of occasional supernova explosions (one every $\sim 5 \times 10^5$ yr).

The enhanced temperature in the ridge probably does not occur because of starlight, since densities $\gtrsim 10^4$ cm^{-3} are needed to efficiently couple the gas to the grains which are directly heated by absorption of starlight. More likely, shock heating results from either stellar winds or supernovae originating in the Orion OB association. Compression of the molecular gas, no matter what the cause, does not directly heat the cloud but may enhance the relaxation of any magnetic field within the cloud (Maddelena 1985). This relaxation provides an additional heat source through ambipolar diffusion, in which viscous dissipation occurs between neutral and ionized particles as the field relaxes.

The density enhancement seen along the ridge could arise from the enhanced pressure of stellar winds, radiation fields, or the hot gas, in the adjacent Orion OB association region. According to Cowie et al. (1979), the pressures are, respectively, $2 \times 10^{-12}, 1 \times 10^{-12}$, and 1×10^{-10} erg cm^{-3}. Thus the hot gas pressure dominates. If it produces the ridge, then the width of the ridge should be $l \approx c_s t \approx (P/\rho)t \approx 10$ pc (c_s is the sound speed, $t \approx 3 \times 10^6$ yr is the lifetime of the OB association). The observed ridge width (Maddelena et al. 1986) is in agreement with this estimate.

(iii) The Distribution of Massive-Star-Forming Cores Within a GMC. Observations of massive cores are important in at least three ways in distinguishing mechanisms of massive-star formation. First, within a GMC, massive cores move supersonically relative to each other, so that collisions between them involve strong shocks, large releases of energy, and highly inelastic interactions. Clump accretion may or may not occur, depending on the details of the collision (Pumphrey and Scalo 1983, Gilden 1984). Second, the trend that increasingly massive stars and higher mean stellar ages are found in increasingly condensed star-forming regions continues from low-mass- to high-mass-star-forming regions. This can be seen within a given star-forming cluster, such as Orion. Here, the most massive stars are clearly localized in a highly compact region, and the most massive star, θ'C Ori, lies at the very center of the cluster. We conclude that within star-forming regions the gas becomes more centrally condensed with time, somehow allowing stars to form in more massive and concentrated clusters with increasingly larger stellar masses.

4.3.4 Intermediate-Scale Structure: Velocity Outflows and Disks

The presence of supersonic outflows of gas from young stellar objects was first recognized in 1976 in the form of broad ($\Delta v \sim 200$ km s^{-1}) wings in CO lines and the presence of molecular H$_2$ quadrupole rovibrational emission toward Ori(BN). This emission comes from hot (~ 2000 K) gas which has been collisionally shocked. The outflow was subsequently found to emanate not from BN, but from the more deeply embedded and highly luminous ($L \simeq 10^5 \, L_\odot$) IRc2 object. More recently, outflow around young stars has been recognized as common. Bally and Lada (1983) find CO outflows in about thirty-five regions of star formation. Some are driven by quite luminous stars, while others are not. For example, Levreault (1983) finds ~ 30 to 50% of low-mass pre-main-sequence T Tauri stars show outflows. Recent data (Myers 1985b) suggest that perhaps all low-mass stars go through an outflow phase. While CO outflows are seen in both high- and low-mass protostellar regions, shocked H_2 emission is common only in massive-star-forming regions (Lane 1984). The majority of flows show evidence for bipolarity, as a redshifted lobe in one direction and a blueshifted lobe in the opposite direction. The classic example is the outflow in the dark cloud L1551 (Snell et al. 1980), apparently driven by an IR source of modest luminosity (30 $L_\odot$). Herbig-Haro objects, often manifestations of outflows, are associated wih the blueshifted CO lobe, and proper motion studies indicate a transverse motion away from the central source at ~ 150 km s^{-1}. Sizes of outflows vary from 0.1 to 3 pc, and velocities from 20 to 2000 km s^{-1}. Surveys of outflows are highly incomplete; about forty cases are currently known.

Outflow properties vary considerably from object to object but exhibit several common features which suggest a common phenomenon. The mechanical luminosity in the flows ($L_m = \frac{1}{2}MV^2/\tau$, where M is the swept-up mass, V the velocity, τ the lifetime of the phenomenon) is an increasing function of L_*, the total stellar luminosity. Although $L_m < L_*$ is found in all cases, radiation pressure is likely not the accelerating agent. The required driving force ($F = \dot{M}V$) always exceeds L_*/c, the radiation pressure, by 10^2 to 10^3 times. Most outflows are accordingly attributed to stellar winds which are not driven by radiation pressure. The bipolarity may be explained as an intrinsic property of the wind, or as an initially isotropic wind which is collimated by circumstellar or interstellar disks.

Are outflows important in the evolution of molecular clouds which contain them? First, consider the conservation of energy. For a typical GMC of mass $1 \times 10^5 \, M_\odot$, the total turbulent kinetic energy is $\frac{1}{2}M\langle v \rangle^2 = 1 \times 10^{49}$ erg for $\langle v \rangle = 3$ km s^{-1}. Outflow kinetic energies range from 4×10^{43} erg for T Tauri stars to 2×10^{47} erg for the three most energetic flows known in massive cores (Mon R2, S140, and NGC 2071). Thus, one requires fifty of the strongest known outflows *per* GMC to yield the turbulent energy. Whether such numbers exist is unknown. A survey by Margulis and Lada (1986) has detected twelve (energetically modest) outflows in the NGC 2264 GMC, while a few square-degree patch of the Ori A GMC has revealed four such outflows (Snell et al. 1986; in progress). The twelve flows in NGC 2264 seem sufficient, on energetic grounds, to support the GMC against gravitation (Snell 1986), although just how the flow energy is distributed throughout the cloud is unclear. For SMCs, the turbulent kinetic energy of the clouds is more typically 1×10^{46} erg ($M = 10^3 \, M_\odot$, $\langle v \rangle = 1$ km s^{-1}, for which 250 T Tauri type outflows *per* SMC would be needed. Among SMCs, only L988 is known to contain multiple (four) flows (Clark 1986); these were selected on the basis of IRAS sources rather than by complete mapping. Although outflow statistics are lacking for both GMCs and SMCs, the required number for support of these clouds does seem rather unlikely in general.

Next, consider the conservation of momentum. In a dense molecular cloud, a wind at < 200 km s^{-1} will be radiative and approximately momentum conserving. Therefore, the mean velocity dispersion acquired by an average volume element in a cloud of mass M_c containing a mass M_* of stars that have lost a fraction ΔM_* of their mass at a typical wind velocity V_w is $\langle \Delta v \rangle = (M_*/M_c)(\Delta M_*/M_*)V_w$. The three factors have representative values of 0.1, 0.1, 50 km s^{-1} for SMCs and 0.05, 0.1, 100 km s^{-1} for GMCs, yielding $\langle \Delta v \rangle \sim 0.5$ km s^{-1} in each case. This is not a negligible fraction of the observed value in SMCs but is negligible for GMCs.

While outflows may not have much effect on molecular clouds as a whole, they clearly have the momentum and energy to disperse core masses. A T Tauri-like wind will drive a shell to a typical low-mass core radius in 4×10^5 yr. The dynamical lifetime of a bipolar flow is $\sim 1 \times 10^4$ yr, but the flow duration and time for the flow to disperse a core could be longer. For massive cores and protostars, the more energetic flows are capable of significant massive-core dispersal, but other more powerful agents such as HII regions and supernovae, which do not attend low-mass stars, will be dominant in the disruption of massive cores (Silk 1985).

The nonsymmetric nature of outflows evidently challenges the notion of spherically symmetric star formation, at least at small enough scales. Recently, dense elongated structures, generally interpreted as disks or toroids, have been observed around a few protostars having outflows. There is little similarity in the sizes and masses of these objects, which range from 0.3 pc and 500 $M_\odot$ for NGC 7538 down to ~ 0.005 pc (1000 AU) and $< 7 \times 10^{-4}$ $M_\odot$ for HL Tau. Only six or seven such objects are presently identified (Wootten 1985), and they are generally $\leqslant 0.1$ pc in diameter, with $n > 10^4$ cm^{-3}, and usually have their long axis perpendicular to the outflow direction. The causal relationship of the "disk" and outflows is unclear: the disk may collimate the outflow on interstellar scales, or the collimation may occur on stellar or circumstellar scales. In the latter cases, the outflow may serve to shape the interstellar disk rather than vice versa.

Rotation and magnetic fields are obvious candidates for producing a favored axis. Some evidence suggests rotation is important in one or two cases. Even more evidence points to magnetic fields, as summarized by Sargent (1985). Outflow and magnetic field directions appear correlated in several cases. In the Taurus molecular cloud, several bipolar outflows are aligned with one another and along the magnetic field. By implication, confining disks must be perpendicular to the magnetic field. The Taurus clouds themselves are apparently elongated perpendicular to the field, so their overall contraction has occurred along the field lines. Thus, the magnetic properties of the parent cloud may continue to dominate even at this relatively advanced stage of evolution. Unfortunately, no direct measures of magnetic field have been obtained at the size scales of cores or confining disks. An intriguing possibility is that the increases in magnetic field strength and rotational velocity that may accompany overall cloud contraction make them likely sources of energy to drive the outflows.

4.3.5 Magnetic Fields

There is qualitative but little quantitative evidence that on all size scales there are correlations between the general structure of star-forming regions and the direction of the magnetic field. In Section 4.2.1.f it was speculated that the Northern and

Southern Filaments in the Orion-Monoceros GMC complex manifest a giant magnetic loop issuing from the galactic plane. The Taurus molecular clouds are another example, although they are elongated perpendicular to the local field (Monetti et al. 1984) so that contraction has taken place preferentially along the field lines. Large-scale filamentary structure in the ρ Oph molecular cloud (Loren and Wootten, 1986; see Section 4.3.6) lies along field lines whose direction has been determined by IR polarization studies (Vrba 1977).

These examples of large-scale alignment with magnetic fields may or may not apply to GMC complexes in general. Blitz (1980) finds a number of the complexes are elongated (e.g., Ser OB1, Cep OB3, Per OB2, Mon OB1, Mon OB2, Ori OB1, and CMa OB1), the majority being nearly parallel to the galactic plane and along the direction the magnetic field is thought to lie. The majority of GMCs not showing elongation lie in the Cyg X region where the line of sight is nearly tangent to the Orion arm of the galaxy.

On the other hand, Hopper and Disney (1974) measured the directions of elongation of over 200 compact dark clouds in Lynds' (1962) catalogue; although the elongation tends to be parallel to the galactic plane, it was *not* correlated with the direction of the magnetic field in the local vicinity as measured from the polarization vectors in nearby stars.

On intermediate and smaller scales, correlation of structure and magnetic field often seems evident. Bipolar outflows are often parallel to the field direction. In those few cases where there are confining disks at the centers of the flows, they are perpendicular to the field [e.g., Cep A (Sargent 1985); L1551 (Kaifu et al. 1984)]. In the Taurus cloud, several outflows are found (Monetti et al. 1984) to be aligned with one another and with the field, while the overall Taurus clouds themselves are elongated perpendicular to the field. The Per OB2 GMC is also elongated perpendicular to the galactic plane (Sargent 1979) while the embedded HH7-11 outflow, and the field, are almost parallel to the plane (Cohen et al. 1984). If confining disks, elongated in the same sense as the flattened molecular clouds, are commonly associated with bipolar outflows, it would imply that the magnetic properties of the parent cloud continue to dominate even at this relatively advanced stage of evolution. However, other examples are not so clear-cut. The major axis of the Cep OB3 GMC lies approximately parallel to the field and to the outflow (Cohen et al. 1984). In the Mon OB2 and Cas OB6 GMCs, containing the CRL961 and W3 outflows, respectively, it is difficult to ascertain any particular major axis (cf. Blitz 1978), although the outflow from W3 IRS5 (Claussen et al. 1984) is quite closely aligned with the field (Dyck and Lonsdale 1979). It appears possible, however, that the contraction of a GMC as a whole, the formation of the member stars, and the evolution of the cloud and of the fragments which become outflow sources may all be affected by the interstellar magnetic field. This field may play a role in determining how star formation progresses in associations from scales as large as those of stellar complexes and clouds, ~ 100 pc, to scales less than 0.1 pc.

Few quantitative measures of magnetic field strengths exist. Current results (Troland and Heiles 1986) suggest strengths of $10–20$ μG, independent of n, over the range $0.1 \leqslant n \leqslant 10^2$ cm^{-3}, consistent with cloud contraction parallel to the local field over this range of n. These Zeeman results involve HI (Heiles and Troland

1982) and OH (Crutcher and Kazes 1983) in a few CO clouds and feature size scales of $\sim 10^{18}$ cm, larger than star-forming cores. Within star-forming cores themselves, there are even fewer measurements. Zeeman splitting of OH masers yields $|B|$ on the order of a few milligauss on scale sizes of 10^{15} to 10^{16} cm (Reid et al. 1980), and it appears that in general the field increases with density in the range $10^2 \leqslant n \leqslant 10^4$ cm^{-3}, although with large variations from cloud to cloud. Application of the flux-freezing relation $B \propto n^k$ ($\frac{1}{3} \leqslant k \leqslant \frac{1}{2}$) (Mouschovias 1978) to these results suggests $|B|$ is in the range of 40 to 160 μG in low-mass cores and 40 to 500 μG in massive cores, but these estimates are very uncertain. These estimates suggest that frozen-in fields would be too weak in massive cores, and possibly too strong in low-mass cores, to be important dynamically, that is, to balance gravitational forces. Since low-mass cores are not expanding, they may typically have less magnetic flux than the frozen-in value.

4.3.6 Small-Scale Structure of Star-Forming Regions

Although IR observations at sufficiently high resolution to distinguish the detailed structure of star-forming cores have been available for some time, comparable radio molecular line observations, necessarily utilizing interferometers, are only very recent. Very few such objects have yet been studied, so it is difficult to say how typical they are of star-forming regions in general. The existence of IR objects and compact continuum sources does not necessarily reflect the presence of protostars. IR objects have often been found to be merely reflection-type phenomena, as occurs for the majority of the IR clusters in Ori(KL) (Werner et al. 1983). Ultracompact continuum sources may be the ionized skins of neutral condensations in clouds (Lacy et al. 1982). Thus only the study of molecular lines can identify star formation regions unambiguously.

(a) The Orion Molecular Core

The KL nebula in Orion is the nearest region of massive star formation (500 pc distant), and the source of a particularly energetic, bipolar outflow. An exhaustive study of this region has been made in many molecular lines (SO, SiO, HCO$^+$, HCN, and ^{29}SiO) and continuum with the Hat Creek interferometer (Plambeck et al. 1985) and has provided the first detailed look at such regions. It is worth summarizing the results in some detail.

Figure 4.3, taken from Plambeck et al., gives the model for Orion (KL). Single-dish spectra of many molecular lines toward KL have been known for many years to consist of "spike," "plateau," and "hot core" components. The narrow ($\Delta v \approx 4$ km s^{-1}) spike arises from the ambient molecular ridge, while the broad "plateau" arises in part from each of the inner disk, doughnut, extended outflow, and high-velocity outflow components. The hot core has $v = 5$ km s^{-1}, differing noticeably from the $v = 8.7$ km s^{-1} of the spike.

The center of activity for the entire system is the IR source IRc2. From IR studies, IRc2 appears to be a blackbody of size $\sim 1''$ ($\sim 1 \times 10^{16}$ cm) and color temperature 700 K, with total luminosity $\sim 10^5 \, L_\odot$, evidently providing the bulk of the luminosity of the KL region (Wynn-Williams et al. 1984). Surrounding IRc2 is a dense ($n \gtrsim 10^7$ cm^{-3}) inner disk of gas oriented at position angle of $\sim 60°$ and tilted with

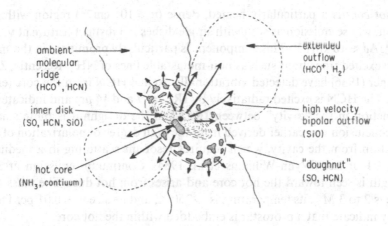

Fig. 4.3. The Orion (KL) region as observed with the Hat Creek interferometer (Plambeck et al., 1985).

respect to the line of sight. A bipolar outflow of high-velocity gas comes from either IRc2 or from the surface of the inner disk. This component is observed in the maser lines of SiO. The inner disk appears to channel this flow into two broad lobes, which expand into lower-density gas to the NW and SE, creating shock fronts which excite the rovibrational transitions of H_2 at 2 μm. A ring or "doughnut" of clumpy, compressed gas is observed in the lines of SO and HCN. It is surmised that the high-velocity outflow also produces the doughnut, by ramming into the dense ridge of the surrounding molecular cloud to the NE and SW. The "hot core" is a particularly dense, warm region within the doughnut, centered about 2" SE of IRc2.

As mapped earlier by single-dish studies (e.g., Turner and Thaddeus 1977), the *ambient molecular ridge* extends for many (~ 6) arcmin (~ 1 pc) along a roughly N-S axis. It appears to rotate, exhibiting a ± 1-km s^{-1} shift about the central velocity of 8.7 km s^{-1}, over the inner 1' extent (0.16 pc). This rotation curve implies a central mass of $\gtrsim 25\ M_\odot$, a lower limit since the inclination angle of the ridge is not known.

The *extended outflow* was first detected in the 2-μm lines of vibrationally excited H_2 (Beckwith et al. 1978), and later in the lines of HCO$^+$ (Olofsson et al. 1982); $\Delta v \approx 20$ km s^{-1}. The H_2 and HCO$^+$ distributions closely resemble each other, extending over a $\sim 1 \times 2'$ region (0.3 pc). It is not clear whether the HCO$^+$ outflow can coexist with the 2000 K H_2 gas. The HCO$^+$ gas may exist in clumps inside the outflow, or in postshock gas which has cooled to $\lesssim 100$ K.

The "*doughnut*" is observed in emission from SO and HCN over the same velocity range as HCO$^+$ is seen in the extended outflow. The doughnut has size $\sim 10'' \times 20''$ (0.03 × 0.06 pc), is centered on IRc2, and is elongated along the ambient ridge. Plambeck et al. (1982) argue that the doughnut is expanding and has been compressed and accelerated by the IRc2 outflow. H$_2$O masers appear to be clustered around the outer edges of the doughnut, at the interface between the outflow and the molecular ridge. The doughnut appears rather clumpy. The HCN abundance seems to be strongly enhanced; it is certainly optically thick, the hot core not being visible in HCN through the doughnut.

The *hot core* is a particularly heated, dense ($n \geqslant 10^7$ cm^{-3}) region within the doughnut, whose emission as seen with single dishes is a distinct feature at velocity 5 km s^{-1}, $\Delta v \approx 10$ km s^{-1}. This component is particularly prominent in the spectra of highly excited transitions such as non-metastable lines of NH$_3$. Recently, Ziurys and Turner (1986) have detected vibrationally excited HCN from the core (energy 2000 K). The HCN is excited radiatively by IRc2 at 7 and 14 μm and indicates that the extinction in the "cavity" between IRc2 and the doughnut is quite small. A similar conclusion was earlier derived from the high degree of polarization of 3-μm IR radiation from the cavity, which requires resonant scattering in a medium of opacity ~ 1 or less (Wynn-Williams et al. 1984). Continuum emission at 3-mm wavelength is seen toward the hot core and arises from hot dust. The mass of the hot core is 2 to 3 $M_\odot$, its temperature is ~ 250 K, and its size is ~ 0.01 pc. The hot dust may indicate that a protostar is embedded within the hot core.

The *inner disk* is deduced by Plambeck et al. (1985) from molecular emission occurring in the velocity range just outside the central 10 km s^{-1} which arises from the ridge, extended outflow, doughnut, and hot core. The inner disk, thus deduced from SO and HCN emission, extends out from IRc2 at least 5″ (2500 AU) along a position angle of $\sim 60°$. OH masers appear to be distributed along a similar position angle and are probably associated with the disk. The velocity seen in the inner disk is too high to indicate either rotation (virial) or collapse (free-fall) alone, and suggests that the disk is participating in the outflow from IRc2. If the disk is thin enough, it is consistent with the evidence that the space between IRc2 and the doughnut is essentially a "cavity."

Other dense "protostellar" clumps exist within the general massive Orion core region. Four arcsec (0.6 pc) north of KL, there are two rapidly rotating clumps seen in emission from NH$_3$ (Harris et al. 1983), each with $n \sim 10^5$ to 10^6 cm^{-3}, size ≈ 0.05 pc, $T \sim 20$ K, and $M \sim 0.2$ to 10 $M_\odot$, but with no known IR sources. In a $2' \times 3'$ region around IRc2 there are four more massive ($\sim 50 \ M_\odot$) clumps, which lie along the edge of the bright central part of the M42 HII region, suggesting that they may have formed as a result of shock compression driven by the HII region (Mundy et al. 1986).

(b) Other Star-Forming Cores

The Orion core is unique as the closest massive-star-forming object. A few other cores studied with similar (7″ to 12″) resolution lie further away and have yielded much less information.

(i) The Cepheus A Massive Core.

The Cepheus A massive core (distance 700 pc) has a CO bipolar flow extending over $\gtrsim 0.2$ pc, whose mass (10 $M_\odot$), velocity (25 km s^{-1}), and expansion age (8×10^3 yr) are all similar to the extended flow in Ori(KL). A possible confining disk in the form of an elongated ^{13}CO molecular cloud of length 0.03 pc has been discovered interferometrically (Sargent 1985). However, elongated structures also seen in NH$_3$ with comparable resolution (0.04 pc; Torrelles et al. 1985) do not coincide with the ^{13}CO structure. It is unclear if any or all of these objects are involved in focusing the flow. The center of activity appears to be two ultracompact HII regions separated by only 2×10^{15} cm and driven by two B1 ZAMS stars (Kassim and Turner 1988).

(ii) *The W3 Massive Core*. The W3 massive core (distance 2.3 kpc) features a bipolar flow of size 0.03 pc in CO centered on IRS5 (Claussen et al. 1984). Each lobe contains $\sim 1\ M_\odot$ and has density $\sim 10^5$ cm^{-3}. Within 10" of IRS5 are three other IR objects in various evolutionary stages, the latest one (IRS3) centering a compact HII region. If Ori(KL) were placed at the distance of W3, the plateau width would appear the same as the CO wings in W3 at the detection level. The derived and observed properties of the W3 flow are quite similar to those of Ori(KL) (smaller and denser than other flows), suggesting that W3 (IRS5) is of comparable age and evolutionary state to Ori(BN/KL). IRS3 would be older, and shows no CO outflow. H$_2$O masers, abundant in the general area, do not clearly associate with any of the IR sources. With current interferometry, more detailed structure as is seen in Ori(KL) cannot be resolved at the distance of W3.

(iii) *The ρ Oph Star-Forming Core*. The ρ Oph star-forming core (distance 160 pc) is an earlier evolutionary phase than those just discussed and appears to be forming only low-mass stars. The three extensively studied cores of this general type are ρ Oph, L1551, and TMC-1. The high star-forming efficiency and proximity of the ρ Oph cloud to Earth optimize the chances of finding the very earliest star-forming cores and the mechanisms that trigger onset of star formation. Figure 4.4 shows the ρ Oph star-forming region, observed at a resolution of 0.06 pc by Loren and Wootten (1986). In addition to the many embedded low-mass stars in the core region (Wilking and Lada 1983), there are two important incipient star-forming regions. One is a dense ($n \approx 5 \times 10^5$ cm^{-3}) massive (35 $M_\odot$), yet cold ($T < 20$ K) core of size 0.26 $\times$ 0.10 pc observed in intense emission lines of DCO$^+$. The other is a region of rarely observed 2-cm H$_2$CO emission at slightly blueshifted velocity bordering the SW tip of the DCO$^+$ region ($n > 1 \times 10^6$ cm^{-3}, $M = 18\ M_\odot$). Both regions lack compact embedded IR sources. The dramatic spectral differences between these adjacent regions suggest that the second region traces a compression front progressing into the DCO$^+$ region. The ridge of C^{18}O gas may represent material swept up by this front. Just behind it is a region of copious recent star formation (HD147889, E16). The magnetic field is in the direction of the long filaments extending to the NE and is the direction along which the shock front will be least impeded. Important questions posed by the morphology of ρ Oph are whether additional stars will form shortly in the DCO$^+$ core or whether the present embedded stars will disperse the core, which is rotating relative to the surrounding region. The DCO$^+$ core apparently has insufficient internal support (thermal plus turbulent pressure) to prevent gravitational collapse. Its future evolution could be to disperse and form no stars (but no dispersive agent is apparent); or, if it is isolated from external influence, it could fragment into a number of low-mass stars; or under the external influence of shock compression and heating, it may form a massive star (Turner 1984). The DCO$^+$ core in ρ Oph is unique at present in being the only known cold, very-high-density core with sufficient mass to form a massive star. Populations of lower-mass stars can form ahead of the shock, in condensations in the filaments, or in the swept-up shell when it becomes massive enough to become gravitationally unstable. The high rate of low-mass star formation apparent in ρ Oph seems to occur when these processes happen together. Formation of OB stars, conversely, is

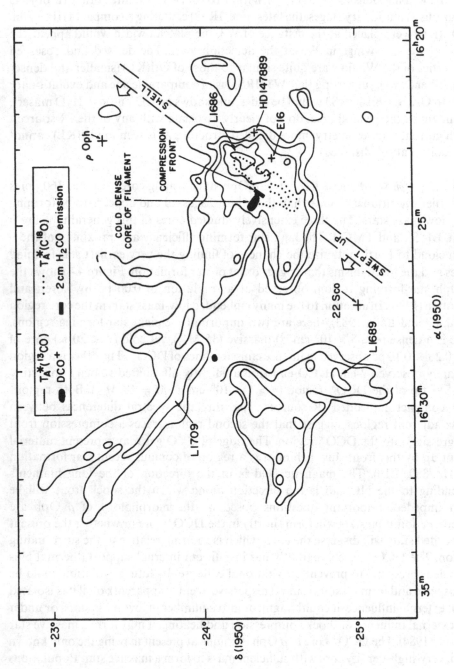

Fig. 4.4. The ρ Oph star-forming region (Loren and Wootten, 1986).

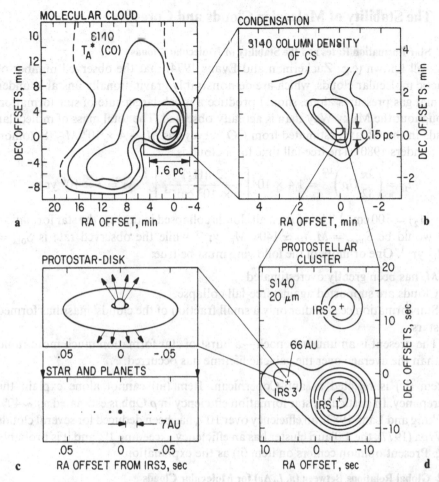

Fig. 4.5. The evolution of a GMC (S140) to form a stellar system. a. Blair et al., 1978; b. Snell 1982; c. Beichmann, 1979; d. Proceedings of the 1st LDR Asilomar Workshop, June 1982, Provided through the courtesy of the Jet Propulsion Laboratory, California Institute of Technology, Pasadena, California.

more episodic and seems to depend on an encounter with a massive condensation or core.

4.3.7 A Summary of Star Formation Morphology

Figure 4.5 depicts the evolution of a GMC, S140, to form a stellar system. The earliest phases (upper left) are observed via the $J = 1-0$ line of CO (Blair et al. 1978); the portion shown here extends ~ 6.3 pc. The box marks the region of star formation. Studies of the star formation region using several transitions of CS, a good densitometer, reveal a dense condensation within the cloud (10^5 to 10^6 cm^{-3}) (upper right; from Snell et al. 1984). Within this dense condensation, a cluster of IR and radio continuum sources has been found (Beichman et al. 1979), shown lower right. This map covers ~ 0.1 pc with a resolution of 4.7×10^{-3} pc. IRS3 is the center of a bipolar flow and of a possible confining toroid seen in NH$_3$ (Torrelles et al. 1985), which is depicted schematically at lower left.

4.4 The Stability of Molecular Clouds and Cores

4.4.1 Star Formation Rates and the Stability of Molecular Clouds

It is well known (e.g., Zuckerman and Evans 1974) that the observed number of galactic molecular clouds, which are demonstrably gravitationally unstable under thermal gas pressures alone, should produce a much larger rate of star formation throughout the Milky Way than is actually observed. The total mass of molecular clouds in the Galaxy, estimated from CO surveys, is $M_c > 4 \times 10^9\ M_\odot$ (Solomon and Sanders 1980). The free-fall time for a cloud is

$$t_{ff} = \left(\frac{3\pi}{32}G\rho\right)^{1/2} = 1.4 \times 10^5 \left[\frac{2n(H_2)}{1 \times 10^5\ cm^{-3}}\right]^{-1/2} yr = 3 \times 10^6\ yr$$

for $n(H_2) = 100\ cm^{-3}$. Therefore, if all clouds collapsed in time t_{ff}, the star formation rate would be $S_{theor} = M_c/t_{ff} \simeq 1400\ M_\odot\ yr^{-1}$ while the observed rate is $S_{Obs} = 3\ M_\odot\ yr^{-1}$. One or more of the following must be true:

(i) M_c has been greatly overestimated.
(ii) Clouds are supported against free-fall collapse.
(iii) Star formation ceases after only a small fraction of the clouds' mass has formed stars.
(iv) The present is an unusual epoch—a burst of star formation much more rapid than the average over the galactic lifetime has occurred.

Item (iv) is rejected as anti-Copernican. Item (iii) cannot alone explain the discrepancy. In particular, star formation efficiency in ρ Oph is estimated as $\sim 40\%$ (Wilking and Lada 1983); an efficiency over 10% has been deduced for several clouds by Vrba (1977); the T Tauri cluster has an efficiency exceeding 1% and it is probably 10%. Present opinion centers on item (ii) as the explanation.

4.4.2 Global Relations Between $(n, L, \Delta v)$ for Molecular Clouds

Although long sought, no direct observationl evidence exists for the gravitational collapse of a molecular cloud. Two or three cases are known of apparent infall of molecular gas onto existing stellar or protostellar objects, but maps of these effects do not exist, and it is not clear that these cases of apparent infall are gravitational in origin.

The question of stability of molecular clouds rests mainly on the demonstration by Larson (1981) of three remarkable relationships between size L, mean density n, and velocity dispersion σ ($\equiv 0.74\ \Delta v_{1/2}$, the FWHP line width for a Gaussian line shape). For fifty molecular clouds with data available at the time, these relationships and their physical meaning have been summarized by Myers (1985) and we follow his discussion here. First, *virial equilibrium* is satisfied:

$$\sigma = (\pi \rho\ G/3)^{1/2}\ n^{1/2}\ L = 0.016\ n^{1/2}\ L\ km\ s^{-1}$$

if n is in cm^{-3} and L is in pc. Second, a *"condensation"* law applies, with best fit

$$n = 3900\ L^{-1.2}\ cm^{-3}\ .$$

Third, there is a *"turbulence"* law, whose best fit is

$$\sigma = 1.2\ L^{0.3}\ km\ s^{-1}\ .$$

These relationships hold for $0.1 \leqslant L \leqslant 100$ pc, and each of the three is consistent, within its uncertainties, with that derived from the other two. Subsequent data have corroborated these relationships at both small scales (49 clouds, $0.05 \leqslant L \leqslant 24$ pc; Leung et al. 1982, Myers 1983, Torrelles et al. 1983) and large scales (34 clouds, $12 \leqslant L \leqslant 384$ pc; Dame et al. 1986, who find $\sigma \sim L^{0.50}$). Further, the three relationships are satisfied not only by a wide variety of molecular clouds, but also by the corresponding smaller range of scale sizes within individual clouds.

Virial equilibrium for molecular clouds is not surprising, since (Section 4.2.3) cloud ages exceed t_{ff} by factors of more than 10. If a cloud consists only of stable, independently moving clumps, then interactions between them will be expected to drive their velocities toward equilibrium values (Scalo and Pumphrey 1982). If a cloud is more a continuous fluid, then virial equilibrium implies that interactions between supersonically moving fluid elements exert a "turbulent" supporting pressure. Probably the latter picture is more accurate for low-mass clouds since the dense clumps contain far less mass than the widespread low-density "continuum." Therefore, the appearance of virial equilibrium implies that nonthermal "turbulent" pressures are central in supporting molecular clouds. These turbulent motions must be continually supplied with fresh energy, since the decay time of supersonic turbulence is much shorter than cloud lifetimes (Field 1978, Fleck 1981).

While virial equilibrium has a physical explanation, the "condensation" and "turbulence" laws are more controversial. Either one might be "fundamental" and the other "derivative," since each can be derived from the other in combination with the virial relation. Larson (1981) noted that the condensation law can result from selection, if the observations are sensitive to a relative range of column density $N = nL$ small compared to the relative range of L. As stated, the condensation law implies that column densities lie in a narrow range between $\sim 4 \times 10^{21}$ and 2×10^{22} cm^{-2}. Certainly this range is exceeded by a few massive, dense cores, but these do not fit the turbulence law well either.

While the turbulence law might also arise solely for statistical reasons, many believe that the law reflects an energy cascade from larger to smaller scales, similar to the process in idealized incompressible fluids, where fully developed subsonic turbulence obeys the Kolmogorov law $\sigma \propto L^{1/3}$ (Larson 1981). Interstellar clouds are both compressible and clumpy, but generalizations taking account of these properties (Fleck 1983) produce a similar law. An alternative picture (Hendriksen and Turner 1984) substitutes an angular momentum cascade (from smaller to larger scales) and incorporates the effects of gravitational forces as well, leading to a similar law $\sigma \propto L^{0.5}$.

An alternative interpretation of the turbulence and condensation laws involves magnetic field support of the clouds (Myers 1986). Again, virial equilibrium is taken as a starting point. In addition, either approximate equality of magnetic energy density with turbulent energy density or approximate equality of mean turbulent velocity with the Alfvén speed must be assumed (either one of these implies the other if virial equilibrium applies). Under these conditions it may be shown that $\sigma \sim L^{1/2}$ and $n \sim L^{-1}$. The size of magnetic field needed to fit these observed laws for Larson's (1981) observed ensemble of clouds is in the range 15 to 150 μG, with median value 50 μG. These values are in agreement with observed field values for the subset of clouds in which such fields have been detected.

The many interpretations of the three laws have yet to be sorted out, but a reasonable picture is that (a) virial equilibrium holds in clouds and cloud cores from 0.1 to 100 pc; (b) supersonic "turbulent" motions, observed in all cloud regions except low-mass cores, are probably supporting the clouds. The origin of the turbulence may be energy or angular momentum cascades, which enter the cloud ensembles at largest scales and which have been hypothesized to arise from the shearing forces acting on clouds by the differential galactic rotation (Fleck 1981). Alternatively, the turbulence could be fed by magnetic field energy, which continuously is transformed into turbulence as virial equilibrium is maintained under (slow) collapse. While such turbulence seems to satisfy the needs of cloud support, we briefly review the possibilities of other supporting agents.

4.4.3 Rotation

Clouds may initially co-rotate with galactic rotation and spin up as they contract. When the spun-up rotation velocity becomes equal to $(3GM/R)^{1/2}$, the cloud is in virial equilibrium against gravity. Such rotational velocities are rarely observed. Several small cores have velocity gradients of 0.1 to 1 km s^{-1} pc^{-1} but these neither equal the turbulent contribution to the line width nor do they demonstrably indicate rotation in most cases. Even fewer massive cores ($\sim$ four) have exhibited unambiguous rotation. Such overall lack of rotation has been speculated to be due to magnetic braking (e.g., Field 1978), which occurs via radiation of Alfvén waves into the external medium on a time scale t_M. Braking ceases when the magnetic field diffuses out of the cloud on a time scale t_D. Braking occurs if $t_M < t_D$; this condition is probably satisfied for diffuse clouds, probably not for SMCs, and definitely not for globules. The time when braking ceases determines the cloud's final rotation velocity. Observationally, it appears that braking, if it occurs, must persist until late in the cloud's evolution (e.g., Vogel and Welch 1983), and thus rotation is not a viable stabilizer against gravitational collapse.

4.4.4 Magnetic Fields

For a smooth, spherically symmetric cloud, the energy of a magnetic field is $E_B \sim (1/R)(R^4 B_0^2) \sim \Phi^2/R \sim R^{-1}$ if the field is frozen in (i.e., Φ, the magnetic flux, is conserved and the field does not drift out of the gas). Since the gravitational energy $E_G \sim GM^2/R$ has the same R dependence, then if a frozen-in field cannot prevent initial contraction, it cannot prevent it at any later time. The collapse is in this case along the field lines, in principle. In practice, this probably is true only up to limited densities; if such collapse proceeds to $n = 100$ cm^{-3}, for example, a collimation of the collapse is already implied over fourteen orders of magnitude in scale size. In actual, nonideal clouds, the magnetic field lines are in any case likely to become tangled, and the magnetic energy may then exceed the dynamical energies. For a frozen-in field, E_B exceeds E_G for cloud masses $M > M_{cr} \sim (10^3 M_\odot)(B/30 \mu G)(R/2$ pc)$^2 = 10^5 M_\odot$ for $R = 20$ pc, $B = 30 \mu G$ (Mouschovias 1976a, b). In actual molecular clouds consisting primarily of neutral particles, the field is not frozen in, so its stabilizing effect can at most only slow, but not prevent, collapse. Present models (Mouschovias 1976a, b) are oversimplified but suggest that only the smallest clouds (globules) can have significantly slowed collapse, consistent with observed upper limits for field strengths. On the other hand, torsional Alfvén waves appear able to spin up cloud fragments as well as brake them (Mouschovias and Morton 1985). Thus even at higher densities, magnetic fields may hinder further contraction as well as aiding it.

Observationally, nine out of twenty clouds studied via the OH Zeeman effect have magnetic fields ranging from 9 μG (current sensitivity limits) to 140 μG (Crutcher et al. 1987). The regions involved are probably the "envelope" gas of both GMCs and SMCs, rather than the denser cores embedded in them. Thus, about half of all molecular clouds may possibly be supported to some degree by magnetic fields, if the diffusion time of the field out of the cloud is sufficiently slow. This latter condition may not apply, however, in view of the low fractional ionization ($\sim 10^{-7}$) observed in these clouds.

4.4.5 Stellar Winds

In Section 4.3.4, we showed that observed outflows probably lack both energy and momentum needed to significantly affect the dynamical evolution of all types of molecular clouds. Their effects may be only local within the cloud; however, they certainly contain sufficient power to disrupt star-forming cores.

4.4.6 "Gravito-Turbulence" or "Star Cloud Turbulence"

The observations are thus consistent with, and differential galactic rotation has been suggested as a source for, a "universal" turbulence whose magnitude is sufficient to support clouds against gravitational collapse. In the angular momentum cascade picture (Henriksen and Turner 1984, Larson 1985), an ensemble of clouds or of substructures within a cloud with a gravitational correlation length l can couple its angular momentum to a larger-scale ensemble with correlation length L. The smaller-scale structures spin down, the larger-scale structures spin up. Simulation of three-dimensional collapse shows development of trailing spiral features which transfer the angular momentum outward. If this process transfers constant angular momentum per unit time per unit volume, $\rho_l v_l^2 = \rho_L v_L^2$, then the clouds will satisfy $\Delta v \propto L^{0.5}$ and $M_{cloud} \propto \Delta v^4$ (Henriksen and Turner 1984). Questions remain as to whether efficient coupling can occur between cloud and intercloud regions, as may be required if galactic rotation is to be an effective source of star cloud turbulence. However, the observed cloud sample, which contains no significant internal heat sources, obeys the relationships $\sigma (km\ s^{-1}) = 0.8R(pc)^{0.5}$ and $\Delta v = 0.37(M/M_\odot)^{0.25}$, lending quantitative support to the notion of star cloud turbulence as the supporting agent for molecular clouds.

In summary, supersonic turbulence is *observed* to be adequate for supporting all clouds from GMCs to small dense cores, and thus is clearly preferred over rotation or the "direct" effects of stellar winds as the general agent supporting molecular clouds of all sizes. Such turbulence may by fed by an external energy or angular momentum cascade, or it may be maintained by magnetic energy under quasi-virial conditions during slow collapse. Observationally, there is some question whether magnetic fields of sufficient strength reside in many clouds. In the picture of star formation we discuss next, the important point is that the origin of the turbulence is unimportant, so long as it is the turbulence itself that directly supports cloud structures of all sizes. This will be the case unless magnetic fields retain fully frozen-in values down to size scales approaching those of protostars, an unlikely situation because of the low ionization and also on recent theoretical grounds (Goodman 1986).

The problem we now confront is not how clouds are supported, but rather how this support is on occasion eliminated so as to allow clouds (or cores) to contract and form stars. This forms the final subject of this review—how stars actually form from dense, stabilized clouds and cores. The problem appears to have a straightforward answer for low-mass stars, in which the turbulence law $\sigma \propto L^p$ plays a direct and fundamental role. As we shall see, multiple mechanisms may be needed to form massive stars.

4.5 The Formation of Stars

4.5.1 Low-Mass Stars from Low-Mass Cores

Any molecular cloud stabilized against gravity by turbulence has supersonic internal motions according to the turbulence law. Such motions in a nonuniform medium will undergo shock compressions, and condensed structures (sheets, filaments, clumps) will form. If these substructures remain large enough, their internal motions will remain supersonic. Additional substructures will then form. Eventually, such structures will be small enough that their internal motions become subsonic. Such subsonic structures suffer no further internal shocks and cannot possess or develop further substructures. Furthermore, within a free-fall time, such subsonic structures can be shown not to grow again to sizes corresponding to supersonic turbulence, by either accretion or by collisions between clumps (Myers 1983, Turner 1984). Thus, a subsonic clump, once formed, is an endpoint, an entity out of which a low-mass star will form. The process is as follows.

First, a subsonic clump will tend to produce large density gradients within itself (Myers 1983), assuming that it is initially in quasi-virial equilibrium. These density gradients may serve to decouple the flow of turbulent energy into the clump from outside, so long as the scale sizes involved are in the *inertial* range of turbulent flow. [The inertial regime is assured, since the microscale size at which turbulence is converted into heat ($\sim 10^{11}$ cm) is much smaller than the clump size.] This means that the coupling of turbulent energy into a given scale size comes predominantly from the next larger scale size. No shocks or other mechanisms exist to dissipate or inject energy on clump size scales. If the clump scale has significantly more mass and inertia than an equal volume of gas at the next larger scale size, then turbulent coupling will be inefficient. Similar arguments apply if the turbulence is torque-driven angular momentum rather than energy.

Once a subsonic core is decoupled, it is free to dissipate its remaining turbulence on a time scale $t_t \approx l/\Delta v \approx (l^3/GM)^{1/2} \approx t_{ff}$, i.e., comparable to the free-fall time. Here, virial equilibrium has been assumed, as observed. If the core is only partly decoupled, collapse would take longer.

This picture leaves unexplained the physical origin of the turbulence and the physics of its maintenance, and assumes that the specific notions of Kolmogorov or of "star cloud" turbulence have some validity; although logical and plausible, these ideas are incomplete and open to question.

An interesting corollary of this picture is the concept of *minimum stellar mass*. The mass of a typical clump when it becomes subsonic turns out to be roughly its Jeans mass. Thus, one expects that the number of (low-mass) stars formed per subsonic clump will be $N \approx M_{c1}/M_J \approx 1$. For a subsonic clump which has lost its turbulence, the internal motions are purely thermal, and for a typical gas temperature of 10 K in such a clump, $\Delta v = \Delta v_{th} = 0.32$ km s^{-1}. The turbulence law then implies a minimum clump size at the subsonic threshold of $R_{min} \approx 0.1$ pc, which is essentially at or just beyond the limits of current observational resolution in nearby low-mass-star-forming regions (0.1 pc = 2.3' in Taurus). Next, if virial equilibrium applies for the subsonic clumps, then we find that $\Delta v = 0.48(M/M_{\odot})^{0.188}$. By setting $\Delta v = 0.32$ km s^{-1}, the minimum "predicted" mass is 0.12 $M_{\odot}$. Although this number is somewhat uncertain, the important idea is that the stellar mass function should begin to turn downward for masses $\lesssim 0.1\ M_{\odot}$. This expectation is consistent with the available data for the local stellar mass spectrum. By contrast, fragmentation theories do not predict a termination of the fragmentation process until masses of order 0.01 $M_{\odot}$ are reached, the limit set by opacity effects.

In summary, low-mass stars should form anywhere that supersonic motions accidentally produce small, self-gravitating, dense, subsonic clumps. There is no fundamental reason why GMCs or SMCs should differ in the efficiency of low-mass–star formation. Small stars should form widely in any type of molecular cloud, as is observed.

This picture of low-mass-star formation is quite simple, mainly because the masses of small stars do not exceed the typical mass of a clump formed as a transient entity by hydromagnetic shocks. The situation is quite different for massive stars, and a more complicated picture for their formation is therefore implied.

4.5.2 Formation of Massive Stars from Massive Cores

(a) A Summary of Observational Constraints

All clouds containing massive stars ($\geq 10\ M_{\odot}$) have $\Delta v > 2\Delta v_{th}$ so that smaller-scale internal structures can be expected to form. In addition, the number of Jeans masses in massive clumps is at least $(\Delta v/\Delta v_{th})^3 \approx 3.4(M/M_{\odot})^{0.56}$ (Larson 1981) so that fragmentation into at least this many smaller clumps is expected. Since the typical mass of condensation formed as transients of hydromagnetic shocks is much less than that of a massive star, it is clear that some form of accretion or coalescence of these clumps or of the supersonically turbulent gas of the parent cloud in general must occur prior to formation of massive stars. Three basic processes may be considered (cf. Turner 1984): (i) accretion or coalescence within a large cloud of initially diffuse gas or of small clumps; (ii) concentration of a lot of gas into a small region by a gravitational potential well caused by an already existing concentration of stellar mass; (iii) external triggers, such as shocks from expanding HII regions, supernovae, etc.

To distinguish these possibilities, some pertinent observational facts may be cited:

1. SMCs do not form large clumps. Accretion of clumps is not important in SMCs. On dynamical and size scale arguments, it can be shown to be even less important in GMCs. Thus "ordinary" clump accretion is not relevant to any kind of star formation.

2. Bright-rim clouds contain no massive stars. Clump masses in these clouds are < 5 $M_\odot$. The observed clouds are SMCs. Thus, presence of nearby HII regions driven by O stars is not a sufficient condition for the triggering of massive star formation, at least within SMCs.

3. Massive stars form near other massive stars. Whether the interaction is sequential or not, it appears that external shocks are a *necessary* if not a sufficient condition.

4. Massive stars form where there are large gravitational potential wells. It has long been recognized that the most massive stars within an association occur in the most concentrated subsystem. It is also suggested (Larson 1982) that older clusters tend to form more massive stars in more condensed regions. Finally, it appears that low-mass-star formation may reach its peak and actually begin to decline before massive-star formation gets under way (Adams et al. 1983; but see criticism by Stahler 1985). These aspects all point to the gas within a star-forming region becoming more centrally condensed with time. This would permit *enhanced* accretion rates for clumps, which is necessary because "ordinary" accretion rates are too slow relative to clump free-fall times.

5. Massive accretion can occur in the absence of strong shocks. The ρ Oph massive, centrally condensed core is the best example. This core contains 550 $M_\odot$ within 1×2 pc, contains copious low-mass-star formation (Wilking and Lada 1983), and does not lie near any strong shocks.

6. The Initial Mass Function (IMF) is close to a power law for massive stars. Such a law arises naturally from an accretion process for clumps (cf. Kwan 1979). Clumps which become sufficiently large accrete most rapidly and undergo a runaway growth relative to the smaller clumps. This produces a hierarchy of a few large clumps, many small ones, and a mass spectrum which is asymptotically a power law.

Items (4) through (6) all imply an "enhanced" accretion rate for clumps where massive stars form. Item (6) in addition suggests that such a process *must* occur.

7. Massive stars form later than small stars. Besides the evidence cited in item (4), there is direct evidence within the Orion cluster (Field 1982). Low-mass stars have recently been seen *within* the ionized Trapezium Nebula for the first time (Field 1982). Their ages, determined from lithium spectral features, are small enough that they must have formed where they are presently observed. The ionized nebula is much younger yet and formed immediately after the massive stars which ionized the nebula were formed. Thus, the massive stars must have formed after the low-mass stars, presumably by a different process.

We conclude that the classical picture of gravitational collapse and fragmentation of a cloud cannot be correct. It provides no mechanism by which stars of different masses form at different times. The observational "rules" for star formation now seem clear. First, if there are no external shocks and no enhanced accretion, then turbulence is eliminated only in subsonic clumps, that is, only in small clumps; so only low-mass stars can form. Second, if there is enhanced accretion but no shocks, then the massive core, which is supersonic, will form substructures which continue to subdivide until subsonic, small clumps occur, which again can form only low-mass stars. Third, various observations indicate that external shocks are necessary, but not sufficient, to form massive stars.

(b) Massive-Star Formation: Enhanced Accretion Plus Shocks
The mechanisms by which massive rather than low-mass stars form when an external shock impinges on a massively accreted (supersonic) clump have been discussed by Turner (1984). They may be summarized as follows:

(i) The shock compresses the clump gas.
(ii) Density gradients resulting from the compression allow the turbulence to decouple.
(iii) The cold postshock (CPS) layer becomes unstable, via Rayleigh-Taylor instabilities, to gravitational collapse (a minimum column density through the CPS layer is required for such an instability to occur).
(iv) Jeans collapse occurs. Initial collapse is isothermal, but later approaches adiabatic conditions when the opacity becomes large. When analyzed in both isothermal and adiabatic conditions, Jeans collapse predicts a *minimum* mass for the resulting star(s) which depends strongly on the gas temperature, in the form $M_{min} = 0.01(M/M_\odot)(T/10)^{2+\delta}$, where δ describes grain composition (really the cooling properties) and has a value of 1 or 2 depending on grain composition (Silk 1985) and T is the initial gas temperature before onset of Jeans collapse but after shock heating. For $T = 100$ K, $M_{min} = 100\ M_\odot$ for $\delta = 2$, 10 $M_\odot$ for $\delta = 1$. The role of the shock in elevating M_{min} through heating is as important to massive-star formation as it is in initiating gravitational instabilities.

The larger the clump mass, the weaker the external shock needed to initiate collapse. Thus, accretion is essential not only because small clumps will simply lose their turbulence and form low-mass stars, but also because HII region shocks and even the sdw shock can initiate massive-star formation when massive clumps are involved.

Figure 4.6 summarizes the mechanisms of star formation as deduced from the large mass of observational data on star-forming regions now collected via molecular spectroscopy. A global source of turbulence (galactic rotation) couples into and supports both SMCs and GMCs. GMCs form from SMCs (or HI clouds) via the sdw shocks. Both SMCs and GMCs form subsonic structures which, in turn, produce low-mass stars. These constitute an "endpoint," a permanent repository of interstellar gas. GMCs also foster massive accretion cores either by means of successive generations of stars in a given region, which attract surrounding gas, or by a central condensation which results from supersonic collisions between clumps. These massive cores contain heated gas (either from nearby HII regions or as a result of an impinging shock) and form only massive stars when hit by a shock. Such stars interact violently with the parent GMC, disrupting it on a time scale of a few tens of millions of years. The likely products of such disruptions are SMCs and some HI. The SMCs are recycled through the sdw on time scales of $\sim 2 \times 10^8$ yr to form more GMCs. Most of the salient observational aspects of star forma-

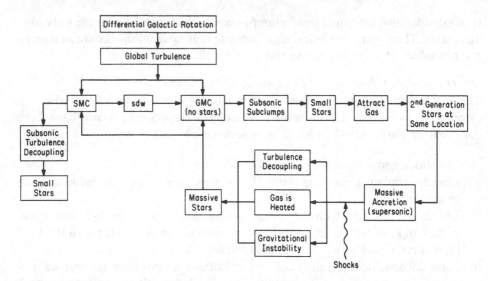

Fig. 4.6. A schematic of the mechanisms of star formation and how they may relate to molecular clouds (GMCs and SMCs) and to galactic dynamics. This picture assumes that the turbulence which supports molecular clouds is replenished by an external source such as galactic rotation; magnetic fields are a less likely but possible alternative.

tion are explained by this picture. Certain details such as bipolar flows are subjects of later stages in protostellar evolution, beyond the scope of radio molecular spectroscopy.

4.5.3 A Stochastic Picture of Star Formation

In the picture of star formation just discussed, additional physics is required to produce massive stars, after low-mass stars are formed. This picture is a "bimodal" theory of star formation, in which massive stars form by different processes and in different regions than low-mass stars. It is motivated by the observation that massive stars are associated only with massive molecular clouds.

An alternative view (e.g., Elmegreen 1985) is that star formation is purely stochastic, with the spectrum of stellar masses simply corresponding to the mass spectrum of cloud fragments. This viewpoint also starts with the observation that massive stars tend to be associated with GMCs while low-mass clouds have only low-mass stars. This observation is now interpreted, however, as just reflecting the probability spectrum of fragmentation. Assume stars form with a single IMF but randomly in a cloud core with some universal probability (independent of cloud mass) of having stellar mass m. Massive clouds will form more stars, and therefore more of the rare massive stars, than will SMCs. A given total mass of SMCs will form the same number of OB stars as the same total mass of GMCs, but an OB star will occur much more rarely in an SMC than in a GMC. With this picture, using a Miller-Scalo IMF, one can easily derive an expected relationship between the mass of a stellar cluster and the mass of its most massive star. The observed relationship between *molecular cloud* masses and the mass of the most massive stars associated with them

(Larson 1982) is found to be very similar and may be brought into agreement if one assumes a mean star formation efficiency of 10%. Such an efficiency seems typical of at least some star-forming regions (see Section 4.4.1). The Miller-Scalo IMF also indicates that ~ 500 low-mass stars (of typical mass 0.6 $M_\odot$) form for every OB star (typical mass 23 $M_\odot$). If star formation has a 10% efficiency, one such OB star should form in every three SMCs of typical mass 10^3 $M_\odot$. Actually, the observed incidence of OB stars in SMCs seems significantly lower than this.

Regardless of which picture of massive star formation applies, once massive stars form, they heat their surroundings and raise the minimum Jeans mass. Formation of massive stars thus inhibits formation of low-mass stars, and this concept alone serves as a basis for "bimodal" star formation.

4.6 Future Prospects

We may envision three directions that future observations will take in order to shed further light upon the subject of star formation and its relation to the molecular component of the ISM.

First, further large-scale mapping of the Milky Way at modest single-dish resolution (one to several arcminutes) will be useful mainly to elucidate further the global morphological aspects of molecular clouds. Higher sensitivity is perhaps most needed, to establish the boundaries of the clouds and thus answer the question whether SMCs, GMCs, and GMC complexes are really well-defined entities. Such studies will necessarily be most fruitful in the solar neighborhood, where the outlines of clouds will not be confused with the ubiquitous molecular emission from the galactic disk. It will probably be difficult to clarify much further the relationship of GMCs with spiral arms in the Milky Way because of the overall blending of the large number of clouds along any line of sight through the inner regions of the galaxy. Further studies of OH in its 1720-MHz transition, particularly in the southern hemisphere, offer most hope. Studies such as those of Maddelena et al. (1986) which in the Orion-Monoceros region have indicated most clearly the morphological relationships of GMC complexes with OB associations, supernovae, and other energetic phenomena, are needed for other regions. Of course, the Orion-Monoceros region is one of few studied in such detail, because of its proximity to the Sun. At present, relatively few solar neighborhood OB associations are studied in sufficient detail to check the star-forming mechanisms which have been proposed. Accurate proper motions and additional photometry and radial velocity measurements (such as the Hipparchos satellite will make) can establish association membership more precisely. This is important to determine whether the original star-forming regions are in general as extended as recent results imply and whether, in associations other than Orion OB1, high mass and youth are correlated. It is also important to ascertain if the formation of the associations is initiated by stochastic processes as well as by mechanisms depending on the effects of earlier generations of stars.

Second, high-resolution studies of the many fine-scale aspects of star formation, of the sort only just begun with the Hat Creek and CIT interferometers, promise to revolutionize our understanding of star formation both in our own galaxy (the small-scale phenomena) and in other galaxies (the large-scale aspects).

Third, new spectral windows, those of the submillimeter and far IR, offer much new information about many aspects of star formation.

The second and third items serve to emphasize current problems and future directions, and we expand upon them in what follows.

4.6.1 High-Resolution Studies of Star Formation

(a) Global Aspects As Seen in Other Galaxies

The first interferometric maps of CO in other galaxies (Lo et al. 1985) show indications of fragmented spiral structure, especially in M51 and IC 342. Current instrumentation is inadequate to yield complete maps containing all of the CO flux, but the results are consistent with the highest-resolution single-dish studies at Nobeyama (14" resolution; Sofue, Nakai, Turner, and Rickard, manuscript in preparation) in M51. Peak brightnesses seem somewhat higher than expected for Milky Way GMCs at the corresponding distances, and currently only large GMCs (~ 300 pc in size) are resolvable in M51. Nevertheless, elongated structures ~ 3 kpc in size are seen with strong velocity gradients along them, and a $\sim 7:1$ arm/interarm contrast. With larger arrays in the future, the prospects of mapping complete spiral arms in CO seem good, with the possibility of determining whether GMCs occur only in the arms. If so, it may also be possible to test the "ballistic" theory of formation of GMCs in spiral arms and its postulates of the way they are launched, by determining whether the GMCs really have postshock velocities, as required on this picture, or preshock velocities, as suggested by Wielen (1978) for M51 on the basis of the relative distribution of the young OB stars and the older star population that defines the spiral arm. Also, if GMCs originate from the sdw shock, they should lie preferentially on the inner edges of spiral arms.

Other intriguing questions have already arisen from preliminary high-resolution studies of extragalactic CO. The brightness temperatures of GMCs in the Milky Way and in M31 are practically identical (Blitz 1981) and less than those observed interferometrically in M51, M82, and IC 342. These latter cases involve clouds within the central few kiloparsecs of these galaxies. These clouds thus appear warmer than the GMCs observed at radii of 8 to 10 kpc in M31 and the Milky Way. Whether this effect should be attributed to excitation or to larger cloud sizes is not known.

With prospective millimeter-wave arrays of the future, a hot-centered GMC core like that in Orion (OMC-1) will be detectable to a distance of ~ 10 Mpc, while more extended and massive GMCs will be detectable as far away as 50 Mpc, well beyond the Virgo cluster.

(b) Star-Forming Cores

The study of cloud cores and their relationship to associated young stars will benefit greatly from the high-resolution facilities coming into use at millimeter, submillimeter, and IR wavelengths in the near future.

The "global" relationships (Section 4.4.2) between $(n, L, \Delta v)$ for molecular clouds and cores on all size scales are basic to the picture of star formation presented here.

Several questions remain. What is the internal density structure of cores over length scales where Δv changes significantly? Does virial equilibrium hold on scales smaller than core sizes? At what scale does the turbulence become uncoupled, and gravitational collapse set in? Do the trends $n(L)$ and $\sigma(L)$ differ between massive cores and low-mass cores? If they do, the details may give clues as to how these different core types are supplied with turbulent energy. Also, do the trends $n(L)$ and $\sigma(L)$ differ between cores with associated stars and those without? It is important to determine the size scale in cores where these trends break down or change, perhaps as a consequence of stellar luminosity and winds. Various considerations suggest that massive cores with internal sources of heating and turbulence will have density gradients which are steeper at smaller length scales, which will bear on the question of fragmentation during the final collapse stage. Perhaps most fundamental is whether the higher turbulence found in massive cores is really due to the requirement of virial support of the core, as discussed in Section 4.4. Since massive cores may contain massive protostars, the increased turbulence could be due solely to the greater luminosity and mechanical energy provided by newly formed massive stars. With high-resolution studies it may be possible to answer this question on the basis of detailed comparison of core velocity dispersions among regions forming low-, intermediate-, and high-mass stars, as well as cores in quiescent clouds and in Bok globules.

High-resolution studies in molecular lines sensitive to high density are also needed to clarify these questions. Resolutions at least equivalent to the Jeans length are desirable. For nearby SMCs such as Taurus and Ophiuchus, $n \sim 5 \times 10^3$ cm^{-3} and $T \approx 10$ K so that the Jeans length is 0.3 pc or 2'. For these regions current single-dish measurements are marginally adequate, though improved sensitivity is needed. For Ori(KL), the nearest massive-star-forming region, $n \sim 1 \times 10^6$ cm^{-3} and the Jeans length is 0.06 pc (25"), requiring interferometry. The higher-density regions are more interesting, since the free-fall time is smaller, and collapse, if it occurs, will be more noticeable. With current limited sensitivity for interferometers, a core region must be very hot to be detectable. Hot regions such as Ori(KL) are very complex and apparently have a large variety of phenomena within a region of a few arcseconds. Thus, it is difficult to unambiguously associate phenomena observed in different wavelength ranges unless the angular resolution is $\sim 1''$. If the Jeans length is any guide, for a given density we expect smaller structures from colder regions, so cold regions are currently not detectable at such scales.

(c) The Question of the Elusive "Protostar"

The current models of protostars imply that they should have energy distributions peaked in the 5- to 20-μm region, but so might other types of objects such as main-sequence stars still embedded in dust shells, or even post-main-sequence objects which have ejected a shell. To distinguish a true protostar, there should be no radio continuum emission or IR recombination lines, and there should be evidence for infall of the surrounding gas. Of the many IR objects and OH or H_2O masers found so far, none shows clear evidence that they are heat sources in which contraction supplies the energy. No conclusive kinematic evidence for gravitational motions of protostar-forming gas has been found either, although in two cases

(IRAS 16293-2422 in ρ Oph: Walker et al. 1986; and G10.6-0.4: Ho and Haschick 1986) the information may be suggestive. It is possible that good candidates are still lacking. The young stellar objects found at optical or near-IR wavelengths may be too evolved, having already completed their accretion. Detection of free-fall motion with speeds greater than a few kilometers per second onto a few-solar-mass star even in the nearby Taurus region requires arcsecond resolution in a strong emission line.

(d) Ori(KL) and Detailed Studies of Star-Forming Regions

It is noteworthy that the intricate model of Orion presented by Plambeck et al. (1985) is based on maps of just six molecular lines and that the different molecular species all show different spatial distributions. Particularly interesting is that SiO, HCN, and HCO$^+$ differ spatially despite having similar dipole moments and hence excitation requirements. This different "chemistry" presumably reflects different physical conditions. HCO$^+$ dominates where the ionization fraction is highest, SiO perhaps where high-velocity outflow shocks have destroyed grains. New information on star formation processes in Orion will therefore come from the study of many additional lines and molecules among the hundreds that are observable toward Orion. At the same time, it is clear that models of this region will become much more complex.

The many different phenomena presently observed in Orion may well mark a very particular evolutionary stage and thus may not characterize other massive-star-forming regions such as Ceph A, NGC 7538, W3, and W31 that we may hope to study in comparable detail with future interferometers. There is already a suspicion that the Ceph A region is different: it may have two perpendicular confining disks on very different size scales (Kassim and Turner, manuscript in preparation; Torrelles et al. 1985). Studies of these other regions comparable to studies of Orion will be difficult, given that their star-forming cores have sizes $\lesssim 0.''5$ (compared to 10$''$ for Orion). Unlike Orion, the W31 and NGC 7538 cores lie in front of HII continuum sources so that absorption techniques, at least at centimeter wavelengths, may be used for several molecular species which would be difficult to detect in emission. Velocities in the absorption features can allow discrimination between infall and outflow on scales very close to those of the forming object.

The role of disk structures around young stars is potentially far-reaching. With high spatial resolution one may determine how common they are; whether they channel the bipolar outflows or are the sources of the outflows; how strongly they affect subsequent star formation by disrupting surrounding material, sweeping up dense shells, compressing preexisting clumps, etc. Observations of dust polarization in the disk can determine the direction of any magnetic field in the disks and cloud cores, and can thus determine whether the fields become more ordered in denser regions and are related to the directions of the outflows. Such studies will determine whether magnetic forces play an important role in the final protostellar contraction phase.

4.6.2 Submillimeter and Far-IR Studies

(a) Far-IR Studies

Infrared continuum studies, although not the subject of this review, have played a major role in the search for star formation, both because IR radiation can escape from dusty regions and because dust around young stellar objects tends to reach temperatures in the range of 20 to 1500 K. Ground-based IR work will continue to make important advances toward the understanding of star formation regions, both through further increases in spatial resolution via speckle interferometry on large (7 to 15 m) telescopes and through increased spectral resolution, which together can rival molecular spectroscopy in providing a probe of protostellar structure in (n, T, v).

The most important advances in IR work will probably come from IRAS, which will provide the major catalogue of not only candidates for massive protostars but also of stars in low-mass cores. Preliminary results already indicate that about one-third of all cores studied by Myers et al. (1983) have associated low-luminosity IRAS sources (Beichman 1984). Recent studies (Beichman et al. 1986) of ninety-five low-mass cores in Taurus, Ophiuchus, and Cygnus indicate that perhaps one-third of the optically invisible cores associated with IRAS sources could be protostars on statistical grounds. The cold IR objects in B335 are also considered good candidates. Finally, several very cold IR sources have also been located in the globule B5 and the Chamaeleon I region (Evans 1984). These appear to be the best candidates so far for true protostars in a very early infall phase. IRAS also provides complete surveys of cloud complexes for dense cores and should reveal the distribution of core masses in each complex and the variation of clumpiness from complex to complex. It has already significantly increased the number of young star candidates in nearby molecular cloud complexes to 10^3–10^4 sources (Rowan-Robinson et al. 1984). Two questions may be answered by these data: (i) Are the "low-mass" and "massive" cores described here distinctly different, or are they part of a continuous spectrum of core masses? (ii) Are core masses and resultant star masses related, and is the mass of the most massive star in a cloud-embedded cluster correlated with the cloud mass, as suggested by Ho et al. (1981a)?

The complete and unbiased sample of young embedded IR sources provided by IRAS can be combined with the high-resolution molecular line observations now becoming available with the new millimeter-wave interferometers to ascertain if there are any differences between the (n, T, v) structures of cloud fragments with and without associated IR emission. Where IR sources exist, their effects on the ambient clouds can be studied, and a better picture of cloud dissipation achieved.

(b) Submillimeter Studies
In the continuum, heated dust produces the bulk of submillimeter emission, the submillimeter spectral region being sensitive to cooler ($T_d \sim 20$ to 50 K) dust than are the currently used IR bands which range from 2 to 400 μm ($T_d \sim 100$ to 1500 K). The submillimeter region will complement the ground-based IR and IRAS studies of the dust by adding spectral information about the dust emissivity, of importance in determining the dust composition. In particular, current preliminary studies (e.g., Schwartz 1980) indicate the dust opacities will be small over the submillimeter range, so that the absorption efficiency may be estimated as a function of wavelength. In addition, dust-to-gas mass ratios may be obtained. Studies to determine if these properties vary from region to region will be important. Finally, as long as the dust temperature T_d is not too low, the brightness temperature is proportional to T_d and the column density. If gradients in T_d induced by internal sources are not

large, submillimeter observations will probe enhancements in column density in shocked regions and in protostellar contraction phases, up to $\sim 4 \times 10^{25}$ cm^{-2} ($\tau_{1mm} \sim 1$). By this means, the evolution of a 1-$M_\odot$ protostar can be traced to radii as small as 165 AU ($\sim 1''$ at the distance of the nearest clouds).

Many new results in the submillimeter region will likely come from spectral line studies. Species which emit uniquely in the submillimeter/far-IR region are the atoms CI, CII, OI, OIII, and NIII via their fine-structure transitions, several hydride molecules (the most important as ISM diagnostics being NH_3, H_2O, and OH), and CO.

Among the atoms, CI has already been seen (Section 4.2.1) to monitor the transition zones between dense molecular clouds and the more diffuse ISM. Because its $^3P_1-^3P_0$ transition has a similar Einstein A coefficient and collision cross section to CO ($J = 1-0$), it may be compared directly with CO and is found apparently to permeate the entire dense ISM as well as the diffuse components (Phillips and Huggins 1981). Because carbon is largely in the singly ionized state in the diffuse ISM, its 158-μm fine-structure $^2P_{3/2}-^2P_{1/2}$ transition is expected to be of prime interest for cooling and possibly for defining the physical conditions of these regions. Nevertheless, first results (Crawford et al. 1985) suggest it is observable from dense regions also. In other galaxies (e.g., M82), the CII distribution is similar to that of CO, but of particular note is the fact that the CII line is very strong, accounting for almost 1% of the total luminosity of M82! The OI, OIII, and NIII lines similarly provide important diagnostics about the diffuse ISM regions, particularly the ionization state. OI should coexist with CII in neutral diffuse regions. Considerable work on the O lines has already been done (e.g., Watson 1984). In interpreting these results, one notes that the column density of HI needed to produce unit opacity in [OI] or [CII] lines is $\sim 10^{21}$ cm^{-2} ($A_v \approx 2$ to 15^m). Also, one would expect the gas producing the fine-structure lines to be less than a UV light penetration depth thick; otherwise, the gas would be molecular. These considerations are puzzling, given the evidence that CI and CII appear to exist in dense molecular clouds. The current interpretation (Tielens and Hollenbach 1985) is that observed lines of CI, CII, OI, and H_2 (2 μm) and even much of the $J = 1-0$ CO arise from the photodissociation region that lies between the HII regions and dense molecular gas in sources such as Orion (KL). Such an interpretation avoids the necessity to invoke nonequilibrium chemistry to explain the large CI abundances toward molecular clouds, but may have difficulty explaining the widespread and strong CII emission from other galaxies. And while this interpretation explains the observed linear relationship between IR CII emission and millimeter-wavelength CO emission (Crawford et al. 1985), it raises serious questions about deriving hydrogen gas column densities from CO, as is commonly done, because such derivations assume the CO samples all of the gas of a cloud while the CII can only arise at the cloud surface. Extensive observations with larger telescopes are needed to settle these questions and to fully exploit the diagnostic potential of the atomic lines for the more diffuse components of the ISM.

Information about star-forming regions from submillimeter molecular emission is preliminary at present, but its potential may be illustrated by the following examples. Many submillimeter/far-IR rotational transitions of CO are seen toward

Ori(KL). These lines are optically thin, and both density and temperature may be obtained from the overall dependence of emission strength on rotation quantum number J. A single-component model for Ori(KL) yields $n \sim 10^6$ cm^{-3}, $T_k = 750$ K, indicating gas which is related to, but cooler than and further "downstream" from, the smaller amount of 2000 K gas which produces the H$_2$ emission at 2μm. These values of (n, T) appear to rule out the possibility of a nondissociative, nonmagnetic shock being responsible for the H$_2$ and CO emission in Ori(KL). Hydromagnetic shocks may thus be indicated.

Far-IR observations of OH at 119 μm also suggest a postshock molecular environment, because elevated temperatures are implied by the overabundance of OH deduced in Sgr B2, the only source studied so far. Two separate far-IR transitions of NH$_3$ seen toward Ori(KL) arise, respectively, from the hot core object (suggesting $T_k = 200$ K) and from the "spike" component. The important 1_{10}–1_{01} fundamental transition of H$_2$O (538 μm), which may be a major coolant in dense molecular clouds, has not yet been detected.

Although submillimeter spectroscopy has not yet radically altered our view of the ISM and of star-forming regions, the abundance and distribution of the atomic and molecular species recently detected in this spectral region have on the whole been greater than anticipated. The fact that the atomic gas appears *not* to be a distinct phase of the ISM as compared to the dense molecular gas is one of great significance and may well cause major modifications to interstellar chemistry schemes. The atoms detected in this region, as well as some of the molecules, will also be important in determining the energy balance of many physical regimes of the ISM.

The far-IR/submillimeter region clearly promises important new information for all scales of the ISM and of star-forming regions, ultimately realized only by space facilities. SIRTF (early 1990s) will provide sensitive photometric studies of star-forming regions, following up IRAS. The Large Deployable Reflector, a $\sim$20-m space telescope planned for the late 1990s, will achieve $1''$ resolution at 100 μm and high-resolution spectroscopy over the entire 50- to 1000-μm region, bringing a whole new clarity to our picture of star formation, and to more extended regions of the ISM, in both the Milky Way and in other galaxies. Together with European facilities like ISO and FIRST, these space telescopes will revolutionize our understanding of the ISM and of star formation.

Recommended Reading

1. M. Piembert and J. Sagaku (eds.). 1986. Star Formation Regions, Int. Astron. Union Symp. No. 115. Dordrecht: Reidel.
2. Turner, B.E. "How Stars Form: A Synthesis of Modern Ideas," Vistas in Astronomy 27:303.

References

Adams, M.T., K.M. Strom, and S.E. Strom. 1983. Astrophys. J. (Suppl.) 53:893.
Bally, J., and C.J. Lada. 1983. Astrophys. J. 265:824.
Bash, F.N., E. Gree, and W.L. Peters. 1977. Astrophys. J. 217:464.
Beckwith, S., S.E. Persson, G. Neugebauer, and E.E. Becklin. 1978. Astrophys. J. 223:464.
Beichman, C. 1984. In D. Black and M. Matthews (eds.), Protostars and Planets II. Univ. of Arizona Press.
Beichman, C.A., E.E. Becklin, and C.G. Wynn-Williams. 1979. Astrophys. J. 232:L47.

Beichman, C.A., P.C. Myers, J.P. Emerson, S. Harris, R. Mathieu, P.J. Benson, and R.E. Jennings. 1986. Astrophys. J. **307**:337.

Benson, P. 1983. Ph.D. thesis, Dept. of Physics, Massachusetts Institute of Technology.

Blaauw, A. 1964. Annu. Rev. Astron. Astrophys. **2**:213.

Blaauw, A. 1984a. Irish Astron. J. **16**:141.

Blaauw, A. 1984b. *In* H. van Woerden, W.B. Burton, and R.J. Allen (eds.), Int. Astron. Union, Symp. No. 106. Dordrecht: Reidel.

Blair, G.N., N.J. Evans, P.A. Vanden Bout, and W.L. Peters. 1978. Astrophys. J. **219**:896.

Blitz, L. 1978, Ph.D. dissertation, Columbia University.

Blitz, L. 1980. *In* P.M. Solomon and M.G. Edmunds, (eds.), Giant Molecular Clouds in the Galaxy. Pergamon Press.

Blitz, L. 1981. *In* L. Blitz and M. Kutner (eds.), Extragalactic Molecules, NRAO Workshop, Green Bank, Nov. 1981.

Blitz, L., and F. Shu. 1980. Astrophys. J. **238**:148.

Blitz, L., and P. Thaddeus. 1980. Astrophys. J. **280**:676.

Blitz, L., L. Magnani, and L. Mundy. 1984. Astrophys. J. (Lett.) **282**:L9.

Bok, B.J. 1977. Publ. Astron. Soc. Pacific **89**:597.

Bok, B.J., C.S. Cordwell, and R.H. Cromwell. *In* B.T. Lynds (ed.), Dark Nebulae, Globules, and Protostars. Tuscon, University of Arizona Press.

Boulanger, F., A.A. Stark, and F. Combes. 1981. Astron. Astrophys. **93**:L1.

Clark, F.O. 1986. Astron. Astrophys. **164**:L19.

Claussen, M.J., G.L. Berge, G.M. Heiligman, R.B. Leighton, K.Y. Lo, C.R. Masson, A.T. Moffet, T.G. Phillips, A.I. Sargent, S.L. Scott, P.G. Wannier, and D.P. Woody. 1984. Astrophys. J. (Lett.) **285**:L79.

Cohen, R.S., H. Cong, T.M. Dame, and P. Thaddeus. 1980. Astrophys. J. (Lett.) **239**:L53.

Cohen, R.J., P.R. Rowland, and M.M. Blair. 1984. Mon. Not. R. Astron. Soc. **210**:425.

Cowie, L.L., A. Songaila, and D.G. York. 1979. Astrophys. J. **230**:469.

Crawford, M.K., R. Genzel, C.H. Townes, and D.M. Watson. 1985. Astrophys. J. **291**:755.

Crutcher, R.M., and I. Kazes. 1983. Astron. Astrophys. **125**:L23.

Crutcher, R.M., I. Kazes, and T.H. Troland. 1987. Astron. Astrophys. **181**, 119; also, private communication.

Dame, T.M., B. Elmegreen, R. Cohen, and P. Thaddeus. 1986. Astrophys. J. **305**:892.

Davies, R.D., and E.R. Cummings. 1975. Mon. Not. R. Astron. Soc. **170**:95.

Dickman, R.L. 1978, Astrophys. J. Suppl. Ser. **37**:407.

Dyck, H.M., and C.J. Lonsdale. 1979. Astron. J. **84**:1339.

Elias, J.H. 1980. Astrophys. J. **241**:728.

Elmegreen, B.G. 1979. Astrophys. J. **231**:372.

Elmegreen, B.G. 1985. *In* R. Lucas, A. Omont, and R. Stora, (eds.), Naissance et Enfance des Etoiles Amsterdam: North-Holland, p. 257.

Elmegreen, B.G., and C.J. Lada. 1977. Astrophys. J. **214**:725.

Encrenaz, P.J., E. Falgarone, and R. Lucas. 1975. Astron. Astrophys. **44**:73.

Evans, N.J. 1978. *In* T. Gehrels (ed.), Protostars and Planets. Tuscon, University of Arizona Press.

Evans, N.J. 1984. Review paper presented at 3rd IAU Regional Astronomy Meeting, Buenos Aires, Dec. 1983.

Evans, N.J., G. Blair, P. Harvey, F. Israel, W. Peters, M. Scholtes, T. de Graauw, and P. Vanden Bout. 1981. Astrophys. J. **250**:200.

Evans, N.J., G. Blair, D. Nadeau, and P. Vanden Bout. 1982. Astrophys. J. **253**:115.

Federman, S.R., and R.F. Willson. 1982. Astrophys. J. **260**:124.

Field, G.B. 1978. *In* T. Gehrels (ed.), Protostars and Planets. Tuscon, University of Arizona Press.

Field, G.B. 1982. *In* A.E. Glassgold, P.J. Huggins, and E.L. Schucking (eds.), Symposium on the Orion Nebula to Honor Henry Draper. New York: New York Academy of Sciences.

Fitzgerald, M.P. 1968. Astrophys. J. **73**:983.

Fleck, R.C. 1981. Astrophys. J. (Lett.) **246**:L151.

Fleck, R.C. 1983. Astrophys. J. (Lett.) **272**:L45.

Frerking, M.A., W.D. Langer, and R.W. Wilson. 1982. Astrophys. J. **262**:590.

Gilden, D. 1984. Astrophys. J. **279**:335.

Goodman, A.A. 1987. *In* Wyoming Summer School on Interstellar Processes. Dordrecht: Reidel.

Harris, A., C.H. Townes, D.N. Matsakis, and P. Palmer. 1983, Astrophys. J. (Lett.) 265:L63.

Hartquist, T.W., and M.A. Snijders. 1982. Nature 299:783.

Heiles, C., and T. Troland. 1982. Astrophys. J. (Lett.) 260:L23.

Henriksen, R.N., and B.E. Turner. 1984. Astrophys. J. 287:200.

Herbst, W., and R.L. Dickman. 1983. Int. Astron. Union, Colloquium No. 76, p. 187.

Ho, P.T.P., and Haschick, A.D. 1986. Astrophys. J. 304:501.

Ho, P.T.P., R. Martin, and A.H. Barrett. 1981a. Astrophys. J. 246:761.

Ho, P.T.P., A.D. Haschick, and F.P. Israel. 1981b. Astrophys. J. 243:526.

Hollis, J.M. 1982. Astrophys. J. 260:159.

Hopper, P.B., and M.J. Disney. 1974. Mon. Not. R. Astron. Soc. 168:639.

Jenkins, E.B. 1978a. Astrophys. J. 219:845.

Jenkins, E.B. 1978b. Astrophys. J. 220:107.

Kaifu, N., S. Suzuki, T. Hasagawa, M. Morimoto, J. Inatani, K. Nagane, K. Miyazawa, Y. Chikada, T. Kanzawa, and K. Akabane. 1984. Astron. Astrophys. 134:7.

Kassim, M.A., and B.E. Turner. 1988. In preparation.

Kazes, I., and J. Crovisier. 1981. Astron. Astrophys. 101:401.

Knapp, G.R., and M. Jura. 1976. Astrophys. J. 209:782.

Kuiper, T.B.H., E.N. Rodriguez, D.F. Dickinson, B.E. Turner, and B. Zuckerman. 1984. Astrophys. J. 276:211.

Kwan, J. 1979. Astrophys. J. 229:566.

Lacy, J.H., S.C. Beck, and T.R. Geballe. 1982. Astrophys. J. 255:510.

Lane, A. 1984. Private communication.

Larson, R.B. 1981. Mon. Not. R. Astron. Soc. 194:809.

Larson, R.B. 1982. Mon. Not. R. Astron. Soc. 200:159.

Larson, R.B. 1985. Mon. Not. R. Astron. Soc. 214:379.

Lazareff, B. 1975. Astron. Astrophys. 42:25.

Leung, C., M. Kutner, and K. Mea. 1982. Astrophys. J. 262:583.

Levreault, R.M. 1983, Bull. Amer. Astron. Soc. 15, 679.

Lo, K.T., G. Berge, M. Claussen, G. Heiligman, J. Keene, C. Masson, T. Phillips, A. Sargent, N. Scoville, S. Scott, D. Watson, and D. Woody, 1985. In International Symposium on Millimeter and Submillimeter Wave Radio Astronomy, Granada, published by URSI.

Loren, R.B. and H.A. Wootten, 1986, Astrophys. J. 306: 142.

Lynds, B.T. 1962. Astrophys. J. (Suppl.) 7, 1.

Maddelena, R.J. 1985. Ph.D. thesis, Columbia University.

Maddelena, R.J. and P. Thaddeus. 1985. Astrophys. J. 294:231.

Maddelena, R.J., M. Morris, J. Moscowitz, and P. Thaddeus. 1986. Astrophys. J. 303:375.

Margulis, M., and C.J. Lada, 1986. Astrophys. J. (Lett.), 309: L87.

Monetti, A., J.L. Pipher, H.L. Helfer, R.S. McMillan, and M.L. Perry. 1984. Astrophys. J. 282:508.

Mouschovias, T.C. 1976a. Astrophys. J. 206:753.

Mouschovias, T.C. 1976b. Astrophys. J. 207:141.

Mouschovias, T.C. 1978. In T. Gehrels, (ed.), Protostars and Planets. Tuscon, University of Arizona Press.

Mouschovias, T.C. and S.A. Morton. 1985. Astrophys. J. 298:190.

Mundy, L. 1984. Ph.D. dissertation, University of Texas, Austin.

Mundy, L., N. Scoville, L. Baath, C. Masson, and D. Woody. 1986. Astrophys. J. (Lett.) 304:L51.

Myers, P.C. 1978. Astrophys. J. 225:380.

Myers, P.C. 1983. Astrophys. J. 270:105.

Myers, P.C. 1985a. In D. Black and M. Matthews (eds.), Protostars and Planets II. Tuscon, University of Arizona Press.

Myers, P.C. 1985b. In G. Serra (ed.), Lecture Notes in Physics. Springer-Verlag.

Myers, P.C. 1987. In Wyoming Summer School on Interstellar Processes. Dordrecht: Reidel.

Myers, P.C., and P. Benson. 1983. Astrophys. J. 266:309.

Myers, P.C., R.A. Linke, and P.J. Benson. 1983. Astrophys. J. 264:517.

Myers, P.C., T.M. Dame, P. Thaddeus, R.S. Cohen, R.F. Silverberg, E. Dwek, and M.G. Hauser. 1986. Astrophys. J. 301:398.

Norman, C.A., and J. Silk. 1980. In B.H. Andrew (ed.), Interstellar Molecules. Dordrecht: Reidel.

Nyman, L.-A. 1983. Astron. Astrophys. **120**:307.

Nyman, L.-A. 1984. Astron. Astrophys. **141**:323.

Olofsson, H., J. Ellder, A. Hjalmarson, and G. Rydbeck. 1982. Astron. Astrophys. **113**:L18.

Phillips, T.G., and P.J. Huggins. 1981. Astrophys. J. **251**:533.

Plambeck, R.L., M.C.H. Wright, W.J. Welch, J.H. Bieging, B. Baud, P.T.P. Ho, and S.N. Vogel. 1982. Astrophys. J. **259**:617.

Plambeck, R.L., S.N. Vogel, M.C.H. Wright, J.H. Bieging, and W.J. Welch. 1985. *In* International Symposium on Millimeter and Submillimeter Wave Radio Astronomy, Granada, published by Union Scientifique Radio Internationale (URSI).

Pumphrey, W., and J. Scalo. 1983. Astrophys. J. **269**:531.

Reid, M., A. Haschick, B. Burke, J. Moran, K. Johnston, and G. Swenson. 1980. Astrophys. J. **239**:89.

Rickard, L.J, B.E. Turner, and P. Palmer. 1985. Astron. J. **90**:1175.

Rowan-Robinson, M., P.E. Clegg, C.A. Beichman, G. Neugebauer, B.T. Soifer, H.H. Aumann, D.A. Beintema, N. Boggess, J.P. Emerson, T.N. Gautier, F.C. Gillett, M.G. Hauser, J.R. Houck, F.J. Low, and R.G. Walker. 1984. Astrophys. J. (Lett.) **278**:L7.

Sargent, A.I. 1979. Astrophys. J. **233**:163.

Sargent, A.I. 1985. *In* W. Boland, H. van Woerden, D.E. Reidel (eds.), The Birth and Evolution of Massive Stars and Stellar Groups Should say Groups.

Sargent, A.I., R.J. van Duinen, H.L. Nordh, C.V.M. Frilund, J.W.G. Aalders, and D. Beintema. 1983. Astron. J. **88**:1236.

Savage, B.D., R.C. Bohlin, J.F. Drake, and W. Budich. 1977. Astrophys. J. **216**:291.

Scalo, J., and W. Pumphrey. 1982. Astrophys. J. (Lett.) **258**:L29.

Schwartz, P.R. 1980. Astrophys. J. **238**:823.

Scoville, N.Z., S. Kleinman, D. Hall, and T. Ridgway. 1983. Astrophys. J. **275**:201.

Shu, F. 1985. *In* D. Black and M. Matthews (eds.), Protostars and Planets II. Tucson, University of Arizona Press.

Silk, J. 1985. *In* R. Lucas, A. Omont, and R. Stors (eds.), Birth and Infancy of Stars (1983 Les Houches Lectures). Amsterdam: North-Holland Press.

Snell, R.L. 1986. *In* M. Peimbert and J. Jugaku (eds.), IAU Symposium # 115 on Star Forming Regions, Dordrecht: Reidel.

Snell, R.L., R.B. Loren, and R. Plambeck. 1980. Astrophys. J. **239**:L17.

Snell, R.L., L.G. Mundy, P.F. Goldsmith, N.J. Evans, and N.R. Erickson. 1984. Astrophys. J. **276**:625.

Snell, R.L., F.P. Schloerb, J. Bally, and J. Morgan. 1986. Private communication.

Solomon, P.M., and D.B. Sanders. *In* P.M. Solomon and M.G. Edmunds (eds.), Giant Molecular Clouds in the Galaxy. New York: Pergamon Press.

Spitzer, L. 1978. Physical Processes in the Interstellar Medium. New York: J. Wiley & Sons.

Stahler, S. 1985. Astrophys. J. **293**:207.

Tielens, A.G., and D. Hollenbach. 1985. Astrophys. J. **291**:722.

Torrelles, J.M., L.F. Rodriguez, J. Canto, P. Carrol, J. Mercaide, J. Moran, and P.T.P. Ho. 1983. Astrophys. J. **274**:214.

Torrelles, J.M., P.T.P. Ho, L.F. Rodriguez, and J. Canto. 1985. Astrophys. J. **288**:595.

Troland, T.H., and Heiles, C., 1986. Astrophys. J. **301**:339.

Turner, B.E. 1974. *In* K.I. Kellermann and G.L. Verschuur (eds.), Galactic and Extragalactic Radio Astronomy. New York, Springer-Verlag.

Turner, B.E. 1978. W.B. Burton (ed.), The Large-Scale Characteristics of the Galaxy. Dordrecht: Reidel.

Turner, B.E. 1979. Astron. Astrophys. Suppl. **37**:1.

Turner, B.E. 1982. Astrophys. J. (Lett.) **255**:L33.

Turner, B.E. 1983. *In* W. Shuter (ed.), Kinematics, Dynamics and Structure of the Milky Way. Dordrecht: Reidel.

Turner, B.E. 1984. Vistas in Astronomy **27**:303.

Turner, B.E., and P. Thaddeus. 1977. Astrophys. J. **211**:755.

Vogel, S.N., and W.J. Welch. 1983. Astrophys. J. **269**:568.

Vrba, F.J. 1977. Astron. J. **82**:198.

Walker, C.K., C.J. Lada, E.T. Young, P.R. Maloney, and B.A. Wilking. 1986. Astrophys. J. (Lett.) **309**:L47

Wannier, P.G., A.A. Penzias, and E.B. Jenkins. 1982. Astrophys. J. **254**:100.
Wannier, P.G., S.M. Lichten, and M. Morris. 1983. Astrophys. J. **268**:727.
Watson, D.M. 1984. In ESLAB Symposium XVI, Galactic and Extragalactic Infrared Spectroscopy Dordrecht: Reidel, p. 195.
Werner, M.W., H.L. Dinerstein, and R.W. Capps. 1983. Astrophys. J. (Lett.) **265**:L13.
Wielen, R. 1978. In E.M. Berkhuijsen and R. Wielebinski (eds.), Structure and Properties of Nearby Galaxies, IAU Symposium No. 77. Dordrecht: Reidel.
Wilking, B.A., and C.J. Lada. 1983. Astrophys. J. **274**:698.
Wootten, H.A. 1985. In P.A. Shaver and K. Kjar (eds.), ESO-IRAM-Onsala Workshop on (Sub)millimeter Astronomy.
Wynn-Williams, C.G., R. Genzel, E.E. Becklin, and D. Downes. 1984. Astrophys. J. **281**:172.
Ziurys, L.M., and B.E. Turner. 1986. Astrophys. J. (Lett.) **300**:L19; also, work in progress.
Zuckerman, B. and N.J. Evans. 1974. Astrophys. J. **192**:L149.

5. Interstellar Molecules and Astrochemistry

BARRY E. TURNER and LUCY M. ZIURYS

5.1 Molecules in Space: An Overview

5.1.1 A Brief History

The quantitative study of the interstellar medium (ISM) began with Hartman's detection in 1904 of the resonance lines of Ca and K, seen in absorption against bright stars by intervening diffuse clouds. Other atoms (Na, Fe, Ti) were subsequently detected, although the overwhelming abundance of H was not recognized until the 1930s. In 1934 came the first discovery that was to suggest the possibility of interstellar molecules—the detection of four diffuse interstellar bands in the visible. Although these bands (now numbering three dozen) have never been identified, Russell in 1935 suggested that they were molecular in nature. In 1937 Swings proposed that the diffuse bands were CO_2 bands, and the same year Swings and Rosenfeld stressed the possibility of interstellar CH, OH, NH, CN, and C_2. In the years 1937 to 1941, the first discoveries of interstellar molecules were made at Mt. Wilson, of CH^+, CH, and CN by means of their optical absorption spectra in the 4230–4300-Å region (e.g., Adams 1941). However, the possibility of a rich chemistry in the ISM, as suggested by such detections, remained unrecognized at the time. Indeed, despite an extensive study by Adams (1949) of the optical spectra of these molecules towards many stars, there remained doubt as to whether these lines were interstellar or circumstellar. The 1951 work by Bates and Spitzer which concluded that the formation of even simple (diatomic) molecules in space was slow and that destruction processes by UV radiation were rapid further cast doubt on the notion of interstellar molecules, although the only formation mechanism considered was radiative association. In view of this study, it was difficult to understand the presence of the simple diatomics CH, CH^+, and CN, let alone more complicated species. Therefore, when both Townes and Shklovsky in the mid-1950s independently published lists of molecular lines of potential interest to radio astronomy, such lists were largely ignored. Even after the first detection of an interstellar molecule at radio wavelengths, that of OH by Weinreb et al. in 1963, the widespread presence of interstellar molecules was not recognized. It was not until two years later, after the observation of strong OH maser emission from HII regions (Weaver et al. 1965, Gundermann 1965), and particularly after the discovery of interstellar NH_3 and H_2O in 1968, that it was finally realized that gas densities were high enough to synthesize considerable abundances of complicated molecules, as well as to shield such molecules from the galactic UV field. With this revolutionized thinking, from 1968 onward, discoveries of new interstellar molecules have occurred rapidly.

Table 5.1. Known interstellar molecules.

No. of atoms	Chemical formula	Molecule	Wavelength region of observation	Type of transition	Objects where observed[a]					
					Diffuse clouds	Envelopes	Dark cloud cores	Cores of GMC's	Circumstellar envelopes	Extra-galactic
2	H_2	Hydrogen	UV	Electronic	✓	✓	✓	✓	✓	✓
			IR	Rovibrational						✓
	CO	Carbon monoxide	UV	Electronic	✓	✓	✓	✓	✓	✓
			IR	Rovibrational, rotational						
			Radio/mm	Rotational			✓	✓	✓	
	CH	Methylidyne	Optical	Electronic	✓	✓	✓	✓		✓
			IR	Rotational						
			Radio/cm	Λ-Doubling						
	OH	Hydroxyl radical	UV	Electronic	✓	✓	✓	✓	✓	✓
			IR	Rotational		✓			✓	
			Radio/cm	Λ-Doubling		✓			✓	
	CH^+	Methylidyne ion	Optical	Electronic	✓					
	C_2	Dicarbide	Optical	Electronic	✓				✓	
	CN	Cyanogen radical	Optical	Electronic*	✓	✓	✓	✓	✓	
			Radio/mm	Rotational						✓
	CS	Carbon monosulfide	IR	Rovibrational		✓				
			Radio/mm	Rotational			✓	✓	✓	
	NO	Nitric oxide	Radio/mm	Rotational				✓		
	NS	Nitrogen sulfide	Radio/mm	Rotational				✓		
	SO	Sulfur monoxide	Radio/mm	Rotational			✓	✓	✓	
	SiO	Silicon monoxide	IR	Rovibrational				✓	✓	
			Radio/mm	Rotational			✓	✓	✓	
	SiS	Silicon sulfide	Radio/mm	Rotational				✓	✓	
	HCl	Hydrogen chloride	Sub-mm	Rotational				✓		
3	HCN	Hydrogen cyanide	IR	Rovibrational		✓	✓	✓	✓	✓
	HNC	Hydrogen isocyanide	Radio/mm	Rotational			✓	✓	✓	

Continued

201

Table 5.1. Known interstellar molecules. (*Continued.*)

No. of atoms	Chemical formula	Molecule	Wavelength region of observation	Type of transition	Diffuse clouds	Envelopes	Dark cloud cores	Cores of GMC's	Circumstellar envelopes	Extra-galactic
					Objects where observed[a]					
	H_2O	Water	Radio/mm, cm	Rotational					✓	✓
	CCH	Ethynyl radical	Radio/mm	Rotational		√?	✓	✓	✓	
	HCO	Formyl radical	Radio/mm	Rotational			✓	✓	✓	
	OCS	Carbonyl sulfide	Radio/mm	Rotational				✓	✓	
	H_2S	Hydrogen sulfide	Radio/mm	Rotational			✓	✓	✓✓	
	SO_2	Sulfur dioxide	Radio/mm	Rotational				✓	✓	
	SiC_2	Silicon dicarbide	Radio/mm	Rotational				✓	✓	
	HCO^+	Formyl ion	Radio/mm	Rotational		✓	✓	✓		✓
	N_2H^+	Protonated nitrogen	Radio/mm	Rotational		✓	✓✓	✓		
	HCS^+	Thioformyl ion	Radio/mm	Rotational			✓	✓		
	H_2D^+	—	Sub-mm	Rotational						
4	NH_3	Ammonia	IR	Rotational		✓	✓	✓	✓	
			Radio/mm, cm	Inversion						
	H_2CO	Formaldehyde	Radio/mm, cm	Rotational		✓	✓	✓		✓
	HCCH	Acetylene	IR	Rovibrational			✓	✓	✓	✓
	H_2CS	Thioformaldehyde	Radio/mm, cm	Rotational			✓	✓		
	HNCO	Isocyanic acid	Radio/mm	Rotational			✓✓	✓		
	HNCS	Thioisocyanic acid	Radio/mm	Rotational			✓✓	✓✓		
	C_3N	Cyanoethynyl radical	Radio/mm	Rotational			✓✓	✓	✓	
	C_3H	Propynyl radical	Radio/mm	Rotational			✓✓	✓✓	✓✓	
	C_3O	Tricarbon monoxide	Radio/cm	Rotational			✓	✓		
	$HOCO^+$	Protonated carbon dioxide	Radio/mm	Rotational				✓		
	$HCNH^+$	Protonated hydrogen cyanide	Radio/mm	Rotational				✓		
5	CH_4	Methane	IR	Rovibrational			✓			
	HC_3N	Cyanoacetylene	Radio/mm, cm	Rotational			✓	✓	✓✓	
	HCOOH	Formic acid	Radio/mm, cm	Rotational				✓	✓	
	H_2CCO	Ketene	Radio/mm	Rotational			✓	✓		

	Formula	Name	Transition	Wavelength
	NH_2CN	Cyanamide	Rotational	Radio/mm
	CH_2NH	Methanimine	Rotational	Radio/mm
	C_4H	Butadiynyl radical	Rotational	Radio/mm
	SiH_4	Silane	IR	IR
	C_3H_2	Cyclopropenylidene	Rovibrational	Radio/mm, cm
6	CH_3OH	Methanol	Rotational	Radio/mm, cm
	CH_3CN	Acetonitrile (methyl cyanide)	Rotational	Radio/mm
	CH_3SH	Methyl mercaptan	Rotational	Radio/mm
	C_2H_4	Ethylene	Rovibrational	IR
	NH_2HCO	Formamide	Rotational	Radio/mm, cm
	C_5H	Pentynylidyne radical	Rotational	Radio/mm
7	CH_3CCH	Methyl acetylene	Rotational	Radio/mm, cm
	CH_3CHO	Acetaldehyde	Rotational	Radio/mm, cm
	CH_2CHCN	Vinyl cyanide	Rotational	Radio/mm
	CH_3NH_2	Methylamine	Rotational	Radio/mm
	HC_5N	Cyanobutadiyne	Rotational	Radio/mm, cm
	C_6H	Hexatriynyl radical	Rotational	Radio/mm
8	CH_3C_3N	Methylcyanoacetylene	Rotational	Radio/mm
	$HCOOCH_3$	Methyl formate	Rotational	Radio/mm, cm
9	CH_3CH_2OH	Ethanol	Rotational	Radio/mm
	CH_3CH_2CN	Ethyl cyanide	Rotational	Radio/mm
	CH_3OCH_3	Dimethyl ether	Rotational	Radio/mm
	CH_3C_4H	Methyl diacetylene	Rotational	Radio/mm
	HC_7N	Cyanohexatriyne	Rotational	Radio/mm, cm
11	HC_9N	Cyanooctatetrayne	Rotational	Radio/cm
13	$HC_{11}N$	Cyanodecapentayne	Rotational	Radio/cm

ᵃIndicates towards what types of objects a species has presently been observed. A blank in any column may only mean that there are no sensitive observations of a species towards such objects at present, not necessarily that the molecule doesn't exist there.

5.1.2 The Presently Known Interstellar Molecules

Table 5.1 lists the sixty-eight interstellar molecules presently accepted as identified in the ISM. Also shown in this table are the types of transitions and wavelengths by which the species have been detected, as well as the regions where they have been observed. As is evident from this list, a wide variety of molecules are present in the ISM, ranging from simple diatomics to complex species with as many as thirteen atoms. Furthermore, in addition to these known molecules, many unidentified lines (U-lines) have been observed in interstellar sources, corresponding to as yet unknown molecular species.

Patterns with respect to chemical type are not easy to generalize from Table 5.1, particularly because techniques for the detection of interstellar molecules embody strong selection effects. All but nine of the sixty-eight species have been initially detected via their rotational transitions that occur at radio/millimeter wavelengths. As Table 5.1 shows, seven remain identified solely on the basis of electronic and/or rovibrational transitions that occur at UV/optical/IR frequencies.

There are several reasons why most of the molecules have been detected by their rotational lines at radio and millimeter wavelengths. First, the generally low-temperature, low-density environment of most molecular sources overwhelmingly favors the population of the rotational energy levels relative to vibrational or electronic states, and the rotational spectra of most common molecules, except for diatomic hydrides, occur in the radio and millimeterwave regions. Instrumental sensitivity and the presence of obscuring atmospheric lines also favors detections at radio/millimeter wavelengths. In addition, the physical conditions of molecular clouds result in generally narrow line widths for molecular spectra, the principal broadening mechanisms often being only turbulent motions of a few kilometers per second. Thus, molecular hyperfine and fine structure, Λ-doubling, and other small rotational splittings can easily be resolved. High-resolution rotational spectra are therefore readily obtained; such spectra constitute "fingerprint patterns" by which individual species are uniquely identified. Such a unique pattern is shown in Figure 5.1, that of the $(J, K) = (1, 1)$ inversion transition of NH_3. Accurate rest frequencies derived from laboratory measurements are thus extremely important for identifying new molecules.

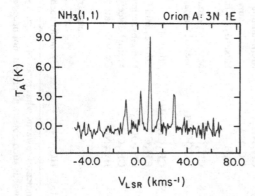

Fig. 5.1. The $(J, K) = (1, 1)$ inversion transition of metastable ammonia, observed toward the $(3N, 1E)$ position in Orion. The line is split into five hyperfine components, which result from the quadrupole interaction of the nitrogen nuclear spin with the molecular rotation. The resulting pattern clearly identifies the spectrum as arising from NH_3.

Because rotational spectra are chiefly involved, certain other selection effects determine what species are detected. A simple linear molecule with only a few populated energy levels will produce stronger lines than a complicated asymmetric top in which the population is spread over hundreds of closely spaced levels. In addition, molecules in general must have a permanent electric dipole moment in order to be detected. Thus, symmetric-top species such as CH_4 and homonuclear molecules such as O_2 and N_2, which are probably fairly abundant in interstellar gas, cannot be observed because they have no dipole moment and thus possess no pure rotational spectrum. Such species can be detected via electronic or rovibrational lines. However, in general, these transitions are not excited in the molecular sources.

In addition to these observational selection effects, detections of new interstellar species rely heavily on the availability of known rest frequencies. Thus, certain types of molecules such as free radicals and molecular ions, the so-called "transient species," may appear to be much less common in the ISM than stable, closed-shell organic molecules, whereas, in fact, these "transient" molecules are extremely difficult to create in the laboratory for the purposes of obtaining rest frequencies. Such is also the case for many silicon- and metal- (in the chemist's sense) containing compounds. Thirteen of the twenty-three "transient" species listed in Table 5.1 have, in fact, been identified without the availability of accurate rest frequencies. Their identification was based on a combination of chemical intuition, quantum-mechanical calculations of approximate molecular constants, and the observation of distinct patterns of lines in molecular sources. Such patterns often corresponded to hyperfine/fine-structure splittings, on the basic of which a molecule could be identified, in conjunction with rough estimates of the splittings from optical and electron spin resonance/argon-matrix laboratory spectra (e.g., Tucker et al. 1974).

Still, two important facts, evident from the species listed in Table 5.1, are largely independent of any selection effects. First, most interstellar molecules are organic, such as is the case on Earth. None of the inorganic species except NH_3 contain more than three atoms, so as on Earth, the carbon bond is apparently the key to synthesis of complex molecules. Second, although many of the species are common "terrestrial" substances such as H_2O, HCOOH, and EtOH, a large fraction of them are extremely reactive "transient" molecules, i.e., free radicals and molecular ions that have only recently been synthesized in the laboratory as a result of their identification in interstellar space. In view of the difficulty of measuring rest frequencies for such species, it is remarkable that so many of them have been identified in the ISM. One of these transient molecules, C_3H_2, is the only ring compound as yet observed in interstellar gas. [The only other ring compound, SiC_2, is seen solely in a circumstellar gas shell, whose chemistry appears quite distinct from interstellar chemistry (Section 5.2)]. Another interesting class of interstellar molecules are the long-chain cyano-polyacetylenes HC_3N, HC_5H, HC_7N, etc., which are rather abundant in certain types of molecular clouds but are not naturally occurring on Earth and can only be made in the terrestrial laboratory by sophisticated synthetic techniques. The existence of these and other species makes it clear that the chemistry of the ISM is not in thermochemical equilibrium. In addition, the molecular species in Table 5.1 are far from those that characterize a highly reducing (hydrogen-rich) chemical environment under conditions of chemical equilibrium at any temperature.

A few more interesting points can be deduced from the list in Table 5.1. First, the NO bond appears to be scarce in comparison with other organic functional groups, being represented only by the NO molecule itself. Second, ring molecules are strongly underrepresented relative to long-chain species. Although C_3H_2, a three-member ring, has been identified, many other three-member rings, and several higher-member rings, have been searched without success, to comparable levels of abundance. Third, no branched-chain species have been identified, despite searches for a few, although these species have large partition function. Fourth, molecules containing second- and third-row elements are not particularly common, although cosmic abundances of such elements as silicon and magnesium are comparable to that of sulfur, and only one order of magnitude less than that of carbon. These facts are discussed in more detail in Section 5.5.

The patterns of chemical species with respect to type of source (Table 5.1) are also subject to strong selection effects. Only recently, as a result of concerted effort, has it become apparent that the chemistry of dark cloud cores is almost as rich as that of the hot GMC cores, and the gap continues to narrow. Nonetheless, Si species and several of the more complex species seem to be absent in dark cloud cores. The apparent absence of all but the simplest species in diffuse clouds and in cloud "envelopes" is likely mostly a result of smaller column depths and of limited excitation of even the millimeter-wavelength rotational transitions. Also, concerted searches for molecules are lacking in these objects, as they are for extragalactic objects.

5.1.3 Interstellar Chemistry: Some Generalities

The wide variety of known species, together with the widely disparate nature of the regions in which they are found, indicates that their chemistry must be highly complex; the formation and destruction of molecules can involve atomic ions, free radicals, molecular ions, and stable species, as well as cosmic rays, energy from shocks, and radiation fields, all involving both gas-phase and solid-state (interstellar grain) processes.

Figure 5.2 illustrates schematically three general ways in which interstellar molecules can, in principle, form. (The dashed pathways indicate unlikely processes.) The simplest idea (Sequence 1) is that molecules are formed "in situ" in molecular clouds. As diffuse material coalesces into a cloud, densities will build up such that two-body collisions readily occur. Then gas-phase reactions and/or reactions on the surfaces of grains can take place, leading to the synthesis of diatomic species, which in turn react to form more complicated molecules. The large molecules may eventually accrete to form further interstellar grains, but this process is highly uncertain.

Another possibility is illustrated by Sequence 2. Here, molecules are formed in the high-density, high-temperature region of young protostars, stars, and pre-solar nebulae embedded in dense clouds. In such regions the densities and temperatures are high enough so that many-body gas-phase reactions can occur and thermo-

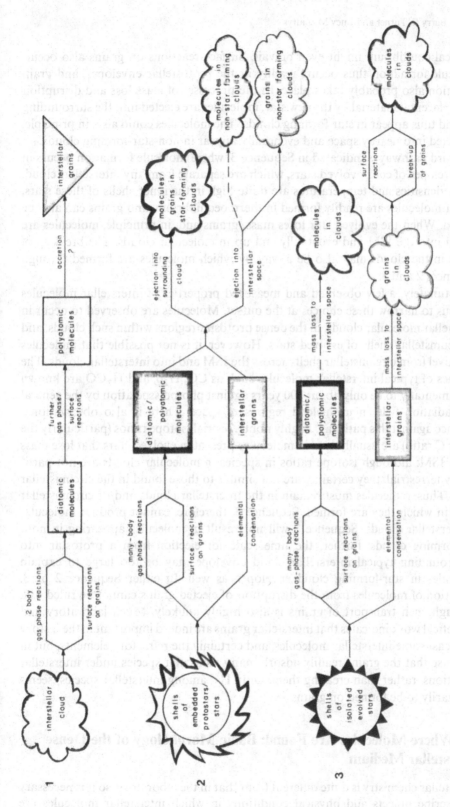

Fig. 5.2. Possible scenarios for molecule formation in the interstellar medium.

207

chemical equilibrium might even prevail. Surface reactions on grains also occur. Molecule formation thus occurs readily in the protostellar envelopes, and grain formation also probably takes place. In the processes of mass loss and disruption of the placental material by the new star, molecules are ejected into the surrounding gas, and thus appear in star-forming clouds. The molecules could also, in principle, be ejected into nearby space and eventually appear in non-star-forming clouds.

A third pathway is indicated in Sequence 3, where molecule formation occurs in the envelopes of cool, evolved stars, which are separate from any interstellar cloud. Again, densities and temperatures are quite high in the inner shells of these stars, so that molecules are readily formed in thermoequilibrium, and grains can also be created. When the evolved star loses mass, grains and, in principle, molecules are ejected into the ISM and eventually end up in molecular clouds. The breakup of grains in the clouds may also be a way by which molecules are formed through Sequences 2 and 3.

Fortunately, a few observed and measured properties of interstellar molecules allow us to narrow these choices at the outset. Molecules are observed to occur in interstellar molecular clouds, in the dense protostar regions within such clouds, and in circumstellar shells of evolved stars. However, it is not possible that molecules can travel from circumstellar shells across the ISM and into interstellar clouds. The lifetimes of typical interstellar molecules such as CO, H_2O, and H_2CO are known experimentally to be only about 100 years against photodissociation by the general UV radiation field in unshielded regions of space. There is also observational evidence against this pathway. Highly nonterrestrial isotope ratios (particularly the $^{12}C/^{13}C$ ratio) are usually seen in molecules present in evolved stars that lose mass to the ISM; although isotope ratios in species in molecular clouds are not particularly terrestrial, they certainly are not similar to those found in the circumstellar shells. Thus, molecules must remain in the interstellar clouds and/or circumstellar shells in which they are formed. Sequence 3, therefore, cannot produce molecules in interstellar clouds. Sequence 2 will not result in molecules appearing in non-star-forming clouds; further, the time scale for ejection from a protostar into a surrounding typical interstellar cloud envelope may be too large to explain molecules in star-forming cloud envelopes as well. In either Sequence 2 or 3, formation of molecules from the disruption of ejected grains cannot be ruled out, although such transport of grains is also highly unlikely. Much laboratory and theoretical work indicates that interstellar grains are indeed important in the history of at least some interstellar molecules, and certainly the refractory elements, but in the sense that the grains readily adsorb many molecular species under interstellar conditions, rather than creating them. Only H_2, among interstellar species, seems necessarily to be formed on grains.

5.2 Where Molecules Are Found: Basic Morphology of the Dense Interstellar Medium

Interstellar chemistry is quite different from that in the laboratory, so it is necessary to describe objects and physical conditions in which interstellar molecules are formed. Molecules are synthesized in a variety of interstellar regimes, namely, diffuse clouds, cloud envelopes, cold cloud cores, warm cores in Giant Molecular Clouds, circumstellar shells, and shocked regions. Figure 5.3 shows representative spectra from each of these regions and emphasizes the fundamental differences that characterize them.

5.2.1 Diffuse Clouds

These objects are defined for present purposes as clouds whose column densities are $N(\text{HI} + \text{H}_2) \lesssim 10^{19}$ cm^{-2}, which is the threshold at which molecules are first observed. Examples of diffuse clouds are ζ Oph, χ Per, and α Cam. Here, $N(\text{CO}) \lesssim 10^{15}$ cm^{-2} is observed optically (Federman and Willson 1982). Such column densities correspond to extinctions $A_v \ll 1^m$ so that UV radiation processes are highly important in the chemistry of these clouds. Independent evidence suggests a density threshold $n < 10^2$ cm^{-3} for onset of molecules. Diffuse cloud densities range from ten to several hundred cm^{-2}. Since typical sizes of diffuse clouds are 1 to 3 pc, it is probable that the higher-density regions are restricted to rather small cores ($\lesssim 0.1$ pc), which are surrounded by extensive envelopes of primarily atomic gas. Temperatures of diffuse clouds are usually 20 to 100 K (cf. Snow 1980). The heating of the gas is thought to occur via its coupling to dust grains, which are warmed by incident UV/visible radiation (Jura 1976, Watson 1976) and via the photoejection of electrons from the grains by the same radiation. Cooling is believed to occur principally via emission in the fine-structure lines of C$^+$, carbon being completely ionized and essentially undepleted onto grain surfaces (e.g., Hobbs et al. 1982), unlike many heavier elements in these clouds (e.g., Morton 1974). The balance of these heating and cooling mechanisms is predicted to result in a $T \propto n^{-1}$ law for these regions, as is observed (Lazareff 1975). The observed electron fractional abundance $x_e \equiv n_e/n$ in these clouds is 3×10^{-4}, the result of ionization of C as well as of Fe, Mg, and Si, and some H.

Such physical conditions mean that the gas in diffuse clouds is chiefly atomic. Only simple molecules (diatomics) are observed, as might be expected both because the densities are relatively low (chemical reactions are infrequent) and because many diatomics have longer lifetimes against photodissociation than polyatomic species. Thus, CH, CH$^+$, OH, CN, and C$_2$ are common constituents of diffuse clouds. These species have been studied almost entirely via their optical/UV electronic spectra which, because the extinction is so low, are readily observed in absorption against bright background stars.

Recently OH, CO, and H$_2$CO have been observed at radio wavelengths in high-galactic-latitude IRAS "cirrus" regions (Magnani et al. 1985, Magnani 1986) and suggest conditions in these regions are very similar to those of the denser parts of diffuse clouds.

Because photodissociation is the only significant destruction mechanism, the molecular chemistry is particularly simple in diffuse clouds.

Apart from photodissociation and ionization processes, the formation of at least some simple molecules (CN: Federman et al. 1984; CH: Federman and Willson 1982) seems to involve only neutral-neutral and ion-molecule reactions between at most diatomic species. Federman (1986) finds that $N(\text{CH}) \propto N(\text{H}_2)$ whereas $N(\text{C}_2) \propto N^2(\text{H}_2)$ and $N(\text{CN}) \propto N^3(\text{H}_2)$. Thus, CH forms in the first stage of reactions between atomic species, while C$_2$ is formed only after two stages, and CN after

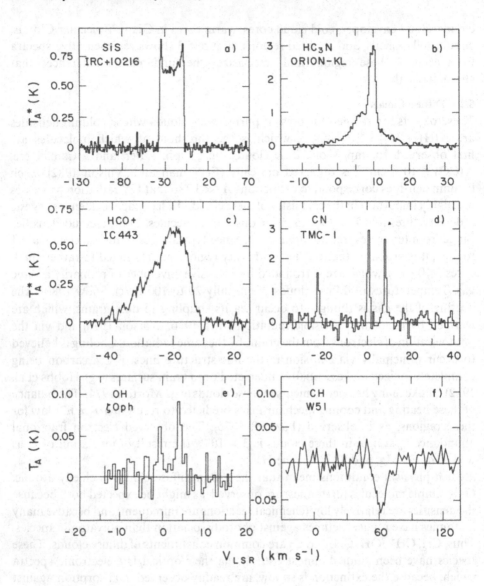

Fig. 5.3. Representative spectra observed toward various types of molecular sources. Spectrum (a) shows the $J = 5-4$ rotational transition of SiS near 90 GHz, arising from the circumstellar envelope of the late-type carbon star IRC + 10216. The nearly square-shaped line profile, topped with "cusps," is characteristic of optically thin transitions originating from outflowing circumstellar shells. Spectrum (b) is the $J = 12-11$ rotational line of HC_3N, observed at a frequency of 110 GHz toward Orion-KL, an active molecular cloud core region where shocked gas and outflows are present (Goldsmith et al., 1986). Three separate velocity components are visible in the HC_3N spectrum: (i) the narrow ($\Delta V_{1/2} \sim 4$ km s^{-1}) "spike" component, at an LSR velocity of 9 km s^{-1} and the most prominent feature in the spectrum; (ii) the broader "hot core" component, which appears as a "shoulder" to the left of the "spike" and has $\Delta V_{1/2} \sim 10$ km s^{-1} and $V_{LSR} \cong 7$ km s^{-1}; and (iii) the "plateau" or "doughnut" feature, which is the very broad underlying "base" of the whole line profile, extending out to ± 15 km s^{-1} with $V_{LSR} \cong 8$ km s^{-1}. These three velocity components are observed in many other molecules toward KL and represent distinct kinematic (and chemical) regions (see Section 5.2). Spectrum (c) is that of the $J = 1-0$ transition of HCO$^+$ at 89 GHz observed toward SNR IC 443 (from DeNoyer and Frerking, 1981). The asymmetric line profile

three stages. CH^+ does not reveal so simple a chemistry. Shocks seem to be involved in its formation (Section 5.4), as suggested both by systematic difference in the velocities of CH^+ relative to other species (Crutcher and Watson 1976) and by the fact that $N(CH^+)$ correlates with the column density of highly rotationally excited H_2, indicating possibly high temperatures. The occurrence of shocks is deduced in several diffuse clouds not only by the presence of velocity shifts, but also by the existence of some cloud cores which have n as high as 10^3 cm^{-3} but $N(H_2) \leqslant 10^{18}$ cm^{-2}. At such high n, H_2 becomes strongly self-shielding and large amounts of H_2 are expected. The high-n, low-$N(H_2)$ combination appears to require passage of a 10- to 15-km s^{-1} shock (from a nearby HII region or supernova remnant).

5.2.2 Envelopes of GMCs and SMCs ("Dark Clouds")

Giant Molecular Clouds (GMCs) and Small Molecular Clouds (SMCs), often termed "dark clouds" (cf. Myers 1985), are generally opaque in the visual ($A_v > 3^m$). Historically, they were discovered via the $J = 1$–0 millimeter-wave transition of CO, and typical column densities in directions not intercepting dense cores may be taken as $N(CO) \sim 10^{18}$ cm^{-2}, or $N(H_2) \sim 10^{22}$ to 10^{23} cm^{-2}. From the chemical standpoint, we may define molecular cloud envelopes as being opaque to UV radiation, although photoejection heating of grains in the envelopes of GMCs appears sufficient to elevate the temperature somewhat ($T \sim 20$ K; e.g., Blitz and Thaddeus 1980) above the "base" level of 10 K that characterizes dense cold molecular cores. Envelopes of molecular clouds have not been extensively studied except in CO, but the limited information suggests the molecular species are similar to those in diffuse clouds. Most important tracers of envelopes are OH and CH, in addition to CO; OH studies suggest $10^2 \leqslant n \leqslant 10^3$ cm^{-3} and $15 \leqslant T \leqslant 50$ K for molecular clouds located in spiral arms, but $T \leqslant 15$ K for clouds in interarm regions (Turner 1982, 1983). Neutral atomic carbon (observed via its fine-structure transitions at 370 and 610 μm) most decisively points to a different chemistry in molecular cloud envelopes and in diffuse clouds. CI is observed throughout envelopes (and apparently in dense cores as well) but is not seen in diffuse clouds (Phillips and Huggins 1981). Clearly, the distinction is due to the ionizing UV fields found in diffuse clouds. Slightly more species may also be seen in molecular cloud envelopes: for example, the molecules HCO^+, N_2H^+, HCN, C_2H, and H_2CO are observed to be widespread in certain molecular cloud complexes. Whether they are

---

results from outflows associated with shocked gas (see Section 5.2). Spectrum (d) shows three main hyperfine lines ($J = \frac{3}{2}$–$\frac{1}{2}$, $F = \frac{5}{2}$–$\frac{3}{2}$, $F = \frac{3}{2}$–$\frac{1}{2}$, $F = \frac{1}{2}$–$\frac{1}{2}$) of the $N = 1$–0 rotational transition of CN at 113 GHz, observed toward the dark cloud core of TMC-1. The extremely narrow lines, which allow resolution of the two stronger hyperfine components, separated by less than 3 MHz, are typical of dark cloud cores. Spectrum (e) is that of the main hyperfine line ($J = \frac{3}{2}$, $F = 2$–2) of the ground state Λ-doubling transition of OH, at a wavelength of 18 cm, taken toward the diffuse cloud ζ Oph. The two different velocity components apparent in the profile are thought to arise from separate regions, one with shocked gas and the other with unshocked material (Crutcher, 1979). Spectrum (f) shows the $F = 2$–2 hyperfine transition of the first rotationally excited Λ-doublet (F_1, $N = 1$, $J = \frac{3}{2}$) of CH, observed in absorption toward W51, another active giant molecular cloud. This transition probably originates in hot, dense gas (see Section 5.4). It appears as absorption rather than emission because of the presence of strong continuum background radiation from the HII region associated with the molecular cloud (Ziurys and Turner, 1985).

as widespread as CO is difficult to determine, as they require higher densities for their excitation. The physical distinctions between envelopes and diffuse clouds may not be well enough defined to ascertain chemical differences at this level of complexity.

Apart from the diminished role of UV in molecular cloud envelopes, there is little to distinguish these envelopes physically from diffuse clouds. Indeed, Combes et al. (1980) find "transition" clouds with properties intermediate between those of diffuse clouds and envelopes [e.g., $N(CO) \sim 3 \times 10^{17}$ to 5×10^{17} cm^{-2}, $300 \leqslant n \leqslant 500$ cm^{-3}, $T \approx 20$ K]. Chemically, the distinction is also small, at least for diatomic molecules. Similar values of density (10^2 to 10^3 cm^{-3}) and similar velocities explain both optical observations of CH and OH in diffuse clouds and radio observations of the same species in cloud envelopes. The $N(CH)$ vs. A_v relationship is observed to be continuous across the diffuse cloud/cloud-envelope boundary, as is predicted if CH is formed by $C^+ + H_2 \rightarrow CH_2^+ + h\nu$ and destroyed by C^+ and by photodissociation. Thus, Federman (1981, 1982) proposes that diffuse clouds are the outer regions of dark molecular cloud envelopes, the diffuse regions being mostly atomic while the deeper envelopes are mostly molecular. However, current counts of dark molecular clouds indicate there are more dark clouds than diffuse clouds, casting doubt on the notion that *all* molecular clouds have extensive atomic outer regions. While there is a general chemical similarity across a range of (n, T) for diffuse clouds and cloud envelopes, the differing role of UV radiation remains important to the chemistry, and we shall keep the two regimes formally distinct.

5.2.3 Cold Cores and Low-Mass Star-Forming Cores in Molecular Clouds

As discussed in detail in Chapter 4 of this volume, both SMCs (commonly known as "dark clouds," and here defined as clouds of $\lesssim 10^3\ M_\odot$ and not associated with OB star formation) and GMCs ($\gtrsim 10^3\ M_\odot$ and usually having associated OB star formation) contain many dense cores which are the sites of star formation. SMC cores, which form only low-mass stars (later than $\sim$A0), are always of modest mass and are never heated much above the "base" temperature of ~ 10 K that characterizes opaque molecular clouds. GMCs, by contrast, form both low-mass and high-mass cores and stars, the massive cores being heated to high temperatures by embedded massive protostars. In this section, we discuss the low-mass cores that occur in SMCs and probably (though not observationally proven) in GMCs.

A few low-mass cores have been recognized and studied for many years in nearby dark clouds such as Taurus and Ophiuchus. Only relatively recently has a survey of ~ 100 such cores by Myers and colleagues (Myers et al. 1983, Myers and Benson 1983, Myers 1983) served to characterize their properties by studies of CO, NH$_3$, HC$_3$N, and HC$_5$N. These cores have temperatures in the range 10 to 20 K, densities of 10^4 to 10^5 cm^{-3}, and sizes of 0.05 to 0.2 pc. Masses range from 0.3 to 10 $M_\odot$. About half of the studied sample contains embedded or nearby low-mass stars, as shown by IRAS observations. Such stars appear to have no observable effect upon the temperatures within the cores, and only a tendency to increase the turbulence of the core gas in some cases; the turbulence is always small, and line widths are often near thermal widths.

These cores seem to comprise a chemically rather homogeneous group in the few extensively studied species. More recent studies of additional species (C_2H, CH_3CN, HCN, HNC, SO, and SO_2; e.g., Irvine et al. 1985) give a similar conclusion. There do appear to be differences in relative abundances among these species from core to core, however. Variations by factors as large as ~ 10 in HC_3N and HC_5N have been noted (Benson and Myers 1983). Similar variations of HC_3N relative to NH_3 have been reported in TMC-1 (Little et al. 1979). In TMC-1, map peaks of NH_3 and HC_3N are displaced by several arcminutes, although no such shift is seen in L1544 or other TMC cores. Differences in map shapes between NH_3, SO_2, and CS occur in L183.

Several nearby SMCs (Taurus, L134, L183, and ρ Oph) contain particularly cold, dense cores, which may represent an earlier evolutionary phase than the protostellar phase. Temperatures of these cores are ~ 10 K or slightly lower in some cases. These are the coldest, densest objects known in the ISM. They also contain the lowest fractional ionization ($< 10^{-8}$) of any known astrophysical regions. Such cores contain the same chemical species as do the slightly warmer cores with two exceptions. (1) The concentration of deuterium in several molecules in these cold cores is extreme, typically $XD/XH \approx 10^{-2}$ to 10^{-1} while $D/H \approx 2 \times 10^{-5}$ cosmically. (2) The remarkable long-chain cyanopolyyne species (HC_7N and higher order) have much higher fractional abundance in cold cores, and the higher-order ones are seen only in such cores, among interstellar as distinct from circumstellar regions. These unique chemical properties seem to be the chemical signature of these extremely cold, dense systems. The first is a prediction of ion-molecule chemistry (Section 5.4). The second has no accepted explanation as yet.

The chemical situation among the cold, low-mass cores in general is complicated, however. Consider the three best-known cores: TMC-1 and 2, L183, and the ρ Oph cores (especially B1 and B2). All show extreme deuteration. Yet the TMC and L183 cores are truly cold, $T \sim 10$ K, while the ρ Oph cores are only cold ($T \sim 15$ K) relative to the rest of the ρ Oph cloud at $T \sim 25$ K. The ρ Oph cores are denser ($n \sim 3 \times 10^5$ cm^{-3}) than the TMC and L183 cores ($n \sim 3 \times 10^4$ cm^{-3}). Further, the B1 and B2 cores in ρ Oph have very different chemical properties, even though (T, n) are the same for each and they lie only 0.25 pc apart (Wootten and Loren 1986). Only B1 shows strong deuteration. Only B2 shows significant SO and SO_2, species usually associated with warm molecular cores. Finally, the L183 core, though cold, resembles the ρ Oph B2 core in exhibiting SO_2, a species which is absent in the cold TMC cores (Irvine and Hjalmarson 1983). All three cores do contain SO, however.

Certain molecular species are conspicuously absent in all cold molecular cores, namely, Si-containing species (SiO, SiS) and more complex species such as EtOH, EtCN, and $(CH_3)_2O$.

The observed variations in abundances among the very coldest cores are at least as great as any differences that occur between these cores and low-mass star-forming cores. Thus, the chemistry is not obviously affected by an increase in temperature from 10 K to 15–20 K. It is as likely that the Taurus cloud chemistry differs from that of other very cold cores as it is that the chemistry shows any trends over the 10–20 K range, or any dependence on the presence or absence of low-mass protostars. For these reasons, we have combined all cold low-mass cores together as one category in Table 5.1.

5.2.4 Warm Massive Cores in GMCs

Warm massive cores form massive (OB) stars and occur only in GMCs. These cores are subject to strong heating by the embedded massive stars or protostars and often to other powerful energy sources (shocks, high-velocity outflows, etc.) which accompany the birth of massive stars. They are a highly diverse class of objects compared with the low-mass cores. Temperatures range from ~ 25 to 2000 K in their inner regions, and densities from 10^4 to $\gtrsim 10^{10}$ cm^{-3}; line widths are always representative of highly supersonic turbulence and in some cases indicate powerful outflows with speeds up to 100 km s^{-1}. The upper ranges of (n, T) are limited only by current observational resolution.

The majority of massive cores have outflows, and seem to fall between two extrema (Wootten 1986). At one extreme are those flows which give rise to broad ^{12}CO lines but which are not as yet observed to have broadened lines of species such as ^{13}CO, HCO$^+$, or CS. Vibrationally excited H$_2$ has not been observed in these sources. The flows in these objects may possibly have broken through the inner, dense cores and are now interacting with the less dense surrounding gas, gas whose column densities are too small to produce observable broad lines of ^{13}CO, HCO$^+$, or CS. The second extreme comprises objects with flows manifested by relatively strong broad lines of all of these species, and also by vibrationally excited H$_2$, which indicates gas at $\gtrsim 2000$ K. In these objects, the flows may still be confined to the inner, dense core, where column densities are high. The properties of warm cores vary smoothly between these extremes for the class of known objects. More sensitive observations may eventually reveal weak broad wings in ^{13}CO, HCO$^+$, and CS in those objects which presently do not show them, and perhaps even vibrationally excited H$_2$.

The temperatures of the core regions are not well correlated with the properties just discussed. As deduced from both IR and molecular studies, temperatures range from only modestly elevated (~ 25 to 100 K) in the majority of warm cores, to several hundred K in the core of Ori(KL). [In Ori(KL), the $\gtrsim 2000$ K gas indicated by vibrationally excited H$_2$ occurs outside the immediate core, where the flow interacts with the surrounding medium. We discuss this very-high-temperature region under the subject of shocks (Section 5.2.5).]

The modestly heated cores (e.g., Ceph A, S140, S255, M17, Ori B, NGC 2264, NGC 2071, and NGC 6334) have not been extensively studied. They certainly exhibit the same simple molecules (HCN, HNC, H$_2$CO, CS, NH$_3$, HC$_3$N, CH$_3$OH, CH$_3$CN, and CH$_3$C$_2$H) as do cold cores (Gottlieb et al. 1979, Loren and Mundy 1984, Mundy 1984) and must surely also contain species such as SO and SO$_2$, which have recently been found in cold clouds (Irvine et al. 1985), although these have not yet been observed. Although the evidence is meager, it appears that the chemistry changes little in passing from cold cores to modestly warm ones. One distinction does occur, however: Si compounds (SiO, SiS) are not seen in cold cores, but make their first appearance in the warm cores (Ori B, NGC 7538, and NGC 2071).

It is likely that warm cores include a group of objects whose temperatures range continuously from ~ 25 K to several hundred K as in the case of Ori(KL). One might expect that the chemistry becomes continuously more complex as the temperature increases. Some evidence that this is so comes from objects such as W51 and the so-called "southern condensation" embedded in the "ambient ridge" component of the Ori(KL) region (Johansson et al. 1984). Besides the simple species seen in modestly warm cores, these sources exhibit spatially confined $(CH_3)_2O$ and CH_3OCHO, and W51 also shows EtCN and VyCN. Both contain SiO. While temperatures in the "southern condensation" probably do not exceed 100 K, those in W51 could well be higher, although this complex source is too distant to resolve such detail at present.

Most of the complex species listed in Table 5.1 have been observed in only two sources, Ori(KL) and Sgr B2. These objects represent the extreme cases of high temperature and of shocks to be found in star-forming regions. Because of its closeness, Ori(KL) has been studied in much finer detail than other sources, so it is not surprising that so many species have been detected there. Yet a few more species have been seen in Sgr B2, despite its large distance (~ 10 kpc). What makes Sgr B2 so special remains unclear, but it appears to be the largest GMC in the Galaxy and may simply have larger column densities or many more highly energized star-forming regions (with slightly differing chemistries) than does Ori(KL). Many star-forming regions in the Galaxy may be similar to Ori(KL), but no other emulates Sgr B2. Only Ori(KL) is close enough, however, so that its highly energetic outflows and shocked regions can be observed in many different molecules for purposes of chemistry. We discuss this source in the next section.

5.2.5 Shocked Regions

Shocks are simply supersonic compressions that occur when the bulk fluid velocity of the gas exceeds the local sound velocity. In neutral interstellar gas, the sound speed is only 0.3 km s^{-1} at $T = 10$ K and 1 km s^{-1} at $T = 100$ K. A variety of interstellar processes produce velocities much in excess of this: expanding HII regions ($\gtrsim 10$ km s^{-1}); stellar winds from OB stars ($\gtrsim 1000$ km s^{-1}); bipolar outflows from protostars (30 to 200 km s^{-1}); supernova explosions (~ 1000 km s^{-1}); cloud-cloud collisons (~ 10 km s^{-1}); galactic spiral density wave (~ 100 km s^{-1}). Thus, shocks are expected to be a common phenomenon throughout the ISM, not only in neutral gas, but also in ionized and in hot coronal gas.

Shocked gas is known to be present in diffuse clouds, in expanding supernova remnants, and in the hot cores of many GMCs. There is little or no observational evidence for shocks in cold dark clouds. A description of shock waves as they affect the chemistry is given in Section 5.4. Here we discuss the physical conditions in shocked material and the current observations.

Shock waves heat and compress the gas. The extent of this depends in part on the shock velocity; observations thus far show that shocked gas can attain temperatures of several thousand K and densities of order 10^{10} cm^{-3}, although these values could be higher. Shocked regions are detected on the basis of broad molecular lines that indicate outflow, seen most readily in CO, and/or by certain variations in observed velocities of spectral lines. Another indicator of shocked gas is the presence of highly excited species, in particular, vibrationally excited H_2 seen at 2 μm. On the basis of such criteria, shocked gas has been observed in many giant molecular clouds (e.g., Beckwith 1980, Bally and Lada 1983), and in many diffuse clouds (e.g., Federman 1982). However, only two regions are simple enough geometrically and contain sufficient column densities to allow as yet meaningful studies of many molecular species in shocked gas, and thus to indicate what occurs chemically to material when a shock wave passes through. These objects are Ori(KL) and supernova remnant IC 443.

Several studies have been done on IC 443 (e.g., De Noyer 1979a, b, Dickinson et al. 1980). Vibrationally excited H_2 has been detected toward this region (Treffers 1979), and studies in OH and CO have shown that there are many cloudlets in the supernova remnant where there is evidence of shocked material. Perhaps the most striking is the "B cloud" (DeNoyer and Frerking 1981), where the CO line profile is characterized by a sharp, narrow "spike" component at a velocity of -5 km s^{-1}, thought to be cold, ambient "preshock" gas, along with a much wider asymmetric line at $V_{LSR} = -10$ km s^{-1} that has a broad tail extending out to -60 km s^{-1} in velocity, the "postshock" gas (see Figure 5.3). This broad feature has been detected in HCO$^+$, HCN, CS, and SiO (DeNoyer and Frerking 1981, Ziurys et al., manuscript in preparation); interestingly, none of these species contain the narrow "preshock" component. Such data have led to the formulation of so-called "shock-enhanced" abundances of such species as HCO$^+$ (Section 5.4). Temperatures and densities for the B cloudlet postshock gas are not particularly high, estimated to be $T \leqslant 100$ K and $n(H_2) \sim 10^4$ to 10^5 cm^{-3}, in contrast to $T \cong 10$ K and a density of a few times 10^3 cm^{-3} in the preshock region.

IC 443 thus is illustrative of the chemical effects of a shock in nearly diffuse gas. The shock in IC 443 has one important distinction with respect to shocks in diffuse clouds. Because it lies at the edge of a supernova remnant, IC 443 is irradiated by X rays as well as UV photons. Thus, the state of ionization of the gas through which the shock passes is significantly higher than is the case for diffuse clouds. The effects of this enhanced ionization on both the "hardness" of the shock and on the chemistry have not yet been investigated.

Ori(KL) is by far the best example where shocks have occurred in dense gas. A more detailed description of this source is given in Chapter 4 of this volume. To summarize, the shock waves in Ori(KL) originate with the infrared object IRc2, a probable protostar located in the KL nebula, which appears to power a massive, high-velocity outflow. To the NE and SW of IRc2, the outflow is believed to have run into ambient cloud material, which has been swept up and accelerated and consequently has formed a sort of "doughnut" or torus of gas centered on the IR source (Plambeck et al. 1982, 1985). The "doughnut" is characterized by molecular spectra with line widths of ~ 20 to 30 km s^{-1} and LSR velocity of ~ 8 km s^{-1}, also called the "plateau" feature (e.g., Johansson et al. 1984). Temperatures in this region are 60 to 100 K, with $n(H_2) \sim 10^6$ to 10^7 cm^{-3}. The doughnut is about $10'' \times 20''$ in size. To the NW and SE of IRc2, perpendicular to the "doughnut," ambient cloud densities are thought to be lower, and the "high-velocity outflow" is observed, characterized by spectra with line wings extending out to velocities of ± 30–100 km s^{-1}. This outflow appears to be bipolar in nature and occupies a region $1' \times 2'$, presumably channeled by the doughnut. Detection of 2-μm H_2 in this region indicates gas temperatures of at least 2000 K, although there appear also to be cooler clumps in the outflow, with $T \sim 100$ K (the "extended outflow"; Vogel et al. 1984). Densities in the high-velocity outflow are probably 10^5 to 10^6 cm^{-3}. The "hot core" is yet another important component, thought to be a clump or clumps of material situated in the inner wall of the doughnut near IRc2. This region is only 6 to 10″ in size with $T \simeq 150$–200 K and densities of at least 10^7 cm^{-3}, although they may be as high as 10^{10} cm^{-3} (e.g., Townes et al. 1983). Line profiles from the hot core have a distinguishing line width of 10 km s^{-1} and an LSR velocity of about 5 km s^{-1}. Finally, surrounding all of these regions is a ridge of relatively undisturbed ambient gas, extending N-S for several arcminutes, known as the "ridge" or "spike" component, identified in spectra by a narrow line width of 3 to 4 km s^{-1} and a velocity of 9 km s^{-1} at KL. Temperatures vary across the ridge, being as high as 70 K near KL and falling off to ~ 20 K several arcminutes north of KL. Densities over this region are $\sim 10^4$ to 10^6 cm^{-3}, but there may be clumps with $n(H_2) \geqslant 10^7$ cm^{-3} (Batrla et al. 1983).

Ori(KL) is thus quite kinematically complex, and its chemistry is also. Because the various components have differing line shapes and velocities, and because of the use of high-spatial-resolution mapping via aperture synthesis techniques, it has been possible to determine exactly in what gas components many molecules are present and what their abundances are. For example, HCN, SO, and SO$_2$ are quite prominent in the doughnut material (Vogel et al. 1984). On the other hand, the high-velocity outflow is traced by SiO close in to IRc2, while HCO$^+$ is abundant in the flow toward its furthest extents. SiO is also very concentrated near the inner edge of the doughnut, near the "hot core," but is not present in the "doughnut" itself. In the "hot core," NH$_3$ is highly abundant, and HCN and HC$_3$N as well, while HCO$^+$ is noticeably absent from the region. EtCN, in contrast, originates exclusively in the "hot core" (Sutton et al. 1985). In the ridge or "spike" gas, most molecules are observed, including HCN, HCO$^+$, HC$_3$N, C$_2$H, and CN but conspicuously absent are SiO and EtCN. In addition, a few species, namely, the free radicals CCH and CN, are *only* observed in the "spike" component. While CH$_3$CN is confined to the inner components, CH$_3$C$_2$H occurs mostly in the outlying ambient ridge.

Most complex molecules have been studied only with single-dish ($\sim 1'$) resolution and thus, on the basis of kinematic information, can be assigned less reliably to the various spatial regions (Johansson et al. 1984, Sutton et al. 1985). Thus, the hot core contains HNCO, EtCN, CH$_3$CN, HC$_3$N, and VyCN (seen nowhere else). HC$_3$N is seen in all regions. HC$_5$N and CH$_3$C$_4$H appear only in the ambient ridge.

An interesting comparison between the extended ridge ($T \sim 20$ to 70 K, $n = 10^4$ to 10^6 cm^{-3}) and the "southern ridge condensation" ($T \simeq 70$ to 140 K, $n \sim 10^7$ cm^{-3}; see Section 5.2.4) can be made reliably by single dish, owing to their large spatial separation. Complex molecular chemistry appears quite different in these two regions: the ambient ridge contains EtCN, CH$_3$C$_4$H, HC$_3$N, and HC$_5$N while the southern condensation contains CH$_3$CN, CH$_3$OH, (CH$_3$)$_2$O, CH$_3$OHCO, H$_2$CCO, and HCO.

Although some of the regional assignments are difficult on the basis of kinematic data only, and may be misleading, rather striking differences in the chemistry do seem to occur between these regions despite similar temperatures and even rather similar densities. Differences in excitation among the different molecular species are not generally an important factor, since many of the species, e.g., SiO, HCN, HCO⁺, and CN, have similar dipole moments and hence similar excitation requirements. The *current* physical conditions may therefore not be the controlling influences. Differences in elemental abundances are one possibility for the differences in spatial distribution. Transient variations in (T, n) in recent history are another possibility. Shocks such as might affect particularly the inner regions may have elevated the temperatures to $\gtrsim 2000$ K and densities to about ten times their current values, at different times. Such shocked regions will typically cool to ~ 100 K in a few hundred years. The chemical imprint of these transient shocks would still be present. There is increasing evidence (Walmsley et al. 1986) that the hot core density in particular may be dominated by the products of shock-sputtered grains.

Though Ori(KL) has been studied in great detail compared with any other source, several complex molecules in Table 5.1 are not found in Ori(KL), but only in Sgr B2 (and in a few cases, in W51). These species include NH_2CN, CH_2NH, CH_3CHO, CH_3NH_2, and EtOH. Because of its great distance, the physical conditions in Sgr B2 are too poorly defined to offer any clues about the uniqueness of these species, but it is worth noting that the fractional abundances of these species are comparable to those of species seen also in Ori(KL). Thus different chemistries are indicated.

5.2.6 Circumstellar Envelopes

Circumstellar envelopes form as a result of mass loss from late-type stars, usually semiregular and regular giants and supergiants. Such stars are generally cool, with surface temperatures of only a few thousand K and typical luminosities of 10^4 to $10^5 L_\odot$ (Glassgold and Huggins 1985). The actual mass loss mechanism is uncertain, but possibilities include radiation pressure on grains and molecules, shock waves, magnetohydrodynamic waves, and sporadic ejection driven by rotation and turbulence (Slavsky and Scalo 1985). Independent of the mechanism, these objects are generally thought to have a constant mass loss rate of typically $\dot{M} \sim 10^{-5}$ to 10^{-4} $M_\odot$/yr, with expansion velocities of 5 to 30 km s^{-1} (Zuckerman 1980). A constant mass loss rate means that the gas density in the envelope should vary as r^{-2}, r being radial distance. Such a radial dependence is reasonably consistent with the shapes of most molecular emission profiles. Envelope sizes vary from object to object, but typical values might be represented by the late-type carbon star IRC + 10216, i.e., an inner shell radius of 6×10^{14} cm, extending out to 10^{18} cm. Most envelopes are not obviously nonspherical, but some may have more of a disklike structure. Densities and temperatures vary throughout the shell, with typical values being $n \sim 10^8$ to 10^{11} cm^{-3} and $T \sim 600$ to 1000 K in the inner regions, decreasing to $n \sim 10^2$ cm^{-3} and $T \sim 10$ K at the outer edge (Scalo and Slavsky 1980). The IRC + 10216 object, the only one studied in detail, consists of an inner hot core $(T = 600$ K, $n \simeq 3 \times 10^7$ cm^{-3} at its outer radius, $r \simeq 6 \times 10^{14}$ cm) surrounded by an "outer core" $(T = 375$ K, $r = 3 \times 10^{15}$ cm), and a cooler extended envelope with $n \propto r^{-2}$ extending out to $r \simeq 10^{18}$ cm; n varies from $\sim 10^7$ to $\sim 10^2$ cm^{-3} and T from ~ 240 to 10 K in this envelope (e.g., Rieu et al. 1984). The gas is heated by collisions with heated grains that are driven by radiation pressure at supersonic speeds through the gas, and is cooled by adiabatic expansion and by line emission by abundant molecules (H_2O for O-rich stars, CO for C-rich stars). These conditions imply that the shells are almost totally molecular in composition, as observations have verified. There is a possibility, however, that shock waves may propagate through the shells, in which case the basic physical properties would be somewhat different (Slavsky and Scalo 1985).

Although the general physical conditions are quite similar for all stellar envelopes, such envelopes differ sharply in elemental composition, being separated into two classes on the basis of their stellar photospheric emission spectra. One class is the oxygen-rich stars, with $O/C \geqslant 1$ (typical of "normal" cosmic abundances). Examples are RX Boo, W Hydra, and IRC + 10011. The other class is the carbon-rich stars ($O/C < 1$), the best studied example being IRC + 10216. This difference fundamentally affects the molecular composition of oxygen-rich vs. Carbon-rich shells, as well as their dust grain composition. There might also be slight chemical differences because carbon-rich stars have somewhat cooler photospheres ($T \sim$ 2000 K) than Oxygen-rich stars ($T \sim$ 2000–3500 K).

In the oxygen-rich envelopes, virtually all the carbon is in the form of CO, leaving oxygen to form the backbone of other molecules. Thus, OH, H_2O, and SiO are species commonly observed at radio wavelengths in these objects. Radio lines of these species usually exhibit strong masering, but thermal transitions are seen as well, including those of CO, SiO, and HCN and IR absorption lines of NH_3 and H_2O. Oxides of many other heavy elements are observed optically. Oxygen-rich stars have not, in the past, been a common target for molecular searches, however, and more species are doubtless present and will be detected in the future. Dust grains in these objects are found to be silicate in composition.

In carbon-rich envelopes, on the other hand, most of the oxygen is thought to be in the form of CO, and carbon-containing molecules dominate. In IRC + 10216, the only extensively studied object of this class, grains account for $\sim 1\%$ of the total mass of the shell and consist mainly of carbon (graphite) and lesser amounts of metals and silicates. Twenty-one species thus far have been detected in these types of shells, primarily at radio/millimeter wavelengths (i.e., CO, CS, SO, HCN, HNC, CCH, SiO, SiS, SO_2, SiC_2, HC_3N, HC_5N, HC_7N, C_4H, C_3N, and CH_3CN), and a few in the IR (C_2H_2, CH_4, HCCH, NH_3, and SiH_4). All of these have been observed toward IRC + 10216, but several have also been seen in other carbon-rich stars such as CIT-6 (Henkel et al. 1983). While simple carbon-containing species such as CCH and HCCH might be expected, it is surprising to find the larger cyano-polyacetylenes (i.e., HC_3N, HC_5N, and HC_7N) in these shells. The cyanopolyynes are also found to be most abundant in cold dark clouds, whose environment is entirely unlike that of circumstellar shells. SO_2 is also surprising; it would be expected only in an oxygen-rich environment. It is also notable that common molecular ions (HCO^+, NNH^+), very abundant species in any dense interstellar cloud, are absent in these shells.

5.3 Molecular Abundances and Their Determination

Any successful model of interstellar chemistry must predict not only what species will form but also, at least semiquantitatively, the relative abundances of those species. Determination of such abundances depends on knowing the excitation of the molecules, which is usually nonequilibrium and requires knowledge of collisional and radiative molecular processes, knowledge which usually does not exist. Here we discuss limitations in abundance determinations.

5.3.1 Methods of Analysis

(a) The Simplest Derivation of Column Densities

Studies of interstellar molecules at radio wavelengths can, in principle, reveal a wider range of physical parameters, in addition to the abundance of the species, than studies at other wavelengths. This is because a large number of molecular energy levels are connected by radio-wavelength transitions and also have sufficiently small energies to be excited under interstellar conditions, either by collisions or by radiation.

A study of the equation of transfer is required to relate the observed quantities to the column density of the given species and to the physical parameters of the source. The observed quantities for a molecular line at frequency v are the excess brightness temperature ΔT_B and the FWHP line width Δv. When observations are meager, as is yet the case for most of the more complex interstellar species, they are interpreted in terms of very simple models, which have only a few free parameters. The simplest model, still much used, is a molecular source of uniform (T, n) for which the equation of transfer at the transition frequency, v_0, may be written

$$\Delta T_B = [J_v(T_{ul}) - J_v(T_{bg})](1 - e^{-\tau_0}) \tag{5.1}$$

where the source function $J_v(T) \equiv (hv/k)[\exp(hv/kT) - 1)]^{-1} \approx T$ in the Rayleigh-Jeans limit $(hv \ll kT)$, T_{ul} is the excitation temperature of the observed transition, and T_{bg} is the background brightness temperature. The excitation temperature T_{ul} is defined by $n_u/n_l = (g_u/g_l)\exp(-\Delta E_{ul}/kT_{ex})$, where ΔE_{ul} is the energy separation between levels and the g's are the degeneracies. The opacity τ_v is the quantity related to the molecular column density in the lower state, N_l, by

$$\tau_v = \frac{\phi_v}{4\pi}\lambda^2 A \frac{g_u}{g_l} N_l[1 - \exp(-hv/kT_{ul})] . \tag{5.2}$$

For a Gaussian line shape, the line profile function $\phi_v = (\ln 2/\pi)^{1/2}/\Delta v$ and $\tau_v = \tau_0$. A is the Einstein coefficient of spontaneous emission for the transition, given by $A = (64\pi^4 v^3/3hc^3)(|\mu_{ul}|^2/g_u)$, where μ_{ul} is the dipole matrix element of the transition. If T_{ul} characterizes all rotational transitions, i.e., $T_{ul} = T_{rot}$, then the total column density N of the molecule can be found from

$$\frac{N_l}{N} = \frac{g_l[\exp(-E_l/kT_{rot})]}{Q} \tag{5.3}$$

where Q is the rotational partition function and E_l is the energy of the lower state.

N may thus be determined by the simple two-step process: (1) determine N_l from τ_0 and T_{ul}, if both can be estimated; (2) determine N from N_l if T_{ul} is known and is unique to all transitions (i.e., $T_{ul} \equiv T_{rot}$).

Equation (5.1) shows that if $\tau_0 \gg 1$, then no information about N_l is available from the single transition, although $T_{ul} - T_{bg}$ and hence T_{ul} is obtained. In this case, τ_0 can often be estimated from multiple transitions. If $\tau_0 \ll 1$, then (in the Rayleigh-Jeans limit) Equations (5.1) and (5.2) yield

$$N_l = \frac{k}{\zeta h \nu} \frac{T_{ul}}{T_{ul} - T_{bg}} \int \Delta T_B \, d\nu \tag{5.4}$$

where $\zeta = (\phi_\nu/4\pi)\lambda^2 A(g_u/g_l)$. At millimeter wavelengths, $T_{ul} \gg T_{bg} (\simeq 3 \text{ K})$ so N_l can be obtained independently of T_{ul}. At centimeter wavelengths, $T_{ul} \ll T_{bg}$ is possible ($T_{bg} = T_c$, the continuum source brightness), and the transition is seen in absorption ($\Delta T_B < 0$). In this case, only N_l/T_{ul} may be obtained.

At this level of interpretation, two or more different transitions are sometimes used to determine the opacity, assuming T_{ul} is the same for each. Hyperfine splitting, if it occurs, is best for this purpose because T_{ul} is most likely to be the same for the hyperfine lines. τ_0 is also often estimated by observing lines of various isotopic species, e.g., ^{12}CO, ^{13}CO, and $C^{18}O$, and assuming an isotopic abundance ratio. Finally, two rotational transitions can be used, but T_{ul} may differ between them, vitiating their use.

Excitation conditions must be known in order to estimate T_{ul} for a given transition and to determine if it applies to the entire rotational ladder of a molecule, thus permitting Q to be evaluated. In our simple model, T_{ul} can be related to T_k, the kinetic temperature, which characterizes collisional excitation, to T_{rad}, the cosmic background excitation, and to T_L, which characterizes any other radiation field, by solving the statistical equilibrium equation for a simple two-level case:

$$n_u(A_{ul} + B_{ul}J_\nu + n\sigma v + r_{ul}^L) = n_l\left(\frac{g_u}{g_l}B_{ul}J_\nu + \frac{g_u}{g_l}n\sigma v e^{-h\nu/kT_K} + r_{lu}^L\right)$$

to yield

$$T_{ul} = \frac{(t_k t_L + t_{rad}t_L + t_{rad}t_k)T_{rad}T_K T_L}{t_{rad}t_k T_{rad}T_K + t_{rad}t_L T_{rad}T_L + t_k t_L T_K T_L} \tag{5.5}$$

where we have assumed $r_{lu}^L/r_{ul}^L \equiv (g_u/g_l)\exp(-h\nu/kT_L)$ and where $t_k = (n\sigma v)^{-1}$, $t_{rad} = [1 - \exp(-h\nu/kT_{rad})]/A_{ul}$, and $t_L = (r_{ul}^L)^{-1}$ are the lifetimes (reciprocal rates) of the various excitation processes. It has been assumed that $h\nu/kT \ll 1$.

In dense interstellar clouds, $t_L \gg t_k$, t_{rad} is typical, and collisions dominate the excitation. In this case, $T_{ul} < T_k$ and $T_{rot} \lesssim T_k$, which at least enables limits to be placed on column densities. Thus, when molecules are observed in emission (as at millimeter wavelengths), $t_k < t_{rad}$ must apply. For an optically thin cloud, a transition will be "thermalized" ($T_{ul} \to T_k$) when $t_k \ll t_{rad}$, i.e., when $n(\sigma v) \gg A_{ul}/[1 - \exp(-h\nu/kT_{rad})]$. Values of n which satisfy this condition range from $\sim 10^3$ cm^{-3} for CO ($J = 1-0$) to several 10^6 cm^{-3} for submillimeter transitions of molecules with large dipole moments. If the cloud is optically thick at the transition frequency, radiation trapping occurs, and thermalization requires $\tau \gg (1 + A_{ul}/C_{ul})^{1/2}$, where C_{ul} is the collision rate. When $T_{ul} = T_k$ for all levels, the condition of local thermodynamic equilibrium (LTE) is met.

At the simple level of analysis discussed here, the $J = 1-0$ line of CO is usually used to determine T_k for our "uniform" cloud. This is because $A_{ul}/C_{ul} \approx 1$ for $n = 10^3$ cm^{-3} (as occurs in all except diffuse molecular clouds), while $\tau(CO, J = 1-0) > 15$ typically. Thus, CO ($J = 1-0$) is thermalized in all dense molecular clouds, and hence $\Delta T_B(CO) = (h\nu/k)[J_\nu(T_k) - J_\nu(T_{rad})]$ from Equation (5.1). This yields T_k.

For cases where radiation dominates the excitation, T_{ul} and T_{rot} can be related in a similar manner to T_{rad}, the temperature characterizing the local radiation field. T_{rad} is often the local dust temperature, or source continuum temperature, and hence can be measured.

In relating ΔT_B to N, we have so far omitted an important observational limitation, that imposed by the "beam filling factor" f, defined as the ratio of source solid angle to the telescope beam solid angle; $f = \theta_s^2/(\theta_s^2 + \theta_B^2)$ for a Gaussian source and beam. Thus Equation (5.1) becomes $\Delta T_B = f[J_v(T_{ul}) - J_v(T_{bg})](1 - e^{-\tau_0})$. If $f < 1$, as for example occurs if the source is "clumpy" on a size scale smaller than the beam, then we observe a reduced excess brightness temperature and erroneously deduce a column density which is smaller than the true value. Effects of clumping or "beam dilution" are major factors in the derivation of column densities and abundances.

In summary, abundances for molecules observed in only one or two transitions are usually derived as follows: (i) the lines are assumed to be optically thin (most often the case), so that Equation (5.4) may be applied; (ii) $T_{ul} \approx T_k$ is assumed, with T_k taken from CO ($J = 1-0$) observations, and the rotational ladder is assumed thermalized at T_k, enabling Q to be estimated. A variant to this approach, applied particularly to species whose transitions are optically thick, is to estimate the opacity from observations of two or more isotopic species, or two or more hyperfine lines, as mentioned previously. Assumption (ii) (the LTE assumption) is probably safe in these cases. The resultant estimates for abundances have actually proven surprisingly consistent with more refined estimates based on observations of many additional transitions and more sophisticated treatment of the radiative transfer.

(b) Elementary Column Density Estimates When Many Transitions Are Observed

When several transitions are observed, one may estimate a column density by solving the statistical equilibrium equations (Section 5.3.1.a) for values of n and T_k which match observed values of ΔT_B for the transitions, and then by summing the populations over all levels.

An even simpler approach is to assume that all transitions of a given molecule are optically thin, and further, that T_{ul} for all transitions is equal to a common value, T_{rot}. Then Equations (5.3) and (5.4) may be combined to yield

$$\log\left[\left(\frac{k}{\xi h v}\right)\int \Delta T_B\, dv\right] = \log\left(\frac{N}{Q}\right) - (\log e)\left(\frac{E_u}{kT_{rot}}\right) \tag{5.6}$$

so that the left-hand side is a linear function of E_u/k with slope $-(\log e)/T_{rot}$ and intercept $\log(N/Q)$ at $E_u = 0$. If the two assumptions are valid, both T_{rot} and N are obtained. Conversely, if the data do not fit the linear relationship of Equation (5.6), the assumptions may be incorrect. Since the value of n needed to thermalize a given transition increases with higher energy levels, inadequate excitation will show up as the value of the left-hand side falling below the straight line representing T_{rot}. This must always occur for sufficiently high energy levels, but if $E_u/kT_{rot} \gg 1$ for these levels, the degree to which N is overestimated will be negligible.

This method has been used in a study of twenty-one molecular species in Sgr B2 (Cummins et al. 1986) and three species in Ori(KL) (Johansson et al. 1984). Such studies are highly important in understanding the chemistry of these regions. For example, if different values of T_{rot} are found for different species, T_{rot} may have been incorrectly determined, due to insufficient excitation for some of the species. Or different values of T_{rot} may signal different spatial distributions among the species within the observing beam, in which case their determined relative abundances are not relevant in chemical modeling.

(c) Column Density Estimates from Radiative Transfer Techniques

When the opacity in any of the molecular transitions is not small, the radiative transfer must be taken into account in determining column densities. This is because the radiation from that transition is trapped within the cloud and contributes to the excitation of that transition along with collisions and other radiation fields. In a simple two-level analysis, the effect of trapped radiation is to multiply the Einstein A coefficient by the escape probability $\beta = (1 - e^{-\tau})/\tau$ so that for large opacity τ, $A_{eff} \to 0$ and $T_{ul} \to T_k$ (Equation 5.5).

In the general case, the equations of radiative transfer and of statistical equilibrium must be solved simultaneously for the multilevel system for specified cloud geometry, spatial distribution of n and T_k, and of fractional abundance X of the molecular species relative to $n \equiv n(H_2)$, collision coefficients, and a specified line-broadening process. The latter is not known but may be systematic motions (e.g., of collapse), microturbulence, or macroturbulence. The choice of these various physical processes has been shown to have little effect on the solution for emergent line intensities (White 1977) as long as they produce similar line profiles ϕ_v.

The large-velocity-gradient model (LVG) is the technique extensively used. It entails a systematic velocity field which allows photons to escape from all layers in the cloud, even in optically thick lines. The observed line widths Δv are equal to the velocity gradient dv/dr times the radius of the cloud R. The LVG approximation is valid if the velocity gradient is sufficiently large that an emitted photon travels only a short distance compared to R before it is a Doppler width away from the velocity of the local gas. Thus $\Delta v/(dv/dr) \ll R$. If the motion is thermal, then values of $dv/dr < 1$ km s^{-1} pc^{-1} satisfy this relation for most molecular clouds.

Such a model explains the striking rarity of self-reversed lines; a cloud with purely turbulent motions should show self-reversal whenever the temperature in the outer regions is less than in the inner regions. An LVG model has no self-reversed lines and also explains the observed similarity in profiles of both optically thick and thin lines, since photons can escape from all parts of the cloud.

In order to use an LVG model, observations of the line intensities of multiple transitions are necessary. The model is then used to reproduce these intensities by varying several input parameters, which include the gas density n, the gas kinetic temperature T_K, and the quantity $X' = X/(dv/dR)$, where the velocity gradient dv/dR is assumed to be constant throughout the cloud. Spherical or slab geometry is generally assumed, and the number of levels for which the calculation is to be done is specified. Once the line intensities are reproduced, column density can be derived. It is simply given by $N = X' n 2R \, dv/dR = X' n \Delta V_{1/2}$, independent of the assumed cloud radius.

5.3.2 Problems in Abundance Determinations

(a) Clumping

As mentioned previously, the effects of beam dilution and of clumping are critical factors in deriving column densities. No method of column depth determination is independent of these factors, since they directly affect measured line temperatures. Unfortunately, the clumping or filling factor is difficult to estimate, as there is little direct information about the small-scale structure of molecular clouds, especially for molecules with low abundances. As the high-resolution studies of Ori(KL) have shown, several kinematically and chemically distinct regions exist within the normal single-dish beam size of about 1'. Similar types of regions are known for several other types of star-forming regions, but in far less detail. Mundy (1984) suggests typical filling factors of $\gtrsim 0.3$ for the GMCs M17, S140, and Orion B. However, studies of dark clouds indicate that clumping on size scales $\lesssim 1'$ may be unimportant for these objects. In general, however, it is not easy to assign a filling factor to a given molecule in a particular source.

The study of twenty-one molecules done in Sgr B2 (Cummins et al. 1986) shows that clumping may be important for a wide variety of species within one source. Using the LTE abundance determination method for multiple transitions described previously in Section 5.3.1.b, Cummins et al. found that all twenty-one species were characterized by consistent rotational temperatures, whose values ranged from 9 to 30 K, depending on the molecule. Such values are much less than the kinetic temperature T_K as estimated from CH_3CN and CH_3CCH to be 54 and 85 K, respectively (Hollis 1982, Churchwell and Hollis 1983). The difference in values for T_K is attributed to the source having a central hot core surrounded by a cooler halo. The only explanation for $T_{rot} < T_K$ for the twenty-one molecules of Cummins et al. is that all the transitions are subthermally excited. It would seem unlikely, however, that under conditions of subthermal excitation, all transitions of a given species would be characterized by a single temperature. Higher transitions would be expected to exhibit increasingly smaller excitation temperatures. Cummins et al. also found that T_{rot} for all twenty-one species has no correlation with permanent dipole moment. Under subthermal conditions, higher dipole moment molecules should have lower T_{rot}.

These results can be explained by clumping. Assume a typical clump is sufficiently dense at its center to thermalize all transitions, but not so toward its outer regions. The higher-lying transitions will become subthermal at larger values of density than the lower-lying lines and hence will have smaller angular size. Thus, ΔT_B for the higher transitions will be more beam-diluted than for the lower-lying transitions, increasing the slope of Equation (5.6) and producing $T_{rot} < T_K$. The independence of T_{rot} on dipole moment is probably explained by a failure of the optically thin assumption within the clumps.

(b) Fractional Abundances

Fractional abundances are subject to far more uncertainties than column densities. The fractional abundance for a given molecular species is defined as $X \equiv N/N(H_2)$, and thus implies establishing the column density $N(H_2)$ of H_2. Independent methods such as the use of IR extinction give quite reliable estimates of $N(H_2)$. A problem with this approach is that one must assume that the molecular species in question has exactly the same volume distribution as H_2. For statistical equilibrium methods, on the other hand (e.g., the LVG approach), the excitation is modeled to fit many observed lines, and the assumption of coexistence is avoided. If $\tau \ll 1$ so that radiative transfer is not important, then $n \equiv n(H_2)$ is derived from the excitation model and N from the resultant derived populations which fit the observed line brightnesses. To get X, one must adopt a size R for the (uniform) cloud. Maps provide an estimate of R. If radiative transfer is included, one again derives n from the excitation, as well as $X/(dv/dr)$, which requires again assuming R to get $dv/dr = \Delta v/R$. Again, $N = [X/(dv/dr)]*\Delta v*n$ is more reliably determined as it is independent of the geometry.

A more serious problem for the determination of X is clumping. A comparison of LVG and LTE methods for the species HCO^+ in twelve warm molecular cores provides an indication of this problem (cf. Wootten et al. 1978). It is found that $X_{LTE} \equiv N_{LTE}/N(H_2) \simeq 10^{-9}$ in these cores, where $N(H_2)$ is independently estimated from IR data, and N_{LTE} according to Section 5.3.1.a. The LVG analysis of the same twelve cores yielded $X_{LVG}/(dv/dr) = 10^{-11}$. Now $N_{LVG} = [X_{LVG}/(dv/dr)]*\Delta v*n$, where n is the density found in the radiative transfer analysis to reproduce observed line brightnesses. It is found that $N_{LVG} \approx N_{LTE}$, indicating that $X_{LVG}/(dv/dr)$ is reliably determined. Estimating dv/dr from $\Delta v/R$ yields $X_{LVG} \approx 2 \times 10^{-11}$, much smaller than X_{LTE}. If the gas is clumped, however, with $R_{clump}/R \ll 1$, then X_{LVG} is proportionately larger, and may be brought into agreement with X_{LTE} if $R_{clump}/R \approx 1/50$. In this picture, the clumps do not shadow one another, and interclump motions are unimportant compared with internal clump motions.

(c) Collisional Cross Sections

Both the effective rate $\langle \sigma v \rangle$ and the selection rules for collisions with H_2, He are poorly known for almost all molecules of astrophysical interest. Since collisional excitation nearly always dominates radiative, uncertainties in $\langle \sigma v \rangle$ directly affect the derived densities n. Analyses of many molecular species tend to average out these uncertainties and provide quite reliable estimates of n. The uncertainties due to clumping completely dominate uncertainties in n. The relative values of $\langle \sigma v \rangle$ as a function of ΔJ have been calculated (e.g., Green 1975) for several species (CO, CS, HC_3N, ...) with widely different dipole and quadrupole moments. These results are thought to apply quite well to other molecules of similar moments, and uncertainties in them probably affect resultant column densities of the molecule in question by at most a factor of 2.

(d) Kinetic Temperature

This quantity is usually derived from observations of the highly saturated $J = 1-0$ CO line. Because this line is so optically thick, it yields temperatures that pertain only to the surface of the cloud, which may be warmer than the interior if exposed to ambient UV heating or cooler than the interior if the latter contains embedded heat sources. Nearby heating sources also produce temperature gradients, which introduce apparent density gradients when the cloud is modeled with a uniform temperature. For uniform-T models, Mundy (1984) finds that changes in T_k by a factor of 2 change the derived N(CS) by 25% and the derived n by a factor of 3.5. Any gradients in T_k probably produce larger uncertainties. Column density uncertainties probably are not much larger when such gradients occur. However, it is not possible to separate unambiguously the effects of temperature and density gradients which may be particularly important near heating sources, and thus physical understanding of the region may be limited. Unfortunately, the chemistry is influenced differently by temperature and density. These ambiguities are quite serious when only one species (e.g., CS) is used as a diagnostic of physical conditions; use of several molecules reduces the uncertainties, and certainly permits the identification of peaks in the maps as due largely either to temperature or to density. Various molecules which are good "thermometers" (NH_3, CH_3CN, and CH_3C_2H) often give discrepant temperatures, most likely because they do not sample the same regions of the cloud. For example, in both Sgr B2 and Ori(KL), CH_3CN indicates higher temperatures than CH_3C_2H; the latter species is known from interferometry to exist in Ori(KL) in the cooler "halo" gas surrounding the hotter core which contains the CH_3CN.

(e) Radiation Fields

These fields can affect the molecular excitation directly, but are poorly known. UV fields are probably unimportant in the interiors of molecular clouds, although this is not certain. IR fields near embedded protostars may be significant in general and are known to dominate the excitation in certain cases [the hot core component of Ori(KL), circumstellar envelopes]. These IR fields provide vibrational excitation of molecules, which can in some cases mimic the effects of collisions quite closely. In order for IR excitation to be important near embedded protostars, small hot sources are required because of solid angle dilution and emissivity factors (cf. Mundy 1984). Extended, cool dust envelopes around many IR sources are probably unimportant in the excitation of nearby molecules. When sources are eventually observed at resolutions comparable to that of Ori(KL) ($\sim 7''$ or 0.015 pc), IR excitation may be expected to dominate, but it does not for most present observations of molecules over more extended regions.

5.3.3 Representative Abundances

Table 5.2 lists typical column densities and fractional abundances for common molecules in various types of objects. These values are subject to the uncertainties discussed in this section, but they give a general view of interstellar abundances.

5.4 The Four Basic Schemes of Interstellar Chemistry

In principle, interstellar molecules can form under nonequilibrium conditions that occur at very low density by gas-phase reactions, by grain surface (catalytic) reactions, by the breakup of interstellar grains, or under equilibrium conditions in the (dense) atmospheres of cool stars. In practice, interstellar molecules are found in objects that have varied physical conditions, and hence different chemical processes. Astrophysically, it is convenient to consider four separate divisions: chemistry that occurs on grain surfaces (e.g., catalytic), chemistry initiated by shock fronts, chemistry centered around ion-molecule gas-phase reactions, and chemistry in the circumstellar shells and atmospheres of late-type stars. Shock fronts can produce molecule formation either through "high-temperature" gas-phase reactions or by breaking up ("sputtering") grains. In interstellar molecular clouds of all types—diffuse clouds, cloud envelopes, cold and warm cores in GMCs and

Molecule	ζ Oph	TMC-1	IRC+10216	Sgr B2	Ori(KL)		
					Spike	Plateau	Hot core
CO	1.2×10^{15} (S)	5.8×10^{17} (FM)	$\sim 10^{17}$–10^{19} (MC)	6×10^{19} (SSP)	$>10^{19}$ (J)	2.5×10^{18} (J)	—
CH	3.4×10^{13} (S)	2×10^{14} (FM)	—	$\geq 2 \times 10^{15}$ (SLG)	↑	$\geq 1.7 \times 10^{14b}$	↑
OH	5.2×10^{13} (S)	6×10^{14} (FM)	—	$\geq 2.7 \times 10^{15}$ (SWT)	1×10^{14} (WTH)	1.5×10^{16} (WTH)	—
CN	4.8×10^{12} (S)	3.0×10^{14} (CCZ)	3×10^{15} (LLO)	3.4×10^{14} (AK)	1.5×10^{16} (SBMP)	—	—
CS	$<2.6 \times 10^{13c}$	2×10^{13} (IH)	1.8×10^{15} (MC)	3.0×10^{15} (G)	2.5×10^{14} (WW)	3.2×10^{14} (WW)	d
SO	—	5×10^{13} (FM)	—	4.4×10^{14} (CLT)	3×10^{14} (J)	3×10^{16} (J)	1×10^{15}–3×10^{15} (J)
SiO	—	$<5 \times 10^{10c}$ (ZFI)	4.1×10^{14} (MC)	$\sim 4 \times 10^{13}$ (WPJKT)	—	2×10^{15} (WPVHW)	2.3×10^{17} (WPVHW)
SiS	—	—	1.6×10^{15} (MC)	1×10^{13}–2×10^{13} (DK)	—	$\sim 10^{14}$ (Z)	—
SiC$_2$	—	—e	1.5×10^{14} (TCL)	—	—	—	—
SO$_2$	—	—	—	8.6×10^{14}, 3.1×10^{15f} (CLT)	3×10^{14}–9×10^{14} (J)	3×10^{16} (J)	2×10^{15}
HCO$^+$	—	8×10^{13} (GLW)	—	2.3×10^{15} (CLT)	5×10^{14} (J)	2×10^{15} (VWPW)	—
N$_2$H$^+$	—	2×10^{12} (IH)	—	$\sim 1 \times 10^{13}$ (TT)	2.0×10^{13} (WW)	—	—
HCNH$^+$	—	—	—	4×10^{14} (ZT2)	—	—	—
HCN	—	1.2×10^{14} (IH)	5×10^{16} (LLO; J)	4×10^{15} (ZT2)	1.2×10^{15}–4.0×10^{15} (J)	2×10^{17} (VWPW)	2.3×10^{18} (ZT1)
HNC	—	8.0×10^{13} (IH)	2×10^{14} (J)	$\sim 10^{14}$	2×10^{13}–8×10^{13} (J)	← 0.3×10^{14}–1.7×10^{14b} → (J)	→
CCH	—	8.4×10^{13} (FM)	6×10^{15} (LLO; J)	$\sim 10^{15}$ (CLT)	2.4×10^{15} (J)	—	—
H$_2$CO	—	1.2×10^{14} (FM)	—	3×10^{15} (FW)	1.2×10^{16} (WW)	—	—

Continued

227

Table 5.2. Representative interstellar abundances.[a] (*Continued.*)

Molecule	ζ Oph	TMC-1	IRC+10216	Sgr B2	Ori(KL)		
					Spike	Plateau	Hot core
NH_3	—	1.0×10^{15} (FM)	1.0×10^{15} (LLO)	$\sim 1 \times 10^{16}$ (W1)	3.2×10^{16} (WW)	1.2×10^{16} (WW)	5×10^{18} (TGWS)
C_3H_2	—	$\gtrsim 10^{14}$ (MIM)	6×10^{12} (TVG)	9×10^{13} (TVG)	7×10^{12} (TVG)	—	—
CH_3OH	—	$\gtrsim 10^{13}$ (I1)	—	2.7×10^{16}, 3.5×10^{15f} (CLT)	5×10^{16} (J)	—	—
CH_3CN	—	7.0×10^{12} (MS1)	$>10^{13}$ (J)	1.3×10^{14} (CLT)	1×10^{14}–2×10^{14} (J;SBMP)	—	2×10^{15} (SBMP)
H_2CCO	—	—	—	1.7×10^{14} (CLT)	5×10^{14} (SBMP)	—	—
$CH_3C \equiv CH$	—	6.0×10^{13} (FM)	—	1×10^{15} (CH)	6×10^{14} (J)	—	—
HC_3N	—	6.0×10^{13} (FM)	1.8×10^{15} (MC)	3.5×10^{14} (CLT)	5×10^{13} (J)	2×10^{14} (J)	0.6×10^{14}–1.5×10^{14} (J)
HC_5N	—	1.0×10^{14} (FM)	4×10^{14} (MC)	1.5×10^{13}, 1.6×10^{14}, 1.1×10^{13g} (A)	7×10^{12} (J)	—	—
HC_7N	—	4.7×10^{12}–5.0×10^{12} (SSYHF)	2×10^{14} (MC)	—	—	—	—
HC_9N	—	3×10^{12}–4×10^{12} (BM)	—	—	—	—	—
$HCOOH$	—	—	—	$\sim 1 \times 10^{14}$ (ZBG)	$\sim 10^{14}$ (SBMP)	—	—
NH_2CN	—	—	—	2.0×10^{13} (CLT)	—	—	—
NH_2CHO	—	—	—	7.2×10^{13} (CLT)	—	—	—
CH_3CHO	—	6×10^{12} (MFI)	—	2.4×10^{14} (CLT)	—	—	—
CH_2CHCN	—	3×10^{12} (MS2)	—	4.7×10^{13} (CLT)	—	—	2×10^{14} (SBMP)

CH_3OCH_3	—	—	—	5×10^{14} (CLT)	3×10^{15} (J;SBMP)	2×10^{15} (SBMP)
$HCOOCH_3$	—	—	—	3.4×10^{13} (CLT)	3×10^{15} (J; SBMP)	—
CH_3CH_2OH	—	—	—	5.3×10^{14} (CLT)	—	—
CH_3CH_2CN	—	—	—	6.2×10^{13} (CLT)	$<1 \times 10^{13}$–4×10^{13c} (J)	—

[a] A blank in any column does not necessarily mean a species is not present in a given object, but that sensitive observations may not yet exist for the molecule in that source.

[b] Column density a blend of components.

[c] Not detected up to this limit.

[d] Detected in the source but column density not available

[e] Detected toward dark cloud L134N with $N_L \simeq 2 \times 10^{13}$–$6 \times 10^{13}$ IGS).

[f] Two-component model.

[g] Three-component model.

References:

S: Snow 1980
FM: Freeman and Millar 1983
MC: McCabe et al. 1979
MS2: Matthews and Sears 1983
SSYHF: Snell et al. 1981
LLO: Lafont et al. 1982
CCZ: Crutcher et al. 1984
IH: Irvine and Hjalmarson 1983
ZFI: Ziurys et al. 1987
IGS: Irvine et al. 1983
GLW: Guelin et al. 1982
MFI: Matthews et al. 1985
MS1: Matthews and Sears 1983
MIM: Madden et al. 1986
BM: Bell and Matthews 1985
TVG: Thaddeus 1985
TCL: Thaddeus et al. 1986
SSP: Scoville et al. 1975
SLG: Stacey et al. 1986
SWT: Storey et al. 1981
AK: Allen and Knapp 1978

G: Gottlieb et al. 1978
WPJKT: Wilson et al. 1971
DK: Dickinson and Rodriguez-Kuiper 1981
W: Woods et al. 1983
TT: Turner and Thaddeus 1977
FW: Fomalont and Weliachew 1973
ZBG: Zuckerman et al. 1971
CH: Churchwell and Hollis 1983
A: Avery et al. 1979
W1: Wilson et al. 1982
WW: Watson and Walmsley 1982
ZT1: Ziurys and Turner 1986a
ZT2: Ziurys and Turner 1986b
J: Johansson et al. 1984
SBMP: Sutton et al. 1985
WPVHW: Wright et al. 1983
Z: Ziurys, manuscript in preparation
WTH: Watson 1982
R: Rydbeck et al. 1976
VWPW: Vogel et al. 1984
TGWS: Townes et al. 1983
CLT: Cummins et al. 1986

SMCs—ion-molecule chemistry is thought to be the principal mechanism of molecule formation for the large majority of species listed in Table 5.1. Diffuse clouds are also subject to UV dissociation/ionization, and there is evidence that shocks influence the chemistry of at least some species (e.g., CH^+) in them, but the basic synthesis is due to ion-molecule and some neutral-neutral gas-phase reactions. Circumstellar envelopes, by contrast, have unique and different physical properties and the chemistry of these objects is in a class by itself. Reactions on grain surfaces possibly occur in all these types of regions to some extent, as may the disruption of grains by shocks and by energetic photons, but gas-phase processes appear to be more important. Grains do, however, play an important and irrefutable role in interstellar chemistry: molecules are lost from the gas phase onto grain surfaces by adsorption, which occurs at a rate $\tau_{gr}^{-1} = n_{gr}n_x\pi r_{gr}^2 \bar{v} S$ cm^{-3} s^{-1} where $\bar{v}$ is the mean speed of the molecule in the gas phase and S the sticking probability per collision (of order unity for many species according to experiment). The lifetime of a given molecule against grain adsorption is $\tau_{gr} \simeq 10^{21}/n\bar{v}$ s $\simeq 10^9$ yr/$n \approx 10^6$ yr for typical values of $\bar{v}$ and n. Two important conclusions follow. First, since clouds have lifetimes longer than depletion times, either all molecules (except H_2) should have disappeared or molecules must somehow be returned from grains to the gas phase. Second, only the more rapid gas-phase reactions (see Table 5.3) can compete with the grain adsorption rate, which suggests ion-molecule reactions as the likely if not the dominant process, at least in cool clouds.

5.4.1 Ion-Molecule Chemistry

At the low temperatures ($T \simeq 10$ to 100 K) and densities ($n \sim 10^2$ to 10^7 cm^{-3}) of typical interstellar molecular clouds, gas-phase reactions must entail two-body collisions only; three-body collisions become important for $n > 10^{11}$ cm^{-3}. Such reactions must also have negligible activation barriers, and must be exothermic except in higher-temperature regions characteristic of shocks. Unless these conditions are satisfied, a given reaction cannot be significant on the time scale of the lifetime of a molecular cloud.

Table 5.3 gives a simplified list of important gas-phase reactions that fit the requirements in general. Also given are typical reaction rates in the absense of

Table 5.3. Reaction types relevant to interstellar chemistry.

Reaction type		Typical rate, k^a	Activation energy?
Radiative association	Diatomic	10^{-17}	Possible
[$A^+ + B \rightarrow AB^+ + h\nu$]	Polyatomic	10^{-17}–10^{-9}	Possible
Neutral-neutral	Neither radical	10^{-13}	Yes
[$A + B \rightarrow C + D$]	One or both radical	10^{-11}–10^{-10}	Usually
Ion-molecule		10^{-9}	50% of all reactions
[$A^+ + B \rightarrow C^+ + D$]			
Dissociative recombination		10^{-7}–10^{-6}	Almost never
[$A^+ + e \rightarrow$ neutral products]			

^{a} Rate (in cm^3 s^{-1}) in absence of activation energy, i.e., $\Delta E_{act} = 0$, where $k = A \exp(-\Delta E_{act}/kT)$.

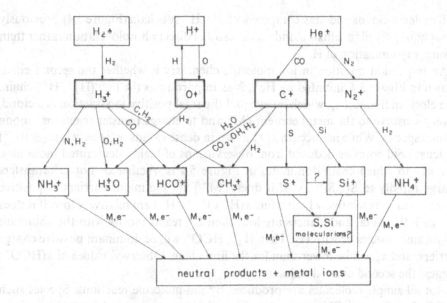

Fig. 5.4. Schematic diagram illustrating some of the major chemical pathways in the ion-molecule scheme of interstellar molecular synthesis, as discussed in Section 5.4 (adapted from Prasad and Huntress, 1980).

activation barriers; at low interstellar temperatures, rates will be much slower if such barriers exist. The given rates are representative only; every chemical reaction is different, and exceptions to standard rates can always exist. An important point is that rates for many reactions may have severe temperature dependences; especially as $T \rightarrow 0$, rates often increase dramatically.

Ion-molecule reactions are rapid because they generally proceed at the Langevin rate (for exothermic cases), which is typically $k \approx 1 \times 10^{-9}$ to 3×10^{-9} cm^3 s^{-1}. The nearly constant value for the Langevin rate independent of the details of any given reaction results because if the ion and neutral are within a minimum distance from one another, the ion sees an infinite potential well and will spiral into the neutral. Every collision thus results in a reaction.

The basic scheme of interstellar ion-molecule chemistry is shown in Figure 5.4. Cosmic rays, which at energies $\gtrsim 2$ MeV easily penetrate even the densest of molecular clouds, initially ionize the most abundant species at a rate of $\sim 10^{-17}$ s^{-1} per molecule (Herbst and Klemperer 1976). In dense clouds, H_2, which is formed on grain surfaces, is ionized to form H_2^+ and H^+. These and He^+ react extremely quickly with the most abundant species (H_2 and CO) to start the two reaction chains shown in Figure 5.4. (The reaction $He^+ + H_2$ is slow.) The species H_3^+, produced by $H_2^+ + H_2$, is particularly important, reacting with CO, N_2, CS, ... to produce HCO^+, N_2H^+, HCS^+ .... Other highly important ions produced directly from H_3^+ such as H_3O^+, NH_3^+, and CH_3^+ (Figure 5.4), go on to produce the hydroxyl, amine, and hydrocarbon families in dense clouds. H_3^+ has recently taken on even greater importance since recent experiments show that its dissociative recombination rate with electrons is slow. Thus, unless photodissociation of H_3^+ is rapid (unlikely), it is probably an important initiator of formation pathways in diffuse as well as dense clouds and may compete with the H^+ subchain (Figure 5.4), previously thought to prevail in diffuse clouds, as a result of direct photoionization rather than cosmic-ray ionization of H.

An important question in ion-molecule chemistry is whether the second chain shown in Figure 5.4, initiated by He^+, is as important as the first (H_2^+, H^+) chain. The electron fraction x_e, which must equal the total positive ion fraction in a cloud, is very sensitive to the metal content [M] and to charge transfer reactions among atomic species. While not accurately known in dense clouds, x_e is very low $(\lesssim 10^{-8})$ in dense cold cores as deduced from observations of highly deuterated molecules. The key to which chain dominates in Figure 5.4 is whether or not C^+ transfers charge rapidly to S^+, Si^+ .... If it does, $[M^+]$ are the major carriers of positive charge, and x_e is relatively high while $x(HCO^+, N_2H^+)$ are relatively low. If it does not, or if the metal abundances are low, then C^+ reacts mostly with the abundant polyatomic species or with H_2. Then H_3^+, HCO^+ will be dominant positive charge carriers, and x_e will be lower than for the first chain. Observed values of $x(HCO^+)$ suggest the second chain dominates.

Not all simple molecules are produced by ion-molecule reactions. Species such as CO, N_2, O_2, and CN are produced largely by many neutral-neutral reactions, which, although slower, result in significant abundances because destruction rates for these species are also usually slow. Most neutral molecules are, however, produced from ions by dissociative recombination and by charge transfer reactions (especially with H_2).

Radiative association reactions are also important; since their reaction rates increase with increasing size of the fragments involved, they may be particularly significant for the synthesis of larger molecules such as CH_3OH. The limitations of these reactions are at present unknown, but it has been speculated that they might eventually build "mushy" grain-sized agglomerations of water, alcohols, aldehydes, etc.*

A complicated network consisting of these kinds of reactions constitutes the ion-molecule scheme of interstellar synthesis. The application of this theory involves complicated computer codes that entail as many as 1500 reactions with corresponding coupled, nonlinear kinetic equations (e.g., Mitchell et al. 1978, Iglesias 1977, Prasad and Huntress 1980, Graedel et al. 1982, Leung et al. 1984) that are solved numerically to derive molecular abundances. In earlier models, only the equilibrium solution was derived, but more recently, time-dependent solutions have been investigated, although with at best rudimentary application to observed clouds.

There are many uncertainties in ion-molecule chemistry. First, the majority of rate constants for ion-molecule reactions have not been measured, especially at low temperatures, so that use of the Langevin rate will often be incorrect. Recently it has been found that ion-molecule reactions can have unusual temperature depen-

---

* On the other hand, the fundamental H_2 molecule cannot be formed by radiative association of two H atoms, because at low interstellar energies the approaching H atoms do not form a bound potential energy manifold, and also because the bound manifold which occurs at higher energies has no permitted radiative transition to the ground state, so that stabilization of the molecule cannot occur.

dence as $T \to 0$. Also, measured rates may not always be reliable, as shown for the H_3^+ dissociative recombination reaction (Smith and Adams 1984) A second problem is that models are sensitive to the input atomic abundances. Calculations have used solar abundances and various "depleted" abundances derived from Copernicus UV observations of *diffuse* clouds. Not only molecular abundances, but also the ionization structure of molecular clouds depend critically upon the metal content; this is unobservable in dense clouds and difficult to predict because it depends strongly on temperature and on grain composition and recycling, properties which cannot be assumed to resemble the conditions found in the few (possibly unusual) diffuse clouds examined by Copernicus. Third, the time-dependent aspect is not treatable in a realistic manner without tying the chemistry self-consistently to the detailed physical evolution of the cloud (cf. Glassgold and Langer 1976), about which little is known. Current time-dependent models largely ignore these larger evolutionary questions and simply follow the chemistry forward from a switch-on time.

Given such uncertainties, it is surprising how successful ion-molecule chemistry has proved to be, at least for the simpler (four atoms or fewer) species. There is, in general, overall agreement between the abundance predictions of the ion-molecule models and observed abundances for most simple compounds containing first-row elements, both in diffuse and dense, quiescent clouds (Watson and Walmsley 1982, Watson 1984). Direct evidence for ion-molecule chemistry exists from the observations as well, namely, the detection of the ions HCO^+, N_2H^+, HCS^+, $HCOO^+$, and $HCNH^+$, predicted by ion-molecule theory to be the direct product of proton transfer to neutral, stable species from H_3^+. HCO^+ is predicted to be particularly abundant, as it is destroyed only by dissociative recombination, which should be slow because of the low concentration of free electrons. A high abundance is indeed observed for HCO^+ (e.g., Turner and Thaddeus 1977). Also encouraging is that the earlier detailed analyses of such species as OH (Turner and Heiles 1974), HCN (Turner and Thaddeus 1977, Huntress 1977), H_2O (Huntress 1977, Herbst 1978), and CN (Turner and Gammon 1975) indicated reasonable quantitative agreement between observations and theory, even though only a few reactions were included in models at the time. In particular, not only the abundance ratios of these fundamental species, but also their predicted dependence on overall particle density, seemed borne out by observations.

A reasonable summary of the current status of ion-molecule models for simple species is found in the models of Graedel et al. (1982). They analyze both a "high metal" (as observed by Copernicus in ρ Oph) and a "low metal" (a factor of 100 in depletion) model as well as time dependence. Carbon-oxygen chemistry seems successfully explained, as exemplified by CO, OH, and H_2CO for either model. Several other important species are predicted well by the low-metal model, such as HCO^+ and CS. The high-metal model explains only a few abundances in cold clouds but appears to explain observed abundances of C, CO, and HCO^+ in warm cores. The conspicuous failures are N_2H^+ and NH_3, observed to be more abundant by a factor of 100 or more than predicted by either model. It is possible that ion-molecule theory incorrectly treats chemistry beginning with atomic N. This may be further indicated by its failure to predict the rarity of $N{=}O$ bonds among polyatomic interstellar molecules or to explain the observed relatively low abundance of NO itself (Liszt and Turner 1978).

Aside from the obvious successes in explaining the observed molecular ions, one other aspect of interstellar chemistry, that of the large concentrations of deuterium observed in several species in cold cloud cores, seems uniquely explained by ion-molecule chemistry. The observed ratios, $DCO^+/HCO^+ \sim 0.01$, $N_2D^+/N_2H^+ \approx NH_2D/NH_3 \approx DCN/HCN \sim 0.001-0.01$, even though $D/H = 2 \times 10^{-5}$, are predicted naturally from the few ion-molecule reactions that govern these species (e.g., Watson 1978, Turner and Zuckerman 1978), provided that the fractional ionization is very low ($\lesssim 10^{-8}$). If $T < 15$ K, where metals freeze onto grains producing an attendant drop in x_e, both the molecular ion abundances and the fraction that contain deuterium are predicted to increase, as observed in cold cores as compared with warm cores. The large ratios observed for DC_3N/HC_3N, DC_5N/HC_5N, and $HDCO/H_2CO$ strongly suggest these species are also governed by ion-molecule reactions (e.g., Langer et al. 1980), regardless of which ion pathway dominates their production. By contrast, no mechanism yet envisioned for reactions on grain surfaces at low temperatures can produce such large deuterium fractions in the gas phase. Even more certain is that surface reactions would not produce a higher fractionation in cooler than in warmer clouds, as observed, because surface deuterium-exchange reactions have measured activation energies of ~ 1000 K (Watson 1978).

Some problems remain for ion-molecule chemistry. An example is the HNC/HCN ratio. Both species are thought to form from dissociative recombination of $HCNH^+$, in approximately equal amounts so that $HNC/HCN \simeq 1$. The observed ratio varies anywhere from 0.03 to 0.4 in warm cores, and may be as high as 4.4 in cold cores. Other processes are probably affecting the HNC/HCN ratio but alternate pathways are yet to be established. Another problem is that of CH^+, observed in diffuse clouds to be far more abundant than predicted by ion-molecule chemistry. There are no recognized ion-molecule reaction chains that can form CH^+ fast enough to compete with its rapid destruction by electron recombination. The endothermic reaction $C^+ + H_2 \rightarrow CH^+ + H$ seems necessary to account for the CH^+ abundance, thus requiring high temperatures and shocks (Watson 1984). In addition, the very large fractional abundance of CH recently observed in several warm and dense cores (Ziurys and Turner 1985) has yet to be accounted for by ion-molecule processes. Perhaps the most striking class of problems is illustrated by the sulfur-bearing molecules such as H_2S and SO, which seem to require high-temperature reactions to reproduce their observed abundances. These problems indicate that ion-molecule chemistry does have its limitations. Further observations, particularly in cold dark clouds where high-temperature processes can be ruled out, are needed to establish the intrinsic limits of applicability.

The discussion so far pertains mostly to dense molecular clouds. Diffuse cloud chemistry, in which starlight is at least as important as cosmic rays in maintaining ionized species, may be a more reliable test of ion-molecule chemistry than is dense cloud chemistry. Many fewer reactions need be considered, because elements other than hydrogen are primarily in atomic form, and the physical conditions are better known. The dependence of abundances of species such as CN, CO, CH, and

OH on total particle density seems well explained (Federman 1986), but actual abundances of CN and CO agree rather poorly with observations. The recently recognized importance of H_3^+ as an initiator of ion-molecule formation even in diffuse clouds has yet to be incorporated in detailed models and may well correct the discrepancies.

5.4.2 Surface Chemistry on Dust Grains

The best evidence for the role of dust grains in interstellar chemistry is the abundance of gaseous H_2 in molecular sources. No gas-phase processes are able to account for this species' abundance. On the other hand, recombination of hydrogen atoms on grains, i.e.,

$$H(gas) + H(gas) + grain \rightarrow H_2(gas) + grain$$

is known to occur readily for many types of surfaces. The high surface mobility of H atoms, combined with the extreme volatility of H_2 molecules, which allows them to pop easily off the grains, makes this process extremely efficient.

It is uncertain, however, whether surface chemistry is important for the synthesis of molecules other than H_2. Facts which suggest surface reactions may play some sort of role are the following. (1) Even in cold clouds, the rate of incidence of gas particles upon grains is comparable, if not faster, than two-body gas-phase collision rates. In fact, it is puzzling why all molecules are not condensed out onto grain surfaces in cold clouds. The probability of the incident particle sticking to the grain is of order unity for "heavy" atoms or radicals, and probably not less than 0.2 for H atoms, even on the most unfavorable surfaces, provided that they contain defect sites. (2) Ample observational evidence exists (Jura and York 1978) that grains selectively adsorb heavy atoms and cause the observed selective and variable depletion of the heavy refractory elements (Si, Mg, Fe, and Ca). (3) Recombination of atoms on surfaces is known to be important in the laboratory, even at quite low (77 K) temperatures. (4) UV photons at interstellar wavelengths may be effective in exploding small grains and releasing their molecules, as shown in the laboratory (e.g., Allamandola 1984, Greenberg 1986).

By contrast with gas-phase reactions, those on grain surfaces are poorly defined, especially since the nature of the grain material and the state of its surface are not well known. Certainly, the four important processes in any model of surface reactions are (1) adsorption of a radical or atom from the gas; (2) migration over the surface via quantum-mechanical tunneling; (3) reaction with a second adsorbed atom or molecule; (4) ejection or evaporation of the product molecules back to the gas phase. The details of each of these processes depend critically on the temperature, composition, and physical properties of the surface. Independent of the composition, extensive laboratory work suggests that in *cold* clouds ($T_{gr} \lesssim 20$ K) the binding energy to grains of saturated molecules, and probably also of atoms and radicals, will be weak, corresponding to physical adsorption. The reason is that even if grains do not initially contain ice mantles or low-binding monolayers of H_2, they will soon accumulate them. On initially more strongly binding surfaces, evaporation is not possible, so condensation of H_2 or of ices of H_2O, NH_3, and CH_4 will occur, on top of which the binding is weak. In *warm* clouds, where IR observations indicate $T_{gr} \sim 40$ to 80 K, ice mantles or H_2 layers will evaporate, surfaces will be chemically active, and atoms or radicals will bind tightly. Saturated molecules will still be only weakly bound.

In order to incorporate these ideas into a meaningful scheme for forming interstellar molecules, we require answers to several questions. (1) Is the heat of formation adequate to eject newly formed molecules from grain surfaces? The answer appears generally to be no. In some cases, experimental evidence allows educated guesses. (2) Are activation barriers generally small enough to permit surface reactions, and are there significant differences in these barriers and in surface mobilities from one species to another? Generally, activation barriers seem sufficiently small between atoms or radicals, but not necessarily between a radical and a saturated species, and not between saturated species. Mobilities definitely depend on the particular species, based on limited data. (3) Are there alternative (nonthermal) ejection mechanisms for newly formed surface molecules? Watson (1978) has considered photodesorption, thermal pulses in small grains, and shock waves. Based on experiment, photodesorption in diffuse clouds would appear to proceed at adequate rates but just what molecules or molecular fragments are desorbed is difficult to predict. Similarly, explosive grain disruption by UV photons (Allamandola 1984, Greenberg 1986) yields unpredictable products. The same statement can be made about desorption by sputtering, as occurs in shocks. There are strong arguments against desorption by thermal pulses, as initiated by photons, cosmic rays, or energy of recombination (Watson 1978).

Detailed models have been made of interstellar molecular abundances expected solely by grain synthesis (Watson and Salpeter 1972, Iguchi 1975, Allen and Robinson 1977), and also as a result of gas-phase reactions together with surface reactions (Millar 1980, Pickles and Williams 1981, Tielens and Hagen 1982). The results of these studies are highly varied, as may be expected from the very different assumptions made. None of the results matches the observational picture very well in general nor appears capable of the degree of specificity needed to describe, for instance, the different relative abundances found in the cold versus the warmer clouds. Iguchi (1975) does present evidence that the mobilities of N and O atoms, or of small molecules containing them, are less than for C atoms or small C-molecules on grain surfaces that contain ice mantles. If so, then molecules containing $N{=}O$ bonds would form less readily than those with C bonds.

(a) Difficulties for Grain-Surface Chemistry: Diffuse and Cold Clouds

The following difficulties for grain-surface chemistry may be cited. (1) Poor abundance predictions: compared with ion-molecule reactions, the relative abundances of the simple species are poorly predicted and are very sensitive to input assumptions. (2) Cold versus warm cloud predictions: largely because of the problem of ejection, abundances of simple species (HCN, H_2CO, and HCO^+) are not predicted to be larger in cold clouds than in warm ones, contrary to observations. (3) The NH/OH ratio: in diffuse clouds, where H is abundant, it impinges on grains at ~ 1000 times the rate of C, N, and O atoms. Under conditions near physical adsorption, where surface mobilities are unrestrained, hydrides should be formed predominantly over other species on grain surfaces. In particular, NH and OH should form about equally per atom (Watson 1978) so that if the optically observed

OH in diffuse clouds is produced on grains, then NH should be present in moderate abundances. A search for NH (Crutcher and Watson 1976) gives NH/OH ⪍ 0.01, which does not augur well for surface chemistry dominating at the diatomic level, although the fact that $N/O \approx 0.14$ as well as other uncertainties do not rule out surface reactions at this level. (4) Enhanced deuteration: surface reactions have been argued as incapable of producing a greater gas-phase deuterium enhancement in cold clouds than in warm clouds as observed, because activation energies for surface deuterium-exchange reactions typically exceed 1000 K (Watson 1978, Tielens 1983). In fact, because HD escapes less readily than H_2 from grain surfaces, the D/HD ratio in the gas phase is higher than the cosmic ratio, so that molecules produced from gaseous D and H on surfaces will subsequently have a higher deuterium fraction than those in the gas phase, if surface formation rates exceed gas-phase formation rates (Tielens 1983). The effect is typically a factor of 2 over the gas phase, and increases at lower temperatures as the escape of HD becomes more difficult. Contrary to observations, this mechanism predicts, if anything, a *decrease* of gas-phase deuterium enhancement at lower temperature as more D is lost to the surface. (5) Long-chain cyanopolyynes are more abundant in cold cores: these species, characterized by highly reactive triple carbon bonds, should adhere firmly to grain surfaces, especially at low temperatures. It has been speculated that long-chain molecules might be the expected products of grain disruption (Webster 1979). This mechanism involves high temperatures behind shocks and, therefore, is contrary to the observed temperature dependence of long-chain abundances.

(b) Grain-Surface Chemistry in Warm Clouds
Evidence for or against surface chemistry in warm clouds is sparse. Here, the grains are not covered with ices, and their basic composition becomes relevant. Absorption features in the IR spectrum in the 2- to 20-μm region have long suggested a composition mainly of silicates. Other traditional candidates are graphite grains or grains with mantles of organic molecules (cf. Hoyle and Wickramasinghe 1976). Combinations of graphite particles and silicates of Fe and Mg have been shown (Mathis et al. 1977) to match the interstellar extinction law over the entire range $0.11 \leqslant \lambda \leqslant 1$ μm. Duley and Williams (1979) suggest that the absence of a C≡H vibrational transition at 3.3 to 3.4 μm in the spectra of several objects implies that $<1\%$ of a cosmic complement of carbon in grains can exist as organic molecules. A third possibility is small (< 100 Å) grains composed of metallic oxides (Millar and Duley 1978). Oxides are well-known laboratory catalysts, their reactivity residing in lattice defects. Such grains appear able to deplete selectively certain elements (Fe, Mg, Ca, and Al), in accordance with the observations (Duley and Millar 1978), and may be particularly effective in catalyzing the formation of interstellar sulfur compounds such as H_2S, OCS, HNCS, and CH_3SH in dense clouds (Millar 1982), so that shock heating does not have to be invoked to drive endothermic reactions (Hartquist et al. 1980).

Some experimental data exist about the types of molecules synthesized on these kinds of grains:

(i) *Silicate surfaces.* Anders et al. (1974) show that the formation of a large variety of organic molecules can be catalyzed on silicate surfaces, at temperatures of ~ 500 K, including probably the HC_nN species. These Fischer-Tropsch reactions are inoperative at the much lower temperatures of even the warm molecular clouds, but may well be productive in the most energetic star-forming cores and in circum-stellar envelopes. These reactions predict a fall-off in abundance with increasing number of carbon atoms that appears much steeper than observed in interstellar species. This, and the failure to detect many other species predicted to be abundant, suggests that at least the energetic star-forming cores cannot have ejected enough material into the surrounding molecular cloud to alter its molecular composition significantly.

(ii) *Graphite grains.* Experiments using graphite surfaces (Bar-Nun 1975; Bar-Nun et al. 1980) show that at low temperatures ($\lesssim 20$ K), impinging H atoms will form methane and other hydrocarbons which form a monolayer on the surface. This mantle could be polymerized by short-UV photons and possibly low-energy cosmic rays; such polymers may be indicated by several IR emission bands in interstellar clouds (Knacke 1977, Greenberg 1986), but it should be noted that once polymers are formed on grains, they would not be readily removed during modest heating of the grains or by other removal mechanisms which are effective for small molecules. Thus the graphite loses its identity at low temperatures. At higher temperatures (>77 K), CO, CO_2, and hydrocarbons are synthesized on bare graphite surfaces from reactions of H, N, O, and S. Now, however, a carbon atom is removed from the graphite for each molecule formed, a process that would soon erode away interstellar grains, which according to observations actually seem to accrete with time (Carrasco et al. 1973). Grain lifetimes would be only $\sim 3 \times 10^5$ yr, much shorter than cloud lifetimes. Moreover, although graphite synthesis might operate in warm cores, the predicted distributions of H_2CO and hydrocarbons are not observed, even in the close vicinity of the cores where diffusion times for the synthesis products are short. Graphite synthesis in circumstellar envelopes may also confront the difficulty that observed species are not predicted.

(iii) *Metallic oxide grains.* Alkaline-earth oxides are well-known catalysts. Based on experiments at 77 K, Duley et al. (1978) have modeled the active sites on $MgO/FeO/SiO/CaO$ grains. In diffuse interstellar clouds, these sites are maintained by UV and cosmic rays, and can apparently form species such as H_2O, HCO, NH_3, and H_2CO with high efficiency. Ejected by the heat of formation, these species are rapidly photodissociated to form CO, OH, and NH. The observed limits on the NH/OH ratio pose a difficulty for this model as it did for the more general considerations mentioned previously. The large observed abundance of H_2CO in diffuse clouds might be possible through formation on oxide surfaces (Millar et al. 1979) but the former difficulty in producing large H_2CO abundances via ion-molecule reactions has been removed (Graedel et al. 1982). In *dense* clouds the active surface sites of oxide grains appear to become poisoned and, at best, might form only molecules with several heavy atoms (Duley et al. 1978). Such molecules apparently cannot be released to the gas. Thus it appears that oxide grains are more likely to account for depletion of the elements than for production of gas-phase molecules.

(c) Summary

As formation sites of interstellar molecules, grains appear to have many difficulties and few positive attributes. With the exception of H_2, grain-surface reactions do not seem to produce any observed molecular abundances that cannot be accounted for by gas-phase processes. Certainly this and additional lines of observational evidence argue against the importance of surface reactions in cold clouds. In warm cores, grain reactions may augment gas-phase reactions.

As a source of interstellar molecules through disruption, grains are more promising. Historically, the enhanced abundances of atomic Na and K in highly energetic regions has been attributed to sputtering of grains by passing shocks. The recent strong observational limits placed on the MgO abundance (MgO/SiO < 10^{-3} Mg/Si) have been interpreted in terms of the breakup of metal oxide grains by shocks or high-temperature volatization (Turner and Steimle 1985). The observations of high HDO and NH_2D concentrations in the hot, energetic star-forming cores of Ori(KL) and W51 have been interpreted in terms of grain breakup, which releases the higher concentrations of deuterium trapped on the grains (Olofsson 1984, Walmsley et al. 1986).

While grains appear to lose out to gas-phase chemistry in cool and even warm core molecular environments, it will be much more difficult to separate the effects of grain disruption and of high-temperature gas-phase chemistry in energetic (shocked) regions.

5.4.3 Shock Chemistry

As discussed in Section 5.2, there are regions in massive star-forming cores of molecular clouds where both temperatures and densities are observed to be unusually high, viz., $T \gtrsim 1000$ K and $n(H_2) \sim 10^7$ to 10^{11} cm^{-3} (e.g., Scoville et al. 1979). Molecular spectra from such regions usually show broad line wings (e.g., Bally and Lada 1983), indicative of high-velocity mass outflow. In diffuse clouds, we saw (Section 5.2.5) that peculiar velocity differences occurred between CH$^+$ and other species, and that CH$^+$ abundances seemed to correlate with regions of high temperature. The unusual physical conditions in each of these cases have been attributed to the passage of interstellar shock waves (e.g., Draine et al. 1983).

The obvious existence of shocks in molecular gas has led to the development of "shock chemistry." The basic principle is that the heating of gas caused by the passage of shock waves overcomes activation energy barriers present in chemical reactions. The result is that endothermic reactions and exothermic ones with activation barriers can now occur, in particular, neutral-neutral processes—reaction pathways that are otherwise inaccessible in cold gas. Consequently, "ion-molecule" abundances become greatly modified. A classic example is oxygen chemistry. At low temperatures, O does not react with H_2, and O$^+$ must react with H_2 to form OH and H_2O. At high temperatures, however, $O + H_2 \rightarrow OH + H$ and $OH + H_2 \rightarrow H_2O + H$ are rapid, and the abundances of both OH and H_2O are enormously enhanced, which probably explains the high abundances necessary in the powerful OH and H_2O masers associated with massive-star formation.

As the shocked gas cools, the pressure tends to remain constant, so the density must rise, possibly by a factor of 100 (in nonmagnetic shocks); this affects the chemistry in additional ways.

Elevated temperatures in interstellar gas can result from processes other than shocks, such as embedded protostars. Such temperatures do not approach those caused by shocks, except very near the protostar, and only temporarily. Such protostars are usually accompanied by shock-producing outflows, so direct heating effects by the IR source are masked. Since shocked gas cools rapidly (a few hundred years), "shock chemistry" products may become frozen for some time in the post-shock gas (e.g. Mitchell 1983). Because of these complicating factors, the terms "shock chemistry" and "high-temperature chemistry" are virtually synonymous in the interstellar context.

Shock chemistry is thus more complex than quiescent cloud chemistry, both because of the many variables associated with the shocks and because of the many additional reactions which operate at the high temperatures. Numerous shock chemistry models are available in the literature (e.g., Iglesias and Silk 1978, Elitzur and Watson 1978, 1980; Hartquist et al. 1980, Mitchell and Deveau 1983, Mitchell 1983, 1984). The models are, in principle, similar to those used in ion-molecule chemistry; they consist of complicated networks of chemical reactions built into extensive computer codes, including endothermic reactions and exothermic ones with otherwise insurmountable activation energies. The coupled kinetic equations are solved for a particular shock velocity, which determines the temperature of the postshock gas. Current models are rudimentary, however. Much of the chemical data needed for modeling is not available. In addition, most models deal only with nonmagnetic, nondissociative shocks. While these may be applicable to some diffuse cloud conditions, line wings are often observed in star-forming regions which indicate shock velocities that well exceed ~ 20 km s^{-1}, the velocity at which molecules will be dissociated in the absence of magnetic effects, which tend to smooth out the shock acceleration and compression. Dissociating shocks ("J type" shocks) reduce all molecules to atoms, and the molecules reform only after H atoms recombine on grain surfaces to form H_2. "C-type" or MHD shocks, which include the smoothing effects of "magnetic precursors," cannot be meaningfully modeled because nothing is known of the magnetic fields in star-forming regions.

Despite the incompleteness of the models, "shock chemistry" has been successful in explaining several, otherwise-puzzling observational results. One of the most striking is the large abundance of CH^+ in diffuse clouds, which is ~ 100 times the abundance predicted on the basis of ion-molecule chemistry. Shock chemistry models can account for the observed CH^+ concentration with the incorporation of a few key endothermic reactions (Elitzur and Watson 1978, 1980). Evidence for the ~ 10-km s^{-1} shock needed in the model is provided by the velocity shift between CH^+ and other molecules observed toward several diffuse clouds (Federman 1982), which suggests that CH^+ occurs in the postshock region, and also by the correlation between $N(CH^+)$ and highly rotationally excited H_2, which Federman (1986) attributes to high temperature.

The peculiar chemical behavior of the highly refractory SiO in molecular clouds is another instance where shock chemistry may be the only viable explanation. Toward several dense GMC cores, SiO emission is found to arise only from spatially confined regions which are associated with high-velocity outflows (e.g., Downes et al. 1982). This behavior is particularly striking towards Ori(KL), where inter-ferometer maps have shown SiO to exhibit a very high column density confined to a small clump of gas (8 to 10 arcsec), centered on infrared source IRc2 (Wright et al. 1983), an apparent protostar undergoing high-velocity mass loss. The small spatial extent of SiO in Ori(KL) could be a result of excitation, but other molecules such as HCO^+, which have comparable dipole moments, are observed to be extended over several arcminutes in Orion (Turner and Thaddeus 1977). The effect is thus purely chemical. In addition, SiO is not seen in dark clouds down to a low level, yet is observed in the more diffuse "postshock" gas of supernova remnant IC 443 (Ziurys et al. 1987). These characteristics of SiO suggest its formation via shock chemistry, although the destruction of silicate-type grains in the shock waves could equally well contribute to its unique abundance and distribution. The quantitative study of grain disruption by shocks has only just begun, and early results suggest that the chemical products may be highly specific. For example, Duley and Boehlau (1986) find that the silicates Mg_2SiO_4 (forsterite) and $MgSiO_3$ (enstatite) both evaporate to yield SiO, but only the latter also yields MgO, a species not observed in shocked regions (Turner and Steimle 1985).

Other molecules also probably show the effects of shock chemistry. HCO^+ seems to be overly abundant in the postshock material of supernova remnant IC 443, relative to dark clouds (Dickinson et al. 1980, DeNoyer and Frerking 1981), as well as in the "high-velocity outflow" gas in Ori(KL) (Vogel et al. 1984). OH is also observed to be several orders of magnitude more abundant towards Ori(KL), as opposed to the quiescent gas of the extended Orion cloud (Watson 1982); the large column densities associated with main-line OH masers may in part be due to shocks (Elitzur 1982). Finally, recent observations have suggested that CH may be formed in shocks towards warm, dense cloud cores. Detection of the first excited rotational state of CH in numerous GMCs has shown that very high column densities of CH must exist in dense ($\gtrsim 10^6$ cm^{-3}) gas (Ziurys and Turner 1985). Such conclusions are in conflict with ion-molecule chemistry models (e.g., Prasad and Huntress 1980), which universally predict small abundances of CH in moderately dense gas ($\sim 10^4$ cm^{-3}) and its complete disappearance as densities become large ($\gtrsim 10^6$ cm^{-3}). The high concentration of CH is explainable by shock chemistry; calculations have suggested that even in dense clouds, diatomic hydrides achieve large postshock abundances (Mitchell 1984).

Finally, sulfur-containing molecules are predicted to have enhanced abundances behind shocked gas (Mitchell and Deveau 1983). Both $S^+ + H_2 \rightarrow SH^+ + H$ (diffuse clouds) and $S + H_2 \rightarrow SH + H$ (dense clouds) are endothermic, but the latter reaction can proceed behind shocks and initiate sulfur chemistry. This can explain the frequent overabundance of S-species relative to O-species relative to the cosmic S/O ratio. One would predict that such an overabundance would not be seen in cold clouds, where there is no evidence for shocks.

Observations are thus quite suggestive that shock chemistry may be the most important source of molecular synthesis in active gas where outflows and shock waves are present. More modest shock chemistry might also occur in dark clouds where low-mass-star formation is occurring, although the flows are not very energetic and ion-molecule chemistry itself seems satisfactory. The overall role of shock chemistry, including the disruption of grains, has yet to be fully evaluated, and thus it remains one of the important frontiers of astrochemistry.

5.4.4 Circumstellar Chemistry

Circumstellar chemistry is traditionally explained in terms of the chemical equilibrium "freeze-out" model (e.g., McCabe et al. 1979, Scalo and Slavsky 1980, Lafont et al. 1982). As previously described (Section 5.2), circumstellar shells exhibit a wide range of physical conditions, from those of the inner envelope ($T \sim 1000$ K; $n(H_2) \sim 10^8$ to 10^{11} cm^{-3}) to those of the shell edges where densities and temperatures are more similar to those in an interstellar molecular cloud ($T \sim 10$ K; $n \sim 10^3$ cm^{-3}). UV photons penetrate the gas at the shell edge. The high temperatures and densities that exist in the inner envelope mean that activation energy barriers can be overcome and three-body collisions can occur. Molecules thus form under conditions of thermochemical equilibrium in the inner-shell regions, and ion-molecule reactions are unimportant for molecule formation. The expansion of the inner shell to its outer edges is thought to be so fast that the abundances of the inner envelope, i.e., the "equilibrium abundances," are "frozen" throughout the shell. This simply mean that on the time scale of the shell expansion, chemical reactions cannot occur quickly enough to noticeably alter abundances determined in the inner regions before the molecules arrive at the shell edge. Thermochemical equilibrium is thought to prevail near the photosphere, where $n \sim 10^{15}$ cm^{-3}, and the "freeze-out" is thought to occur at densities less than 10^{10} cm^{-3}. At this point, grains are believed to start forming as well, at least in some models. Because of the freeze-out, molecular abundances in the shell should be primarily stable species, and not radicals and molecular ions. At the shell edges, however, UV photons will penetrate the shell gas and photodestruction will occur. The chemical composition of the shell is then thought to be changed to chiefly photodestruction fragments of stable molecules, and many free radicals should be present. It may also be the case that ion-molecule reactions play some role at the shell edge as well. In such cases, however, photoionization probably creates the ions (e.g., Nejad et al. 1984). Indeed, recent studies of shell chemistry suggest that $C_2H_2^+$, produced from photoionization of acetylene, is an important reactant leading to the formation of the long-chain carbon species (Glassgold and Huggins 1985).

Of course, the details of the chemistry will depend on whether the shell involved is O-rich or C-rich. O-rich circumstellar envelopes have been modeled (using the "freeze-out" assumptions) in detail by Slavsky and Scalo (1985), and a review of chemical processes important for C-rich shells is given by Huggins and Glassgold (1985). To summarize briefly the reactions and species relevant to each type of shell, O-rich envelopes are characterized by principally neutral-neutral reactions involving mostly H, O, S, Si, and N. Two important reactions are $H_2 + O \rightarrow OH + H$ and $H_2 + OH \rightarrow H_2O + H$. Other principal formation chains are those concerning sulfur species, which synthesize SH, H_2S, SO, and SO_2. Silicon-containing compounds comprise another reaction cycle, producing SiS, SiO, and SiH. NH_3 is formed through additions of H_2 to N. Important reactions in C-rich shells are those synthesizing CO, HCCH, HCN, SiS, and SiO, which again are chiefly neutral-neutral reactions (Lafont et al. 1982). CO is formed through C + O, while HCCH is made by the reaction $CCH + H_2 \rightarrow HCCH + H$, and HCN via $CN + H_2 \rightarrow HCN + H$. The long-chain species are thought to be formed, on the other hand, by ion-molecule reactions involving $C_2H_2^+$, at the shell edge.

There are several problems with the freeze-out model, however, the first of which is that most models use a single freeze-out location for all molecules (Glassgold and Huggins 1985). This assumption is unrealistic, as less reactive species will tend to freeze out close to the inner envelope, while reactive molecules may not become "frozen" until well into the shell. Also, some molecules may heavily condense out on grain surfaces, which models have not yet included. In addition, thermochemical data are not always available for all species possibly present in the shells, and assumptions must be made. Nonetheless, predictions of the equilibrium "freeze-out" model have been reasonably successful, as we discuss next.

(a) Carbon-Rich Envelopes

Almost all observations of molecules in C-rich shells have been made for one object, IRC + 10216. Thus, comparison of observation with theory means comparison with IRC + 10216, in general. A half-dozen or so molecules have been detected towards two other C-rich shells, namely, CIT-6 and AGL 2688, and their abundances do not significantly differ from those in IRC + 10216 (Glassgold and Huggins 1985). Hence, IRC + 10216 may be representative of these types of shells.

Several thermochemical equilibrium models have been made for IRC + 10216, including those of McCabe et al. (1979) and Lafont et al. (1982). These models do not consider the effects of UV radiation at the envelope edge. In general, the abundances of the simplest molecules containing C, N, O, and H are well predicted by the models, including species such as CO, HCCH, HCN, and CH_4 (e.g., Glassgold and Huggins 1985). The abundance of HC_3N is also predicted accurately, as well as that of HNC, such that the HNC/HCN ratio, predicted under conditions of thermoequilibrium, is $\leqslant 100$, very different from the value of about 1 found in interstellar clouds. Several other molecules have abundances that are not so well reproduced, including NH_3, HC_5N, and most radicals, especially CN and CCH. All of these species are observed to be far more abundant than the calculations predict. In contrast, SiS and SiO are calculated to have abundances 10^3 or so greater than observed.

The discrepancy concerning silicon species may be due to the fact that most of the silicon in the shell is contained in grains, consistent with the 11-μm spectral feature observed toward IRC + 10216; alternatively, the molecules may simply be highly condensed onto grain surfaces (McCabe et al. 1979). The observed high radical abundance (versus prediction) can be explained by photodissociation at the shell edges, which is not considered in the models. However, the discrepant predictions for other species probably represent definite failures of the "freeze-out" model for C-rich shells.

(b) Oxygen-Rich Envelopes

Chemical models of O-rich stars are not as common as for C-rich stars, although recently a few have appeared (Scalo and Slavsky 1980, Slavsky and Scalo 1985). In addition, not very many chemical species have been observed in O-rich shells, and those that have are often masering, making abundance estimates difficult (e.g., OH, SiO, and H_2O). The presence of these masers at least indicates sufficient column densities are present to produce significant maser gain. OH may be linked to the chemistry of the outer shell. OH is thought to be quickly converted to water in the shell envelope. If present with the considerable column densities required to produce the observed maser gains, it would be expected to occur in the outer shell, where it is formed from UV dissociation of H_2O. Indeed, observations have shown that OH masers arise in the shell at a radius of about 10^{16} to 10^{17} cm, in agreement with calculations of sizes of photoproduced shells (Glassgold and Huggins 1985). A few other qualitative comparisons can be made between theory and observation. Freeze-out calculations have shown that SiO should be the principal Si-containing molecule in O-rich shells. Observations of SiO in the shells, and the failure to detect SiS to low levels, support this conclusion (e.g., Zuckerman 1980). Other molecules observed in absorption in the IR and therefore deep in the envelope (NH_3 and CO) appear to be satisfactorily explained by the freeze-out model. Over a wide range of temperatures and pressures, most of the oxygen is predicted to occur in CO, H_2O, and OH, and indeed these molecules are observed with large abundances. These few simple conclusions at present exhaust the comparison of theory with observation. So few observational tests of O-rich shell equilibrium chemistry exist that the models have not yet posed any definite problems.

(c) Summary

There is little reason to doubt that equilibrium "freeze-out" chemistry dominates molecule formation throughout at least a good fraction of both O-rich and C-rich circumstellar shells. However, the effects of photodissociation at the shell outer edge, as well as differential condensation onto grains, are not currently understood and may influence the chemistry considerably. The role of shocks in the chemical processes of the envelopes will also eventually have to be evaluated, as observational evidence of such shocks has been found (Hinkle et al. 1982).

5.5 Current Dilemmas and Future Directions

There are many areas of astrochemistry for which current understanding is at best rudimentary, and it is impossible to discuss them all here. Many have been mentioned already. Here are described a few of the older, but as yet unsolved, dilemmas and some of the most recent questions.

5.5.1 The Composition of Interstellar Dust Grains

The composition of dust grains is a long-standing problem in astrochemistry and astrophysics. There is some observational evidence for graphite grains (the 2200-Å dust absorption feature: e.g., Draine 1984), as well as for a silicate composition (the ~ 9.7- and 18-μm dust features: e.g., Willner 1984). The 2200-Å feature has also been attributed to magnesium oxide grains (Millar and Duley 1978). In addition to these features, many other interstellar absorption/emission bands seen at IR wavelengths are thought to arise from dust. Some are postulated to be due to SiC grains, and some are thought to arise from ice mantles composed of NH_3 and H_2O or from H_2CO (see Willner 1984, for a review). While such identifications are based on laboratory data (e.g., Allamandola 1984), these identifications and the inferred composition of dust grains must be treated with caution. First, the bands measured in the laboratory (and in the dust features) correspond to stretching and bending modes of groups of atoms in a bond in the solid state and are not directly attributable to a specific molecule. In addition, the exact frequency of a vibration within a bond depends highly on what the bond's environment is. Slight changes in composition of a solid-state substance can significantly shift its vibrational band frequencies (e g , by 15% for O—H stretch in H_2O/NH_3 mixtures of 1 : 1 vs. 3 : 1 ratios). Unambiguous identification of bands from dust grains is extremely difficult if not impossible.

Recent studies (Puget et al. 1985, Allamandola et al. 1985) have suggested a completely new identification for certain dust features, namely, that they arise from polycyclic aromatic hydrocarbons, or PAHs, lending credence to the possibility that large molecules aggregate into grains. More evidence is needed to substantiate this theory. The recent laboratory results of Kroto et al. (1985) pose an even more interesting suggestion for dust grain composition: on vaporizing graphite with a laser, much of the gaseous carbon assumes the form of a remarkable sixty-atom structure shaped like the pentagon and hexagon outlines of a soccer ball. This C_{60} structure is extremely stable and could, in principle, itself constitute dust grains. Alternatives to graphite composition of carbon-based grains may be indicated. Recent carbonaceous material extracted from meteorites is found to be mostly amorphous macromolecular organic matter rather than well-crystallized graphite (Nuth 1985). The stringent observational limits for the gas-phase abundance ratio MgO/SiO ($<10^{-3}$) have been interpreted as suggesting a magnesium silicate composition for grains (Turner and Steimle 1985). While such results are to be pondered, there still exists little definitive evidence, observational or experimental (cf. Greenberg 1986), for the composition of interstellar dust grains.

5.5.2 Diffuse Bands

The interstellar diffuse bands, which consist of thirty-nine spectral-like features lying between 4400 and 6850 Å in wavelength, remain an astrochemical enigma after more than fifty years of study (see Herbig 1975, for a review). Recently, further diffuse bands have been detected between 6500 and 8900 Å (Sanner et al. 1978). To date, these bands defy identification. Among solid-state candidates have been grain impurities (e.g., Weltner and Savage 1977) and S_2^- and S_3^- molecules in lattices (Exarhos et al. 1982). Gas-phase candidates include the bending/internal conversion modes of such species as C_3 and benzene (Smith et al. 1977), cyanopolyynes (Douglas 1977), and even bands arising from the radical C_4H, which were suggested by the detection of this species in the gas near Cas A (Bell et al. 1983). A recent exhaustive study of the varying properties of four of the strongest diffuse bands over eighty lines of sight by McNally et al. (1986) suggests that they may result from four different carriers, or at least from four different excitation or ionization states. These observations do not differentiate between gas-phase and solid-phase carriers, although McNally et al. favor a gas-phase carrier. They conclude that identification is unlikely to come from astronomical studies, although once laboratory spectroscopic evidence is available, the astronomical evidence will be crucial for a selection among the contenders. Whatever the carrier, there is no doubt that the diffuse interstellar absorption lines represent an important unidentified component of the ISM. If the features are not saturated, it is easy to show that the abundance of their carriers is $\sim 10^{-8\pm1}$ (McNally et al. 1986), comparable with the abundance of interstellar diatomic molecules. Until such a common constituent of the ISM is identified, our knowledge of the ISM must remain incomplete, perhaps seriously so.

5.5.3 Synthesis of Complex Molecules

Accounting for the presence of such large species as HC_9N and CH_3CH_2CN is one of the more challenging questions for astrochemistry (e.g., Watson 1984). It has been suggested that radiative association reactions may be the main pathways for the synthesis of such species (e.g., Bates 1983a, Herbst 1982). Most rates for these reactions producing large molecules are not well known, and generally have been estimated only theoretically (e.g., Bates 1983b). A combination of radiative association and ion-molecule reactions may also be significant in the formation of the long-chain carbon molecules (Freed et al. 1982, Herbst 1983, Suzuki 1983). Leung et al. (1984) have written a complex ion-molecule computer model using such reactions, with the specific goal of predicting the abundances of complex molecules. For the most part, however, the model cannot reproduce the relatively large abundances of such species as CH_3CH_2OH, HC_5N, and $(CH_3)_2O$ found in sources such as Ori(KL), TMC-1, and Sgr B2. Towards Ori(KL), it is interesting that CH_3CH_2CN and $(CH_3)_2O$ appear at LSR velocities that suggest the species arise from the "hot core" gas, which may be as dense as 10^{10} cm^{-3}. At such densities, three-body collisions are becoming significant. Perhaps complex molecules are formed only in very dense regions where three-body collisions can aid in their formation (see Sequence 2 in Fig. 5.2). More observation and theory are clearly needed to establish the synthetic routes to complex interstellar molecules.

5.5.4 Ring Molecules

Since the early days of molecular radio astronomy, a variety of aromatic and aliphatic ring molecules have been unsuccessfully sought in interstellar space, including pyrrole (Myers et al. 1980), imidazole (Irvine et al. 1981), and benzene derivatives such as benzaldehyde (Fertel and Turner 1975). Only recently has the picture changed, with the detection of the first two ring compounds, SiC_2 (Thaddeus et al. 1984; circumstellar) and C_3H_2 (Thaddeus et al. 1985; interstellar). Unlike species previously searched for, both of these molecules are highly reactive radicals. Indeed, the microwave spectrum of SiC_2 has yet to be obtained in the laboratory, as the species is so unstable. Thus far, SiC_2 has been detected only in one object, the circumstellar shell of the late-type star $IRC + 10216$; it is not observed in any of the usual molecular clouds such as Ori(KL) or Sgr B2. In contrast, C_3H_2 seems to be present everywhere, from dark clouds to GMCs to stellar envelopes (Matthews and Irvine 1985). However, unlike SiC_2, whose rotational temperature in $IRC + 10216$ may characterize the species as a "high-temperature" molecule, C_3H_2 appears to favor the colder regions, as evidenced by its high abundance in TMC-1 and its presence in Ori(KL) as only the "spike" component.

The detection of such molecules raises the question of how many more simple ring compounds are present in interstellar gas. There is a good possibility that other such rings are radicals or molecular ions (e.g., $C_3H_3^+$) and hence accurate laboratory frequencies are not available for searches. It is interesting that searches for larger carbon rings have never been successful, although long-chain carbon species are quite abundant in certain sources. Thermodynamically, carbon rings are sometimes energetically favored over their long-chain analogues. Interstellar chemistry is, however, for the most part kinetically rather than thermodynamically controlled.

Therefore, it is likely very difficult to create a low-entropy situation in space in which a long, linear molecule can close into a ring or two conjugated species can come together and create a ring in a Diels-Alder type reaction. On the other hand, ring molecules often have many low energy internal degrees of freedom, and thus their populations can easily be spread over a large manifold of energy levels. Detecting any one transition is then difficult because of the "dilution" effect of the partition function. C_3H_2 and SiC_2, being three-membered rings, are not subjected to so much dilution. If more rings are detected in the future, it will be intriguing to see if any are aromatic.

5.5.5 NO Bonds

A great variety of heteronuclear bonds exist in presently known interstellar molecules (Table 5.1). For example, many interstellar species contain the OH bond, in the form of the hydroxyl group OH (i.e., CH_3OH, CH_3CH_2OH) or the carboxyl group COOH (i.e., HCOOH). Many also contain the CO bond in the form of the carbonyl ($C{=}O$) (e.g., H_2CO, H_2CCO) or the aldehyde (CHO) group (CH_3CHO, NH_2CHO) and in esters (CH_3OCHO). There does, however, appear to be a lack of species containing the NO bond (Turner 1980). The only interstellar molecule that has been observed with this bond is NO itself (Liszt and Turner 1978), although HNO (Guelin 1984), NH_2OH (Turner and Rubin, unpublished), and N_2O (Wilson and Snyder 1981) have been sought, as well as nitric acid (Giguere et al. 1973). Nitrogen bonds to hydrogen and carbon, on the other hand, are quite common in interstellar species, as evidenced by the variety of molecules containing the CN group (CH_3CN, CH_2CHCN, CH_3CH_2CN, and HCCCN, etc.) and the amine groups NH and NH_2 (i.e., NH_2CN, NH_2CHO, HNCO, and CH_2NH, etc.). Certainly the presence of N_2H^+ indicates that a reasonably large abundance of N_2 must exist in the interstellar medium, so at least one NN bond is also common. The overwhelming abundance of hydrogen can explain the predominance of NH-type bonds, but the frequent appearance of CN (and NN?) bonds versus the obvious rarity of NO bonds must be a result of chemistry, as oxygen is more cosmically abundant than carbon and nitrogen by greater than a factor of two and a factor of ten, respectively. Additionally, none of carbon, nitrogen, or oxygen is thought to be subject to much depletion onto grains (e.g., Guelin 1984). The lack of NO bonds therefore requires a chemical explanation. One possibility involves the apparent lack of mobility of N and O atoms on grain surfaces, as compared with C and H atoms (Iguchi 1975), although this would explain the lack of NO bonds only if most molecular formation occurred catalytically on grains, a situation we have discussed as unlikely (Section 5.4).

On the other hand, it could well be that the most abundant interstellar species containing the NO bond have never been sought in the ISM. It is interesting that in terrestrial chemistry, the nitro group, NO_2, is typically more common in compounds than the nitroso group, NO, especially in organic species. Extra bond stability is gained in the delocalization of electrons in the NO_2 group. As a major fraction of known interstellar molecules are "terrestrial," it might well be more productive to search for NO_2-containing species than NO-containing compounds.

5.5.6 Chemistry Involving Second- and Third-Row Elements

At present, the abundances of most simple interstellar molecules that contain only elements from the first row of the periodic table (i.e., C, N, O, as well as H) can be explained fairly well by gas-phase ion-molecule reactions (Watson and Walmsley 1982, Watson 1984). However, little is understood concerning the synthesis of species that contain second- and third-row elements. The only molecules of this class that have been observed thus far are those containing sulfur, silicon, and possibly chlorine, but even with these limited data, there is strong evidence that their chemistry is quite different from that of species composed of first-row elements. Ion-molecule reactions simply cannot account for the measured abundances of these species. For example, sulfur-containing compounds, especially H_2S, have observed abundances that are significantly larger than those predicted by ion-molecule chemistry (e.g., Prasad and Huntress 1982). Also, PO is predicted to be quite prevalent in interstellar gas (Thorne et al. 1984), but searches thus far are negative (Matthews et al. 1986). Silicon-containing species, in addition, seem to be present only in hot or shocked gas, or in circumstellar envelopes (Downes et al. 1982, Ziurys et al. 1987). The fact that interstellar HCl may have been detected (Blake et al. 1985), but not such molecules as MgO (Turner and Steimle 1985), FeO (Merer et al. 1982), and NaH (Plambeck and Erickson 1982), all of which contain heavy elements at least an order of magnitude more abundant than Cl, is also peculiar. Elemental depletions onto surfaces of dust grains and the formation/destruction of such grains probably play a major role, together with shock waves and high temperatures, in determining what molecules of this class are formed.

Because so little is really understood concerning "shock" or "high-temperature" chemistry in the interstellar medium, as well as the chemical significance of grains, studies of molecules containing second- and third-row elements is important. A few theoretical studies do exist concerning the gas-phase chemistry of these species, including models of sulfur (Oppenheimer and Dalgarno 1974, Duley et al. 1980, Prasad and Huntress 1982) and silicon chemistries (Turner and Dalgarno 1977, Millar 1980, Clegg et al. 1983). More observational data are needed for this class of molecules, however, to make such chemical formulation meaningful.

5.5.7 "Biological" Molecules

The chemistry of the interstellar medium is basically organic in nature, and hence it might be expected that biologically significant species such as amino acids or purines and pyrimidines should be present in space. Amino acids, for instance, have been found in meteorities (Kvenvolden et al. 1971). Glycine, the simplest amino acid (NH_2CH_2COOH), has been searched unsuccessfully in the ISM in both conformer forms by Hollis et al. (1980) and Snyder et al. (1983). Glycine is a large asymmetric-top molecule with low-lying vibrational satellites. It therefore has innumerable energy levels that can be populated at interstellar temperatures. While it seems chemically likely that glycine is present at some concentration in sources where complex organic species have been observed, its detection will be difficult, because of the partition-function dilution effect. One other biologically significant molecule, urea [$(NH_2)_2CO$], has been searched unsuccessfully. All such species, however, are structurally complicated with many low-lying states, so that finding them in the

ISM will be difficult, from the viewpoint both of detecting the weak signals and of identifying the many spectral features.

5.5.8 Unidentified Lines

One of the most exciting problems in astrochemistry is the question of unidentified lines (U-lines). Recent spectral surveys by Johansson et al. (1984), Sutton et al. (1985), Cummins et al. (1986), and Turner (manuscript in preparation) have shown that even with modest limits of sensitivity, many U-lines are present in sources such as Sgr B2, Ori(KL), and IRC + 10216. Recent observations have also shown that strong U-lines exist in dark clouds such as TMC-1 (e.g., Suzuki et al. 1984). The number of lines per frequency interval increases exponentially with decreasing flux, indicating that at a brightness level of perhaps 0.01 K, molecular lines will become completely overlapped in objects that have broader lines such as Sgr B2. It may be estimated that perhaps 150 to 200 molecular species are identifiable at levels above the "confusion" level. Many of the U-lines probably correspond to "nonterrestrial"-type molecules; otherwise, they would have been identified. They remain interesting puzzles and clear tests of chemical intuition and spectroscopic knowledge.

5.5.9 A Final Perspective: What Has Astrochemistry Revealed?

The goal in studying astrochemistry is to attain from molecular observations an understanding of the physical nature of the environment in which these molecules are found, the consequences of the chemical composition and how it may vary and affect the evolution of the environment, and finally to place the objects in question with other objects in evolutionary sequences.

To date, we have learned that the chemical composition varies on astrophysically significant time scales and is sensitive to density, temperature, radiation, and elemental abundances. We have probably included by now most of the crucial gas-phase reactions in existing models, as well as probably many that are unimportant. Nevertheless, the rates of these reactions are largely unknown, and these uncertainties, together with the still greater uncertainties about the contribution of grains to formation and destruction of molecules, preclude at present any definitive conclusion that might be drawn from a comparison of models and observations. Only if these uncertainties are smaller than the modifications caused by dynamical events such as shocks may we conclude anything of astrophysical significance.

Diffuse cloud chemistry is probably best understood, because of the large amount of observational information available from both optical and radio methods, and the relatively simple chemistry involved. From diffuse clouds we can predict with confidence that H_2 is indeed the major constituent of photon-free dense clouds, and we can probably set a conservative upper limit of 10^{-16} s^{-1} for the cosmic-ray ionization rate. Almost certainly, warm regions are needed (to produce CH^+).

Dense cloud chemistry is more uncertain. While the simple molecules can be explained by ion-molecule reactions (with a few exceptions), the more complex species obviously involve so many possible reactions and so many uncertainties that it is unclear whether much is learned from the continued detection of such species, or from either the successes or failures of the models. Limited knowledge about physical conditions on small scales and no direct knowledge about elemental abundances are major handicaps. Cloud age is an important aspect of the chemistry, and this raises the possibility that chemistry might be used eventually to determine cloud ages. Depletion of the molecules onto grains is a major process, given the long time scales needed for the gas-phase formation of complex molecular species. Thus, their return to the gas phase must somehow occur, and such a release has not yet been explored in any realistic model, but its effects will surely be profound.

By focusing attention on molecular ions, several of the uncertainties in the chemistry are diminished, and interesting astrophysical conclusions can result. For example, the recent analysis of deuterium fractionation in the molecular ions (Dalgarno 1986) confirms several earlier studies that indicate a fractional ionization of $\leqslant 2 \times 10^{-7}$ in cold dense cores and that the cosmic ratio $D/H \approx 1 \times 10^{-5}$. The high ratio indicates that baryonic matter cannot close the universe.

While some useful conclusions have been reached, much must be done to develop realistic scenarios before astrochemistry is of general utility. The paramount needs are for more relevant data on reaction rates and greater understanding of the evolution of the various regimes of the ISM, to which the chemistry is tied.

Recommended Reading

Gammon, R.H. 1978. "Chemistry of Interstellar Space," Chem. Eng. News. **56**:21.
Solomon, P.M. 1973. "Interstellar Molecules," Physics Today **26**(3):32.
Thaddeus, P. 1981. "Radio Observations of Molecules in the Interstellar Gas," Phil. Trans. R. Soc. London **A303**:469.

References

Adams, W.S. 1941. Astrophys. J. **93**:11.
Adams, W.S. 1949. Astrophys. J. **109**:354.
Allamandola, L.J. 1984. In M.F. Kessler and J.P. Phillips (eds.), Galactic and Extragalactic Infrared Spectroscopy. Dordrecht: Reidel, p. 69.
Allamandola, L.J., A.G.G.M. Tielens, and J.R. Barker. 1985. Astrophys. J. (Lett.) **290**:L25.
Allen, M., and G.R. Knapp. 1978. Astrophys. J. **225**:843.
Allen, M., and G.W. Robinson. 1977. Astrophys. J. **212**:396.
Anders, E., R. Hayatsu, and M.H. Studier. 1974. Astrophys. J. (Lett.) **192**:L101.
Avery, L.W., T. Oka, N.W. Broten, and J.M. MacLeod. 1979. Astrophys. J. **231**:48.
Bally, J., and C.J. Lada. 1983. Astrophys. J. **265**:824.
Bar-Nun, A. 1975. Astrophys. J. **197**:341.
Bar-Nun, A., M. Litman, and M.L. Rappaport. 1980. Astron. Astrophys. **85**:197.
Bates, D.R. 1983a. Astrophys. J. (Lett.) **267**:L121.
Bates, D.R. 1983b. Astrophys. J. **270**:564.
Bates, D.R., and L. Spitzer. 1951. Astrophys. J. **113**:441.
Batrla, W., T.L. Wilson, P. Bastien, and K. Ruf. 1983. Astron. Astrophys. **128**:279.
Beckwith, S. 1980. In B.H. Andrew (ed.), Interstellar Molecules, Int. Astron. Union Symp. No. 87. Dordrecht: Reidel.
Bell, M.A., and H.E. Matthews. 1985, Astrophys. J. (Lett.) **291**:L63.
Bell, M.A., P.A. Feldman, and H.E. Matthews. 1983, Astrophys. J. **273**:L35.
Benson, P.J., and P.C. Myers, 1983. Astrophys. J. **270**:589.
Blake, G.A., J. Keene, and T.G. Phillips. 1985. Astrophys. J. **295**:501.
Blitz, L., and P.M. Thaddeus. 1980. Astrophys. J. **241**:676.
Carrasco, L., S.E. Strom, and K.M. Strom. 1973. Astrophys. J. **182**:95.
Chernhoff, D.F., Hollenbach, D.J., and McKee, C.F. 1982, Ap. J. (Lett.), **259**:L97.
Churchwell, E., and J.M. Hollis. 1983. Astrophys. J. **272**:591.
Clegg, R.E.S., L.J. van Ijzendoorn, and L.J. Allamandola. 1983. Mon. Not. R. Astron. Soc. **203**:125.
Combes, F., E. Falgarone, J. Guibert, and Nguyen-O-Rieu. 1980. Astron. Astrophys. **90**:88.
Crutcher, R.M. 1979. Astrophys. J. (Lett.) **231**:L151.

Crutcher, R.M., and W.D. Watson, 1976. Astrophys. J. **220**:124.

Crutcher, R.M., E. Churchwell, and L.M. Ziurys. 1984. Astrophys. J, **183**:668.

Cummins, S.E., R.A. Linke, and P.M. Thaddeus. 1986. Astrophys. J. (Suppl.), **60**:819.

Dalgarno, A. 1986. Quarterly J. R. Astron. Soc. **27**:83.

Deguchi, S., and P.F. Goldsmith. 1985. Nature **317**:336.

DeNoyer, L.K. 1979a. Astrophys. J. (Lett.), **228**:L41.

DeNoyer, L.K. 1979b. Astrophys. J. (Lett.), **237**:L43.

DeNoyer, L.K., and M.A. Frerking. 1981, Astrophys. J. (Lett.) **246**:L37.

Dickinson, D.F., and E.N. Rodriguez-Kuiper. 1981. Astrophys. J., **247**:112.

Dickinson, D.F., E.N. Rodriguez-Kuiper, A. St. Claire-Dinger, and T.B.H. Kuiper. 1980. Astrophys. J. (Lett.) **237**:L43.

Douglas, A.E. 1977. Nature **269**:130.

Downes, D., R. Genzel, A. Hjalmarson, L.A. Nyman, and B. Ronnang. 1982, Ap. J. (Lett.) **252**:L29.

Draine, B.T. 1980, Astrophys. J., **241**:1021.

Draine, B.T. 1984. Astrophys. J. (Lett.) **277**:L71.

Draine, B.T., and W.G. Roberge. 1982. Astrophys. J. (Lett.) **259**:L91.

Draine, B.T., W.G. Roberge, and A. Dalgarno. 1983. Astrophys. J. **264**:485.

Duley, W.W., and E. Bochlau. 1986. Mon. Not. R. Astron. Soc. **221**:659.

Duley, W.W., and T.J. Millar. 1978. Astrophys. J. **220**:124.

Duley, W.W., and D.A. Williams. 1979. Nature **227**:40.

Duley, W.W., and D.A. Williams. 1981. Mon Not. R. Astron. Soc. **196**:269.

Duley, W.W., T.J. Millar, and D.A. Williams. 1978. Mon. Not. R. Astron. Soc. **185**:915.

Duley, W.W., T.J. Millar, and D.A. Williams. 1980, Mon. Not. R. Astron. Soc. **192**:945.

Elitzur, M. 1982. Rev. Mod. Phys. **54**:1225.

Elitzur, M., and W.D. Watson. 1978. Astrophys. J. (Lett.) **222**:L141.

Elitzur, M., and W.D. Watson. 1980. Astrophys. J. **236**:172.

Exarhos, G.H., J. Mayer, and W. Klemperer. 1982. Phil. Trans. R. Soc. Lon. **A303**, 463.

Federman, S.R. 1981. Astron. Astrophys. **96**:198.

Federman, S.R. 1982. Astrophys. J. **257**:125.

Federman, S.R. 1986. In M.S. Vardya and S.F. Tarafdar (eds.), Astrochemistry, Int. Astron. Union, Symp. No. 120. Dordrecht: Reidel.

Federman, S.R., and R.F. Willson. 1982. Astrophys. J. **260**:124.

Federman, S.R., A.C. Danks, and D.L. Lambert. 1984. Astrophys. J. **287**:219.

Fertel, J.H., and B.E. Turner. 1975. Astrophys. Lett. **16**:61.

Fomalont, E.B., and L. Weliachew. 1973. Astrophys. J. **181**:781.

Freed, K.F., T. Oka, and H. Suzuki. 1982. Astrophys. J. **263**:718.

Freeman, A., and T.J. Millar. 1983. Nature, **301**:402.

Giguere, P.T., F.O. Clark, L.E. Snyder, D. Buhl, D.R. Johnson, and F.J. Lovas. 1973. Astrophys. J., **182**:477.

Glassgold, A.E., and W.D. Langer. 1976. Astrophys. J. **204**:403.

Glassgold, A.E., and P.J. Huggins. 1986 In H.R. Johnson and F. Querci (eds.), The M-Type Stars. NASA/CNRS.

Gottlieb, C.A., E.W. Gottlieb, M.M. Litvak, J.A. Ball, and H. Penfield. 1978. Astrophys. J. **219**:77.

Gottlieb, C.A., J.A. Ball, E.W. Gottlieb, and D.F. Dickinson. 1979. Astrophys. J. **227**:422.

Graedel, T.E., W.D. Langer, and M.A. Frerking. 1982. Astrophys. J. (Suppl.), **48**:321.

Green, S. 1975. Astrophys. J. **201**:221.

Greenberg, J.M. 1986. In M.S. Vardya and S.F. Tarafdar (eds.), Astrochemistry, Int. Astron. Union, Symp. No. 120. Dordrecht: Reidel.

Guelin, M. 1984, In G.H.F. Dierksen et al. (eds.), Molecular Astrophysics. Dordrecht: Reidel, p. 23.

Guelin, M., W.D. Langer, and R.W. Wilson. 1982. Astron. Astrophys., **107**:107.

Gundermann, E. 1965. Ph.D. dissertation, Harvard University.

Hartquist, T.W., M. Oppenheimer, and A. Dalgarno. 1980. Astrophys. J. **236**:182.

Henkel, C., H.E. Matthews, and M. Morris. 1983. Astrophys. J. **267**:184.

Herbig, G.H. 1975. Astrophys. J. **196**:129.

Herbst, E. 1978. Astrophys. J. **222**:508.

Herbst, E. 1982. Astrophys. J. **252**:810.

Herbst, E. 1983. Astrophys. J. (Suppl.) **53**:41.

Herbst, E., and W. Klemperer. 1976. Physics Today **29**:32.

Hinkle, K.H., D.N.B. Hall, and S.T. Ridgway. 1982. Astrophys. J. **252**:678.

Hobbs, L.M., D.G. York, and W. Oegerle. 1982. Astrophys. J. (Lett.) **252**:L21.

Hollis, J.M. 1982. Astrophys. J. **260**:159.

Hollis, J.M., L.E. Snyder, F.J. Lovas, and B.L. Ulich. 1980. Astrophys. J. **241**:158.

Hoyle, F., and Wickramasinghe, N.C. 1976. Nature **264**:45.

Huntress, W.T. 1977. Astrophys. J. (Suppl.) **33**:495.

Huntress, W.T., L.R. Thorne, and V.G. Anicich. 1983. Bull. Amer. Astron. Soc. **15**:955.

Iglesias, E.R. 1977. Astrophys. J. **218**:697.

Iglesias, E.R., and J. Silk. 1978. Astrophys. J. **226**:851.

Iguchi, T. 1975. Publ. Astron. Soc. Jpn. **27**:515.

Irvine, W.M., and A. Hjalmarson. 1983. *In* C. Ponnamperuma (ed.), Cosmochemistry and the Origin of Life. Dordrecht: Reidel, p. 113.

Irvine, W.M., J. Ellder, A. Hjalmarson, E. Kollberg, O.E.H. Rydbeck, G.O. Sorenson, B. Bak, and H. Svanholt. 1981. Astron. Astrophys. **97**:192.

Irvine, W.M., J.C. Good, and F.P. Schloerb. 1983. Astron. Astrophys. **127**:L10.

Irvine, W.M., F.P. Schloerb, A. Hjalmarson, and E. Herbst. 1985. *In* D.C. Black and M.S. Matthews (eds.), Protostars and Planets II. Tucson: University of Arizona Press, p. 579.

Irvine, W.M., et al. 1986. Private communication.

Johansson, L.E.B., C. Andersson, J. Ellder, P. Friberg, B. Hjalmarson, B. Hoglund, W.M. Irvine, H. Olofson, and G. Rydbeck. 1984, Astron. Astrophys. **130**:227.

Jura, M. 1976. Astrophys. J. **204**:12.

Jura, M., and D.G. York. 1978. Astrophys. J. **219**:861.

Knacke, R.F. 1977. Nature **269**:132.

Kroto, H.W., J.R. Heath, S.C. O'Brien, R.F. Curl, and R.E. Smalley. 1985, Nature **318**:162.

Kvenvolden, K.A., J.G. Lawless, and C. Ponnamperuma. 1971. Proc. Natl. Acad. Sci. USA **86**:486.

Lafont, S., R. Lucas, and A. Omont. 1982. Astron. Astrophys. **106**:210.

Langer, W.D., P.F. Goldsmith, E.R. Carlson, and R.W. Wilson. 1980. Astrophys. J. (Lett.) **235**:L39.

Lazareff, B. 1975. Astron. Astrophys., **42**:25.

Leung, C.M., E. Herbst, and W.F. Huebner. 1984. Astrophys. J. (Suppl.) **56**:231.

Liszt, H.S., and B.E. Turner. 1978. Astrophys. J. (Lett.) **224**:L73.

Little, L.T., G.H. MacDonald, P.W. Ripley, and D.N. Matheson. 1979. Mon. Not. R. Astron. Soc. **189**:539.

Loren, R.B., and L.E. Mundy. 1984. Astrophys. J. **286**:232.

Madden, S.C., W.M. Irvine, and H.E. Matthews 1986, Astrophys. J. (Lett.) **311**:L27.

Magnani, L. 1987, Ph.D. thesis, University of Maryland.

Magnani, L., L. Blitz, and L. Mundy. 1985. Astrophys. J. **295**:402.

Mathis, J.S., W. Rumpl, and K.H. Nordsieck. 1977. Astrophys. J. **217**:425.

Matthews, H.E., and T.J. Sears. 1983a. Astrophys. J. (Lett.) **267**:L53.

Matthews, H.E., and W.M. Irvine. 1985. Astrophys. J. (Lett.), **298**:L61.

Matthews, H.E., and T.J. Sears. 1983b. Astrophys. J. **272**:149.

Matthews, H.E., P.F. Friberg, and W.M. Irvine. 1985. Astrophys. J. **290**:609.

Matthews, H.E., P.A. Feldman, and P.F. Bernath. 1987. Astrophys. J. **312**:358.

McCabe, E.M., R. Connon-Smith, and R.E.S. Clegg. 1979. Nature **281**:263.

McKee, C.F., D.F. Chernhoff, and D.J. Hollenbach. 1984. *In* M.F. Kessler and J.P. Phillips (eds.), Galactic and Extragalactic Infrared Spectroscopy. Dordrecht: Reidel, p. 357.

McNally, O., M. Ashfield, D. Baines, S. Fossey, P. Rees, W. Somerville, and D. Whittet. 1986. *In* S.F. Tarafdar and M.S. Vardya (eds.), Astrochemistry, Int. Astron. Union, Symp. No. 120. Dordrecht: Reidel.

Merer, A.J., C.M. Walmsley, and E. Churchwell. 1982. Astrophys. J. **256**:151.

Millar, T.J. 1980. Astrophys. Space Sci. **72**:509.

Millar, T.J. 1982. Mon. Not. R. Astron. Soc. **149**:309.

Millar, T.J., and W.W. Duley. 1978. Mon. Not. R. Astron. Soc. **183**:177.

Millar, T.J., W.W. Duley, and D.A. Williams. 1979. Mon. Not. R. Astron. Soc. **186**:685.
Mitchell, G.F. 1980. Mon. Not. R. Astron. Sci. **205**:765
Mitchell, G.F. 1984. Astrophys. J. (Suppl.) **54**:81.
Mitchell, G.F., and T.J. Deveau. 1983. Astrophys. J. **266**:646.
Mitchell, G.F., J.L. Ginsburg, and P.J. Kuntz. 1978. Astrophys. J. (Suppl.), **38**:39.
Morton, D.C. 1974. Astrophys. J. (Lett.) **252**:L21.
Mundy, L.G. 1984. Ph.D. thesis, University of Texas, Austin.
Myers, P.C. 1983. Astrophys. J. **270**:105.
Myers, P.C. 1985. In D.C. Black and M.S. Matthews (eds.), Protostars and Planets II. Tucson: University of Arizona Press, p. 81.
Myers, P.C., and P.J. Benson. 1983. Astrophys. J. **266**:309.
Myers, P.C., P.M. Thaddeus, and R.A. Linke. 1980. Astrophys. J. **241**:155.
Myers, P.C., R.A. Linke, and P.J. Benson. 1983. Astrophys. J. **264**:517.
Nejad, L.A.M., T.J. Millar, and A. Freeman. 1984. Astron. Astrophys. **134**:129.
Nuth, J.A. 1985. Nature **318**:166.
Olofsson, H. 1984. Astron. Astrophys. **134**:36.
Oppenheimer, M., and A. Dalgarno. 1974. Astrophys. J. **187**:321.
Phillips, T.G., and P.J. Huggins. 1981. Astrophys. J. **251**:537.
Pickles, J.B., and D.A. Williams. 1981. Mon. Not. R. Astron. Soc. **197**:429.
Plambeck, R.L., and N.R. Erickson. 1982. Astrophys. J. **262**:606.
Plambeck, R.L., M.H.C. Wright, W.J. Welch, J.H. Bieging, B. Baud, P.T.P. Ho, and S.N. Vogel. 1982. Astrophys. J. **259**:617.
Plambeck, R.L., S.N. Vogel, M.C.H. Wright, J.H. Bieging, and W.J. Welch. 1985. In International Symposium on Millimeter and Submillimeter Wave Radio Astronomy, Granada, published by URSI.
Prasad, S.S., and W.T. Huntress. 1980. Astrophys. J. **239**:151.
Prasad, S.S., and W.T. Huntress. 1982. Astrophys. J. **260**:590.
Puget, J.L., A. Leger, and F. Boulanger. 1985. Astron. Astrophys. (Lett.) **142**:L19.
Rieu, N-Q., V. Bujarrabal, H. Olofsson, L.E.B. Johansson, and B.E. Turner. 1984. Astrophys. J. **286**:276.
Rydbeck, O.E.H., E. Kollberg, A. Hjalmarson, A. Sume, J. Ellder, and W.M. Irvine. 1976. Astrophys. J. (Suppl.) **31**:333.
Sanner, F., R.L. Snell, and P. Vanden Bout. 1978. Astrophys. J. **226**:460.
Scalo, J.M., and D.B. Slavsky. 1980. Astrophys. J. (Lett.) **239**:L73.
Scoville, N.Z., P.M. Solomon, and A.A. Penzias. 1975. Astrophys. J. **201**:352.
Scoville, N.Z., D.N.B. Hall, S.G. Kleinman, and S.T. Ridgway. 1979. Astrophys. J. (Lett.) **232**:L121.
Slavsky, D.B., and J.M. Scalo. 1985. Preprint.
Smith, D., and N.G. Adams. 1984. Astrophys. J. (Lett.) **284**:L13.
Smith, W.H., T.P. Snow, and D.G. York. 1977. Astrophys. J. **218**:124.
Snell, R.L., F.P. Schloerb, J.S. Young, A. Hjalmarson, and P. Friberg. 1980. Astrophys. J. **244**:45.
Snow, T.P. 1980. In B.H. Andrews (ed.), Interstellar Molecules, Int. Astron. Union, Symp. No. 80. Dordrecht: Reidel, p. 247.
Snyder, L.E., J.M. Hollis, R.D. Suenram, F.J. Lovas, L.W. Brown, and D. Buhl. 1983. Astrophys. J., **268**:123.
Stacey, G.J., J.B. Lugten, and R. Genzel. 1986. Bull. Amer. Astron. Soc. **17**:871.
Storey, J.W.V., D.M. Watson, and C.H. Townes. 1981. Astrophys. J. (Lett.) **244**:L27.
Sutton, E.C., G.A. Blake, C.R. Masson, and T.G. Phillips. 1985. Astrophys. J. (Suppl.) **58**:341.
Suzuki, H. 1983. Astrophys. J. **272**:579.
Suzuki, H., N. Kaifu, T. Miyaji, and M. Morimoto. 1984. Astrophys. J. **282**:197.
Thaddeus, P.T., S. Cummins, and R. Linke 1984. Astrophys. J. (Lett.) **283**:L45.
Thaddeus, P.M., J.M. Vrtilek, and C.A. Gottlieb. 1985. Astrophys. J. (Lett.) **299**:L63.
Thorne, L.R., V.G. Anicich, S.S. Prasad, and W.T. Huntress. 1984. Astrophys. J. **280**:139.
Tielens, A.G.G.M. 1983. Astron. Astrophys. **119**:177.
Tielens, A.G.G.M., and W. Hagen. 1982. Astron. Astrophys. **114**:245.
Townes, C.H., R. Genzel, D.M. Watson, and J.W.V. Storey. 1983. Astrophys. J. (Lett.) **269**:L11.
Treffers, R.R. 1979. Astrophys. J. (Lett.) **233**:L17.
Tucker, K.P., M.L. Kutner, and P.M. Thaddeus. 1974. Astrophys. J. (Lett.) **193**:L115.

Turner, B.E. 1980. J. Mol. Evol. **15**:79.

Turner, B.E. 1982. Astrophys. J. (Lett.) **255**:L33.

Turner, B.E. 1983. *In* W.L. Shuter (ed.), Kinematics, Dynamics, and Structure of the Milky Way. Dordrecht: Reidel.

Turner, B.E., and R.H. Gammon. 1975. Astrophys. J. **198**:71.

Turner, B.E., and C.E. Heiles. 1974. Astrophys. J. (Lett.) **187**:L59.

Turner, B.E., and T.C. Steimle. 1985. Astrophys. J. **299**:956.

Turner, B.E., and P.M. Thaddeus. 1977. Astrophys. J. **211**:758.

Turner, B.E., and B. Zuckerman. 1978. Astrophys. J. (Lett.) **225**:L75.

Turner, J.L., and A. Dalgarno. 1977. Astrophys. J. **213**:386.

Vogel, S.N., M.C.H. Wright, R.L. Plambeck, and W.J. Welch. 1984. Astrophys. J. **283**:655.

Walmsley, C.M., W. Hermsen, C. Henkel, R. Mauersberger, and T.L. Wilson. 1986. Astron. Astrophys. **172**:311.

Watson, D.M. 1982. Ph.D. thesis, University of California at Berkeley.

Watson, W.D. 1976. Rev. Mod. Phys. **48**:513.

Watson, W.D. 1978. Proceedings of the 21st Liège Astrophysical Symposium, Université de Liège.

Watson, W.D. 1984. *In* M.F. Kessler and J.P. Phillips (eds.), Galactic and Extragalactic Infrared Spectroscopy. Dordrecht: Reidel, p. 5.

Watson, W.D., and E.E. Salpeter. 1972. Astrophys. J. **174**:321.

Watson, W.D., and C.M. Walmsley. 1982. *In* R.S. Roger and P.E. Dewdney (eds.), Regions of Recent Star Formation. Dordrecht: Reidel, p. 357.

Weaver, H., D.R.W. Williams, N.H. Dieter, and W.T. Lum. 1965. Nature **208**:29.

Webster, A. 1979. IAU Circular No. 3425.

Weinreb, S., A.H. Barrett, M.L. Meeks, and J.C. Henry. 1963. Nature **200**:829.

Weltner, G.L., and B.D. Savage. 1977. Astrophys. J. **215**:788.

White, R.E. 1977. Astrophys. J. **211**:744.

Williams, D.A. 1984. *In* M.F. Kessler and J.P. Phillips (eds.), Galactic and Extragalactic Infrared Spectroscopy. Dordrecht: Reidel, p. 66.

Willner, S.P. 1984. *In* M.F. Kessler and J.P. Phillips (eds.), Galactic and Extragalactic Infrared Spectroscopy. Dordrecht: Reidel, p. 37.

Wilson, R.W., A.A. Penzias, K.B. Jefferts, M.L. Kutner, and P.M. Thaddeus. 1971. Astrophys. J. (Lett.) **167**:L97.

Wilson, T.L., K. Ruk, C.M. Walmsley, R.N. Martin, T.A. Pauls, and W. Batrla. Astron. Astrophys. **115**:185.

Wilson, W.J., and L.E. Snyder. 1981. Astrophys. J. **246**:86.

Woods, R.C., C. Gudeman, R. Dickman, P. Goldsmith, G. Huguenin, W. Irvine, A. Hjalmarson, L.-A. Nyman, and H. Olofsson. 1983. Astrophys. J. **270**:583.

Wootten, H.A. 1986. Private communication.

Wootten, H.A., and Loren, R.B. 1986. Astrophys. J. **306**:142.

Wootten, H.A., N.J. Evans, R.L. Snell, and P. Vanden Bout. 1978. Astrophys. J. (Lett.) **225**:L143.

Wright, M.C.H., R.L. Plambeck, S.N. Vogel, P.T.P. Ho, and W.J. Welch. 1983. Astrophys. J. (Lett.) **267**:L41.

Ziurys, L.M., and B.E. Turner. 1985. Astrophys. J. (Lett.) **292**:L25.

Ziurys, L.M., and B.E. Turner. 1986a. Astrophys. J. (Lett.) **300**:L19.

Ziurys, L.M., and B.E. Turner. 1986b. Astrophys. J. (Lett.) **302**:L31.

Ziurys, L.M., P.F. Friberg, and W.M. Irvine. 1987, manuscript in preparation.

Zuckerman, B. 1980. Annu. Rev. Astron. Astrophys. **18**:236.

Zuckerman, B., J.A. Ball, and C.A. Gottlieb. 1971, Astrophys. J. (Lett.), **163**:L41.

6. Astronomical Masers

MARK J. REID and JAMES M. MORAN

6.1 Introduction

The study of astronomical maser sources has been an exciting and rapidly grow-ing area of research over the past twenty-five years. In addition to providing a fascinating astrophysical laboratory for the study of maser processes, the intense emissions have been used as probes of the dynamics and physical conditions of their surroundings on scales as small as one astronomical unit throughout the Galaxy and in nearby galaxies. Since masers are found in regions of star formation and are also associated with evolved stars, their study provides unique information on the birth and death of stars. Finally, because maser sources are extremely small in angular size, they can be used to study interstellar scattering and as astrometric tools, for example, to measure distances by methods of trigonometric and statistical parallax.

The first detection of what was later understood to be astronomical maser emission came in 1965. While attempting to study the characteristics of the hydroxyl (OH) radical in absorption against regions of hot, ionized hydrogen (HII regions), Weaver et al. (1965) and Gundermann (1965) surprisingly found intense emission. These and subsequent observations posed several difficult problems: the relative strengths of different hyperfine components were incompatible with thermodynamic equilibrium; the emission was often highly polarized (Weinreb et al. 1965); and while linewidths indicated a molecular excitation temperature of about 10 K, the lower limits on the brightness temperature of the radiation (i.e., the temperature of a blackbody required to emit radiation of the same spectral intensity) implied by the pointlike nature of the sources was greater than 10^3 K (McGee et al. 1965). Experiments with connected element interferometers in the U.S. and Great Britain showed that spectral features were spatially separated, and in 1968 a very long baseline interferometric (VLBI) experiment involving telescopes across the U.S. by Moran et al. (1968) established that the OH emission originated from many spots, each of which was milliarcseconds in size. At this point the conclusion that the "mysterious" emission was due to a maser radiation process became almost inescapable. The first theoretical models for maser action, which involved ultraviolet pumps, were published by Perkins et al. (1966) and Litvak et al. (1966).

Hydroxyl maser emission was found toward many galactic HII regions, establish-ing it as a landmark for sites of active star formation (Shklovskii 1969). However, in 1968, OH maser emission was discovered toward many infrared (IR) sources listed in the 2-micron Sky Survey. Most of these IR sources were long-period variable (LPV) stars, objects well studied by optical astronomers and known to be evolved stars in the red giant phase of evolution (see Wilson and Barrett 1972). Thus, before 1970 it was known that astronomical maser emission was associated with both the early and late stages of stellar evolution.

Water vapor (H_2O) was the second molecule identified as a masing species (Cheung et al. 1969). Similar to OH, H_2O was also found to be associated with sites of active star formation and with evolved stars. H_2O masers are the most spectacular of the presently known sources. Provided the radiation in our direction is not substantially stronger than in other directions, the most powerful source in the Milky Way emits nearly a solar luminosity in a single spectral line only 50 kHz wide!

Several other molecular species have to date been observed as strong masers. Silicon monoxide (SiO) was discovered in maser emission by Snyder and Buhl (1974) toward an active star-forming region in the Orion Nebula. However, with the notable exception of the Orion maser, most SiO masers are found associated with IR stars. Maser emission from the methanol (CH_3OH) molecule toward the Orion Nebula was discovered by Barrett et al. (1971). Other methanol transitions exhibit maser emission. Recently, a strong HCN maser has been detected (Guilloteau et al. 1987), and many other molecules exhibit weak maser emission.

In Section 6.2, we outline a simplified radiative transfer theory as developed for astronomical masers. This section includes a discussion of saturation, the apparent brightness distribution of the emission, the source of input radiation that is amplified, constraints on pump mechanisms, and several other topics of "maser physics."

Strong maser activity in astronomical settings occurs in regions where the dust and gas density are considerably higher than those usually detected in giant molecular clouds (e.g., hydrogen densities from 10^5 to 10^{11} cm^{-3}). In addition, a highly luminous ($\geqslant 10\ L_\odot$) source seems required to provide the energy to pump the masers. For masers associated with star-forming regions, a newly formed or forming O- or B-type star probably supplies the pump energy. The masing molecules most likely are contained in condensations of interstellar material in the vicinity of these luminous objects, and hence we will refer to them as "interstellar masers." In Section 6.3, we will review observations and theories that describe the characteristics of interstellar masers.

For maser sources associated with IR stars, an M-type red giant or supergiant is responsible for the excitation. These masers form in regions just above the turbulent photosphere and in the outward-flowing circumstellar envelope. In contrast to "interstellar masers," these masing molecules come from stellar material that is being returned to the interstellar environment, and hence we will refer to these sources as "stellar masers." In Section 6.4, we review the observations and theories pertaining to stellar masers.

H_2O and OH masers can be so intense that some have been detected in other galaxies. The most luminous H_2O maser yet detected is in the spiral galaxy NGC 3079 and its "isotropic" luminosity exceeds $500\ L_\odot$! The exceedingly high luminosity of such a source may indicate that we are observing a different type of maser activity than seen in the Milky Way. In Section 6.5, we discuss the phenomenon of extragalactic maser sources.

In Section 6.6, the use of masers as probes of the electron distribution in the interstellar medium via scattering is reviewed. Finally, in Section 6.7, we describe measurements of proper motions of H_2O masers and the use of masers as direct indicators of galactic and extragalactic distances via trigonometric and statistical parallax measurements.

6.2 Maser Theory

In order to understand the physical conditions in masers, suitable models for the molecular excitation and radiative transfer must be developed. Radiative transfer in masers is complicated because stimulated emission rates in maser transitions can exceed the rates by which population can be "recycled" by collisions and other radiative processes. Thus, the intensity of the maser emission can significantly affect the level populations. When such a condition exists, astronomical masers are said to be "saturated."

There are several basic questions any theory of masers attempts to answer. What is the source of the "input" radiation that is amplified? What processes cause the population inversion or "pump" the maser? Are astronomical masers saturated? Also, a maser theory should allow the observed properties of apparent brightness, size, line width, and polarization to be interpreted in terms of physical parameters of the masing region. In this section, we outline the most important aspects of maser theory using an elementary approach to the radiative transfer problem. More detailed treatments of the radiative transfer in astronomical masers have been carried out by Litvak (1973), Goldreich and Keeley (1972), Kegel (1975), Western and Watson (1984), and Alcock and Ross (1985).

6.2.1 Radiative Transfer

In this section, attention is focused on a simple maser cloud model that can describe the observed emission from a single feature or spot. The model can explain several characteristics of a maser amplifier: exponential gain, saturation, and line narrowing. The sources of the energy of masers are also discussed.

Consider a cloud of molecules having two energy levels, denoted 1 and 2 for the lower and upper levels, respectively. The transition frequency between them is v_0. The statistical weights of the levels are taken to be the same to simplify the presentation. In exact calculations, it is necessary to consider the difference in the statistical weights. If n_1 and n_2 are the total population densities of the two levels, then the number of molecules in a given velocity interval dv along the line of sight can be defined as $n_1 f(v) dv$ and $n_2 f(v) dv$. For the case where thermal motion dominates, the distribution function $f(v)$ will be a Gaussian function,

$$f(v) = \frac{1}{\sqrt{2\pi}} \frac{1}{u} \exp\left[-\frac{v^2}{2u^2}\right] \tag{6.1}$$

where $u = (kT_k/M)^{1/2}$, M is the mass of the molecule, k is Boltzmann's constant and T_k is the kinetic temperature. The distribution can be written as a function of frequency, v, through the Doppler effect,

$$\frac{v - v_0}{v_0} = \frac{v}{c} \tag{6.2}$$

where c is the speed of light. Hence

$$f(v) = \frac{1}{\sqrt{2\pi}} \frac{1}{w} \exp\left[-\frac{(v - v_0)^2}{2w^2}\right] \tag{6.3}$$

where

$$w = \left(\frac{v_0}{c}\right)\left(\frac{kT_k}{M}\right)^{1/2} \tag{6.4}$$

If Δv_D is the full width at half-maximum of $f(v)$, then to a good approximation

$$f(v) = \frac{1}{\Delta v_D} \exp\left[-\frac{4\ln 2(v - v_0)^2}{\Delta v_D^2}\right] . \tag{6.5}$$

The intensity of the radiation propagating along a given ray path will be decreased by absorption and increased by emission. The equation describing the radiation transfer, that is, the change of specific intensity, I_v, with distance, ℓ, can be written as

$$\frac{dI_v}{d\ell} = -\kappa_v I_v + \eta_v \tag{6.6}$$

where

$$\kappa_v = (n_1 - n_2)B\frac{hv}{4\pi}f(v) \tag{6.7}$$

$$\eta_v = n_2 A\frac{hv}{4\pi}f(v) ; \tag{6.8}$$

κ_v and η_v are the volume absorption and emission coefficients, respectively. A and B are the Einstein coefficients, which are related by the equation

$$A = \frac{2hv^3}{c^2}B . \tag{6.9}$$

There is only one Einstein B coefficient since the statistical weights are assumed to be equal. Hence, with the substitution of Equations (6.5), (6.7), (6.8), and (6.9) into Equation (6.6), the radiative transfer equation *at the line center* becomes

$$\frac{dI}{d\ell} = -(n_1 - n_2)B\frac{hv}{4\pi\Delta v_D}I + n_2 A\frac{hv}{4\pi\Delta v_D} . \tag{6.10}$$

Note that we have made the customary assumption that $f(v)$ is the appropriate profile function for the absorption and emission coefficients (see Goldreich and Keeley 1972).

The population inversion can be calculated from the steady-state rate equations. The rate equations can be written in different ways. For example, the energy levels

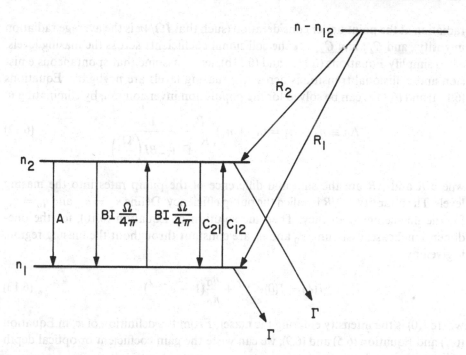

Fig. 6.1. Schematic energy level diagram for a maser with relevant transition rates shown. The masing levels have population densities n_1 and n_2. Statistical weights are assumed equal for these levels. These levels are linked by transitions at rates per molecule of $BI(\Omega/4\pi)$ for stimulated emission and absorption, A for spontaneous emission, and C_{12} and C_{21} for collisions. A and B are the Einstein coefficients of spontaneous emission and absorption, respectively, and I is the radiation intensity. The masing levels are also linked to other molecular energy levels, depicted by one other level. The pump rates per molecule into the masing levels are characterized by R_1 and R_2, and decays out of the masing levels at rate per molecule Γ. Equations (6.11a) and (6.11b) in the text give the steady-state rates of transition explicitly.

can be represented by a pseudo two-level system with pump rates *between* the levels to describe the pumping action that actually involves many other levels. Alternatively, we can describe the system by the two maser levels but include net pump rates into the masing levels from all other levels plus decay rates out of the masing levels to all other levels. We adopt the latter approach (see Goldreich and Keely 1972).

In steady state, the time rates of change of the upper and lower level populations ($\dot{n}_2$ and $\dot{n}_1$, respectively) are zero (see Figure 6.1):

$$\dot{n}_2 = R_2(n - n_{12}) - (n_2 - n_1)BI\left(\frac{\Omega}{4\pi}\right) - n_2 A - n_2 C_{21} - n_2\Gamma = 0 \quad (6.11\text{a})$$

$$\dot{n}_1 = R_1(n - n_{12}) - (n_1 - n_2)BI\left(\frac{\Omega}{4\pi}\right) - n_1 C_{12} - n_1\Gamma = 0 \quad (6.11\text{b})$$

where n is the total population density, $n_{12} = n_1 + n_2$ is the total population in the maser levels, R_1 and R_2 are the pump rates into levels 1 and 2, respectively, from all other states, Γ is the rate at which molecules are transferred from the masing levels to other levels (assumed equal for the two levels), Ω is the solid angle of the radiation at the point under consideration (such that $I\Omega/4\pi$ is the average radiation intensity), and C_{12} and C_{21} are the collisional coefficients across the masing levels.

To simplify Equations (6.11a) and (6.11b), let us assume that spontaneous emission and collisional transitions across the masing levels are negligible. Equations (6.11a) and (6.11b) can be solved for the population inversion, Δn, by eliminating n:

$$\Delta n \equiv n_2 - n_1 = (n_2 + n_1)\frac{\Delta R}{R}\frac{\Gamma}{\Gamma + 2BI\left(\dfrac{\Omega}{4\pi}\right)} \tag{6.12}$$

where R and ΔR are the sum and difference of the pump rates into the masing levels. The quantity $\Delta R/R$ is called the pump efficiency. Define $\kappa_0 = \kappa_{v_0}$ and $\eta_0 = \eta_{v_0}$ for the line-center frequency. Then the solution of Equation (6.10), for the one-dimensional case, assuming κ_0 and η_0 are constant throughout the masing region, is given by

$$I(\ell) = I(0)e^{-\kappa_0\ell} + \frac{\eta_0}{\kappa_0}(1 - e^{-\kappa_0\ell}) \tag{6.13}$$

where $I(0)$ is the intensity entering the maser. From the definition of κ_v in Equation (6.7) and Equation (6.5) and (6.9), we can write the gain coefficient or optical depth at line center, $\kappa_0\ell$, as

$$\kappa_0\ell = -\frac{\Delta n\lambda^2 A\ell}{8\pi\Delta v_D} . \tag{6.14}$$

For a maser, $n_2 > n_1$ and it is clear from Equation (6.7) that $\kappa_0 < 0$. Thus, the optical depth ($\kappa_0\ell$) is negative and exponential amplification, rather than absorption, occurs. As is obvious from Equation (6.12), the condition that κ_0 and η_0 are constant requires that $\Gamma \gg 2BI(\Omega/4\pi)$. In this case the maser is said to be unsaturated. (Note that the astronomical nomenclature is the opposite of that used for laboratory masers, where if $\Gamma \gg 2BI(\Omega/4\pi)$, the maser is called saturated because the pump is sufficiently strong to maintain exponential amplification.)

It is convenient to convert intensity to brightness temperature, T_B, of an equivalent blackbody. Using the Rayleigh-Jeans approximation to the blackbody law, $I_v = 2kT_Bv^2/c^2$, Equation (6.13) can be expressed as

$$T_B(\ell) = [T_B(0) - T_x]e^{-\kappa\ell} + T_x . \tag{6.15}$$

In Equation (6.15), T_x is the excitation temperature for the transition defined by

$$\frac{n_2}{n_1} = e^{-hv/kT_x} . \tag{6.16}$$

Note that T_x is negative under conditions of population inversion (see Figure 6.2). The relative importance of background emission and spontaneous emission (from the maser itself) as sources of input signal to the maser depends on the magnitudes of $T_B(0)$ and T_x, respectively.

Maser emission occurs when $n_2 > n_1$ so that the absorption coefficient and hence the opacity and excitation temperature are negative. Equation (6.14) or (6.15) will

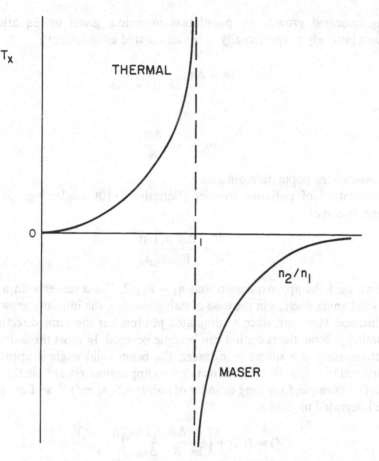

Fig. 6.2. Excitation temperature, T_x, for a molecular transition as a function of the ratio of densities (per statistical weight) in the upper to lower levels, n_2/n_1. Under conditions of thermal equilibrium, n_2/n_1 varies from 0 to 1 as the excitation temperature varies from 0 to ∞. When the level population densities are inverted to support maser action $n_2/n_1 > 1$. As the maser pumping efficiency increases, n_2/n_1 increases, and T_x goes toward zero (from the negative side). Thus, *weakly pumped* masers, or those that are *saturated*, have *large negative* excitation temperatures, whereas *strongly pumped* masers have *small negative* excitation temperatures.

describe a masing cloud as long as the intensity or brightness temperature, which grows exponentially, is small enough not to disturb the population distribution. Eventually, however, the growth will be limited as the upper level becomes depleted, thus changing T_x, and the astronomical maser is said to be saturated.

The effect of the radiation on the level populations will now be considered. From Equation (6.12) one can see that the population inversion is reduced by stimulated emission when $2BI\Omega/4\pi$ (the stimulated emission rate averaged over 4π steradians) exceeds Γ. Under these conditions the maser is saturated. Using the Rayleigh-Jeans approximation, Equation (6.9), and setting Γ equal to $2BI\Omega/4\pi$, we derive the brightness temperature at which the maser saturates to be

$$T_s = \frac{h\nu}{2k} \frac{\Gamma}{A} \frac{4\pi}{\Omega} \ . \tag{6.17}$$

During saturated growth the population inversion, given by Equation (6.12), decreases inversely proportionally to the stimulated emission rate:

$$\Delta n = \Delta n_0 \frac{\Gamma}{2BI(\Omega/4\pi)} \tag{6.18}$$

where

$$\Delta n_0 = n_{12} \frac{\Delta R}{R} \tag{6.19}$$

is the *unsaturated* population difference.

The equation of radiative transfer (Equation 6.10), neglecting spontaneous emission, becomes

$$\frac{dI}{d\ell} = \frac{\Delta R}{R} \frac{n_1 \Gamma h\nu}{\Delta v_D \Omega} \tag{6.20}$$

where we used the approximation that $n_1 \sim n_{12}/2$. For a maser with a constant beam solid angle, such as in the case of slab geometry, the intensity grows linearly with distance. However, since a stimulated photon has the same direction as the stimulating photon, the radiation can become beamed. In most three-dimensional geometries, such as a sphere or cylinder, the beam solid angle is approximately proportional to ℓ^{-2}, so that the intensity grows approximately as ℓ^3 in the saturated region. For example, for a long cylinder of radius r, $\Omega \approx \pi r^2/\ell^2$, so Equation (6.20) can be integrated to yield

$$I(\ell) = I(\ell_s) + \left(\frac{1}{3\pi} \frac{\Delta R}{R} \frac{n_1 \Gamma h\nu}{\Delta v_D} \right) \frac{(\ell - \ell_s)^3}{r^2} \tag{6.21}$$

where ℓ_s is the path length at which the maser saturates as given by Equation (6.17).

6.2.2 Input Sources

There are two sources of input signal to a maser as described by Equation (6.15): background radiation from discrete radio sources or the galactic and cosmic background, characterized by $T(0)$, and spontaneous emission from the *boundary* of the masing region, characterized by T_x. Thus, in order to determine which source is more important, we must know which temperature is greater (in magnitude). While background source brightness temperatures may be easily estimated (e.g., about 10^4 K for optically thick HII regions and 3 K for the cosmic background), maser excitation temperatures depend upon details of the pump and saturation conditions, which are not well known. The excitation temperature of a maser with a strong pump capable of maintaining a one percent population inversion (i.e., $\Delta R/R = 0.01$) would be -5 K for OH and -50 K for H_2O. Masers probably have *lower* population inversions at their boundaries than at their centers because of decreased pump efficiencies or saturation. Since the *magnitude* of the excitation temperature *increases* as population inversion *decreases*, it is likely that masers have excitation temperatures at their boundaries that are larger in magnitude than those given above.

While there are considerable uncertainties in determining excitation temperatures of masers and brightness temperatures of some background sources, there are some cases where one can identify the most likely source of input radiation that is amplified by the maser. As discussed in Section 6.3.2, OH masers associated with star-forming regions are usually found projected against HII regions. Since the free-free emission from the ionized hydrogen behind the masers is usually optically thick at 18-cm wavelength, it has a brightness temperature of $\sim 10^4$ K. This is substantially larger than the expected excitation temperatures for these masers, even accounting for the effects of saturation. Thus, OH masers associated with HII regions probably amplify this background emission. On the other hand, stellar OH masers generally show no signs of amplifying stellar 18-cm photons from their underlying 2500 K stellar photospheres. The OH emission is generally observed at velocities both red- and blueshifted with respect to the stellar velocity and from a region on the sky about two orders of magnitude larger than the stellar disk. This observation suggests that the OH masers from the circumstellar envelopes of evolved stars generally amplify spontaneous emission and have excitation temperatures at their boundaries of < -200 K (Reid et al. 1977). Recently, however, a few maser features that appear to be amplifying their stellar photospheres have been found (Norris et al. 1984).

6.2.3 Geometry and Apparent Sizes

It has long been recognized that the observed sizes of individual maser spots are likely to be considerably smaller than the size of the cloud as defined by the length over which amplification occurs. It is difficult to achieve the required gain in models where the gain path, ℓ, is as small as the observed dimension of a cosmic maser spot. For example, for $\Delta n = 0.03$ (e.g., $n_{H_2} \sim 10^5$, $n_{OH}/n_{H_2} \sim 10^{-4}$, pump efficiency $\Delta R/R \sim 0.01$, and $\approx 30\%$ of the OH molecules in the ground state), a maser of observed diameter 10^{14} cm would have a gain of only ~ 3. Two different geometric models of cosmic masers alleviate this difficulty. If, for example, the masers are long tubes or filaments of length ℓ and radius r, where $\ell \gg r$, the gain length can be increased by l/r, or up to about 100, over the observed spot size ($\approx 2r$). The maser radiation would be beamed into the solid angle $\Omega \approx \pi(r/\ell)^2$, and only those filaments pointed towards the earth would be observed. The time variability might be explained by rotation of the filaments or by turbulent motions that change the path length over which the velocity is constant to within the line width.

The second model is a spherical maser whose appearance is not immediately obvious. Consider a spherical maser of diameter d that is unsaturated. Rays passing through the sphere and reaching the observer will be nearly parallel since the angular size of the maser is very small, as shown in Figure 6.3. A ray passing through the center has a gain path d, while a ray at projected distance x has a gain path of only $(d^2 - 4x^2)^{1/2}$. Hence, the brightness distribution as a function of x would be

$$T_B(x) = T_c \exp\left[\kappa_0 d\left(1 - \frac{4x^2}{d^2}\right)^{1/2} \right] \tag{6.22}$$

where T_c is the input temperature (assumed greater than the magnitude of the excitation temperature). The observed width at half-maximum brightness, d^*, will

512-081

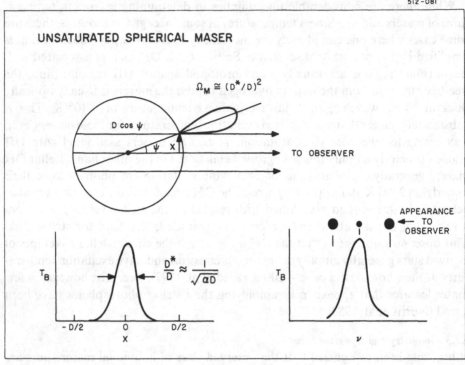

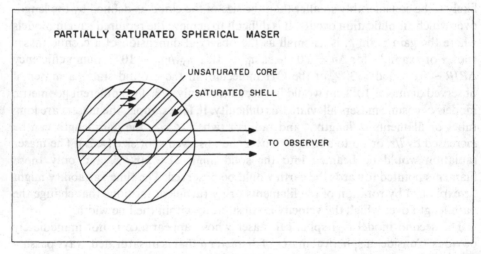

Fig. 6.3. *Top*: The rays passing through an unsaturated and uniformly pumped spherical maser. Off-center rays will have less gain and the sphere will appear to be smaller than it is by the amount given by Equation (6.23). *Bottom*: The rays passing through a partially saturated spherical maser. The rays that pass through the unsaturated core will be very much stronger than those that do not. From Moran 1976.

be given by the relation

$$\frac{d*}{d} \approx \frac{1}{\sqrt{\kappa_0 d}} .$$ (6.23)

Molecules at the surface will see radiation coming primarily from the diametric direction, and the beam angle will be

$$\Omega \simeq \left(\frac{d*}{d}\right)^2 \approx \frac{1}{\kappa_0 d} .$$ (6.24)

The maser will, of course, appear isotropic to any observer, provided that spontaneous emission, not background photons, is predominantly amplified.

In general, the beam solid angle is approximately equal to the square of the ratio of the apparent diameter, as measured by the interferometer, to the true amplification path length. A rough estimate of the true amplification path length is the transverse distance between maser spots in tight clusters. Beam solid angles estimated in this way are $\sim 10^{-2}$ steradians. For a spherical maser, however, the emission as seen by all observers is the same even though the radiation at any spot on the surface is highly beamed. Note that the spatial coherence of astronomical masers is very small, being on the order of 10λ in order to produce the modest beam solid angles discussed above.

6.2.4 Saturation

In the previous example of a spherical maser, assume that the pump source weakens so that the pump rate decreases uniformly throughout the sphere. Also, assume that spontaneous emission drives the maser so that there is no preferred direction. The maser will first saturate at the outer edge where the intensity is greatest because of rays that have been amplified over the total diameter d. This situation happens when the radiation intensity (brightness temperature) reaches the value given by Equation (6.17). The maser will then have an unsaturated core of diameter d' surrounded by a saturated shell. The observer will see only those rays that pass through the unsaturated core and are exponentially amplified. Radiation originating in the saturated shell and not passing through the core will not grow exponentially and will be very weak at the edge of the maser. Hence, as the pump rate continually decreases, the maser appears to get smaller down to some critical size, after which it would appear to increase in size as the entire maser becomes saturated. In many cases, most of the power comes from the unseen saturated shell. The microwave emission rate per molecule, W, is given by

$$W = BI\frac{\Omega}{4\pi} .$$ (6.25)

In terms of observable quantities, Equation (6.25) becomes

$$W = \frac{\lambda^2 A}{8\pi h\nu}S_m\frac{\Omega}{\theta^2}$$ (6.26)

where S_m is the observed flux density, λ is the wavelength, and θ is the observed

Table 6.1. Typical parameters of strong masers.[a]

Type	Γ (s^{-1})	T_s^b (K)	S_m (Jy)	θ (")	T_B (K)	W^c (s^{-1})
OH-stellar (1612 MHz)	0.03	1×10^{11}	2×10^2	1	10^8	10^{-5}
OH-interstellar (1665 MHz)	0.03	2×10^{10}	2×10^2	10^{-2}	10^{12}	1
H$_2$O-stellar	1	3×10^{11}	4×10^3	10^{-2}	10^{11}	10^{-1}
H$_2$O-interstellar	1	3×10^{11}	4×10^4	10^{-3}	10^{14}	10^2
SiO-stellar (43 GHz)	5	2×10^9	1.5×10^3	10^{-2}	10^{10}	10

[a] Ω equal to 10^{-2} in all cases.
[b] T_s given by Equation (6.17).
[c] W given by Equation (6.27).

source size. Saturation occurs where $2W = \Gamma$ [see Equation (6.12)]. For H$_2$O masers

$$W = 0.42 \frac{S_m \Omega}{\theta^2} \tag{6.27}$$

where S_m is in Janskys, θ is in milliarcseconds (mas), and Ω is in steradians. Hence, interstellar H$_2$O masers, for which $\theta \sim 1$ mas and Γ is thought to be ~ 1 s^{-1}, are saturated when $S_m > 100$ Jy for $\Omega = 10^{-2}$. The strongest H$_2$O masers, which have flux densities from 10^3 to 10^6 Jy, are probably saturated, at least near their surfaces. Estimates of microwave emission rates and saturation brightness temperatures, calculated from Equations (6.26) and (6.17), are given in Table 6.1.

A fully saturated maser requires the least pump power for a given output intensity because every pump photon results in a maser photon, whereas in an unsaturated maser the radiation field is not strong enough to take full advantage of the available population inversion. The total luminosity, L, or power from a fully saturated *cylindrical* maser is

$$L = 2I\Omega A_m \Delta v \tag{6.28}$$

where Δv_m is the maser linewidth and where the dominant emission comes from the ends of the cylinder of cross-sectional area A_m. The radiation is primarily beamed into two cones of emission of solid angle $\Omega \approx A_m/\ell^2$. Substitution of Equations (6.21) and (6.19) into Equation (6.28) yields $L = \frac{1}{3}\Delta n_0 \Gamma h v V$, where $V = A_m \ell$ is the volume of the maser and where we assumed $\ell_s = 0$ in Equation (6.21). In this case, one-third of the total luminosity comes out of the "ends" of the cylinder and two-thirds of the luminosity emerges as low-brightness emission from the sides of the cylinder. The total luminosity of the maser is

$$L = \Delta n_0 \Gamma h v V . \tag{6.29}$$

Equation (6.29) is general and does not depend on the maser geometry. Since the rate of transfer of population into the masing levels must equal the rate of transfer out of the masing levels, $(n - n_{12})(R_1 + R_2) = (n_1 + n_2)\Gamma$. Then, from Equation (6.19), it follows that $n\Delta R \approx \Delta n_0 \Gamma$ (for $n \gg n_{12}$), and Equation (6.29) can be rewritten as

$$\tilde{L} = n\Delta R h v V . \tag{6.30}$$

The flux density at the Earth is

$$S_m = \frac{L}{4\pi D^2 \Delta v} .$$ (6.31)

Hence, from Equations (6.30) and (6.31) and the equation above equation (6.30), the required population inversion is

$$\Delta n_0 = \frac{4\pi D^2 S_m \Delta v}{h\nu \Gamma V} .$$ (6.32)

The total gas density, which is predominately molecular hydrogen, n_{H_2}, can be estimated for an H_2O maser by the equation

$$n_{H_2} = \Delta n_0 \frac{n_1}{\Delta n_0} \frac{n_{H_2O}}{n_1} \frac{n_{H_2}}{n_{H_2O}}$$ (6.33)

where n_{H_2O} is the density of water molecules. For a strong H_2O maser, $S \sim 10^{-19}$ erg s^{-1} cm^{-2} Hz^{-1} (10^4 Jy), $\Delta v \sim 50$ KHz, $V \sim \frac{4}{3}\pi 10^{45}$ cm^3, $\Gamma = 1$ s^{-1}, and $D = 10^{22}$ cm (≈ 3 kpc). From Equation (6.32) we obtain $\Delta n_0 \sim 100$ cm^{-3}. If the pump efficiency is 1%, then from Equation (6.19), $n_1/\Delta n_0 = 50$. For $n_{H_2O}/n_2 = 50$, appropriate for temperatures of ~ 200 K, and $n_{H_2}/n_{H_2O} = 10^4$, we calculate from Equation (6.33) that $n_{H_2} \sim 2 \times 10^9$ cm^{-3}. The luminosity is 6×10^{31} erg s^{-1} and the microwave photon rate, $L/h\nu$, is 4×10^{47} photons s^{-1}.

It is often argued that one can infer the state of saturation from studies of the time variation of masers. The reasoning behind this argument is that small changes in the pump, and hence in κ_0, can result in dramatic changes in an unsaturated maser since the output is proportional to $\exp(\kappa_0 \ell)$. On the other hand, since the output of a fully saturated (spherical) maser is only linearly proportional to $\kappa_0 \ell$, the saturated maser should be far less sensitive an indicator of pump variations than the unsaturated maser. However, we suspect that most saturated masers grow proportional to $(\kappa_0 \ell)^3$ (see Section 6.2.1). Also, without a fairly detailed understanding of the pump mechanism and its variations, it is usually impossible to quantify such relationships. Thus, time variations of masers are probably unreliable indicators of saturation.

6.2.5 Line Widths

Most astronomical masers have line widths that are smaller by factors of up to 10 than those expected from thermal motions in the masing regions. Some OH masers have line widths that would correspond to kinetic temperatures of <5 K were thermodynamic equilibrium to hold [see Equation (6.4)].

Line narrowing in masers is expected since, during unsaturated growth, the half-power width of the spectral profile will decrease. Consider a case in which the input signal to the maser is dominated by a background continuum source, characterized by brightness temperature T_c, so that the brightness temperature, with frequency dependence is, from Equation (6.15),

$$T_B(v) = T_c e^{\kappa \nu \ell} .$$ (6.34)

Substitution of Equations (6.5) into (6.7) gives the equation

$$\kappa_\nu = \kappa_0 e^{-4 \ln 2 (\nu - \nu_0)^2 / \Delta \nu_D^2} .$$ (6.35)

Hence, the brightness temperature at any frequency is given by

$$T_B(\nu) = T_c \exp(\kappa_0 \ell e^{-4 \ln 2 (\nu - \nu_0)^2 / \Delta \nu_D^2}) .$$ (6.36)

The profile of T_B is approximately Gaussian for $\kappa_0 \ell \gg 1$ and has a width, $\Delta \nu$, of

$$\Delta \nu \approx \frac{\Delta \nu_D}{\sqrt{\kappa_0 \ell}} .$$ (6.37)

The line width decreases until the center of the line begins to saturate. The line width then broadens as the wings of the line continue to experience exponential growth until the line width again equals $\Delta \nu_D$. For many masers, a value of $\kappa_0 \ell$ of approximately 16 to 25 is required to provide enough gain to explain the observed brightness temperatures. Thus, the line widths of unsaturated masers should be narrower, by a factor of 4 to 5, than the thermal line width.

Goldreich and Kwan (1974a) suggested that the rebroadening of the line during saturated growth can be inhibited if trapping of radiation at the (infrared) wavelengths involved in the pump cycle is significant. Provided that the trapped photons can cycle population from the line wings (through some other molecular energy levels) to the line center at a rate faster than the stimulated emission rate in the masing transition, the intensity of the line center can continue to grow more rapidly than that of the line wings. Thus, although the maser is saturated and the intensity does not grow exponentially, the line will remain narrow. From the estimates of the infrared transition and stimulated emission rates, this process appears likely to affect OH masers significantly and may affect H_2O masers to a lesser extent.

6.2.6 Polarization

OH maser sources display a variety of polarization characteristics. Zeeman splitting of the spectral lines is the most likely explanation for the polarization. When the magnetic field in the masing region is well ordered and strong enough to split the lines by amounts comparable to or greater than the maser line widths, one would expect to observe polarized emission. Stellar OH masers, with a few notable exceptions (see Section 6.4), are unpolarized. The absence of polarization from the stellar sources is expected, because at distances from the central star of 10^{16} to 10^{17} cm, where the masers usually form, the magnetic field is probably too weak to split the lines by more than the line widths. Also, these maser sources often contain dozens of components that are spectrally blended. Therefore, even if the Zeeman splitting were large, any single-telescope polarization study would not find much net polarization.

In contrast to the stellar sources, interstellar OH masers are highly polarized. However, complete Zeeman patterns in these masers are not observed. The strongest and best-studied masers seem deficient in linearly polarized emission; this situation can be attributed to the combined effects of Faraday rotation inside the masing region, which will slow the growth of linearly polarized components, and resonant trapping of infrared radiation, which allows a cross-relaxation among magnetic sublevels.

Another problem in the identification of Zeeman pairs in interstellar OH maser spectra is that in most sources the relative strengths of the left and right circularly polarized Zeeman components are not equal. Differences in position and relative intensity of Zeeman components may be caused by gradients in the magnetic field and velocity field, which can make the gain path lengths of a Zeeman pair unequal (Cook 1968). However, observations of ground-state OH transitions have shown some pairs of oppositely polarized components that are separated by less than about 10^{15} cm on the sky. Since the amplification path length for these masers is also about 10^{15} cm, these pairs are probably from within the same condensation in the masing region. The most convincing evidence supporting the Zeeman interpretation of OH maser polarization comes from a study of an *excited-state* OH transition (Moran et al. 1978). The spectrum of W3(OH) at 6035 MHz can be decomposed into twelve Zeeman pairs whose RCP and LCP components are separated on average by less than 3×10^{15} cm. The magnetic field strengths are ~ 5 mG.

Observations of H_2O sources show that about 25% of them have polarized features. The H_2O molecule is nonparamagnetic, and a magnetic field strength of about 50 gauss is required for the Zeeman splitting to equal the Dopper widths of the lines. Scaling from the parameters of interstellar OH sources, where $B = 5$ mG, $n \sim 10^6$ cm^{-3}, to the H_2O regions, where $n < 10^{11}$ cm^{-3}, for $B \propto n^{1/2}$ implies that $B < 1$ G. Under these conditions, no detectable circular polarization is expected, and linear polarization can be generated only if the maser is saturated. Polarized emission will occur when the stimulated emission rate, W, exceeds both the decay rate, Γ, from the maser levels and the cross-relaxation rate, γ, and also when the Zeeman splitting is greater than $(W\Gamma)^{1/2}$. Γ and γ are probably set by spontaneous transition rates in the far IR and are about 1 s^{-1}. W can be deduced from observed source sizes with assumptions about the source geometry [see Equation (6.26)] and is probably between 10^{-1} and 10^2 for most masers (see Table 6.1). Hence, since the Zeeman splitting is about 1 kHz G^{-1} for H_2O, the magnetic fields must only be greater than 0.3 to 10 mG to yield observable polarization. A detailed study of polarization versus source size could improve the limits on the magnetic fields.

Most SiO masers in stars contain linearly polarized features (Troland et al. 1979, Barvanis and Predmore 1985). Since SiO is also nonparamagnetic, the polarization theory of Goldreich et al. (1973) predicts that the SiO masers are saturated and form in regions where $B > 0.1$ mG. The magnetic fields are likely to be many orders of magnitude greater than this value for these masers which form close to red gaint (or supergiant) stars.

6.2.7 Pump Models

One of the challenges in the development of a theory of masers is to explain how different masers are pumped. Pump models require consideration of a large number of molecular transitions. Often the evaluation of a model requires knowledge of physical conditions (e.g., temperature and density) and molecular properties (e.g., collisional cross sections) beyond that which is presently available. In spite of these difficulties, many viable pump models can be found in the literature, demonstrating that relatively simple processes can yield sufficient population inversion to account for astronomical masers.

There are thermodynamic constraints that can be used to understand some basic requirements for maser pumps and to provide some means to discriminate among models. Masers cannot exist under conditions of thermodynamic equilibrium. In order to allow for population inversion, there must be a flow of energy between two "heat reservoirs" at different temperatures. In many pump models, radiation in certain key (usually IR) transitions must be optically thin in order to couple molecules to the low-temperature reservoir (e.g., cold sky).

The pumping of astronomical masers usually requires the solution of rate equations involving many levels in the molecule. By setting Γ, R_1, and R_2 equal to zero, Equations (6.11a) and (6.11b) describe an isolated two-level system. If we assume that the radiation field is isotropic ($\Omega = 4\pi$) and characterized by a radiation temperature $T_R = Ic^2/2kv^2$, and we note that

$$\frac{C_{12}}{C_{21}} = e^{-hv/kT_k} . \tag{6.38}$$

One can derive the relation for the excitation temperature:

$$T_x = T_k \frac{T_0 + T_R}{T_0 + T_k} \tag{6.39}$$

where

$$T_0 = \frac{hv}{k} \frac{C_{12}}{A} . \tag{6.40}$$

T_x is bounded between T_R and T_k and cannot be negative. Hence a two-level cosmic maser is not possible. The influence of other levels is essential in establishing the population inversion. Two-level laboratory masers can be built in which the molecules in the lower level are spatially separated from those in the upper level by magnetic or electric fields.

Pumps are usually classified in terms of their source of excitation—with collisional and radiative pumps being the most widely discussed. Pumps in which the excitation is from chemical reactions are thought not to be effective. However, since no pump models have as yet been widely accepted, it would be premature to eliminate any class of models.

There is an energy constraint that can be used to rule out some pumping schemes. Most proposed pumping mechanisms require at least one pump photon per maser photon. The number of pump photons per second entering the maser cloud, N_p, is given by

$$N_p = S_p 4\pi D^2 \frac{\Delta v_p}{hv_p} \frac{\Omega_m}{4\pi} \tag{6.41}$$

where S_p is the flux density of the pump measured by the observer, Δv_p is the line width of the pump transition, Ω_m is the solid angle subtended by the maser cloud seen from the pump source, D is the distance between the observer and the pump source, and v_p is the pump frequency. The number of maser photons per second is

$$N_m = S_m 4\pi D^2 \frac{\Delta v_m}{h v_m} \frac{\Omega_0}{4\pi} \qquad (6.42)$$

where the m-subscripts apply to the maser source and specifically where Ω_0 is the solid angle of the radiation pattern of the maser. $\Omega_0 = \Omega$ for a cylindrical maser and 4π for a spherical maser. The condition that $N_p > N_m$ yields

$$S_p > S_m \frac{\Omega_0}{\Omega_m} \frac{\Delta v_m}{v_m} \frac{v_p}{\Delta v_p} . \qquad (6.43)$$

Often the pump source is not visible due to absorption by intervening dust, and S_p cannot be observed directly. However, limits on the total pump luminosity can be obtained by assuming that at most a small number of stars provide the pump energy. In this case, we wish to express Equation (6.43) in terms of the luminosity and temperature of the pump source. Assuming a blackbody-like spectrum for the pump source,

$$S_p = B_{v_p}(T_p)\Omega_p , \qquad (6.44)$$

where $B_{v_p}(T_p)$ is the Planck function at the pump frequency for a pump source at temperature T_p, and Ω_p is the solid angle of the pump source as seen from the Earth. Ω_p can be replaced by using the relation

$$L_p = 4\sigma T_p^4 \Omega_p D^2 \qquad (6.45)$$

where σ is the Stephan-Boltzmann constant. Finally, substitution of Equations (6.44) and (6.45) into Equation (6.43) yields

$$L_p > S_m 4\pi D^2 \frac{\sigma T_p^4}{B_{v_p}(T_p)} \frac{\Omega_0}{\Omega_m} \frac{\Delta v_m}{v_m} \frac{v_p}{\Delta v_p} \qquad (6.46)$$

Since the Planck function is

$$B_v(T) = \frac{2hv^3}{c^2} \frac{1}{e^{hv/kT} - 1} , \qquad (6.47)$$

the pump luminosity is minimized for a fixed pump frequency by minimizing the expression $T^4(e^{hv/kT} - 1)$. This expression has a minimum when $hv/kT \approx 3.9$, and substitution of this value into Equation (6.46) gives

$$L_p > 20.5 v_p S_m D^2 \frac{\Omega_0}{\Omega_m} \frac{\Delta v_m}{v_m} \frac{v_p}{\Delta v_p} . \qquad (6.48)$$

For the strongest H_2O maser sources, one can probably rule out ultraviolet or optical pumps, as well as most radiative pumps in which the exciting source is outside of the maser cloud. For example, consider UV pumping for a strong H_2O source such as in W49 by an O-type star embedded in the maser cloud. If we assume $S_m = 10^{-19}$ erg s^{-1} cm^{-2} Hz^{-1} (10^4 Jy), $\lambda_p = 0.1$ μm, $D = 3 \times 10^{22}$ cm (10 kpc), equal fractional bandwidths for the maser and pump lines, and $\Omega_0 \approx \Omega_m$, then Equation (6.48) requires $L_p > 10^9$ $L_\odot$, for an optimum pump temperature of

37,000 K. This luminosity is about four orders of magnitude greater than an O-type star can produce. Thus, unless all strong H_2O maser lines in W49 are highly beamed and the pump is internal (i.e., $\Omega_m = 4\pi$), UV pumping cannot work. In the above example, the maser photon rate is $\sim 10^{48}$ photons s^{-1}, and thus masers with photon rates greater than $\sim 10^{44}$ photons s^{-1} are probably not pumped by UV photons from a continuum source. The required pump luminosity is proportional to ν_p, and thus infrared pumping seems more attractive, especially given the proximity of nearly every maser to an infrared source. However, even infrared pumping cannot explain the strongest H_2O masers unless they are highly beamed.

Nonradiative pumping is possible, and for the strongest H_2O masers is likely. Collisional pumping may be significant in H_2O masers where the gas density is high. At first, collisional pumping seems implausible since collisional processes suggest thermal equilibrium. However, inversion can occur, for example, because there are more spontaneous decays from a high-lying state, populated by collisions, to the upper level of the maser transition than to the lower level. The inversion will be maintained if the photons emitted during the spontaneous decay are removed from the gas. For example, they can escape from the gas if the maser is thin in one dimension, or they can be absorbed on cold grains. Note that the flux density in a saturated maser increases as the square of the hydrogen density, since the inversion density and number of collisions per second are both proportional to n_{H_2}.

In many sources the maser emission is thought to come from condensations flowing in a stellar wind. In the Orion-KL region, the H_2O masers appear to be moving ballistically away from a compact infrared source (designated IRc 2). Most of the masers are substantially removed from IRc 2, as well as any other radiation source, and hence would require an external pump were a radiative pump invoked. This seems unlikely based upon the discussion above. The most likely source for pumping the masers is the kinetic energy of the flow itself (or of the clouds in the flow). Suppose a maser cloud or "bullet" plows into a fragment of the ambient cloud material which surrounds the newly formed star which drives the wind (e.g., Norman and Silk 1979, Tarter and Welch 1986). The rate of dissipation of kinetic energy is

$$L_k = \frac{\pi}{2} r^2 n_a m_{H_2} v^3 \tag{6.49}$$

where r and v are the radius and velocity of the maser bullet, n_a is the number density of gas in the ambient cloud, and m_{H_2} is the mass of a hydrogen molecule. For $r = 3 \times 10^{14}$ cm, $n_a = 10^8$ cm^{-3}, and $v = 40$ km s^{-1}, Equation (6.49) indicates that $L_k \sim 1\ L_\odot$. The apparent isotropic maser luminosity, L_m, is given by the equation

$$L_m = \frac{4\pi}{\Omega_0} \frac{\nu_m}{\nu_p} L_k \varepsilon \tag{6.50}$$

where ε is the pump conversion efficiency, i.e. $L_p = L_k \varepsilon$. Hollenbach and McKee (1979) showed that the cooling in molecular lines, and for H_2O in particular, is very efficient in a postshocked regime. Hence it is reasonable that a few percent of the cloud kinetic energy may be converted to IR lines of H_2O, which could pump the maser. We adopt $\varepsilon = 0.02$. If the pump operates at $\sim 40\ \mu$m, then $\nu_m/\nu_p \sim 0.003$.

Thus if $\Omega_0 = 0.1$, then $L_m \sim 10^{-2} L_\odot$, which corresponds to about 2×10^{47} photons s^{-1}. This is adaquate to pump all but the strongest H_2O masers.

Conventional pumping mechanisms involving radiative and collisional processes cannot satisfactorily explain the emission from the strongest masers (i.e., those that have photon rates greater than $\sim 10^{47}$ photons s^{-1}). The basic problem is that the maser output is limited by the pump rate which is limited by (1) the Einstein A coefficients linking the masing levels to other levels in the molecule and (2) the total hydrogen density which cannot exceed $\sim 10^{11}$ cm^{-3}, at which point the maser would be thermalized by collisions. Models that avoid these limitations have been proposed by Strelnitskii (1984) and Kylafis and Norman (1986) and involves collisional pumping in a region where the free electrons and neutral particles have different temperatures. Such a two-temperature gas is inherently out of thermodynamic equilibrium, and a population inversion can be maintained regardless of the molecular density.

(a) Interstellar Maser Pumps

Both collisional and radiative pumps have been proposed for the strong interstellar H_2O masers. For example, de Jong (1973) proposed a pump model in which collisions with hydrogen molecules excite H_2O molecules. Deexcitation is by spontaneous emission with some of the resulting IR photons escaping from the masing cloud. Inversion occurs because the lowest level of a given rotational ladder (e.g., the 6_{16} level) is the last to become optically thin. Hence this level remains in thermodynamic equilibrium the longest, while other levels (e.g., the 5_{23} level) can have populations lower than for thermodynamic equilibrium. Alternatively, Goldreich and Kwan (1974b) suggested that some H_2O masers are radiatively pumped. Their model requires that H_2O molecules be excited by 6-μm photons emitted by hot dust grains. Inversion occurs because there are more absorptions from the lower than from the upper level in the masing transitions. In this model, the "heat sink" is cool hydrogen gas that collisionally deexcites the H_2O molecules.

(b) Stellar Maser Pumps

Stellar water masers may be pumped by the collisional excitation of the ν_2 band and radiative decay at 6 μm. The effectiveness of the pump mechanism is unclear since the collision cross sections are poorly known, but probably there is adequate 6-μm emission to account for the maser emission.

There have been several pump models proposed to explain the appearance of strong 1612-MHz and often significant 1665- and 1667-MHz (main-line) OH emission from stellar sources. Elitzur et al. (1976) proposed that the 1612-MHz masers could be pumped by far-IR (35-μm) radiation from circumstellar dust. The correlation between the 35-μm flux density and the maser flux density supports this model. The main-line transitions from stellar sources may also be pumped via far-IR radiation from dust. Alternatively, infrared radiation from the central star excites the H_2O molecules, which might invert the main lines via the subsequent radiation from certain H_2O transitions that overlap in frequency with OH lines.

A pump mechanism for SiO masers was proposed by Kwan and Scoville (1974) that can explain maser emission in the rotational transitions of the first excited vibrational state of the molecule. The masing level is populated by indirect radiative routes through higher vibrational states and depopulated primarily by radiative decays. Population inversion can occur if radiation from higher vibrational levels into the masing level is optically thin, and if radiation out of the masing level to lower vibrational levels is optically thick. This scheme can only account for maser emission from one vibrational level in a given volume of gas if conditions in that volume are homogeneous. However, VLBI observations (Lane 1984) show that some maser spots in the $v = 1$ and $v = 2$ levels are coincident and probably arise from the same gas. A model in which the SiO is pumped by a cycle involving collisional excitation and radiative deexcitation (Elitzur 1980) can account for maser action in more than one vibrational level at a time but will work only in regions of high density close to the stellar surface.

6.3 Interstellar Masers

Interstellar masers are generally found in regions of star formation and are excellent signposts of massive-star formation. Newly formed O and B stars heat and ionize their environments, creating sources of infrared and radio emission. The sources appear as: (1) masers, (2) molecular hot spots, (3) compact HII regions, (4) near- and far-IR sources. The molecular cloud in NGC 6334 is an active star-forming region with dimensions of 2×20 pc that has been probed in great detail and found to contain five water-vapor masers; two OH masers; four CO peaks; three compact HII regions; one ultracompact HII region; four 1-mm peaks; five far-IR peaks; and one near-IR peak. These twenty-five sources are excited by at least six objects or clusters.

Many attempts have been made to order objects in an evolutionary sequence based on the types of emission observed. Firm observational constraints are (1) OH masers are almost always associated with compact HII regions smaller than 3×10^{17} cm (0.1 pc); (2) most H_2O masers are located near, but not exactly coincident with, compact HII regions; and (3) H_2O masers show high-velocity features due to high-velocity mass motions. Most evolutionary models have H_2O masers appear in the earliest phase of stellar life when the region of ionized gas is too small to be detected. The masers persist for about 10^5 years as the ionization region expands until it has reached 3×10^{17} cm, at which point the molecular density and excitation have decreased below the limits necessary for maser action. Whether or not H_2O maser sources become OH maser sources is unclear. There are several observational problems that limit our ability to determine source evolution. Firstly, the sampling in most wavelength regions is poor. Since the beamwidths of typical telescopes at the H_2O maser transition of 1 cm are about one arcminute, a survey of the sky is hopeless and a complete survey of even small molecular clouds is laborious. Hence, searches for H_2O masers are generally made in regions where success seems likely, i.e., near compact HII regions. Secondly, the observations are sensitivity limited. As sensitivity increases, more and more masers are found. Thirdly, subarcsecond angular positions for most compact infrared and submillimeter sources are lacking. Hence there is a fundamental problem of establishing whether a group of sources are coincident and excited by some object or are a cluster of objects at slightly different stages of development. For example, if the Orion Nebula were at 5 kpc instead of 0.5 kpc, the relation between the Trapezium stars and the BN and KL objects would probably be unclear.

6.3.1 H₂O Masers

H_2O maser emission arises from the 6_{16}–5_{23} rotational transition at 1.3-cm wavelength due to a fortuitous near-coincidence in energy states in adjacent rotational ladders of ortho-water. The 6_{16} level is 447 cm^{-1} above the ground state, and in thermal equilibrium at a temperature of 400 K about one percent of the H_2O molecules are in this level. Emission and absorption from H_2O in the Earth's atmosphere in this transition are pressure broadened to a width of about 5 GHz, and in the zenith direction on clear days cause an absorption of about 10% and an excess propagation delay corresponding to a distance of about 30 cm.

Most H_2O masers have been found during searches of limited fields with antennas having beamwidths of 30 to 120 arcsec and limiting sensitivities of 1 to 100 Jy. Sources are generally named in reference to groups that are not resolved by a single antenna (e.g., separated by less than about 10^{18} cm). Many of these sources undoubtedly contain a number of stars. There are presently about 200 known H_2O maser sources in regions of star formation. However, arguments based upon the extrapolation of log N–log S plots for known masers, the assertion that a newly formed O or B star has a maser for a period of about 10^5 years, and the results of a very sensitive search for H_2O masers in large molecular clouds suggest that there are about 10^3 masers in the Galaxy with flux densities greater than 1 Jy. Finding all them will be tedious work.

The spectra of interstellar H_2O masers range from simple ones containing a single feature to very complex ones containing hundreds of features. Figure 6.4 shows the spectrum of a complex source, W51M, along with a VLBI map of the relative locations of individual maser components. Maps such as this one show that a spectral feature may be composed of one or more individual maser components or spots, each of which arises from a spatially distinct position.

Maser components cluster over a very wide range of scale lengths. Compact clusters, sometimes called "centers of activity," persist for times much longer than the lifetimes of individual components. For example, in W51M there are very compact clusters, about 10^{14} to 10^{15} cm in diameter, which are spread over a region of about 10^{16} cm. Clustering on the scale of 10^{16} cm is probably due to masers that are dynamically associated with a single star. Clusters that are separated by 10^{17} cm, as found in W49, may be associated with a single star or a cluster of stars that are related to each other. Finally, the largest clustering scale of $<10^{20}$ cm refers to masers associated with different sites of star formation in the same giant molecular cloud.

The clustering on the smallest scales mentioned above is difficult to explain. Genzel et al. (1979) considered whether nonkinematic effects could cause the velocity dispersion of typically 10 to 30 km s^{-1} in very compact knots. The hyperfine structure of the transition has six components spread over 5.9 km s^{-1} (Kukolich 1969). Although some observed spots have velocity separations close to that of some hyperfine components, the velocity structure is usually more complex than that of

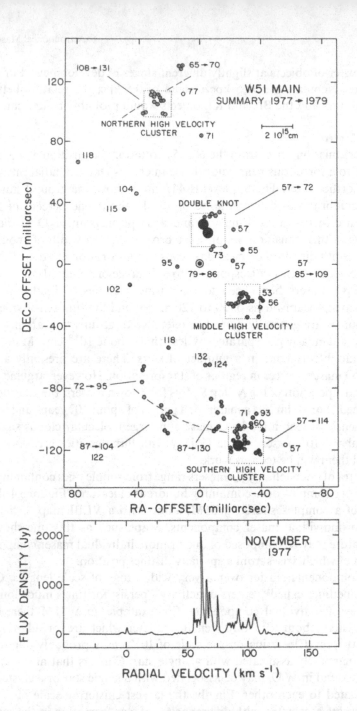

Fig. 6.4. H_2O maser emission from the central part of W51(Main) reported by Genzel et al. (1981b). *Bottom*: Spectrum of maser emission with the radial velocity axis referred to the local standard of rest for a rest line-frequency of 22,235.080 MHz for the 6_{16}–5_{23} transition. A vertical bar indicates the velocity of the molecular cloud. *Top*: A VLBI map of the maser emission. Dots indicate individual maser spots, the filled circles indicate clusters of masers, and the encircled dot indicates the maser that served as a reference for the map. The regions of highest concentration of high-velocity masers are marked with dashed lines. The linear distance scale is based upon a statistical parallax distance (see Section 6.7) of 7 kpc. (From Reid and Moran, 1981. Reproduced, with permission, from the Annual Reviews of Astronomy and Astrophysics, Vol. 19. © 1981 by Annual Reviews Inc.)

the hyperfine pattern alone. The Zeeman effect requires a magnetic field strength of about 700 gauss to produce splittings of 10 km s^{-1}, and hence this effect should be negligible. The resonant Stark effect requires brightness temperatures of about 10^{18} K to produce a splitting of 10 km s^{-1} and thus cannot explain the observations. Other possible mechanisms for inducing large frequency shifts in the emission, such as Raman or Langmuir scattering, do not appear to be applicable. The most likely explanation for the velocity dispersion is a combination of hyperfine splitting and of turbulence caused by a strong stellar wind impacting on dense molecular cloud material (Walker 1984).

There is much discussion in the literature concerning the nature of high-velocity features ($|v - \bar{v}| \gtrsim 30$ km s^{-1}). Strelnitskii and Syunyaev (1973) first proposed the idea of high-velocity ballistic flows around newly formed stars to explain the H_2O maser spectra. The high-velocity features do not appear to be intrinsically different from low-velocity features. The proper-motion results (see Section 6.6) show that transverse velocities are as high as radial velocities and, therefore, both the transverse and radial velocities are primarily kinematic in origin. In well-studied sources, such as W49 and Orion, the distribution of the high-velocity features suggests that they are part of a global flow from a single object. Models for wind-driven mass outflow in such sources can account for the high-velocity features.

The spectral appearance of H_2O masers changes rapidly. Features can vary in strength significantly on a daily basis, and the lifetime of features typically ranges from months to years. In contrast to OH masers, where the spectra have remained easily recognizable over two decades, most H_2O maser spectra have changed completely over that time period. In general, the temporal sampling of observations has been inadequate to follow the variations of maser features. In one case where data were acquired daily, the flux density of a feature in W3OH rose for eight days and then decayed slowly over the following few weeks (Haschick et al. 1977). This burst may have been caused by a sudden application of pump energy from a heat pulse that diffused through the maser medium. Detailed modeling of the intensity curves may provide a way of selecting the proper pump model.

The Orion molecular cloud (OMC-1) is one of the best-studied regions of star formation. In 1979, a spectacular maser component at 8 km s^{-1} "flared," recently reaching a flux density of over 10^7 Jy. This feature is the brightest maser even seen, having a brightness temperature in excess of 10^{15} K. It has a high degree of linear polarization ($>60\%$), suggesting magnetic field strengths approaching 1 Gauss (Garay, Moran, and Haschick, in preparation).

The intense H_2O maser emissions from Orion have been mapped to an accuracy of 0.001 arcsec in a series of six VLBI observations spanning a two-year period (Genzel et al. 1981a). The spectrum contained low-velocity features within ± 20 km s^{-1} of the molecular cloud velocity as well as high-velocity features extending out to ± 100 km s^{-1}. During this period the positions of many features changed linearly with time at rates of up to 0.02 arcsec yr^{-1}, which corresponds to transverse velocities of up to 50 km s^{-1}. These proper motions, combined with radial velocities, provided three-dimensional velocity vectors of many components. The pattern of velocities clearly indicated a symmetric outflow. Modeling these data yielded the position of the expansion center, the expansion speed, and the distance to the Orion source (see Section 6.7 for more details on distance estimates). The center of expansion was found to lie between IRc 2 and IRc 4, and the expansion speed was found to be 18 km s^{-1}. Because the H_2O expansion speed matches that observed in other molecules, Genzel et al. suggested that a single "low-velocity" flow, probably driven by IRc2, can explain most molecular observations.

H_2O masers may also be associated with lower-mass stars than the O and B stars discussed above. In dark clouds containing Herbig–Haro (HH) objects, weak H_2O masers can be found. These masers are displaced from compact HII regions that are powered by B1- to B4-type (ZAMS) stars. These H_2O masers may be from high-velocity gas fragments accelerated by the stellar wind from the B stars. As the fragments move farther away from their origin and subsequently collide with the surrounding molecular cloud material, they may radiate optically as HH objects. Alternatively, the H_2O masers, compact HII regions, and HH objects may be associated with separate lower-mass stars in a forming cluster.

6.3.2 OH Masers

Interstellar OH maser emission has been detected from several lambda-doublet, hyperfine split transitions in the $^2\Pi_{3/2}$ and $^2\Pi_{1/2}$ rotational ladders. The strongest emissions are from the ground state, particularly in the $^2\Pi_{3/2}$, $J = \frac{3}{2}$, $F = 1-1$ transition at rest frequency 1665.402 MHz (see Figure 6.5). Weaker emission is usually observed from the other main-line transition ($F = 2-2$) at 1667.359 MHz, and sometimes detectable emission is observed from the two satellite-line transitions ($F = 1-2$) at 1612.231 MHz and ($F = 2-1$) at 1720.530 MHz. There are cases where the 1665-MHz transition is not dominant. For example, the 1720-MHz transition is strongest in some sources; however, these sources have not received sufficient observational scrutiny to be discussed further.

Typical interstellar OH maser spectra appear to consist of many spectral components that span a range of about 5 to 10 km s^{-1} in radial velocity. High-velocity emission such as seen in H_2O masers is rare. Most spectral components are strongly (up to 100%) circularly polarized. Linear-polarized components do not seem as common as circular ones, although this may only be true for the strongest and best-studied sources.

Compact ($\ll 1$ pc) HII regions and sources of IR emission are almost without exception found at nearly the same position on the sky from which interstellar OH masers are detected (see Figure 6.6). OH masers are not found associated with HII regions that are more than 3×10^{17} cm in extent. In all cases where sufficiently accurate (better than 1″) OH and radio continuum maps exist, the masers appear projected very close to regions of HII emission. Thus, there is little doubt that interstellar OH masers are physically associated with very young, hot stars that create compact HII regions.

Individual maser components appear as "spots" about 10^{14} cm in size that are spread over regions 10^{16} to 10^{17} cm in diameter. Several interstellar OH sources have been mapped with highly sensitive VLBI aperture synthesis techniques. These studies indicate that the maser components tend to cluster together on a scale size of 10^{15} cm. Interstellar OH masers vary in intensity on time scales of years or longer. Since present observations span only about two decades, we do not know their

Fig. 6.5. Emission spectra for the four ground-state Λ-doublet transitions in the $^2\Pi_{3/2}$, $J = \frac{3}{2}$ state of hydroxyl (OH) from Fouquet and Reid (1982). The solid-line spectra were measured in right circular polarization and the dashed-line spectra in left circular polarization. Measured antenna temperatures were multiplied by Boltzmann's constant and divided by the effective antenna collecting area to convert to flux density.

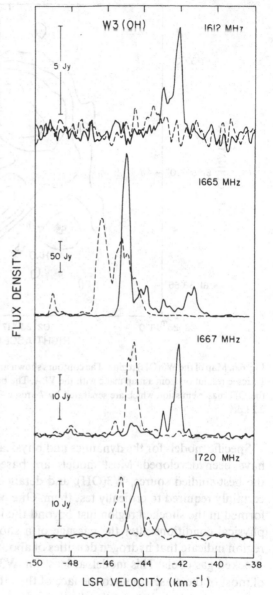

lifetimes. In contrast, H_2O masers from the same general regions have lifetimes of as short as weeks and typically months. The masing molecules are part of an envelope of neutral gas and dust surrounding a compact HII region. The dust emission from the envelope is responsible for the IR emission, implying that the temperature of the dust is between 10^2 and 10^3 K. Hydrogen densities are probably between 10^5 and 10^9 cm^{-3} in order to provide sufficient column density for maser amplification, but not so high as to thermalize the level populations (see Section 6.2). Magnetic field strengths, which are inferred from Zeeman splitting, in the envelope appear to be in the range of 2 to 10 mG.

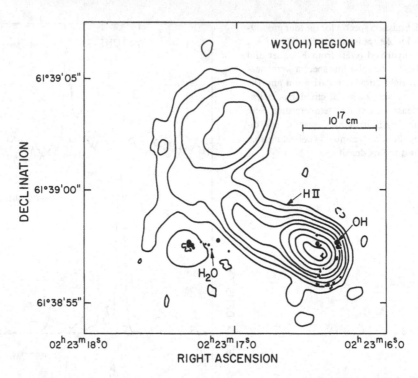

Fig. 6.6. Map of the W3(OH) region. The contours shown are the 18-cm-wavelength emission with about 1-arcsec resolution from a map made with the VLA. The back dots show the locations of the H_2O and the OH maser emission which are separated by 7 arcsec. The linear scale is based upon a distance of 2.2 kpc.

Specific models for the dynamics and physical conditions in the masing envelope have been developed. Most models are based primarily upon observations of the best-studied source, W3(OH), and detailed observations of more sources are certainly required to critically test them. One possible model is that the masers are formed in the shocked region just beyond the ionization front. Calculations of the physical conditions after the passage of a shock outside of a dust-bounded HII region indicate that hydrogen densities of about 10^7 cm^{-3} can be achieved after the shocked gas cools. This model, at least for W3(OH), can explain the confinement of most of the maser spots to the face of the HII region.

The data do not support the shock model as outlined above. A synthesis map of the W3(OH) maser emission indicates that the OH masers occur in clumps, almost all of which appear to be redshifted with respect to the underlying compact HII region. The HII region velocity can be estimated from high-frequency recombination lines where the optical depth of the emission is low. The HII region is optically thick at the OH maser frequency of 1665 MHz; this implies that the masing clumps are in front of the HII region and are therefore falling inward relative to it. A similar pattern of molecular material redshifted with respect to center-of-mass velocity estimates is seen in many other OH maser sources and by using NH_3 line velocities instead of OH/HII line velocities. Thus, OH masers may be part of a remnant accreting envelope that has not (yet) encountered an expanding shock front (Reid et al. 1980).

The question as to whether shock chemistry is required to achieve high OH densities is unclear. The answer may be that the envelope is clumpy, giving rise to some high-density regions. Alternatively, it is possible that further from the central star than the ionization-induced shock, another shock may be associated with the infall of material and the shock (or high-temperature) chemical calculations might apply there. In any event, observations of maser emission can provide crucial checks on theories of the formation and early evolution of massive stars.

One point of great interest in star formation theory concerns how angular momentum and magnetic fields affect the collapse of interstellar material to form a star. Observations of masers can provide maps of the velocity and magnetic field conditions on scales of 10^{13} to 10^{17} cm that fill in a significant gap between those conditions found in molecular clouds ($>10^{19}$ cm) and in stars ($<10^{12}$ cm). Maps of magnetic fields across an OH source require the identification of Zeeman pairs, which is best accomplished with long baseline radio interferometric maps. The most successful mapping of magnetic field structure is by Moran et al. (1978). They identified ten Zeeman pairs in an excited-state OH transition toward W3(OH) and found that the magnetic field varies from 2 to 9 mG across the masing region. No field reversals were found, indicating a coherent magnetic field structure over the entire source. Davies (1974) points out that the magnetic field direction, as inferred from a general shifting of the spectrum in one sense of circular polarization with respect to the other sense of circular polarization, is parallel to the direction of rotation of the Galaxy at the positions of the OH sources. Thus, the magnetic field probed by OH maser observations appears to have originated in the general interstellar field that has been compressed during the formation of the central star. It is interesting to note that the magnetic field strength appears to scale proportionally to the square root of the density during compression from interstellar conditions ($B \sim 1\ \mu G$ and $n \sim 1$ cm^{-3}) to those found in masers ($B = 6$ mG and $n \sim 10^7$ cm^{-3}). This is somewhat different from the relationship between magnetic field strength and density expected for homologous contraction with a frozen-in magnetic field (B proportional to $n^{2/3}$) and appears to support the theory of nonhomologous contraction.

Observations of maser sources can provide maps of the velocity field across a source and hence can contribute to our understanding of the rotational characteristics of the material from which stars are forming. There have been claims that some OH maser spot maps show a large-scale rotation across a source. However, in most cases the data are susceptible to other interpretations, because (1) maps do not give three-dimensional positional and velocity information and (2) Zeeman splitting can give "apparent velocities" of several kilometers per second. Proper-motion studies seem essential to determine unambiguously the three-dimensional dynamics of most sources (see Section 6.7). In W3(OH), any large-scale rotation is less than about 2 km s^{-1}. W75N is perhaps the only source in which OH maps clearly indicate both an elongated ($>5:1$ axial ratio) spot distribution and a velocity gradient (about 10 km s^{-1}) across the source (Haschick et al. 1981).

6.4 Stellar Masers

In addition to regions of active star formation, molecular masers are found associated with late- (usually M) type stars. These are evolved stars in the red giant or supergiant phase of evolution. The most numerous of these stellar masers are long-period variable (LPV) stars. However, molecular masers are also found with semiregular variable stars and many optically unidentified sources. At present about 400 stellar masers have been discovered (see Engels 1979).

Long-period variable stars have photospheric temperatures just above 2000 K and radiate most of their 10^4 $L_\odot$ in the red and infrared. Optical emission lines of hydrogen and metallic absorption lines are copious and exhibit a complex and time-varying velocity distribution. Maser emission seems to favor stars with strong infrared excesses above that attributable to stellar emission, which indicates that an extensive envelope of gas and dust surrounds the star.

Stellar masers usually display emission from SiO, H_2O, and OH molecules. Figure 6.7 displays maser spectra from the supergiant star VY CMa. The energy levels of the masing states in the SiO, H_2O, and OH molecules expressed in temperature units are 1800 K, 600 K, and ~ 0 K, respectively. This suggests that the SiO masers reside close to the stellar photosphere, the H_2O masers occur farther out in the circumstellar envelope, and the OH masers exist at the greatest distance from the central star. Thus, the study of these three masing species can add a great deal to our understanding of the photosphere and circumstellar envelope of these luminous stars.

The OH masers are the best studied of the stellar maser sources. The emission is usually dominated by the satellite-line transition at 1612 MHz, although significant radiation from the main lines is often observed. The OH spectra are characterized by two complexes of emission lines which are separated by about 10 to 50 km s^{-1} in radial velocity (see Figure 6.7). As a general rule, the OH emission is only slightly polarized. The intensity of the OH lines oscillates fairly smoothly with the light of the (usually) variable star, suggesting that the masers are at least partially saturated and that they may be radiatively pumped.

VLBI observations have established that the OH maser components appear to be much larger than those associated with HII regions. In LPV sources, both the low- and the high-velocity emissions extend for more than about 10^{16} cm on the sky. Both emission complexes contain many maser components whose apparent sizes seem to differ by at least a factor of 10. Interferometer amplitudes are generally low for fringe spacings less than 1 arcsec (Reid et al. 1977), and thus connected element interferometers and short baseline VLBI systems are best suited to study these sources.

The very regular double-peaked OH emission spectra suggest a simple dynamical explanation. For example, such spectral characteristics could be explained if the masing molecules are part of a circumstellar envelope that is expanding, contracting, or rotating about the star. The large sizes of the OH masers, coupled with the velocity separations of the two emission peaks, rule out contraction or rotation as the dominant dynamical characteristic of the circumstellar envelope. Either

Fig. 6.7. Spectra of maser emission toward the star VY CMa. From *top* to *bottom*: OH, H_2O, SiO ($v = 1, J = 1-0$), and SiO ($v = 0, J = 2-1$) transitions. The parabolic profile (dashed line) in the bottom spectrum fits the thermal SiO emission whose velocity centroid indicates the stellar velocity; a weak maser feature appears near 18 km s^{-1} and is one of very few ground-vibrational-state SiO masers. (From Reid and Moran, 1981). Reproduced, with permission, from the Annual Reviews of Astronomy and Astrophysics, Vol. 19. © 1981 by Annual Reviews Inc.

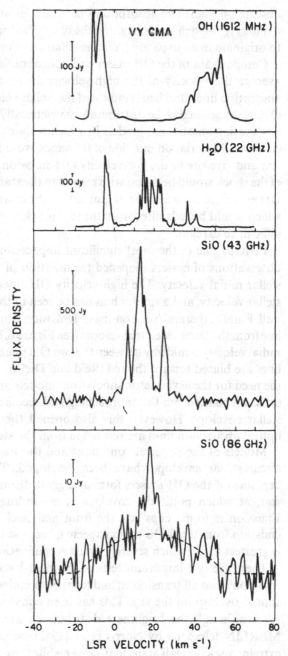

condition requires gravitational energy to support the velocity separation, and would imply exceedingly massive (e.g., $> 100\ M_\odot$) central stars.

Expansion is the most likely dynamical condition of the circumstellar envelope in LPV stars. Evidence for expansion has long been cited by optical spectroscopists. In LPV stars, emission lines of hydrogen appear blueshifted with respect to metallic absorption lines. The absorption-line velocities at maximum light have been used to indicate the stellar velocity, and the blueshifted emission lines have been suggested to originate in an expanding circumstellar envelope.

Comparisons of the OH maser and optical radial velocities indicated that on the average the velocity of the high-velocity OH peak matched that of the optical absorption lines, and hence indicated the stellar velocity. Thus, the highly symmetric OH peaks seemed to be distributed assymetrically about the stellar velocity. These observations motivated models in which a shock front exists in the circumstellar envelope. Material on one side of the shock would be expanding about the central star and give rise to the low-velocity OH emission, while material on the other side of the shock would be at rest with respect to the star and give rise to the high-velocity OH emission. Such a model required that the emission from the far side of the star, which would be redshifted from the stellar velocity, not be observable. This proved difficult to explain.

Perhaps one of the most significant impacts on the study of LPV stars is that observations of masers reopened the question of which spectral lines indicate the stellar radial velocity. The high-velocity OH lines seem to be redshifted from the stellar velocity, and a similar bias can be seen in the optical absorption-line data as well. Finally, thermal (i.e., non-maser) emission from a ground-vibrational-state SiO line from the circumstellar envelopes (see Figure 6.7) clearly indicated that the stellar radial velocity is midway between the two OH peaks and that the optical absorption lines are biased toward the red (Reid and Dickinson 1976). These results removed the need for the asymmetric shock-front models and clearly suggested that the OH emission comes from the approaching and receding sides of an expanding circumstellar envelope. However, this also opened the problem of explaining why the optical absorption lines are redshifted from the stellar velocity.

Models of the physical conditions and the maser-line formation in expanding circumstellar envelopes have been developed. The observational data are best explained if the OH masers form at a great distance ($> 10^{16}$ cm) from the central star, at which point the envelope is expanding at a nearly constant velocity. Emission is from "caps" on the front and back sides of the star. Strong maser emission from the limbs is not expected, because the line-of-sight velocity gradient is greatest there, which severely limits amplification lengths.

There are two important tests of the front-back emission model. First, the OH emission from all transitions, and from both emission complexes in any transition, should overlap on the sky. This has been convincingly verified for many sources. Also, although the brightness distributions are complex, maps made with the MERLIN telescope by Norris et al. (1982) indicate that the emissions from the extreme velocities fall at the map center while the emissions from the central velocity components appear as portions of rings that are concentric with the map center (see Figure 6.8). This is the structure expected for emission from two "caps" from the approaching and receding sides of an expanding spherical shell.

The second test of the front-back emission model was suggested by Schultz et al. (1978). Assuming that both sides of the circumstellar envelope are excited simultaneously by the central star, the maser emission from the two sides should vary in intensity in a similar manner, except that the high-velocity emission from

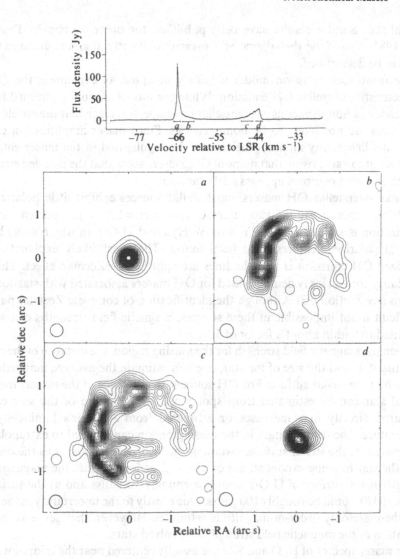

Fig. 6.8. *Top*: The 1612-MHz spectrum of the stellar source OH127.8. The bracketed velocity intervals *a*, *b*, *c*, and *d* are the intervals for which the spatial distributions shown below have been determined. *Bottom*: Maps of the spatial distribution of OH maser emission for the velocity ranges *a*, *b*, *c*, and *d*. The contour interval is 5% of the peak intensity, and the lowest contour is at 10%. These maps clearly show the shell-like structure of the OH maser emission region. These data are from Booth et al. (1981). (Reprinted by permission from Nature Vol. 290, No. 5805, pp. 382, Copyright(c) 1981, Macmillan Magazines Ltd.

the far side of the star should be delayed by the light travel time across the envelope relative to the low-velocity emission from the near side of the star. A convincing detection of this delay was published by Jewell et al. (1980) for IRC + 10011. They found that the variations of the high-velocity OH emission lag those of the low-velocity emission by 25 ± 5 days, strongly supporting the front-back model and indicating that the strongest OH emission occurs at about 3×10^{16} cm from the central star. Similar results have been published for other sources by Diamond et al. 1985. Also, if the shell diameter is measured directly, then the distance to the star can be determined.

The front-back emission model is widely accepted as explaining the general characteristics of stellar OH emission. While the model is usually outlined for the ideal case of a homogeneous circumstellar envelope, it is clear that the envelopes of most stars are not likely to be homogeneous. Since maser amplification can be highly nonlinear, any inhomogeneities will be highlighted in the maser emission, and it is not too surprising that maps of OH masers show that the detailed structure of each emission complex appears fairly random.

Unlike interstellar OH masers, most stellar sources exhibit little polarization. Notable exceptions are the supergiant OH sources where a few percent circular polarization is seen and the LPV stars W Hya and U Ori, in which some highly ($>50\%$) polarized features have been found. The most likely explanation for polarized OH emission is that the lines are split by the Zeeman effect. This has been fairly convincingly demonstrated for OH masers associated with star-forming regions (see Section 6.3). Although the identification of complete Zeeman patterns is difficult if not impossible in these sources, magnetic field strengths can still be estimated to within about a factor of 3.

Given the magnetic field strength for the masing region, the distance of the region from the star, and the size of the star, one can estimate the average magnetic field strength at the stellar surface. For OH sources, the distance of the masers from the central star can be estimated from spot maps, and the size of the star can be measured directly in some cases or estimated from the star's luminosity and temperature. The field strength in the masing region can be used to extrapolate to the strength at the stellar surface, assuming the field strength varies as the distance from the star to some exponent, α. For $\alpha = -2$, as for the Sun, the magnetic field strength at the surface of U Ori would be roughly 10 gauss and at the surface of IRC + 10420 would be roughly 100 gauss. Due mostly to the uncertainty in the value of α, these are only order-of-magnitude estimates. However, they serve as the first indication of the magnetic field strength for evolved stars.

The maser spectra of H_2O and SiO are usually centered near the midpoint of the OH emission but typically have a narrower range of radial velocity compared to OH. Double-peaked spectra, indicative of shell-type structure as for OH masers, are observed in some cases of H_2O sources. The H_2O emission varies with the stellar light; however, the variation is more extreme and erratic than that observed for the OH masers. Long-term studies comparing SiO and visible or IR radiation generally show a strong correlation between the IR and maser flux densities (e.g., Nyman and Olofsson 1986).

VLBI observations show that H_2O maser components are typically 10^{13} cm in size and that these components are spread over about 10^{14} cm on the sky. This spread is much smaller than the OH emission in the same objects. Also, VLBI observations indicate that the SiO maser components have characteristic dimensions comparable to H_2O masers in these stars. Both H_2O and SiO masers are spread on the sky over a region which is only slightly larger than the stellar diameter, and thus they are probably formed within about ten stellar radii of the star. In this region of the circumstellar envelope, grains are just forming and the mass motions may be very complex. Pulsation of the central star might cause oscillations in this portion of the envelope with shock waves propagating outward and with material falling inward. Conditions such as these may explain why the optical absorption lines appear redshifted with respect to the central star as discussed above. Clearly, there is great potential for learning about stellar pulsation from more detailed observations of H_2O and SiO masers.

One very important astrophysical problem concerns the mechanism and rate by which mass is lost from red giant stars. These stars are responsible for the return of a significant fraction of all material to the interstellar medium. Observations of maser emission, in particular of OH, yield the expansion speed of the molecular material. Also, if progress is made in quantifying the pumping mechanisms, information about the molecular density in the outflow region, the depletion of silicon into silicate grains, and the radius of grain formation can be obtained. Hopefully, the study of masers will continue to provide unique and useful information in this area.

6.4.1 Supergiants

One of the best-studied classes of stellar masers are the supergiant sources. The central stars in these sources are much more luminous that typical red giants; the luminosities of the supergiants typically are $10^5 \, L_\odot$ and may in some cases even approach $10^6 \, L_\odot$. Although there are not a large number of identified supergiant sources, their maser emission can be very intense, and detailed observations of a handful of these sources have been carried out. Apart from their strength, the maser spectra of the supergiants are similar to those of the LPVs. However, it is dangerous to ascribe the properties deduced for the supergiants to other sources. In fact, it is probable that the supergiants do not form a homogeneous group and that some may not even be evolved objects.

The best-studied supergiant masers are VY CMa, NML Cyg, VX Sgr (Chapman and Cohen 1986), and IRC + 10420. In a very general way, the observations indicate that, in comparison to the LPV sources, the supergiants are larger, more luminous stars with more extended and thicker circumstellar envelopes and higher expansion velocities. The OH emission appears to be composed of dozens of components spread over 10^{16} to 10^{17} cm on the sky. As in the case of the LPV stars, the H_2O and SiO masers are more compact than the OH masers, but in these sources they extend from about 10^{15} to 10^{16} cm from the central star. These properties indicate that supergiant sources appear to be losing mass at a substantially greater rate than the LPV sources.

There are two very unusual supergiants which deserve special attention: VY CMa and IRC + 10420. VY CMa has a spectral type of M3-5 and appears buried in a small reflection nebula at the tip of an arrow-shaped HII region. Herbig (1969) suggested that it is a pre-main-sequence object, surrounded by an accretion disk from which the OH maser emission emanates. There is a molecular cloud associated with VY CMa, supporting the interpretation of this object as a very young, although not necessarily a pre-main-sequence, object. Also, the "triple-peaked" spectrum of the SiO maser can be very well explained by a disk geometry for the circumstellar envelope. The OH and H_2O maser emission from VY CMa has been mapped with VLBI techniques, but the data do not clearly determine either the geometry of the circumstellar envelope or the evolutionary status of the central object.

IRC + 10420 is unique among stellar maser sources in that the central star is spectral type F, indicating that it is much hotter than all other (M-type) stellar masers. Only OH maser emission has been detected in this source. The lack of H_2O and SiO maser emission may be explained by the absence of these molecular species close to such a hot star. It is possible that IRC + 10420 may be a rapidly evolving star which has left behind a "fossil remnant" of a circumstellar shell from a recent M-type stellar phase.

6.5 Extragalactic Masers

Several nearby spiral galaxies have been surveyed for H_2O masers, and a small number have been found (Churchwell et al. 1977). They are located near HII regions and have characteristics similar to those of galactic H_2O masers. A much more powerful class of H_2O masers has been found in the nuclei of active galaxies (Claussen and Lo 1986). The H_2O masers in NGC 3079 and NGC 4258 have luminosities (assuming isotropic emission) of $500\,L_\odot$ and $120\,L_\odot$, respectively, more than two orders of magnitude more powerful than the strongest galactic maser, W49. In principle, the luminosity of a maser can be made arbitrarily large by having a suitably large volume. Hence, these nuclear masers could conceivably be pumped by collections of O-stars.

We can calculate the smallest volume required for a maser that radiates $100\,L_\odot$ by use of Equation (6.29) or (6.30). Assume that the maser emission arises in a spherical cloud of gas with density less than 10^{11} cm^{-3} (to avoid collisional quenching). If we assume a pump efficiency of 1%, then from Equation (6.19) we find $\Delta n_0/n_1 \approx 0.02$; if $n_{H_2O}/n_{H_2} \sim 10^{-4}$ and for $n_1/n_{H_2O} \approx 0.02$, then Equation (6.33) implies that $\Delta n_0 \sim 4000$. Using this value for Δn_0, which is probably too large for most H_2O masers, and $\Gamma \sim 1$ s^{-1}, we obtain from Equation (6.29) a maser volume of 6×10^{47} cm^3 and a maser diameter of $\sim 10^{16}$ cm. However, VLBI observations show that for NGC 3079 and NGC 4258, most of the emission originates within a region of diameter less than $\sim 10^{15}$ cm. Clearly, standard pumping schemes cannot account for these strong masers, provided that the masers are not highly beamed primarily in our direction.

Strelnitskii (1984) and Kylafis and Norman (1986) have described a pump mechanism involving hydrogen and electrons at different kinetic temperatures. The maximum pump rate occurs when the fractional ionization is $\sim 10^{-5}$. There is no collisional quenching of the pump action in this model. Thus, hydrogen densities might greatly exceed 10^{11} cm^{-3} and give rise to more luminous maser sources.

Very powerful OH masers have been found in other galaxies. In the case of the nucleus of the peculiar galaxy IC 4553, the maser luminosity is $\sim 10^8$ times greater than that of strong galactic OH masers like W3(OH) (Baan and Haschick 1984). In contrast to the strong extragalactic H_2O masers, this maser has an angular extent of several arcseconds and exactly coincides with the continuum radio source in the nucleus. This maser may be a low gain, unsaturated maser that is amplifying the background (high-brightness) radio source.

6.6 Interstellar Scattering

Astronomical masers may be useful probes of the interstellar medium. Masers are both intense and extremely small in angular extent, and it is likely that scattering effects due to irregularities in the electron density distribution of the interstellar medium may be observable. Also, maser emission from OH and H_2O molecules could be observed from many regions in the Galaxy, enabling one to study the frequency dependence of the scattering from 1.6 to 22 GHz. These measurements complement those of pulsars and extragalactic sources (e.g., Cordes et al. 1985).

At present, evidence suggests that interstellar scattering broadens many maser spots. There is a tendency for H_2O and OH sources that are at great distances from the Sun to have larger apparent sizes than those that are close to the Sun. For example, the masers in the source W3(OH), which is at a distance of about 2 kpc, have apparent spot sizes less than 0.005 arcsec for OH and about 0.08 mas for H_2O, while those in Sgr B2 and W49 that are at distances greater than about 8 kpc have apparent spot sizes greater than 0.010 arcsec for OH and about 0.3 mas for H_2O. Also, the ratio of the apparent sizes of the H_2O and OH masers from several sources is approximately that expected from a wavelength-squared scattering law. [The data are not accurate enough to distinguish between a scattering size that varies as λ^2, as predicted by a screen with a Gaussian correlation function, and one that varies as $\lambda^{2.2}$, as predicted for a scattering medium with a (Kolmogorov) power law distribution of irregularities.]

It seems unlikely that the more distant maser sources would be intrinsically larger in size than the closer ones. Although the more distant sources are probably more luminous than the nearer ones, maser theory predicts that (in the absence of scattering) apparent sizes should be a weak function of luminosity, and that apparent-to-real-size ratios should decrease with increasing source strength. Thus, the simplest interpretation of the observations is that some of the more distant OH masers are scattered by intervening diffuse HII regions in the Galaxy.

The low-frequency variability of some extragalactic sources is thought to be due to refractive effects in the interstellar medium caused by the large-scale ($> 10^{14}$ cm) electron density fluctuations. These fluctuations should also cause the image of a radio source to wander. This wander should be on the order of the scattered image size, and the time scale is on the order of the time required for the line of sight to the source to move through the scattering material by the scattered image size (Rickett 1986). At the present time, there is no evidence for refractive wandering in masers (Gwinn et al., 1988).

To date, very little effort has been expended on the use of masers as probes of interstellar scattering. With the application of VLBI aperture synthesis techniques, it should be possible to determine accurate spot sizes for many more sources than we have at present. Since masers can be seen throughout the galaxy and over a wide range of radio frequencies, they can be used to extend our knowledge of the interstellar medium.

6.7 Distance Measurements

Individual maser spots are extremely small (i.e., 0.001 arcsec), and many are clustered together in any one source. Thus, in some ways, maser sources resemble star clusters, suggesting that proper-motion measurements can be used to estimate distances to them. The proper motions of many H_2O spots can be measured with VLBI techniques, and these motions can be used to determine distances to H_2O sources at any distance in the Galaxy and possibly to other galaxies with accuracies approaching 10%.

Genzel et al. (1981a) measured proper motions of H_2O maser spots in the Orion-KL source and found that the spots are expanding about a central point (presumably a massive young star). Using the two dimensions of spatial information (i.e., relative positions) and the three dimensions of velocity information (i.e., radial and proper motions) available for the maser spots, they constructed a three-dimensional kinematic model of the source. The distance of the entire source which is needed to relate the angular (proper) motions to true velocities was then estimated by least-squares fitting techniques to be 480 ± 80 pc, in accord with optical luminosity distances from associated O-stars.

Genzel et al. (1981b) and Schneps et al. (1981) measured proper motions of H_2O masers in W51(M) and W51(N), respectively. The motions in these sources appeared largely random, and the technique of statistical parallax was employed to determine distances to the sources. The distance to these sources was estimated to be 7 ± 1.5 kpc. This value is consistent with the far kinematic distance to the W51 region.

Studies of the water masers in the galactic-center region (Sgr B2) indicate a phenomenon similar to the Orion case (Reid et al. 1987). The H_2O maser spectrum of Sgr B2(North) is composed of ~ 100 maser spots with an LSR velocity range from 15 to 115 km s^{-1}. Spectral-line VLBI maps made between 1980.9 and 1982.5 clearly identify up to 50 spots stronger than 1 Jy per epoch. With four telescopes spanning the United States, a synthesized beam of 2.4×0.4 mas was obtained. The maser spots in Sgr B2(North) were slightly resolved and the location of peak brightness was estimated by fitting an elliptical Gaussian model. Theoretically, the center of a Gaussian brightness distribution can be estimated to an accuracy given by

$$\sigma_\theta \approx \frac{0.5\theta_{app}}{\text{SNR}}$$
(6.51)

where θ_{app} is the apparent size (FWHM) of the maser spot (dominated by the synthesized beam) and SNR is the peak signal-to-noise ratio for the spot in the map. Plots of position versus time for features at the same LSR velocity were made. These plots indicated that the maser spots typically appeared to move in straight lines on the sky to within a factor of 2 of the expected (theoretical) position uncertainties.

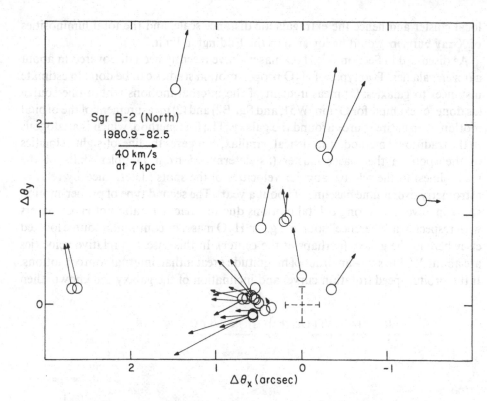

Fig. 6.9. Proper motions of the H_2O masers in the star-forming region Sgr B2(North). Circles indicate the positions of H_2O maser spots. The arrows indicate the directions and speeds of maser features. The masers are expanding from a newly formed star whose position, indicated by the error bar, can be inferred by tracing back the arrows. Modeling these data yields an estimate of the distance to the center of the Milky Way of about 7 kpc.

A map displaying the proper motions of the H_2O maser spots in Sgr B2(North) is shown in Figure 6.9. The maser spots appear to be expanding outward from a position indicated by the origin of the map. A least-squares fit of a uniformly expanding spherical source resulted in an estimate of the distance to the source of 7.1 ± 1.5 kpc. The estimated expansion velocity is 45 ± 7 km s^{-1}, and the velocity residuals were about 15 km s^{-1}. These residuals are much greater than the measurement errors, suggesting that the source contains "turbulent" motions of about 15 km s^{-1}.

The star-forming region Sgr B2 is almost certainly within 0.3 kpc of the galactic center. Thus, the Sgr B2 distance can be used directly as an estimate of R_0, the distance between the Sun and the galactic center. The H_2O maser distance estimate of 7.1 kpc to the center of the Galaxy is significantly lower than the previous standard value of 10 kpc and in better agreement with the revised (1985) IAU value of 8.5 kpc. A decrease in the value of R_0 has widespread impact on astrophysics. For example, all kinematic distances would be decreased; the mass of the Galaxy and the galactic center would also be decreased; recent revisions of the absolute magnitudes of RR Lyrae variables would be supported, affecting distances to the local cluster and hence the extragalactic distance scale; and the total luminosities of X-ray bursters would be lowered to the Eddington limit.

As discussed in Section 6.5, H_2O masers have recently been discovered in about a dozen galaxies. Two types of H_2O proper-motion studies can be done to estimate distances to galaxies: (1) measurement of the internal motions within one source (as done for example for Orion, W51, and Sgr B2) and (2) measurement of the orbital motion of an entire source around the galaxy. The first method, which is analogous to the traditional method of statistical parallax, compares the line-of-sight velocities of the spots in the maser cluster (as determined from Doppler shifts of the maser lines) to the relative angular velocities of the spots (determined by relative astrometry over a time baseline of about a year). The second type of proper-motion study involves measuring orbital motions due to galactic rotation of H_2O masers with respect to a "reference" source, e.g., an H_2O maser or continuum source located elsewhere in the galaxy (perhaps at the center). In this case, the relative velocities are about 250 km s^{-1}—an order of magnitude greater than internal source motions. If the orbital speed (rotation curve) and inclination of the galaxy are known, then

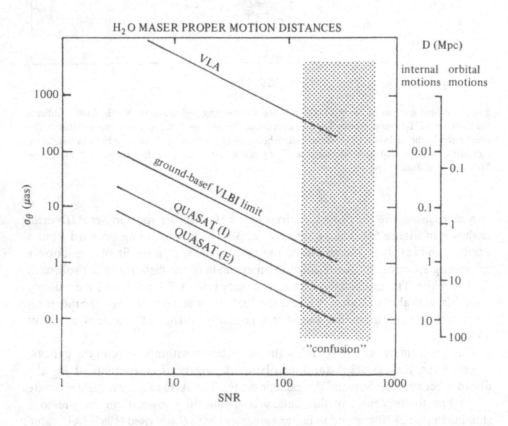

Fig. 6.10. Positional measurement accuracy (σ_θ) as a function of interferometer signal-to-noise ratio (SNR) for various interferometers. The approximate distance (D) to which a source can be measured is indicated, as well as the SNR at which "confusion" is likely to limit measurements. Confusion, in the sense used here, occurs when two maser features with overlapping spectral lines appear within the synthesized beam of the interferometer. The lines labeled QUASAT are for two sizes of orbits being considered for a satellite-to-ground VLBI system by NASA and the European Space Agency.

measurement of the proper motion directly yields the galaxy's distance. A maser located away from the center of a face-on galaxy 1 Mpc from the Sun moves about 50 μas per year, a motion comparable to the synthesized beam of proposed space-based VLBI observations.

Distance estimates similar to those described above are of great value to astronomy and cosmology since they are potentially as accurate as those provided by most other methods and, most importantly, they are independent of the complex hierarchy of distance indicators required for all but the nearest objects to the Sun. The measurement accuracy needed to extend the present studies throughout the Galaxy and to other galaxies is indicated in Figure 6.10. For example, in order to apply the statistical parallax technique to M33, relative positional accuracies of about one *micro*arcsecond are needed. A one-mircoarcsecond displacement corresponds to one degree of interferometer fringe phase on intercontinental base-lines. A measurement of this accuracy may be best achieved with space-to-ground VLBI aperture synthesis observations.

Recommended Reading

Downes, D. 1985. *In* R. Lucas, A. Omant, and R. Stora (eds.), Birth and Infancy of Stars, Les Houches Sess. XLI. Amsterdam: North-Holland, p. 560.

Elitzur, M. 1982. Rev. Mod. Phys. **54**:1225.

Goldreich, P. 1974. *In* R. Balian, P. Encrenaz, and J. Lequeux (eds.), Atomic and Molecular Physics and the Interstellar Matter, Les Houches Sess. XXVI. Amsterdam: North-Holland, p. 413.

Litvak, M.M. 1974. Annu. Rev. Astron. Astrophys. **12**:97.

Moran, J.M. 1976. *In* E.H. Avrett (ed.) Frontiers of Astrophysics. Cambridge, Mass.: Harvard U. Press, p. 385.

Moran, J.M. 1982. *In* M. Weber (ed.), Handbook of Laser Science and Technology, Vol. 1. Boca Raton, Florida: CRC Press, p. 483.

Reid, M.J., and J.M. Moran. 1981. Annu. Rev. Astron. Astrophys. **19**:231.

References

Alcock, C. and R.R. Ross 1985. Astrophys. J. **290**:433.

Baan, W.A., and A.D. Haschick. 1984. Astrophys. J. **279**:541.

Barrett, A.H., P.R. Schwartz, and J.W. Waters. 1971. Astrophys. J. (Lett.) **168**:L101.

Barvainis, R. and C.R. Predmore 1985. Astrophys. J. **288**:694.

Booth, R.S., A.J. Kus, R.P. Norris, and N.D. Porter. 1981. Nature **290**:382.

Chapman, J.M., and R.J. Cohen 1986. Mon. Not. R. Astron. Soc. **220**:513.

Cheung, A.C., D.M. Rank, C.H. Townes, D.D. Thornton, and W.J. Welch. 1969. Nature **221**:626.

Churchwell, E., A. Witzel, W. Huchtmeier, I. Pauliny-Toth, J. Roland, and W. Suben. 1977. Astron. Astrophys. **54**:969.

Claussen, M.J., and K.Y. Lo. 1986. Astrophys. J. **308**:592.

Cook, A.H. 1968. Mon. Not. R. Astron. Soc. **140**:299.

Cordes, J.M., J.M. Weisberg, and V. Boriakoff. 1985. Astrophys. J. **288**:221.

Davies, R.D. 1974. *In* Galactic Radio Astronomy, Int. Astron. Union Symp. 60. Dordrecht: Reidel, p. 275.

Diamond, P.J., R.P. Norris, P.R. Rowland, R.S. Booth, and L-A. Nyman. 1985. Mon. Not. R. Astron. Soc. **212**:1.

de Jong, T. 1973. Astron. Astrophys. **26**:297.

Elitzur, M. 1980. Astrophys. J. **240**:553.

Elitzur, M., P. Goldreich, and N. Scoville. 1976. Astrophys. J. **205**:384.

Engels, D. 1979. Astron. Astrophys. Suppl. Ser. **36**:337.

Fouquet, J.E., and M.J. Reid. 1982. Astron. J. **87**:691.

Genzel, R., D. Downes, J.M. Moran, K.J. Johnston, J.H. Spencer, L.I. Matveyenko, L.R. Kogan, V.I. Kostenko, B. Rönnäng, A.D. Haschick, M.J. Reid, R.C. Walker, T.S. Giuffrida, B.F. Burke, and I.G. Moiseev. 1979. Astron. Astrophys. **78**:239.

Genzel, R., M.J. Reid, J.M. Moran, and D. Downes. 1981a. Astrophys. J. **244**:884.

Genzel, R., D. Downes, M.H. Schneps, M.J. Reid, J.M. Moran, L.R. Kogan, V.I. Kostenko, L.I. Matveyenko, and B. Ronnang. 1981b. Astrophys. J. **247**:1039.

Goldreich, P., and D.A. Keeley, 1972. Astrophys. J. **174**:517.

Goldreich, P., D.A. Keeley, and J. Kwan. 1973. Astrophys. J. **182**:55.

Goldreich, P., and J. Kwan. 1974a. Astrophys. J. **190**:27.

Goldreich, P., and J. Kwan. 1974b. Astrophys. J. **191**:93.

Guilloteau, S., A. Omont, and R. Lucas. 1987. Astron. Astrophys. (submitted).

Gundermann, E. 1965. Ph.D. thesis, Harvard University, Cambridge, Massachusetts.

Gwinn, C.R., J.M. Moran, M.J. Reid, and M.H. Schneps. 1988. Astrophys. J. *in press.*

Haschick, A.D., B.F. Burke, and J.H. Spencer. 1977. Science **198**:1153.

Haschick, A.D., M.J. Reid, B.F. Burke, J.M. Moran, and G. Miller. 1981. Astrophys. J. **244**:76.

Herbig, G.H. 1969. Mem. Soc. R. Sci. Liège **8**:13.

Hollenbach, D., and C.F. McKee. 1979, Astrophys. J. Suppl. Ser. **41**:555.

Jewell, P.R., J.C. Webber, and L.E. Snyder. 1980. Astrophys. J. (Lett.) **242**:L29.

Kegel, W.H. 1975. *In* B. Bascheck (ed.), Problems in Stellar Atmospheres and Envelopes. Berlin: Springer, p. 257.

Kukolich, S.G. 1969. J. Chem. Phys. **50**:3751.

Kwan, J., and N. Scoville. 1974. Astrophys. J. (Lett.) **194**:L97.

Kyafis, N.D. and C. Norman 1986. Astrophys. J. (Lett.) **300**:L73.

Lane, A.P. 1984. *In* R. Fanti, K. Kellermann, and G. Setti (eds.), VLBI and Compact Radio Sources, Int. Astron. Union Symp. 110. Dordrecht: Reidel, p. 329.

Litvak, M.M. 1973. Astrophys. J. **182**:711.

Litvak, M.M., A.L. McWhorter, M.L. Meeks, and H.J. Zeiger. 1966. Phys. Rev. Lett. **17**:821.

Mcgee, R.X., B.J. Robinson, F.F. Gardner, and J.G. Bolton 1965. Nature **208**:1193.

Moran, J.M., B.F. Burke, A.H. Barrett, A.E.E. Rogers, J.C. Carter, J.A. Ball, and D.D. Cudaback. 1968. Astrophys. J. (Lett.) **152**:L97.

Moran, J.M., M.J. Reid, C.J. Lada, J.L. Yen, K.J. Johnston, and J.H. Spencer. 1978. Astrophys. J. (Lett.) **224**:L67.

Norman, C., and J. Silk. 1979. Astrophys. J. **228**:197.

Norris, R.P., P.J. Diamond, and R.S. Booth. 1982. Nature **299**:133.

Norris, R.P., R.S. Booth, P.J. Diamond, L-A. Nyman, D.A. Graham, and L.I. Matveyenko, 1984. Mon. Not. R. Astron. Soc. **208**:435.

Nyman, L-A., and H. Olofsson. 1986. Astron. Astrophys. **158**:67.

Perkins, F., T. Gold, and E.E. Salpeter. 1966. Astrophys. J. **145**:361.

Reid, M.J., and D.F. Dickinson. 1976. Astrophys. J. (Lett.) **209**:505.

Reid, M.J., D.O. Muhleman, J.M. Moran, K.J. Johnston, and P.R. Schwartz. 1977. Astrophys. J. **214**:60.

Reid, M.J., A.D. Haschick, B.F. Burke, J.M. Moran, K.J. Johnston, and G.W. Swenson, 1980. Astrophys. J. **239**:89.

Reid, M.J., M.H. Schneps, J.M. Moran, C.R. Gwinn, R. Genzel, D. Downes, and B. Ronnang. 1987. *In* M. Peimbert and J. Jugaku (eds.) Star Formation, Int. Astron. Union Symp. 115. Dordrecht: Reidel.

Rickett, B.J. 1986. Astrophys. J. **307**:564.

Schneps, M., A.P. Lane, D. Downes, J.M. Moran, R. Genzel, and M.J. Reid. 1981. Astrophys. J. **249**:124.

Schultz, G.V., W.A. Sherwood, and A. Winnberg. 1978. Astron. Astrophys. (Lett.) **63**:L5.

Shklovskii, I.S. 1969. Sov. Astron. **13**:1.

Snyder, L.E., and D. Buhl. 1974. Astrophys. J. (Lett.) **189**:L31.

Strelnitskii, V.S. 1984. Mon. Not. R. Astron. Soc. **207**:339.

Strelnitskii, V.S., and R.A. Syunyaev. 1973. Sov. Astron. **16**:579.

Tarter, J., and W.J. Welch. 1986. Astrophys. J. **305**:467.

Troland, T.H., C. Heiles, D.R. Johnston, and F.O. Clark 1979. Astrophys. J. **232**:143.

Walker, R.C. 1984. Astrophys. J. **280**:618.

Weaver, H., D.R.W. Williams, N.H. Dieter, and W.T. Lum. 1965. Nature **208**:29.

Weinreb, S., M.L. Meeks, J. Carter, A.H. Barrett adn A.E.E. Rogers 1965. Nature **208**:440.

Western, L.R. and W.D. Watson 1984. Astrophys. J. **285**:158.

Wilson, W.J., and A.H. Barrett. 1972. Astron. Astrophys. **17**:385.

7. The Structure of Our Galaxy Derived from Observations of Neutral Hydrogen

W. Butler Burton

7.1 Observations of Galactic Neutral Hydrogen

Although insight into the structure of our Galaxy can be sought in various ways, study of the radio emission from the neutral hydrogen component is particularly suitable. Neutral atomic hydrogen, HI, is the main observed constituent of the interstellar medium; its physical properties are closely related to the properties of other Galactic constituents, both stellar and interstellar. The interstellar medium is transparent enough to hydrogen emission at the 21-cm radio wavelength that investigation of the entire Galaxy is possible, with the exception of a few directions along the Galactic equator. This transparency allows investigation of regions of the Galaxy which are too distant to be studied optically. HI is particularly important for the information it provides on the form and overall mass in the outer Galaxy, where it is the only directly observable constituent. Interstellar neutral hydrogen is so abundant and is distributed in such a general fashion throughout the Galaxy that the 21-cm hyperfine transition line has been detected in emission in every direction in the sky at which a suitably equipped radio telescope has been pointed. No time variation of a neutral hydrogen line has been found.

The 21-cm line of atomic hydrogen results from a hyperfine transition in the ground state of the atom. This state is characterized by two possible relative orientations of the spins of the electron and proton; when the spins are parallel the energy of the system is slightly greater than when they are antiparallel. The energy separation between the two hyperfine sublevels corresponds to a quantum of radiation with a natural frequency of 1420.406 MHz. The probability that this spin-flip transition from the $F = 1$ to the $F = 0$ state will occur spontaneously is given by the Einstein emission coefficient, $A_{10} = 2.85 \times 10^{-15}$ s^{-1}. Thus a spontaneous electron spin flip would occur after some eleven million years. In fact, atomic encounter collisions largely determine the relative populations of the two energy levels. At the typical HI densities of about 0.4 cm^{-3} which characterize much of interstellar space, encounters result in reorientation of the spins of a hydrogen atom about once every three or four hundred years. Effects of ambient interstellar radiation field on the spin states are less important than those of encounters, except in certain unusual interstellar environments such as found near recent supernovae or near regions of intense star formation.

The 21-cm line was first observed in 1951. The detection at Harvard by H.I. Ewen and E.M. Purcell was confirmed within a few weeks in the Netherlands by C.A. Muller and J.H. Oort and in Sydney by J.L. Pawsey; all three of these initial sets of observations were announced in the September 1, 1951, issue of *Nature*.

H.C. van de Hulst (1945) had predicted that the line might well be observable despite the long intervals between spontaneous transitions because of the large number of hydrogen atoms expected along a line of sight through Galactic interstellar space. van de Hulst had been encouraged in the study which led to his prediction by Oort, who realized the potential of a narrow spectral line from the most abundant component of the interstellar medium and at a wavelength to which the entire galactic interstellar medium is quite transparent.

Since the detection, the emphasis in 21-cm-line work on the global aspects of Galactic hydrogen has involved three stages. Throughout the 1950s and until the late 1960s many efforts were directed toward mapping the large-scale structure of the Galaxy. During that period, HI was considered largely representative of the interstellar medium in general, with physical properties more or less constant over the entire Galactic disk. The expectation that these efforts would reveal the distribution of gas in spiral arms was largely based on the optical appearance of external galaxies. The Dutch and Australian work culminating in the Leiden-Sydney map (see Oort et al. 1958) did fulfill these expectations, but many details of the HI Galactic morphology have proven difficult to unravel. Many of the questions addressed in our own Galaxy twenty or thirty years ago are now being addressed in nearby external galaxies; the angular resolution offered by radio interferometers is sufficient for these studies, and the external perspective is in many regards favorable.

The embedded, close-up perspective which we have of our own Galaxy is, however, advantageous for studies of details of the gas distribution. In a second stage beginning in the mid- or late 1960s, increased emphasis was placed on the smaller-scale aspects and on the physical properties of the interstellar hydrogen, motivated by the realization that the interstellar environment is so complicated that a wide range of temperatures and densities will naturally occur. This stage is being continued especially in the attention being given to the relationship of HI structures to those seen in emission from molecules and from dust. A third stage in galactic-structure 21-cm investigations began in the late 1970s, when it became clear that among the constituents of the interstellar medium, atomic hydrogen is unique in its large-scale distribution. Especially important in this regard is the study of the HI distribution in the outer parts of the Milky Way; the Galactic disk as defined by atomic hydrogen has a diameter fully twice as large as that defined by the ionized and molecular states of hydrogen. The rotation velocity measured in the outer Galaxy does not decrease with increasing distance as would be expected from Kepler's laws. Investigation of the outer-Galaxy HI distribution is important in the context of understanding the nature of the dark matter whose existence is implied by the flat rotation curve at large radii and which evidently is the dominant Galactic mass component at these large distances. HI is the only Galactic constituent widely observable in the far outer Galaxy, where its motions and distribution are largely governed by the dark matter.

Table 7.1 lists the comprehensive, current surveys of the HI 21-cm line available for Galactic structure studies. Earlier and more specialized surveys are summarized by Kerr (1968), Burton (1974), Burton and Liszt (1983), and Bajaja (1983). The table shows that most of the sky has been surveyed at 21 cm, with single-dish telescopes.

Table 7.1. Surveys of the galactic distribution of HI and of CO, a surrogate for H_2.

Author	Date	Beam (arcmin)	l-Coverage (degrees)	b-Coverage (degrees)	v-Coverage* (km s^{-1})	Sensitivity* (K)
Surveys of HI emission:						
Weaver and Williams	1973, 1974a	36	10–250; $dl = 0.5$	−10–10; $db = 0.25$	$v_0 \pm 100$; $dv = 2.1$ (6.3)	1.7 (1.0)
Weaver and Williams	1974b	36	12–249.5; $dl = 2.5$	$10 < [b] < 30$; $db = 0.5$	$v_0 \pm 100$; $dv = 2.1$ (6.3)	1.7 (1.0)
Heiles and Habing	1974	36	All l at $[b] > 10$; $d > -30$; $dl = 0.3/\cos b$	All $[b] > 10$ at $d > -30$; $db = 0.6$	± 50 at $dv = 2.1$; to $-92, +75$ at $dv = 6.3$	1.2
Colomb et al.	1977, 1980	30	All l at $[b] > 10$; $d < -25$; $dd = 1$	All $[b] > 10$ at $d < -25$; $dd = 1$	$-38 < v < 38$; $dv = 2$	2
Cleary et al.	1979	48	All l at $[b] > 10$, $b > -25$, $d < -30$; $dd = 1$	$[b] > 10$, $b > -25$, $d < -30$; $dd = 1$	$-148 < v < 300$; $dv = 7$	0.3
Kerr et al.	1981	16	230–350; $dl = 0.1$	$-2 < b < 2$; $db = 0.1$	$v_0 \pm 160$; $dv = 2$	2
Westerhout and Wendlandt	1982	13	$11 < l < 235$; $dl = 0.2$	$-2 < b < 2$; $db = 0.1$	$v_0 \pm 125$; $dv = 2$	3
Stark et al.	1985	150	All l at $d > -40$; $dl = 2$	All b at $d > -40$; $db = 2$	$v_0 \pm 300$; $dv = 5.3$	<0.1
Burton	1985	21	All l at $d > -46$, $-20 < b < 20$ (33) at $d > -46$; $dl = 1$	$-20 < b < 20$ (33) at $d > -46$; $db = 1$	$v_0 \pm 250$; $dv = 1$ (2.1)	0.2 (0.1)
Kerr et al.	1986	48	$240 < l < 350$; $dl = 0.5$	$[b] < 10$; $db = 0.25$	$v_0 \pm 150$; $dv = 2$	0.8
Surveys of ^{12}CO emission:						
Cohen et al.	1986	8	$12 < l < 60$; $dl = 0.125$ (0.25)	$-1.0 < b < 1.0$; $db = {>}125$ (0.25)	$-13 < v < 153$; $dv = 1.3$	1.4
Sanders et al.	1984	0.8	$8 < l < 90$; $dl = 0.05$ (0.1)	$-1.05 < b < 1.05$; $db = 0.05$ (0.1)	$-50 < v < 150$; $dv = 1$	1.2
Bronfman et al.	1986	8	$300 < l < 348$; $dl = 0.125$ (0.25)	$-2 < b < 2$; $db = 0.125$ (0.25)	$v_0 \pm 166$; $dv = 1.3$	0.3

* The velocity-coverage and sensitivity values in parentheses refer to low intensity wings of the spectra, which have been smoothed.

The principal limitation of the data, especially important for investigations which involve comparisons of the HI data with molecular or infrared data, lies in the relatively coarse angular resolution. One may expect improvements in the velocity coverage and resolution, and in the sensitivity. Such improvements will require a substantial investment of telescope time because of the large number of beam areas in the sky. Improvement in sensitivity will have to be made with the realization that suitable single-dish telescopes [excepting the small Bell Labs horn reflector used in the Stark et al. (1985) survey] are plagued by stray radiation entering the antenna-pattern sidelobes which cover the entire sky. Lockman et al. (1986) have emphasized that the stray radiation can contribute spurious signals of order 0.5 K, which is a significant level compared to the currently attainable survey sensitivity of order 0.1 K, and describe a procedure for correcting for this spurious emission. Improvement in angular resolution will be even more difficult; high-resolution radio interferometers are not suitable for large-scale surveying because of their limited response to emission distributed as ubiquitously at that from galactic HI.

In the temperature range from a few tens of degrees Kelvin to a few thousands, the interstellar gas is almost entirely in the form of atomic hydrogen. Hydrogen in the molecular form predominates over all gaseous material in regions colder than a few tens of degrees Kelvin. The CO molecule serves as a surrogate for the much more abundant, but generally unobservable, hydrogen molecule. Table 7.1 also summarizes survey observations of the CO tracer. Having no permanent electronic dipole moment, H_2 molecules at the characteristic interstellar temperatures do not emit in the radio or optical windows. The H_2 Lyman absorption bands in the ultraviolet suffer from extinction due to interstellar dust, and so do not carry information from transgalactic paths. Transgalactic paths are largely transparent to the millimeter-wavelength emission from rotational transitions of CO. The CO molecule traces molecular hydrogen because the most important source of excitation of the CO rotational transitions involves collisions with H_2; furthermore, CO is stable at low temperatures and is some 100 times more abundant than any molecule other than H_2.

A number of other widely distributed components of the interstellar medium with generally observable spectral lines reveal different aspects of Galactic structure. Hydrogen recombination lines trace the hot, ionized gas. Lines from the OH radical are useful for a number of purposes, but the usefulness for morphological purposes of lines usually measured in absorption is limited by the distribution of the sources of absorbed background radiation.

The deuterium isotope of hydrogen produces a spectral line from a hyperfine spin-flip transition in its ground state which is analogous to the transition producing the 21-cm line, but with a wavelength of 92 cm. In the standard cosmological picture, most of the deuterium formed during the Big Bang reacted to form helium, but some remained after the first few minutes. The amount left is thought to depend sensitively on the baryon density during helium production (e.g., Wagoner 1973). Not more than a quarter of the matter in the Galaxy is accounted for by direct observation. Because detection of the 92-cm DI emission line could constrain inferences on the amount of matter in familiar, baryonic form in the early Universe, substantial efforts to detect the very weak line have been made for the past several decades. Results so far are rather ambiguous.

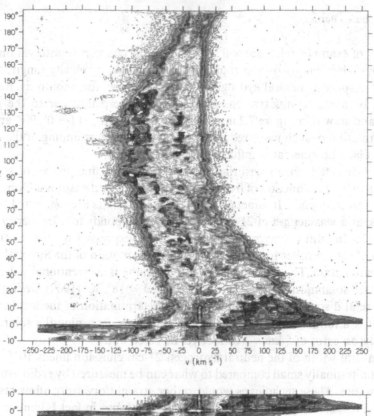

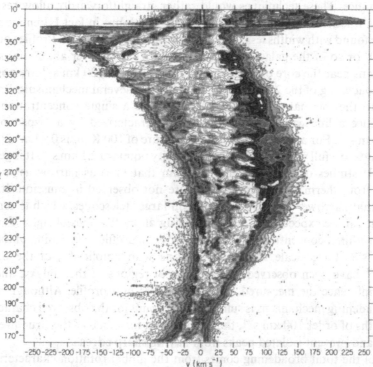

Fig. 7.1. Neutral hydrogen emission intensities displayed in longitude-velocity coordinates along the Galactic equator, $b = 0°$. The contours represent brightness temperatures at levels of $T_b = 0.4, 0.8, 1.5, 2.5, 4, 7, 10, 15, 20, 25, 35, 45, \ldots$ K. The grey-scale levels follow the same divisions, with shading darkening in steps until the $T_b = 20$ K level and resuming at the 60 K contour. Velocities are expressed with respect to the local standard of rest. The observations in the range $-10° \leqslant l \leqslant 240°$ are from Burton (1985); those in the range $240° < l < 350°$, from Kerr et al. (1976).

By way of example, reference will be made in this chapter to observations made along the Galactic equator and shown in Figure 7.1 in a velocity-longitude map, at $b = 0°$, displaying neutral hydrogen brightness temperature contours. Reference will also be made to observations made in strips perpendicular to the galactic equator and shown in Figure 7.2 in velocity-latitude maps, at $l = 0°, 90°$, and $180°$.

Interpretation of such maps requires first of all an understanding of the manner in which Galactic kinematics influence the observations.

What is detected when measuring the 21-cm line is a line profile, or spectrum, giving intensity as a function of frequency. The frequency measures a Doppler shift from the rest frequency. In practice, these frequency shifts are converted to radial velocities; at a wavelength of 21 cm, 1 km s^{-1} corresponds to a frequency shift of -4.74 kHz. In Milky Way studies, velocities are expressed relative to the local standard of rest, in order to correct for the peculiar motion of the Sun with respect to the nearby stars. The local standard of rest frame is conventionally defined by the standard solar motion of 20 km s^{-1} toward $\alpha, \delta = 18^h, 30°$ (1900). Delhaye (1965) gives a useful discussion of the observational determination of the local standard of rest. It is a fundamental but not simple matter to determine the solar motion accurately; an error of 5 km s^{-1} cannot yet be ruled out.

The natural width of the neutral hydrogen 21-cm emission line is 10^{-16} km s^{-1} and is infinitesimally small compared to what can be measured by radio astronomical techniques. These techniques would rather straightforwardly allow resolution of a line a fraction of a kilometer per second broad, but in fact HI emission lines are rarely found with widths less than a few kilometers per second. Profiles observed within 10° or so of the galactic equator typically extend over about 100 km s^{-1}; in directions near the core of the Galaxy, total widths of 500 km s^{-1} are observed.

The broadening of the 21-cm line occurs through several mechanisms. Broadening due to the thermal velocities of atoms within a single concentration of gas will produce a line with a Gaussian shape characterized by a dispersion $\sigma_v = 0.09\sqrt{T}$ km s^{-1}. For a realistic kinetic temperature of 100 K, σ_v is 0.9 km s^{-1}, which corresponds to a full width between half-intensity points of 2.1 km s^{-1}. It is generally true in HI studies of the interstellar medium that lines as narrow as would be expected from thermal broadening alone are not observed in emission. It is also the case that narrower lines are measured on larger telescopes, which is consistent with what can be expected for severe blending along the lines of sight. Turbulent motions within a concentration of HI gas produce profile broadening of the order of 5 km s^{-1}. Large-scale streaming motions with amplitudes of the order of 10 km s^{-1} have been observed in a number of regions of the Galaxy, and these motions influence the measured width of features in a profile. Although none of these broadening mechanisms is sufficient to account for the observed characteristic total widths of order 100 km s^{-1}, they do account for much of the structure within a profile and prevent isolated peaks from having sharp edges.

Most of the total broadening comes from the global rotation characteristics of the Galaxy. This is of particular importance because it means that 21-cm profiles can give information about galactic kinematics and, in particular, about differential Galactic rotation.

The observed intensity of an HI line profile is usually expressed as a brightness

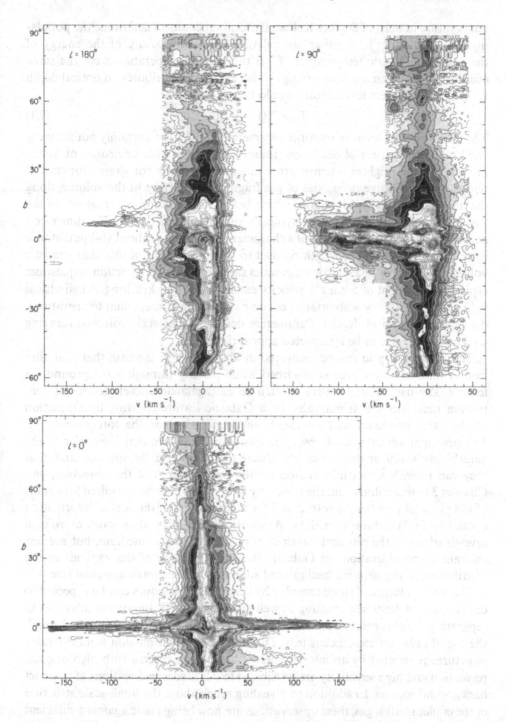

Fig. 7.2. Neutral hydrogen emission intensities displayed along three cuts perpendicular to the Galactic equator, in the cardinal directions $l = 0°$, $90°$, and $180°$. The intensities are indicated in the same ways as in Figure 7.1. The observations are from Heiles and Habing (1974) and from Burton (1985). The cuts show the signatures of several large-scale Galactic phenomena.

temperature in units of degrees Kelvin. If atomic encounters determine the populations in the energy levels, then the relevant physical measure of the energy of the gas is the kinetic temperature, T_k. If the kinetic temperature were the same everywhere along an entire line of sight where the gas contributes an optical depth τ, then the brightness temperature would be

$$T_b = T_k(1 - e^{-\tau}) . \tag{7.1}$$

The assumption of a gas of uniform parameters is, however, certainly not formally valid under the variety of conditions found in the interstellar environment. Interpretation of the brightness temperature measure is usually not straightforward; it refers to the collective properties of gas fragments occurring in the volume along the line of sight sampled by the telescope beam. For emission observations of the sort listed in Table 7.1, this beam typically subtends a half degree. The dimension perpendicular to the line of sight of a characteristic volume at about 10-kpc distance is thus about 100 pc. The length parallel to the line of sight of this characteristic volume depends on the Galactic kinematics sampled along the direction in question; typically, a segment of 5-km s^{-1} velocity extent might be 1 kpc long. A cylindrical volume 100 pc $\times$ 1 kpc will certainly encompass a range of interstellar temperatures, densities, and optical depths. Parameters derived under such volume-averaging circumstances have to be interpreted appropriately.

The 21-cm line also can be measured in absorption, in the case that radiation from a background source is absorbed while passing through a foreground HI feature. The background source is often an extragalactic discrete source of continuum radiation, but it may also be a Galactic feature emitting line radiation at the same frequency and on the same line of sight as the foreground one. The principal advantages of absorption measurements are that they refer to the vanishingly small angles which the discrete sources usually subtend and that they can provide direct information on the optical depth of the absorbing gas. Chapter 3 in this volume and the review by Kerr (1968) may be consulted for details of the physical processes involved and for a description of the sort of information contained in HI absorption data. Absorption data play an important role in investigations of the physical parameters of the interstellar medium, but are less relevant to considerations of Galactic structure, because of the vagaries of the distribution of the required background sources and their small apparent size.

The large volumes of space sampled by single-dish telescopes can be expected to contain, at low latitudes, emitting as well as absorbing HI. Important advances in separating the absorption from the emission contributions are being made through the use of radio interferometers (e.g., Dickey et al. 1983). Emission from extended structures is rejected by an interferometer. Absorption spectra with high angular resolution and high sensitivity are possible at low latitudes in directions of compact background sources. In addition to revealing much about the small-scale structure of the cooler neutral gas, these observations are now being made against a sufficient number (several hundred) of low-latitude background sources to begin to provide useful data on the Galactic distribution of HI temperatures and optical depths. Background emission is widely enough distributed that an interferometer could, in principle, be used to measure the optical-depth properties in most regions of the Galactic gas layer.

The total number of hydrogen atoms in the line of sight at all velocities, N_{HI}, is given in terms of the integral $\int T_k \tau(v)\, dv$, which is, in general, not a measured quantity. Only if the gas distribution is optically thin at all velocities does N_{HI} become measurable through the profile integral $\int T_b(v)\, dv$, because in this case the observed brightness temperature profile given by Equation (7.1) becomes $T_b(v) \approx T_k \tau(v)$ and

$$N_{HI} = 1.823 \times 10^{18} \int T_b\, dv \; \text{cm}^{-2} \; . \tag{7.2}$$

The distribution on the plane of the sky of HI column densities derived in this way is shown in Figure 7.3. The velocity range of the integrals $\int T_b\, dv$ entering Figure 7.3 are sufficient to include most of the HI associated with the Galaxy. Burton and te Lintel Hekkert (1985) have published a series of maps showing the HI column densities in 2.5-km s^{-1} bins, in which individual features can be more easily identified.

The validity of the assumption of optical thinness is, of course, important for the validity of densities derived from the profile integrals. At Galactic latitudes more than $5°$, say, from the Galactic equator, the volumes sampled by HI emission spectra are typically transparent, although evidence is commonly found for isolated, partly saturated features (see Dickey et al. 1983). Along lines of sight lying closer than a few degrees to the Galactic equator, optical thinness may pertain over only portions of the profile; optical depths are substantial in the general directions of the Galactic center and anticenter, where rotational motions are largely perpendicular to the lines of sight and thus where the motions are crowded together near $v = 0$ km s^{-1}. Densities derived under an assumption of optical thinness will generally underestimate the true amount of HI.

7.2 Kinematics of Galactic Neutral Hydrogen

7.2.1 Velocities Due to Differential Galactic Rotation

In order to interpret HI observations in terms of large-scale structure, it has been common to assume that the motions of the gas about the center of the Galaxy are everywhere circular and that the angular velocity, $\Omega(R)$, is a decreasing function only of distance, R, from the center of the system. These assumptions are not, strictly speaking, true, but they are valid to first order and especially useful when dealing with global characerics of Galactic structure. Under these assumptions, radial velocities in the Galactic-equator plane $b = 0°$ which are due to differential Galactic rotation may be expressed simply, making use of the construction shown in Figure 7.4. We write l for Galactic longitude and θ for Galactocentric azimuth, both angles being measured as in the figure in the sense that they become equal for a point at infinity. The distance from the observer to the emitting concentration of hydrogen is r. The observer is, of course, embedded within the system, at a distance R_o from the center, and is rotating about this center with angular velocity Ω_o. The radial velocity, v, which is measured (by convention positive in sign if the emitting gas is moving away from the observer) is the Doppler-shifted velocity of the gas along the line of sight:

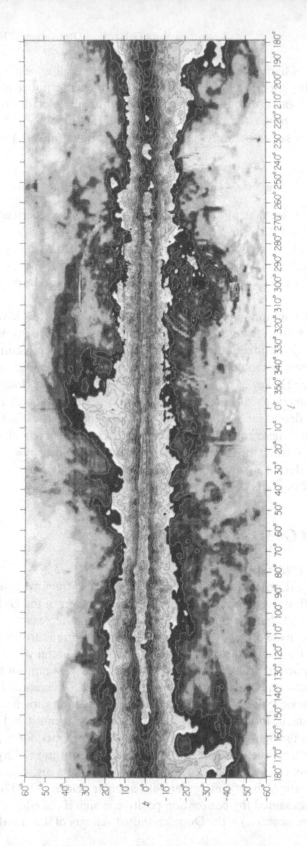

Fig. 7.3. Arrangement in Galactic coordinates of the integrals $\int T_b \, dv$ which correspond to HI column densities if the gas is optically thin. The composite data set was generated by E. R. Deul from the surveys of Weaver and Williams (1973), Heiles and Habing (1974), Cleary et al. (1979), Westerhout and Wendlandt (1982), Burton (1985), and Strong et al. (1982). The range of integration is the total velocity coverage listed for each survey in Table 7.1.

Fig. 7.4. Diagram illustrating the construction used in the derivation of Equation (7.4)

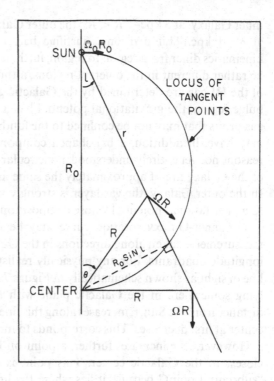

v = component of ΩR along line of sight minus component of $\Omega_0 R_0$ along line of sight

$$= \Omega R \cos(90° - l - \theta) - \Omega_0 R_0 \cos(90° - l)$$

$$= \Omega R(\sin\theta\cos l + \cos\theta\sin l) - \Omega_0 R_0 \sin l \ . \tag{7.3}$$

Because $r\sin l = R\sin\theta$ and $R\cos\theta = R_0 - r\cos l$, this becomes

$$v = R_0[\Omega(R) - \Omega_0]\sin l \ . \tag{7.4}$$

Equation (7.4) is the fundamental equation of 21-cm Galactic structure analysis. If the function $R_0[\Omega(R) - \Omega_0]$ is known, then, in principle, distances along the line of sight in the Galactic plane can be attributed to each measured velocity. It is worth emphasizing that kinematic distances cannot be determined directly, but require, in advance, knowledge of the velocity field throughout the Galaxy. Practical application of Equation (7.4) raises a number of problems, some of which are discussed in this chapter.

The first of these problems involves the accuracy with which the angular-velocity rotation curve $\Omega(R)$ can be determined. Different methods have to be employed to determine the rotation curve for different regions of the Galaxy; the accuracies of the different methods vary.

7.2.2 The Galactic Rotation Curve

It is convenient to discuss Galactic kinematics separately for different regions of the Milky Way, which we can define roughly as the bulge region, at $R < 3$ kpc; the inner Galaxy, at $3 \text{ kpc} < R < R_o$; the outer Galaxy, at $R > R_o$; and the local region, at $r < 1$ kpc. This division is justified because the methods used to derive the kinematics differ from region to region; furthermore, the dynamical situation might be rather different in the different regions, although in all cases the global motions of the HI gas are determined by the Galactic gravitational potential. Thus, in the bulge region, the gravitational potential has a spheroidal component, with stable gas orbits that may not be confined to the fundamental plane $b = 0°$. The potential may have, in addition, a bar-shaped component of dynamical importance. For reasons not yet entirely understood, noncircular motions in the inner few kiloparsecs of the Galaxy are of approximately the same amplitude as the rotational motions. In the outer Galaxy, the gas layer is strongly warped and responds in its motions to a gravitational potential whose composition and shape is still largely unknown.

The inner-Galaxy rotation curve may be determined as follows from 21-cm measurements taken along directions in the Galactic equator in the first and fourth longitude quadrants. For any physically realistic rotation law, v will vary along a line of sight as shown schematically in Figure 7.5. Consider a particular line of sight lying somewhere in the Galactic plane with l in the range $0° < l < 90°$. As the distance from the Sun, r, increases along this line of sight, the distance to the Galactic center at first decreases. This corresponds to increasing $\Omega(R)$ and thus to increasing v. However, as r increases further, a point on the line of sight is reached which is closest to the Galactic center. This point is called the "tangent point," or the "subcentral point" because it lies where the line of sight is tangent to a Galactocentric circle. Here $R = R_{min} = R_o \sin l$; $\Omega(R)$ and thus v reach maximum values. For still larger r, R increases so that v decreases from the value it reached at the tangent point. By suitably defining the cutoff value of v on a profile observed at each longitude along the Galactic equator and attributing this "terminal velocity" to the distance from the center, $R_o \sin l$, corresponding to that longitude's tangent point, one obtains $\Omega(R)$, provided R_o and Ω_o are known from other methods.

The terminal velocity is, in general, $v_t = R_o[\Omega(R_o|\sin l|) - \Omega_o] \sin l$. In determining this velocity from the observations, it must be taken as a carefully chosen point on the high-velocity edge of each profile; the edges are in practice not sharp but are blurred by various broadening mechanisms. The linear-velocity rotation curve, giving the circular velocity $\Theta(R) = \Omega R$ as a function of v_t, is then obtained using

$$\Theta(R_o|\sin l|) = |v_t| + \Omega_o R_o|\sin l| . \tag{7.5}$$

Note that the terminal-velocity method formally refers to the kinematics only along the locus of tangent points, which defines a circle passing through the Sun and the Galactic center. It is also necessary, of course, to assume that hydrogen is actually present at the tangent point. This assumption is plausible. Indeed, it is the ubiquity of HI that makes it more suitable than some of the other tracers of transgalactic kinematics for rotation curve determination. Over a range of radii $4 \text{ kpc} < R < 7$ kpc, interstellar molecular clouds occur so commonly that observations of CO can be subject to a terminal-velocity analysis; over that range, CO observations have a marginal advantage over HI observations of a somewhat smaller intrinsic velocity dispersion, so that the cutoff velocities can be more accurately determined. Outside that range, however, molecular clouds are too

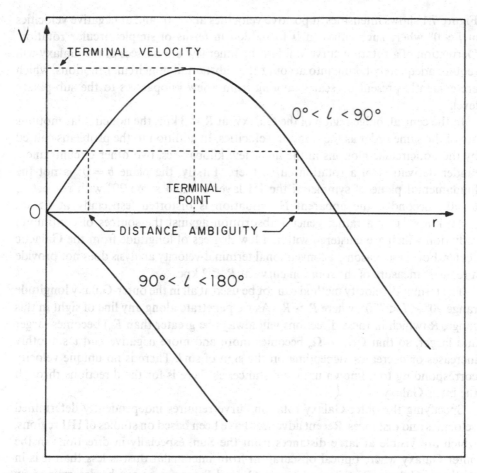

Fig. 7.5. Run of velocities with respect to the local standard of rest as a function of distance from the Sun, appropriate to a simple, axisymmetric velocity field. For the longitude range $0° < l < 90°$, the diagram illustrates the distance ambiguity, the terminal velocity, and the tangent-point distance. The schematic situation in the quadrants $270° < l < 360°$ and $180° < l < 270°$ is similar to that illustrated here, except for a reversal of the sign of the velocity. For a less well-behaved, but more realistic, velocity field, the variation of radial velocities is substantially more complicated than illustrated here, with important consequences for the shapes of Galactic spectral lines.

sparsely distributed to gaurantee emission from the locus of tangent points. The terminal-velocity analysis based on CO data confirms the general kinematic conclusions derived from the HI data.

The terminal-velocity procedure is weak at longitudes $75° < l < 90°$ and $270° < l < 295°$. Because of the geometry in these directions, R and thus v change very slowly with increasing distance, r, along the line of sight. Where the kinematic parameter $|dv/dr|$ is small, a small velocity uncertainty corresponds to a large uncertainty in distance from the Sun.

In practice, the terminal-velocity procedure fails at directions within about $15°$ of the direction of the Galactic center, where the assumption of circular rotation is clearly invalid. The reference map of HI emission from the Galactic disk in

Figure 7.1 shows intensities at positive velocities at $l < 0°$ and at negative velocities at $l > 0°$ where such emission is forbidden in terms of simple circular rotation. Derivation of a rotation curve valid in the inner few kiloparsecs of the Galaxy will require adequately taking into account the substantial noncircular motions, which are evidently present on scales ranging from a few kiloparsecs to the sub-parsec level.

In the central, bulge region of the Galaxy, at $R < 3$ kpc, the noncircular motions are of the same order as the rotation velocities. In addition to the problems caused by the noncircular motions in the inner few kiloparsecs, two other circumstances hinder derivation of a rotation curve there. Firstly, the plane $b = 0°$ is not the fundamental plane of symmetry: the HI layer is tilted some 20° with respect to $b = 0°$. Secondly, the apparent HI situation is distorted (especially at $l > 0°$, $v < 0$ km s^{-1}) by a rather general absorption against the sources of continuum radiation which are clustered within a few degrees of longitude from the Galactic center. For these reasons, a conventional terminal-velocity analysis does not provide a realistic measure of the rotation curve at $R < 2$ kpc.

The terminal-velocity method cannot be used at all in the outer-Galaxy longitude range $90° < l < 270°$, where $R > R_o$. As we penetrate along any line of sight in this range, R (which in these directions will always be greater than R_o) becomes larger and larger, so that $\Omega(R) - \Omega_o$ becomes more and more negative and v smoothly increases or decreases, depending on the sign of $\sin l$. There is no unique velocity corresponding to a known unique distance, as there is for the directions through the inner Galaxy.

Specifying the outer-Galaxy rotation curve requires independently determined velocities and distances. Recent advances have been based on studies of HII regions, which are visible at large distances from the Sun, especially in directions in the outer Galaxy where optical obscuration from interstellar dust is less than it is in the inner Galaxy. Distances to young O- and B-type stars can be determined by spectrophotometric methods. Velocities can be determined from observations of the spectral lines (in particular from CO molecules) which are invariably found associated with HII regions or from the stellar spectral lines. This approach has been emphasized by Jackson et al. (1979), Blitz (1979), Blitz et al. (1980, 1982), and Brand and Blitz (1987). Because the method is not restricted to a particular locus, it has the particular advantage over the terminal-velocity method that it can, in principle, lead to a Galactic velocity field, rather than to an azimuthially smoothed rotation curve. Uncertainties in the general velocity field remain substantial, however, principally because of the limited number of optically visible HII regions at large Galactic radii.

For the reasons mentioned above, the rotation curve of our Galaxy can be determined from 21-cm HI observations only over the range 3 kpc $< R < R_o$. The Galactic rotation curve shown in Figure 7.6 is based on a terminal-velocity analysis of HI data at $R < R_o$ and on combined optical/radio data on HII regions at $R > R_o$. The signatures of several important aspects of Galactic structure are present in this figure. $\Theta(R)$ changes much more slowly with respect to R than would be the case for solid-body rotation; this strong differential rotation indicates a strong increase in total Galactic mass toward the center of the system. The rotation curve also

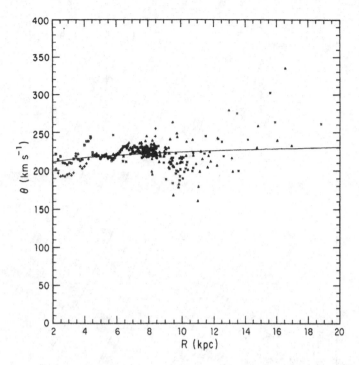

Fig. 7.6. Galactic rotation curve, $\Theta(R)$, showing the variation with distance from the Galactic center of the linear velocity of differential rotation (From Brand and Blitz 1988, in preparation). The plot corresponds to the galactic-constant values $R_0 = 8.5$ kpc and $\Theta_0 = 220$ km s^{-1}. The information at $R < R_0$ pertains to an analysis of the HI terminal velocities, with the northern data indicated by pluses, the southern by boxes. At $R > R_0$ the distance information comes from optically studied HII regions, whose kinematics can be determined by radio observations of the molecular clouds with which they are always associated; the northern data are indicated by crosses, the southern by triangles. Interpreting the observed data points as a generally valid rotation curve involves a number of important assumptions. The smooth-line curve represents the power law given in the text. As is generally the case for spiral galaxies, the rotation curve is not observed to fall off at large distances; this implies substantial invisible mass in our Galaxy.

indicates, however, that a large mass remains in the Galaxy at distances well beyond the last measured point on the rotation curve. If a distance on the rotation curve has been reached exterior to most of the Galactic mass, then the rotational velocities would fall off as $R^{-1/2}$ in the manner described by Kepler for the motions of the planets, which reside exterior to most of the mass of the Solar System. The total mass of the Galaxy is much larger than was realized before the flat, or even rising, nature of the outer-Galaxy rotation curve was known. The irregularities in this rotation curve can be attributed to systematic large-scale streaming motions. Over much of the Galaxy, the deviations from circular velocity are of the order of 5% of the rotation velocities. We will see below that, although rather modest in amplitude, these noncircular motions considerably complicate analysis of Galactic spectral-line data.

The rotation curve for our Galaxy resembles those of other spiral galaxies of more or less similar type. Spirals of widely different morphology have rotation curves

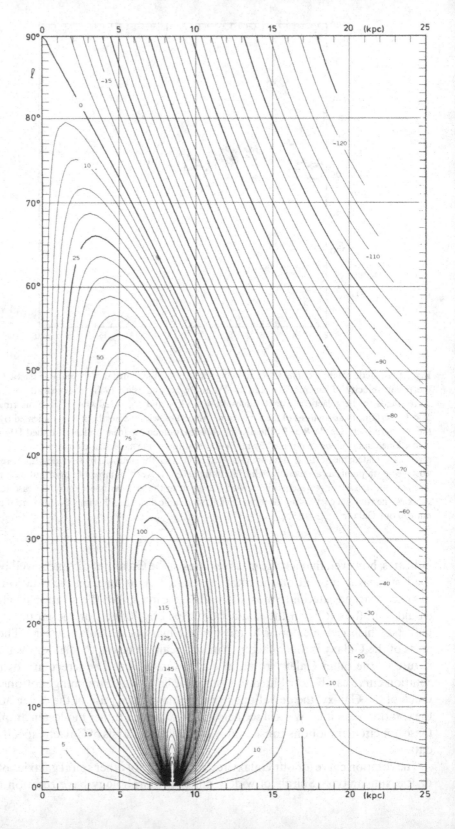

Fig. 7.7

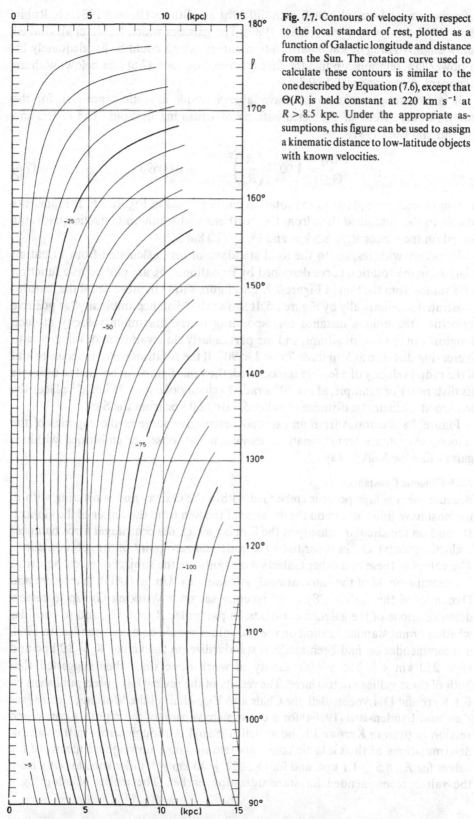

Fig. 7.7. Contours of velocity with respect to the local standard of rest, plotted as a function of Galactic longitude and distance from the Sun. The rotation curve used to calculate these contours is similar to the one described by Equation (7.6), except that $\Theta(R)$ is held constant at 220 km s^{-1} at $R > 8.5$ kpc. Under the appropriate assumptions, this figure can be used to assign a kinematic distance to low-latitude objects with known velocities.

of quite similar form, although with different amplitudes (Bosma 1981a, b, Rubin et al. 1985, Sancisi and van Albada 1987). The galaxies which Rubin et al. studied show slightly rising or flat outer rotation curves, which could be fit adequately by a power law relationship like the first term in Equation (7.6) (see below) with an exponent in the range 0.0 to 0.2.

It is sometimes useful to have available a simple analytic expression for the rotation curve freed of the perturbations of streaming motions. The expression (Brand and Blitz 1987)

$$\frac{\Theta(R)}{\Theta_o} = 1.0074 \left(\frac{R}{R_o}\right)^{0.0382} + 0.00698 \qquad (7.6)$$

is a satisfactory simple fit to the rotation curve plotted in Figure 7.6. It represents the fit to the combined data from the northern and southern Galactic quadrants based on the values $R_o = 8.5$ kpc and $\Theta_o = 220$ km s^{-1}.

Velocities with respect to the local standard of rest, calculated from Equation (7.4) using the rotation curve described by Equation (7.6), are plotted as a function of distance from the Sun in Figure 7.7. This figure shows in more detail the features illustrated schematically by Figure 7.5. It shows the "distance ambiguity" at positive velocities, the unique distance corresponding to the maximum velocity at each longitude in the first quadrant, and the particularly slow variation of velocity with increasing distance at longitude $75° < l < 90°$. If the rotation curve is accurate and if the radial velocity of a feature is measured, then this figure can be used to estimate its distance. For example, at $l = 30°$ a radial velocity of $v = +75$ km s^{-1} places the feature at a kinematic distance of either 5.3 or 12.0 kpc from the Sun.

Figure 7.8 illustrates from an external perspective some of the vagaries of the velocity-to-distance transformation relevant for an observer embedded within a galaxy like the Milky Way.

7.2.3 Galactic Constants

Because our vantage point is embedded within the Galaxy and is rotating with it, we must have information on the distance of the Sun from the center of the Galaxy, R_o, and on the circular velocity at the Sun (at which the centrifugal force balances Galactic gravity), Θ_o, in order to fix the scale and zero point of the rotation curve. The values of these and other Galactic constants are the subject of recent scrutiny.

Commission 33 of the International Astronomical Union (IAU), "Structure and Dynamics of the Galactic System," recently set up a Working Group to review determinations of the galactic constants, in particular R_o and Θ_o, and to consider whether or not standardization on specific values should be recommended. In 1964, a recommendation had been made to standardize on the values $R_o = 10$ kpc and $\Theta_o = 250$ km s^{-1}, but a wide variety of work done since then suggested that both of these values are too large. The results of the review have been published by F.J. Kerr and D. Lynden-Bell, the Chair and Vice-chair of the Working Group [see Kerr and Lynden-Bell (1986a) for the complete version of the report; a condensed version is given in Kerr and Lynden-Bell (1986b)]. Although errors in the various determinations of the Galactic constants remain large, unweighted means of the values for R_o, 8.5 ± 1.1 kpc, and for Θ_o, 222 ± 20 km s^{-1}, differ substantially from the values recommended for standardization earlier. The merits of adopting a

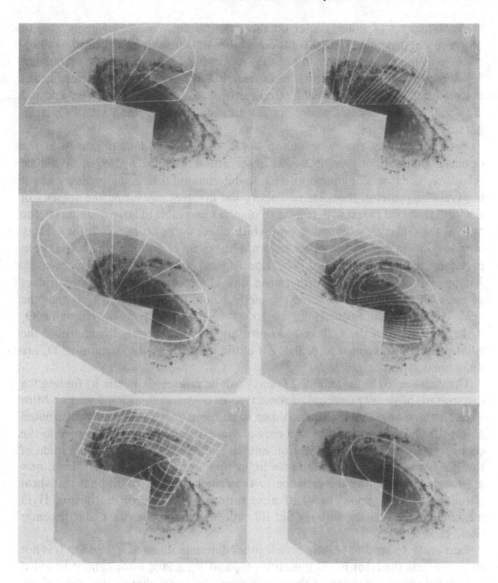

Fig. 7.8. Schematic diagram illustrating the transformation of velocity to distance for an observer embedded within a galactic system. The location of the observer is indicated by the small circle; the representative galaxy is rotating counterclockwise. (a) Spatial plane perpendicular to the galaxy, oriented along $l = 330°$, $150°$, and with lines drawn at $30°$ increments in b. (b) Equal-velocity contours in the plane portrayed in panel (a). The dashed line separates positive radial velocities, in the outer galaxy, from negative ones, in the inner galaxy. (c) Surface formed by a cone at $b = 30°$, with lines drawn at $30°$ increments in l. (d) Equal-velocity contours on the surface portrayed in panel (c). For rotation as in our own Galaxy, the velocity contours indicated occur at intervals of about 25 km s^{-1}. The dashed line refers to $v = 0$ km s^{-1}. (e) Spatial surfaces corresponding to a single observed velocity. The inner-galaxy distance ambiguity is shown. (f) Representation of a cylinder at constant galactocentric radius and of a plane at constant azimuth. These surfaces correspond to the data displays in Figures 7.18 and 7.21. If the observations can be transformed to such surfaces, the analysis can proceed in a more straightforward way than when the display remains in the velocity domain. This diagram was prepared by Vincent Icke.

conventional set of values center on the practical convenience of being able to compare observational and theoretical results of different investigators. Consequently, at the General Assembly of the IAU in New Delhi in 1985, Commission 33 adopted a resolution in which it recommends, "... on the basis of the conclusions of its Working Group on "Galactic Constants," use of the values $R_0 = 8.5$ kpc and $\Theta_0 = 220$ km s^{-1} in cases where standardization on a common set of Galactic parameters is desirable."

Determination of the basic quantities R_0 and Θ_0 has lain primarily in the optical domain, although in both cases observations of neutral hydrogen have some relevance. The most common approach to determining the circular velocity about the Galactic center of the solar neighborhood involves measuring the velocity of the local standard of rest with respect to objects observable at large distances in the halo of the Galaxy, or beyond it, and which either do not rotate with the Galaxy or whose kinematics are somehow known. The system of globular clusters has been especially useful in this regard. The Local Group of galaxies can also be used. Knapp et al. (1978) describe, furthermore, how HI observations at the outer fringes of the Galaxy can be used to determine Θ_0; note from Equation (7.4) that the velocity observed from the outer boundary of the Galaxy could determine Θ_0 if the dimension of the Galaxy is known and if, for example, the shape of the rotation curve is assumed to be flat. All of the methods used to determine Θ_0 are unfortunately rather indirect.

The distance to the center of the Galaxy can be measured directly by finding the distance to the density centroid of some type of object, such as RR Lyrae or Mira variable stars, or globular clusters, taken to be more or less spherically distributed about the Galactic center. The accuracy of these determinations is sensitive to the correction for interstellar absorption and to the adopted absolute magnitude of the RR Lyrae stars or other distance indicators. A promising, rather direct, new approach to measuring the distance to the center involves traditional statistical parallax methods applied to VLBI observations of the proper motions of H_2O maser clumps associated with the Sgr B2 molecular cloud near the Galactic center (Reid et al. 1987).

Kerr and Lynden-Bell (1986a) review other determinations of R_0 and Θ_0. It is not easy to assess the error in the quantities Θ_0 and R_0; a 20% error cannot be ruled out. A subsequent revision of these two quantities may change the scale of the rotation curve and its overall slope, but many of the general kinematic conclusions drawn from it would not be affected.

The observations relevant to setting R_0 and Θ_0 pertain mostly to Galactic constituents at great distances from the Sun. It is interesting for many purposes to consider the characteristics of Galactic kinematics in the neighborhood of the Sun. Oort (1927) developed the following description of the local kinematics resulting from differential Galactic rotation.

Equation (7.4) for the radial velocity is a general one, derived assuming only purely circular rotation. We may approximate the difference $(\Omega - \Omega_0)$ in a first-order Taylor expansion

$$\Omega - \Omega_0 = \left(\frac{d\Omega}{dR}\right)_{R_0} (R - R_0) + \cdots ,$$

assuming that $(R - R_o)$ is small, and may then write the derivative of the angular velocity in terms of the linear velocity as

$$\frac{d\Omega(R)}{dR} = \frac{d(\Theta/R)}{dR} = \frac{1}{R}\frac{d\Theta}{dR} - \frac{\Theta}{R^2} .$$

Then Equation (7.4) for the radial velocity becomes

$$v_R = \left[\left(\frac{d\Theta}{dR}\right)_{R_o} - \frac{\Theta_o}{R_o}\right](R - R_o)\sin l .$$

Near the Sun, $(R - R_o) \approx r \cos l$, so that

$$v_R = \left[\frac{\Theta_o}{R_o} - \left(\frac{d\Theta}{dR}\right)_{R_o}\right]r \cos l \sin l .$$

Use of the trigonometric identity $\sin 2l = 2 \sin l \cos l$ yields

$$v_R = \frac{1}{2}\left[\frac{\Theta_o}{R_o} - \left(\frac{d\Theta}{dR}\right)_{R_o}\right]r \sin 2l .$$

Defining the Oort constant A as

$$A \equiv \frac{1}{2}\left[\frac{\Theta_o}{R_o} - \left(\frac{d\Theta}{dR}\right)_{R_o}\right] \tag{7.7}$$

allows rewriting the expression for the radial component due to differential rotation of objects near the Sun as

$$v_R = Ar \sin 2l . \tag{7.8}$$

Note that the definition of A may be written as

$$A = -\frac{1}{2}R_o\left(\frac{d\Omega}{dR}\right)_{R_o} , \tag{7.9}$$

showing that A is a measure of the local rate of shear of galactic rotation.

An approximate expression analogous to Equation (7.8) can be derived for the tangential component of differential rotation. In general, assuming circular rotation, the tangential velocity of an object is

$v_T = $ component of Θ perpendicular to the line of sight minus
component of Θ_o perpendicular to the line of sight

$$= \frac{\Theta}{R}[R_o\cos(l) - r]R_o\cos(l) - \Theta_o\cos l$$

or

$$v_T = [\Omega(R) - \Omega_o]R_o\cos(l) - \Omega(R)r . \tag{7.10}$$

After a first-order Taylor expansion, and maintaining the restriction to the solar neighborhood so that $\Omega r \approx \Omega_o r$, this equation becomes

$$v_T = \left[\frac{\Theta_o}{R_o} - \left(\frac{d\Theta}{dR}\right)_{R_o}\right]r \cos^2 l - \frac{\Theta_o}{R_o}r .$$

Use of the trigonometric identity $\cos^2 l = \frac{1}{2}(1 + \cos 2l)$ yields

$$v_T = \left[\frac{\Theta_o}{R_o} - \left(\frac{d\Theta}{dR} \right)_{R_o} \right] \frac{r}{2}(1 + \cos 2l) - r\frac{\Theta_o}{R_o}$$

$$= Ar\cos(2l) - \frac{r}{2}\left[\frac{\Theta_o}{R_o} + \left(\frac{d\Theta}{dR} \right)_{R_o} \right]$$

Defining the Oort constant B as

$$B \equiv -\frac{1}{2}\left[\frac{\Theta_o}{R_o} + \left(\frac{d\Theta}{dR} \right)_{R_o} \right] \qquad (7.11)$$

allows rewriting the expression for the tangential component due to differential rotation of objects near the Sun as

$$v_T = r[A\cos(2l) + B] . \qquad (7.12)$$

Because both radial and tangential velocities can be described in terms of A and B, the Oort constants describe the consequences of circular differential Galactic rotation observable within about a kiloparsec from the Sun. Oort's (1927) demonstration that Equations (7.8) and (7.12) are consistent with observation firmly established the differential nature of Galactic rotation. The observed radial velocities of nearby objects show a characteristic double sine-wave variation with Galactic longitude in accordance with Equation (7.8); measurement of the amplitude of this variation for objects of known distances gives a measure of A. Observations of proper motions show, in accordance with Equation (7.12), a double-cosine variation with Galactic longitude as well as a longitude-independent term. Measurement of the amplitude of the first term provides an estimate of A; the longitude-independent offset term provides an estimate of B. Because proper motions are more difficult to observe than radial velocities, B is known with less certainty than A. The rounded-off means of the Oort-constant determinations summarized by Kerr and Lynden-Bell (1986a, b) are $A = 14$ km s^{-1} kpc^{-1} and $B = -12$ km s^{-1} kpc^{-1}.

It is useful to note that determinations of A and B can constrain the local values of the angular rotation rate and of the radial gradient of the linear rotation velocity, because

$$\frac{\Theta_o}{R_o} = \Omega_o = A - B \qquad (7.13)$$

and

$$-\left(\frac{d\Theta}{dR} \right)_{R_o} = A + B . \qquad (7.14)$$

Some caution is required when comparing the locally derived values of A and B with values for R_o and Θ_o which are based on global averages, because of the possibility of global anomalies.

We note that although 21-cm observations of neutral hydrogen cannot provide A or R_o separately, they can provide a determination of the product AR_o. Measurement of this product gives a valuable independent check of the values of A and R_0 determined separately, by other methods. The 21-cm observations provide AR_0 as follows. The angular velocity $\Omega(R)$ is expanded in a Taylor approximation:

$$\Omega(R) = \Omega_0 + (R - R_0)\left(\frac{d\Omega}{dR}\right)_{R_0}. \tag{7.15}$$

The first-order approximation is applicable where $(R - R_0)$ is small. Using the definition of the Oort constant A,

$$\Omega(R) = \Omega_0 - (R - R_0)\left(\frac{2A}{R_0}\right) \tag{7.16}$$

so that substitution into Equation (7.4) gives

$$v = -2A(R - R_0)\sin l . \tag{7.17}$$

The assumption of circular rotation places the terminal velocity, v_t, at the Galactocentric distance $R = R_{min} = R_0 \sin l$. Consequently, measurements of the terminal velocities give

$$AR_0 = \frac{v_t}{2\sin l(1 - |\sin l|)} \tag{7.18}$$

The product should be determined from observations at longitudes not too much less than $90°$ in order for the assumption $R \approx R_0$ to remain valid. The variations observed in v_t, which are of the order of 5 km s^{-1} and are attributed to systematic streaming motions, will introduce large errors in AR_0, because the denominator in Equation (7.18) is less than 1. In principle, the product AR_0 could be determined in a similar fashion using stellar objects. It is difficult, however, to be sure that the stellar objects really are at R_{min}.

The period of revolution of the solar neighborhood around the Galactic center is $2\pi R_0/\Theta_0 = 2.4 \times 10^8$ yr, for the standard values $R_0 = 8.5$ kpc and $\Theta_0 = 220$ km s^{-1}. The length of the Galactic year in our vicinity is thus only 1 or 2% of the age of the Galactic disk. The period of revolution at about 4 kpc from the center is about half of the local value. One of the classic questions of Galactic structure research is the winding dilemma, which asks how structure in the disk can be maintained against the shearing forces of such strong differential rotation for times longer than one Galactic revolution. Substantial, continuing efforts have been directed toward a theoretical understanding of the common phenomena of spiral structure. These efforts may conveniently be grouped under three headings, in which spiral structure is viewed respectively as a stable dynamic mode of a rotating galaxy (Kalnajs 1983a, Lin and Bertin 1985); as a phenomenon which is continuously excited, for example, by a central bar structure (Norman 1983); and as the response to stochastically occurring events percolating through galactic disks (Seiden 1985). (The references given here refer to recent reviews of quite extensive fields of theoretical research.)

7.2.4 Deviations from Circular Symmetry and Circular Motions
The discussion above has been based on the assumption of an axisymmetric Galaxy. There are a variety of indications that this assumption is not strictly

valid. Unambiguous proof of noncircular motions in the Galaxy is given by profiles observed in the galactic plane in the cardinal directions $l = 0°$, $90°$, and $180°$. Figure 7.2 shows 21-cm observations made in these directions. At longitudes $0°$ and $180°$ all circular motions would be perpendicular to the line of sight, and in the presence of only such motions, the profiles would have one peak, centered on zero velocity, and broadened symmetrically about $v = 0$ km s^{-1} by the intrinsic gas dispersion of several kilometers per second. Actually, significant radial motions are observed: the peak emission in these directions is not centered at $v = 0$ km s^{-1}, and it is not symmetric. For $l = 90°$, the circular-motion assumption permits no positive-velocity peak in the profiles. Actually, there is a peak at $v \approx +6$ km s^{-1}.

The deviations from circular motion pointed out above are less than a few percent of the rotational velocities. The situation is fundamentally different in the inner few kiloparsecs of the Galaxy, where noncircular motions of the same amplitude as the rotational velocities are observed. The most dramatic example of forbidden-velocity emission is given by the 3-kpc arm (van Woerden et al. 1957, Rougoor and Oort 1960, Rougoor 1964). This feature is evident in the reference map in Figure 7.1 as the ridge of intensities extending from $l \approx 338°$ at $v \approx -140$ km s^{-1}, across the Sun-center line $l = 0°$ at $v = -53$ km s^{-1}. There is a deep absorption feature at $l = 0°$, $v = -53$ km s^{-1}, indicating that this branch of the arm is located between the Sun and the central continuum source Sagittarius A. The negative radial velocity of the absorption dip indicates that the feature has a net expansion away from the center. Because of this expansion motion, the distance scale of the arm cannot be measured by the standard kinematic procedure. Instead, the distance scale of the feature has been determined geometrically: it can be traced to $l \approx 338°$, where it is seen tangentially. Here $R = R_0 \sin 22° \approx 3$ kpc (hence the feature's name). Admittedly, the longitude at which the feature becomes tangential to the line of sight is not very easy to determine. This is, nevertheless, the only point on the arm at which a distance can be estimated. Its distance from the center at $l = 0°$ is therefore uncertain, except that it must pass between the Sun and the center. The 3-kpc arm can also be traced in the CO lines. The total molecular gas mass involved is $\approx 3 \times 10^7 M_\odot$; the HI mass is about $4 \times 10^7 M_\odot$ (Bania 1980). The mechanism responsible for the 3-kpc arm remains unknown (see Oort 1977). It is, in particular, not clear if the feature is a transient and perhaps rare phenomenon, or if, on the other hand, it represents a permanent flow of gas perhaps due to a barlike potential.

The forbidden-velocity emission feature seen on the $l = 180°$ map near $b = 6°$, $v = -120°$ is an example of the high-velocity cloud phenomenon. Although clearly indicative of noncircular motions, high-velocity clouds are patchy, albeit extended, features with velocities deviating by 100 km s^{-1} or more from the differential rotation motions and thus are not to be confused with the smoothly streaming flows, with length scales coherent on the order of a kiloparsec and velocities deviating by 10 km s^{-1}, that evidently are characteristic of the entire galactic disk. We note that high-velocity clouds do not occur in the part of Galactic phase space that would cause them to contaminate the terminal-velocity analysis.

The rotation curve plotted in Figure 7.6 does, in fact, show localized irregularities in the velocities of the observed points plotted against distance from the center. In the original determination of the rotation curve by Kwee et al. (1954), this sort of irregularity was attributed to extended regions that did not contain enough gas near the tangent point to determine the profile cutoff there. The idea was that if there was not much gas at the tangent point, the observed cutoff velocity would be due to lower velocity gas at $R > R_0 \sin l$. If this was the case, then the rotation curve should be drawn using the upper envelope of the observed terminal velocities. Circular motion could in this way be retained. Shane and Bieger-Smith (1966) showed, however, that the irregularities could better be attributed to large-scale streaming motions, i.e., to systematic deviations from circular rotation. They rejected the possibility of large empty regions because low-intensity extensions on the ends of profiles are not observed at longitudes corresponding to the dips in the run of observed terminal velocities. Indeed, one of the most pronounced characteristics of the observations in Figure 7.1 is the persistent ridge of high intensities near the terminal-velocity locus in both the first and fourth longitude quadrants. In order for the whole maximum-velocity part of the profile to shift to lower velocities, there would have to be essentially no hydrogen (a factor of 100 or so less than the average hydrogen density) along the whole region of the line of sight with velocities within approximately 10 km s^{-1} of the velocity corresponding to the tangent point. Inspection of Figure 7.7 shows that for such a velocity range, regions near the tangent point of 3- or 4-kpc extent would have to be essentially empty of gas and that these regions would have to have a preferential orientation with respect to the observer. Such an artificial density distribution is implausible. Thus, we are left with the important conclusion that the observed irregularities in the run of the terminal velocities reflect corresponding irregularities in the Galactic velocity field. There is no reason to expect these irregularities to be axially symmetric. The characteristic length scale of the systematic irregularities in the rotation curve of our Galaxy is about one or two kiloparsecs (see Burton and Bania 1974).

Because of the kinematic irregularities, the data points derived using the terminal-velocity method and plotted in Figure 7.6 describe an apparent rotation curve, because they refer to motions along one particular locus, and not over the entire Galactic plane. This locus is not necessarily the locus of tangent points, because in the presence of deviations from circular motion, the terminal velocity does not necessarily originate at the tangent point. Thus, it is not surprising that the apparent rotation curve derived from observations of the terminal velocities in the longitude quadrant $270° < l < 360°$ has irregularities somewhat differently placed, although of about the same amplitude. What is more disturbing is that the unperturbed or basic rotation curve, obtained by drawing a smooth curve through the irregular apparent one, is different when determined solely from fourth-quadrant data than when determined solely from first-quadrant data. There is a systematic difference of about 7 km s^{-1} between the two curves over the region between $R = 4$ and 8 kpc, which requires us to accept kinematic asymmetries of this amplitude on a very large scale or correspondingly large errors in the determination of the local standard of rest (cf. Kerr 1962, Shuter 1981, Brand 1986).

The method of combining optical distances to HII regions with velocities from the associated radio-wavelength molecular-line emission, used especially in the outer Galaxy, has the favorable aspect of not being restricted to a particular locus, so that a two-dimensional velocity field can be found if the observational material

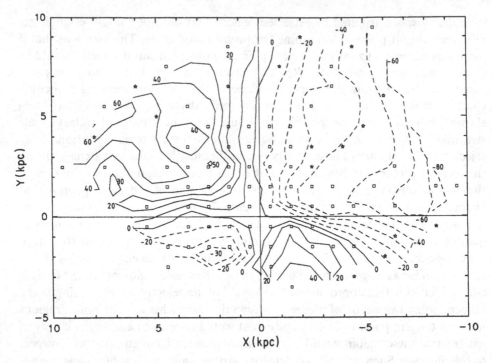

Fig. 7.9. Two-dimensional display of radial velocities derived from optical distances combined with radio velocities (From Brand 1986), projected onto the Galactic plane. Azimuthal smoothing is not intrinsic to this method, as it is for the terminal-velocity method. Because noncircular motions are included, this diagram is useful for kinematic distance determinations. The Galactic velocity field is, however, generally symmetric about the Sun-center line. Boxes indicate HII-region data averaged in 1-kpc bins; stars indicate interpolated points. The Sun is at $(X, Y) = (0, 0)$; the Galactic center in the direction $(0, -\infty)$.

is rich enough. Brand and Blitz (see Brand 1986) found in this way the velocity field shown in Figure 7.9. The residuals between this plot and one derived from the same data but azimuthally smoothed have amplitudes of order 10 km s^{-1}, similar to the situation in the inner Galaxy.

Deviations from circular symmetry in our Galaxy may be spatial as well as kinematic. This can be demonstrated in a general way by comparing the cutoffs on both the positive-velocity and negative-velocity wings of profiles observed at complementary North-South Galactic longitudes. Such a comparison is shown in Figure 7.10. The cutoffs plotted in this figure provide information from different parts of the Galaxy, as suggested by the labels "local," "subcentral region," and "outskirts" on the figure. The differences in the profile extends are especially large and systematic in the outskirts of the Galaxy. It seems more plausible to explain these outer-Galaxy asymmetries as spatial asymmetries than as kinematic ones, because the difference between the smoothed outer-Galaxy rotation curve derived only from data at $l > 180°$ and the one derived only from data at $l < 180°$ is not large enough to account for the difference in the profile extents. The HI distribution is known to be similarly lopsided in a number of nearby spiral galaxies (Baldwin et al. 1980).

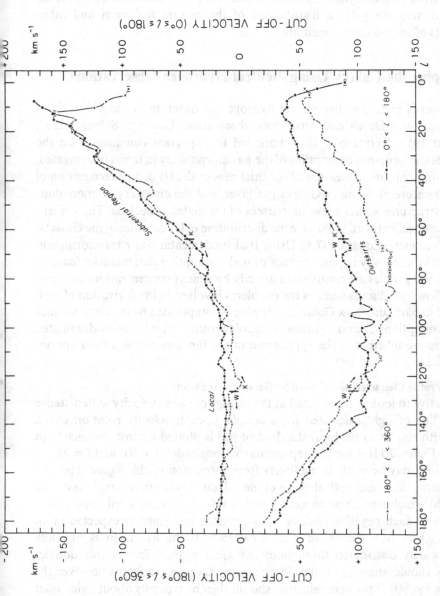

Fig. 7.10. Comparison of cutoffs on positive- and negative-velocity wings of profiles observed in the longitude range $0° < l < 180°$ (dots, right-hand scale) with those in the corresponding longitudes in the range $180° < l < 360°$ (crosses, left-hand scale). The data are from the surveys of Westerhout (1969) and Kerr and Hindman (1970), as indicated. For the purposes of this figure, the cutoff was defined as the velocity at which the intensity at a particular antenna temperature: this was $T_A = 5$ K for the "W" points and 6 K for the "K" ones. The systematic differences in the cutoff velocities imply large-scale deviations from axial symmetry. For the local and subcentral (tangent-point) regions, these deviations are largely kinematic, but recent work on the Galactic velocity field indicates that the very large deviations in the outer Galaxy are probably of a structural nature.

Velocity-field determinations in many external spiral galaxies show irregularities of the same characteristic size and amplitude as found in the Milky Way (see Bosma 1981a,b). The kinematic variations from strictly circular motion amount to only some few percent of the overall motions, over most of the Milky Way. From some points of view, this is a remarkably regular situation. The general phenomenon of deviations from circular symmetry do, however, have important consequences for attempts to map the galactic distribution of the neutral hydrogen and other components of the interstellar medium.

7.3 Mapping the Galaxy Using Neutral Hydrogen Observations

The first overall picture of the neutral hydrogen distribution in our Galaxy was derived from early Dutch and Australian observations (see, e.g., Schmidt 1957, Westerhout 1957, Kerr 1962). That work led to important conclusions on the Galactic rotation curve, a description of the gas morphology in terms of elongated, spiral-arm-like features, a measure of the thickness of the HI disk, a recognition of the warped nature of the outer-Galaxy gas layer, and the discovery of anomalous kinematic structures within a few kiloparsecs of the Galactic nucleus. The Leiden-Sydney map (see Oort et al. 1958) gave the distribution of HI densities in the Galactic plane, derived using Equation (7.4). Using that fundamental v-to-r transformation in a straightforward way requires the additional assumption that intensity features in the observed profiles are contributed directly by density concentrations in space. In what follows, we discuss some of the problems involved in the derivation of such a map, and in particular how Galactic mapping is complicated by the circumstance that the assumptions of axial symmetry, circular rotation, and density-dominated profile structure inherent in the application of the fundamental equation are not strictly valid.

7.3.1 Line Profile Characteristics Caused by Geometrical Effects

It is instructive to look in more detail at the change of radial velocity with distance along the line of sight calculated for a simple, circular-velocity rotation curve. Velocity with respect to the local standard of rest is plotted against distance from the Sun in Figure 7.11(a), for two representative longitudes, $l = 50°$ and $l = 75°$.

Two points may be made immediately from inspection of this figure. The first point concerns the global optical depth of the HI gas. There are two regions on the line of sight which contribute to each positive velocity, whereas only one region contributes to each negative velocity. This distance ambiguity is expected in all directions where $0° < l < 90°$ or $270° < l < 360°$. If the hydrogen is optically thin and widely distributed throughout the Galaxy, the effect of this double-valuedness should show up in the observations, because the intensities over the sector $0° < l < 90°$ at positive velocities should then be typically about twice what they are at negative velocities. The same expectation holds, with the signs of the velocities reversed, in the fourth Galactic quadrant. The situation observed conforms to this expectation. The actual existence of sharp drops in intensities near zero velocity in the ranges $20° < l < 70°$ and $290° < l < 330°$ is evident in the reference maps in Figure 7.1. These intensity drops are consistent with those

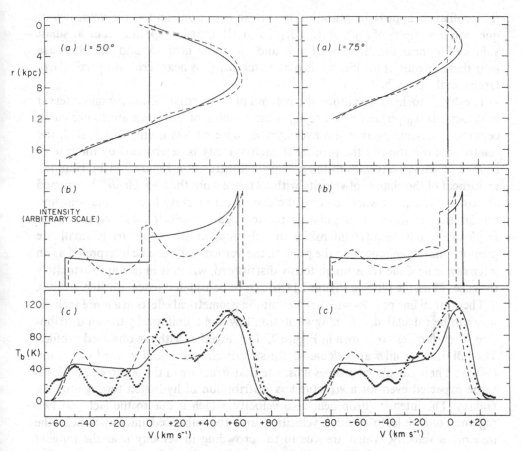

Fig. 7.11. Diagram illustrating for two representative longitudes the importance of the geometrical velocity-crowding effects for the case of purely circular rotation (full-drawn lines) and for motions appropriate to a density-wave velocity field, involving noncircular motions of about 5 km s^{-1}. (a) Velocity with respect to the local standard of rest as a function of distance from the Sun. (b) Schematic line profiles, each considered as the sum, for the near and far sides of the tangent point, of the slopes $(dv/dr)^{-1}$ of the corresponding curve in panel (a). These profiles show the relative geometrical enhancement at each velocity. (c) Synthetic line profiles calculated with the velocity fields in panel (a) and a completely uniform distribution of HI density, temperature, and dispersion. Structure in the synthetic profiles corresponds to regions along the lines of sight where the velocity changes relatively slowly with distance. The heavy dots represent the observations.

predicted in Figure 7.11; their existence indicates that the hydrogen gas at these longitudes and located at small values of $|v|$ is indeed globally optically thin.

The second point which can be made from inspection of Figure 7.11(a) concerns velocity crowding of material near the tangent points. The run of v with r plotted in the figure shows that the velocity observed from regions near the tangent point changes relatively slowly along a length of path several kiloparsecs long. Consequently, the profiles contain, near the terminal velocity, a contribution from an especially long path length. This crowding in velocity results in the high-velocity ridge pattern, which is a striking characteristic of the observations in both the first and fourth quadrants of the Figure 7.1 reference map. Velocity crowding can cause the profiles to approach saturation in certain directions and at the velocities in question. Examples of high optical depths in HI emission spectra occur at small values of $|v|$ near the directions $l \approx 0°$ and $180°$, and near $75°$ and $285°$, because near these directions the kinematic parameter $|dv/dr|$ is near zero over particularly large lengths of path.

The density of hydrogen atoms in a column of 1-cm^2 cross section per unit interval of velocity is $n_{HI}(|dv/dr|)^{-1}$, where n_{HI} is the number of hydrogen atoms per cubic centimeter. Assuming that the hydrogen is more or less evenly distributed, the relative contribution to the profiles at each velocity is determined by the rate of change of the velocity with distance. This is illustrated in Figure 7.11(b). Here the reciprocal of the change of velocity with distance from the Sun, $|dv/dr|^{-1}$, summed at positive velocities where there is the distance ambiguity between the near and the far side of the tangent point, is plotted against velocity. The vertical scale is $|dv/dr|^{-1}$, but because intensities at velocities at which $|dv/dr|$ is small are proportionately enhanced on the profiles, the vertical scale can be interpreted as an intensity scale if the HI is ubiquitously distributed, which is in fact approximately the case. Thus, the plots in Figure 7.11(b) can be considered schematic line profiles.

Theoretical line profiles which represent the geometrical effects in a more realistic way can be calculated, assuming a rotation law and a uniform hydrogen distribution. Such profiles are shown in Figure 7.11(c), together with the observed profiles. The full-draw profiles are calculated, for strictly circular rotation, for $l = 50°$ and $l = 75°$. These calculated profiles illustrate that structure in the observed profiles is to be expected even for a structureless distribution of hydrogen throughout the Galaxy. The intensity drop near zero velocity, which is due to the fact that two regions contribute to positive velocities, and the enhanced intensities near the maximum velocity, which are due to the crowding in velocity near the tangent point, are geometrical effects which are model independent in the sense that structure of this sort would be present in the profiles for any reasonable rotation law and density distribution. Obviously, this sort of profile structure must be satisfactorily accounted for in the subsequent analysis, and not attributed to spurious characteristics of the spatial distribution of the hydrogen. Although there are numerous cases in the literature where this has not been done, it should not be too difficult to take the model-independent effects into account.

Figure 7.12 shows the arrangement of the velocity-crowding parameter $|dv/dr|$ in an idealized Galaxy. The behavior of this parameter along individual lines of sight determines important aspects of the basic appearance and degree of saturation of 21-cm profiles. Regions in Figure 7.12 which are shaded show where the kinematic resolution of distances is low and where the optical depths will be relatively high; mapping the shaded regions of the Galaxy will be therefore especially difficult.

7.3.2 Profile Characteristics Caused by Kinematic Irregularities

Allowing for the effects of systematic streaming motions is more complicated than dealing with the model-independent effects inherent in the overall viewing geometry. The full-drawn curves in Figure 7.11 are calculated for simple circular rotation. The dashed curves in the figure, however, are calculated for a Galactic kinematic situation in which large-scale streaming motions are present. At this stage it is not important that the rotation law used is one derived using the kinematics

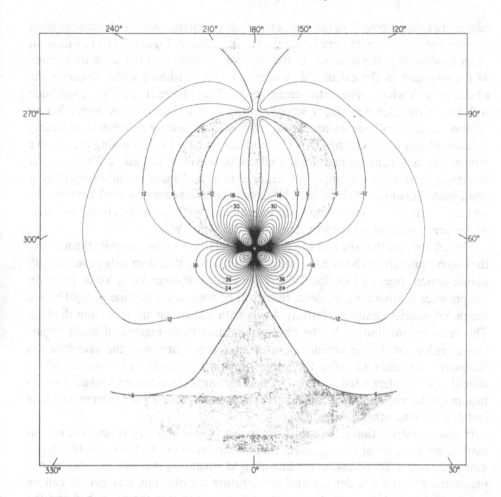

Fig. 7.12. Galactic distribution of the kinematic parameter dv/dr (Burton 1976. Reproduced, with permission, from the Annual Review of Astronomy and Astrophysics, Vol. 14. © 1976 by Annual Reviews Inc.). The behavior of this parameter along individual lines of sight determines, through velocity crowding, important aspects of the basic appearance and degree of saturation of the 21-cm profiles. Thus, the model-independently enhanced intensities and optical depths in the directions $l \approx 0°$, $180°$, $75°$, and $285°$ and in the tangent-point region in general are attributable to low values of $|dv/dr|$. A monotonically varying rotation curve enters this plot; in the more realistic case of a perturbed velocity field, the arrangement of dv/dr becomes correspondingly more complex and so does the analysis of the HI observations. The contours are labeled in units of km s^{-1} kpc^{-1}.

appropriate to the density-wave theory for the dynamical maintenance of spiral structure (see Toomre 1977); what is important is that in the presence of deviations from circular motion, the variation of v along the line of sight will not be as regular as in the circular rotation case, but will show the sort of structure illustrated by the dashed lines in Figure 7.11(a). Irregularities in the plot of v against r will show up as structure in the schematic profile constructed from the $|dv/dr|^{-1}$ against r relation.

The dashed-line synthetic profiles plotted in Figure 7.11(c) were calculated using the density-wave-theory velocity field illustrated by Figure 7.11(a) and a uniform hydrogen distribution. The profiles show the model-independent effects, but these effects are superimposed on the geometrical effects attributable to deviations from circular rotation. The observed profiles are included in Figure 7.11(c) as dots, in order to show that the structure in the observed profiles is of the same magnitude as the structure in the calculated profiles. The dashed-line profiles illustrate the efficiency with which systematic streaming motions of about 5-km s^{-1} amplitude will distort the observations; it would require density contrasts of about 3:1 to achieve intensity differences equivalent to those obtained by systematic streaming motions of only a few kilometers per second (see Burton 1971). Judging the relative importance of kinematic and density effects is, however, not eacy. We note that streaming motions of the amplitude required to be dominant are observed to exist (see Figure 7.6) in our Galaxy; high HI density contrasts are observed between the arm and interarm regions of some external galaxies, but there is little direct evidence on the arm:interarm HI density contrast in the Milky Way.

Thus, it appears that the sort of velocity irregularities known to exist throughout the Galaxy probably influence the line profiles in a way that dominates over density manifestations (see, e.g., Burton 1971, Tuve and Lundsager 1972, Yuan 1969). In the presence of streaming motions, there will be regions on the line of sight where the radial velocity changes relatively slowly with increasing distance from the Sun. The relative contribution to the observed profiles from regions of small $|dv/dr|$ will be enhanced. In the absence of saturation, irregularities in the velocities are themselves sufficient to cause the appearance in the profiles of structure which should not be interpreted in terms of density concentrations on Galactic structure maps derived assuming purely circular rotation, unless such interpretation is justified by some other information.

In order to stress that line profiles are so sensitive to velocity variations, we can assume for the sake of the argument that *all* structure in the 21-cm profiles has a kinematic origin. By adjusting the line-of-sight streaming motions in a model, but maintaining a uniform density and temperature distribution, any profile can be reproduced to within a scale factor. The difference between the perturbed and the basic circular-velocity line-of-sight velocity fields gives information on the spatial distribution of the streaming motions. Burton and Bania (1974) applied this method to HI profiles observed in the outer Galaxy and showed that the observations could be mimicked with line-of-sight streaming amplitudes of the order of 7 km s^{-1}. The sense and magnitude of the kinematic variations are consistent with gas motions which are in fact observed in a number of regions and with motions predicted by the density-wave theory. The kinematic variations found by the perturbation procedure for the gas are consistent with those found directly by Brand and Blitz (1987) in their combined optical/radio study of the outer Galaxy kinematics. It is, of course, not physically realistic to expect temperature or density to be uniform, but a purely kinematic approach does serve to emphasize and exploit the extreme sensitivity of the observations to streaming motions.

7.3.3 Some Remarks on the Spiral Structure of the Galaxy

It is clear from Figure 7.11(c) that any deviations from the adopted rotation law which are systematic over a few degrees on the sky, and are of a few km s^{-1} amplitude, would cause serious errors in the interpretation of the profiles unless adequately accounted for. Because the kinematics dominate the appearance of the profiles, interpretation of the observed profile structure in terms of a map of the density distribution requires knowledge of the velocity field throughout the Galaxy. This knowledge must be rather detailed: for example, if the motions are those predicted by the linear density-wave theory, then the resulting spectral feature will occur at a velocity different from the one corresponding to the center of the mass of the structural feature inducing the motions; if the motions are those predicted by the nonlinear density-wave theory, a single structural feature can contribute multiple peaks to the profile (Roberts 1972).

It is, of course, only plausible to accept that fluctuations in velocity and density will accompany each other; determining the relative importance of the various physical parameters remains problematic. The technique of model fitting using synthetic profiles is a useful way to gauge the relative importance of the various physical parameters under controlled conditions, and to account for the vagaries of the convolution from Galactic space to the observed position-velocity co-ordinates. The relative importance of the parameters may first be established in the outer Galaxy, because there the velocity field is being directly measured and the profile structure is not complicated by distance ambiguities (see Blitz 1983, Blitz et al. 1983).

It is clear that structure in the temperature distribution will also result in some structure in the observed profiles. Consequently, it is also necessary to know the temperature distribution on a large scale. There is abundant evidence on the temperature characteristics of individual interstellar features with characteristic scales of the order of 1 pc. Chapter 3 of this volume deals with this sort of information. As we have stressed above, however, the large telescope beams with which the sky has been surveyed in the 21-cm emission line sample the collective behavior along long lengths of path through the Galactic gas layer. Isolation of individual structural features in the heavily blended low-latitude profiles is generally very difficult. In any case, much information at length scales less than 100 pc, say, is lost by the blending. The observational evidence for a Galactic-scale variation in what is perceived as a sort of harmonic mean temperature of the neutral hydrogen is rather shaky so far. Maps of the HI distribution have generally been based on the assumptions that the kinetic temperature of the gas is constant and that the optical depth is low. These assumptions represent gross oversimplifications, but they have been necessary in order for the analysis to proceed.

It does not seem possible to determine directly from the observations the relative importance of the variables for each peak or valley in an HI emission profile. Near the Galactic plane, the interstellar environment includes HII regions, supernovae and their expanding shells, regions of star formation, gravitational effects from mass concentrations, and the largely unknown effects of magnetic fields, which are probably coupled to the neutral gas by collisions between the plasma component and the neutral one. Each of these mechanisms can produce differences in the motions, densities, and temperatures on a scale large enough to affect the appearance of the profiles. Only in the special case of gas concentrations (1) with no systematic streaming, (2) free from the shearing effects of differential rotation, (3) with constant temperature throughout, and (4) isolated from the model-independent geometrical effects can the relative intensities of peaks and valleys in the profiles be interpreted directly in terms of a density contrast between "cloud" and "intercloud" or "spiral arm" and "interarm" regions. Similarly, only in such a case will the measured dispersion be a direct indication of the random cloud velocities. Such a situation may pertain locally at large distances from the Galactic plane, but it certainly does not commonly pertain in the plane itself. These procedural difficulties do not mean that the structural characteristics of low-latitude HI are inaccessible, although it does seem important not to overrate the definiteness of the picture of our Galaxy's HI density distribution.

It is safe to conclude that 21-cm observations indicate structure on a large scale in the Galactic plane. Some of the features observed define extended regions of some kiloparsecs length situated, roughly, along Galactocentric arcs. We recognize this structure as the spiral structure which is present to some degree in almost all galaxies with sufficient gas in a rotating-disk distribution. Examination of external spiral galaxies in photographs such as those in the *Hubble Atlas of Galaxies* shows that the structure is more often than not very irregular, although in many galaxies a general spiral "grand design" emerges from a more or less tangled background structure. But it does not seem that the 21-cm observations have demonstrated that our Galaxy does, or does not, belong to the galaxies which display such a grand design.

Galaxies which do exhibit such a grand design can be expected to provide much of the information necessary for confronting theories of the origin and maintenance of spiral structure. Observations are necessary which will provide answers to questions, so far not entirely answered for our own Galaxy, which include: Does our Galaxy exhibit a "grand design" of spiral structure? If our Galaxy does have more or less regular arms, are these arms trailing or winding with respect to Galactic rotation? How many such arms are there? What pitch angle and what spacing between arms characterize the structure? What is the density distribution across an arm? What are the motions of the gas between arms? What are the motions and distribution of the gas relative to the stars?

The weakness of our answers to these questions as far as our Galaxy is concerned is largely due to our vantage point embedded within the system. There is also much which has been learned about the larger-scale aspects of the neutral hydrogen distribution for which the embedded vantage point is advantageous. This is discussed in the following sections.

7.4 The Inner-Galaxy Gas Layer

Observations of the hydrogen distribution in the z-direction, perpendicular to the Galactic plane, have shown that the gas is confined to a thin and rather flat layer at distances $R < R_o$. This morphology contrasts sharply with the situation in the outer parts of the Galaxy, where the HI layer is thicker than in the inner parts and is systematically distorted from the plane defined by $b = 0°$. Furthermore, many components of the Galaxy which occur commonly in the inner parts are quite scarce in the outer parts.

The Galactic morphology of atomic hydrogen is unique. The Galactic disk as defined by hydrogen in the atomic state has a diameter fully twice as large as that defined by hydrogen in the molecular and ionized states. In general, the radial scale length of HI is considerably larger than that of all other known tracers of the disk. The HI Galaxy is about twice as large as the Galaxy traced by Population I components, including interstellar molecules, supernova remnants, pulsars, gamma radiation, synchrotron radiation, and dust emission. It is also considerably larger than the distributions traced by Population II components. We focus attention here on comparing the morphology of galactic HI with that of CO, taken as representative of the cold, high-density interstellar gas, and with that revealed by the 100-μm infrared emission from interstellar dust.

Figure 7.13 shows schematically the distribution of emission from ^{12}CO. The two panels of the figure show the arrangement on the plane of the sky of the total flux, integrated over velocity, and the arrangement in longitude-velocity space of the emission from near the Galactic equator (Dame et al., 1987). Figure 7.14 shows a map in Galactic coordinates of IRAS measurements of emission at 100 μm contributed by interstellar dust. Before the IRAS satellite was lauched, important exploratory information was available from balloon-mounted telescopes with limited angular and spectral resolution, sky coverage, and sensitivity; but the IRAS mission provides access to dust properties on a global scale.

7.4.1 Radial Distribution of the Inner-Galaxy Gas and Dust Layer

If one compares the sky map of integrated HI emission shown in Figure 7.3 with the one in Figure 7.13 showing the integrated CO emission and with the one in Figure 7.14 showing the 100-μm dust radiation, one sees directly that the HI is much less confined than the other tracers to the inner Galaxy. Unlike the HI situation, the CO and dust emission occur sparsely in the second and third longitude quadrants, representing $R > R_0$. The impression that the dust and CO maps give is of an external galaxy viewed edge-on. This impression in not an unrealistic one, because the Sun is situated exterior to all but a few percent of the CO and 100-μm emission. The situation is different for the atomic hydrogen distribution, because the Sun is embedded within an HI gas layer extending well beyond our local neighborhood.

Compared to the HI longitude-velocity map shown in Figure 7.1, the paucity of CO emission seen in the lower panel of Figure 7.13 at velocities (negative in the first quadrant, positive in the fourth) contributed from the outer Galaxy also shows the relative confinement of the molecular material.

The radial distributions of these tracers is shown in Figure 7.15. For the cases of HI and CO, conversion of the longitude-velocity distributions at $b = 0°$ can be done using the kinematic information on distances available in spectral-line data. For the case of the continuum emission from dust, such information is lacking, so a geometrically based unfolding process is used. The radial distributions plotted in the figure are consistent with the conclusion which holds generally: all interstellar tracers accessible on a Galactic scale, except HI, show a morphological confinement to the inner Galaxy.

Putting this qualitative conclusion on a quantitative foundation involves different

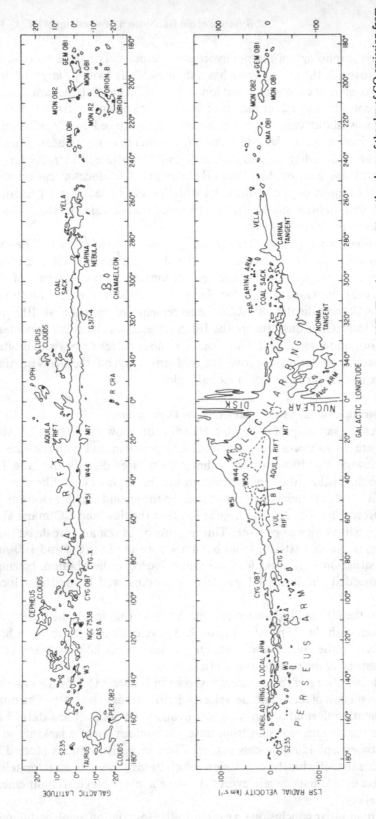

Fig. 7.13. Synoptic maps of CO emission from the Galactic molecular cloud ensemble (Dame et al. 1987). *Top:* Arrangement on the plane of the sky of CO emission from the Milky Way, integrated over the velocity range $|v| < 15$ km s^{-1}. *Bottom:* Longitude-velocity map of CO emission vertically integrated over the range $-3°25 < b < 3°25$. The principal molecular complexes are labeled.

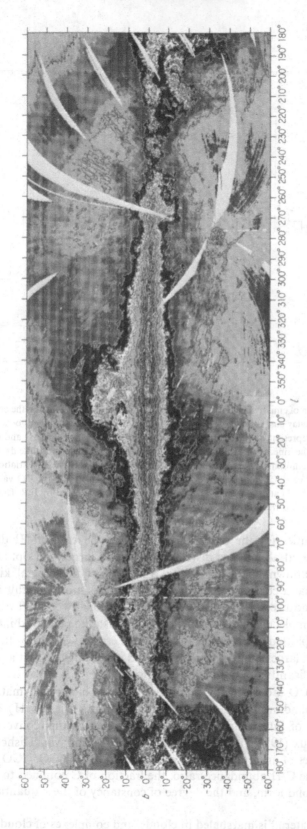

Fig. 7.14. Distribution in Galactic coordinates of emission from interstellar dust observed by the 100-μm IRAS detector. A band of faint emission from dust in the Solar System follows the sinusoidal path of the zodiac, but most of the emission comes from dust diffusely distributed along transgalactic paths. Dust whose emission dominates at 100 μm is distributed more like HI than like CO, except regarding the uniquely large radial extent of HI.

Radial Emissivity Distributions

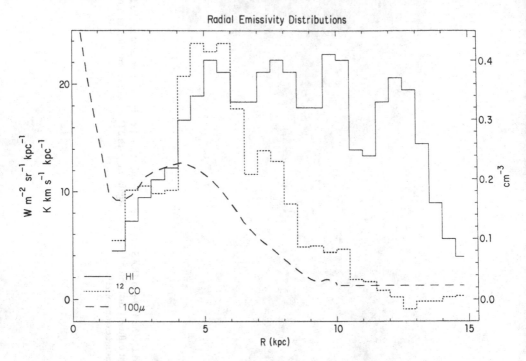

Fig. 7.15. Radial variation of emissivities in the Galactic equator from HI, representing the cool, diffuse component of the interstellar gas; from CO, representing the cold, clumped component of the gas; and from the 100-μm band, representing the diffuse component of interstellar dust. The HI and CO radial emissivities incorporate the kinematic information on distances available in spectral-line data; the dust emissivity is based on an unfolding algorithm appropriate to continuum data. The HI variation is based on observations by Kerr et al. (1976) and Westerhout (private communication); the CO variation, on observations by Dame and Thaddeus (1985). The dust variation was calculated by E.R. Deul (private communication) from 100-μm IRAS data.

sources of uncertainties for the different tracers. For the case of HI data, uncertainties in the radial abundance distribution depend largely on optical-depth effects. Enough is known about the global variation of the crucial kinematic parameter $|dv/dr|$ that a correction can be attempted. It is more difficult to know how to correct for localized self-absorption from exceptionally cold HI residual in and near clumps of molecular gas, which would diminish the perceived intensities. The HI abundance plotted in Figure 7.15 might be in error by 20%. For the case of CO data, there are a larger number of uncertainties which suggest that a plausible error might be significantly more than in the HI situation.

Observations of CO are especially interesting because of the information they can give on H_2 abundances, but conversion of CO intensities to H_2 densities involves a number of intervening assumptions. Aspects of this conversion are reviewed by Lequeux (1981), Liszt (1984), Israel (1985), and van Dishoeck and Black (1987). Sources of uncertainty include the values of the $^{12}CO/^{13}CO$ isotopic abundance ratio, the C/H abundance ratio, the fraction of C confined to CO, the fraction of CO in solid form, and the degree of constancy of these quantities over the Galaxy.

The molecular material is marshaled in clouds, and complexes of clouds, whose number, typical size and separation, and relative speeds can be measured. References to this general subject include Dronfman et al. (1986), Burton and Gordon (1978), Clemens (1985), Sanders et al. (1984), and Solomon et al. (1985). The ensemble is largely confined to an annulus extending over Galactic radii 4 to 7 kpc. In this annulus only about 1% of the volume is filled by the clouds of compressed material. The filling factor for the diffuse HI structures is a subject of current scrutiny, but it is certainly much larger than for the molecular clouds; the range 0.2 to 0.8 is realistic. Molecular clouds display a range of sizes and are often arranged in larger complexes. A substantial complex might be 50 pc or more across and might comprise ten or more recognizable subclouds. The degree of small-scale structure in the subclouds is a current subject of study. The clouds in the molecular annulus move with respect to each other with speeds characterized by a dispersion of about 4 km s^{-1} (Liszt et al. 1984, Clemens 1985); there are indications that the velocity dispersions of clouds situated exterior to the annulus are somewhat larger (Stark 1984, Magnani et al. 1985). Molecular complexes may have H_2 masses of 10^6 $M_\odot$, but a typical mass for an individual cloud is two or three orders of magnitude less. The total molecular mass in the interstellar medium is about 1×10^9 to 3×10^9 $M_\odot$; this is about the same as the total HI mass, but all but a few percent of the molecular material is confined within the solar circle, whereas less than half of the total amount of HI is found there (see Sanders et al. 1984, Bloemen et al. 1986, Dame et al. 1987).

It is natural to ask how the global properties of the dust component of the interstellar medium compare with those of the gas. Several groups are currently working on aspects of this subject. Comparison of the map in Figure 7.14 of the dust emission with the map in Figure 7.3 of total HI emission reveals that details of their distribution are shared by the two tracers.

At latitudes more than, say, 10° from the Galactic equator, where individual structures can be identified, there is a tight, general correlation between the 21-cm HI features and the 100-μm dust ones. These rather filamentary structures, some of which resemble loops or parts of shells, have been studied for some time in the HI line and are the same structures called "cirrus" in the infrared data. The atomic hydrogen and dust seem to be well mixed in these structures (e.g., Burton et al. 1986). Some of the cirrus structures contain isolated concentrations from which weak CO emission has been observed (Blitz et al. 1984, Magnani et al. 1985, Weiland et al. 1986). The HI/dust cirrus structures have not, however, been traced in their entirety in the molecular lines.

At lower latitudes, lines of sight through the inner Galaxy are evidently sampling the collective properties of the dust. Like the HI, the dust is distributed so ubiquitously that separate features cannot easily be isolated; also in this regard the dust emission sampled at 100 μm does not resemble the cold clumps traced by CO.

7.4.2 Thickness and Flatness of the Inner-Galaxy Gas and Dust Layer

The linear thickness of the gas layer can be studied at distances $R < R_0$ by measuring at each longitude the distribution in latitude of intensities observed near the terminal velocity, where there is no distance ambiguity. The deviation of the centroid of the gas distribution from the Galactic plane can similarly be measured at the tangent-point locus. Studies of this sort have shown that although latitude variations in various parameters are present, most of the inner-Galaxy HI and molecular emission structures are confined to a remarkably thin and flat layer. The average full thickness of the HI layer to half-density points is about 220 pc at $R < 9$ kpc (Lockman 1984). This value is an upper limit, because it is obtained from observations which contain at the tangent-point velocity a contribution from an extended line-of-sight region, along which the mean z may fluctuate, thereby increasing the apparent thickness.

Although most of the hydrogen is confined to the central layer, there is a rather structureless component extending at least several hundred parsecs beyond the main concentrations. The distribution of densities in the central layer in the direction perpendicular to the Galactic plane is approximately Gaussian. However, the distribution does deviate from Gaussian in the form of low-intensity wings extending to higher z-distances. Decomposition of individual line profiles into Gaussian components shows some evidence for a background envelope of hydrogen with a larger characteristic velocity dispersion, a correspondingly higher temperature, and a larger thickness in latitude to half-density points (~ 500 pc; Lockman 1984) than exhibited by the more intense structure in the central layer. Emission from the diffuse background shows a much smoother distribution than that shown by the main concentrations of hydrogen near the Galactic equator and evidently fills much of the volume within a kiloparsec or so from the Galactic equator.

The thickness of the molecular annulus, containing the cold, compressed gas, is about half that of the more diffuse HI gas layer (e.g., Sanders et al. 1984). The molecular gas confined to dense clouds is available for star formation; the youngest stars are also confined to a layer substantially thinner than the HI layer (see Blaauw 1965). Consistent with the molecular-cloud ensemble being thinner than the diffuse gas layer are the lower temperatures and the smaller velocity dispersions which pertain both within a single molecular cloud structure and to the members of the ensemble of clouds.

The layer thickness of the widely distributed interstellar dust can be derived by radially unfolding the longitudinal distribution to get characteristic distances and by modeling the latitudinal distribution. The dust observed at 60 and 100 μm has a z-thickness characterized by a Gaussian dispersion of 120 pc. This thickness is equivalent to that of the pervasive HI gas and twice that of the molecular layer.

Over much of the region within $R < R_0$, the deviation of the center of mass of the HI gas layer from the equator defined by $b = 0°$ is less than 30 pc (Gum and Pawsey 1960, Bania and Lockman 1984, Lockman 1984). Such deviations can be measured in detail for the molecular clouds, because their velocity dispersion is low and because the layer is so thin. The deviations, which refer to the tangent-point locus, are typically a few tens of parsecs (see Cohen et al. 1979, Sanders et al. 1984). The dust-layer deviations from the flat equator are of the same amplitude.

Figures 7.16 and 7.17 summarize information on the thickness and flatness of the inner-Galaxy interstellar components, between the bulge region and the solar distance. It is interesting to compare the measures of the thickness and flatness with the overall diameter of the HI gas layer, which is about 50 kpc. Thus, the inner gas layer is flat to within about 0.1%; its thickness is about 1% of its diameter.

The gas layer in the innermost few kiloparsecs of the Galaxy, roughly the extent

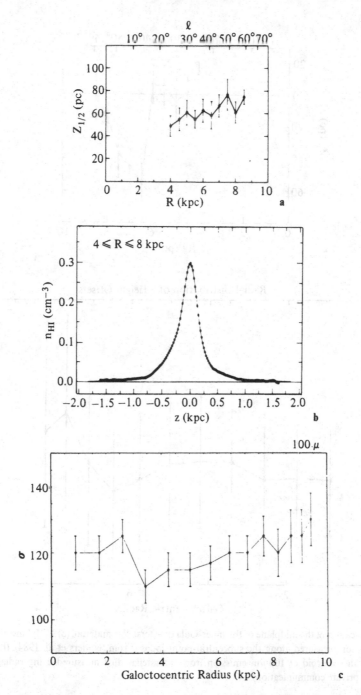

Fig. 7.16. Measures of the thickness of the inner-Galaxy interstellar material. (a) Thickness (half-width at half-maximum) of the layer of CO emission as a function of Galactocentric radius (Sanders et al. 1984). (b) z-Distribution of HI densities characterizing the region $4 < R < 8$ kpc (Lockman 1984); note the low-level extensions at larger $|z|$. (c) Radial distribution of Gaussian half-widths of 100-μm emissivity from dust (E.R. Deul, private communication). The cold, compressed gas has a layer thickness of about 60 pc, approximately half of the thickness of the less cold, more diffusely distributed interstellar material.

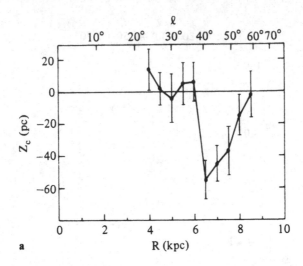

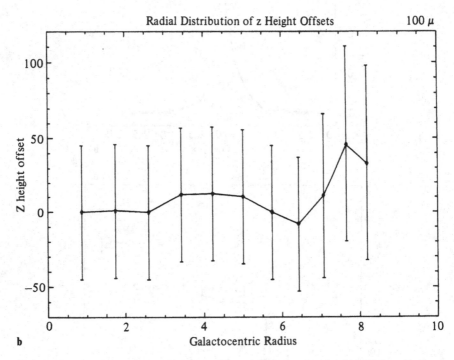

Fig. 7.17. Location of the midplane of the inner-Galaxy interstellar material. (a) Midplane displacement of CO emission measured along the subcentral-point locus (From Sanders et al. 1984). (b) Midplane location of the centroid of 100-μm emission from interstellar dust measured using radial unfolding (E.R. Deul, private communication).

included within the Galactic bulge, is anomalous in terms of both its z-extent and its confinement to $z = 0$ pc. There and especially at $R \lesssim 100$ pc, the III thickness is quite a bit less than the 220 pc measured at large R (see Rougoor and Oort 1960, Brown and Liszt 1984, Liszt and Burton 1980). Most of the gas between a few hundred parsecs from the nucleus and two or three kiloparsecs lies in a disk which is tilted some $20°$ with respect to the plane $b = 0°$. This tilt is shared by the atomic gas (see, e.g., Kerr 1967, Cohen 1975, Cohen and Davies 1976, Burton and Liszt 1978, Sinha 1979, Liszt and Burton 1980) and by the molecular gas (e.g., Lizst and Burton 1978, Sanders et al. 1984). The tilted disk is the fundamental plane of symmetry of the kinematics and distribution of gas in that region. Pervasive non-circular motions are observed here. The dynamical situation is not yet clear. If the motions are along closed, elliptical streamlines, no net flux of gas is involved; otherwise, a net outward flux of about $4\ M_\odot$ per year has to be accounted for.

7.4.3 Other Parameters of the Inner-Galaxy Gas and Dust Layer

It is difficult to measure the temperature of HI emission structures near the Galactic disk because of confusion due to blending of unrelated features in the large volumes of space sampled by the single-dish telescope beams and because of the generally low optical depths. Higher-optical-depth atomic hydrogen coexists with molecular gas in the dense molecular clouds at temperatures of order 10 or 15 K; HI at temperatures of several thousand degrees and higher occurs throughout much of the Galactic volume but contributes little to the mass budget and little to the observed emission profiles. The emission profiles are dominated by gas at temperatures between these extremes. This gas shows a hierarchy of physical properties determined by the mechanical and radiative influences of the local environment.

The temperature of the molecular gas clumped into clouds of high density follows in a straightforward way from the high opacity of the ^{12}CO line. In clouds without special sources of heat, for example from just-formed stars, an excitation temperature of 12 K is representative. There seems to be some variation of cloud properties across the molecular annulus. Solomon et al. (1985) show that the inner annulus contains more effective heat sources than the outer annulus: the radial distribution of cloud cores characterized by kinetic temperatures less than 10 K peaks near $R = 7$ kpc, whereas the distribution of clouds with hotter cores peaks near $R = 5$ kpc. This conclusion is supported by the very narrowly confined radial distribution of diffuse HII emission measured by Lockman (1976) in the H166α recombination line, which traces heat sources.

The temperature of the interstellar dust is an important parameter whose derivation involves a number of problems. Higher observed dust emissivity can indicate higher temperatures, but it can also indicate higher dust densities. The spectral coverage of the four detectors on board IRAS helps distinguish between the relative importance of density and temperature. Being hotter than the diffuse Galactic dust because of its proximity to the Sun, the zodiacal dust is more intense in the 60-μm band than in the 100-μm one, unlike the situation with the diffuse dust. The effective color temperature of the galactic dust may be derived from the ratio of intensities in these two bands. A map of the resulting temperatures (see Burton et al. 1986) shows discrete sources of hot dust, strongly confined to the

Galactic equator and almost always associated with known HII regions; their distribution and number are also quite like those of the subset of molecular clouds which are observed to have embedded heat sources. But most of the emission observed at 100 μm is contributed by dust which is distributed rather diffusely and which can be characterized by a single temperature of about 22 K.

From the general flatness of the central layer inside the solar distance, it is reasonable to expect that systematic motions in the z-direction are smaller than a few kilometers per second. The random gas velocities in the z-direction are expected to increase toward the Galactic center in order to maintain dynamically the constant layer thickness against the total mass density (see Mihalas and Binney 1981). The actual random gas velocities are difficult to measure near the Galactic plane. Furthermore, it is possible to get information on the z-components of the dispersion only for the immediate vicinity of the Sun. The question naturally arises, although it is not satisfactorily answered so far, whether the Θ-, R-, and z-components are the same. Observations of the z-component of the dispersion are better made in external galaxies, viewed face-on. Such observations have been made for the stellar component of a number of face-on galaxies by van der Kruit and Freeman (1986), who found that the vertical velocity dispersion of disk stars increases inward in a manner dynamically consistent with the thickness of the stellar disk which, like the gas layer in the inner Milky Way, is observed to have a constant thickness (van der Kruit and Searle 1982).

The density of the neutral hydrogen gas remains roughly constant over the major part of the Galactic disk, in contrast to the stellar density, which increases strongly toward the center (see, e.g., Gordon and Burton 1976).

Some additional remarks are appropriate on the filamentary nature of the HI gas distribution. The filamentary structure is seen rather directly at latitudes greater than, say, 10°. Although it could well be the dominant sort of structure at lower latitudes too, the large amount of blending along the long low-latitude paths of sight blurs much of the information on the details of the distribution. As mentioned above, the HI filaments generally have counterparts in the cirrus dust features detected in the infrared. Much of the observational work on the HI filaments has been done by Heiles and his collaborators (see, e.g., Heiles 1976, 1978, 1979, 1983, Hu 1981). Many of the filaments form portions of circular arcs. The kinematics of the arcs suggest that the features are parts of shells which are generally expanding. The shells which can be delineated with the most confidence are all at rather high latitudes, and thus are quite nearby. The angular sizes of these shells may be several tens of degrees or more; their linear sizes are typically less than 100 pc, and the expansion motions of order 20 km s^{-1}. An expansion energy of up to about 10^{49} ergs is involved for shells clearly delineated at the higher latitudes. Energies of this magnitude could be produced by stellar winds from massive stars and supernovae. The energies ejected by novae are substantially less.

A class of "supershells" has been identified at lower latitudes (see Heiles 1979). The features identified with supershells occur at radial velocities corresponding to large distances from the Sun. The energies derived for the supershell features are of order 10^{52} ergs; some would require 10^{53} ergs. Such energies are greater even than those produced by supernova explosions. Although there can be no doubt about

Table 7.2. Simplified description of properties of cool components of the interstellar medium.

Property	Cool atomic gas	Cold molecular gas	Diffuse dust
Structure	Hierarchy of structures: filaments, loops, shells	Highly clumped; isolated complexes	Like cooler HI
Density	$n(HI) \sim 0.1-1 \ cm^{-3}$	$n(H_2) \sim 10^3-10^5 \ cm^{-3}$	
Temperature	20–few hundred K (collective mean)	3–30 K	~ 22 K
Filling fraction	0.2–0.8	0.01	>0.2
Layer radial extent ($R_o = 8.5$ kpc)	3–25 kpc; warped at $R > 11$ kpc	3–7 kpc	3–7 kpc & bulge?
Layer z extent	120 pc; thicker at $R > R_o$	60 pc	120 pc

the reality of the higher-latitude shells, the supershell phenomenon at low latitudes requires further observational confirmation. Attempts within our group, using HI data of higher angular resolution, higher velocity resolution, and higher sensitivity, have not confirmed the low-latitude supershell parameters with any confidence. It should be noted, however, that giant filamentary, ionized shells are found in several nearby galaxies, including the Magellanic Clouds, where the viewing perspective is more favorable than in our own; these shells have diameters ranging to more than 1 kpc and clearly involve large energies (see Meaburn 1978, 1983). The class of HI holes identified in M31 (see Brinks and Bajaja 1986) may also be relevant in this regard.

Further detailed investigation of the features identified as holes, as supershells, and, most convincingly, as shells is important because of the information it is likely to reveal on the energy processes globally ordering the interstellar medium (see, e.g., Cox and Smith 1974, McKee and Ostriker 1977, Bruhweiler et al. 1980, Tenorio-Tagle 1980).

Table 7.2 summarizes, in a highly simplified form, some of the physical parameters of the cool components of the interstellar medium.

7.5 The Outer-Galaxy Gas Layer

The early 21-cm surveys revealed that the layer of atomic hydrogen in the outer parts of the Milky Way is systematically warped from the Galactic equator plane $b = 0°$ (Burke 1957, Kerr 1957, Westerhout 1957, Oort et al. 1958). Since then, it has become clear that large-scale, systematic deviations from a flat disk are a common characteristic of the HI morphology in the outer regions of spiral galaxies. The nearest large spirals, M31 and M33, are both warped: the hydrogen layer in M31 resembles the one in the Milky Way (Brinks and Burton 1984); the warp in M33 is more severe (Rogstad et al. 1974, 1976). Sancisi (1976, 1983) found an integral-sign shape to characterize the projected HI surface density of many edge-on spirals, including NGC 5907, NGC 4244, and NGC 4565. Recognition of the integral-sign warp signature requires a very favorable viewing geometry, nearly edge-on, with the line of nodes of the warp oriented more or less along the line of sight. Bosma (1981a,b) showed on the basis of twists in velocity of their line of nodes that many other, not necessarily edge-on, galaxies are warped, including NGC 2841, NGC 5055, and NGC 7331, in their outer regions.

7.5.1 The Shape of the Warped, Flaring, Outer-Galaxy Gas Layer

The accumulated evidence that systematic deviations from flatness are a common aspect of spiral-galaxy morphology provides part of the motivation for investigation of the shape of the gas layer in the outer Milky Way. Additional motivation is provided by the realization that the unobserved dark-matter component dominating the Galactic potential can be traced most effectively by study of the motions and form of the outer-Galaxy HI layer. Although the perspective from our position embedded within the Galaxy complicates many studies, it is an ideal perspective from which to study the bent gas layer. The Milky Way can serve as a prototype of a warped galactic system.

An important obstacle to quantitative improvement of the description made in the late 1950s of the HI gas layer in the Galactic outskirts was the lack of adequate observational material covering the part of the Milky Way only visible from the southern hemisphere. Kerr et al. (1986; see Table 7.1) used the Parkes 18-m telescope to effectively extend the coverage provided by the Weaver and Williams (1973) survey in the range $-10° < b < 10°$ in the north to include all Galactic longitudes. Henderson et al. (1982) used the combined $|b| < 10°$ material to produce the first global description of any detail of the warped HI layer.

A second development important in this regard is the increased confidence in the outer-Galaxy distance scale. The earlier investigations made use of a steeply falling rotation curve, whereas it has since become clear that our Galaxy, like essentially all spirals, has a rotation curve which does not fall in a Keplerian fashion, even for the largest radii for which the curve can be measured. Henderson et al. (1982) used a velocity-to-distance transformation appropriate to a flat rotation curve. Kulkarni et al. (1982) reanalyzed the Weaver and Williams material using the slightly rising rotation curve found by Blitz et al. (1980) from CO velocities of emission nebulae with optically determined distances. The insights into the scale and form of the warped gas layer resulting from these studies are reviewed by Blitz et al. (1983).

An additional lingering observational constraint was that the signature of the Galactic warp extends beyond the 10° latitude limit of the surveys used in the analyses mentioned above. The Leiden-Green Bank HI survey (Burton 1985, Burton and te Lintel Hekkert 1985) was carried out with the NRAO 140-ft telescope in order to broaden the latitude coverage, and deepen the sensitivity, of the earlier surveys. The data were combined with existing material [most importantly the surveys of Weaver and Williams (1973) and Kerr et al. (1986)] in order to give the highest possible density of coverage of the outer Galactic disk.

The composite material contains HI emission intensities in l, b, v coordinates. In the outer Galaxy, radial velocities correspond to galatocentric distances in a single-valued manner. (Primarily because of this one-to-one correspondence, the parts of the profiles contributed by the outer regions of the Galaxy are somewhat simpler in appearance than the parts contributed by the interior regions; this apparent difference, evident in Figure 7.1, does not by itself imply that the *physical* structure is simpler in the outer Galaxy than in the inner, although in fact it probably

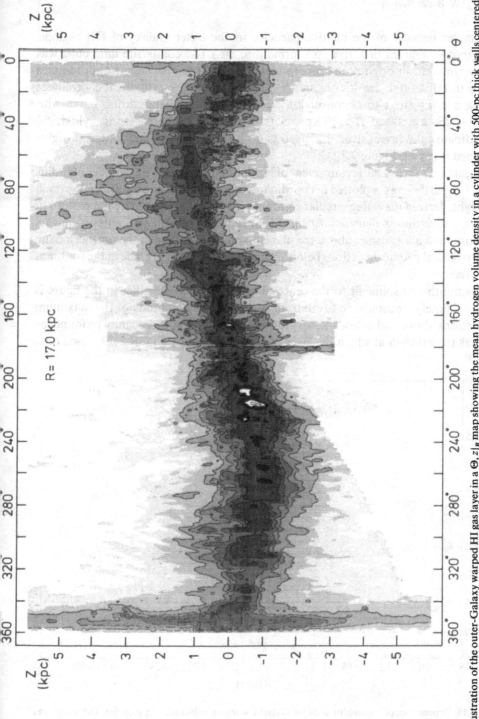

Fig. 7.18. Illustration of the outer-Galaxy warped HI gas layer in a Θ, $z|_R$ map showing the mean hydrogen volume density in a cylinder with 500-pc thick walls centered at $R = 17$ kpc. At this distance, the azimuthal variation of the warp is approximately sinusoidal, although the amplitude of the warp is somewhat larger at $\Theta < 180°$ than at $\Theta > 180°$. At larger distances, this asymmetry is more pronounced (From Burton and te Lintel Hekkert 1986). The anomalies near $\Theta = 0°$, $180°$, and $360°$ stem from the weakness of the v-to-R transformation in these directions, and may be ignored.

is simpler because of the relative pacuuity in the outer Galaxy of HII regions, supernovae, regions of active star formation, etc.) The composite data cube was transformed from brightness temperatures in an array of heliocentric coordinates l, b, v to HI volume densities in an array of galactocentric cylindrical coordinates R, Θ, z using the v-to-R conversion consistent with a flat rotation curve with $\Theta(R) = 250$ km s^{-1} at $R > 10$ kpc. Cuts through this galactocentric, cylindrical coordinate cube reveal structural aspects of the outer Galaxy in a rather straightforward way [see Figure 7.8(f)].

Figure 7.18 shows the arrangement of HI volume densities on a cylinder of radius $R = 17$ kpc; the map is plotted in coordinates of galactocentric azimuth and vertical z-height, derived assuming circular rotation. The warp is revealed by the pattern in the Θ, z variation of densities. At this distance from the Galactic center, z-heights of more than a kiloparsec above the plane $b = 0°$ are reached in the northern data; approximately equal deviations below the galactic equator are made in the southern material.

A smooth-line spline-fit to the centroid of the band of densities in the figure is approximately sinusoidal in form; from such a fit the amplitude of the maximum excursions above and below the equator occurring at $R = 17$ kpc may be found, as well as the azimuth at which the layer crosses the equator. Figure 7.19, based on a

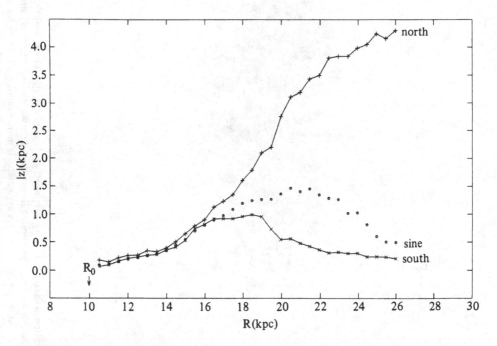

Fig. 7.19. Dependence of the amplitude of the Galactic warp on galactocentric distance. The pulses and crosses represent the absolute values of the extreme positive and negative values respectively, of the z-height found in spline-curves fit to $\Theta, z|_R$ maps of the sort shown in Figure 7.18. The amplitude of a simple sine-curve of period 2π is shown by the open circles. The gas layer becomes significantly warped at $R \sim 13$ kpc. The amplitude of the warp grows approximately linearly, and equally in the two Galactic hemispheres, until $R \sim 16$ kpc; at larger R, the warp amplitude increases strongly in the northern hemisphere but decreases in the southern hemisphere.

similar analysis carried out at other R-values, shows how the amplitude of the warped HI layer varies with distance from the Galactic center. The details of the dependence of layer amplitude on R depend on the particular rotation curve adopted, but the global characteristics of that dependence are well established. The layer remains quite flat until about 12 or 13 kpc from the Galactic center. The amplitude of the warp grows linearly, and approximately equally in the two galactic hemispheres, until about $R = 16$ kpc. At larger radii, the amplitude of the warp in the northern material continues to increase, until the sensitivity limits of the data are reached at $R \sim 25$ kpc, where the emission centroid lies some 4 kpc above the plane $b = 0°$. The behavior of the warp at large distances in the southern-hemisphere material is different: after reaching a maximum excursion of about 1 kpc below the equator at $R \sim 18$ kpc, the gas layer folds back again towards the equator at the largest distances.

The shape of the warped gas layer may be further specified by measuring the azimuthal angles, at various distances, at which the excursions of the emission centroids from the equator are greatest and the angles at which the gas layer crosses the equator. This information is plotted in Figure 7.20. The azimuthal location of the plane crossings defines the line of nodes of the Galactic warp; the regularity of the warp is reflected by the location of the maxima of the excursions at about 90°

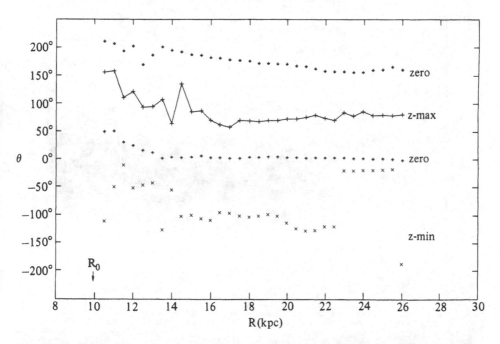

Fig. 7.20. Radial dependence on azimuth of the positions in the Galaxy where the HI gas layer reaches its maximum height above (+ symbols) the plane $z = 0$ kpc and its maximum distance below (× symbols) that plane, as well as the positions (↑ and ↓ symbols) where the gas layer crosses the plane $z = 0$ kpc. The zero crossings, separated by approximately 180°, give the position angle of the line of nodes of the Galactic warp. The azimuth of the line of nodes varies remarkably little with R: evidently, the line of nodes does not precess under the influence of differential Galactic rotation.

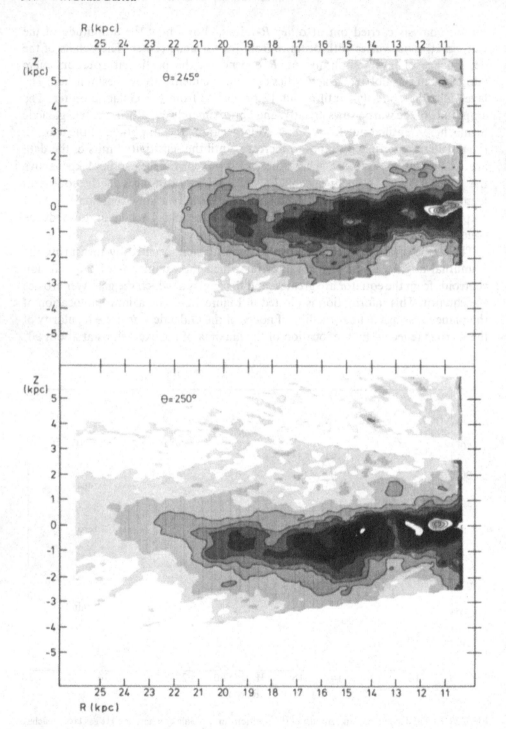

Fig. 7.21. Illustration of the outer-Galaxy warped HI gas layer in a $R, z|_\Theta$ map showing the arrangement of hydrogen volume densities in a sheet through the Galactic center perpendicular to the equator at the indicated Galactocentric azimuths. The approximate 2π-symmetry of the Galactic warp makes the

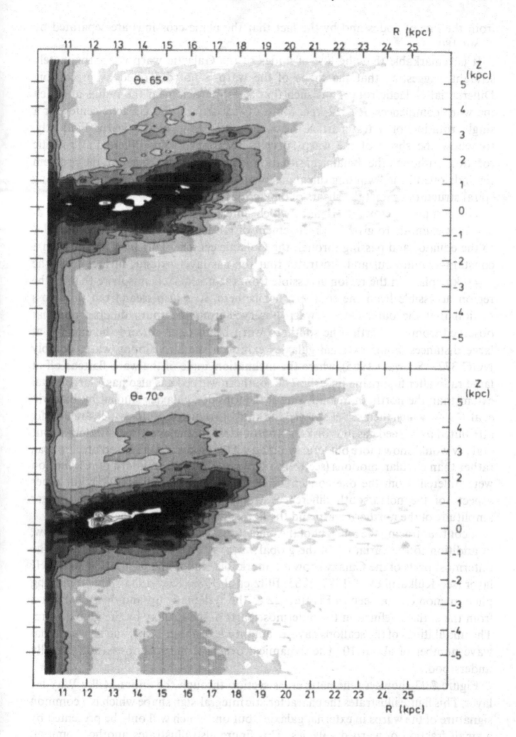

pairing of maps separated by 180° in azimuth appropriate (From Burton and te Lintel Hekkert 1986). The finite velocity dispersion of the gas causes the band near $R < 11$ kpc, which, in the context of the warp, may be ignored.

from the line of nodes and by the fact that the plane crossings are separated by about 180°.

It is remarkable that the line of nodes of the Galactic warp is approximately straight, suggesting that the shape of the warp is not deformed by precession. Differential Galactic rotation is such that a test particle near the radius at which the warp commences, $R \sim 12$ kpc, rotates twice around the Galactic center for a single rotation of a test particle near the outer edge of the warp. Evidently, somehow the shape of the warp survives shearing due to differential galactic rotation. Although the dynamical situation is no doubt very different, the problem reminds one of the "winding dilemma," which poses the question of maintaining spiral structure against the shears of differential rotation.

The composite cube of HI data which enters Figure 7.18 can also be cut at constant azimuth, to give the arrangement of HI densities in sheets perpendicular to the equator and passing through the Galactic center. Figure 7.21 shows such a constant-azimuth cut and illustrates that the gas layer extends further from the $z = 0$ kpc plane in the region accessible from the northern hemisphere than in the region accessible from the southern hemisphere. It is also clear from this cross section that the outer-Galaxy warp observed from the south differs from that observed from the north. The southern warp bends back toward the equator at large distances. Some external galaxies exhibit a similarly floppy warp, notably NGC 3729 (Schwarz 1985), where the inclination angle of the warp flattens off at large radii after first rising linearly. The southern warped gas also has a larger total extent than the northern, reminiscent of the lopsided galaxies studied by Baldwin et al. (1980), which themselves showed the kind of HI distributions which are usually attributed to warped gaseous disks. A contour tracing the outer radial extent of our Galaxy would show more bilateral symmetry if the analysis were based on elliptical, rather than circular, motions (see Kerr 1983) or if the local standard of rest motion were different from the one conventionally adopted (see Shuter 1981); but other aspects of the north/south differences would remain, in particular, the larger amplitude of the northern warp and the floppy aspect of the southern one.

A contour tracing the z-height of the Galactic gas layer at large radii would show, in addition to the variation of the global warp, structure on a smaller scale. The outermost parts of the Galaxy show a remarkable scalloping, or pleating, of the HI layer (see Kulkarni et al. 1982, 1983, Blitz et al. 1983, Kerr 1983). The scalloping phenomenon can be seen in Figure 7.22 as the systematic up-and-down variation from the surface defined in the outermost parts of the Galaxy by the global warp. The undulations of the scallops have an amplitude of about 1 kpc, and an azimuthal wave number of about 10. The dynamical origin of the scallops is not yet fully understood.

Figure 7.23 shows a schematic cross section through the entire Milky Way HI layer. This figure illustrates the characteristic integral-sign shape which is a common signature of the warps in external galaxies, but one which will only be presented by a small fraction of warped galaxies. This figure also illustrates another common aspect of warped outer galaxies: the gas layer flares to larger thickness at larger radii. The scale height in the warped layer at $R \sim 25$ kpc is about an order of magnitude greater than it is near R_o. This circumstance provides an important

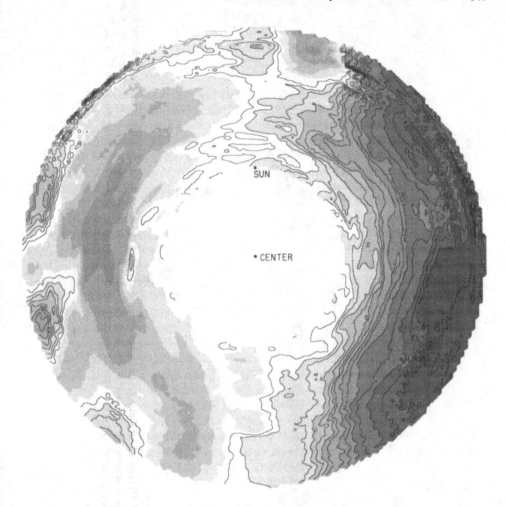

Fig. 7.22. Plan view showing the z-heights of the maximum density of the HI gas layer as determined from smooth lines spline-fit to the centroid of HI volume densities in cuts of the composite data cube at constant R (see Figure 7.18). The grey-scale shading gives the z-heights; contours enclose regions of $z > = 0$ kpc. The diagram illustrates the scalloping, or pleating, in the outermost regions. This diagram was prepared by K.K. Kwee.

constraint on the vertical distribution of dark matter (see van Albada and Sancisi 1986, Kulkarni et al. 1982), when used with the observations of external, face-on galaxies which show that the vertical velocity dispersion varies little with R (e.g., van der Kruit and Shostak 1984). Although the detailed conclusions are not yet firm, it does seem indicated from the flaring gaseous disks that the large amount of dark matter implied by the flattish rotation curves cannot be confined to a thin disk.

7.5.2 Some Characteristics of the Warped Gas Layers in Other Galaxies

The hydrogen layer in external systems commonly starts to bend at radii where the stellar luminosity is terminating. The conclusion that warps generally occur in the regime dominated by dark matter requires, in addition to the nonfalling rotation

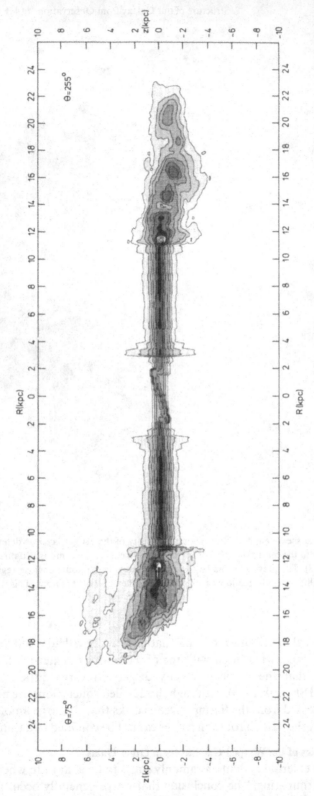

Fig. 7.23. Schematic, but not arbitrary, representation of the HI layer of our Galaxy. The cross section represents a cut through the Galactic center perpendicular to $b = 0°$ in a sheet sampling strong deviations from a flat layer. The representation at $R > 11$ kpc is from R, z maps, like the ones in Figure 7.21, at $\Theta = 90°$ and $270°$. Over the range $2 < R < 10$ kpc, the parameters of the layer are those given by Lockman (1984) for the subcentral-point region; at $R < 2$ kpc, they are those given by Liszt and Burton (1980).

curve, the observation that the warp exists beyond the optical disk (see Kalnajs 1983b). Although accurate photometry of the Milky Way disk is difficult (but see van der Kruit 1986), it does seem to be the case that the Galactic warp begins near the edge of the stellar distribution. The optical tracers which are found rather sparsely distributed at large radii in the Milky Way evidently do participate in the warp (Brand 1986).

Although many other galaxies have been recognized to be warped, the details of the shape of the warp are available only for a few nearby galaxies. It is interesting to compare the situation pertaining in the outer Milky Way with that pertaining in some of the other warped systems. NGC 3718 (Schwarz 1985) and M33 (Rogstad et al. 1974, 1976) are both examples of gaseous disks more severely warped than the one in the Milky Way, with tilts of the layer to almost 90°. But as in the Milky Way, these warps are global phenomena and show no evidence for precession of the line of nodes of the bent layer.

The nearest large spiral, M31, exhibits a warp in its outer regions which is quite similar to the Galaxy's (Henderson 1979, Brinks and Burton 1984). The morphology of the M31 HI distribution and kinematics can be modeled by a thin, flat disk extending over the region at $R < 16$ kpc and a flaring, warped gas distribution at larger radii. The thickness of the HI gas in the warped part of M31 increases approximately linearly with radius to a maximum of about 1.7 kpc. The maximum deviation of the warp from the plane of the extended inner disk is about 3.8 kpc. Figures 7.24 and 7.25 show in a schematic way how similar the warp of the M31 gaseous disk is to the one in the Milky Way.

If the warped outer-Galaxy gas disk does not represent a persistent, relaxed dynamical state, it is natural to ask how the material reached such large distances into the halo and, once there, how it is maintained, without evaporating into the halo, at the same modest velocity dispersions as characterize gas features confined to low z-distances. The inference is that some replenishment or maintenance mechanism is necessary.

Burke (1957) and Kerr (1957) considered that the bending might be a tidal distortion by the Magellanic Clouds but doubted that the Clouds could be massive enough to account for the observed warping. Hunter and Toomre (1969) also favored a tidal distortion, but one which arose from a single close transit of the Large Cloud. Hunter and Toomre's model requires a passage of the Large Cloud at a minimum distance of about 20 kpc from the center of the galaxy. Alternative interpretations of the observed warping have been given by Kahn and Woltjer (1959), who suggested that the distortion might be due to a flow of intergalactic gas past the Galaxy, and by Lynden-Bell (1965), who considered a free oscillation mode of the spinning Galactic disk.

An important general characteristic which the Milky Way shares with other warped galaxies is that the warp is a global phenomenon, involving not just a single quadrant, but a full 2π of azimuth. This symmetry, and the general regularity of the warps, seems to argue against the warps being the consequence of tidal interactions or mergers with companion galaxies. Detailed modeling of tidal interactions of M32 with M31 by Byrd (1978) showed that the disturbance to M31 would be confined to a small portion of that galaxy, and would not be global in nature. Furthermore,

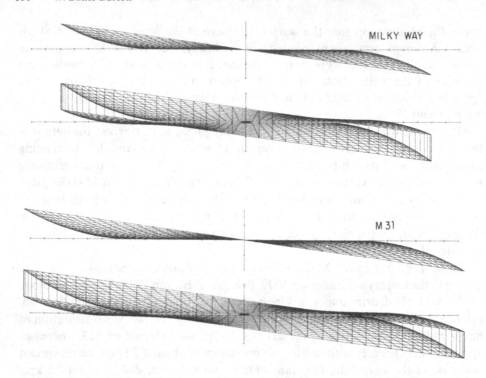

Fig. 7.24. *Upper pair*: Simple geometrical model of the warped Galaxy, showing an external view looking along the line of nodes. The uppermost panel shows the appearance of the midplane of the warped gas layer; the second panel shows the surfaces at one scale height above, and one below, the midplane. *Lower pair*: Representation as in the upper pairs of the warped, flaring gas layer in M31, using the model parameters determined by Brinks and Burton (1984). Here M31 is viewed from the same distance as in the representation of our Galaxy, and under an orientation with the line of nodes pointing toward the viewer.

some warped galaxies, noticeably NGC 5301 (Krumm and Shane, 1982), are quite isolated.

Because warps occur so commonly in spiral galaxies, including isolated ones, it seems likely that they are *intrinsic* to galaxies. Merritt and de Zeeuw (1983) and Sparke (1986) considered out-of-plane families of gas orbits which are stable in the presence of a thick triaxial gravitational potential. Mark (1983) and Bertin and Mark (1980) have developed a theory of bending waves that provides for the self-generated warps in spiral galaxies; their work also requires a substantial amount of mass in a spherical distribution. Higher-order modes might be relevant to the observed scalloping.

7.5.3 Brief Remarks on the Phenomenon of High-Velocity Clouds

High-velocity clouds (HVCs) are fragments of hydrogen emission occurring at radial velocities which clearly do not fall within the purview of circular differential Galactic rotation. These fragments were first detected in the Dwingeloo surveys (see Muller et al. 1963) and have subsequently been found to cover about 10% of the surface area of the sky. Because the anomalous velocities rule out derivation of

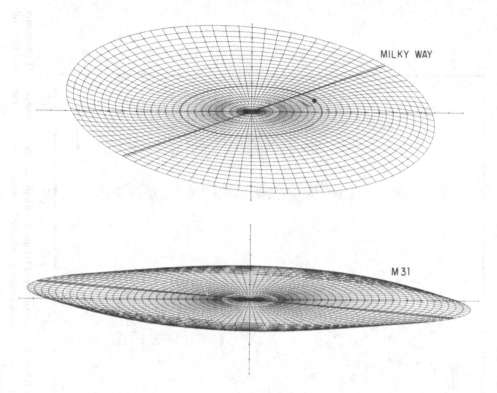

Fig. 7.25. *Top*: Midplane of the simple model of the warped Galaxy with the $z = 0$ plane inclined 20° from the line of sight and viewed from a position at an azimuth of 120°. This orientation gives the view of our Galaxy seen from the Andromenda galaxy. Heavier lines indicate $R = 10$ kpc and the line of nodes of the warp along $\theta = 165°, 345°$. The dot gives the position of the Sun. *Bottom*: Midplane of the M31 gas layer drawn in accordance with the model determined from observations. The $z = 0$ plane is inclined 13°, and the line of nodes of the warp is twisted from the line of sight through an azimuth of 20°. These representations suggest that even rather severely warped Galactic gas layers might be difficult to recognize as such from an external perspective.

kinematic distances based on the circular-motion assumption and because no other distance measure is known of which one can be confident, the nature of the HVCs remains problematic. Recent reviews of the phenomenon are given by van Woerden et al. (1985), Giovanelli (1980), Oort and Hulsbosch (1978), and Verschuur (1975).

The distribution of HVCs on the portion of the sky accessible to the Dwingeloo telescope is shown in Figure 7.26 (Hulsbosch 1985). The emission fragments are labeled according to their radial velocities with respect to the local standard of rest. It has been common to define as intermediate-velocity clouds those fragments with velocities differing by up to 80 km s^{-1} from velocities consistent with circular differential Galactic rotation, and as high-velocity clouds those with greater speeds. This criterion has proved useful but it is somewhat arbitrary and difficult to apply, especially in regions where the warped gas layer is detectable quite far from the Galactic equator. There can be little doubt, however, that most of the features recorded in Figure 7.26 are truly anomalous. The observed velocities differ not only by more than 100 km s^{-1} from those permitted by circular rotation, but they are

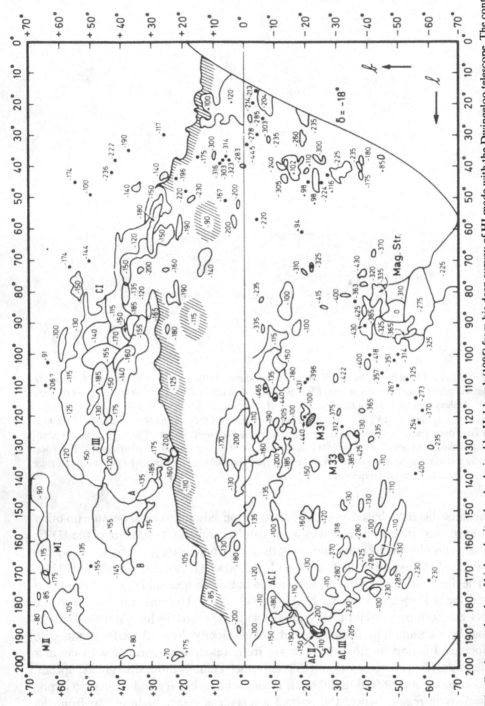

Fig. 7.26. Distribution on the sky of high-velocity clouds, derived by Hulsbosch (1985) from his deep survey of HI made with the Dwingeloo telescope. The contours enclose regions of emission brightness temperature ≥ 0.05 K. The characteristic radial velocity of each region is indicated, in units of km s^{-1} with respect to the local standard of rest. The principal complexes and streams of high-velocity clouds are labeled.

also more than an order of magnitude greater than either the random velocities or the systematic streaming motions observed in the gas layer.

The sky distribution of HVCs shows certain kinematic and spatial patterns. Negative radial velocities predominate in the region $l < 210°$; positive velocities predominate in the higher-longitude region. There are many more HVCs in the first and second Galactic quadrants than in the third and fourth. Thus, there are in general many more negative-velocity HVCs than positive-velocity ones. Although the kinematics of the ensemble of HVCs do show a component resembling the effect of Galactic rotation (see Giovanelli 1980, Verschuur 1975), the negative-velocity features prevalent in the directions of both Galactic poles as well as in the direction of the anticenter show that there is in addition a strong component of motion toward the Galactic disk.

The HVCs often, but not always, occur in elongated, patchy, filaments showing some continuity in velocity as well as in position. The most obvious example of such continuity is given by the Magellanic Stream (see Mathewson et al. 1979), which extends from the direction of the Clouds through the Galactic south pole and beyond. Other ribbons of anomalous-velocity HI gas extending over large arcs on the sky are also indicated in Figure 7.26. Despite the general continuity in position and velocity in these elongated features, they all show a great deal of small-scale structure, both as density concentrations and as jumps in velocity (see Schwarz and Oort 1981).

High-velocity clouds have only been observed in HI emission from the 21-cm line. Attempts to measure HI in absorption against distant extragalactic radio sources have not given positive results. Substantial efforts have been directed toward detection of a molecular component in the clouds, but despite long integrations in the CO lines by a number of observers, the efforts have failed to reveal molecular material. Searches in the IRAS data for emission from dust in the HVCs have also been negative. Particularly important are the ongoing attempts to measure optical or ultraviolet absorption lines against early-type stars (see de Boer 1985), because a detection would not only give the first proof that HVCs contain interstellar material processed to elements heavier than hydrogen, but it would also give at least an upper limit to the distance to the cloud. There has not yet been a definite detection of a bona-fide HVC in absorption against stars.

Interpretation of the origin and fate of the HVC phenomenon has been particularly stymied by lack of definite information on their distances. Without this information, substantial doubt remains regarding such crucial physical parameters as mass, size, density, and flux of the infall.

The reviews referenced above summarize various interpretations of the HVC phenomenon; it is not ruled out that more than one interpretation will be required to account for the variety of properties of the anomalous-velocity hydrogen. We mention here two different sorts of models. The kinematic and positional coincidence of the Magellanic Stream with the Clouds supports the tidal models which envisage the Stream as debris pulled out of the Clouds by the tidal action of the Milky Way and its massive halo (see Mathewson et al. 1979, Murai and Fujimoto 1980, Lin and Lynden-Bell 1982). The galactic fountain model proposed by Bregman (1980) accounts for some of the properties of other HVC features.

According to Bregman's model, gas in the Galactic disk is heated by supernovae, causing it to bubble up into the halo. The gas radiatively cools in the halo and condenses into neutral clouds which then fall back into the Galactic disk. Shearing by Galactic rotation in the lower halo can account for the elongated nature of the infalling features. The fountain model is successful in accounting for the predominance of negative-velocity HVCs at high latitudes; the model does not, however, address the substantial asymmetries in the overall distribution of HVCs.

Acknowledgments. I am indebted to several Leiden colleagues for providing un-published material: E.R. Deul's work on the morphology of Galactic dust is represented in Figures 7.15 and 7.17, and he generated the data displays shown in Figures 7.3 and 7.14; Vincent Icke designed the schematic diagram shown in Figure 7.8 illustrating the vagaries of the velocity-to-distance transformation; K.K. Kwee determined the Galactic-warp parameters shown in Figures 7.19, 7.20, and 7.22. Leo Blitz, of the University of Maryland, kindly commented on a draft of the manuscript.

Recommended Reading

Berendzen, R., R. Hart, and D. Seeley. 1976. Man Discovers the Galaxies. New York: Science History Publications.
Mihalas, D., and J. Binney. 1981. Galactic Astronomy: Structure and Kinematics. San Francisco: W.H. Freeman.
Athanassoula, E. (ed.). 1983. Internal Kinematics and Dynamics of Galaxies. Dordrecht: Reidel.
Shuter, W.L.H. (ed.). 1983. Kinematics, Dynamics and Structure of the Milky Way. Dordrecht: Reidel.
van Woerden, H., R.J. Allen, and W.B. Burton (eds.). 1985. The Milky Way Galaxy. Dordrecht: Reidel.

References

Bajaja, E. 1983. *In* W.B. Burton and F.P. Israel (eds.), Surveys of the Southern Galaxy. Dordrecht: Reidel, p. 49.
Baldwin, J.E., D. Lynden-Bell, and R. Sancisi. 1980. Mon. Not. R. Astron. Soc. **193**:313.
Bania, T.M. 1980. Astrophys. J. **242**:95.
Bania, T.M., and F.J. Lockman. 1984. Astrophys. J. **54**:513.
Bertin, G., and J.W-K. Mark. 1980. Astron. Astrophys. **88**:289.
Blaauw, A. 1965. *In* A. Blaauw and M. Schmidt (eds.), Stars and Stellar Systems, Vol. 5, Galactic Structure. Chicago: University of Chicago Press, p. 167.
Blitz, L. 1979. Astrophys. J. **231**:L115.
Blitz, L. 1983. *In* W.B. Burton and F.P. Israel (eds.), Surveys of the Southern Galaxy. Dordrecht: Reidel, p. 117.
Blitz, L., M. Fich, and A.A. Stark. 1980. *In* B.H. Andrew (ed.), Interstellar Molecules. Dordrecht: Reidel, p. 213.
Blitz, L., M. Fich, and A.A. Stark. 1982. Astrophys. J. Suppl. **49**:183.
Blitz, L., M. Fich, and S. Kulkarni. 1983. Science **228**:1233.
Blitz, L., L. Magnani, and L. Mundy. 1984. Astrophys. J. **282**:L9.
Bloemen, J.B.G.M., A.W. Strong, L. Blitz, R.S. Cohen, T.M. Dame, D.A. Grabelsky, W. Hermsen, F. Lebrun, H.A. Mayer-Hasselwander, and P. Thaddeus. 1986. Astron. Astrophys. **154**:25.
Bosma, A. 1981a. Astron. J. **86**:1791.
Bosma, A. 1981b. Astron. J. **86**:1825.
Brand, J. 1986. Dissertation, University of Leiden.
Brand, J., and L. Blitz. 1987. In press.
Bregman, J.N. 1980. Astrophys. J. **236**:577.
Brinks, E., and E. Bajaja. 1986. Astron. Astrophys. **169**:14.

Brinks, E., and W.B. Burton. 1984. Astron. Astrophys. **141**:195.
Bronfman, L., R.S. Cohen, H. Alvarez, J. May, and P. Thaddeus. 1987. Astrophys. J. (in press).
Brown, R.L., and H.S. Liszt. 1984. Annu. Rev. Astron. Astrophys. **22**:223.
Bruhweiler, F.C., T.R. Gull, M. Kafatos, and S. Sofia. 1980. Astrophys. J. **238**:L27.
Burke, B.F. 1957. Astron. J. **62**:90.
Burton, W.B. 1971. Astron. Astrophys. **10**:76.
Burton, W.B. 1974. In G.L. Verschuur and K.I. Kellermann (eds.), Galactic and Extragalactic Radio Astronomy. New York: Springer-Verlag, p. 82.
Burton, W.B. 1976. Annu. Rev. Astron. Astrophys. **14**:275.
Burton, W.B. 1985. Astron. Astrophys. Suppl. **62**:365.
Burton, W.B., and T.M. Bania. 1974. Astron. Astrophys. **34**:75.
Burton, W.B., and M.A. Gordon. 1978. Astron. Astrophys. **63**:7.
Burton, W.B., and H.S. Liszt. 1978. Astrophys. J. **225**:815.
Burton, W.B., and H.S. Liszt. 1983. Astron. Astrophys. Suppl. **2**:63.
Burton, W.B., and P. te Lintel Hekkert. 1985. Astron. Astrophys. Suppl. **62**:645.
Burton, W.B., and P. te Lintel Hekkert. 1986. Astron. Astrophys. Suppl. **65**:427.
Burton, W.B., E.R. Deul, H.J. Walker, and A.A.W. Jongeneelen. 1986. In F.P. Israel (ed.), Light on Dark Matter. Dordrecht: Reidel p. 357.
Byrd, G.G. 1978. Astrophys. J. **226**:70.
Cleary, M., C. Heiles, and G. Haslam. 1979. Astron. Astrophys. Suppl. **36**:95.
Clemens, D.P. 1985. Astrophys. J. **295**:422.
Cohen, R.J. 1975. Mon. Not. R. Astron. Soc. **171**:659.
Cohen, R.J., and R.D. Davies. 1976. Mon. Not. R. Astron. Soc. **175**:1.
Cohen, R.S., G.R. Tomasevich, and P. Thaddeus. 1979. In W.B. Burton (ed.), The Large-Scale Characteristics of the Galaxy. Dordrecht: Reidel, p. 53.
Cohen, R.S., T.M. Dame, and P. Thaddeus. 1986. Astrophys. J. Suppl. **60**:695.
Colomb, F.R., W.G.L. Poppel, and C. Heiles. 1977. Astron. Astrophys. Suppl. **29**:89.
Colomb, F.R., W.G.L. Poppel, and C. Heiles. 1980. Astron. Astrophys. Suppl. **40**:47.
Cox, D.P., and B.W. Smith. 1974. Astrophys. J. **189**:L105.
Dame, T.M., and P. Thaddeus. 1985. Astrophys. J. **297**:751.
Dame, T., H. Ungerechts, R.S. Cohen, E. de Geus, I. Grenier, J. May, D.C. Murphy, L.-A. Nyman, and P. Thaddeus. 1987. Astrophys. J. **322**.
de Boer, K.S. 1985. In H. van Woerden, R.J. Allen, and W.B. Burton (eds.), The Milky Way Galaxy. Dordrecht: Reidel, p. 415.
Delhaye, J. 1965. In A. Blaauw and M. Schmidt (eds.), Stars and Stellar Systems, Vol. 5, Galactic Structure. Chicago: University of Chicago Press, p. 61.
Dickey, J.M., S.R. Kulkarni, J.H. van Gorkom, and C.E. Heiles. 1983. Astrophys. J. Suppl. **53**:591.
Giovanelli, R. 1980. Astron. J. **85**:1155.
Giovanelli, R., G.L. Verschuur, and T.R. Cram. 1973. Astron. Astrophys. Suppl. **12**:209.
Gordon, M.A., and W.B. Burton. 1976. Astrophys. J. **208**:346.
Gum, C.S., and J.L. Pawsey. 1960. Mon. Not. R. Astron. Soc. **21**:150.
Heiles, C. 1976. Astrophys. J. **208**:L137.
Heiles, C. 1978. Scientific American, January, p. 74.
Heiles, C. 1979. Astrophys. J. **229**:533.
Heiles, C. 1983. In W.B. Burton and F.P. Israel (eds.), Surveys of the Southern Galaxy. Dordrecht: Reidel, p. 195.
Heiles, C., and H.J. Habing. 1974. Astron. Astrophys. Suppl. **14**:1.
Henderson, A.P. 1979, Astron. Astrophys. **75**:311.
Henderson, A.P., P.D. Jackson, and F.J. Kerr. 1982. Astrophys. J. **263**:116.
Hu, E. 1981. Astrophys. J. **248**:119.
Hulsbosch, A.N.M. 1985. In H. van Woerden, R.J. Allen, and W.B. Burton (eds.), The Milky Way Galaxy. Dordrecht: Reidel, p. 409.
Hunter, C., and A. Toomre. 1969. Astrophys. J. **155**:747.
Israel, F.P. 1985. In J.-L. Nieto (ed.), New Aspects of Galaxy Photometry. p. 101. Berlin: Springer-Verlag.
Jackson, P.D., M.P. FitzGerald, and A.F.J. Moffat. 1979. In W.B. Burton (ed.), The Large-Scale Characteristics of the Galaxy. Dordrecht: Reidel, p. 221.

Kahn, F.D., and L. Woltjer. 1959. Astrophys. J. **130**:705.

Kalnajs, A.J. 1983a. *In* E. Athanassoula (ed.), Internal Kinematics and Dynamics of Galaxies. Dordrecht: Reidel, p. 109.

Kalnajs, A.J. 1983b. *In* E. Athanassoula (ed.), Internal Kinematics and Dynamics of Galaxies. Dordrecht: Reidel, p. 87.

Kerr, F.J. 1957. Astron. J. **62**:93.

Kerr, F.J. 1962. Mon. Not. R. Astron. Soc. **123**:327.

Kerr, F.J. 1967. *In* H. van Woerden (ed.), Radio Astronomy and the Galactic System. London: Academic Press, p. 239.

Kerr, F.J. 1968. *In* B.M. Middlehurst and L.H. Aller (eds.), Stars and Stellar Systems, Vol. VII, Nebulae and Interstellar Matter. Chicago: University of Chicago Press, p. 575.

Kerr, F.J. 1983. *In* W.B. Burton and F.P. Israel (eds.), Surveys of the Southern Galaxy. Dordrecht: Reidel, p. 113.

Kerr, F.J., and J.V. Hindman. 1970. Aust. J. Phys. Astrophys. Suppl. **18**:1.

Kerr, F.J., and D. Lynden-Bell. 1986a. Mon. Not. R. Astron. Soc. **221**:1023.

Kerr, F.J., and D. Lynden-Bell. 1986b. *In* J.P. Swings (ed.), Highlights of Astronomy. Dordrecht: Reidel, p. 889.

Kerr, F.J., and G. Westerhout. 1965. *In* A. Blaauw and M. Schmidt (eds.), Stars and Stellar Systems, Vol. 5, Galactic Structure. Chicago: University of Chicago Press, p. 167.

Kerr, F.J., R.H. Harten, and D.L. Ball. 1976. Astron. Astrophys. Suppl. **25**:391.

Kerr, F.J., P.F. Bowers, and P.D. Henderson. 1981. Astron. Astrophys. Suppl. **44**:63.

Kerr, F.J., P.F. Bowers, P.D. Jackson, and M. Kerr. 1986. Astron. Astrophys. Suppl. **66**:373.

Knapp, G.R., S.D. Tremaine, and J.E. Gunn. 1978. Astron. J. **83**:1585.

Krumm, N., and W.W. Shane 1982. Astron. Astrophys. **116**:237.

Kulkarni, S.R., L. Blitz, and C. Heiles. 1982. Astrophys. J. **259**:L63.

Kulkarni, S.R., L. Blitz, and C. Heiles. 1983. *In* W.L.H. Shuter (ed.), Kinematics, Dynamics and Structure of the Milky Way. Dordrecht: Reidel, p. 97.

Kwee, K.K., C.A. Muller, and G. Westerhout. 1954. Bull. Astron. Inst. Neth. **12**:117.

Lequeux, J. 1981. Comments Astrophys. **9**:117.

Lin, C.C., and G. Bertin. 1985. *In* H. van Woerden, R.J. Allen, and W.B. Burton (eds.), The Milky Way Galaxy. Dordrecht: Reidel, p. 513.

Lin, D.N.C., and D. Lynden-Bell: 1982, Monthly Not. Roy. Astron. Soc., **198**:707.

Liszt, H.S. 1984. Comments Astrophys. **10**:137.

Liszt, H.S., and W.B. Burton. 1978. Astrophys. J. **226**:790.

Liszt, H.S., and W.B. Burton. 1980. Astrophys. J. **236**:779.

Liszt, H.S., W.B. Burton, and D.-L. Xiang. 1984. Astron. Astrophys. **140**:303.

Lockman, F.J. 1976. Astrophys. J. **209**:429.

Lockman, F.J. 1984. Astrophys. J. **283**:90.

Lockman, F.J., K. Jahoda, and D. McCammon. 1986. Astrophys. J. **302**:432.

Lynden-Bell, D. 1965. Mon. Not. R. Astron. Soc. **129**:299.

Magnani, L., L. Blitz, and L. Mundy. 1985. Astrophys. J. **295**:402.

Mark, J.W-K. 1983. *In* W.L.H. Shuter (ed.), Kinematics, Dynamics and Structure of the Milky Way. Dordrecht: Reidel, p. 289.

Mathewson, D.S., V.L. Ford, M.P. Schwarz, and J.D. Murray. 1979. *In* W.B. Burton (ed.), The Large-Scale Characteristics of the Galaxy. Dordrecht: Reidel, p. 547.

McKee, C.F., and J.P. Ostriker. 1977. Astrophys. J. **218**:148.

Meaburn, J. 1978. Astrophys. Space Sci. **59**:193.

Meaburn, J. 1983. *In* R.M. West (ed.), Highlights of Astronomy, Vol. 6. Dordrecht: Reidel, p. 665.

Mebold, U.: 1972. Astron. Astrophys. **19**:13.

Merritt, D., and T. de Zeeuw. 1983. Astrophys. J. **267**:L19.

Mihalas, D., and J. Binney. 1981. Galactic Astronomy: Structure and Kinematics. W.H. Freeman.

Murai, T., and M. Fujimoto. 1980. Publ. Astron. Soc. Jpn. **32**:581.

Muller, C.A., J.H. Oort, and E. Raimond. 1963. C.R. Acad. Sci. Paris **257**:1661.

Norman, C. 1983. *In* E. Athanassoula (ed.), Internal Kinematics and Dynamics of Galaxies. Dordrecht: Reidel, p. 163.

Oort, J.H. 1927. Bull. Astron. Inst. Neth. **3**:275.

Oort, J.H. 1977. Annu. Rev. Astron. Astrophys. **15**:295.

Oort, J.H. 1978. *In* L. Mirzoyan (ed.), Problems of Physics and Evolution of the Universe. Yerevan: Armenian Acad. Sci., p. 259.

Oort, J.H., and A.N.M. Hulsbosch. 1978. *In* A. Reiz and T. Andersen (eds.), Astronomical Papers Dedicated to Bengt Stromgren. Copenhagen: Copenhagen University Observatory, p. 409.

Oort, J.H., F.J. Kerr, and G. Westerhout. 1958. Mon. Not. R. Astron. Soc. **118**:379.

Reid, M.J., M.H. Schneps, J.M. Moran, C.R. Gwinn, R. Genzel, D. Downes, and B. Ronnang. 1987. *In* M. Peimbert and J. Jugaku (eds.), Star Forming Regions. Dordrecht: Reidel, p. 554.

Roberts, W.W. 1972. Astrophys. J. **173**:259.

Rogstad, D.H., I.A. Lockhart, and M.C.H. Wright. 1974. Astrophys. J. **193**:309.

Rogstad, D.H., M.C.H. Wright, and I.A. Lockhart. 1976. Astrophys. J. **204**:703.

Rougoor, G.W. 1964. Bull. Astron. Inst. Neth. **17**:381.

Rougoor, G.W., and J.H. Oort. 1960. Proc. Natl. Acad. Sci. USA **46**:1.

Rubin, V.C., D. Burstein, W.K. Ford, Jr., and N. Thonnard. 1985. Astrophys. J. **289**:81.

Sancisi, R. 1976. Astron. Astrophys. **53**:159.

Sancisi, R. 1983. *In* E. Athanassoula (ed.), Internal Kinematics and Dynamics of Galaxies. Dordrecht: Reidel, p. 55.

Sancisi, R., and T.S. van Albada. 1987. *In* J. Kormendy, and G.R. Knapp (eds.), Dark Matter in the Universe. Dordrecht: Reidel, p. 67.

Sanders, D.B., D.M. Solomon, and N.Z. Scoville. 1984. Astrophys. J. **276**:182.

Schmidt, M. 1957. Bull. Astron. Inst. Neth. **13**:247.

Schwarz, U. 1985. Astron. Astrophys. **142**:273.

Schwarz, U., and J.H. Oort. 1981. Astron. Astrophys. **101**:305.

Seiden, P.E. 1985. *In* H. van Woerden, R.J. Allen, and W.B. Burton (eds.), The Milky Way Galaxy. Dordrecht: Reidel, p. 551.

Shane, W.W., and G.P. Bieger-Smith. 1966. Bull. Astron. Inst. Neth. **18**:263.

Shuter, W.L.H. 1981. Mon. Not. R. Astron. Soc. **199**:109.

Sinha, R.P. 1979. *In* W.B. Burton (ed.), The Large-Scale Characteristics of the Galaxy. Dordrecht: Reidel, p. 341.

Solomon, P.M., D.B. Sanders, and N.Z. Scoville. 1985. Astrophys. J. **292**:L19.

Songaila, A., D.G. York, L.L. Cowie, and J.C. Blades. Astrophys. J. **293**:L15.

Sparke, L. 1986. Preprint.

Stark, A.A. 1984. Astrophys. J. **281**:624.

Stark, A.A., J. Bally, R.A. Linke, and C. Heiles. 1985. Circulated to individuals.

Strong, A.W., P.A. Riley, J.L. Osborne, and J.D. Murray. 1982. Mon. Not. R. Astron. Soc. **201**:495.

Tenorio-Tagle, G. 1980. Astron. Astrophys. **88**:61.

Toomre, A. 1969. Astrophys. J. **158**:899.

Toomre, A. 1977. Annu. Rev. Astron. Astrophys. **15**:437.

Tuve, M.A., and S. Lundsager. 1972. Astron. J. **77**:652.

van Albada, T.S., and R. Sancisi. 1987. Phil. Trans. R. Soc. (in press).

van Dishoeck, E., and J. Black. 1987. Astrophys. J. (in press).

van de Hulst, H.C. 1945. Ned. Tijd. Natuurkunde **11**:201.

van der Kruit, P.C. 1986. Astron. Astrophys. **157**:230.

van der Kruit, P.C., and K.C. Freeman. 1986. Astrophys. J. **303**:556.

van der Kruit, P.C., and L. Searle. 1982. Astron. Astrophys. **110**:61.

van der Kruit, P.C., and G.S. Shostak. 1984. Astron. Astrophys. **134**:258.

van Woerden, H., G.W. Rougoor, and J.H. Oort. 1957. C.R. Acad. Sci. Paris **244**:1961.

van Woerden, H., U.J. Schwarz, and A.N.M. Hulsbosch. 1985. *In* H. van Woerden, R.J. Allen, and W.B. Burton (eds.), The Milky Way Galaxy. Dordrecht: Reidel, p. 387.

Verschuur, G.L. 1975. Annu. Rev. Astron. Astrophys. **13**:257.

Wagoner, R.V. 1973. Astrophys. J. **179**:343.

Weaver, H.F., and D.R.W. Williams. 1973. Astron. Astrophys. Suppl. **8**:1.

Weaver, H.F., and D.R.W. Williams. 1974a. Astron. Astrophys. Suppl. **17**:1.

Weaver, H.F., and D.R.W. Williams. 1974b. Astron. Astrophys. Suppl. **17**:251.

Weiland, J.L., L. Blitz, E. Dwek, M.G. Hauser, L. Magnani, and L.J. Rickard. 1986. Astrophys. J. **306**:L101.

Westerhout, G. 1957. Bull. Astron. Inst. Neth. **13**:201.

Westerhout, G. 1969, "Maryland-Green Band Galactic 21-cm Line Survey:, 2nd ed., University of Maryland.

Westerhout, G., and H-U. Wendlandt. 1982. Astron. Astrophys. Suppl. **17**:251.

Yuan, C. 1969. Astrophys. J. **158**:871.

8. The Galactic Center

Harvey Liszt

8.1 Introduction and Apologia

Deciphering the morphology and kinematics of the galactic center constitutes one of the enduring programs of the study of galactic structure. This is especially true for radio astronomy, which was the first of the astronomical sciences to penetrate the mysteries of the galactic center, but radio-frequency observations alone cannot provide more than a fraction of the knowledge which is required. Nowadays, the galactic center is accessible and commonly observed in all wavebands above 1 μ, and is accessible but less commonly observed in the X-and gamma-ray regimes.

One difficulty which impinges on this article is the sheer scope of material which may be associated with the galactic center; in the astronomical literature, the rubric "Galactic Center" appears associated with regions whose size varies from 10^{14} cm to 10^{22} cm. The former is the smallest size found for the compact source Sgr A*; the latter is the galactocentric radius at which gas in the galactic disk takes on the characteristics of the disk at large and no longer exhibits the high noncircular motions and tilts associated with nuclear material.

Such difficulties are mitigated by the presence of more general reviews of recent vintage, most recently Oort (1977, 1985), Brown and Liszt (1984), and Genzel and Townes (1987). Here, the scope of discussion will be severely limited to presentation of radio astronomical results, with some emphasis on regions immediately surrounding Sgr A on a scale of 0 to 50 pc or so. This chapter is neither a review nor a summary. Rather, the intention is to highlight some of the contributions, insights, and conundrums provided by radio astronomical research.

We begin with a pictorial survey of the galactic-center radio continuum, displaying each of the prominent Sagittarius sources and discussing the outlying regions in cursory fashion [more detailed discussion of many of these sources, observed at lower resolution, can be found in the work of Downes et al. (1978)]. This is followed by a somewhat more detailed exposition of Sgr A itself, and of the environment of the compact source at (or next-door to) the actual galactic center. At the end, we present a brief discussion of the large-scale (100 to 3000 kpc) kinematics and structure of the immense reservoir of neutral gas which pervades the inner galactic regions.

Throughout this article, we adopt the "old" value $R_0 = 10$ kpc (see Chapter 7) for the Sun-Center distance, whereby an angle of 1' projects to a linear dimension of 2.9 pc ($1° = 175$ pc).

8.2 Radio Continuum Emission from the Provinces of Sagittarius

Figure 8.1 is a panoramic view of the strongest galactic-center continuum sources, made by the author with the VLA at a wavelength of 18 cm; the angular resolution is approximately 17″. This view is a mosaic of maps from five separate fields which overlap (approximately) at the half-power points of the individual VLA telescopes; the well-defined interstices between prominent emission regions are real, as is the confinement of strong emission to a narrow band around the galactic equator. However, this is not the view which would occur to us if we had radio-frequency eyes. The peak intensity in each field has been adjusted to produce one point of equal brightness in all regions. Only Sgr B has intensities innately comparable to those of Sgr A; in all the other cases, the adjustment is upward by a factor of approximately 3 to 5. From left to right, the sources are usually called Sgr D (longitude 1.1°), B (0.6° to 0.8°), A, C (−0.6°), and E ($l \leqslant -1.0°$), respectively. Excepting Sgr A for last, they are presented in that order below.

8.2.1 Sgr D

As shown in Figure 8.2, Sgr D is composed of two components. Emission from the brighter of these, an HII region complex, peaks at $(l, b) = (1.13°, -0.11°)$. The complex's emission is sharply bounded on most sides. Diffuse emission, seen at the head of the "wishbone" or y-shaped feature above the bright core, is probably leakage out of a cavity formed within the neutral ambient material.

The lower-latitude component of Sgr D has been known (Little 1974) to be a supernova remnant (SNR); it is seen in the VLA image to be composed of a series of wisps or fossil shells with maximum dimensions rather similar to those of the HII region above. At its top, nearest the HII region, the SNR exhibits a bright rim which is actually connected to the thermal source over a region 1′ to 2′ wide. This osculating pair of thermal and nonthermal sources is highly unusual in its morphology.

Sgr D is deserving of more attention than it has received heretofore. There is a concentration of molecular gas in its vicinity (see Liszt and Burton 1978 or Figure 8.14) and the region as a whole would seem to offer an unusually detailed opportunity for simultaneous study of several distinct stages in the evolution of a star-forming region.

8.2.2 Sgr B

Sgr B (Figure 8.3) is a heavily studied source because of its multitude of compact HII regions (and their associated H_2O and OH masers) and as a result of the copious quantity of molecular gas which surrounds them. It is among the three to four wellsprings of our knowledge of the complex processes of dense-gas interstellar chemistry (see Chapters 4 and 5). The bright compact objects in Sgr B2—the higher longitude regions of Sgr B—are among the densest HII regions in the Galaxy. They are discussed in detail by Downes et al. (1978) and references given there.

What is revealed for the first time in the VLA observations is the network of filamentary structures in and among Sgr B1 and B2. These structures, which give the appearance of a mesh in their complexity, undoubtedly arise from the interaction between stars and molecular gas. They must be thermal in origin, because Sgr B as

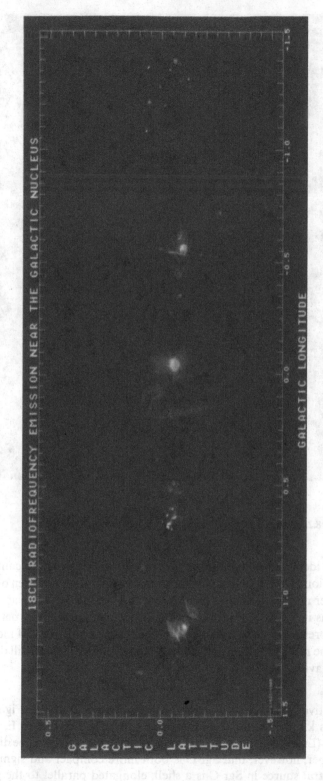

Fig. 8.1. A mosaic of $\lambda 18$ cm continuum emission from the galactic center, observed by the VLA at 17″ resolution. From left to right, the sources are usually called Sgr D ($l = 1.1°$), B, A, C and E ($l \le -1.0°$).

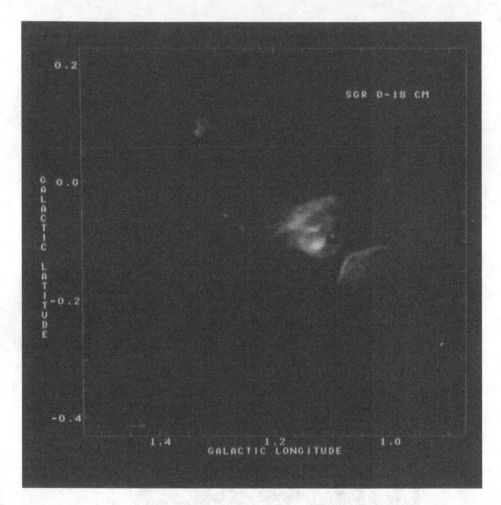

Fig. 8.2. Expanded view of λ18 cm continuum emission from Sgr D.

a whole is clearly identified as thermal and because the mesh structure contributes a substantial portion of the total flux. Similar morphologies have not been observed heretofore in other nominally similar regions, but this could be entirely accidental. Few if any regions in the disk of the Galaxy have been mapped in any detail over the linear scales present in these radiographs. The recent availability of molecular observations at the resolution afforded by the VLA continuum image will do much to clarify this behavior.

8.2.3 Sgr C

Situated at negative longitude $l \approx -0.6°$ is the source Sgr C (see Figure 8.4). Relatively little is known of this region, as is the case for Sgr D and E, but the structure of Sgr C (Liszt 1985) presages much of the behavior which will be discussed in Sgr A (remember, however, that Sgr C is both more compact and significantly weaker). The central source in Sgr C is a shell, elongated parallel to the galactic plane. Upon this, to the North, is superposed a bright patch of thermal emission

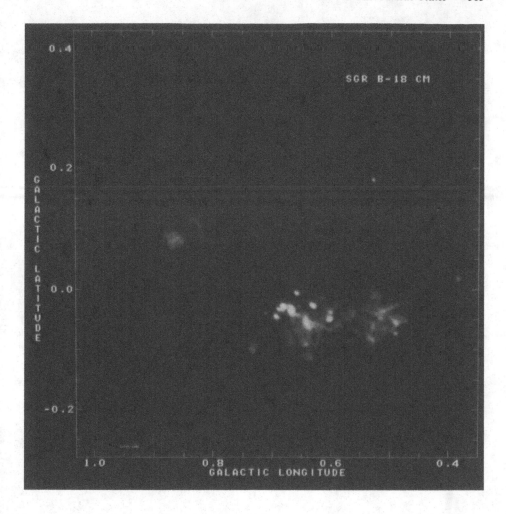

Fig. 8.3. Expanded view of $\lambda 18$ cm continuum emission from Sgr B.

which produces an H110α recombination line at -60 km s^{-1} (unpublished VLA observations by the author).

More important is the filament to the East of the shell, extending in this picture over the region $-0.1° \leqslant b \leqslant 0.1°$. To this end, it seems wise to refer the reader to larger-scale, single-dish continuum maps such as are contained in the work of Downes et al. (1978) and Sofue and Handa (1984). Both this filament and those seen elongated across the galactic plane in Sgr A are the brightest portions of structures which extend over several hundred parsecs of galactic latitude. These large-scale filaments (or arcs or lobes as they are called) will be discussed in more detail with reference to Sgr A, but Figure 8.4 gives the unmistakable impression that there is a physical association between the shell and adjacent quasi-linear structure.

8.2.4 Sgr E
Least well studied and understood of all the galactic center complexes is Sgr E at $l \leqslant -1°$ in Figure 8.1. Readily obvious is the fact that in this region alone is the

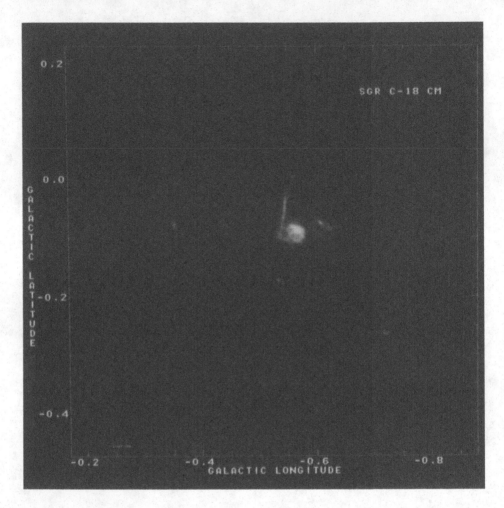

Fig. 8.4. Expanded view of λ18 cm continuum emission from Sgr C.

flux concentrated in discrete sources; even in Sgr B the fraction of flux in extended emission is about 0.5.

None of these discrete sources has an extent exceeding 2′ nor do they have structures which merit closer examination here. Perhaps because of their relatively large projected distance from the center, there is little prior information on which to gauge their character. However, it seems most probable that they are actually HII regions near the galactic center. They have the characteristic distribution (in latitude) of sources which we have come to associate with the center, and Caswell and Haynes (1987) have recently detected a recombination line from this region at -209 km s^{-1}. Inspection of the galactic-center CO distribution (see Section 8.5) shows a concentration of molecular gas at the pertinent position and velocity.

8.2.5 Taking Stock of the Situation in the Provinces

Considered individually, the outlying sources in the Sagittarius complex are for the most part examples of phenomena—active star-forming regions, HII region–

molecular cloud complexes—which are well known in the galactic disk (Sgr C is the outstanding exception to this generality). At most they are extreme (Sgr D) or interesting (Sgr D) variants on common themes. Less attention has been focused on some of them than on powerful sources in the galactic disk, an unfortunate result of Sagitarrius' low declination.

Less common is the appearance of so much activity in such a limited region of the sky. Even in this regard, however, such disk sources as W51 are about equally complex and angularly somewhat more compact (W51 is usually understood as a coincidental projection of the line of sight along a spiral arm). What then is the intrinsic interest of the Sagittarius sources? It is the matrix and environment in which they are embedded, and perhaps their proximity and relationship to Sgr A. Sgr A will be discussed immediately below. The large-scale galactic-center environment, as viewed in neutral atomic and molecular gas, will be considered at the end of this chapter.

8.3 Sgr A and Its Immediate Environment

Sgr A is probably the most intensively and frequently studied region in the Galaxy. It does not lend itself to cursory discussion; even enumerating its named parts requires some patience, and detailed exegesis of any of them would require a breadth and degree of detail which is inappropriate in the present context. The approach taken here is to "build" Sgr A in steps, moving outward from the smallest (0.001″ or 1.5×10^{14} cm) to intermediate (2′ or 6 pc) to larger (50 pc) scales. For a more detailed observational discussion of a fantastic wealth of recent continuum VLA results, see Yusef-Zadeh (1987).

An enlarged radiographic view of Sgr A is shown in Figure 8.5; this figure is itself a mosaic of observations in two fields which have been patched together (the quality of the observations in the northern inset is not high). A schematic view of the emission, with nomenclature superposed, is given in Figure 8.6.

8.3.1 The Compact Source Sgr A*

Balick and Brown (1975; see also Downes and Martin 1971) drew attention to a compact, time-variable radio source near what is believed to be the precise center of the Galaxy, the $\lambda 2$ μm peak IRS 16 of Becklin and Neugebauer (1968). The compact source is known as Sgr A*, and a recent summary of its characteristics by Lo et al. (1985) quotes a radio luminosity of 100 $L_\odot$ (the X-ray luminosity is several times higher) and spectral index $\alpha \approx -0.2$ (the flux varies as $S_\nu \propto \nu^{0.2}$). Moderate variations in the flux density of Sgr A* occur on time scales of about 1 day; at $\lambda 6$ cm, $S_\nu \approx 1$ Jy. IRS 16 is commonly interpreted as a stellar cluster massing some 3×10^6 $M_\odot$, but the nature of the radio source and its precise degree of coincidence with IRS 16 continue to be matters of active debate (see Brown and Liszt 1984). On the scale of Figures 8.5 and 8.6, both IRS 16 and Sgr A* sit squarely in the middle of Sgr A West. The position of the radio source is $\alpha(1950) = 17^h42^m29.335^s$, $\delta(1950) = -28°59'18.6''$.

Soon after detection of the compact source, Davies et al. (1976) pointed out that its apparent angular size varies rapidly with observing wavelength. Collating the

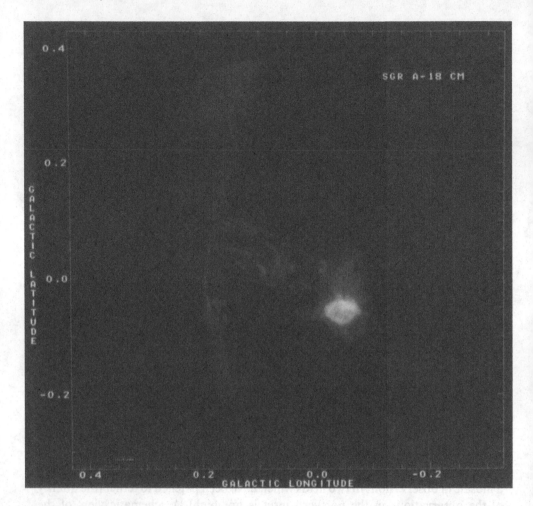

Fig. 8.5. Expanded view of $\lambda 18$ cm continuum emission from Sgr A. This is a mosaic of observations in two separate fields.

available interferometric results (see Lo et al. 1985) produces Figure 8.7; variation in the fitted Gaussian FWHM continues over a factor of about 25 in wavelength, from $\theta_{1/2} = 1.5''$ at $\lambda 31.25$ cm to $\theta_{1/2} = 0.0021''$ at $\lambda 1.35$ cm. A weighted fit to the data produces a power law dependence $\theta_{1/2}(\lambda) \propto \lambda^{2.05 \pm 0.05}$. Despite the asymmetry seen by Lo et al. (1985) at one intermediate wavelength, there is no conclusive evidence that anything more than an upper limit to the true source dimensions has been established; this upper limit is $0.0021'' \approx 10^{-4}$ pc $\approx 3 \times 10^{14}$ cm ≈ 3 light-hours.

As discussed by Davies et al. (1976) and by Lo et al. (1985), a λ^2 variation in apparent source diameter could result either from the presence of a scattering screen in the vicinity of the compact source or from a variation of source electron density with radius ($n_e \propto r^{-1}$). In the latter case, the variation of size results from progres-

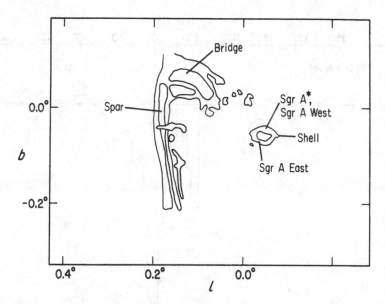

Fig. 8.6. Schematic finding chart for named sources within the Sgr A complex, complementary to Figure 8.5.

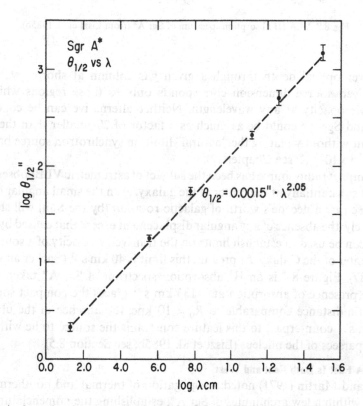

Fig. 8.7. Derived angular size versus wavelength for the compact source Sgr A*. Data were gathered from Davies et al. (1976) and Lo et al. (1985).

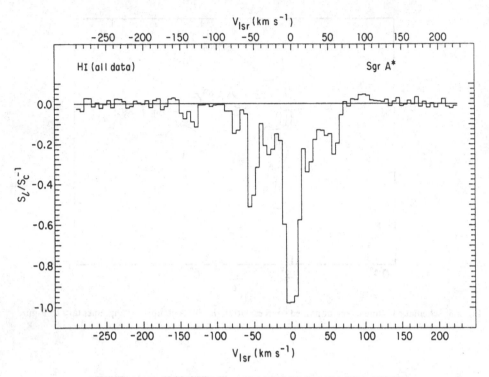

Fig. 8.8. VLA HI absorption spectrum of Sgr A* (from Liszt et al. 1985b).

sively lower optical depth through a given gas column at shorter wavelength; the perceived source dimension corresponds only to those regions which have appreciable opacity at any wavelength. Neither alternative can be conclusively rejected and Sgr A* could be as much as a factor of 20 smaller than the present upper limit without violating the "natural" limit on synchrotron source brightness ($T_b \leqslant 10^{11}$ to 10^{12} K; see Chapter 12).

The compact radio source has been the subject of astrometric VLBI observations. If it is not substantially at the center of the galaxy, even the small angular displacement caused by a decade's worth of galactic rotation (by the Sun) will show this; alternatively, the absence of any angular displacement except that caused by galactic rotation can be used to establish limits on the transverse velocity of a source at the precise center of the Galaxy. At present, this limit is 40 km s^{-1} (Backer and Sramek 1982, 1987). Figure 8.8 is an HI absorption spectrum of Sgr A* taken with the VLA; the presence of absorption at -135 km s^{-1} places the compact source at a heliocentric distance comparable to $R_0 \approx 10$ kpc; the absence of the ubiquitous $+165$ km s^{-1} counterpart to this feature constrains the source to lie within a few hundred parsecs of the nucleus (Liszt et al. 1985b; see Section 8.5).

8.3.2 Sgr A East Is Both East and West

Downes and Martin (1971) noted a segregation of thermal and nonthermal radio emission within a few arcminutes of Sgr A*, establishing the nomenclature "Sgr A West" for the thermal emission and "Sgr A East" for the nonthermal component. In Figure 8.5, Sgr A West is the bright patch at the top of the Sgr A shell (i.e., it is

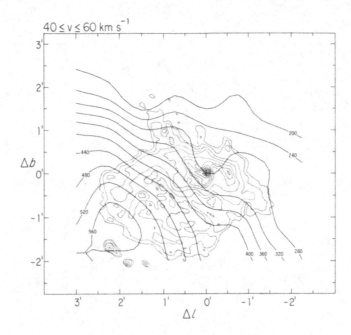

$40 \leq v \leq 60$ km s^{-1}

Fig. 8.9. Comparison of λ20 cm radio continuum (shaded contours; Ekers et al. 1983) and integrated λ2.6 mm CO emission at 40 to 60 km s^{-1}. (From Liszt et al. 1985a.)

more properly called "North") and Sgr A East refers to the brightest and most southerly portions of the shell emission. Here, that entire structure will be referred to as the Sgr A East shell.

Ekers et al. (1983) established the structure of Sgr A East as being a full shell stretching along the galactic equator; it passes through the position of Sgr A West, obfuscating the thermal gas behavior at sufficiently long wavelengths (above perhaps λ10 cm). The shell is usually believed to be a supernova remnant (Chapter 10), and consideration of a Sedov model suffices to show that an age in excess of ≈ 5000 years is unlikely (the shell radius, some 5 pc, could be attained after this time with an average expansion rate below 1000 km s^{-1}). To support a great age it would be necessary to suppose either a very small mechanical energy input to the surrounding medium or too large a quantity of confining gas.

Figure 8.9 presents a comparison of 40 to 60-km s^{-1} CO emission and the 20-cm radio continuum observations of Ekers et al. (1983; see Liszt et al. 1985b). Located just to the southeast of the shell are several compact HII regions whose velocities have been established as 40 to 50 km s^{-1} (van Gorkom et al. 1985). Superposed on the HII region positions is a strong peak in the 40 to 60-km s^{-1} gas cloud in Sgr A. The visual impression gleaned from this figure is of star formation induced by interaction between Sgr A and the molecular cloud. This view must be tempered by the symmetry and elongation of the shell, which indicate that it has not yet been distorted by interactions with the lumpy ambient material, and by the fact that the shell, if it is a supernova remnant, is not old enough to have caused the formation of main-sequence OB stars. The HII regions outside the shell could have resulted from phenomena long since vanished or unrelated to an SNR, or the shell might be

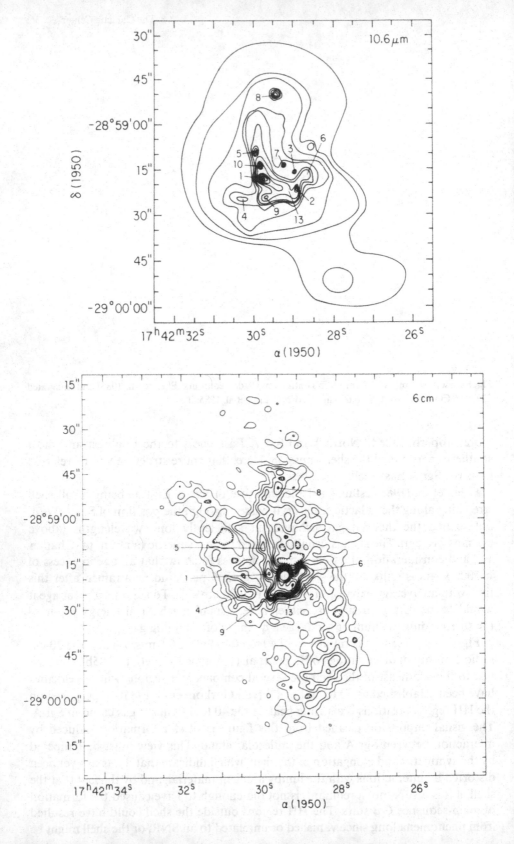

a peculiar and unique object. Lest such suppositions seem out of place, we note that a similar comparison of CO and radio continuum emission on a slightly larger scale is sufficient to place the 50-km s^{-1} molecular cloud in intimate contact with Sgr A (Section 8.4).

8.3.3 Sgr A West

VLA observations by Brown et al. (1981) and Ekers et al. (1983), especially the latter, established the structure of the thermal radio emission around Sgr A*, Sgr A West; it is the same as found previously at $\lambda 10\,\mu$ in the infrared, albeit with some significant extension and elaboration. This structure is shown in Figure 8.10; the compact source Sgr A* is the brightest point in the radio map.

Our current understanding of this material is informed by the fact that two distinct kinematic and thermodynamic patterns are present (Lo and Claussen 1983, Brown and Liszt 1984). The armlike extensions seen to the northeast and southwest of Sgr A* are cooler and exhibit motions which are dominated by galactic rotation; high (≈ 100 km s^{-1}) speeds are viewed at their extremities and low ones occur closest to the center of symmetry. In the material just below Sgr A*, the temperature is higher and the velocities vary oppositely; high speeds ± 250 km s^{-1} are seen nearest the center of symmetry. The cooler material will be discussed first.

(a) A Massive Rotating Ring

The far-IR observations of Becklin et al. (1982), indicated that most of the gaseous material within $\approx 45'' \approx 2$pc of Sgr A* lies in a ring which is highly inclined to our line of sight and noticeably tilted away from the galactic equator; the line of nodes is close to North-South in Figure 8.10. Coupled with the rotational motions found in the arm features, it is now inferred that there is a massive (perhaps $10^4\,M_\odot$) ring of gas in orbit about the galactic center. On the largest scales ($\approx 3'$; Liszt et al. 1985a), the ring material is neutral and molecular. Nearer the center (Figure 8.10), it is presumably ionized by a central source of radiation.

There are some difficulties in constructing detailed models of this phenomenon. One stumbling block is that the central sources IRS 16 and Sgr A* are not symmetrically located with respect to the "arms" or rotating material (see the model outlines of Brown and Liszt 1984); a second is that the velocity gradient is larger in the more northerly arm, suggesting that it is closer to the center or has noncircular components of motion (or both). If the ring is a full structure, the radiation ionizing it must leave the center so anisotropically that whole quadrants are missing in our maps. Such anisotropy could be an important clue to deciphering the correct placement of the structures seen in Figure 8.10. A consistent picture should fit the various parts together in such a way that blockage of the central source is explained.

(b) A Small "Bar"

The second kinematic pattern seen in the Sgr A West material occurs in the 1' "bar" which Figure 8.10 exhibits just below the compact source; not evident in this diagram is the fact that the long axis of the bar is parallel to the galactic rotation axis. The bar material is appreciably hotter than that in the arms (Brown and Liszt 1984) and likely lies closer to the center of the Galaxy.

◁

Fig. 8.10. Comparison of $\lambda 10.6\,\mu$m (Rieke et al. 1978) and $\lambda 6$ cm continuum emission in Sgr A West. The labeled points refer to infrared source numbers.

The highly symmetric velocity pattern seen in the bar could, in principle, arise from outflow (with the caveat that no readily apparent centrally located object is present) or infall (small problem), or it could arise in a rotating disk (not a ring) if the disk axis lies projected in the galactic plane. The displacement of the bar from Sgr A* is a substantial embarrassment in each case. Nonetheless, the combination of higher temperature and velocity closer to Sgr A*-IRS 16 suggests that the gas motion is controlled by gravity.

8.3.4 Sgr A* Redux

The compact radio source has received much attention as a black-hole candidate; its size and brightness clearly render it an unusual object and the presence of a positron annihilation line (Leventhal et al. 1978, Lingenfelter and Ramaty 1982) is usually taken as indirect evidence of a black hole. In the case where no substantial mass in Sgr A West resided in anything except a black hole, the equilibrium (circular) velocity field would be Keplerian, varying as $r^{-1/2}$; freely infalling material would possess the same behavior although the motion would project onto the line of sight quite differently. However, a substantial distributed mass is present in the form of IRS 16, whose stellar density varies as $\rho_*(r) \propto r^{-1.7}$ (Allen et al. 1983). For IRS 16 considered alone, the stellar mass inside radius r is $M_*(r) \propto r^{1.3}$ and rotation or infall velocities vary as $v^2(r) \propto M_*(r)/r \propto r^{0.3}$.

The velocities observed at positions within the bar material are approximately 250 km s^{-1} at $\approx 2''$ separation and 160 km s^{-1} at $7''$; a model fit to the ring yields 110 km s^{-1} at 65$''$ separation. Clearly, the qualitative characteristics of the motion do not fit the pattern expected for the cluster mass alone, nor do they show a pattern of Keplerian decline beyond the bar. The run of velocities can be crudely approximated by superposing a 2×10^6-$M_\odot$ point mass on IRS 16, but this is little proof that Sgr A* is a black hole. The kinematics are simply not well enough understood to constitute a definitive probe of the inner-galaxy mass distribution.

8.3.5 All the King's Horses

At the present time, it is unclear how the various pieces of Sgr A West and East actually fit together. Sgr A East, long believed to lie behind Sgr A West because it (East) alone showed a strong 50-km s^{-1} absorption feature, can really only be shown to lie behind some portion of the 50-km s^{-1} molecular cloud; Sgr A West is simply off to one side of that cloud. To confuse matters, Sgr A* *does* exhibit a 50-km s^{-1} feature (Figure 8.8) but this must be attributed to a noncircular component of motion in the "rotating" ring (Liszt et al. 1985a). As noted above, no entirely satisfactory model exists for the material in Sgr A West which is clearly dominated by galactic rotation. The bar motions are subject to varying interpretations, and any relationship between the ring and bar material is only very dimly perceived.

8.4 Material in Sgr A Observed on 10 to 50-pc Scales

In Section 8.3.2, we alluded to the fact that the 50-km s^{-1} molecular cloud can be shown to be intimately related to Sgr A. This is illustrated in Figure 8.11, where the integrated intensity of CO emission from this cloud is projected against a single-dish

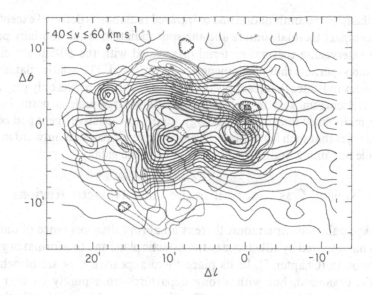

Fig. 8.11. Comparison of radio frequency continuum (shaded; Downes et al. 1978) and 40- to 60-km s^{-1} CO emission over a larger region in Sgr A. (From Liszt et al. 1985a.)

continuum image. The molecular gas nests entirely within the continuum contours and the extraordinary filamentary structures crossing the plane at $l \approx 0.2°$ (Figures 8.5 and 8.6; Yusef-Zadeh 1987, Yusef-Zadeh et al. 1984) are projected against an abrupt boundary in the molecular material. Bally et al. (1987) have pointed out an association between some portions of the Sgr A continuum and material in the velocity range -30 km s$^{-1} \leqslant v \leqslant 0$ km s^{-1}.

The portions of the radio continuum observed above and left of the molecular cloud in Figure 8.11 are collectively known as the Arc. Here these portions are separately referred to (see Figure 8.6) as the Bridge and the Spur. As shown by Liszt et al. (1985b), and as recombination-line observations suggested earlier, the Bridge is largely thermal in origin and the Spur nonthermal. There is no clear understanding at present of the manner in which the Bridge either emerges from Sgr A or joins onto the Spur, nor have models been advanced which account for the intricately detailed structure which is seen in the Spur (cf. Yusef-Zadeh 1987). Only a small fraction of these details are visible in Figure 8.5.

The Spur is one part of a structure which spans nearly 2° of galactic latitude (some 350 pc), as is remarked in Sofue and Handa (1984) and is clearly evident in Figure 1 of Downes et al. (1978). It must also be a physical near-relative of the filament remarked earlier in Sgr C, which is again only the brightest part of a structure extending far above the galactic plane. Sofue and Handa (1984) suggested that the Sgr A and C spurs (filaments) form a single lobe over the galactic center, but they do not have the symmetries expected of a single structure and may well be separately powered.

The Spur in Sgr A does not have a steep spectral index but is mostly nonthermal emission nonetheless (the filament in Sgr C does have a fairly steep spectral index near the shell). No recombination line or infrared radiation is detected except nearby in a few compact thermal sources and the spur radiation is substantially polarized wherever intervening thermal material (associated with the 50-km s^{-1} cloud, or other features) permits that phenomenon to be observed. The relative flatness of the spectral index indicates that resupply of energy is occurring or has only very recently stopped. It seems unlikely in the extreme that any large-scale structural integrity would be maintained long after the energy supply or causative agent had ceased. It is unlikelier yet that such features could simply have propagated outward in a single event while remaining undistorted in the plane of the galaxy.

8.5 The Neutral Gas Reservoir in the Inner Galactic Regions

Inside 4-kpc galactocentric radius there is an abrupt disappearance of the neutral gas disk characterized by adherence to the galactic plane and to a format of generally circular motion (Chapter 7). In its place there appears a new set of behavior— systematic, organized, but with strong departures from purely circular motion and/or the plane of the outer galaxy. The distribution of neutral gas in the Milky Way is at least mildly unusual in having both an annular distribution in the disk and a strong peak around the nucleus. Other galaxies with copious HI in the inner regions show no discontinuity between the inner and outer regions; conversely, when the disk HI in external galaxies is annular, it does not peak again in the innermost regions.

8.5.1 The "Expanding 3-kpc Arm"

Moving inward from the point at which the "normal" disk is replaced by aberrant structures, one probably first encounters a feature known as the (expanding) 3-kpc arm (see Rougoor 1964, Bania 1980); its name arises from a seeming tangent point at longitude 338°, at which angle the subcentral point was taken as 3 kpc at the epoch of the original discussion. This feature appears in absorption at -53 km s^{-1} toward all of Sgr A (for example, see Figure 8.8); because it lies between the Sun and nucleus, its noncircular motion is outward-directed. The 3-kpc arm can be observed continuously over the range $-22° \leqslant l \leqslant 10°$ in regions not far removed from the galactic equator. At higher longitude it blends more or less irretrievably with unrelated material and a positive-longitude terminus is not easily discerned.

The 3-kpc arm is a difficult feature to interpret because it has no counterpart on the far side of the nucleus. It is apparently not a wholly symmetrically distributed feature, nor is it a typical spiral arm because little if any star formation is viewed within it. This is perhaps less of a conundrum now that similar, one-sided distortions have been observed in other galaxies but a full explanation is lacking (see, however, Mulder 1985).

8.5.2 A Tilted Disk—The Galaxy's Bar?

For material closer to the center than ≈ 3 kpc (the 3-kpc arm is not nearly as well located as its name suggests!), a more complicated phenomenon emerges. On scales of 200 to 1500 pc, the gas distribution lies in a plane which makes a $\approx 30°$ angle with that of the "normal" disk observed at galactocentric radii of 4 to 12 kpc.

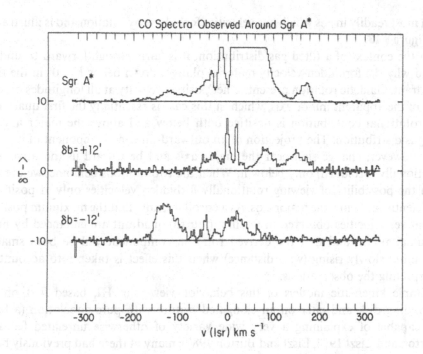

Fig. 8.12. $\lambda 2.6$ mm CO spectra taken at the longitude of Sgr A*. The upper and lower spectra are displaced by 12' from Sgr A* itself.

Perceived velocities usually have substantial components of both circular and noncircular motions even when they are entirely consistent with rotation or expansion alone.

The -135-km s^{-1} feature observed toward Sgr A (Figure 8.8) is an example of these phenomena and can be straightforwardly employed to illustrate one aspect of the tilt in the galactic center gas; the orientation is such that the nearer edge of the distribution is tipped down below the galactic plane and the far side, above. Figure 8.12 is a plot of CO spectra observed above, below, and toward Sgr A*. To the South, the -135-km s^{-1} feature is prominent, to the North, that at $+165$ km s^{-1}. In between, toward Sgr A*, both are present (in emission; compare the absorption structure of Figure 8.8). Clearly, these two features form a pair representing strong, outward-directed motion on both sides of the galactic nucleus. Toward the galactic center, circular motions are projected across the line of sight; the 300-km s^{-1} spread in velocity between the two features results almost solely from outward-directed motion. At higher and lower longitudes, both these features show prominent velocity gradients attributable to rotational motions.

As for external galaxies, the tilt of the gas distribution is specified by two angles; the paragraph immediately preceding identified only one of these, analogous to inclination. The sense of the second component of the tilt is that material appears progressively further below $b = 0°$ at higher longitudes (the position angle of the gas is about 112° measured counterclockwise from $l = 0°$). This phenomenon is also seen most readily in gas which has velocities forbidden by rotation and is illustrated in Figure 8.13.

In the context of a tilted gas distribution, it is fairly straightforward to understand why the forbidden-velocity material plunges down below $b = 0°$ in the first quadrant. Galactic rotation presents a net positive velocity at all longitudes on one side of the apparent minor axis, which in this case *is* essentially the first quadrant; the rotational contribution is positive both below and above the major axis of the gas distribution. The projection of an outward-directed component of motion will, however, change sign near the major axis and be toward us (negative and rotationally forbidden) only below it. When the two motions combine, we are left with the possibility of viewing rotationally forbidden velocities only at positions sufficiently far below the major axis. As a corollary, note that the maximum positive (negative) velocities observed in the first (fourth) quadrant are enhanced by noncircular motion and that the derived rotation component will be both smaller and more slowly rising (with distance) when this effect is taken into account in interpreting the observations.

Simple kinematic models of this behavior viewed in HI, based first on an expanding disk and later on an elongated feature with pure circulation (a bar), are capable of explaining a very wide variety of otherwise unrelated features (Burton and Liszt 1978, Liszt and Burton 1980); many of these had previously been interpreted as ejecta from the nucleus (Burton and Liszt 1983a). The tilt and prevalence of noncircular motions have been given a firmer dynamical basis in the work of Vietri (1986) and Mulder (1985). If the central mass distribution in the Galaxy is triaxial—a figure with three unequal moments of inertia or an asymmetric rotor in the language of interstellar molecules—equilibrium gas orbits exist which do not lie parallel to any of the (three) principal planes. Triaxial mass distributions can produce both the noncoplanarity and peculiar velocity fields observed, and the kinematic model advanced above indicates that a bar of gas is present in the Milky Way. The 3-kpc arm and other phenomena such as warping of the galactic plane at radii ≥ 12 kpc might well be driven by the peculiarities of the innermost material.

8.5.3 More on the Central Gas Reservoir

Figure 8.14 is a position-velocity diagram of CO emission over the central few hundred parsecs, made at the latitude of Sgr A* (Liszt and Burton 1978). The previously mentioned features at -135 and 165 km s^{-1} are apparent toward Sgr A*, as is a portion of the 50-km s^{-1} cloud toward Sgr A, the 60-km s^{-1} cloud in Sgr B2, gas at -60 and -120 km s^{-1} in Sgr C, and gas at -200 km s^{-1} in the region of Sgr E. The 3-kpc arm is easily discerned as a prominent ridge around -50 km s^{-1}. The kinematic patterns seen in Figure 8.14 are perplexing, and there is a good deal of poorly understood kinematic structure in galactic latitude as well (Burton and Liszt 1983b). Note the difficulty incurred in separating any "clouds" from the general distribution.

Figures 8.11 and 8.14 show that the Sagittarius continuum sources are immersed in and associated with copious quantities of molecular material. Moreover, Figure 8.1 shows that the continuum sources in the provinces of Sagittarius all lie rigidly along a locus drawn parallel to the galactic equator. The absence of tilts in the

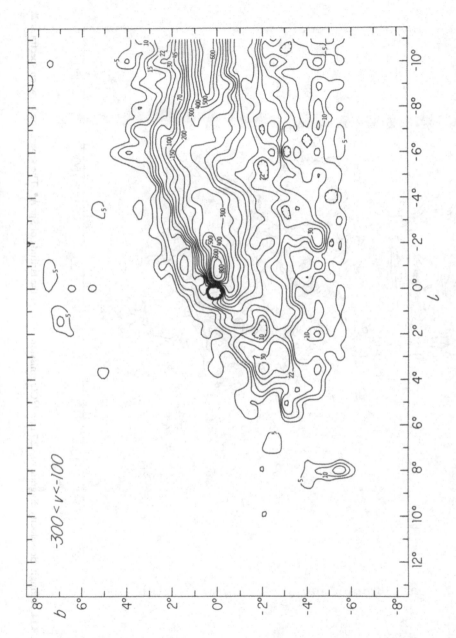

Fig. 8.13. A map of forbidden-velocity ($-300\ \mathrm{km\ s^{-1}} \leqslant v \leqslant -100\ \mathrm{km\ s^{-1}}$) HI emission from the galactic center, taken from **Liszt and Burton (1980)**.

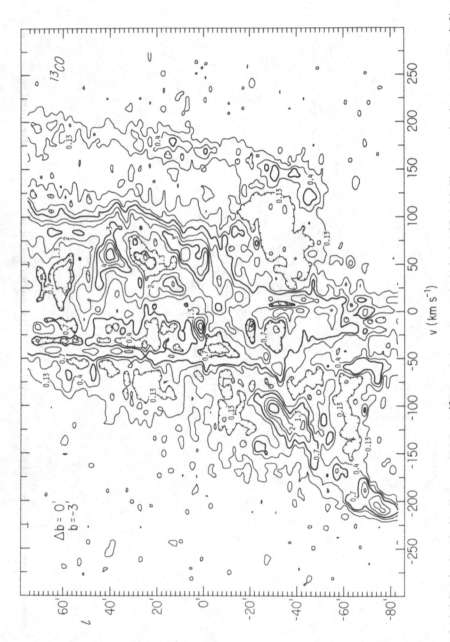

Fig. 8.14. A longitude-velocity diagram of λ2.7 mm ^{13}CO emission taken at the latitude of Sgr A*. The separation between spectra is 2′.

strong, innermost sources is a complication, but a useful one. It implies that the unusual geometry is limited in radial extent, and no model has yet been advanced which accounts for this phenomenon, but it also constrains the Sagittarius sources to reside in a confined region (inside the tilt). They are probably not vastly further apart than they appear in Figure 8.1.

When the quantity of neutral hydrogen is summed over the ≈ 3-kpc diameter of the tilted inner-galaxy gas distribution, the result is $M_{HI} \approx 4 \times 10^7\ M_\odot$; because the gas is usually optically thin, this mass must be fairly accurately determined. When the quantity of molecular material is approximated, much larger masses and controversy ensue. For the tilted distribution at large, Liszt and Burton (1978) attempted to gauge the mean density required to produce the observed, widespread CO lines and found $M_{H_2} \approx 4 \times 10^9\ M_\odot$. Perhaps some 10% of this total could be found within the more restricted longitude region sampled in Figure 8.14.

Such high masses result whenever gas densities are estimated or when the usual factor is applied to convert integrated CO intensity to H_2 column density. In fact, many of the galactic center clouds (i.e., at 50 km s^{-1} around Sgr A and 60 km s^{-1} toward Sgr B) must be massive (10^6 to $10^7\ M_\odot$) or they would be ripped apart by the galactic tide on very short time scales. However there appears to be a conflict between the gamma-ray emissivity observed over the inner galactic regions and that calculated on the basis of a "normal" cosmic-ray flux (Blitz et al. 1985). This discrepancy amounts to about an order of magnitude and it implies that the molecular masses have been overstated by such a factor. If the larger amounts of gas are truly present, our galaxy is among the most copiously endowed. If not, then perhaps some future discussion of the galactic-center gas distribution can be made a few pages shorter than even this one.

Recommended Reading

The Galactic Center, Riegler, G.R. and R.D. Blandford (eds.), New York: American Institute of Physics.

"Sagittarius A and Its Environment", Brown, R.L., and H.S. Liszt 1984. Annu. Rev. Astron. Astrophys. **22**:223.

"The Galactic Nucleus". J.H. Oort 1985. in H. van Woerden, R.J. Allen, and W.B. Burton (eds.), The Milky Way Galaxy. Dordrecht Reidel, p. 345.

"Physical Conditions, Dynamics, and Mass Distribution in the Center Of The Galaxy", Genzel, R. and C.H. Townes. 1987. Annu. Rev. Astron. Astrophys. **25**:377.

References

Allen, D.A., A.R. Hyland, and T.J. Jones. 1983. Mon. Not. R. Astron. Soc. **204**:1145.

Backer, D.C., and R. Sramek. 1982. Astrophys. J. **260**:512.

Backer, D.C., and R. Sramek. 1987. In D.C. Backer (ed.) The Galactic Center. New York: American Institute of Physics, p. 163.

Balick, B., and R.L. Brown. 1974. Astrophys. J. **194**:265.

Bally, J., A. Stark, R. Wilson, and C. Henkel. 1987. Astrophys. J. (in press).

Bania, T.M. 1980. Astrophys. J. **242**:95.

Becklin, E.E., and G. Neugebauer. 1968. Astrophys. J. **151**:145.

Becklin, E.E., I. Gatley, M.W. Werner. 1982. Astrophys. J. **258**:134.

Blitz, L., J.B.G.M. Bloemen, W. Hermsen, and T. Bania. 1985. Astron. Astrophys. **143**:267.

Brown, R.L., and H.S. Liszt. 1984. Annu. Rev. Astron. Astrophys. **22**:263.

Brown, R.L., K.Y. Lo, and K.M. Johnston. 1981. Astrophys. J. **250**:155.

Burton, W.B., and H.S. Liszt. 1978. Astrophys. J. **225**:815.

Burton, W.B., and H.S. Liszt. 1983a. Astron. Astrophys. Suppl. Ser. **52**:63.

Burton, W.B., and H.S. Liszt. 1983b. in W.B. Burton and F.P. Israel (eds.), Surveys of the Southern Galaxy. Dordrecht: Reidel, p. 149.

Caswell, J.L., and R.F. Haynes. 1987. Astron. Astrophys. **171**:161.

Davies, R.D., D. Walsh, and R. Booth. 1976. Mon. Not. R. Astron. Soc. **177**:319.

Downes, D., and A.H.M. Martin. 1971. Nature **233**:112.

Downes, D., W.M. Goss, U.J. Schwarz, and J.G.A. Wouterloot. 1978. Astron. Astrophys. Suppl. Ser. **35**:1.

Ekers, R.D., J.H. van Gorkom, U.J. Schwarz, U.J., and W.M. Goss. 1983. Astron. Astrophys. **12**:143.

Genzel, R., and C.H. Townes. 1987. Annu. Rev. Astron. Astrophys. **25**:377.

van Gorkom, J.H., U.J. Schwarz, and D. Bregman. 1985. *In* H. van Woerden, R.J. Allen, and W.B. Burton (eds.), The Milky Way Galaxy. Dordrecht: Reidel, p. 371.

Leventhal, M., C.J. McCallum., and P.D. Stang. 1978. Astrophys. J. (Lett.) **225**:L11.

Lingenfelter, R.E., and R. Ramaty. 1982. In G.R. Riegler and R.D. Blandford (eds.), The Galactic Center. New York: American Institute of Physics, p. 148.

Liszt, H.S. 1985. Astrophys. J. (Lett.) **293**:L65.

Liszt, H.S., and W.B. Burton. 1978 Astrophys. J. **226**:790.

Liszt, H.S., and W.B. Burton. 1980. Astrophys. J. **236**:779.

Liszt, H.S., W.B. Burton, and J.M. van der Hulst. 1985a. Astron. Astrophys. **142**:237.

Liszt, H.S., W.B. Burton, and J.M. van der Hulst. 1985b. Astron. Astrophys. **142**:245.

Little, A.G. 1974. *In* F.J. Kerr and S.C. Simonson (eds.), Galactic Radio Astronomy. Dordrecht: Reidel, p. 491.

Lo, K.Y., and M.J. Claussen. 1983. Nature **306**:647.

Lo, K.Y., D.C. Backer, R.D. Ekers, K.I. Kellermann, M. Reid, and J. Moran. 1985. Nature **315**:124.

Mulder, W. 1985. Astron. Astrophys. **156**:354.

Oort, J.H. 1977. Annu. Rev. Astron. Astrophys. **15**:295.

Oort, J.H. 1985. *In* H. van Woerden, R.J. Allen, and W.B. Burton (eds.), The Milky Way Galaxy. Dordrecht: Reidel, p. 345.

Rieke, G.H., C.M. Telesco., and D.A. Harper, D.A. 1978. Astrophys. J. **220**:556.

Rougoor, G.W. 1964. Bull. Astron. Soc. Neth. **17**:381.

Sofue, S., and T. Handa. 1984. Nature **310**:568.

van Gorkom, J.H., U.J. Schwarz, and O. Bregman. 1985. *In* H. van Woerden, R.J. Allen, and W.B. Burton (eds.), The Milky Way Galaxy. Dordrecht: Reidel, p. 371.

Vietri, M. 1986. Astron. Astrophys. **306**:48.

Yusef-Zadeh, F. 1987. Ph.D thesis, Columbia University.

Yusef-Zadeh, F., M. Morris, and D. Chance 1984. Nature **310**:557.

9. Radio Stars

ROBERT M. HJELLMING

9.1 The Early Years

Nearly all of the important work on radio emission from non-solar stars and stellar systems has occurred since 1970. During this period the wide variety of stellar radio emission phenomena has been made clear, and in some cases there are striking similarities to solar radio emission. Therefore, a review of earlier stellar work, and solar radio phenomena, provides useful perspective. In this chapter, we will consider only continuum radio emission, leaving circumstellar spectral line emission to other chapters.

9.1.1 The Search for the Radio Sun

Soon after Heinrich Hertz detected, in 1888, the electromagnetic waves predicted by Maxwell's electromagnetic theory, Sir Oliver Lodge made the first attempt to detect radio emission from a star—the Sun. This attempt was made in his laboratory in Liverpool, England, using a crude detector, without even an "antenna" wire; however, only the radio interference from the industrial city of Liverpool was detected. Soon after that, a graduate student at the University of Paris, Charles Nordmann, came to some conclusions that were far ahead of his time, and attempted an experiment that might have been successful if he had been more patient. Nordmann realized that: (1) an antenna was necessary to provide collecting area, even if it was only a long wire; (2) a quiet site far away from industrial radio noise was important; (3) the radio emission from the Sun might vary with the solar cycle; and (4) he might find what we now call "nonthermal" emission from solar noise storms related to sunspots. Nordmann carried his apparatus to the top of a high glacier in France, tried to detect the Sun with a 175-m antenna wire on a single day (19 September 1901), but obtained negative results. It is surprising that he did not try the experiment again, because he expected variations with the solar cycle, and it was a time of minimum solar activity. His detection apparatus was probably capable of detecting the stronger bursts of solar radio emission, so it has been speculated that he would have detected solar radio bursts if he had been more persistent. It was not until the 1940s that solar radio emission was finally detected, almost ten years after Jansky and Reber had already detected (whether they knew it or not) radio emission from the galactic center, a supernova remnant, a radio galaxy, and ionized hydrogen complexes.

One of the major causes of the forty-year delay was the effect of a theory. In 1902, Planck announced his revolutionary radiation theory. Unfortunately, one result was that too many people then thought they *knew* solar radio emission would not be detected with available equipment. It is amusing to speculate whether a successful experiment by Nordmann would have been used to argue that Planck's theory was "wrong." In any case, after decades during which radio operators listened unknowingly to radio static from the Sun, Hey, Southworth, and Reber all independently recognized that they had detected solar radio emission.

9.1.2 Early Flare Star Observations

The early thinking about radio emission from non-solar stars was guided by the fact, as we will shortly discuss, that most solar radio emission occurs in a large variety of flares of varying lifetimes. Since certain red dwarf stars were known to exhibit flares at optical wavelengths, they were obvious candidates for searches for radio flares. Lovell (1969) at Jodrell Bank pioneered observations of this class of objects, reporting a large variety of flaring events, some of which seemed to be coincident with optical flares.

Unfortunately, many of the early Jodrell Bank results may have been only measurements of interference. The first indisputable observations of radio flare star emission from Jodrell Bank were reported by Davis et al. (1978); they used long baseline interferometers with techniques to discriminate amongst flare star events, interference, and confusion from other radio sources. In their paper, they noted that "previous results were not trustworthy" and unambiguously observed a 408-MHz radio flare in YZ CMi which lasted three hours and contained 2.8×10^{25} ergs in a 4-MHz bandwidth. This is about 10^6 times the energy in a large solar flare of importance 3; they also observed a flare in UV Ceti which was a factor of six weaker than the YZ CMi event. Fortunately, we now have a variety of exciting flare star results obtained by methods similar to those used by Davis et al. (1978); unfortunately, the very strong events discussed by Lovell (1969), which are thousands of times stronger, have not yet been confirmed with new interferometric techniques or when independent monitoring of interference is done. As will be discussed later, observations with Arecibo and the Very Large Array (VLA) have established tantalizing similarities between stellar flares and solar flare events, although the former are considerably more energetic.

9.1.3 When Is a Star a Radio Star?

Some very simple considerations will tell us the gross properties of any star likely to be observed as a radio source. We will consider any radio star to be represented by an average brightness temperature, T_B, and an average solid angle for the emitting source region(s), Ω_s; then the flux density at a radio frequency v is given by

$$S_v = \left(\frac{2kv^2}{c^2}\right) T_B \Omega_s \tag{9.1}$$

where k is Boltzmann's constant, c is the speed of light, and we have used the Rayleigh-Jeans (low-frequency) approximation for the Planck radiation formula. It is useful to define an equivalent disk diameter, θ, by $\Omega_s = \pi\theta^2/4$; we can then write Equation (9.1) in the following useful form:

$$S_v = \frac{(T_B\theta^2)}{(1970\lambda^2)} \tag{9.2}$$

where S_ν is in janskys ($1 \text{ Jy} = 10^{-26} \text{ W m}^{-2} \text{ Hz}^{-1} = 10^{-23} \text{ erg s}^{-1} \text{ cm}^{-2} \text{ Hz}^{-1}$), T_B is in kelvins, θ is in arcseconds, and λ is in centimeters.

If S_{min} is the lowest flux density that can be measured with a particular radio telescope, it can detect a particular radio star only if

$$T_B \theta^2 \geq 1970 \lambda^2 S_{min} \ . \tag{9.3}$$

The parameters of a radio star that govern its detectability are the brightness temperature and angular diameter for the emitting region(s). Of course, this is true for any radio source, but Equation (9.3) is very important in the radio star business because, of the two parameters, the angular diameter of an emitting region affects detectability the most, and almost by definition, θ is very small for a star. In 1970, detection thresholds were 0.01 Jy or higher, requiring brightness temperatures considerably higher than the effective temperatures of almost all stars in order to produce a detectable radio source.

One conclusion from this obvious arithmetic is that if any radio stars are detectable, particularly at high flux density levels, there must be either sizes much larger than the optical diameters or physical processes involving unusual amounts of energy and unusually high surface brightnesses. To see this in another way, let us consider the maximum distance at which a star like the Sun would be observable at, say, $\lambda = 6$ cm. If 1 mJy ($=0.001$ Jy) is taken to be a typical minimum flux density, which is valid for multihour observations with the VLA, the quiet Sun could be seen out to 0.2 pc; the slowly varying component ot the Sun could be seen at 0.3 pc; at very strong solar radio bursts could be seen out to 7.6 pc. The nearest star, Proxima Centauri, is 1.3 pc away, and our galaxy is $\sim 20{,}000$ pc in size, so it would seem that solarlike emission would be seen from only a few nearby stars. However, since 1970 a wide variety of stars and stellar systems, even at distances of kiloparsecs or more, have proven to have either unusually high brightness temperatures or unusually large angular sizes for their radio-emitting regions, or both. Thus, studies of stars and stellar systems have become a major field of radio astronomy.

As we shall see, there are many combinations of large size scale and/or high brightness temperatures that have been found in the first two decades of radio star work. Figure 9.1 shows a diagram of stellar magnitudes plotted against stellar spectral types in which active nonthermal radio stars, both single stars and late-type components of binary systems, are plotted (Gibson 1985). In Figure 9.1, the symbol sizes are proportional to the radio luminosities, and the crosses indicate coherent emission, which is necessary to compensate for the small sizes of lower main-sequence stars.

9.1.4 The "Twilight" Years
From the initial observations of red dwarf flare stars until about 1970, the radio astronomy of stars was in sort of a "twilight" phase. Very few radio astronomers were interested in observing the red dwarf flare stars because of the enormous investment in observing time on the biggest available instruments needed to observe what seemed to be relatively few events.

In addition, there were a number of reports of detections which, both then and now, are difficult to assess fairly. As pointed out earlier, the angular diameter of a

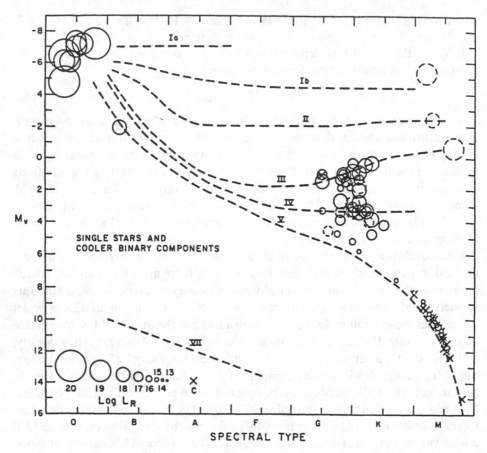

Fig. 9.1. The H-R diagram for normal nonthermal radio stars, including both single stars and the late-type components of active binaries. The range of radio luminosities is indicated by the size of each circle, with the scale indicated in the lower left part of the diagram. Coherent emitters are indicated with a cross, and where the radio luminosity is for frequencies other than 5 GHz, the circles are dashed.

star is so important in its detectability that early efforts to observe radio stars concentrated on the nearby red supergiants. The first report of a detection of radio emission from a red supergiant was made by Kellermann and Pauliny-Toth (1966), who reported an apparent signal of 0.11 ± 0.03 Jy while observing α Orionis (Betelgeuse) at $\lambda = 2$ cm on February 21, 1966. They found no signal, however, on the next eleven nights. Seaquist (1967) reported the possible detection at 2.8 cm of radio emission from α Ori and another red supergiant, π Aurigae, at the 0.023 ± 0.006 and 0.031 ± 0.009 Jy levels, respectively. Although α Ori has been reliably detected since then, neither star has been observed at anywhere near these flux levels. The problem with evaluating these reports is simple, but insoluble: as with most scientific experiments, results that are not reproducible are assumed (sometimes by the experimenters and nearly always by everybody else) to be the results of unknown problems with the equipment. Because most radio stars are variable on some time scale, or range of time scales, it can be unclear when different results mean some measurements are "bad" and when a difference is "proof" of time variability, unless the same differences are seen in independent data. Further, at weak levels, the problem of confusion with background sources can arise, particularly with single antennas with resolution of some minutes of arc. Both of these effects produce serious skepticism when reported detections are near the noise or confusion levels—even when such skepticism is unwarranted.

An example of unwarranted skepticism about results obtained during the twilight years is the early history of the radio emission from the peculiar binary star identified with the strong X-ray source Sco X-1. In 1968, Andrew and Purton (1968) found weak radio emission from the vicinity of Soc X-1 at the level of 0.021 ± 0.007 Jy at 4.8 cm. Soon after that, Ables (1969a) reported time variations about a mean level of 0.022 ± 0.002 Jy at 6 cm. It is fair to say that most radio astronomers did not believe the evidence for variability, and many felt the detection of a source was doubtful because of confusion effects. We now know that in some sense, nearly everybody was correct. As we shall discuss in connection with the radio counterparts of X-ray binaries, the single-dish measurements were seeing three radio sources, only one of which coincides with the X-ray star, and the star's radio emission can be very variable. Because of this, the early work on the radio counterpart of Sco X-1 can be properly described as part of the twilight years.

9.1.5 The Dawn of Stellar Radio Astronomy
There are three main reasons why one can say that the systematic study of radio stars dawned in 1970. These are the three reasons why variable radio stars are best observed with multiantenna, phase-stable, tracking interferometers, and observations of radio stars with such interferometers began in 1970.

The first reason for the importance of radio interferometers is the capability they provide to determine very accurate positions of radio sources. It is very important to establish that the position of a radio source coincides with that of a star to within an arcsecond or better. The second reason is that high sensitivity can be obtained because *all the data* collected for long observing times can be added together to improve the signal-to-noise ratio of the image of a radio star. Sometimes this cannot be done reliably with continuum measurements by a single antenna, even if it has comparable noise characteristics, bandwidth, and collecting area, because of systematic atmospheric and instrumental effects. This problem does not exist for an array of three or more radio telescopes, except when there is *both* a very high frequency of observation and a very inhomogeneous atmosphere over the array; hence, the useful sensitivity almost always improves as the square root of the time spent observing a particular field. The third advantage a multielement interferometer has for the radio star business is now obvious, but only after it has been shown that most radio star emission is highly variable on some time scale: *the proof that a time-variable or transient event is real is most certain if independent subsets of an array are detecting the same time-variable event.*

The previous discussion indicates both why the systematic study of radio stars began in 1970 and why the NRAO VLA became a dominant instrument for radio star studies in the 1980s. In 1970 the NRAO three-element interferometer in Green Bank, West Virginia, and the Westerbork interferometer in Holland were first used to observe radio stars. Through most of the 1970s these instruments dominated radio star work, until the 27-antenna VLA, with much greater sensitivity and instrumental flexibility, was available to dominate the field by the end of that decade. Hundreds of different stellar systems are now known to be radio sources. In some cases, such as for the flare stars and active binaries, the radio phenomena are analogous, but usually more powerful, versions of solar phenomena. In other cases, the objects are radio sources because of massive stellar winds. Some stars eject shells or features frequently called jets. Binaries with strong X-ray star components are prodigious sources of synchrotron emission in many forms, including jets, lobes, and extended synchrotron-radiating nebulosities. However, before discussing the variety of the new stellar components of radio astronomy, let us briefly discuss some more fundamentals.

9.2 Fundamental Radio Emission Processes and Stellar Radiative Transfer

9.2.1 Radiative Transfer in Stellar Environments

Because the radio emission from a stellar system can begin in the spherically symmetric environment near the surface, it is more worthwhile to consider the emission and absorption of radiation in spherically symmetric layers than is common in the rest of radio astronomy. In particular, spherically symmetric models have been very successful in interpreting the radio emission from many stellar winds and novae, largely because they are simple outflow phenomena. In more complicated cases, the spherically symmetric model is still the starting point from which more complicated geometries are considered. For this reason, and because some cases lead to simple solutions, let us consider the radiative transfer of radiation in spherically symmetric layers around stars. Let us assume that the microphysics of radio emission processes for frequency v can be represented by a mass emission coefficient j_v and a mass absorption coefficient κ_v. If I_v is the specific intensity, then the equation of radiative transfer along a one-dimensional path with coordinate s is

$$\frac{dI_v}{ds} = -\kappa_v \rho I_v + j_v \rho \tag{9.4}$$

where ρ is the mass density at position s. Defining the optical depth at frequency v by

$$d\tau_v = \kappa_v \rho \, ds \, , \tag{9.5}$$

Equation (9.4) has the well-known "formal solution" for the specific intensity seen by a distant observer. It can be written as

$$I_v(\tau_v) = \int \left(\frac{j_v}{\kappa_v}\right) e^{-(\tau_v - t_v)} \, dt_v \tag{9.6}$$

where we have assumed a negligible contribution due to the radio emission beyond the star and τ_v is the total optical depth through a particular line of sight. One sees

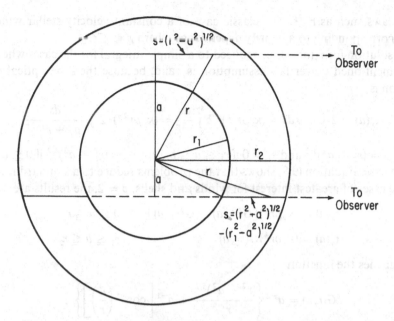

Fig. 9.2. A schematic diagram showing the cross-section geometry for both spherical shells and cones, where r is a radius with r_1 and r_2 identifying the inner and outer boundaries of a shell, a is a coordinate identifying a line of sight for an observer, and s is one-half of the line-of-sight path through the shell.

that the problem of calculating specific intensities is partly a problem of integrating Equation (9.5) to obtain the optical depth variable (t_ν) for all necessary combinations of s and lines of sight and partly a problem of integrating Equation (9.6) for each line of sight. For the cases where j_ν/κ_ν is constant, Equation (9.6) can be integrated immediately to obtain

$$I_\nu(\tau_\nu) = \left(\frac{j_\nu}{\kappa_\nu}\right)[1 - e^{-\tau_\nu(a)}] \qquad (9.7)$$

so for this case we need only calculate the optical depth for each line of sight (coordinate a).

In Figure 9.2, we define the geometry of successive spherically symmetric layers in which each line of sight passes a distance a from the center of the star and r is the radial coordinate with respect to the stellar center, with r_1 and r_2 identifying the inner and outer radii of some layer of interest. Inspection of the two lines of sight in Figure 9.2, with one passing through the inner "hole" and one not, indicates that the solutions of the radiative transfer problem for these cases can be applied to infinite stellar winds (r_1 = stellar radius, $r_2 \to \infty$), finite stellar winds [r_1 = stellar radius, $r_2 = v_{\mathrm{wind}}(t - t_0)$, where t is the time and t_0 is when the ejection begins], shell ejecta such as nova shells [$r_1 = v_1(t - t_0)$ and $r_2 = v_2(t - t_0)$ for inner and outer radii with velocities v_1 and v_2, respectively], *or to any combination of multiple layers due to successive winds, shells, etc.* The simplifying assumption that allows us to write both moderately general and useful solutions for all of these cases is that each layer has dependencies of both $\kappa_\nu\rho$ and $j_\nu\rho$ on r which can be expressed as power laws such as r^{-2p}. The classic case of a constant-velocity stellar wind has $p = 2$, corresponding to a density dependence where $\rho \propto r^{-2}$.

The solution for $\tau_v(a)$ can be reduced to a simple integral for the cases where the above-mentioned power law assumption is valid, because then the optical depth equation is

$$\tau_v(a) = 2 \int_{s_1}^{s_2} \kappa_v \rho \, ds = (\kappa_v \rho r^{2p}) \cdot 2 \int_{s_1}^{s_2} \frac{ds}{r^{2p}} = (\kappa_v \rho r^{2p}) \cdot 2 \int_{s_1}^{s_2} \frac{ds}{(a^2 + r^2)^{2p}} \qquad (9.8)$$

where $s_2 = (r_2^2 - a^2)^{1/2}$ and $s_1 = 0$ if $r_1 \leq a \leq r_2$, but $s_1 = (r_1^2 - a^2)^{1/2}$ if $0 \leq a \leq r_1$. In each case, Equation (9.8) shows that the problems reduce to a standard integral. For the cases of greatest interest for winds and shells, $p = 2$, the results are

$$\tau_v(a) = (\kappa_v \rho r^4)[G(r_2, a) - G(r_1, a)] \qquad 0 \leq a \leq r_1$$
$$\tau_v(a) = (\kappa_v \rho r^4) G(r_2, a) \qquad\qquad\quad r_1 \leq a \leq r_2 \qquad (9.9)$$

if one defines the function

$$G(r, a) = a^{-2} \left\{ \frac{(r^2 - a^2)^{1/2}}{r^2} + a^{-1} \left[\cos^{-1}\left(\frac{a}{r}\right) \right] \right\} . \qquad (9.10)$$

For the important limit where $r_2 \to \infty$, $G(\infty, a) = \pi/(2a^3)$, so

$$\tau_v(a) = (\kappa_v \rho r^4)\left(\frac{\pi}{2a^3}\right) . \qquad (9.11)$$

For the cases where (j_v/κ_v) is constant along the line of sight, the solution is now complete. If it is not constant, different integrals must be evaluated either analytically or numerically.

The specific intensity is, of course, never an observed quantity. At best, one averages over a small solid angle corresponding to an antenna beam or synthesized beam shape. This calculation is basically the convolution of a grid of calculated specific intensities with a function representing the beam shape. However, there are two other "observables" that are derived from such a grid: the source flux density (S_v) and the observed visibility function (V_v) if one observes with an interferometer. By definition,

$$S_v = \int_{\text{source}} I_v \, d\Omega_s = \frac{2k}{\lambda^2} \int_{\text{source}} T_b \, d\Omega_s \qquad (9.12)$$

and V_v is the two-dimensional Fourier transform of the product of I_v and an antenna pattern function, $f(x, y)$, assuming we can use the tangent plane approximations $x = (\alpha - \alpha_0) \cdot \cos \delta$ and $y = \delta - \delta_0$ to relate the angular offsets (x, y) to the (assumed) antenna pointing and interferometer tracking position (α_0, δ_0). Expressing the projection of the interferometer basline upon the sky as (u, v), this means

$$V_v(u, v) = \iint I_v(x, y) \cdot f(x, y) \cdot e^{-2\pi i(u \cdot x + v \cdot y)} \, dx \, dy , \qquad (9.13)$$

which "simplifies" even further in the case of a spherically symmetric object, taking $q = (u^2 + v^2)^{1/2}$ and $\theta = (x^2 + y^2)^{1/2}$, to the Hankel transform

$$V_v(q) = \int I_v(\theta) \cdot f(\theta) \cdot J_0(2\pi q\theta) \cdot 2\pi\theta \cdot d\theta \qquad (9.14)$$

Table 9.1. The observables for simple brightness temperature formulas.

Model	$I_b(\theta)/I_{b,\max}$	$S_\nu(\text{Jy})$	$V_\nu(q)/S_\nu$
Gaussian	$\exp[-4\ln 2(\theta/\theta_{\rm H})^2]$	$\dfrac{T_b({\rm K})\theta_{\rm H}^2(\text{arcsec}^2)}{1360\lambda^2(\text{cm}^2)}$	$\exp[-(\pi^2/4\ln 2)(q\theta_{\rm H})^2]$
Uniform disk	$1, \theta \leq \theta_{\rm H}/2$ $0, \theta > \theta_{\rm H}/2$	$\dfrac{T_b({\rm K})\theta_{\rm H}^2(\text{arcsec}^2)}{1961\lambda^2(\text{cm}^2)}$	$2J_1(\pi q\theta_{\rm H})/(\pi q\theta_{\rm H})$
Limb-brightened (shotglass)	$2\theta_{\rm H}/[(\theta_{\rm H})^2 - \theta^2]^{1/2}, \theta \leq \theta_{\rm H}/2$ $0, \quad \theta > \theta_{\rm H}/2$	$\dfrac{T_b({\rm K})\theta_{\rm H}^2(\text{arcsec}^2)}{981\lambda^2(\text{cm}^2)}$	$\sin(\pi q\theta_{\rm H})/(\pi q\theta_{\rm H})$
Thin ring	$\delta(\theta - \theta_{\rm H}/2)$	$\dfrac{420T_b({\rm K})\theta_{\rm H}(\text{arcsec})}{\lambda^2(\text{cm}^2)}$	$J_0(\pi q\theta_{\rm H})$

in which J_0 is a Bessel function of zeroth order. If x, $y = 0, 0$ corresponds to the center of the object, then $\theta = a/d$, where d is the distance to the object. In Table 9.1, we present the results for the observables for a number of simple formulas for stellar surface brightnesses, all of which can correspond to some simple extreme of a model.

In a later section, we will discuss the observables involved in ionized and partially ionized stellar winds with spherical symmetry, constant velocity, and an r^{-p} density distribution.

9.2.2 Fundamental Radiation Processes

(a) Thermal Bremsstrahlung
In the previous sections we noted that for thermal bremsstrahlung radiation, the quantities $j_\nu\rho$ and $\kappa_\nu\rho$ are proportional to ρ^2 and that j_ν/κ_ν can often be assumed to be independent of position, leading to simple solutions to the radiative transfer equations in spherically symmetric environments. Partly to justify this and partly to lead into a contrast with another major radiation process in radio astronomy, synchrotron radiation, let us summarize the formulas for these quantities. Thermal bremsstrahlung or free-free radiation processes are caused by interactions between free electrons and positive ions in a partially or fully ionized plasma. The emission coefficient for free-free emission is given by

$$j_\nu\rho = 5.4 \times 10^{-39} N_e^2 T_e^{-1/2} g_{\rm ff} e^{-h\nu/kT_e} \cong 7.45 \times 10^{-39} N_e^2 T_e^{-0.34} \nu_{\rm GHz}^{-0.11} \qquad (9.15)$$

where we have used

$$g_{\rm ff}(\nu, T_e) = \left(\frac{3^{1/2}}{\pi}\right)\left[17.7 + \ln\left(\frac{T_e^{3/2}}{\nu}\right)\right] \cong 1.38 T_e^{0.16} \nu_{\rm GHz}^{-0.1} \qquad (9.16)$$

for the free-free Gaunt factor with an approximation valid at radio wavelengths. In these equations, N_e is the electron concentration, T_e is the electron temperature, and h is Planck's constant. The Planck function for blackbody radiation relates the emission and absorption coefficients, i.e.,

$$\frac{j_\nu}{\kappa_\nu} = \frac{2h\nu^3/c^2}{e^{h\nu/kT_e} - 1} \cong 2kT_e\left(\frac{\nu}{c}\right)^2 \qquad (9.17)$$

where the approximation is again valid for radio wavelengths. For $v > 10^{12}$ Hz, there are additional terms (Hjellming et al. 1979) in Equation (9.15) due to free-bound transitions that make a significant difference at infrared wavelengths.

Equations (9.15)–(9.17) show that j_v/κ_v is constant if the temperature in the region of absorption/emission is constant and that both $j_v\rho$ and $\kappa_v\rho$ are proportional to N_e^2 so, with $N_e \propto \rho$, they become proportional to r^{-2p} when $\rho \propto r^{-p}$.

(b) Magneto-Bremsstrahlung
The other radiation process of dominant importance is due to the interaction between fast-moving electrons and magnetic fields, which we will generally call magneto-bremsstrahlung. There are three major varieties of magneto-bremsstrahlung emission. When the moving electrons are very relativistic, the emission process is the very efficient, highly beamed synchrotron emission which dominates the emission from many galactic and most extragalactic radio sources. However, in some stars two other magneto-bremsstrahlung processes occur when the moving electrons are non-relativistic or only mildly relativistic. When mildly relativistic ($\gamma \approx 2$ to 3) electrons undergo interactions with magnetic fields, the emission process is called gyrosynchrotron, which is much less beamed and much less efficient at producing radio emission; however, it tends to be highly circularly polarized, a characteristic commonly found in stellar radio emission. Even less efficient is the magneto-bremsstrahlung resulting from less relativistic particles, with $\gamma \lesssim 1$, which is called cyclotron or gyroresonance emission.

The details of magneto-bremsstrahlung radiation processes are straightforward but very complicated for the nonrelativistic (cyclotron) and mildly relativistic (gyrosynchrotron) cases; we will not discuss them in this chapter. They are summarized by Dulk (1985) and extensively discussed by Kundu (1965) and Zheleznyakov (1970).

Emission from expanding synchrotron-emitting layers or bubbles of relativistic electrons is important for some radio stars; because of this, let us summarize the most commonly used form of the synchrotron radiation emission and absorption coefficients. We assume that relativistic electrons can be described by power law energy distribution density $N(E) = KE^{-\gamma}$ in the energy range E to $E + dE$, where γ is a constant. If these electrons are mixed in with a uniform distribution of magnetic fields of strength H, which can be described as having uniformly random directions on a large scale, then (Ginzburg and Syrovatskii 1965) the emission and absorption coefficients are

$$j_v\rho = 1.35 \times 10^{-22} a(\gamma) K H^{(\gamma+1)/2} \left(\frac{6.36 \times 10^{18}}{v} \right)^{(\gamma-1)/2} \tag{9.18}$$

and

$$\kappa_v\rho = 0.019 g(\gamma)(3.5 \times 10^9)^\gamma K H^{(\gamma+2)/2} v^{-(\gamma+4)/2} \tag{9.19}$$

in CGS units with H in gauss so that

$$\frac{j_v}{\kappa_v} = 2.84 \times 10^{-12} \frac{a(\gamma)}{g(\gamma)} 2^{-\gamma/2} H^{-1/2} v^{5/2} . \tag{9.20}$$

Table 9.2 gives some of the values of the functions $a(\gamma)$ and $g(\gamma)$.

Table 9.2. Incoherent synchrotron emission constants

γ	$a(\gamma)$	$g(\gamma)$	$\alpha_{opt.\ thin}$
1.0	0.283	0.96	0.0
1.5	0.147	0.79	0.25
2.0	0.103	0.70	0.50
2.5	0.0852	0.66	0.75
3.0	0.0742	0.65	1.0
4.0	0.0725	0.69	1.5
5.0	0.0922	0.83	2.0

We see from Equation (9.20) that the important case where j_v/κ_v is constant occurs if the magnetic field is uniform. Equations (9.18)–(9.20) also tell us that for the optically thick and thin limits, I_v is proportional to $H^{-1/2}v^{5/2}$ and $KH^{(\gamma+1)/2}v^{-(\gamma-1)/2}L$, respectively, where L is a line-of-sight path length. Defining the spectral index α by $S_v \propto v^{-\alpha}$, then $\alpha = -2.5$ and $(\gamma - 1)/2$ in the optically thick and thin limits, respectively. The values of α in the latter case for various values of γ are listed in Table 9.2.

Spherically symmetric layers around stars that emit or absorb via synchrotron radiation processes have simple solutions for each layer's contribution to the optical depth as long as K and H are constant inside that region. As can be seen from Figure 9.2 and Equation (9.19), the optical depth for lines of sight through an inner spherical region is equal to a constant for each frequency times the path length $L = 2(r^2 - a^2)^{1/2}$. If the region is a spherically symmetric shell with inner radius r_1 and outer radius r_2, then $L = 2(r_2^2 - a^2)^{1/2}$ for $a \geq r_1$ and $L = 2[(r_2^2 - a^2)^{1/2} - (r_1^2 - a^2)^{1/2}]$ for $a \leq r_1$. We will use these facts later in discussing the evolution of synchrotron "events," particularly in X-ray binaries.

(c) Thermal Versus Nonthermal and Incoherent Versus Coherent
Emission processes are frequently described as thermal or "nonthermal." In many cases, the authors mean to identify thermal bremsstrahlung or magneto-bremsstrahlung processes. Technically, a thermal emission process is one in which the electrons involved in the emission process have a Maxwellian velocity distribution. A nonthermal process either has a non-Maxwellian velocity distribution function or has both a Maxwellian component and one or more non-Maxwellian components. One also distinguishes between coherent and incoherent emission processes. In an incoherent emission process, the velocity distribution is uniform in all directions whereas a coherent process has a velocity distribution that depends upon the direction of motion.

9.3 Solar Radio Emission

9.3.1 Types of Solar Radio Emission
The Sun has been and always will be the most extensively studied star exhibiting radio emission. During the last few decades, a wide variety of solarlike radio characteristics have been discovered in other stars, so it is not possible to rationally discuss radio stars without comparing and contrasting with solar radio phenomena.

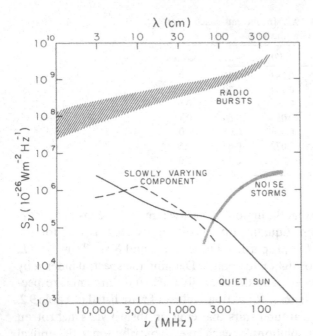

Fig. 9.3. A schematic representation of radio flux density as a function of frequency for four types of radio emission from the Sun. The slowly varying component corresponds to that for sunspot maximum.

For our brief discussion, we will divide solar radio events into four major groups: (1) quiet Sun radio emission; (2) the slowly varying component; (3) noise storms; and (4) bursts. The flux densities typical of these groups are shown schematically in Figure 9.3. Since the Sun is 4.85×10^{-6} pc in distance, while the next nearest star is about 1 pc, a simple scaling with distance would indicate that exact equivalents of the solar events in Figure 9.3 should range from 2×10^{-8} to 0.1 Jy. In practice, radio emission events 10^5 to 10^{10} times these levels are seen in other stars.

9.3.2 Quiet Sun Emission and the Slowly Varying Component

The radio emission from the quiet Sun is a relatively stable component that is always present. At a particular frequency, v, this emission is produced by thermal bremsstrahlung in or near the atmospheric levels where $v = v_p = 0.008 \, N_e^{1/2}$ MHz. N_e is the electron concentration in cm^{-3}. At meter wavelengths, the radiation is emitted by the hot, tenuous corona. At centimeter wavelengths, it comes mainly from the much denser corona and chromosphere, seen optically in certain emission lines. Only at short millimeter waves does it come from the still denser and cooler photosphere that is the surface of the star seen optically in continuum and absorption lines. Other types of radio events are strongly affected by the thermal absorption characteristics of the extended atmosphere responsible for quiet Sun emission. This is because of the simple physics of the propagation of radio waves through an ionized plasma. The index of refraction of radio waves of frequency v in a completely ionized plasma is given by

$$n = \left[1 - \left(\frac{v_p}{v} \right)^2 \right]^{1/2} \qquad (9.21)$$

so that at any level in a stellar atmosphere where $v_p > v$, n will be imaginary and the radio emission will be suppressed by absorption. Since radio waves propagate freely only when $v > v_p$, radiation of frequency v escapes the atmosphere only when it comes from heights above the level where $v \approx v_p$, and the atmosphere is opaque below the density level where $v \approx v_p$. Therefore, since N_e usually increases monotonically with increasing distance from the surface, observations at different frequencies tend to "see" different heights in the atmosphere. The higher the frequency, the deeper and more dense the layer at which the observed radiation can be produced. In the case of an optically thick, thermally emitting stellar atmosphere, it is exactly the level at which $v \approx v_p$ that is observed at frequency v, and the observed brightness temperature will be the electron temperature at that level. As we will discuss further later in this chapter, the closest equivalent of quiet Sun emission observed in other stars is the thermal emission from very extended, fully or partially ionized winds. For $v \geq 10$ GHz, the solar surface is largely a uniform disk with brightness temperatures $\leq 10^4$ K, with occasional brighter areas corresponding to active regions. For lower frequencies, the brightness and dominance of the active regions increases much more rapidly than the disk emission.

9.3.3 Enhanced Radio Emission Associated with Sunspots and Active Regions

The so-called *slowly varying component* is probably due to thermal magneto-bremsstrahlung and comes from bright regions with high-density coronal condensations, at 1×10^6 to 2×10^6 K electron temperatures, existing over sunspots and plage regions. At frequencies of about 1 GHz, this emission dominates on the Sun because of its 1×10^6 to 2×10^6 K brightness temperatures, compared to the 5×10^4 K of the disk component.

In Figure 9.4, we show images of the entire surface of the Sun as obtained by Dulk and Gary (1983). The upper left image shows the radio Sun at 1.4 GHz; the upper right shows the circularly polarized image at 1.4 GHz, with dark and light regions representing LH and RH polarization from $-$ and $+$ magnetic fields, respectively; the lower left image shows the emission at He $\lambda 10,830$ Å; and the lower right image shows a magnetogram of the photospheric emission. The clear association of radio emission from active regions with circular polarization and magnetic fields is related to the magnetic field dependence of polarized absorption processes for the thermalized high-energy electrons attached to high magnetic fields in loops or other field structures.

At frequencies below 1 GHz, the active region decreases in apparent brightness relative to the "disk," until at ≤ 100 MHz the disk is mainly coronal in nature, with a roughly uniform brightness temperature of $\sim 10^6$ K, upon which one sees bright features corresponding to coronal streamers and, occasionally, dark features corresponding to coronal holes and filamentary cavities.

At even lower frequencies, the solar radio emission is dominated by various types of bursts or flares, the strongest of which are intimately associated with noise storms.

Figure 9.5 shows a four-panel cartoon of various stages of solar events associated with various kinds of radio emission. The upper left panel shows that the roots of everything except quiet Sun disk emission are in regions of magnetic heating, somewhere in the photosphere or chromosphere, where there is an active region.

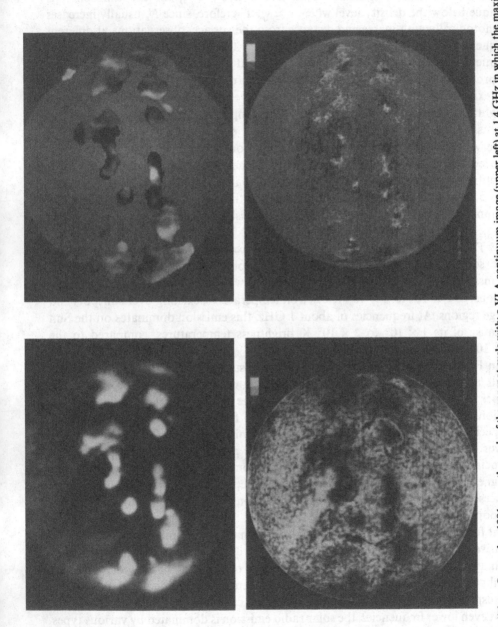

Fig. 9.4. The solar disk on September 1981, near the peak of the sunspot cycle with: a VLA continuum image (upper left) at 1.4 GHz in which the maximum surface brightness is 2.2×10^6 K; a polarization image (upper right) in which dark and light indicate LH and RH circular polarization, respectively; an optical image (lower left) in the He $\lambda 10{,}830$ Å line; and a magnetogram (lower right) for the photospheric magnetic field. (From Dulk and Gary 1983.)

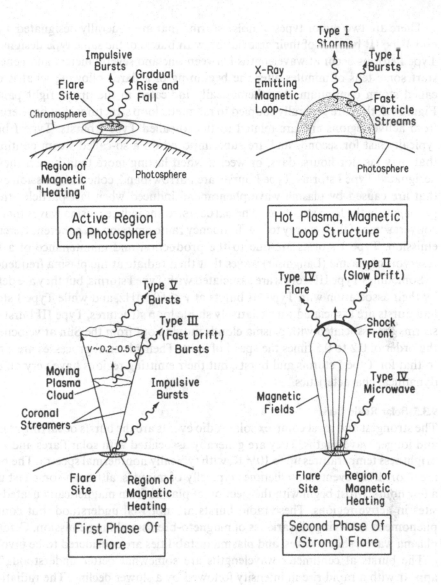

Fig. 9.5. A schematic representation of the basic environments in which various types of solar radio emission are related to other solar phenomena and events.

Such regions can exist without ever developing into flares; however, their pre-existence is necessary for the development of flares.

9.3.4 Solar Noise Storms

Solar noise storms frequently contribute the major part of the solar radio emission at 1- to 10-m wavelengths. They consist of a slowly varying broadband continuum enhancement lasting from a few hours to a few days. Near the maximum of the 11-year sunspot cycle, noise storms are in progress about 10% of the time. These storms originate between 0.1 and 1 solar radii about the photosphere, and they are generally beamed almost radially outward from the Sun.

There are two major types of noise storms that are generally designated Type I and Type III because of their association with bursts of the same type designation. Type I storms occur at wavelengths between one and several meters and generally start some tens of minutes after the beginning of a flare, following what is often called storm continuum. As schematically indicated in the upper right panel of Figure 9.5, they are generally coupled to magnetic loop structures that have erupted from active regions and are related to the so-called Type I bursts. Type I bursts typically last for seconds and are superimposed on a so-called storm continuum that may last for hours, days, or weeks; when lasting more than hours, they are designated Type I storms. Type I bursts are narrow-band, coherent emission events that are caused by plasma wave phenomena induced when fast particle streams propagate in the loop structures. The particle streams induce plasma waves that then convert some of their energy to low-frequency radio emission via coherent Cerenkov emission. Type I storms are due to the production and maintenance of a large reservoir of plasma (Langmuir) waves that then radiate at the plasma frequency.

Sometimes Type III storms are associated with Type I storms, but they are defined by their association with Type III bursts at $v \leq 50$ MHz, and while Type I storms and bursts are associated with relatively stable loop structures, Type III bursts and storms are associated with plasma clouds moving out from the Sun at velocities of the order of 0.2 to 0.5 times the speed of light. Their physical processes are similar to that for Type I storms and bursts, but their emitting regions have very different dynamical characteristics.

9.3.5 Solar Radio Bursts

The strongest and most complex solar radio events are the bursts occurring at meter and longer wavelengths. They are generally associated with solar flares and attain brightness temperatures up to 10^{12} K, with typically nonthermal spectra. The bursts are short-lived events with lifetimes typically of minutes, although some last up to a few hours. Most begin with the ejection of plasma from magnetically heated flare sites in active regions. These radio bursts are not well understood, but complex phenomena involving all varieties of magneto-bremsstrahlung emission, Cerenkov plasma waves, shock waves, and plasma instabilities are considered to be involved.

The bursts at centimeter wavelengths are somewhat better understood. They appear with a rapid rise in intensity followed by a slower decline. The radiation is frequently partially circularly polarized and appears as a smooth continuum. Three major categorizations of centimeter bursts are: (1) the *impulsive bursts*, in which a rapid rise in times of the order of a minute is followed by a decline lasting a few minutes; (2) the *postburst*, lasting up to a few tens of minutes, in which there is a slow decay of intensity to a preburst level; and (3) bursts showing a *gradual rise and fall* of intensity, lasting up to a few tens of minutes. The centimeter bursts are empirically well correlated with X-ray flare events.

(a) Radio Emission Associated with First-Stage Flares
While radio burst phenomena can show a bewildering variety of characteristics, there are some major features that can be briefly described. The lower left panel of Figure 9.5 shows some of the major features of a simple view of the so-called first phase of both weak and strong flares. In this case, one of the consequences of a

region of magnetic heating is not only the usual impulsive bursts and gradual rise and fall, but in addition a cloud of plasma is ejected at high velocities, typically $0.2c$ to $0.5c$, with radial boundaries that can sometimes be identified with coronal streamers seen at optical wavelengths. During just a few minutes, often at the beginning of a major (second-phase) flare, one sees Type III or fast-drift bursts which change in frequency from the order of 500 MHz to tens of MHz and which are highly circularly polarized; the fast frequency drift occurs because of the rapid motion of emitting material through different density layers and hence different plasma frequencies. The emission process is radiation at both fundamental and higher harmonics from plasma waves in the rapidly moving ejecta. Frequently following Type III bursts, by one to a few minutes, are Type V bursts which consist of continuum emission over a wide frequency range (10 to 100 MHz). Type V bursts are usually of the opposite circular polarization to the preceding Type III bursts; they are related to more slowly moving electron streams exciting the same type of plasma waves, but second-harmonic radiation dominates and the angular distribution of Langmuir waves is larger. As indicated in the lower left panel in Figure 9.5, one can think of Type III bursts as related to higher-energy, more collimated "beams" of electrons and Langmuir waves, while the Type V bursts develop later when the electron streams are slower and the angular distribution is broadened.

(b) Radio Emission Associated with Second-Stage Flares

The strongest and most long-lasting types of flares occur when the previously discussed first-stage flares are followed by second-phase events. The ejection of very large plasma clouds leads to major stretching of magnetic field lines and the formation of a more slowly moving (few thousand kilometers per second) shock in front of decelerated (by interaction with external material) plasma cloud ejecta, with considerable leakage of high-energy particles back down the field lines to denser levels and the original flare site itself. In association with this second phase of a flare is a second phase of acceleration of particles to higher energies, e.g., ~ 1 GeV for protons and ~ 10 MeV for electrons, because of some combination of shock and turbulent acceleration processes. Electron streams in the shock-front region generate slow drift bursts due to coherent Cerenkov emission of the Langmuir waves that are induced, and the emission takes a few tens of minutes to drift from frequencies of ~ 300 MHz to 10 MHz, since everything is moving more slowly outward from the solar surface.

There are many varieties of radio emission associated with second-stage flares that are all called Type IV. The so-called Type IV microwave continuum is emission at higher frequencies than normal for bursts, that is, $3 \leq \nu \leq 30$ GHz, lasting for a few tens of minutes, due to a combination of synchrotron and plasma radiation, which occurs because of electron streams that return to near the solar surface and the original flare site. Many other varieties of Type IV emission—moving Type IV, Type IV flare continuum, and Type IV storm continuum—result from emission related to high-energy particles in outward-moving or stationary portions of very extended loops of plasma and magnetic field.

As we will be discussing later, there is a probable association between strong shocks and the very-high-energy relativistic electrons seen in radio emission from some other stars and stellar systems, particularly the X-ray binaries.

9.3.6 The Solar Wind

Although the solar wind is not generally directly observable from ground-based radio telescopes, its existence and well-observed characteristics are important indicators of winds that become observable at radio and other wavelengths around non-solar stars. As discussed by Parker (1963), the solar wind results because of heating and conduction at the base of the solar corona that results in a supersonic outflow of several hundred kilometers per second in the vicinity of the Earth. We now know that the solar wind is highly asymmetric near the surface of the Sun, with the high-velocity wind that is eventually seen by interplanetary probes originating from the low-coronal-density region over coronal holes (low-intensity X-ray emission regions) and very-low-velocity expansion originating from higher-density, X-ray-emitting, coronal loops.

Indeed, two of the main categories of radio stars are stellar wind objects, mainly hot, ionized stellar winds, probably driven by radiation pressure, and cool giant and supergiant winds, whose driving energy sources are some mixture of mechanical waves and Alfven waves. However, let us discuss these objects only after we have discussed the very variable radio emission from flare stars and active binaries.

9.4 Flare Stars and Active Binaries

9.4.1 Flare Star Radio Emission

As mentioned earlier, both the first reported and the first indisputable observations (Davis et al. 1978) of radio flares from nearby red dwarf flare stars occurred in the 1960s at Jodrell Bank. Since that time, the large collecting areas of the VLA and the Arecibo antenna have made possible significant strides in detailed observations of flare stars.

Solar flares are often highly circularly polarized and contain very impulsive events. In Figure 9.6, we show a radio flare star event in AD Leo that was observed by Lang et al. (1983) and which shows both characteristics. In the upper part of the figure one sees that over a period of about thirty minutes a complete flare event occurred that was subdivided into three components that were flares in right, then left, and then right circular polarization. The overall flare event can be interpreted in terms of a flare which accelerates electrons to energies up to 1 MeV and these electrons then interact with magnetic fields to radiate as gyrosynchrotron radiation with brightness temperatures of the order of 10^9 K. In the lower part of the figure, the second left circularly polarized (LCP) event, which lasts just a few minutes, is shown with 200-ms time resolution. One can see very "spiky" characteristics reminiscent of solar impulsive events. The rise times for some spikes are less than 200 ms. The spiky emission may be due to either a cyclotron maser or plasma radiation from Langmuir waves generated by the high-energy electrons.

As of this writing, some tens of red dwarf flare stars have been observed, including objects like UV Ceti, YZ CMi, EQ Peg, AU Mic, and AD Leo that have shown radio flares a large number of times. UV Ceti, also known as L726-8 A and B, and EQ Peg A and B are binary red dwarf systems in which both quiescent and/or flare emission have been observed. Because their binary components are separated in

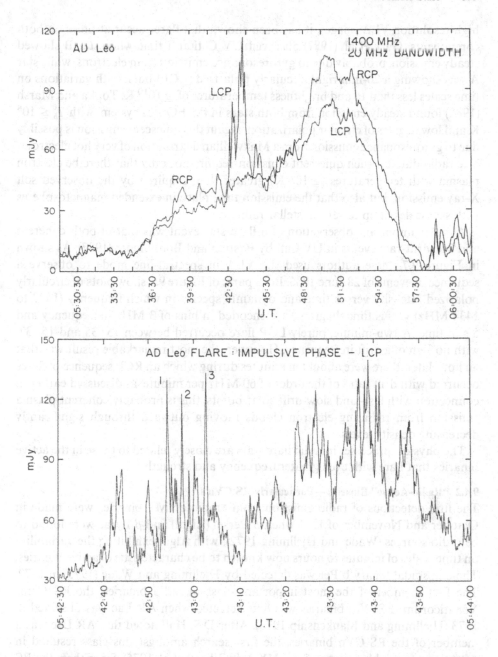

Fig. 9.6. A radio burst event at 1.4 GHz in the flare star AD Leo. The top portion shows thirty minutes of an event with moderate circular polarization while the bottom portion shows a spiky portion of the LH-polarized event with a time resolution of 200 ms in which the rise times of some spikes are still not resolved. (From Lang et al. 1983.)

high-resolution VLA images, it has been shown that flares occur at times on both components. Gary et al. (1983) observed UV Ceti at a time when star B showed steady emission, probably due to gyroresonance emission from electrons, while star A was showing a 20-min right circularly polarized (RCP) burst with variations on time scales less than 10 and brightness temperatures of $\geq 10^{10}$ K. Topka and Marsh (1982) found steady emission from both stars in the EQ Peg system, with $T_b \leq 10^8$ K and low degrees of circular polarization. Again this quiescent emission is possibly due to gyroresonance emission from a Maxwellian distribution of very hot electrons. The radio data on such quiescent emission require not only that there be electron plasma with temperatures ($\geq 10^7$ K) greater than required by the observed soft X-ray emission, but also that the emission take place in extended magnetospheres with size scales of up to several stellar radii.

Another important observation of a flare star event was that of both coherent and fast-drift-rate events in UV Ceti by Bastian and Bookbinder (1987). As shown in Figure 9.7, these authors used the VLA in spectral-line mode to observe a sequence of events of 28 June 1986. Both parts of Figure 9.7 show plots of circularly polarized intensity versus time and dynamic spectra in which frequency (1393 to 1434 MHz) versus time diagrams are encoded in bins of 3 MHz in frequency and 5 s in time. A two-minute, purely LCP flare occurred between 15 : 35 and 15 : 37, with no sign of a drift in frequency. However, the most remarkable result was that an hour later, there were about ten minutes during which an RCP sequence of flares occurred with drift rates of the order of 60 MHz per minute. as discussed earlier in connection with fast- and slow-drift solar bursts, this is probably coherent plasma emission from radiating electron clouds moving outward through significantly decreasing density levels.

The physical processes in radio flare stars are closely related to those in the active binaries that flare with even greater frequency and strength.

9.4.2 "Radio-Active" Binaries—Particularly RS CVn's

The first detections of radio emission from "radio-active" binaries were made in October and November of 1971 when β Persei (Algol) and β Lyrae were found to be radio sources (Wade and Hjellming 1972), with Algol exhibiting the variability on time scales of minutes to hours now known to be characteristic of active binaries. The ellipsoidal binary b Per was detected by Hjellming and Wade (1973) in 1972. The first member of the most important class of radio binaries, the RS Canis Venaticorum (RSCVn) binaries was first detected when AR Lac was observed in 1973 (Hjellming and Blankenship 1973). After D.S. Hall noted that AR Lac was a member of the RS CVn binaries, the first search amongst this class resulted in detection of variable emission from UX Ari (Gibson et al. 1975). Since then, the RS CVn's have constituted the largest group of variable radio stars, with more than forty known at the end of 1986.

In Figure 9.8 a dynamic spectrum display of three successive days of observation of the RS CVn binary UX Ari is shown (Hjellming and Brown 1987). These data were obtained by a combination of VLA observations at 1.4, 5, and 15 GHz and Green Bank interferometer observations at 2.7 and 8.1 GHz. When RS CVn binaries like UX Ari, HR1099, and HR5110 are "active," the radio emission processes are continuous series of "events" that frequently merge together. One of the most

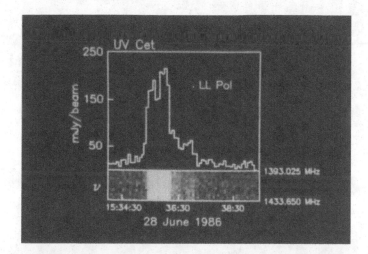

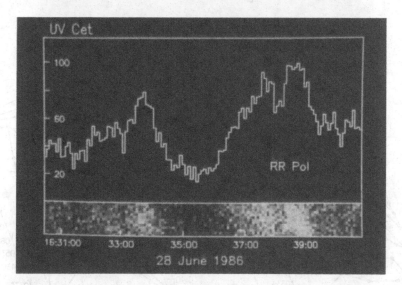

Fig. 9.7. LH (top) and RH (bottom) circularly polarized events in the flare star UV Ceti in which both the total intensity at 1.4 GHz and the frequency drift are shown as a function of time; no drift is seen in the LH event, but the later portions of the RH event show drifts over 40 MHz in about thirty seconds. (From Bastian and Bookbinder 1987. Reprinted by permission from Nature, Vol. 326, pp. 678, Copyright(c) 1987 Macmillan Magazines Limited.)

important "events" is the circularly polarized flare. This was first found by Brown and Crane (1978) in the event shown in Figure 9.9 for HR1099. In this case, an initial event of flaring only in LCP at 2.7 GHz was followed by a joint RCP and LCP event at 8.1 GHz.

Transcontinental and intercontinental radio telescope arrays have the potential to image the flaring in active binary systems if the radio-emitting regions are of the size scale of the binary system rather than being small regions on stellar surfaces. The pioneering VLBI observations of Clark et al. (1975) and Clark et al. (1976)

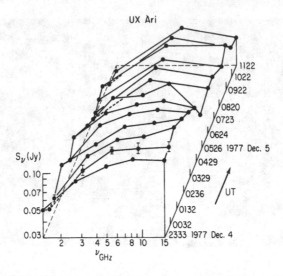

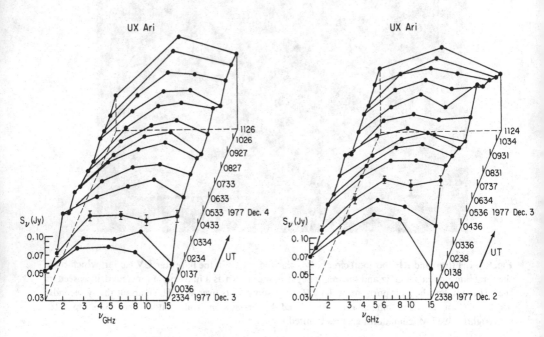

Fig. 9.8. Three days of observations of the RS CVn binary UX Ari at five frequencies are plotted in the form of spectra displaced upward as time increases.

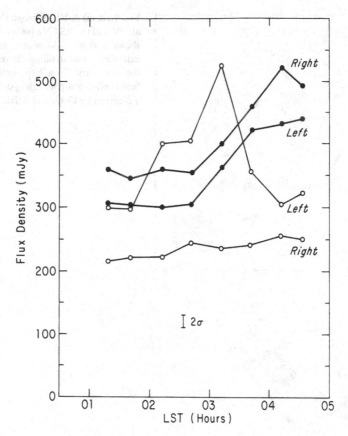

Fig. 9.9. An LH circularly polarized event is seen at 2.7 GHz in the RS CVn binary HR 1099. It is followed by an increase in flux in both polarizations at 8.1 GHz. (From Brown and Crane 1978.)

showed that β Per (Algol) has resolvable radio flaring events with size scales of the binary system ($\sim 10^{12}$ cm). The same VLBI techniques have been used by Mutel et al. (1985) to detect an unresolved "core" and an extended "halo" in UX Ari. Given the assumption that the unresolved component is associated with the more active K0 IV star, while the halo component extends over both magnetospheres since it is on the scale of the binary system, they obtained the image shown in Figure 9.10(a). In Figure 9.10(b) a schematic diagram, based upon the geometry discussed by Uchida and Sakurai (1983), shows a possible concept for the extended stellar magnetospheres that merge into the interacting field region between the stars; the relative orientation and size scale of Figure 9.10 illustrate the probable correspondence of stars, radio emission, and both magnetospheres. On one extreme, one can view all the high-energy electrons as coming from stellar surfaces, with radio emission possibly coming from the source star magnetosphere, the interacting field region between the stars, or the "target" star's magnetosphere. On the other extreme, one can view the continually interacting, reconnecting, etc. intrastar region as playing a role in particle acceleration processes. Because of the diversity of both the known observations and the lessons from the Sun about the many types of solar

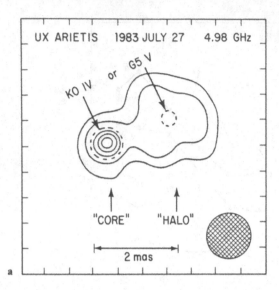

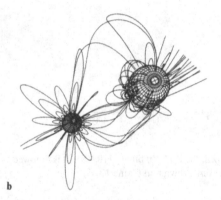

Fig. 9.10. (a) A VLBI image (from Mutel et al. 1985) of the RS CVn binary UX Ari which shows a core radio source and a halo of extended emission filling the region between the two stars; (b) a hypothetical magnetic field configuration for this system based upon a diagram by Uchida and Sakurai (1983).

phenomena, it is likely that all possibilities and combinations occur in some stellar systems at some times.

It is not within the scope of this chapter to discuss the details of the reasons for the relationship between stellar activity, of which time-variable radio emission is an observable symptom, and the generation of magnetic flux loops and starspots. However, this connection is believed to be the reason some stars in binary systems are more active, and hence exhibit more radio emission, than they would have as single stars. The generation of magnetic flux loops in the Sun is not well understood but is believed to be due to magnetic dynamos in the upper convective regions of the star. When these flux tubes protrude from the surface of the star, the magnetic fields produce a cooler region seen as a sunspot. Active binaries, particularly the RS CVn binaries, are known to have cool stars with very large fractions of their stellar surfaces covered by darker and cooler "starspots." Observations of line profiles in these stars indicate they are more rapidly rotating, in synchronism with the orbit of their companion star, than normal for single stars of the same type. It is therefore believed that binary synchronism has forced a higher rotation rate on the star and that this drives stronger convective based magnetic dynamos. The result is larger spot areas and more, and perhaps stronger, magnetic flux loops protruding from the stellar surface. As seen in our brief survey of solar radio emission, the magnetic heating in active regions is fundamental to producing ejected plasmoids, magnetic loops, particle acceleration, and associated radio phenomena.

9.4.3 Pre-Main-Sequence Stars

Beginning with the detection of radio emission from T Tauri by Spencer and Schwartz (1974), a number of cool stars that are in the process of evolving to the main sequence have been found to be continuum radio sources, often in connection with intense infrared emission, and always with complex molecular gas emission. The optical designations of these pre-main-sequence radio stars are usually T Tauri stars or Herbig-Haro objects. The best-known pre-main-sequence objects that are interesting radio sources are the north component of T Tauri, DG Tau, and V410 Tau. The emission from these objects is often resolved, and sometimes shows asymmetries and changes in structure (Cohen and Bieging 1986). The radio emission mechanism is free-free emission, probably due to jet-like ionized outflows (Schwartz et al. 1986). Although detailed models have not yet been worked out, the general idea is that infalling gas associated with star formation plays a role in making hot, ionized circumstellar structures that are the sources of, or become, windlike outflows. Physically, the physics of these objects is related to the general topic of bipolar outflows.

9.5 Stellar Winds

9.5.1 Ionized Stellar Winds

Another large group of radio stars contains the OB stars with ionized stellar winds. The massive O-star binary CC Cas (Gibson and Hjellming 1974) was the first in this class, but more than fifty were known by the end of 1986 (Abbott 1985, Abbott et al. 1986b).

9.5.2 Stellar Wind Observables

In an earlier section, we discussed the general problem of the observables involved in ionized and partially ionized stellar winds with spherical symmetry. Stellar winds are particular examples where one can assume constant velocity, and an r^{-2} density distribution, so that one gets the following simple expression for the flux density:

$$S_v = 130\{[\dot{M}/(M_\odot \ yr^{-1})][(km \ s^{-1})/V]/V\}/\mu\}^{4/3} T_e^{0.11} v_{Hz}^{0.59} \gamma^{2/3}/d_{kpc}^2 \ Jy \ . \quad (9.22)$$

The size of ionized stellar winds is important because they can be resolved with high-resolution aperture synthesis arrays like the VLA. At any frequency v, the angular diameter corresponding to $\tau_v = 1$ is given by

$$\theta_1 = 0.053''\{[\dot{M}/(10^{-5}M_\odot \ yr^{-1})][(10^3 km \ s^{-1})/V]/\mu\}^{2/3}(T_e/10^4)^{-0.42}$$

$$(5 \times 10^9 v_{Hz}^{0.7})\gamma^{1/3}/d_{kpc} \ . \quad (9.23)$$

Newell (1981) and White and Becker (1982) first showed that the ionized wind from

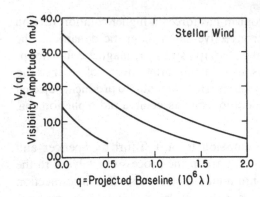

Fig. 9.11. The visibility function, $V_\nu(q)$, is plotted as a function of q for a spherically symmetric, ionized stellar wind with $\theta_1 = 0.3'' \nu_{GHz}^{0.703}$.

P Cygni was resolved with the highest resolution of the VLA, and the latter authors have systematically used measurements of stellar wind visibility functions to determine sizes and temperatures of stellar winds.

For an isothermal stellar wind the brightness temperature at radius θ can be written as

$$T_b = T_e[1 - \exp(-\tau_\nu)] = T_e\{1 - \exp[-(\theta_1/\theta)^3]\} \qquad (9.24)$$

so the visibility function becomes

$$V_\nu(q) = (4\pi k T_e/\lambda^2) \int \{1 - \exp[-(\theta_1/\theta)^3]\} \cdot J_0(2\pi q\theta) \cdot 2\pi\theta \cdot d\theta \qquad (9.25)$$

which can be further transformed, by changing the integration variable to $\theta\nu^{0.703}$, into a frequency-dependent coefficient and an integral describing the dependence of the visibility upon q and the parameters T_e and $\theta_1 \nu^{0.703}$. Figure 9.11 shows this functional dependence for a case where $\theta_1 = 0.3'' \nu_{GHz}^{0.703}$.

The observed stellar winds from P Cygni and γ^2 Vel are the best cases where most of the observations fit the expected theory for ionized stellar winds. White and Becker (1982) showed that the observations fit the theory for a $\sim 18,000$ K isothermal wind while Hogg (1985) was able to fit visibility function measurements for γ^2 Vel to a model with a mass loss rate of 8×10^{-5} solar masses per year and a temperature of 5800 K.

Unfortunately for the simple wind theory, but fortunately for those who like a variety of challenges, many radio-emitting OB or Wolf-Rayet stars show phenomena that do not fit the simple theory (Abbott 1985). Cyg OB2 No. 9 was the first to be widely recognized as deviant because of short time scale variability (Abbott et al. 1984) and the fact that it was unresolved by the VLA (White and Becker 1983) when it should have been resolved. Both a nonthermal radio spectral index and a high degree of variability were indications that the 9 Sgr radio source was not just a simple ionized stellar wind. Phenomena inconsistent with free-free emission from ionized winds are now known to be common amongst OB star radio sources (Abbott 1985). A sample of forty Wolf-Rayet stars was studied by Abbott et al. (1986b), who showed that while nonthermal phenomena are rarer than amongst the OB stars, only thirty-three showed a match of optical and radio characteristics that was consistent with simple free-free emission from ionized winds. Synchrotron emission from relativistic electrons generated by turbulence or terminating shocks in the outer parts of winds has been suggested as a source of the anomalous radio emission, but as of this writing, definitive matches between more complex models and the observations have not been established.

9.5.3 Cool, Weakly Ionized Winds

The first detection of the red supergiant α Ori (Betelgeuse), at what is now known to be its typical level, was by Altenhoff and Wendker (1973). The first extensive observations of many types of stars with the NRAO interferometer at Green Bank carried out in June 1970 included the detection of radio emission from α Sco (Wade and Hjellming 1971a, Hjellming and Wade 1971a); however, the emission from α Sco was resolved into two barely detected components by Gibson (1979), which were shown by Hjellming and Newell (1983) to consist of emission associated with both stars in the α Sco binary. The red supergiant component, Antares, showed radio emission from its partially ionized wind, and Antares' B2.5V binary companion, 2.9" away, was surrounded by a resolved ionized subregion of the Antares wind. The true Antares and Betelgeuse radio sources were the first of a small class of giants and supergiants with largely neutral but partially ionized stellar winds with weak bremsstrahlung emission. The nebulosity around the B-star companion of Antares makes α Sco the radio prototype of the VV Cephei radio binaries.

The general theory of spherically symmetric stellar winds with electron densities with power law dependences like r^{-p} are applicable to both the supergiants Antares and Betelgeuse and the cool giants α^1 Her, α Boo, ρ Per, α Tau, and μ Gem (Drake and Linsky 1986). The equations for the observables are modified because of the different power law. For example, in general a power law of r^{-p} in electron concentration will produce an optically thick radio source with a spectral index $\alpha = (4p - 6.2)/(2p - 1)$ (Wright and Barlow 1975). The data for the radio spectra of Betelgeuse (Newell and Hjellming 1982) and Antares (Hjellming and Newell 1983) are fit by $S_\nu = 0.24\nu_{GHz}^{1.32}$ and $0.47\nu_{GHz}^{1.05}$, respectively. This is consistent with $p = 3.59$ and 2.71, respectively, indicating that while the cool supergiant winds may have a density dependence of r^{-2}, the fractional ionization is also decreasing with radius so that N_e is proportional to $r^{-3.59}$ and $r^{-2.71}$ for the two stars.

The red giants α^1 Her, α Boo, ρ Per, and α Tau have been found (Drake and Linsky 1986) to have spectral indices of 0.80, 0.84, 0.95, and ≥ 0.87, which are steeper than the 0.6 expected for a r^{-2} stellar wind but less than that for the supergiants.

9.5.4 VV Cephei Binaries—Ionized Subregions of Cool Supergiant Winds

VV Cep binaries are basically systems consisting of a red supergiant star and an OB star. The thing that produces a different type of radio source in these systems is the fact that the OB star orbits inside a dense, cool, and nearly neutral wind, ionizing a subregion of the wind that is the major source of free-free radio emission. The nebulosity around the B2.5V companion of Antares is the prototype of these ionized subregions. Figure 9.12 shows both a VLA 4.9-GHz image of the Antares radio emission and the emission from the companion nebulosity surrounding the B-star (marked with an X) and a model for a point source on Antares and the HII region produced when the cool supergiant wind, with an r^{-2} density distribution, flows through the HII region that can be maintained by the Lyman continuum from the

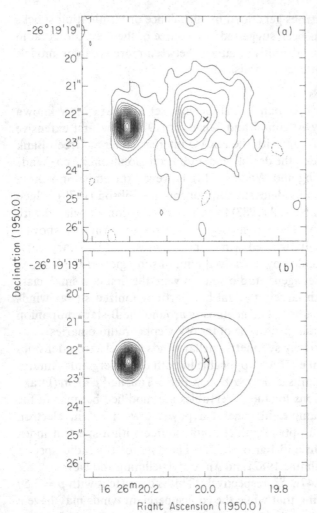

Fig. 9.12. A VLA image at 4.9 GHz (a) and a theoretical model (b) for the radio emission of Antares (left, unresolved point source) and the HII region produced by the companion B2.5V star (marked with a cross) in the otherwise unionized wind of the red supergiant. (From Hjellming and Newell 1983.)

B2.5V star (Hjellming and Newell 1983). Most of the VV Cep binaries have periods of the order of tens of years rather then the > 3000-year period of Antares and its companion. If one scales from the $\sim 10^4$ cm^{-3} densities in the Antares companion nebulosity, one would expect densities of the order of $\geq 10^8$ for the other VV Cep binary nebulosities, and the possibility exists that the HII region would change in flux density because an asymmetric HII region orbits the supergiants in short time scales. In addition to the Antares companion nebulosity, six other VV Cep binaries have been found (Hjellming 1985b) to be surprisingly strong radio sources (VV Cep, KQ Pup, HR 8164, FR Sct, WY Gem, and HD237006), apparently because the HII region gas is much hotter, $\geq 10^5$ K, due to collisional deexcitation of the normal coolants in HII regions. In addition, at least one system, HR8164, has shown changes in radio flux on time scales of years, probably due to changes in the apparent cross section of the optically thick HII region orbiting in the wind of the supergiant.

9.5.5 Symbiotic Stars—Interacting Winds and Ionized Subregions of Winds

The first of the radio-emitting symbiotic stars, V1016 Cyg, was detected by Purton et al. (1973). More than twenty were known by the end of 1986. They exhibit varieties of steady radio emission, extended nebulosities, and even moving jets (CH Cyg). Complex interactions between extended winds, outflows, and ionization of subregions of winds are involved (Taylor and Seaquist 1984). The changes in the CH Cyg radio source on the time scale of 75 days, as found by Taylor et al. (1986) and shown in Figure 9.13, make it clear that asymmetric ejecta, with proper motions of $\sim 1'' \, y^{-1}$, occur. AG Peg is another object that has variously been called a symbiotic star or a slow nova; it shows unresolved emission, asymmetric ejecta, and an extended nebulosity (Hjellming 1985c). Most of the other tens of currently known symbiotic stars (Seaquist et al. 1984) are unresolved with spectral indices ranging from 0.6 to 1.05. Most of these radio symbiotics can be interpreted in terms of partial ionization of cool winds, interaction between cool and hot star winds, and ionization phenomena (Taylor and Seaquist 1984).

A number of emission-line stars that have been found to be radio sources (MWC349, Vy2-2, Hb12) are representative of objects that involve interacting stellar winds, but which are generally categorized as proto-planetary nebulae. While there is probably no clear dividing line between the symbiotic stars, slow novae, stellar winds, and interacting winds, we will not discuss these massive examples of ejecta any further in this chapter.

9.6 Cataclysmic Variables

9.6.1 Classical Novae

The first extensive observations of many types of stars with the NRAO interferometer at Green Bank carried out in June 1970 resulted in the detection of radio emission from the novae FH Serpentis 1970 and HR Delphini 1967 (Hjellming and Wade 1970). These two novae and Nova V1500 Cygni 1975 have been extensively studied and interpreted (Hjellming et al. 1979, Seaquist and Palimaka 1977) in terms of shell ejection with linear velocity gradients or time-dependent winds (Kwok 1983). Classical novae occur in binary systems with a white dwarf, and usually a cool companion dwarf, in which an explosion event, due to either an accretion instability or thermonuclear flash on the surface of the white dwarf, expels a complicated shell of gas from the white dwarf and/or its accreting atmosphere.

Figure 9.14 shows the extensive data on Nova V1500 Cygni 1975 obtained at radio and infrared wavelengths (Hjellming et al. 1979, Ennis et al. 1977). Theoretical curves passing through the data were computed by Hjellming et al. (1979) using a model of a finite, spherically symmetric shell with a linear velocity gradient across the shell. The theory developed in Section 9.2.1 is used with the usual $\rho = (\dot{M}/V)$ $(1/4\pi r^2)$ for a stellar wind replaced by

$$\rho(r, t) = \frac{M/(4\pi r^2)}{r_2 - r_1} \tag{9.26}$$

where M is the mass in the shell and r_1 and r_2 are inner and outer shell radii

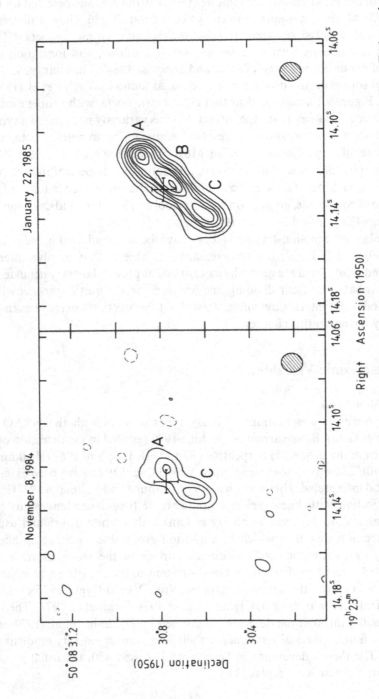

Fig. 9.13. Two epochs of 15-GHz VLA images of the symbiotic star CH Cyg showing jets moving at a rate of 1.1 arcseconds per year. (From Taylor et al. 1986. Reprinted by permission from Nature, Vol. 319, pp. 38, Copyright(c) 1986. Macmillan Magazines Limited.

specified by (different) velocities and initial radii. In order to fit the infrared data, it is necessary to include both free-bound and free-free radiative transitions. The difference between the dashed and solid curves for the infrared data is the necessary inclusion of finite initial radii effects in the latter in order to fit the data. The beautiful fit to all the radio and infrared data, due to this simple model, is marred only when after about 200 days from outburst, one sees the first signs of excess infrared due to the formation of dust in the nova shell. The parameter combinations of the model are $M(T_e/10^4)^{0.58} = 2.4 \times 10^{-4}(d/1400 \text{ pc})^{5/2}$, $v_1/v_2 = 0.036$, and $v_2(T_e/10^4)^{1/2} = 5600(d/1400 \text{ pc})$, $r_{10} = 1.5 \times 10^{13}$ cm, and $r_{20} = 3.5 \times 10^{14}$ cm. The optically thin decay at all frequencies evolves, for all three novae, to the simple analytic solution given by

$$S_v = \frac{4\pi(j_v \rho r^4)}{d^2(r_2 - r_1)r_2 r_1} . \tag{9.27}$$

A perfectly symmetric shell should have an optically thin decay of $(t - t_0)^{-3}$, whereas the three novae just mentioned had decays with power laws slightly less steep because the shell had departures from spherical symmetry.

Since the initial three detections, a number of other classical novae have been detected as radio sources, with some of them fitting the simple shell theories just discussed: Nova V1370 Aql 1982 and Nova PW Vul 1984. However, Nova Vulpecula #2 1984 produced a complex radio source in which interaction between ejecta, winds, etc. produced at least two distinct types of radio sources (Taylor et al. 1987).

One old nova shell is known to have produced a nonthermal radio source as a result of the interaction of the ejected shell with the interstellar medium. That this should occur, because of turbulence generated during the interation process, was originally suggested by Chevalier (1977). As part of a search for such radio emission in association with nova shells, Reynolds and Chevalier (1984) found a partial shell of extended, linearly polarized emission associated with Nova GK Persei 1901; in Figure 9.15 their contour map of the radio-emitting shell is shown with respect to the original center of nova shell ejection (marked with a plus). They estimate that about 1% of the original kinetic energy ($\sim 10^{45}$ ergs) in the shell has gone into relativistic electrons and magnetic fields ($\sim 6 \times 10^{-5}$ gauss). One important aspect of this result is that it indicates that all explosive ejecta and expansive winds have the potential to produce a component of synchrotron or gyrosynchrotron emission due to interaction with the external medium.

9.6.2 Recurrent Novae

The first strong radio event from a recurrent nova was detected by Padin et al. (1985) as a result of the fifth known outburst of RS Ophiuchi. Their observations of the initial rise of the event, and extensive multifrequency observations of the peak and decay by Hjellming et al. (1986), provide nearly complete information about the rise and decay of the first year of this recurrent nova radio event. A recurrent nova occurs in binary systems involving a white dwarf orbiting a red giant companion. The wind from the red giant is responsible for more extensive accretion on the white dwarf and regular occurrence of explosive instabilities in the accretion environment due to mass transfer "events."

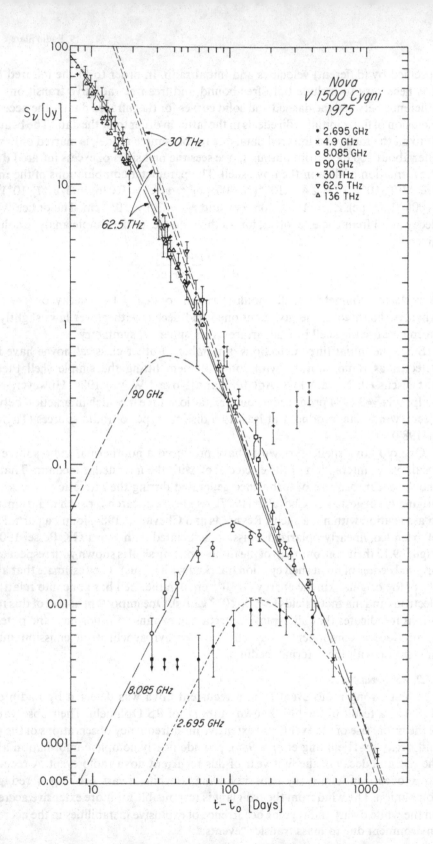

Nova
V 1500 Cygni
1975

- • 2.695 GHz
- × 4.9 GHz
- ○ 8.085 GHz
- □ 90 GHz
- + 30 THz
- ▽ 62.5 THz
- △ 136 THz

S_ν [Jy]

30 THz

62.5 THz

90 GHz

8.085 GHz

2.695 GHz

$t-t_0$ [Days]

Fig. 9.15. The nonthermal radio source produced by the interaction of the ejected shell from Nova Persei 1901 and the surrounding interstellar medium. (From Reynolds and Chevalier 1984.)

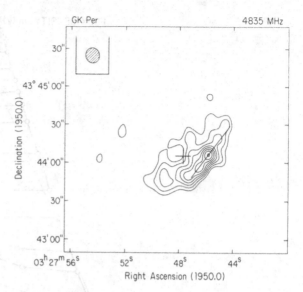

Figure 9.16 shows the first year of evolution of the RS Oph radio source from a slightly optically thick source, through a stage with two radio components, one dominant at high frequencies and another contributing a component with a nonthermal spectrum at low frequencies. The latter is almost certainly a gyrosynchrotron or synchrotron source component due to particle acceleration in regions where the ejected material interacts with the unusually dense external medium provided by the wind. The high-frequency component had a complicated decay behavior, but to first order decayed as $(t - t_0)^{-1.1}$ during early stages and somewhat more steeply at later phases. Shortly after the evolutionary sequence seen in Figure 9.16, the RS Oph radio source settled into an approximately stable source with a radio spectrum described by $\sim 1.0v^{-0.2}$ mJy. At its peak, brightness temperatures in the range of 10^5 to 10^7 K were observed, because the radio source was resolvable a month after outburst. It is noteworthy that this time scale of decay is common in the radio supernovae analyzed by Weiler et al. (1986), where it is almost certain that an expanding synchrotron-emitting source is interacting with an external, compressed shell of hot thermal plasma. The latter is produced when the expanding blast wave compresses overrun portions of the previously existing wind of the star that went supernova.

Beyond simple models of an optically thin synchrotron source with external free-free absorption, detailed models for the evolution of radio components under these conditions have not yet been worked out for either strong supernova explosions or the weaker equivalents in recurrent novae. However, in all cases it is believed that explosive ejecta produced strong shocks when the ejecta were slowed by external gas, setting up conditions that generate relativistic particles, and

◁ ──

Fig. 9.14. Both data and models for the variation of radio and infrared flux densities with time for the shell ejection event in Nova V1500 Cyg 1975. (From Hjellming et al. 1979.)

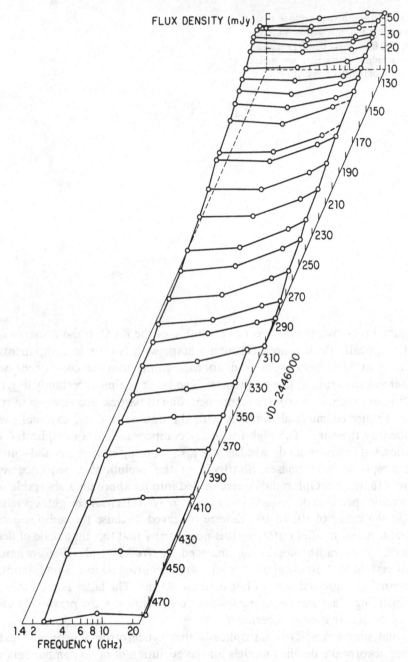

Fig. 9.16. The spectra (log flux density versus log frequency) observed after the January 1985 outburst of the recurrent nova RS Ophiuchi are plotted with displacements in time for each different epoch of measurement, with time increasing from the upper right to the lower left. (From Hjellming et al. 1986.)

possibly magnetic field dynamos. Once layers or bubbles of mixed relativistic electrons and magnetic fields exist, they radiate copious synchrotron emission and evolve as the radiating particles lose energy by adiabatic expansion, synchrotron radiation losses, or other loss mechanisms.

9.6.3 Dwarf Novae

Dwarf novae, which have small but very frequent optical flares, have been reported as radio sources, but their characteristics are not definitively known. Benz et al. (1983) reported 1.3-mJy radio emission from SU Uma at 4.75 GHz using the Bonn 100-m telescope at a time when this dwarf nova was in optical outburst; however, subsequent VLA observations at 4.9 GHz, also during optical outburst, by some members of the same group (Furst et al. 1986) yielded an upper limit of 0.1 mJy. Furst et al. (1986) also obtained negative results for YZ Cnc, Z Cam, V603 Aql, EM Cyg, and RZ Sge. Turner (1985) has reported detection of radio emission from the dwarf novae TY Psc and UZ Boo at levels of 10 and 2.4 mJy, respectively, at 2.5 GHz with the Arecibo interferometer system. These reports make it likely that dwarf novae can produce radio emission during optical outbursts, but without information on time dependence and spectrum, little can be said about these types of radio sources as yet.

9.6.4 Magnetic Cataclysmic Variables

Magnetic cataclysmic variables are divided into two classes: the AM Her type in which a white dwarf has a large magnetic field ($\sim$ few $\times$ 10^7 gauss), a magnetosphere that reaches to a companion star, no accretion disk, and presumed synchronous rotation in the white dwarf; and the DQ Her type where the magnetosphere of the white dwarf does not extend to the other star either because the field is too weak (10^5 to 10^6 gauss) or because of greater binary separation.

Radio emission from AM Her was first detected in 1981 by Chanmugam and Dulk (1982). Since that time, AM Her has exhibited a unique variety of behavior. Originally detected with the VLA at 4.9 GHz at a level of 0.67 mJy, one could infer a brightness temperature of 3×10^9 K for a source radius of 10^{11} cm and a distance of 100 pc, which led to the argument that it was gyrosynchrotron emission from $\sim$0.5-MeV electrons radiating up to the fortieth harmonic in a region of the white dwarf magnetosphere with a field of $\sim$40 gauss. By the middle of 1982, it declined to 0.55 mJy, was 0.52 mJy in early 1983, and had declined below detection limits of about 0.3 mJy at the end of 1983. This decline coincided with a change in the X-ray emission from AM Her that indicated that a second accretion column on the other magnetic pole of the white dwarf had appeared (Heise et al. 1985), and may have been due to an opening of the magnetosphere and escape of the particles responsible for the radio emission. In the middle of 1984, the radio emission was detected at the 0.32-mJy level at 4.9 GHz. A few months later, it peaked at levels of 0.4 and 0.7 mJy at 4.9 and 15 GHz, respectively, then underwent a steady decline. The pre-1984 radio source was not circularly polarized; however, when it reappeared in 1984 it was circularly polarized at the 25% level, confirming the probable gyrosynchrotron nature of the radio emission.

However, the most interesting "event" in AM Her was the occurrence of a flare in July 1982 (Dulk et al. 1983) during a several-minute period when the 0.55-mJy source increased to a peak level of 9.7 mJy before decaying to the previous quiescent level. This increase by a factor of ~ 20 and the fact that the flare was 100% right circularly polarized indicate a coherent emission process that may be due to an electron-cyclotron maser operating near the surface of the red dwarf companion, at a region where the magnetic field is ~ 1000 gauss and the red dwarf has a modest corona. Thus, AM Her seems to exhibit at least two different varieties of magnetospheric radio emission. While the origins of the energetic electrons are unclear, one of the possibilities is acceleration during reconnection processes in intrastellar field regions.

The first detection of radio emission from a DQ Her-type magnetic cataclysmic variable was the detection of AE Aquarii by Bookbinder and Lamb (1987) with flux levels of 16 and 5 mJy at 4.9 and 1.4 GHz, respectively. The AE Aqr binary contains a magnetized white dwarf, with a spin period of 33 seconds, that orbits a red dwarf with a period of 9.8 hours. Because the magnetosphere of the white dwarf does not extend to the red dwarf, Bookbinder and Lamb argue that the emission implies a 10^3-gauss field on the red dwarf, with the radio emission occurring (for both AM Her and AE Aqr) on or near the red dwarf, where the energetic electrons are produced according to their arguments.

These two magnetic cataclysmic variables join the active binaries in the sense that magnetospheric phenomena are involved in the radio emission process, but are unique in the size of the fields and the existence of accretion columns at one or both of the magnetic poles of the white dwarfs.

9.7 Radio Emitting X-Ray Binaries

9.7.1 Summary of Major Radio-Emitting X-Ray Binaries

Amongst the stellar systems first detected with the NRAO interferometer in June 1970 were the radio emission from Sco X-1 and two companion radio sources 1.3 and 2.0 arcmin NE and SW on the same position angle (Hjellming and Wade 1971b). The discovery of variable radio emission from Sco X-1 was a prelude to the discovery that X-ray binaries could produce remarkable varieties of radio emission. Cyg X-1 and GX17 + 2 were the next to be found (Hjellming and Wade 1971c, Braes and Miley 1971), with the radio detection of Cyg X-1 playing a major role in identifying the X-ray source with the binary star HDE 226868, which is one of the most likely candidates for being a black hole in orbit around an OB star. The next remarkable radio-emitting X-ray binary was associated with Cyg X-3 (Braes and Miley 1972). This object is a distant binary system (> 10 kpc) which exhibits flaring to levels of 20 Jy or more. These flares are nearly ideal examples of radiation events of the type most often thought of in connection with quasars. Amongst the radio-emitting X-ray binaries is SS433, discovered to be a radio source by Ryle et al. (1979) and Seaquist et al. (1979), which is a spectacular case where there is a binary system which is continuously ejecting radio-emitting plasma in observable twin-corkscrew patterns with proper motions of 3.0 arcsec per year, or 0.26 times the speed of light (Hjellming and Johnston 1981a). SS433 is a rare example where we can directly measure the motion and evolution of synchrotron-radiating plasma regions in high Mach number ($M \sim 40$) conical jets.

One common denominator links the radio characteristics of these X-ray binaries. This is the fact that in each case the radio emission process is almost certainly synchrotron emission from highly energetic relativistic electrons. This is quite different from the bulk of radio stars discussed earlier, where thermal bremsstrahlung, plasma radiation, or magneto-bremsstrahlung in the cyclotron or gyrosynchrotron regime predominates. As we will discuss, SS433 and Cyg X-3 clearly show radio emission from expanding bubbles or cones of synchrotron-emitting relativistic electrons. The best working hypothesis for the radio emission from X-ray binaries is the continuous, periodic, aperiodic, or transient production of relativistic electron plasma regions behind strong shocks. It is noteworthy that solar events involving very-high-energy electrons, generally designated with Type IV labels, occur when ejected material is decelerated by external media so that the decelerated ejecta are preceded by strong shocks (cf. lower right panel of Figure 9.5). The equivalent phenomenon is initiated by continuously produced shock regions in conical jet environments. In some cases, there is both steady ejection in the form of conical jets and flares due to sudden increases in the production of relativistic plasma.

9.7.2 Relativistic Jets and Synchrotron Flaring in Cyg X-3

(a) Cyg X-3 as a Synchrotron Flaring Source
The X-ray source Cyg X-3 has an extensive history of observed radio emission [cf. review by Hjellming (1973b)] at flaring (> 20 Jy) and quiescent (~ 0.1 to 0.3 Jy) levels since its initial discovery as a radio source by Braes and Miley (1972). Because of the nearly unique and extensive observing of Cyg X-3 during the August-September 1972 radio flaring event, twenty-one papers appeared in a the October 23, 1972, issue of *Nature Phys. Sci.* (**239**, No. 95). Cyg X-3 is associated with a binary system with a 4.8^h period that is well observed at X-ray (Parsignault et al. 1972) and infrared (Becklin et al. 1972) wavelengths. Figures 9.17 and 9.18 show examples of the radio behavior of Cyg X-3 during the 1972 period when it was first "caught" flaring above 20 Jy, including the first of its events that are amongst the best-known examples of observed expanding synchrotron-emitting sources, and during a 1973 period when it was exhibiting a mixture of quiescent and erratically flaring source behavior.

Most of the time, Cyg X-3 has a low-level radio source at flux levels of a few tenths of a Jansky, spectral indices that indicate internal or external absorption of the emitted synchrotron radiation, and variation on time scales of a few hours, as can be seen from Figure 9.18 for the times when obvious flaring episodes are not occurring. Most of the time it is difficult to establish periodic effects, because of the frequent flaring; however, Molnar et al. (1983, 1984, 1985) have used multifrequency VLA data taken during times when Cyg X-3 was relatively quiescent to reveal a periodicity close to the X-ray and infrared periodicities in Cyg X-3. We will discuss this again when we discuss other X-ray binaries that always show perodic variations in their radio emission.

The model of a binary with white dwarf and red dwarf components proposed by

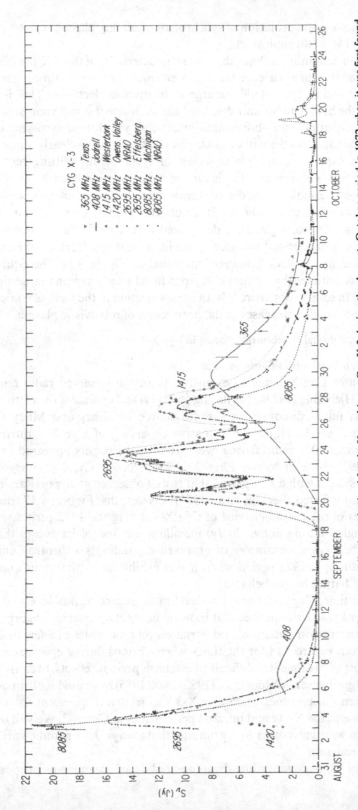

Fig. 9.17. A summary of some of the multifrequency radio data obtained for the X-ray binary Cyg X-3 during the August–October period in 1972 when it was first found to have strong synchrotron-flaring events.

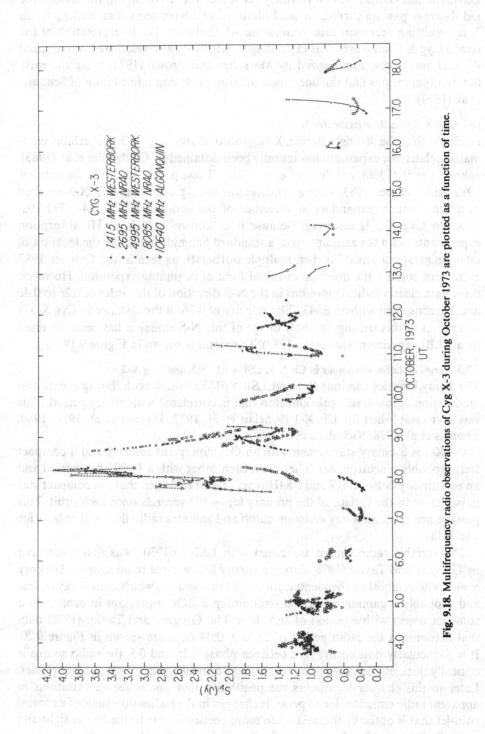

Fig. 9.18. Multifrequency radio observations of Cyg X-3 during October 1973 are plotted as a function of time.

Davidsen and Ostriker (1974) is widely accepted, and according to this model, the red dwarf is passing through a brief phase of extensive mass loss, leading to an X-ray-emitting accretion disk around the white dwarf. The interpretation of the strong Cyg X-3 radio events as expanding synchrotron sources, mixed with ionized thermal gas, is best summarized by Marscher and Brown (1975) using the early flaring observations and the linear polarization evolution information of Seaquist et al. (1974).

(b) Cyg X-3 as a Relativistic Jet Source

Evidence that the strongly flaring X-ray/radio source Cyg X-3 can exhibit colli-mated, relativistic expansion has recently been obtained by Geldzahler et al. (1983), Johnston et al. (1986), and Spencer et al. (1986). These papers show that in October 1982 and October 1983, strong synchrotron flaring events in Cyg X-3 resulted in a radio source expanding at velocities of the order of 0.2 to 0.4c. The dis-tance to Cyg X-3 is uncertain because it is known only from HI absorption experiments, with the assumption of a standard Schmidt model for the location of intervening spiral arms. Further, multiple outbursts, as seen in the October 1983 event, can confuse the question of initial time of beginning expansion. However, these data clearly indicate motions in the N-S direction of the order of 0.2c to 0.4c and are consistent with an SS433-like velocity of 0.26c if the distance to Cyg X-3 is 11 kpc. As of this writing, the only image of this N-S emission has been obtained by a VLBI experiment (Johnston 1987). The result is shown in Figure 9.19.

9.7.3 Periodic Radio Variations in Cir X-1, LSI +61°303, and Cyg X-3

The X-ray binaries Circinus X-1 and LSI +61°303 show both flaring events and modulation of low-level radio emission that is correlated with binary period. This was first established for Cir X-1 (Whelan et al. 1977, Haynes et al. 1978, 1980, Thomas et al. 1978, Nicholson et al. 1980).

Cir X-1 is a binary star system with an OB supergiant primary and a compact star, probably a neutron star. They orbit each other with a 16.595-day period and an eccentricity between 0.7 and 0.8 (Haynes et al. 1980) such that the compact star passes close to the surface of the primary for $\sim 10^5$ seconds once each orbit. This passage produces an X-ray emission cutoff and initiates radio flaring that lasts for 1 to 3 days.

The variable radio source associated with LSI +61°303 was first discovered by Gregory and Taylor (1978) during a survey for variable radio sources. Gregory et al. (1979) established the identification with this system, which is an X-ray source and probably a gamma-ray source, containing a BOe supergiant in orbit with a compact object with a period of 26.5 days. The Gregory and Taylor (1978) data that determined the radio period of 26.52 ± 0.04 days are shown in Figure 9.20. It is particularly noteworthy that between phase 0.25 and 0.5, the radio source is optically thick but becomes a flat-spectrum or optically thin source at other phases. Later in this chapter we discuss the possibility that one is seeing variations in apparent radio emission due to periodic changes in the inclination angle of a conical twin-jet that is optically thinnest when more perpendicular to the line of sight and optically thickest when least perpendicular to the line of sight.

We have already mentioned that at low flux levels, Cyg X-3 exhibits flux

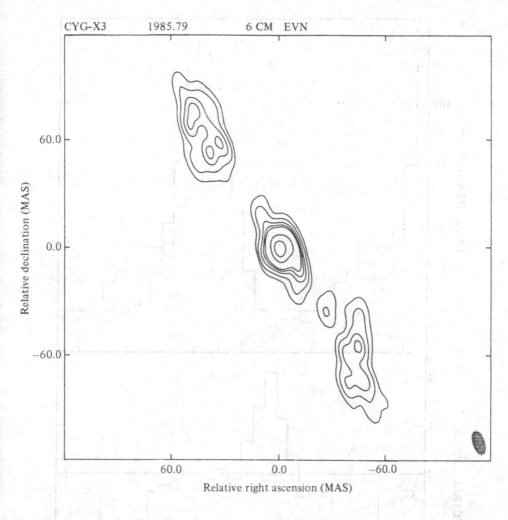

Fig. 9.19. A high-resolution radio image of the X-ray binary Cyg X-3 with the European VLBI network. (From Johnston 1987.)

modulation that matches the infrared and X-ray periodicities; however, unlike LSI +61°303, the modulated radio emission always remains optically thick except when a flare event occurs and the dominant emitting plasma expands enough to become optically thin. We will also discuss this source later, after we have established the radio emission characteristics of a simple model for conical twin-jets.

9.7.4 Relativistic Jets in SS433

The SS433 star system (V1343 Aql) has optical emission lines of HI and HeI (Margon 1984) which change wavelength in a manner that can be interpreted as twin Doppler shifts with a range of 80,000 km s^{-1} and period of 162.5 days. The twin-jet model for the optical data indicates an absolute velocity of ejection of 0.26c, a jet axis either 80° or 20° to the line of sight, and an ejection vector that rotates around the jet axis every 164 days at an angle of 20° or 80°.

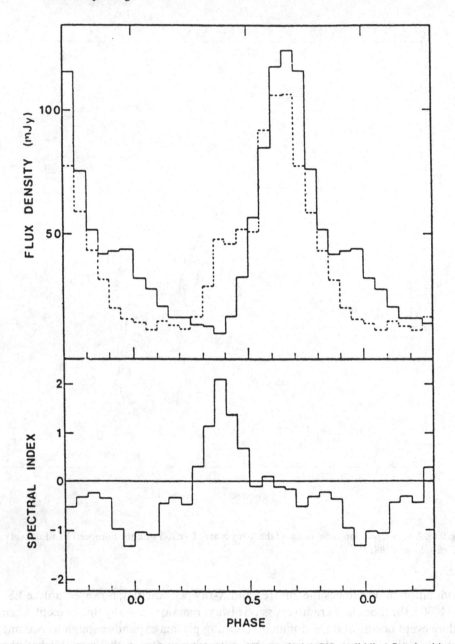

Fig. 9.20. The periodic variations of the 5-GHz (dashed line) and 10.5-GHz (solid line) flux densities of the X-ray binary LSI + 61°303 are plotted as a function of binary phase, together with the implied spectral index variations. (From Gregory and Taylor 1978.)

The radio emission of SS433 was independently discovered by Ryle et al. (1979) and Seaquist et al. (1979), and was shown in the large-scale map of Geldzahler et al. (1980), to be a compact source inside the 1° by 2° "supernova remnant" W50. Hjellming and Johnston (1981a, 1981b, 1982) showed that a series of high-resolution maps made in 1979–1980 indicated that extended portions of the radio emission had oppositely directed proper motions of 3.0" per year and that these proper motions could be used to determine all the remaining parameters in the twin-jet kinematic model that were either ambiguous or unknown from the optical data. The model that fits the 1979–1986 proper-motion data has: a jet axis inclined 80° ($= i$) to the line of sight at position angle 100° ($= \phi$); an angle between the jets and the jet rotation axis of 20° ($= \psi$); the jets rotated in a clockwise (left-handed) sense about the jet axis with a period of 162.5 days ($= P$); the oppositely directed eastern and western (on the sky) jet rotation axes on the near the far sides, respectively, of the central object; and a ratio of constant jet velocity to distance to SS433 of 3.0" per year. The measurable differences in paired features on the east and west sides, due to time delay effects, allow one to determine that the absolute jet velocity is $0.26c$. The velocity and proper-motion determinations yield a value for the distance to SS433 of 5.5 kpc.

The rotating ejection vector for the twin-jets of SS433 has a time-dependent angle, Θ_{LOS}, between the vector and the observer's line of sight (LOS) that is given by (Hjellming and Johnston 1981b)

$$\cos \Theta_{LOS} = \cos i \cos \psi + \sin i \sin \psi \cos 2\pi(t - t_{ref})/P \qquad (9.28)$$

where t is the time of ejection, t_{ref} is a reference time for the kinematics, and the other parameters were defined in the previous paragraph. This result for Θ_{LOS} will be important later when we discuss the evolution of twin-jets.

Between 1979 and 1986 Hjellming and Johnston (1986) made observations at 4- to 6-week intervals during all the 3- to 4-month periods that the VLA was in its largest (35-km) configuration. These data provide an ongoing test of the validity of the previously published kinematic parameters that we have just summarized. They also provide a wealth of data on the evolution of twin radio jets, since the ejecta exhibit a wide range of behavior. Figure 9.21 shows an excerpt of results for a sequence of events in 1980–1981 (left) and 1982 (right). The first two maps on the left are for 15 GHz, while the bottom three maps at the left are for 4.9 GHz. The four maps on the right are for 15 GHz. All maps have superimposed "corkscrews" (with filled circles identifying ejection intervals of 20 days) indicating the predicted proper-motion paths for the kinematic solution of Hjellming and Johnston (1981b), with the modification that the best-fit period is 162.5 rather than 164 days.

The SS433 radio jet data from 1979 through 1986 provide a wealth of detailed information about the evolution of ejected radio sources. One striking characteristic of the SS433 radio source is the omnipresence of a nearly $v^{-0.66}$ radio spectrum. From this result we conclude that the SS433 radio source seldom exhibits any self- or external absorption effects. This behavior is contrary to the behavior of virtually all stellar radio sources, which show self- or external absorption because of the typical compactness of the radio source in the relatively high-density gas environment of star systems.

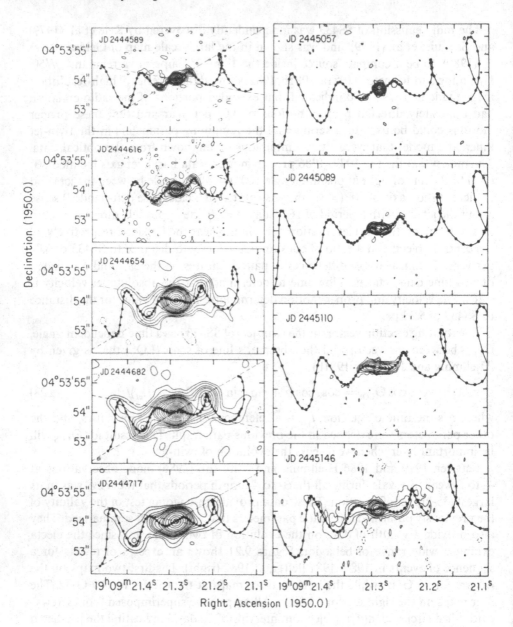

Fig. 9.21. Two sequences of VLA radio images of SS433 are shown with the superimposed proper-motion path predicted from the kinematics of the jets. On the left are images from 1980–81 in which two epochs at 15 GHz are placed above three later epochs at 4.9 GHz, and on the right there is a time sequence of 15-GHz images from 1982 (Hjellming and Johnston 1986, Hjellming 1985).

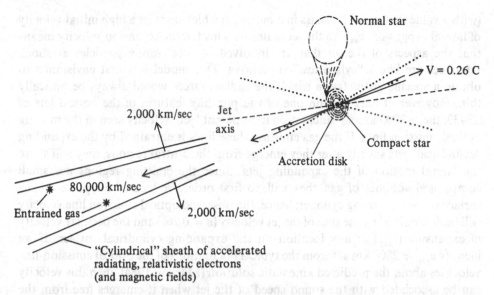

Fig. 9.22. A schematic diagram illustrating the working hypothesis whereby the radio and optical jets of SS433 arise from conical gas jets emerging from the axis of a precessing accretion disk, with a Mach number of ~40, because of lateral expansion imposed by the sound speed when the jets escape the accretion environment. The radio emission occurs in an outer sheath around the main jet material. (From Hjellming and Johnston 1986.)

The 1980–1981 sequence on the left side of Figure 9.21 shows an apparent double radio source moving outward at a rate of 3″ per year. It is one of the best examples of cases where one can derive conclusions about the evolution of the ejected material. The velocity of 0.26c, determined roughly by the east-west (time delay) asymmetries seen in most pairs of ejecta, and the observed proper motion of 3.0″ per year then determine the distance to be 5.5 kpc. The data for the decay of intensity of the identifiable double in Figure 9.21 indicate an exponential decay constant of about 80 days, with *no change* in the spectral index of ~ −0.6 during this observed expansion. These and other data for the extended emission indicate that the extended structure of SS433 shares the tendency to always have a spectral index of −0.6 to −0.7. The reason that an identifiable double is clearly observed in the 1980–1981 sequence is the fact that its decay constant of 80 days is much longer than the 30 days found to be typical for most ejected jet material in SS433 (Hjellming and Johnston 1986).

The 15-GHz sequence of maps on the right in Figure 9.21 shows how a long-surviving ejection of material can be seen to spread out *exactly* on the predicted proper-motion corkscrews, becoming resolved at 15 GHz about 100 days after ejection.

Hjellming and Johnston (1986) have discussed a model for the geometry of the expanding SS433 radio jets which explains much of the data, including the relatively large and optically thin characteristics of the radio emission—and the residual "jitter" of the optical jets (Margon 1984). Figure 9.22 is a schematic diagram of this model in which the jets emerge from a thick accretion disk with a high temperature (with a value that we will discuss in a moment) which imparts a high initial velocity of lateral expansion, v_{exp}, to the jet material. This lateral expansion velocity means that the aspects of the jet that are involved in accelerating particles at shock interfaces are like a "cylindrical" supernova. This model was first envisioned to obtain a specific geometry in which the radio sources would always be optically thin. However, it may explain one of the puzzling features of the optical jets of SS433: the ± 2000-km s^{-1} fluctuation in apparent "jet" velocity seen in the moving optical emission lines. If the gas emitting these lines is entrained by the expanding "cylindrical" jets shortly after they emerge from the central regions, they will share the lateral motion of the expanding jets. Since the emitting regions are small compressed volumes of gas, they will, to first order, be located anyplace on the surface of the expanding cylinder; hence, the observed optical emission line velocity will be determined by the sum of the jet velocity ($v = 0.26c$) and the peculiar velocity of expansion (v_{exp}) at any location on the expanding cylindrical surface. If we identify $v_{exp} \approx 2000$ km s^{-1} from the typical fluctuations of the optical emission-line velocities about the predicted kinematic solution (Margon 1984), then this velocity can be associated with the sound speed of the jet when it emerges free from the accretion disk environment. This velocity corresponds to a temperature of 3×10^8 K. Detailed models of X-ray sources and associated jets have in common a simple association of $kT \approx GM/R$ (G is the gravitational constant, M is the mass of the compact star, and R is the star's radius), that is, kT is the order of the depth of the gravitational potential well from which the X-rays or gas flows are escaping, so it is interesting to note that 3×10^8 K corresponds to the potential well typical of a neutron star.

As can be seen from Figure 9.22 and the previous discussion, a useful working hypothesis for SS433 is a jet model with a conical geometry having a cone angle of $\sim 4°$ and a Mach number of the order of 30 to 40. In order to maintain the completely optically thin characteristics of the source, even during flaring events, the emitting regions are probably in a relatively thin sheath around the laterally expanding, "kinetic energy" jets.

9.7.5 Evolving Bubbles, Layers, and Cones of Relativistic Electrons

(a) Type of Geometry for Synchrotron-Radiating Regions
The least-understood step in the evolution of a synchrotron radiation source in the environment of a stellar system is the generation of the initial relativistic electron plasma. Despite this, it is believed that highly energetic ejection of gas leads to shocks and it is in association with these shocks that particles are accelerated. However, if we can assume sudden, periodic, aperiodic, or continuous production of relativistic plasma, then we can consider the physics of the evolution of a radio source.

In order to have a geometric framework for discussing the synchrotron radiation regions in Cyg X-3, SS433, and other X-ray binaries, let us consider three possible working hypotheses for the geometry of these synchrotron-radiating regions of relativistic electrons: a "spherical bubble" model; ejection of "twin-spherical bubbles" in opposite directions; and "conical twin-jets."

The spherical bubble model has been used for the interpretation of Cyg X-3 events by Marscher and Brown (1975) and Seaquist et al. (1979), with considerable success in fitting major aspects of the data. The N-S relativistic expansion of Cyg X-3 and the continuous radio jet production in SS433 require us to consider nonspherical geometries. It is clear that two bubbles of relativistic plasma ejected in opposite directions form one simple geometry in which the evolution of the bubbles is indistinguishable from that in single-bubble models unless there are significant asymmetric expansion effects. Another model with a simple geometry is based upon the idea that twin-jets are ejected continuously from the stellar system, possibly because of acceleration along the axis of the accretion disk responsible for the X-ray emission, and the relativistic plasma is generated in a conical environment around the laterally expanding jet. It is resonable to suppose that gas jets emerging from an accretion disk environment will begin a lateral expansion at a velocity set by the sound speed. According to this argument, temperatures of the order of 10^7 to 10^8 K, which are reasonable because of the observed X-ray emission, would give lateral expansion velocities of the order of thousands of kilometers per second, certainly sufficient to create laterally expanding shock structures in a conical geometry with a cross section that grows with time.

(b) Radio-Emitting Bubbles and Layers with Spherical Symmetry
Let us consider the basic elements of the evolution of volumes of relativistic plasma in these three geometries. van der Laan (1966) and Kellermann (1966) developed the first simple treatment of the evolution of expanding spheres or bubbles of relativistic electrons. Let us summarize the basic treatment of such adiabatically expanding spheres; then we can apply a similar treatment to conical geometries.

Let us assume a geometry as defined in Figure 9.2, in which a bubble at time t has a radius r, an angular radius $\theta = r/d$ (where d is the distance), a uniform magnetic field (with random directional distribution) of strength H, and a power law spectrum $N(E) = KE^{-\gamma}$ for the relativistic electrons at some early time in the evolution of the object. The parameters of the model are then the values of the following at a time t_0: r_0 or θ_0, H_0, and K_0. Another option is to define $j_v/\kappa_v = j_0/\kappa_0 = 2kT_0/\lambda^2$, and then view the brightness temperature parameter, T_0, and the optical depth parameter, $\tau_0 = 2\kappa_0\rho_0 r_0$, as the variables of the model. Assuming conservation of magnetic flux $H = H_0(r/r_0)^2$ describes the evolution of the magnetic field, the course of evolution of the radio source, which we can assume is expanding as a function of time, then depends upon the predominant energy loss mechanisms. A general formulation of the problem would require a complicated evolutionary equation for $N(E)$, but let us begin with the simple case of adiabatic losses due to source expansion, which means $E = E_0(r/r_0)^{-1}$ for each particle. This form of energy loss plays an important role when relativistic plasma regions are produced in small volumes; hence, it has greatest applicability for compact objects like stars and quasars. Large-volume radio sources tend to be dominated by other mechanisms such as synchrotron losses. With adiabatic losses, if one then assumes conservation of the number of particles in the radiating volume, one can assume $r^3 K \int E^{-\gamma} dE =$ constant, so if one assumes a finite spectrum from E_1 to E_2 with evolution of the spectrum between these energies one finds $K(t) = K_0(r/r_0)^{-(\gamma+2)}$, a result which is independent of the upper and lower limits on the energy spectrum. We can now use both the mass absorption and emission coefficients and the radiative transfer results from earlier in this chapter to describe the optical depth for the line of

Table 9.3. $\xi_3(\tau, u_1)$—A correction function for spherical geometries.

u_1	$\xi_3(\tau, u_1)$ for $\tau =$											
	$\ll 1$	0.04	0.08	0.16	0.32	0.64	1.28	2.56	5.12	10.24	20.48	$\gg 1$
0	2/3	0.670	0.673	0.680	0.693	0.718	0.766	0.844	0.931	0.980	0.993	1.00
0.1	0.666	0.669	0.672	0.679	0.692	0.718	0.766	0.844	0.931	0.980	0.993	1.00
0.2	0.661	0.665	0.668	0.675	0.688	0.714	0.763	0.842	0.931	0.980	0.993	1.00
0.3	0.649	0.652	0.656	0.663	0.677	0.704	0.755	0.838	0.931	0.980	0.993	1.00
0.4	0.624	0.628	0.632	0.640	0.655	0.684	0.739	0.829	0.928	0.980	0.993	1.00
0.5	0.583	0.588	0.592	0.600	0.617	0.649	0.710	0.810	0.921	0.980	0.993	1.00
0.6	0.523	0.527	0.532	0.541	0.558	0.593	0.660	0.773	0.906	0.978	0.993	1.00
0.7	0.438	0.443	0.447	0.456	0.474	0.511	0.582	0.708	0.869	0.970	0.993	1.00
0.8	0.325	0.330	0.334	0.342	0.359	0.393	0.462	0.592	0.782	0.937	0.991	1.00
0.9	0.181	0.184	0.186	0.192	0.204	0.228	0.279	0.396	0.573	0.794	0.944	1.00

sight a as

$$\tau_\nu(a) = 0.019 g(\gamma)(3.5 \times 10^9)^\gamma K H^{(\gamma+2)/2} \nu^{-(\gamma+4)/2} L$$

$$= \tau_0 \left(\frac{r}{r_0}\right)^{-(2\gamma+3)} \left(\frac{\nu}{\nu_0}\right)^{-(\gamma+4)/2} \left[1 - \left(\frac{a}{r}\right)^2\right]^{1/2} = \tau \left[1 - \left(\frac{a}{r}\right)^2\right]^{1/2} \quad (9.29)$$

where ν_0 is a reference frequency. Since the magnetic field in the volume is assumed to be uniform, this means we can write the solution for the source flux density as

$$S_\nu = (2\pi\theta_0^2)\left(\frac{2kT_0}{\lambda_0^2}\right)\left(\frac{\nu}{\nu_0}\right)^{5/2}\left(\frac{r}{r_0}\right)^3\left[1 - \exp\left[-\tau_0\left(\frac{r}{r_0}\right)^{-(2\gamma+3)}\left(\frac{\nu}{\nu_0}\right)^{-(\gamma+4)/2}\right]\right]\xi_{\rm sph}(\tau) \quad (9.30)$$

where

$$\xi_{\rm sph}(\tau) = \frac{\int u\{1 - \exp[-\tau(1 - u^2)^{1/2}]\}\,du}{\int u[1 - \exp(-\tau)]\,du} \quad (9.31)$$

which can be considered as a correction function for the differences between a spherical geometry for the line of sight and a constant line-of-sight path length corresponding to the diameter of the source. The values of $\xi_{\rm sph}(\tau)$ for various τ correspond to the entries for $\xi_3(\tau, u_1)$ for $u_1 = 0$ in Table 9.3. With the exception of the factor $\xi_{\rm sph}(\tau)$, Equation (9.30) is identical to the so-called van der Laan model (van der Laan 1966). We see that differences up to 33% will affect the relative fluxes between optically thick and optically thin stages.

The main importance of the discussion of the geometrical correction function ξ is that we now can consider a spherically symmetric shell with outer radius r and finite inner radius r_i. The line-of-sight path length becomes $(r^2 - a^2)^{1/2}$ for $a \geq r_1$ and $[(r^2 - a^2)^{1/2} - (r_1^2 - a^2)^{1/2}]$ for $a < r_1$. This leads to a more general solution for a geometry correction function, ξ_3, which is given by

$$\xi_3(\tau, u_1) = \frac{\int u(1 - \exp\{-\tau[(1 - u^2)^{1/2} - (u_1 - u^2)^{1/2}]\})\,du}{\int u[1 - \exp(-\tau)]\,du} \quad (9.32)$$

where $u_1 - r_1/r$, which we assume to be a constant. The geometry correction function is now a function of two parameters, with values listed in Table 9.3. The use of the subscript 3 indicates this ζ is a result of a three-dimensional geometry. The function $\zeta_3(\tau, u_1)$ varies between a value of $\frac{2}{3}(1 - u_1^3)$ in the optically thin limit and a value of 1 in the optically thick limit.

Equation (9.30) has $S_\nu \propto (r/r_0)^3 (\nu/\nu_0)^{5/2}$ for an optically thick limit and $S_\nu \propto (r/r_0)^{-2\gamma}(\nu/\nu_0)^{-(\gamma-1)/2}$ for an optically thin limit. This rapid decay of the flux density with r is the reason spherically expanding adiabatic radio sources tend to have short lifetimes.

The solution for S_ν as a function of r does not yet specify the time dependence of the radio source. If the sphere is freely expanding at a constant velocity, one scales between distance and time using $r/r_0 = (t - t_0)/t_{scale}$. If the expansion is decelerated by external gas, one uses scaling laws like $r/r_0 = (t/t_{scale})^{1/2}$ or $(t/t_{scale})^{2/5}$, depending upon whether the external medium is an r^{-2} windlike density distribution or a constant-density medium.

(c) Evolution of Conical Radio Jets

The geometry of conical expanding jets is a simple one that is useful for jets like those of SS433. The solution for the radiative transfer for this case is only slightly more complicated than that for the spheres and spherical layers discussed in the last section. We assume a geometry in which twin-jets are ejected in opposite directions, with the z-axis being the axis of ejection. The velocity of motion along the z-axis divided by the velocity of motion in the perpendicular r-coordinate is the Mach number M. For simplicity, we specify initial conditions at time t_0 when the jet has a radius r_0 at a distance z_0 from the origin, because then at time t the conically expanding jets have $r/r_0 = z/z_0$ and $z = M \cdot r$. We also assume parameters at t_0 such as $\theta_0 = r_0/d$, a magnetic field H_0, and K_0 (for the electron energy spectrum). As with the previous discussion of the spherically symmetric case, we can include the case where the synchrotron-emitting region is between radii r_1 and r, with r/r_1 being constant, leading to another two-parameter geometry correction function $\zeta_2(\tau, r_1/r)$. We denote this ζ with the subscript 2 because it corresponds to a two-dimensional geometry.

In this conical geometry, conservation of magnetic flux requires $H = H_0(r/r_0)^{-1}$, and applying the adiabatic energy loss formula, one obtains $E = E_0(r/r_0)^{-2/3}$. Conservation of the number of relativistic electrons then gives $K = K_0(r/r_0)^{-2(\gamma+2)/3}$. Given these scaling laws, one then has

$$\tau_\nu(a) = \tau_0 \left(\frac{r}{r_0}\right)^{-(7\gamma+8)/6} \left(\frac{\nu}{\nu_0}\right)^{-(\gamma+4)/2} \frac{(L/2r)}{\sin\Theta_{LOS}} \tag{9.33}$$

where $L = (r^2 - a^2)^{1/2}$ if $a \leq r_1$ and $L = (r^2 - a^2)^{1/2} - (r_1^2 - a^2)^{1/2}$ if $a > r_1$. We have used the angle Θ_{LOS} which describes (cf. Equation 9.28) the angle between the jets and the line of sight. It is important because by changing the angle of inclination with respect to the observer, by either orbital motion of the source of the jets or by precession, one introduces the possibility of changes with time of the appearance of the jets. By a calculation analogous to that used for the spherical case, for each line of sight

Table 9.4. $\xi_2(\tau, u_1)$—A correction function for conical geometries.

u_1	$\xi_2(\tau, u_1)$ for $\tau =$											
	$\ll 1$	0.04	0.08	0.16	0.32	0.64	1.28	2.56	5.12	10.24	20.48	$\gg 1$
0.0	0.785	0.788	0.790	0.795	0.804	0.822	0.855	0.907	0.962	0.990	0.996	1.00
0.1	0.778	0.780	0.782	0.787	0.797	0.816	0.851	0.905	0.962	0.990	0.997	1.00
0.2	0.754	0.757	0.760	0.765	0.777	0.798	0.837	0.898	0.960	0.990	0.997	1.00
0.3	0.715	0.718	0.721	0.728	0.741	0.767	0.813	0.885	0.957	0.990	0.997	1.00
0.4	0.660	0.664	0.668	0.675	0.691	0.720	0.775	0.861	0.950	0.989	0.997	1.00
0.5	0.589	0.593	0.598	0.606	0.624	0.657	0.721	0.824	0.936	0.988	0.997	1.00
0.6	0.503	0.507	0.512	0.521	0.540	0.576	0.646	0.765	0.908	0.984	0.997	1.00
0.7	0.401	0.405	0.410	0.419	0.437	0.473	0.545	0.676	0.852	0.969	0.997	1.00
0.8	0.283	0.287	0.290	0.298	0.314	0.346	0.412	0.541	0.740	0.919	0.990	1.00
0.9	0.149	0.152	0.154	0.159	0.169	0.191	0.236	0.332	0.511	0.740	0.921	1.00

$$I_\nu(a, z) = \left(\frac{2kT_0}{\lambda_0^2}\right)\left(\frac{r}{r_0}\right)^{1/2}\left(\frac{v}{v_0}\right)^{5/2}\left\{1 - \exp\left[\frac{-\tau(L/2r)}{\sin\Theta_{\mathrm{LOS}}}\right]\right\} \tag{9.34}$$

where

$$\tau = \tau_0\left(\frac{r}{r_0}\right)^{-(7\gamma+8)/6}\left(\frac{v}{v_0}\right)^{-(\gamma+4)/2}. \tag{9.35}$$

The final result for the flux density from the two jets from the inner z_0 point out to a distance z, where we make use of $z = r \cdot M$, is

$$S_\nu = 2 \cdot M\left(\frac{2kT_0}{\lambda_0^2}\right)\theta_0^2\left(\frac{v}{v_0}\right)^{5/2}\sin\Theta_{\mathrm{LOS}}\int_0^{r/r_0}\left(\frac{r}{r_0}\right)^{3/2}\left[1 - \exp\left(\frac{-\tau}{\sin\Theta_{\mathrm{LOS}}}\right)\right]$$
$$\cdot \xi_2(\tau, u_1)d(r/r_0) \tag{9.36}$$

which has the interesting optically thin limit of

$$S_\nu \propto \left(\frac{r}{r_0}\right)^{-7(\gamma-1)/6}\left(\frac{v}{v_0}\right)^{-(\gamma-1)/2}. \tag{9.37}$$

This is a very important result because it indicates that the flux density contribution from the outer portions of adiabatically expanding conical jets declines much less rapidly with r (or z) when compared with the spherical geometry decrease of $r^{-2\gamma}$. The function $\xi_2(\tau, u_1)$ is analogous to Equation (9.31) for spherically symmetric systems. It is defined by

$$\xi_2(\tau, u_1) = \frac{\int (1 - \exp\{-\tau[(1 - u^2)^{1/2} - (u_1^2 - u^2)^{1/2}]\})\,du}{\int [1 - \exp(-\tau)]\,du} \tag{9.38}$$

which varies between a value of $\pi(1 - u_1^2)/4$ in the optically thin limit to 1 in the optically thick limit. In Table 9.4, we list the values of ξ_2 for a range of values of τ and u_1 appropriate to intermediate optical depths.

In this discussion of the evolution of conical jets, we have not addressed the cases where synchrotron or other energy loss mechanisms play a major role, because

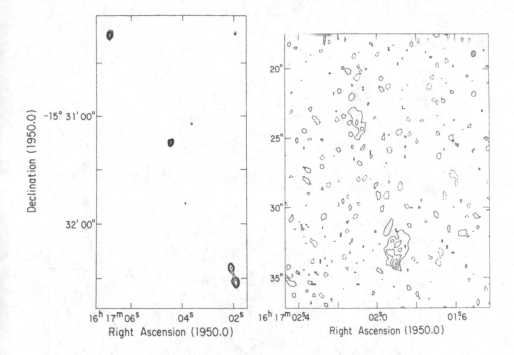

Fig. 9.23. A 1.465-GHz VLA image of the Sco X-1 ratio triplet, where the X-ray binary is coincident with the central source, is shown on the left and an expanded, higher-resolution image of the SW source is shown on the right (From Fomalont et al. 1983.)

the equations become much more intractable. Instead, we have emphasized the tractability of adiabatic evolution in both spherical and conical geometries because such treatments may explain many cases and allow essentially analytic expressions for the evolution of the radio emission components.

9.7.6 Sco X-1—Radio Variable with Companion Double Radio Source
Figure 9.23 shows images of both the Sco X-1 radio triplet and the SW component at higher resolution. The central radio source on the X-ray binary is highly variable (Wade and Hjellming 1971c, Bradt et al. 1975). The NE component is unresolved. The SW component contains a "hot spot" at its outer edge, a characteristic very common in extragalactic double radio sources. The similarity of the Sco X-1 radio triplet to a common type of extragalactic radio source may be because there is considerable similarity in the way in which compact objects produce local, highly variable radio emission and extended radio lobes.

9.7.7 The Very Large Remnants Around SS433 and Cir X-1
The radio lobes of Sco X-1 are weak manifestations of large-size-scale radio emission compared to the synchrotron-emitting nebulosities associated with SS433 and Cir X-1. SS433 is at the center of a 2° by 1° synchrotron radio source called W50 (Geldzahler et al. 1980). Originally thought to be a supernova remnant, W50 has perturbed "ears" exactly in the cone of precession of the SS433 jets. The X-ray

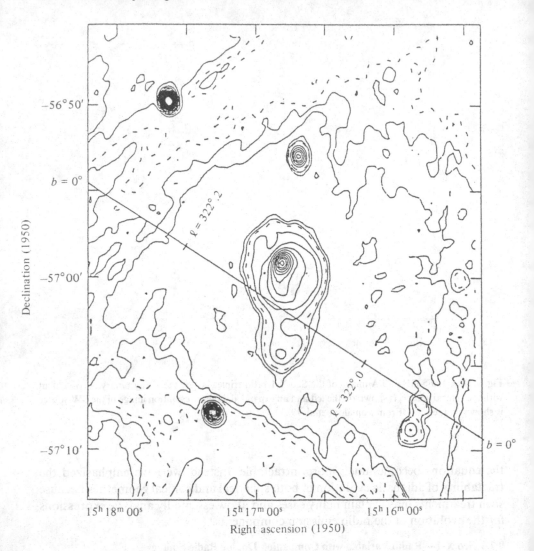

Fig. 9.24. A 843-MHz image of the region around Cir X-1, obtained with the Molonglo Observatory Synthesis Telescope, which shows Cir X-1 radio emission in a compact source and extended emission emanating from Cir X-1 to form a large-scale synchrotron-emitting radio source. (From Haynes et al. 1986. Reprinted by permission from Nature, Vol. 324, pp. 233. Copyright(c) 1986, Macmillan Magazines Limited.

lobes of SS433, on size scales of the order of a degree, are also inside the precession cone of SS433 (Watson et al. 1983). SS433 clearly has affected the W50 remnant because of the energetics of its jets, and may have supplied a considerable part of the energy budget for W50.

Cir X-1 also has a large complex radio nebulosity that is probably produced because of the characteristics of the central object. The image of this extended nebulosity produced by Haynes et al. (1986) is shown in Figure 9.24. The most intense emission is at the position of Cir X-1, but there is an extension from Cir X-1 into a large nonthermal nebulosity.

9.7.8 Other X-Ray Binaries

There are two X-ray binaries whose radio emission is in some sense most remarkable because they are relatively constant in flux level and spectrum. The best-known of these is Cyg X-1. The identification of Cyg X-1 with the star system HDE 226868 began with the detection of its radio emission (Hjellming and Wade 1971c, Braes and Miley 1972) inside the relatively large error region of an X-ray source position. This rapidly led to the accumulating evidence that the dominant 0.9.5Ib star has a companion that may be so massive that it must be a black hole. Early observations of Cyg X-1 (Hjellming 1973a) established that between March 22 and March 31, 1972, the radio source turned on, in the sense that it went from below a detection limit of 10 mJy to a radio flux level that remained relatively constant until May 1975. During this period, the radio flux of Cyg X-1 remained mostly constant at a level of 15 mJy, with a spectral index of ~ 0. The "turn-on" in March 1971 occurred exactly during the time of a change of state of the Cyg X-1 X-ray source (Tananbaum et al. 1972). After four years of nearly stable behavior, Cyg X-1 again, on or before May 1975 (Hjellming et al. 1975), underwent a change of state of both X-ray and radio source, with radio "flaring" to a peak level of 45 mJy and then decaying back to normal levels. Since that time, Cyg X-1 has been only sporadically observed, but when observed, the source was the normal 15 mJy.

The other radio-emitting X-ray binary that is relatively quiescent is Cyg X-2 (Hjellming and Blankenship 1973); however, it has been observed only a few times.

Two other radio-emitting X-ray binaries that have occasionally been found to flare are GX17+2 (Hjellming and Wade 1972c) and GX5-1 (Braes et al. 1972, Geldzahler 1983). However, they have not yet been extensively studied, largely because of their typically weak flux levels.

9.7.9 Transient X-Ray/Radio Stars

Owen et al. (1976) found the first case where a transient X-ray source exhibited transient radio emission when they observed the decay phase of highly variable emission from A0620-03. Hjellming (1979) found a similar transient radio source associated with an X-ray event in Cen X-4. In the latter event, the decaying radio source maintained a constant spectral index of ~ -0.6, which is typical of synchrotron radiation sources. The transient production of strong shocks, generating an expanding bubble or layer of mixed relativistic electrons and magnetic fields, is almost certainly the origin of the radio emission in these binary systems.

9.7.10 A Unified View of Radio-Emitting X-Ray Binaries

The principal difference between radio emission associated with strong X-ray sources and radio emission from "other" radio stars is the clear dominance of strong synchrotron emission sources. This is likely to be associated with the high energies of shocks expanding in spherical or cylindrical geometries. This expansive acceleration phenomenon has long been used to explain the production of the radiating electrons of supernova remnants and has recently been found to produce an extended radio source around the old nova (Nova Persei 1901) shell associated with

GK Per (Reynolds and Chevalier 1984). White (1985) has discussed an equivalent phenomenon for the production of nonthermal emission in stellar winds. For SS433, and possibly many other X-ray binaries, the expansive acceleration occurs in basically conical environments.

In this section, we have emphasized the working hypothesis that all of the radio-emitting X-ray binaries have SS433-like jets. The evolution of conical jets expanding adiabatically is simple to quantify, leading to a wide variety of radio source characteristics. Variability of radio emission in this context is obtained in two basic ways. Periodic variations occur when the kinematics of jets, with substantial radio emission from the optically thick inner portions, causes changes in both apparent solid angle and the line-of-sight optical depths for the jets. The short-lived flaring events are identified with episodes of such efficient production of relativistic plasma, in the acceleration regions, that the pressure of the relativistic particles causes sudden expansion of the plasma. These "plasmoids" can be described as twin bubbles moving along the path of motion of normally conical jets or bubbles that quickly merge into a single expanding sphere of radiating plasma. The combination of conical jets and spherical bubbles is described by using two types of evolutionary equations for different parts of the same system. In some cases, the energy losses in the spherically or conically expanding geometries may not be dominantly adiabatic, in which case the simple evolutionary equations derived in this chapter must be replaced with more complicated numerical computations.

Rees (1982) has shown that jets such as those found in extragalactic radio sources, and SS433, have structures that scale with $\dot{M}/M$, and hence are similar except for scaling in this parameter. Thus, both the accretion and the expansion environment of X-ray binaries and cores of active galaxies may have qualitative similarities and common scaling factors.

9.8 Future Work on Radio Stars

The field of stellar radio astronomy is still undergoing a rapid growth as of this writing. There are a large number of areas that should be very important in coming years.

The imaging of nearby active binaries using trans- and intercontinental VLBI techniques should become very important with the use of the U.S. VLBA, the growing European VLBI network, and antennas on other continents. Imaging the stellar and intrastellar magnetospheric emission as a function of time will provide unique information on magnetospheric physics on scales that make the Sun and Jupiter insignificant by comparison. The other principal observational frontier is that of shorter time resolution and more simultaneous frequency spectra. The bursts and drifts seen in UV Ceti by Bastian and Bookbinder (1987) are indicators that detailed $S(v, t)$ measurements will be very important when done with instruments with sufficient sensitivity for the interesting domains of changes in frequency and time to be explored. Coordinated observations of radio, X-ray, infrared, optical, and UV data have been attempted only a very few times, so in some sense the organizational and political domains, which provide the biggest barriers for this type of data taking for unpredictable, short-time-scale events, are the source of the major problems.

On the theoretical side, the problems of radio stars are in many cases formidable. Complex plasma processes, time dependence, and complex geometries that cannot be observationally resolved in any detail make theoretical work in complex objects extraordinarily difficult. However, while a pessimist might say that another field is easier to work in, an optimist would say that the challenges of radio stars will last for a long time.

Recommended Reading

Hjellming, R.M., and D.M. Gibson (eds.). 1985. Radio Stars. Dordrecht: Reidel.
Mason, K.O., M.G. Watson, and N.E. White (eds.). Physics of Accretion onto Compact Objects. 1986. New York: Springer-Verlag.
Kundu, M.R. 1965. Solar Radio Astronomy. New York: Interscience.
Zheleznyakov, V.V. 1970. Radio Emission of the Sun and Planets. Oxford: Pergamon.

References

Abbott, D.C. 1985. *In* R.M. Hjellming and D.M. Gibson (eds.), Radio Stars. Dordrecht: Reidel, p. 61.
Abbott, D.C., J.H. Bieging, and E. Churchwell. 1984, Astrophys. J. **280**:671.
Abbott, D.C., J.H. Bieging, and E. Churchwell. 1986a. Astrophys. J. **303**:239.
Abbott, D.C., J.H. Bieging, E. Churchwell, and A.V. Torres. 1986b. Astrophys. J. **303**:239.
Ables, J.G. 1969a. Proc. Astron. Soc. Aust. **1**:237.
Ables, J.G. 1969b. Astrophys. J. **155**:L27.
Altenhoff, W.J., and H. Wendker. 1973. Nature **241**:27.
Andrew, B.H., and C.R. Purton. 1968. Nature **218**:855.
Bastian, T.S., and J. Bookbinder. 1987. Nature **326**:678.
Bastian, T.S., G.A. Dulk, and G. Chanmugam. 1987. Astrophys. J. (in press).
Becklin, E.E., G. Neugebauer, F.J. Hawkins, K.O. Mason, P.W. Sanford, K. Mathews, and C.G. Wynn-Williams. 1972. Nature Phys. Sci. **239**:130.
Benz, A.O., E. Furst, and A.L. Kiplinger. 1983. Nature **302**:45.
Bookbinder, J.A. and D.Q. Lamb. 1987. Astronphys. J. (Lett.) (in press).
Bradt, H.V., L.L.E. Braes, W. Forman, J.F. Hesser, W.A. Hiltner, R.M. Hjellming, E. Kellogg, W.E. Kunkel, G.K. Miley, G. Moore, J.W. Pel, J. Thomas, P. Vanden Bout, C.M. Wade, and B. Warner. 1975. Astrophys. J. **197**:443.
Braes, L.L.E., and G.K. Miley. 1971. Nature **232**:246.
Braes, L.L.E., and G.K. Miley. 1972. Nature **237**:507.
Braes, L.L.E., G.K. Miley, and A.A. Schoenmaker. 1972. Nature **236**:392.
Brown, R.L., and P.C. Crane. 1978. Astron. J. **83**:1504.
Chanmugam, G., and G.A. Dulk. 1982. Astrophys. J. **255**:L107.
Chevalier, R.A. 1977. Astron. Astrophys. **59**:289.
Clark, B.G., K.I. Kellermann, and D. Shaffer. 1975. Astrophys. J. **198**:L127.
Clark, T.A., L.K. Hutton, Ma., C., W. Webster, C.A. Hinteregger, C.A. Knight, A.E.E. Rogers, A.E. Whitney, I.I. Shapiro, J.J. Wittels, A.E. Niell, and G.M. Resch, 1976, Astrophys. J. **206**:L107.
Cohen, M., J.H. Bieging, and P.R. Schwartz. 1982. Astrophys. J. **253**:707.
Davidsen, A., and J.P. Ostriker. 1974. Astrophys. J. **189**:331.
Davis, R.J., B. Lovell, H.P. Palmer, and R.E. Spencer. 1978. Nature **273**:644.
Drake, S.A., and J.L. Linsky. 1986. Astron. J., **91**:602.
Dulk, G.A. 1985. Annu. Rev. Astron. Astrophys. **23**:169.
Dulk, G.A., and D.E. Gary. 1983. Astron. Astrophys. **124**:103.
Dulk, G.A., T.S. Bastian, and B. Chanmugam. 1983. Astrophys. J. **273**:249.
Ennis, D., E.E. Becklin, S. Beckwith, J. Elias, I. Gatley, K. Mathews, and G. Neugebauer. 1977. Astrophys. J. **214**:478.
Fomalont, E.B., B.J. Geldzahler, R.M. Hjellming, and C.M. Wade. 1983. Astrophys. J. **275**:802.

Furst, E., A.O. Benz, W. Hirth, A.L. Kiplinger, and M. Geffert. 1986. Astron. Astrophys. **154**:377.

Gary, D.E., J.L. Linsky, and G.A. Dulk. 1983. Astrophys. J. (Lett.) **263**:L79.

Geldzahler, B.J. 1983, Astrophys. J. (Lett.) **264**:L49.

Geldzahler, B.J., T. Pauls, and C. Salter. 1980. Astron. Astrophys. **84**:237.

Geldzahler, B.J., K.J. Johnston, J.H. Spencer, W.J. Klepczynski, F.J. Josties, P.E. Angerhofer, D.R. Florkowski, D.D. McCarthy, D.N. Matsakis, and R.M. Hjellming. 1983. Astrophys. J. (Lett.) **273**:L65.

Gibson, D.M. 1979. Bull. Am. Astron. Soc. **10**:631.

Gibson, D.M. 1985. *In* R.M. Hjellming and D.M. Gibson (eds.), Radio Stars. Dordrecht: Reidel, p. 225.

Gibson, D.M., and R.M. Hjellming. 1974. Publ. Astron. Soc. Pacific **86**:652.

Gibson, D.M., R.M. Hjellming, and F.N. Owen. 1975. Astrophys. J. **200**:L99.

Ginzburg, V.L., and S.I. Syrovatskii. 1965. Annu. Rev. Astron. Astrophys. **3**:297.

Gregory, P.C., and A.R. Taylor. 1978. Nature **272**:704.

Gregory, P.C., P.P. Kronberg, E.R. Seaquist, V.A. Hughes, A. Woodsworth, M.R. Viner, D. Retalleck, R.M. Hjellming, and B. Balick. 1972. Nature Phys. Sci. **239**:114.

Gregory, P.C., A.R. Taylor, D. Crampton, J.B. Hutchings, R.M. Hjellming, D.E. Hogg, H. Hvatum, E.W. Gottlieb, P.A. Feldman, and S. Kwok. 1979. Astron. J. **84**:1030.

Haynes, R.F., D.L. Jauncey, P.G. Murdin, W.M. Goss, A.J. Longmore, L.W.J. Simons, D.K. Milne, and D.J. Skellern. 1978. Mon. Not. R. Astron. Soc. **185**:661.

Haynes, R.F., I. Lerche, and P. Murdin. 1980. Astron. Astrophys. **87**:299.

Haynes, R.F., M.M. Komesaroff, A.G. Little, D.L. Jauncey, J.L. Caswell, D.K. Milne, M.J. Kesteven, K.J. Wellington and R.A. Preston. 1986. Nature **324**:233.

Heise, J., A.C. Brinkman, E. Gronenschild, M. Watson, A.R. King, L. Stella, and K. Kleboom. 1985. Astron. Astrophys. **148**:L14.

Hjellming, R.M. 1973a. Astrophys. J. (Lett.) **182**:L29.

Hjellming, R.M. 1973b. Science **182**:1089.

Hjellming, R.M. 1976. X-Ray Binaries, NASA publication SP-389, p. 233.

Hjellming, R.M. 1979. Int. Astron. Union Circular No. 3369.

Hjellming, R.M. 1985a. *In* R.M. Hjellming and D.M. Gibson (eds.), Radio Stars. Dordrecht: Reidel, p. 97.

Hjellming, R.M. 1985b. *In* R.M. Hjellming and D.M. Gibson (eds.), Radio Stars. Dordrecht: Reidel, p. 151.

Hjellming, R.M. 1985c. *In* R.M. Hjellming and D.M. Gibson (eds.), Radio Stars. Dordrecht: Reidel, p. 301.

Hjellming, R.M., and L.C. Blankenship. 1973. Nature Phys. Sci. **243**:81.

Hjellming, R.M., and R.L. Brown. 1988. Astron. J. (in preparation).

Hjellming, R.M., and K.J. Johnston. 1981a. Nature **290**:100.

Hjellming, R.M., and K.J. Johnston. 1981b. Astrophys. J. (Lett.) **246**:L141.

Hjellming, R.M., and K.J. Johnston. 1982. *In* D.S. Heeschen and C.M. Wade (eds.), Extragalactic Radio Sources. Dordrecht: Reidel, p. 197.

Hjellming, R.M., and K.J. Johnston. 1986. *In* K.O. Mason, M.G. Watson, and N.E. White (eds.), Physics of Accretion onto Compact Objects. Berlin: Springer-Verlag.

Hjellming, R.M., and R.T. Newell. 1983. Astrophys. J. **275**:704.

Hjellming, R.M., and C.M. Wade. 1970. Astrophys. J. **162**:L1.

Hjellming, R.M., and C.M. Wade. 1971a. Astrophys. J. (Lett.) **168**:L115.

Hjellming, R.M., and C.M. Wade. 1971b. Astrophys. J. (Lett.) **164**:L1.

Hjellming, R.M., and C.M. Wade. 1971c. Astrophys. J. (Lett.) **168**:L21.

Hjellming, R.M., and C.N. Wade. 1971d. Astrophys. J. **164**:L1.

Hjellming, R.M., and C.M. Wade. 1973. Nature **242**:250.

Hjellming, R.M., C.M. Wade, N.R. Vandenberg, and R.T. Newell. 1979. Astron. J. **84**:1619.

Hjellming, R.M., D.M. Gibson, and F.N. Owen. Nature **256**:111.

Hjellming, R.M., J.H. van Gorkom, A.R. Taylor, E.R. Seaquist, S. Padin, R.J. Davis, and M.F. Bode. 1986. Astrophys. J. (Lett.) **307**:L71.

Hjellming, R.M., R.L. Brown, and L.C. Blankenship. 1974. Astrophys. J. (Lett.) **194**:L13.

Hogg, D.E. 1985. *In* R.M. Hjellming and D.M. Gibson (eds.), Radio Stars. Dordrecht: Reidel, p. 117.

Johnston, K.J. 1987 (Private communication).

Johnston, K.J., B.J. Geldzahler, J.H. Spencer, E.B. Waltman, F.J. Klepczynski, P.E. Josties, P.E. Angerhofer, D.R. Florkowski, D.D. McCarthy, and D.N. Matsakis. 1984a. Astron. J. 89:509.

Johnston, K.J., J.H. Spencer, F.J. Klepczynski, P.E. Josties, P.E. Angerhofer, D.R. Florkowski, D.D. McCarthy, D.N. Matsakis, and R.M. Hjellming. 1984b. Astrophys. J. (Lett.) 273:L65.

Johnston, K.J., J.H. Spencer, R.S. Simon, E.B. Waltman, G.G. Pooley, R.E. Spencer, R.W. Swinney, P.E. Angerhofer, D.R. Florkowski, F.E. Josties, D.D. McCarthy, D.N. Matsakis, D.E. Reese, and R.M. Hjellming. 1986. Astrophys. J. 309:707.

Kellermann, K.I. 1966. Astrophys. J. 146:621.

Kellermann, K.I., and I.I.K. Pauliny-Toth. 1966. Astrophys. J. 146:953.

Kundu, M.R. 1965. Solar Radio Astronomy. New York: Interscience.

Kwok, S. 1983. Mon. Not. R. Astron. Soc. 202:1149.

Lang, K.R., J. Bookbinder, L. Golub, and M. Davis. 1983. Astrophys. J. (Lett.) 272:L15.

Lovell, B. 1969. Nature 222:1126.

Margon, B. 1984. Annu. Rev. Astron. Astrophys. 22:507.

Marscher, A.P., and R.L. Brown. 1975. Astrophys. J. 200:719.

Mason, K.O., E.E. Becklin, L.C. Blankenship, R.L. Brown, J. Elias, R.M. Hjellming, K. Matthews, P.G. Murdin, G. Neugebauer, P.W. Sanford, and S.P. Willner. 1976. Astrophys. J. 207:78.

Molnar, L.A. 1984. Nature 310:662.

Molnar, L.A. 1985. In R.M. Hjellming and D.M. Gibson (eds.), Radio Stars. Dordrecht: Reidel, p. 329.

Molnar, L.A., M.J. Reid, and J.E. Grindlay. 1983. Int. Astron. Union Circular No. 385.

Mutel, R.L., J.F. Lestrade, R.A. Preston, and R.B. Phillips. 1985. Astrophys. J. 289:262.

Newell, R.T. 1981. Ph.D. thesis, New Mexico Institute of Mining and Technology.

Newell, R.T., and R.M. Hjellming. 1982. Astrophys. J. (Lett.) 263:L85.

Nicholson, G.D., M.W. Feast, and I.S. Glass. 1980. Mon. Not. R. Astron. Soc. 191:293.

Owen, F.N., T.J. Balonek, J. Dickey, Y. Terzian, and S. Gottesman. 1976. Astrophys. J. (Lett.) 293:L15.

Padin, S., R.J. Davis, and M.F. Bode. 1985. Nature 315:306.

Parker, E.N. 1963. Interplanetary Dynamical Processes. New York: Wiley Interscience.

Parsignault, D.R., H. Gursky, E.M. Kellogg, T. Matilsky, S. Murray, E. Schreier, H. Tananbaum, R. Giacconi, and A.C. Brinkman. 1972. Nature Phys. Sci. 239:123.

Preston, R.A., D.D. Morabito, A.E. Wehrle, D.L. Jauncey, M.J. Batty, R.F. Haynes, and A.E. Wright. 1983. Astrophys. J. 268:L23.

Purton, C.R., P.A. Feldman, and K.A. Marsh. 1973. Nature Phys. Sci. 245:5.

Rees, M. 1982. In D.S. Heeschen and C.M. Wade (eds.), Extragalactic Radio Sources. Dordrecht: Reidel, p. 211.

Reynolds, S.P., and R.A. Chevalier. 1984. Astrophys. J. (Lett.) 281:L33.

Ryle, M., J.L. Caswell, G. Hine, and J. Shakeshaft. 1979. Nature 276:571.

Schwartz, P.R., T. Simon, and Campbell, R. 1986. Astrophys. J. 303:233.

Seaquist, E.R. 1967. Astrophys. J. 148:L23.

Seaquist, E.R., and J. Palimaka. 1977. Astrophys. J. 217:781.

Seaquist, E.R., P.C. Gregory, R.A. Perley, R.H. Becker, J.B. Carlson, M.R. Kundu, R.C. Bignell, and J.R. Dickel. 1974. Nature 251:394.

Seaquist, E.R., R.E. Garrison, P.C. Gregory, A.R. Taylor, and P.C. Crane. 1979. Astron. J. 84:1037.

Seaquist, E.R., A.R. Taylor, and S. Button. 1984. Astrophys. J. 284:202.

Spencer, J.H. and Schwartz, P.R. 1974. Astrophys. J. 188:L105.

Spencer, R.E., R.W. Swinney, K.J. Johnston, and R.M. Hjellming. 1986. Astrophys. J. 309:694.

Tananbaum, H., H. Gursky, E. Kellog, R. Giacconi, and C. Jones. 1972. Astrophys. J. 177:L5.

Taylor, A.R., and P.C. Gregory. 1982. Astrophys. J. 255:210.

Taylor, A.R., and E.R. Seaquist. 1984. Astrophys. J. 286:263.

Taylor, A.R., E.R. Seaquist, and J.A. Mattei. 1986. Nature 319:38.

Taylor, A.R., E.R. Seaquist, J.M. Hollis, and S.R. Pottasch. 1987. Astron. Astrophys. (in press)

Thomas, R.M., M.L. Duldig, R.F. Haynes, and P.G. Murdin. 1978. Mon. Not. R. Astron. Soc. 185:29.

Topka, K., and K.A. Marsh. 1982. Astrophys. J. 254:641.

Turner, K. 1985. In R.M. Hjellming and D.M. Gibson (eds.), Raido Stars. Dordrecht: Reidel, p. 283.

Uchida, Y., and T. Sakurai. 1983. In M. Rodono and P. Byrne (eds.), Activity in Red Dwarf Stars, Int. Astron. Union Coll. 71. Dordrecht: Reidel, p. 629.

van der Laan, H. 1966. Nature **211**:1131.

Wade, C.M., and R.M. Hjellming. 1971a. Astrophys. J. **163**:L105.

Wade, C.M., and R.M. Hjellming. 1971b. Astrophys. J. **163**:L65.

Wade, C.M., and R.M. Hjellming. 1971c. Astrophys. J. **170**:523.

Wade, C.M., and R.M. Hjellming. 1972. Nature **235**:270.

Watson, M.G., R. Willingale, J.E. Grindlay, and F.D. Seward. 1983. Astrophys. J. **273**:688.

Weiler, K.W., R.A. Sramek, N. Panagia, J.M. van der Hulst, and M. Salvati. 1986. Astrophys. J. **301**:790.

Whelan, J.A.J., S.K. Mayo, D.T. Wickramasinghe, P.G. Murdin, B.A. Peterson, T.G. Hawarden, A.J. Longmore, R.F. Haynes, W.M. Goss, L.W. Simons, J.L. Caswell, A.G. Little, and W.B., McAdam, 1977, Mon. Not. R. Astron. Soc., **181**:259.

White, R.L. 1985. *In* R.M. Hjellming and D.M. Gibson (eds.), Radio Stars. Dordrecht: Reidel, p. 45.

White, R.L., and R.H. Becker. 1982. Astrophys. J. **262**:657.

White, R.L., and R.H. Becker. 1983. Astrophys. J. **272**:L19.

Wright, A.E., and M.J. Barlow. 1975. Mon. Not. R. Astron. Soc. **170**:41.

Zheleznyakov, V.V. 1970. Radio Emission of the Sun and Planets. Oxford: Pergamon.

10. Supernova Remnants

STEPHEN P. REYNOLDS

10.1 Introduction

Supernovae and their remnants play a central part in modern astrophysics. Supernovae end the lives of massive stars, form neutron stars and pulsars, synthesize heavy elements, and inject energy into the interstellar medium which can maintain turbulent cloud motions, heat a large fraction of the volume of the galaxy, and accelerate particles. Supernova remnants with their shock waves can also generate cosmic rays and compress clouds, perhaps starting another cycle of star formation.

The remnants of supernova explosions have also played an important role in the history of radio astronomy, and radio avenues of investigation remain very productive tools for the study of supernova remnants (SNR). Recent years have seen, in addition to higher-resolution observations of "normal" shell remnants, the growth of a class of SNR resembling the Crab Nebula, the routine study of SNR in other galaxies, and the detection and explication of radio emission from supernova outbursts themselves. This chapter will describe a basic observational and theoretical framework in which to discuss SNR. Review articles addressing various topics in more depth are cited throughout the text and listed at the end of the chapter.

10.1.1 History

The concept of a supernova as an event vastly more powerful and catastrophic than a normal nova dates only from the 1930s, when Baade and Zwicky (1934a, b) produced a series of prescient papers drawing together earlier speculations by various astronomers. They set forth a picture, essentially the current view, of supernovae as the explosions of massive stars at the end of their evolutionary lifetimes, releasing 10^{51} ergs or more, primarily in the form of kinetic energy of ejected material moving at thousands of kilometers per second. Among their more striking assertions were the claims (Baade and Zwicky 1934a) that a superdense stellar remnant might remain and that supernovae might produce cosmic rays. Baade (1942) later proposed the identification, not verified until 1969, of the neutron star remnant of the supernova of 1054 A.D. (the Crab pulsar). See Trimble (1982) for a fuller account of the history of the idea of supernovae.

Lundmark (1921) was the first to associate the Crab Nebula (M1) with the historical reports in Chinese and Japanese records of the "guest star" of 1054 A.D. This apparition may have reached apparent magnitude -4, or about as bright as Venus at its peak; it was visible in daylight for twenty-three days and disappeared only after twenty-one months. Its association with the Crab Nebula made the latter the first SNR to be identified, and it is still the optically most prominent. Before the development of radio astronomy, identifications of other optical SNR were conjectural; while the Cygnus Loop was nominated by Hubble (1937), the position of Kepler's (1604 A.D.) supernova yielded only feeble wisps, bearing little resemblance to the brilliant Crab. Radio astronomy entered the field with the identification of the strong source Taurus A with the Crab Nebula (Bolton et al. 1949). The positions of Tycho's and Kepler's supernovae soon yielded radio sources; in the former case, the radio position led to the discovery of optical filaments even fainter than those in Kepler's remnant. The extremely powerful source Cas A was found to be accompanied by faint, peculiar optical nebulosity, remarkable for showing extremely high velocities, up to 6000 km s^{-1}. Thus it was the first SNR identified without a known supernova.

Multifrequency radio observations of these objects showed power law spectra, $S_\nu \propto \nu^\alpha$, where α was found to be -0.6 to -0.8 except for the Crab where $\alpha \cong -0.3$. While the galactic plane is full of discrete radio sources, most are characterized by thermal spectra: $\alpha \sim 2$ (opaque) or, more commonly, $\alpha \cong 0$ (optically thin). Thus, a galactic-plane source with a steep or "nonthermal" spectrum came to be identified as a supernova remnant; with the general acceptance of Shklovskii's (1953) proposal that the radiation process was synchrotron emission, polarization evidence was taken to clinch the argument. Higher-resolution radio observations made it possible to add the additional requirement that the source be fairly large, to discriminate against the population of background radio galaxies and quasars. Such observations also led to the general realization by 1960 that the Crab Nebula, with its center-brightened blobby structure (Figure 10.1), was not a typical SNR in morphology or spectrum: most SNR were found to have shell-like structure. Of the 145 galactic SNR listed in a recent catalogue by Green (1984), only ten or so depart from the norms of relatively steep spectrum ($-0.3 \geqslant \alpha \geqslant -0.8$) and at least a hint of shell structure. The Crab Nebula now leads a little band of five to ten socalled "Crab-like SNR" or "plerions" (Weiler 1985) distinguished by center-brightened morphology, higher than typical polarization, and flatter spectrum ($0 \geqslant \alpha \geqslant -0.3$); these are presumably an early stage in the life of an SNR powered by a pulsar.

The successful identification of the Crab Nebula led to attempts to identify remnants of other "guest stars" mentioned in Chinese records. A great deal of effort has gone into interpreting the Chinese descriptions and separating the small number of suspected supernovae from the much larger number of observations of comets, ordinary novae, and assorted other transient phenomena. Clark and Stephenson (1977) review the subject, concluding that in the last two thousand years, seven events are reasonably reliably to be identified with supernovae: those of A.D. 185, 393, 1006, 1054, 1181, 1572, and 1604, of which the laggard Europeans managed to record only 1006, 1572, and 1604. Remnants are securely identified for 1006, 1054, 1181 (3C58), 1572, and 1604; possible identifications of varying plausibility exist for the others. The observations by Tycho (1572) and Kepler (1604) were sufficiently careful to allow the supernovae to be classified as Type I (see below).

Once a large sample of SNR was made available through radio investigations, optical investigations became common enough for generalizations to be made. Optically, a supernova remnant can be distinguished from other types of emission nebulosity by emission lines characteristic of collisional ionization, such as [SII]

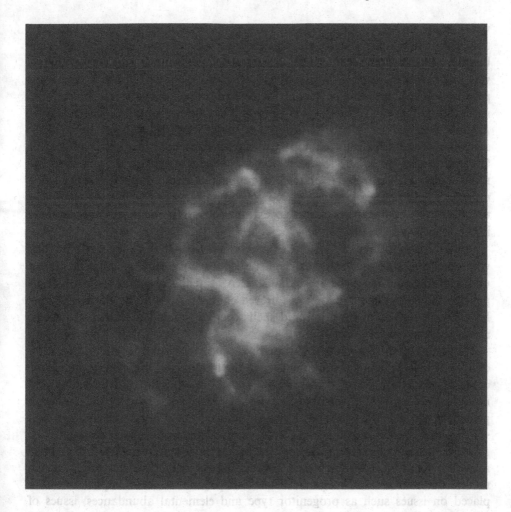

Fig. 10.1. The Crab Nebula at 1.4 GHz, observed with the Very Large Array (Courtesy NRAO/AUI; Observations by A.S. Wilson, D.E. Hogg and H.H. Samarsinha).

$\lambda\lambda$ 6717, 6731: a ratio of (Hα + [NII]) to [SII] less than 2.5 generally selects SNR. A few remnants show strong [OII] and [OIII] emission without much [SII], after the example of Cas A, while Tycho and SN 1006 show only hydrogen lines. These discoveries led to surveys of external galaxies with interference filters, to identify objects based on the [SII] criterion. Samples of candidate SNR in M31 (D'Odorico et al. 1980, Blair et al. 1981) and in M33 (D'Odorico et al. 1976, Blair and Kirshner 1985) were produced in this manner. Such samples provide valuable homogeneous collections of SNR at the same distance, avoiding the problems of disparate discovery criteria and poorly known distances that plague the collection of galactic remnants.

The Crab Nebula was one of the first extrasolar X-ray sources to be discovered, and even in the early days of rocket flights it soon became clear that as a class, SNR were frequently X-ray emitters. A fortunate lunar occultation led to the discovery that the Crab's X-ray source was extended, long before routine X-ray imaging became possible; and yet another pioneering field, extrasolar X-ray polarimetry, was born with the report (Weisskopf et al. 1978) that the Crab's X rays were polarized. (The Crab remains the only detected example of a polarized source of X rays outside the Sun.) The development of X-ray imaging and in particular the *Einstein Observatory* (HEAO-2) satellite revealed that many SNR were X-ray sources, generally showing line spectra and thermal continua characteristic of temperatures of 10^6 to 10^7 K and shell morphologies similar to those seen in the radio. Again, the Crab-like remnants are anomalous, exhibiting center-brightened morphology and power law spectra. X-ray techniques made possible the identification of another homogeneous sample of extragalactic SNR, the thirty-odd remnants in the Magellanic Clouds (Mathewson et al. 1983, 1984).

Infrared observations of SNR are in their infancy. Aside from the Crab Nebula, whose near-infrared spectrum has been studied for some time, most remnants are unobservably faint from the ground. The IRAS satellite observations promise to provide the first systematic look at infrared emission from SNR; a few remnants have been observed to have infrared shells roughly coincident with radio shells, attributable to heated dust (Braun 1985).

The study of radio emission from supernovae themselves began with the detection of SN 1970g in M101 as a weak increase in the total flux of a region in the galaxy containing an HII region as well as the supernova. No other supernova was detected until the Very Large Array became operational, because of sensitivity and confusion problems. Since 1979, five more radio supernovae have been well studied, and sources have been found coincident with historical supernovae in M83.

The subsequent discussion will concentrate on those features of SNR of greatest relevance to radio astronomy: the nonthermal emission and the origin of the relativistic electrons and magnetic fields that produce it. Less emphasis will be placed on issues such as progenitor type and elemental abundances, issues of considerable interest but primarily involving optical and X-ray data. Section 10.2 contains a brief sketch of supernovae: types, rates, etc. The observed properties of SNR are described in Section 10.3, which includes a discussion of radio supernovae. Section 10.4 comprises an outline of the hydrodynamic evolution of SNR and their interaction with the interstellar medium, and of the theory of generation of relativistic particles and magnetic fields in SNR. Theories of radio emission from supernovae are included. Section 10.5 discusses collective effects of SNR on the Galaxy as a whole, and Section 10.6 sketches future prospects in the study of supernova remnants.

10.2 Supernovae

10.2.1 Type I and Type II

The several hundred supernovae observed in other galaxies in the last fifty years or so can be classified mainly into two types, on the basis of spectroscopic evidence. Type I included those supernovae whose optical spectra near maximum light show no obvious hydrogen lines; Type II spectra include strong, wide hydrogen emission lines. CaII H and K can be tentatively identified in both types. For both types, emission features are accompanied by blueshifted absorption (P Cygni profiles), allowing the deduction of expansion velocities (after a few weeks) of about 10,000 km s^{-1} for Type I and 4000 km s^{-1} for Type II. At maximum light, both types show a continuum in addition to the lines; at late times, the continuum in Type I disappears, leaving primarily a complex of lines around 4600 Å, probably [FeII].

A small number of events seem to fit neither Type I nor Type II. Zwicky (1965) defined Types III, IV, and V, but few later events have been assigned to these latter categories. Odd or somewhat deviant events continue to represent a small fraction of supernovae. Cas A appears to be the result of an unobserved supernova in about 1660 A.D.; this event could have belonged to a theoretically expected class of anomalously faint supernovae (Chevalier 1976). The nearby supernova 1987a in the Large Magellanic Cloud was also unusually faint. While possessing hydrogen lines, it showed unusually high expansion velocities compared to normal Type II supernovae, and its spectrum evolved about five times faster than normal. The visual light curve was unprecedented for a Type II event, taking several months to rise to maximum light. It may have been the result of the explosion of a metal-poor, massive blue supergiant (Hillebrandt et al. 1987).

The most significant recent developement in supernova classification has been the recognition of a potentially important subset of Type I supernovae, typified by two recent radio supernovae, 1983n and 1984ℓ. These lacked hydrogen lines but were fainter than normal Type I supernovae, and also showed spectroscopic distinctions such as a lack of silicon features at early times. Branch (1986) has designated these Type Ib, and a considerable fraction of earlier Type I supernovae may also turn out to have been Type Ib. The remaining, traditional Type I supernovae have been retroactively renamed Type Ia. The origin of the Type Ib events is still not clear; they may have Population I progenitors.

The two basic types differ in the characteristic evolution of optical luminosity with time (light curve). Type Ia supernovae have quite similar light curves, rising steeply to a peak absolute blue magnitude of about -18.2 (for a Hubble constant of 100 km s^{-1} Mpc^{-1}) (Tammann 1982). This peak is remarkably constant from one event to another; the observed dispersion of about half a magnitude is small compared to the uncertainty due to the Hubble constant. Type I light curves then decline by about 3 magnitudes in the first three or four weeks after maximum light. Thereafter, they decay more slowly, following an exponential law with e-folding time 70 to 90 days. They can typically be followed for half a year to three years; the total optical radiation emitted is of order 10^{49} to 10^{50} erg. Barbon et al. (1973) display average light and color curves derived from observations of over eighty Type I supernovae. Type I supernovae are found in both spiral and elliptical galaxies, suggesting a connection with an old population of stars.

Type II supernovae have similar peak magnitudes, but with more scatter. The light curves are more disparate than Type I light curves as well, generally showing some type of structure. Plateaus (periods of slower luminosity decay) are seen in about two-thirds of Type II supernovae. [See Barbon et al. (1979) for discussion of a large number of Type II supernova light curves.] However, the total energy radiated optically and the typical observable duration of a year or two are about the same as for Type I supernovae. Type II events are found only in spiral or irregular galaxies and are predominantly concentrated in spiral arms, whereas the Type I supernovae that occur in spirals seem more distributed throughout the disk. Thus, Type II seems connected (in the normal, perverse astronomical fashion) with Population I (young, massive) progenitors.

Both types, in spite of what appear to be totally different origins, release about the same total amount of energy: 10^{51} ergs, to within a factor of 3 or so; a few percent of this emerges as photons, while the rest is deposited as kinetic energy of the expanding ejecta. The energy estimates arise from the observed photospheric velocities and plausible guesses for the ejected mass: 0.5 $M_\odot$ for Type I and 8 $M_\odot$ for Type II. Models of the total X-ray flux from SNR also produce estimates of order 10^{51} erg. There appears to be a fair degree of consensus on the origin of the two types: Type II are the traditional death throes of massive single stars, while (with somewhat less agreement) Type I are thought to arise from the explosion of an accreting white dwarf in a binary as it is pushed over the Chandrasekhar limit. Type I explosions presumably involve the total disruption of the white dwarf, though this is not unanimously accepted; Type II explosions hold out the possibility of formation of a neutron star or black hole, though total disruption is also a possibility. These ideas account for the absence of hydrogen lines in the Type I supernovae, the similarity in Type I light curves (all progenitors have the same mass, 1.4 $M_\odot$), the presence of Type I supernovae in galaxies devoid of young massive stars, the greater scatter in Type II light curves along with the presence of hydrogen lines, and the Population I distribution of Type II. The exponential decay of Type I light curves at late times is ingeniously explained as continuing energy input from radioactive decay of ^{56}Ni, though that should result in a solar mass or so of iron produced by each Type I supernova, a requirement that may be at odds with observations. Thus Types I and II tap different energy sources: for Type I, it is the nuclear energy resulting from burning a carbon-oxygen white dwarf up to iron; for Type II, the gravitational energy released in forming a neutron star or black hole.

At late times, then, one might expect to see little systematic difference in the remnants of Type I and Type II supernovae, since the energy input into the interstellar medium is similar and the dynamics at late times are dominated by the swept-up interstellar material rather than the ejected material. At earlier times, however, the presence or absence of a neutron star perhaps functioning as a pulsar could make an observable difference, while the ejecta could dominate the X-ray or optical spectra. Since the total ejected mass is thought to be about an order of magnitude larger for Type II supernovae, while composed primarily of hydrogen, this could cause observable differences. However, it must be admitted that no SNR is unambiguously identified with a normal Type I or Type II supernova. The most plausible identifications remain the remnants of Tycho's and Kepler's supernovae as Type I, based on historical light curves.

10.2.2 Rates

Zwicky originally estimated the rates of supernovae in our galaxy as one per thousand years or so, but large-scale surveys soon made it clear that this was a considerable underestimate. While no supernova has been seen from Earth in our galaxy since the seventeenth century (probably Kepler's of 1604, though Flamsteed may have recorded the anomalously faint event associated with the formation of Cas A; Ashworth 1980), the existence of five undoubted supernovae in a thousand years within a small part of the galaxy near the Earth suggests a much higher rate, perhaps one per thirty to sixty years (Shklovskii 1960a). Careful accounting for incompleteness and scaling for galaxy luminosity gives an estimate comparable to Shklovskii's; Tammann (1982) gives figures for different galaxy types from 0.1 per hundred years per 10^{10} $L_\odot$ for S0 galaxies to 1.4 per hundred years per 10^{10} $L_\odot$ for late-type spirals (Sbc to Sd). The expected rate for the Milky Way then works out to be one per twenty-two years (both Type I and Type II); for M31, one per twenty-four years. Since no supernova remnant younger than 300 years has been unambiguously identified in our galaxy, while no supernova has been seen in M31 since S Andromedae of 1885, there may be some cause for concern; this will be discussed below.

10.3 Observed Properties of Supernova Remnants

10.3.1 Radio Characteristics

Galactic supernova remnants are overwhelmingly radio objects. Fewer than thirty percent have been detected at optical or X-ray wavelengths, primarily because of obscuration in the galactic plane. Radio morphology then serves to classify them in the two broad categories of shell-like and filled-center or Crab-like, though most "shells" are incomplete or confused, and a substantial fraction of catalogued SNR have not been examined at high enough resolution to allow classification.

Galactic shell remnants are generally large compared to extragalactic sources; Kepler, with a mean diameter of 3 arcminutes, is one of the smallest identified, while the median diameter in large surveys is of the order of half a degree. Thus, it has sufficed that an object have a nonthermal spectrum, be polarized, and be fairly large to be classified as an SNR. While well-defined, almost complete shell structure is rare, some form of edge brightening is common [see Caswell et al. (1975) for a large sampling of moderately high-resolution maps]. Most shell remnants are polarized at the level of 5% to 15%, though one of the defining characteristics of the class of Crab-like remnants is their considerably higher fractional polarization. A few remnants appear to be shells containing small Crab-like cores (Weiler 1985). Young shell remnants such as Cas A, Tycho, Kepler, SN 1006, and a few others show electric vectors parallel to the remnant edge, indicating a predominance of radial magnetic field. By contrast, older and fainter remnants generally show a tangential or complex field structure (Dickel and Milne 1976); the Cygnus Loop is an example. High-resolution images of Cas A, SN 1006, IC 443, and 3C58 are shown in Figures 10.2 to 10.5; several more anonymous but representative older remnants are shown in Figures 10.6 to 10.9.

A fairly complete list of remnants so far identified is given by Green (1984), who lists 145 SNR along with basic properties such as size, flux at 1 GHz, and spectral index, and including references to optical and X-ray data. The mean value of spectral index is -0.45; the vast majority of values lie between -0.3 and -0.7, with those

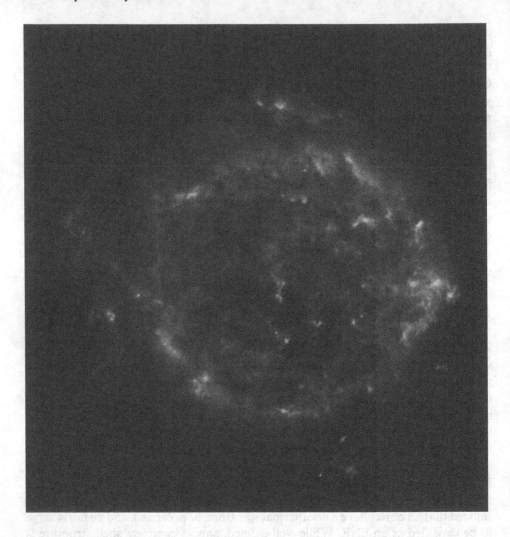

Fig. 10.2. Cas A at 5 GHz, observed with the Very Large Array (Courtesy NRAO/AUI; observations by P.E. Angerhofer, R. Braun, S. Gull, R. Perley, and R. Tuffs).

with flatter spectra generally sharing the other properties of Crab-like remnants: center-filled morphology and higher polarization. For neither class of remnant is there much evidence for appreciable spatial variations of spectral index. The observational data on SNR observed to contain neutron stars are reviewed by Seward (1985). It is likely that the galactic census of SNR is far from complete, even given a generous surface-brightness limit. Large, faint SNR could be misplaced in the general galactic background, or be unidentifiably distorted. Confusion with HII regions might also have hindered the detection of some SNR. Crab-like objects, with their flat spectra, might easily be taken for thermal sources. Even young objects such as Tycho's SNR would be difficult to detect across the Galaxy. This incompleteness can impede statistical analyses, as discussed below.

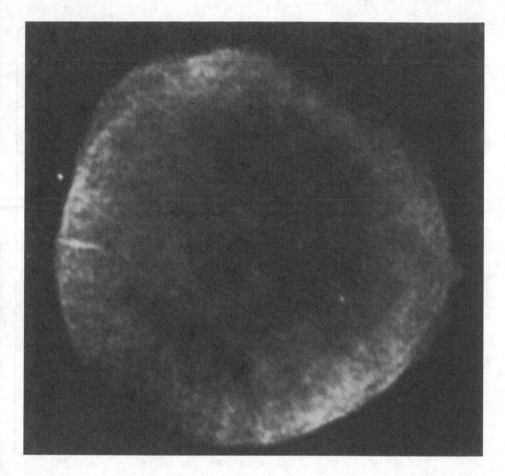

Fig. 10.3. SN 1006 at 1.4 GHz, observed with the Very Large Array (National Radio Astronomy Observatory; observation by author and D. Gilmore).

Unfortunately, supernova remnants as a class probably have some of the most poorly determined distances of any astronomical objects. Often distances are uncertain by factors of 2 or 3, and dramatic shifts in the properties of remnants based on new distance information are common. Clark and Caswell (1976) quote distances for twenty-four SNR that they regard as reliably determined, typically by HI absorption measurements. However, Green (1984) reconsiders these data, proposing different distances for several remnants. For the historical remnants, rough knowledge of the peak optical brightness of the supernova is an aid, and proper motions of optical filaments can sometimes be determined. In a few other cases, SNR are assumed to be associated with a nearby object of known distance. However, the reader is cautioned to regard typical SNR distances with an open mind.

10.3.2 Optical Characteristics
As mentioned above, except for the Crab, SNR tend to be inconspicuous optically. Faint filaments are commonly all that can be observed in young remnants; in Tycho,

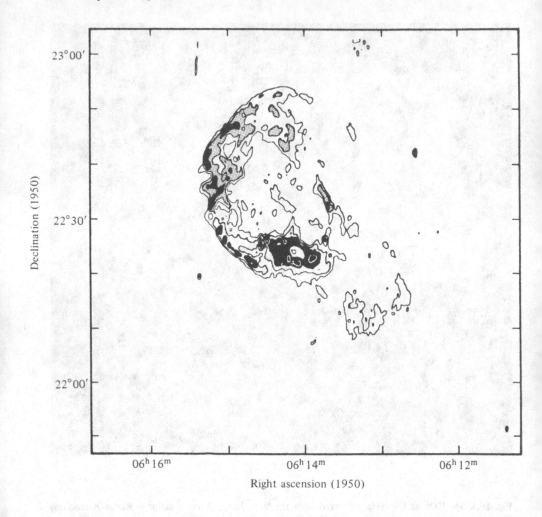

Fig. 10.4. IC 443 at 327 MHz, observed with the Westerbork Synthesis Radio Telescope (from R. Braun).

Kepler, and SN 1006, one or two linear filaments can be identified, generally near the radio edges of the remnants. Cas A shows a multitude of knots and flocculi, some of which may correspond to radio knots; these vary in brightness on time scales of a few years. Older remnants can be more impressive; the Vela remnant and the Cygnus Loop present complex traceries vaguely forming circular shapes (Figure 10.10). In these objects, radio and optical emission correspond quite closely (Straka et al. 1986).

Spectroscopy of these faint filaments yields velocity and abundance information crucial for the understanding of remnants. Prominent lines are those characteristic of collisional ionization: [OII], [OIII], [SII], and [NII] are typical. While Balmer lines are frequently seen, they are generally much weaker than [SII] $\lambda\lambda$ 6717, 6731, so a low value of that ratio is often taken as evidence that an optical nebulosity is an SNR. Velocities seen range from 6000 km s^{-1} in some optical knots in Cas A to values as low as 50 to 60 km s^{-1} as in IC 443. Densities of optically emitting gas

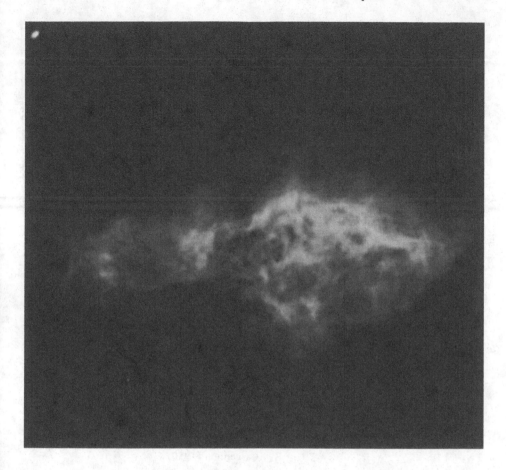

Fig. 10.5. 3C58 at 1.4 GHz, observed with the Very Large Array (National Radio Astronomy Observatory; observations by H. Aller and author).

can be obtained from the [SII] $\lambda\lambda$ 6717, 6731 doublet ratio, and temperatures (or at least lower limits) from [OIII] lines; the latter can range up to 60,000 K, characteristic of shock-heated gas. A few remnants have deviant optical properties: Tycho and SN 1006 reveal only Hα, with the line in Tycho showing broad wings corresponding to velocities of several thousand kilometers per second. This circumstance has been explained by Chevalier et al. (1980) as being due to partially neutral material overtaken by the shock, radiating some line photons before being ionized. The broad wings arise from high-velocity neutral H atoms created by charge exchange from fast protons on slow neutral atoms. Since young remnants are not expected to show cooling shock waves (that is, they are younger than the radiative time scale for shocked interstellar gas), it is not surprising that the spectra are less rich than those of larger, fainter remnants. Several remnants in the Large Magellanic Cloud fall into the Balmer-dominated category.

A substantial class of remnants has appeared with a strong predominance of oxygen lines. The prototype for this class is Cas A, at least the knots in the latter

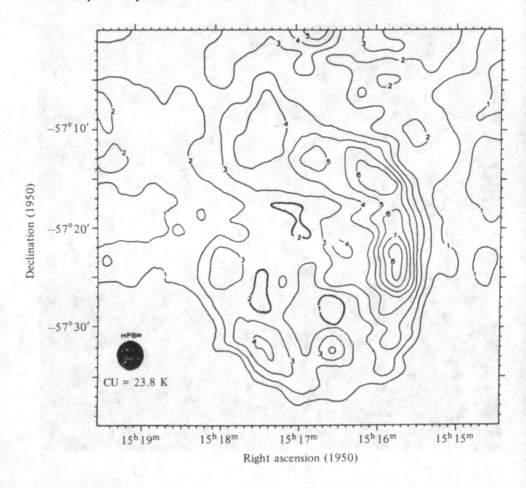

Fig. 10.6. G 321.9-0.3, observed at 408 MHz with the Molonglo Observatory Synthesis Telescope (from Caswell et al. 1975).

remnant that show high velocities and spectra of almost pure oxygen. Several objects in other galaxies, the remarkable remnant in NGC 4449 and three remnants in the Magellanic Clouds, also seem overabundant in oxygen compared to the typical SNR.

Unfortunately, the class of Crab-like remnants is not large enough for generalizations to be made about optical characteristics. The Crab Nebula, of course, is extremely well studied optically, showing both lines and continuum. Helium is overabundant by a factor of several, though with gradients across the remnant; strangely, nitrogen seems depleted with respect to carbon and oxygen, though it ought to dominate them. The analysis is complicated since photoionization from the nebular continuum dominates the ionization structure, rather than shock heating. Extensive discussion of this and other issues related to the Crab Nebula can be found in the *The Crab Nebula and Related Supernova Remnants* (Kafatos and Henry 1985). The other Crab-like remnants show very faint or no optical

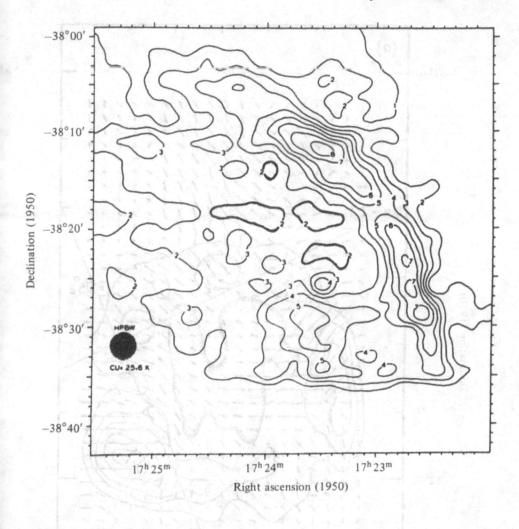

Fig. 10.7. G 350.0-1.8, observed at 408 MHz with the Molonglo Observatory Synthesis Telescope (from Caswell et al. 1975).

emission, some being at distances greater than 10 kpc and hopelessly obscured. Only recently have lines of high velocity been detected in 3C58, as might be expected given its young age (Fesen 1983).

The interpretation of these optical data can be highly complex. For older remnants, radiative-shock calculations do a fairly good job of making sense of the line ratios. Observed optical shock velocities reassuringly drop for larger, presumably older remnants; in the Cygnus Loop, velocities of 200 km s^{-1} are typical of many filaments. While difficulties of interpretation dash any early hopes of identifying Type I and Type II remnants based on their abundances, the oxygen-rich remnants are almost certainly the results of Type II supernovae, in particular, of progenitors with masses above 25 $M_\odot$ or so. Another point that seems clear is that no remnant shows the quantity of iron that one might expect if it results from the

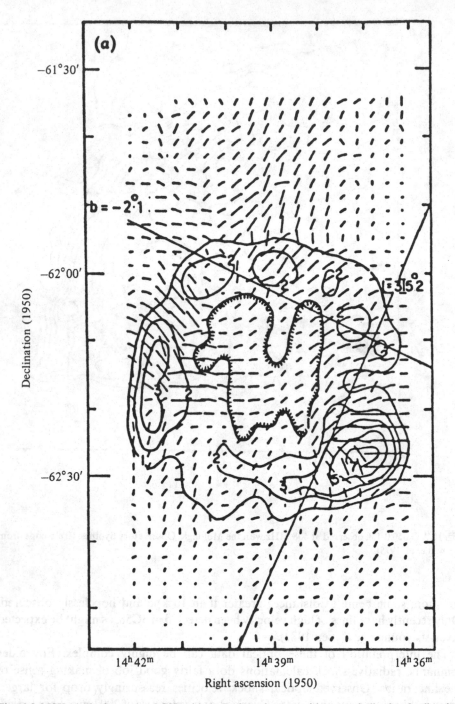

Fig. 10.8. MSH 14-63, observed at 5 GHz with the Parkes telescope, with vectors indicating the direction of the projected magnetic field (from Dickel and Milne 1976).

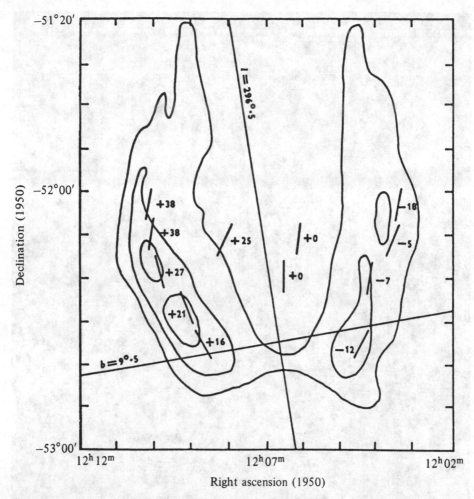

Fig. 10.9. G 296.5 + 9.7 (MSH 1209-51/52), observed at 5 GHz with the Parkes telescope, with vectors indicating the direction of the projected magnetic field (from Dickel and Milne 1976).

conversion of half of a white dwarf into iron, though one might expect to see this most readily in the ejecta-dominated X-ray spectra of young remnants.

The Crab, of course, possesses a novel optical feature shared by only one other known object: in addition to the line-emitting filamentary gas, an amorphous component of polarized optical continuum. The level of polarization (up to 60% near the edges of the nebula) implies synchrotron radiation. Until the discovery of a very similar optical nebulosity in the remnant 0540-693 in the Large Magellanic Cloud (Chanan et al. 1984), the Crab was the only stellar-scale example of an optical synchrotron nebula. Both of these objects contain observed pulsars; no other Crab-like remnant shows optical synchrotron emission, indicative of a young age and very powerful pulsar. The Crab's entire spectrum, from some tens of MHz, where the integrated flux of the pulsar drops below that of the nebula, to 100 keV, where the pulsar spectrum again seems to dominate the nebula, is apparently synchrotron radiation. Except for a bump near 10^{12} Hz attributed to dust, the

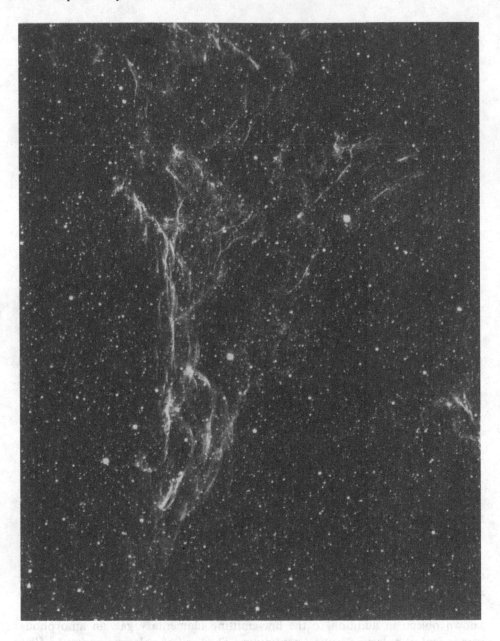

Fig. 10.10. Optical image of a portion of the Cygnus Loop supernova remnant (Hale Observatories; reprinted with permission from the National Optical Astronomy Observatories).

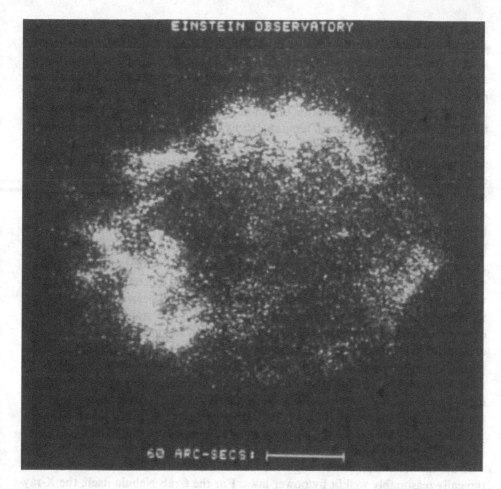

Fig. 10.11. X-ray image of Cas A, from *Einstein Observatory* (Center for Astrophysics; courtesy S. Murray).

spectrum smoothly steepens from the radio index of -0.26 through the optical, where $\alpha \sim -0.4$ to -0.7 depending on the adopted extinction, to the X-ray where $\alpha = -1.1$. The remnant 0540-693 is confused at radio frequencies, but the optical and X-ray indices are both -0.8.

10.3.3 X-Ray Characteristics

While several SNR were known to be X-ray emitters before 1978, the detailed X-ray study of SNR essentially began with the HEAO series of satellites, most importantly with HEAO-2 (*Einstein*). An excellent summary of the significance of this satellite for X-ray studies of SNR is the volume of proceedings of IAU Symposium 101 on *Supernova Remnants and Their X-ray Emission* (Danziger and Gorenstein 1983). Imaging of SNR revealed that in most cases, X-ray morphology resembled radio morphology. Shell remnants prove to be shells in X rays; Cas A (Figure 10.11), Tycho, Kepler, and SN 1006 all have this property. In Kepler, the X-rays and radio are fairly well correlated (Matsui et al. 1984), while this is less true for Tycho (van Breugel et al., manuscript in preparation) and SN 1006 (Reynolds, manuscript in preparation). For older remnants such as the Cygnus Loop and IC 443, the X-ray–radio correspondence is fairly good, though X rays may be relatively more prominent in remnant interiors.

Spectroscopic X-ray information on remnants has led to dramatic changes in our thinking about SNR. Most shell remnants show spectra dominated by line emission, with complex continua typically not well fit by thermal bremsstrahlung at a single temperature. Two-temperature fits do rather better; the low-temperature component in such fits is typically about 0.5 keV (Holt 1983), while the higher can range up to 7 keV or above. The two components can be associated with the blast wave moving into the interstellar medium (ISM) and a reverse shock wave moving back into the ejecta as they begin to decelerate; this picture will be discussed at some length in Section 10.4. Lines typically seen include silicon, sulfur, argon, calcium, and iron. However, abundance determinations depend on the ionization state of the hot gas; early reports of large excesses of sulfur and silicon have disappeared with the realization that ionization equilibrium has not yet been reached for young remnants and that the fractional abundances of He-like states of Si and S far exceed their equilibrium values (Shull 1983). Current models must invoke modest departures from solar abundances; the most dramatic effect, however, is in comparing the expected iron from Type I supernova models, which are expected to manufacture 0.5 to 1 $M_\odot$ of iron, with only a slight overabundance of iron deduced from Tycho spectra, corresponding to less than 0.05 $M_\odot$ of iron heated to X-ray temperatures (Hamilton et al. 1986). For some remnants the reverse shock may not yet have reached the iron in the ejecta, so that it should be looked for as [FeII] instead of much higher ionization states (see below).

Crab-like SNR show center-brightened X-ray emission and lineless X-ray spectra typically reasonably well fit by power laws. For the Crab Nebula itself, the X-ray spectrum has $\alpha = -1.1$; the emission is known to be synchrotron, confirming Shklovskii's daring extrapolation (1965), as a result of rocket and satellite experiments detecting X-ray polarization at the level of 19% ± 1% (Weisskopf et al. 1978). For other Crab-like SNR, the power law spectrum is assumed to be due to synchrotron radiation, though thermal emission can under some circumstances mimic such a spectrum.

10.3.4 Variability
The remnant of a supernova less than a thousand years old might be expected to vary in flux by a fractional amount of order a tenth of a percent per year. Using rather more sophisticated arguments than this (see Section 10.4.4), Shklovskii (1960b) predicted that the radio flux from Cas A should be decreasing at a rate of order one percent per year. As Cas A is the brightest source in the sky at some frequencies, and as it was being used as a fundamental calibrator for radio observations, this prediction received some attention and was soon verified (Högbom and Shakeshaft 1961, Scott et al. 1969, Baars and Hartsuijker 1972). In fact, it seems that the decay rate is frequency dependent and the spectrum is flattening in time (Dent et al. 1974), a circumstance explicable in terms of a model involving turbulent shock acceleration of particles (Scott and Chevalier 1975; see below).

Unfortunately, most supernova remnants are not nearly as bright as Cas A at radio frequencies, and observations of flux variations are much more difficult. Strom et al. (1982) attempted a determination of the change in time of Tycho's radio structure as well as flux, using interferometer observations. They found an expansion rate ($d \ln R/d \ln t$) of 0.47 and a marginally significant decay of flux at the rate of $(0.23 \pm 0.19)\%$ yr^{-1}, perhaps consistent with Tan and Gull's (1985) result of $(0.7 \pm 0.5)\%$ yr^{-1}. Aller and Reynolds (1985a) determined a decay rate of the Crab Nebula at 8 GHz of $(0.17 \pm 0.02)\%$ yr^{-1}, consistent with theoretical predictions (see below), and almost identical to the value of $(0.18 \pm 0.01)\%$ yr^{-1} at 927 MHz reported by Vinyajkin and Razin (1979), indicating that the Crab Nebula's decay, unlike that of Cas A, is independent of frequency. Aller and Reynolds (1985b) also reported that 3C58, another Crab-like remnant, was *brightening* at 8 GHz at the rate of $(0.28 \pm 0.05)\%$ yr^{-1}, a result inexplicable by any current theory.

Direct observations of expansion have been made optically for Tycho (Kamper and van den Bergh 1978), Cas A (Kamper and van den Bergh 1976, 1983), and SN 1006 (Hesser and van den Bergh 1981); for Kepler, van den Bergh and Kamper (1977) find little evidence for expansion. For Tycho and SN 1006, the fractional expansion rates are 0.38 and 0.47, respectively, consistent with straightforward theoretical expectations (Chevalier 1982c; see below). Cas A is a complicated case; optical features divide into fast-moving knots and quasi-stationary flocculi. The fast knots appear undecelerated and indicate an expansion lifetime of about 300 years, giving a date for the supernova of 1658 ± 3 years. The radio results on expansion of unresolved features are somewhat confused, though it seems clear that the rapid expansion of the fast optical knots is not shared by the radio features, which show random proper motions with perhaps some slower expansion.

The Crab has shown optical variability on 10-year time scales, associated with "wisps" (Scargle 1969) which appear to change brightness near the center of the nebula. No systematic motion has been observed, and the phenomenon may be due to irregularities in the interaction of the pulsar wind of particles and magnetic flux with the body of the nebula. The knots in Cas A appear, brighten, and fade with 20-year time scales, to be replaced by new knots.

10.3.5 Statistics of Supernova Remnants
It is a fairly trivial observation that small SNR are bright, while large ones are faint. Similarly, it might be expected that as remnants age, they both expand and fade. Thus, one might be led to attempt to study the evolution of SNR by plotting luminosity versus diameter. In fact, the distance uncertainties are such that a safer procedure is to plot the mean surface brightness Σ (total flux divided by solid angle) versus the mean diameter, reducing distance-related uncertainties to one power of distance for the abscissa. This is the famous $\Sigma-D$ diagram, a traditional way to describe SNR. [Milne (1970) appears to have been the first to produce an observational $\Sigma-D$ plot, though Poveda and Woltjer (1968) plotted Σ versus angular diameter, and the whole idea of the usefulness of such a diagnostic apparently originated with Shklovskii (1960b) in his theoretical discussion of the evolution of the radio emission from SNR. Later detailed discussions appear in Woltjer 1972, Clark and Caswell 1976, Caswell and Lerche 1979, Green 1984.] Figure 10.12, from Green (1984), illustrates the $\Sigma-D$ plot for a subsample of SNR

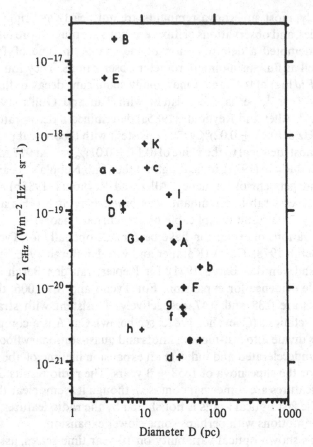

Fig. 10.12. $\Sigma-D$ diagram for twenty-four SNR whose distances are assumed known (Green 1984; with permission from the Royal Astronomical Society).

whose distances were regarded as known. One might expect two kinds of results from such a procedure: the elucidation of evolutionary tracks of SNR, and the possible use of such a relation as a distance determination for remnants of unknown distance. In practice, however, incompleteness of catalogues, intrinsic variation in the properties of remnants and their surroundings, and uncertainties in the distance even of "calibrator" remnants have rendered the $\Sigma-D$ diagram almost useless for either of these purposes (Green 1984). It seems clear that pulsar-driven remnants should be excluded since their presumed energy source is not thermalization of kinetic energy of the supernova explosion. In addition, with Green's (1984) distances, the remnants of known age define a rather steeper track than the mean relation of $\Sigma \propto D^{-3}$, suggesting that a spread in initial conditions can mask real evolutionary trends. As a means of determining distances, the $\Sigma-D$ diagram is a suggestion at best; there are enough anomalies in the remnants of known distance that one is inclined to be less than confident that a remnant of unknown distance is average in its radio properties. As an example, note the anomalously low surface brightness of SN 1006, whose distance is fairly well agreed to be of order 1 kpc, though its $\Sigma-D$ distance would be 6.3 kpc, implying a much larger linear size and requiring absurd expansion velocities. An attempt has been made to take into account the distribution of remnants in z (distance from the galactic plane), on the principle that high-latitude objects encounter lower ambient densities and will be larger and fainter at a given age than low-latitude ones (Caswell and Lerche 1979), but it has not been unqualifiedly accepted. In fact, Green (1984) questions several of Clark and Caswell's distances and finds a $\Sigma-D$ diagram with even more scatter. He maintains that it is useless for determining SNR distances.

An even simpler diagnostic of SNR evolution is the integral number-diameter, or $N(<D)$ relation. The number of remnants smaller than diameter D, $N(<D)$, should simply be given by the time a remnant takes to evolve to diameter D times the supernova rate. For galactic remnants, Clark and Caswell (1976) find that $N(<D)$ obeys $N = 8 \times 10^{-3} D^{5/2}$, consistent with most SNR being in the adiabatic phase of evolution (see below). However, the galactic sample of SNR is heterogeneous and plagued with distance uncertainties; Mills (1983) recalibrates distances and obtains $N \propto D^{1.15}$. For M31, the $N(<D)$ relation is similar to that in the Milky Way; but in the Large Magellanic Cloud and M33, the relation has the form $N \propto D$, strongly at variance with the galactic remnants and suggesting a different phase of evolution for those SNR or, equally plausibly, increasing incompleteness of catalogues for faint remnants. These issues will be discussed further below.

Given a secure $N(<D)$ relation, one could infer the rate at which observed SNR appear. Since supernovae certainly produce SNR, and probably produce most pulsars, the supernova rate ought to show crude agreement with pulsar and SNR birthrates. In view of the above-mentioned problems with $N(<D)$ relations, however, it seems premature to regard any of the reported relations for the Milky Way or nearby galaxies as above criticism; thus, a reliable estimate of the birthrate of SNR is beyond reach at present. Published estimates for the Milky Way range from one per 150 years (Clark and Caswell 1976) to one per 80 years (Caswell and Lerche 1979), reasonably close to the pulsar birthrate of one per 40 to 120 years inferred by Lyne et al. (1985), but much slower than the rate for supernovae of one per 22 years found by Tammann (1982). Rates derived for other galaxies lie within a factor of 2 of the inferred supernova rates—agreement at least as good as could be expected, given the uncertainties and assumption dependence of this kind of analysis.

One might expect light to be cast on some of these issues by searching for correlation of various quantities with spectral index α. However, except for the division of remnants into Crab-like and shell types, α is uncorrelated with any obvious property of SNR, including D, z, or luminosity.

Associations of SNR with their presumed progenitor objects (molecular clouds, OB associations) or their presumed products (pulsars) are distressingly less than compelling. No completely convincing case has been made for the association of an SNR with an OB association, though the Gem OB1 association is probably related to IC 443. This SNR also represents one of the handful of firm associations with a molecular cloud; to the NE the remnant appears to be interacting with molecular gas. However, Huang and Thaddeus (1985) find fourteen positional coincidences of SNR with molecular clouds and use the distances to the clouds to calibrate a $\Sigma-D$

relation. Only four remnants contain more or less normal observed pulsars: the
Crab, Vela, MSH 15-52, and 0540-693 in the Large Magellanic Cloud. Several
others contain neutron stars manifested as X-ray binaries or unresolved X-ray
sources (see Helfand and Becker 1985), and the Crab-like remnants are taken to be
evidence for a pulsar otherwise unobservable because of beaming, high dispersion,
low luminosity, or too fast a period. While one might expect half of SNR to contain
pulsars, most pulsars will have outlived their SNR, since they remain visible some
10 to 100 times as long as a typical SNR.

10.3.6 Radio Supernovae

Before the advent of the VLA, only one supernova had been observed to be
accompanied by radio emission: 1970g in M101. However, the routine observation
of weak ($\sim$ mJy) sources with the VLA has resulted in the discovery of several more
in only a few years. Weiler et al. (1986) exhaustively report the extant data on radio
supernovae 1979c in M100, 1980k in NGC 6946, 1981k in NGC 4258 (discovered
by its radio emission!), and 1983n in M83. The Type Ib supernova 1984ℓ in NGC
991 has also been detected in the radio (Panagia et al. 1986). All have luminosities
at 1 GHz greater than Cas A by factors of 10 to 100; all appeared at shorter
wavelengths first; and all eventually develop nonthermal spectral indices (though
the data for 1970g are scanty). All but 1970g, which abruptly disappeared, are fading
regularly as a function of time, $S_v \propto t^{-0.6}$ to $t^{-0.7}$ for the Type II supernovae and
$S_v \propto t^{-1.6}$ for 1983n (Weiler et al. 1986). Supernova 1987a in the Large Magellanic
Cloud made a brief appearance at radio wavelengths a few days after its discovery,
and disappeared soon thereafter (Turtle et al. 1987). It was far fainter than the other
radio supernovae, with a peak total radio luminosity only about 10^{-4} of that of
Supernova 1979c, and was detectable only because of its proximity. In addition,
Cowan and Branch (1985) have reported faint emission from the locations of SN
1950b and 1957d in M83, but its behavior with time is not known. Thus, magnetic
field and relativistic particles, evidently present in SNR, can appear immediately
after the supernova. These issues will be discussed further below.

10.4 Theory

10.4.1 Evolution

The details of the explosion of a star are complex and less than perfectly understood
at present. The discussion here will ignore all complications and consider the
explosion to deposit some 10^{51} ergs in the form of kinetic energy of ejected material
into the ISM (which may or may not have been modified by the star's presupernova
evolution), in a spherically symmetric manner. If binary collisions formed the only
coupling between the ejected material and interstellar material, the ejecta would
simply interpenetrate the ISM, heating it throughout, since the mean free path
for individual-particle collisions is of the order of hundreds of parsecs. However,
astrophysical shock waves frequently form under similar circumstances; the coupling is not via binary collisions but through the magnetic field. The field energy is
nowhere near sufficient to decelerate the supersonic material, but it can effectively

transmit the momentum of the ejecta to the ISM, forming a collisionless shock wave. The earth's bow shock in the solar wind is an example of such a shock; the shock thickness is not the binary mean free path, but the ion cyclotron radius, which is smaller by orders of magnitude under typical interstellar conditions.

The ejecta thus begin sweeping up interstellar material, driving a shock wave into the ISM and heating it to very high temperatures. Initially, the ejecta are separated from shocked ISM by a contact discontinuity, across which the pressure is constant. Processes mixing ISM and ejecta act on a slower time scale than the dynamical time, so the two media can be considered separate initially. When a mass comparable to the ejected mass has been swept up, the shock begins to decelerate; as ejecta from the interior pile into the decelerated ejecta, an inward-facing shock (the reverse shock) is formed, which shocks the ejecta. The outer shock continues to decelerate until the radiative time scales become comparable to the dynamical time, and the shock becomes effectively isothermal, with much higher compression ratios than before. Eventually the shock becomes weak, then disappears into sound waves in the ISM. See Spitzer (1978) for a discussion of the basic physics of shock waves, in particular the "jump conditions" giving the postshock density, pressure, and temperature in terms of the preshock values. Let v_s be the shock velocity and $\bar{m}$ the mean mass per particle behind the shock; ρ, P, and T are mass density, pressure, and temperature, respectively, and k is Boltzmann's constant. Subscripts 1 and 2 label pre- and postshock quantities, respectively. For strong shocks ($\rho_1 v_s^2 \gg P_1$) and an adiabatic index of 5/3,

$$\rho_2 = 4\rho_1 , \tag{10.1}$$

$$P_2 = \frac{3}{4}\rho_1 v_s^2 \tag{10.2}$$

$$T_2 = \frac{3}{16}\frac{\bar{m} v_s^2}{k} . \tag{10.3}$$

A simple picture of the development of an explosion in a uniform medium was described by Shklovskii (1968) and Woltjer (1972) and remains a useful framework in which to discuss the evolution of SNR. Four phases are divided by the onset of deceleration, the dominance of radiative cooling, and the transition to subsonic expansion. Label the shock radius R_s and the supernova energy E_0; assume the ambient density ρ_1 to be constant.

Phase I: Free expansion. The ejected mass M_0 dominates the swept-up mass ($M_0 \gg 4\pi R_s^3 \rho_1/3$). The ISM has no effect on the expansion; the initial properties of the explosion dominate.

Phase II: Sedov or adiabatic phase. The swept-up ISM now dominates the mass, but its pressure is negligible (the shock is still strong), and cooling times are long compared to the age of the remnant. Energy is conserved; the expansion is governed by Sedov's (1959) result for the dynamics of a point explosion in a uniform medium. For a constant-density medium, the equation of conservation of energy can be trivially integrated to obtain $R_s \propto t^{2/5}$. Sedov used dimensional analysis to show that the expansion was self-similar and to derive, for a gas with ratio of specific heats 5/3, the relations

$$R_s = 1.17 \left(\frac{E_0}{\rho_1}\right)^{1/5} t^{2/5} \tag{10.4}$$

$$v_s = 0.4 R_s/t \tag{10.5}$$

where the shock has been assumed strong.

The self-similarity means that the solutions for $P(r)$, $\rho(r)$, and $T(r)$ scale with time and radius so that when written as $P(r)/P_2$, etc., they are functions only of the dimensionless "similarity variable" $\xi \equiv r(\rho_1/E_0 t^2)^{1/5}$. In the Sedov solution, the density drops to zero as $r \to 0$, while the temperature becomes infinite; the pressure, proportional to ρT, stays finite. Most of the mass in the interior is thus fairly close behind the shock.

The total kinetic energy is $1/2(4\pi R_s^3 \rho_1/3)v_s^2 = $ const.; since the total energy is also conserved, the thermal energy is conserved separately. The transition to this decelerated phase from Phase I involves the deceleration of the ejecta and the attainment of pressure equilibrium throughout the remnant interior via a reverse shock, which thermalizes the ejecta.

Phase III: Radiative (snowplow) phase. Eventually the remnant's age becomes comparable to radiative cooling time scales near the shock. (The cooling time in the remnant interior is much longer.) The expansion energy is no longer conserved; the deceleration becomes rapid and the shock compression becomes large. The remnant consists of a dense, cool shell with a hot interior. Radial momentum is conserved; again, this statement can be trivially integrated to obtain $R \propto t^{1/4}$ (Oort 1946). This phase begins roughly when half the thermal energy of the SNR has been radiated, at time t_{rad}; Woltjer (1972) derives

$$t_{rad} = 3.5 \times 10^4 \left(\frac{E_0}{10^{51} \text{ erg}}\right)^{4/17} n_1^{-9/17} \text{ yr} \tag{10.6}$$

$$v_{rad} = v_s(t_{rad}) = 230 \, n_1^{2/17} \left(\frac{E_0}{10^{51} \text{ erg}}\right)^{1/17} \text{ km s}^{-1} \tag{10.7}$$

where n_1 is the preshock number density; thus, a shock velocity of about 200 km s^{-1} marks the beginning of Phase III.

Phase IV: Dissipation. The shock disappears as the expansion velocity drops below the local speed of sound, and the remnant dissipates in the interstellar medium. This is thought to occur after times of order 10^6 years.

As Woltjer (1972) points out, this discussion neglects a number of important effects: the importance of cosmic-ray (relativistic particle) pressure on the shock, dynamical effects of the magnetic field, pressure forces in Phase III, and, most importantly perhaps, inhomogeneity of the ISM. As well, the discussion of Phase I above assumes that the ejecta are uniformly expanding with a constant density in space, whereas recent work suggests a more complicated density law, with concomitant changes in the time dependence of the shock radius (Chevalier 1982a). Most glaringly, Crab-like remnants seem not to fit this picture at all. They will be discussed separately; a discussion of the other effects follows.

Numerical work (Gull 1973, 1975, Jones et al. 1981) has shown that the transition to pure Sedov evolution is rather slow, and that this transition is not complete until several times the ejected mass has been swept up. Details of the transition can then be important for the interpretation of observations of young remnants. Chevalier (1982a) has derived similarity solutions for the expansion of ejecta with a power law density profile into media with constant density or with density declining as a power of radius. Recent hydrodynamic calculations have suggested that both Type I and Type II supernovae produce ejecta in which part of the mass has uniform density, preceded by material with a steep density gradient (for instance, $\rho \propto r^{-7}$ for Type I supernovae). A reasonable model for a Type I supernova has ejecta with constant density for the inner four-sevenths of the mass, and the outer three-sevenths obeying $\rho \propto r^{-7}$. For such a density profile moving into a constant-density ambient medium, the shock radius obeys $R_s \propto t^{4/7}$ until $t = 1.7 \, (M_0^5/E_0^3 \rho_1^2)^{1/6}$ when the Sedov phase is reached. A characteristic feature of all such models in which the outer material does not have constant density is that the shock radius does not rise proportionally to time, even in the earliest phases, since the shock radius moves in with respect to ejecta unlike the simple picture of a shock driven in front of a spherical "piston."

Cosmic rays accelerated at the shock front can constitute a significant fraction of the postshock pressure. Their effect on global SNR evolution is not thought to be large in the early stages but can lead to reduced deceleration in Phase III. If they constitute a large fraction of the postshock energy density, the compression in the shock may be slightly increased since the effective ratio of specific heats of the combined thermal and relativistic gas is less than 5/3. More important for the study of SNR is the effect of the shock on the cosmic rays, as this is thought to be related to the origin of nonthermal emission from SNR and of the galactic cosmic rays. See below.

If the pressure forces from the gas in the interior of the SNR in Phase III cannot be neglected, the dynamics are modified. Such a "pressure-driven" radiative shock is described by

$$R_s = 0.48 \left[\frac{[R(t_{\rm rad})]^2 (E_0/10^{51} \, {\rm erg})}{n_1} \right]^{1/7} t^{2/7} \, {\rm pc} \qquad (10.8)$$

(e.g., McKee and Ostriker 1977); here $R(t_{\rm rad})$ is in parsecs and t in years. Numerical work confirms that the shock radius increases with time somewhat faster than $t^{1/4}$, indicating that pressure forces are still important in the remnant interior.

The magnetic field is unlikely to be dynamically important in the early stages of evolution of an SNR, while the shock is adiabatic and the compression ratio ρ_2/ρ_1 only 4. When the shock becomes radiative, however, and the compressions are of order 100, the magnetic pressure can come to dominate in the postshock gas. While this does not affect the shock dynamics, it can have an effect on the interpretation of optical spectra from cooling-shock filaments. If shock compressions are governed by magnetic rather than gas pressure, estimates of the latter from line ratios will underestimate the total pressure. Effects of this may have been seen in samples of remnants in M31, M33, and the Large Magellanic Cloud (see, e.g., Blair et al. 1981) where the thermal energy of an SNR inferred from line ratios increases with SNR diameter, suggesting that magnetic pressure is more highly dominant in smaller remnants.

Even if the field is dynamically unimportant, it can influence the evolution of an SNR by impeding heat conduction and slowing the evaporation of clouds in the remnant interior. This effect is discussed further below.

Possibly the most extensive modification of the simple four-phase evolution sketched here occurs in dealing with an inhomogeneous ISM. The idea that the ISM consists of several components of differing characteristics in pressure balance dates back at least to the work of Field et al. (1969), who imagined a two-component medium made up of cold clouds and a warm intercloud medium, with a thermal instability responsible for manufacturing the former from the latter. The effects of supernovae on such an arrangement were first examined by Cox and Smith (1974), who realized that SNR have hot, low-density interiors that last a long time and that for a sufficient supernova rate, their interiors might come to overlap, creating a third component of the ISM, which might eventually occupy a significant fraction of the volume of the galactic disk. This idea was extended and elaborated by McKee and Ostriker (1977), who added the physics of mass exchange between the hot component and clouds via cloud evaporation. They concluded that most of the volume of the galaxy was occupied by material of density ~ 0.003 cm^{-3} at a temperature $T \sim 5 \times 10^5$ K, created and maintained by supernova injection of energy. Thus, in this picture the typical ambient interstellar medium is not warm neutral gas with a density of ~ 0.2 cm^{-3} but mainly hot gas with much lower density, with embedded clouds containing most of the mass. The warm component then consists only of edges of clouds. These ideas require substantial modification of the simple four-stage picture.

McKee (1982) summarizes the chief effects of SNR evolution in a cloudy medium. One effect, recognized early, resolves a long-standing difficulty in understanding X-ray and optical observations of the Cygnus Loop and other SNR. The shock velocity as inferred from optical spectra was of order 100 km s^{-1}, while the X-ray temperature through Equation (10.3) indicated a considerably higher value. This discrepancy can be understood if the ambient medium contains clouds with considerably higher than average density. The shock moves more slowly into these clouds, and the higher density causes the cloud shocks to become radiative much sooner than the blast wave as a whole. Thus, the SNR may be globally still in an adiabatic phase but filled with lower-velocity cooling shock waves in denser clouds producing optical emission. The impression one obtains from optical photographs of the Cygnus Loop and the Vela SNR supports such a picture of inhomogeneity (though difficulties appear in attempting to construct detailed models for arclike filaments). This effect is generally accepted.

Another effect of a cloudy medium appears when evaporation of clouds in the remnant interior becomes important (at typical radii of 20 pc or so or ages of about 10^4 yr, for expansion into a low-density medium). The effective mass interior to the shock begins to increase, resulting in an alteration of the expansion law to $R \propto t^{3/5}$. The mass contributed by evaporating clouds is most influential for SNR expanding into hot-phase material (very low-density); if the ambient density is greater than 0.1 or so, evaporation may never become very important. As well, magnetic fields can act to inhibit thermal conduction and hence cloud evaporation. However, since conduction along field lines is unaffected, this effect is highly geometry dependent and its importance is difficult to estimate quantitatively.

Though subshocks in denser clouds are immediately radiative in this scheme, the blast wave as a whole becomes radiative at a much later time given by the cooling time in the lower-density ambient intercloud medium (hot phase, according to McKee and Ostriker). Thus, the onset of Phase III is very much delayed compared to an estimate based on an ambient density of order unity (see Equation 10.6). In terms of radius, $R(t_{rad}) \sim 150$ pc instead of 40 as implied by Equations (10.4) and (10.6). Thus, in this picture, the only radiative remnants should be extremely large objects such as the North Polar Spur. Once a remnant is radiative, this model predicts different structure than does the simple picture: for example, the dense shell forms some distance inside the shock instead of right behind it. Again, little observational evidence can be brought to bear on this prediction.

10.4.2 Relation of Evolutionary Theory to Observations

Observations of SNR for the most part fit well into this theoretical framework. Cas A seems to be a good example of a remnant still in the free expansion stage, or just beginning the transition to Sedov evolution. The highest velocities measured spectroscopically are of order 6000 km s^{-1}; the original ejection velocity was probably no higher. Thus, the fast-moving knots probably represent undecelerated ejecta, consistent with their extremely enriched composition. The quasi-stationary flocculi, on the other hand, show more normal composition and are most likely preexisting clouds resulting from presupernova mass loss. The radio image of Cas A demonstrates that the remnant edge is not sharp; most of the emission probably originates near the contact discontinuity between ejecta and shocked ISM (see below).

While no other young SNR seems to be in a free expansion stage, it is possible that the nova shell of GK Per (Nova Cygni 1901) represents, at a much smaller scale, a stage of evolution comparable to Cas A (Reynolds and Chevalier 1984b). The same physics should obtain for the mass ejected in a nova outburst; there seems to have been little deceleration in GK Per's shell, and the radio emission is quite strong relative to the total energy of the outburst, again resembling Cas A.

The other historical shell SNR seem to have begun significant deceleration. The optical velocities are much less than the presumed original velocities of ~ 5000 to 10,000 km s^{-1}, and the proper-motion observations described above are consistent with an expansion law close to $R \propto t^m$ with $m \simeq 0.4$, though perhaps not quite as decelerated. The result quoted above for the expansion of ejecta with a power law density profile into a medium of constant density seems able to account well for the observed deceleration of Tycho and SN 1006, as it predicts m in the range 0.4 to 0.57, consistent with observations. This is interesting in that a model involving expansion into a medium with $\rho \propto r^{-2}$, as might be expected for a stellar wind from the progenitor's companion, does not fit the observations as well (Chevalier 1982c).

X-ray spectral observations of the young remnants are difficult to interpret because of the need for at least two components to fit the data. One of these is presumably to be identified with the ejecta, for which the relative abundances are free parameters. Two-temperature fits also require a hot component which is identified with the shocked ISM, at a temperature of $\sim 10^7$ to 10^8 K, consistent with Equation (10.3). The morphology of Tycho in X rays suggests that emission from the reverse shock moving into the ejecta can be identified spatially (Seward et al. 1983); interior emission in SN 1006 may be due to the same cause. Kepler's SNR shows a less symmetric shell than either of these remnants; as well, no high-velocity optical filaments have been found, consistent with the lack of observed proper motions. The much larger size and lower surface brightenss of SN 1006 compared to Tycho seem due to a considerably lower ambient density, as well as a somewhat more advanced age.

Older SNR such as Puppis A, the Cygnus Loop, and IC 443 show the effects of a cloudy ambient medium, as discussed above. The low velocities inferred from optical observations and radiative-shock models are almost certainly not characteristic of the blast wave velocities, which are considerably higher as implied by the observed X-ray temperatures. Objects indisputably in the radiative phase are hard to find; several investigations agree that there are insufficient cool shells in the nearby ISM compared to the number expected based on a supernova rate of one per twenty or thirty years. However, extremely large, low-surface-brightness objects such as the North Polar Spur and the Gum Nebula are likely ancient SNR.

In order to discuss the observed $\Sigma-D$ diagram in the light of evolutionary models, it is necessary to understand the physics of the nonthermal emission. This will be discussed below. However, the $N(<D)$ distribution is more straightforward if one assumes the observed remnant diameters accurately reflect the shock location (or, at least, are proportional to the shock radius). The predictions of the simple four-phase picture are clear. For similar E_0 and ρ_1, Phase I has $R \propto t$ (unless the models with nonconstant-density ejecta are correct), so that $N \propto D$. In Phase II, since $R \propto t^{2/5}$, $N \propto D^{5/2}$. In Chevalier's (1982a) solution for expansion with nonconstant density, $R \propto t^m$ with $m \sim (0.4-0.57)$, implying $N \propto D^{7/4}-D^{5/2}$. The galactic remnants seem to fit the adiabatic phase (II) prediction, more or less (Clark and Caswell 1976), as do the M31 remnants. But M33 and the Large Magellanic Cloud show distributions well fit by $N \propto D$. This cannot be simply interpreted as indicating that remnants in these galaxies are still in free expansion (out to radii of tens of parsecs), though such a situation might conceivably arise in a three-phase picture in which the remnants all expand into hot-phase, very-low-density material. Several groups (Hughes et al. 1984, Fusco-Femiano and Preite-Martinez 1984) have shown that a collection of remnants with a large range of values of E_0 and ρ_1 will, in the adiabatic phase, form a distribution with $N \cong D$, mimicking free expansion. Furthermore, incompleteness at the faint end lowers $N(<D)$ for large D, flattening the relation (Green 1984). Thus, the data are currently unable to discriminate between free expansion and very inhomogeneous conditions for these remnants.

10.4.3 Crablike Supernova Remnants

The above picture of SNR evolution clearly does a very poor job of describing the original SNR, the Crab Nebula. To summarize its anomalous properties: it shows no evidence of shell structure; its X-ray spectrum is lineless and nonthermal; its optical velocities are only 1200 km s^{-1} or so at maximum and indicate, if anything, an acceleration rather than a deceleration since 1054 A.D.; and the total mass in optical filaments is only 1 $M_{\odot}$ or so, indicating a total kinetic energy in the remnant of only about 10^{49} ergs. Its scanty lightcurve information seems consistent with a more or less normal Type II supernova, as does the presence of a pulsar; but if this was the case, the remnant needs a great deal of explaining. Chevalier (1977b) sought to rationalize the properties of the nebula with an origin in a normal Type II supernova by asserting that most of the ejecta, moving at thousands of km s^{-1} and containing the canonical 10^{51} ergs of energy, have yet to be observed. The expected shell from this material would be found at several times the radius of the observed nebula but could be faint at radio or X-ray wavelengths if the ambient density is sufficiently low. Then the observed nebula represents a small fraction of the mass and energy of the supernova; its peculiar characteristics are explained as due to the interaction of the material with energy input from the pulsar in the form of relativistic particles and magnetic field. The core of a presupernova star has been found in numerical hydrodynamic calculations to collide with the stellar envelope, ejecting the latter at normal supernova speeds but leaving the one or two solar masses of core material moving at much lower velocities. This material is identified with the observed nebula.

Recently, Nomoto (1987) has proposed an alternative origin for the Crab Nebula. He accepts the observed nebula with its 10^{49} ergs as all there was to the supernova and explains the recorded two years of visibility of the supernova by invoking an unusual progenitor: a white dwarf with an extended helium envelope. This may be able to produce an optical display not unlike a normal Type II supernova, but releasing only a few times 10^{49} ergs. In this case, the Crab Nebula should be interacting with interstellar or previously ejected material at its periphery.

Given such pictures, one then must calculate the evolution of an expanding cloud of gas containing a pulsar (Reynolds and Chevalier 1984a). For a pulsar like the Crab, a bubble of relativistic material is inflated at the center of the ejecta. After a few hundred years, the ionized ejecta become transparent at radio frequencies, allowing the very bright synchrotron emission from the bubble to be seen. The pulsar input accelerates the thermal material to velocities over 1000 km s^{-1}; the resulting adiabatic expansion losses on the relativistic material are heavy, and after a few thousand years the bubble becomes unobservably faint. At this point the Chevalier and Nomoto pictures diverge. In the former, at ages of a few thousand years, the outer, fast material begins to decelerate as it enters the Sedov phase; the reverse shock moves in to the remnant center, compressing and rebrightening the synchrotron bubble which thereafter expands slowly in pressure equilibrium with the remnant interior. One might expect a remnant following this prescription to appear first as a Crab Nebula-like object for the first $\sim 10^3$ years, then to develop a shell outside the Crablike center as the kinetic energy is thermalized, and finally to lose the central bubble and appear as a normal shell SNR. In a Nomoto model, a shell might never form. Independently of the detailed calculations, observed statistics of Crablike SNR indicate that their lifetime must be short if they are an inevitable consequence of pulsar formation and if their pulsars are like the Crab pulsar. The pulsar birthrate is such that most or all Type II supernovae must make pulsars; then about half of all SNR should contain young pulsars, and unless the

Crablike phase is short, some fifty or sixty SNR should contain Crab Nebulas, even if their pulsars are themselves invisible for one reason or another.

This picture describes the Crab Nebula itself quite well and explains the decay of its radio flux mentioned above. Several objects have been observed which appear to be shell remnants surrounding Crablike cores, as would be expected in the Chevalier picture. However, attempts to find shells of fast-moving material around the Crab Nebula and its near-twin 3C58 have failed, at radio, optical, and X-ray wavelengths. This requires either that the interstellar density is quite low, as might be expected if the remnants are expanding into the hot phase of the ISM, or that the Nomoto picture is correct and SN 1054 and 1181 were very-low-energy events. The former explanation carries with it the implication that other supernovae of order a thousand years old have yet to produce observable remnants. If the supernova rate is really one per twenty years or so, this might be desirable, since the Galaxy contains nothing like the population of twenty to fifty objects brighter than the Crab Nebula or Cas A that one might want to associate with even younger SNR. While a low density for the ISM can cause the problem of absence of shells to vanish, the absence of younger Crab Nebulas can only be ascribed to properties of the newborn pulsars, assuming the pulsar birthrate is correct. If all young pulsars are like the Crab pulsar, their synchrotron bubbles would be extremely bright and conspicuous; but it is possible that the typical pulsar is in fact born with a considerably slower rotation period than the Crab pulsar's inferred 19 ms, so that its initial luminosity in particles and magnetic flux and the brightness of its bubble are much lower than those of the Crab pulsar.

10.4.4 Nonthermal Emission from Supernova Remnants

For most of the history of the field, the defining characteristic of SNR has been the presence of nonthermal radio emission. The power law spectra and polarization indicate synchrotron radiation from distributions of electrons (or positrons) described by $N(E) = kE^{-p}$ radiating in magnetic fields. Initially, then, the understanding of the origin of the emission involves specifying the sources of electrons and magnetic field. Four possibilities have been envisioned for the source of fast particles: the original supernova event, a pulsar within the remnant, ambient cosmic rays, and acceleration of particles at the shock wave or at the (Rayleigh-Taylor unstable) contact discontinuity between shocked ejecta and shocked interstellar medium. Similarly, the field could be generated by a pulsar as seems to be the case for the Crab Nebula, swept up along with ambient ISM, or turbulently amplified in the shock or at the contact discontinuity.

Several of these possibilities seem fairly unlikely and have received little theoretical attention. Particles produced in the original supernova now occupy a volume larger by an enormous factor; adiabatic expansion losses would seem to be a prohibitive drain, if the particles remain confined by the remnant. (If they escape, they could still account for cosmic-ray electrons but would be useless for explaining SNR.) While a pulsar source for particles in Crab-like remnants seems eminently reasonable, for shell remnants the problem of silently transporting the particles out from the pulsar to radiate in a distant shell seems quite difficult. In fact, the scanty evidence for pulsars in shell remnants suggests that both field and particles are due to other processes.

For Crab-like remnants, then, a pulsar source for both electrons and magnetic field seems most reasonable, while for shell remnants, field and particles are either locally generated or borrowed from the ambient medium. For old remnants with radiative, high-compression shocks, compression of ambient magnetic field and cosmic rays, boosting the energies of the latter ("adiabatic gains"), can represent a powerful source of synchrotron emissivity (van der Laan 1962). This may explain emission from remnants such as the Cygnus Loop; even if the blast wave is dominantly adiabatic, slower, high-compression shocks in the clouds could produce the radio emission, which is in fact quite well correlated with optical filaments. van der Laan found that for the Cygnus Loop, most of the radio flux in fact could be explained with only a modest increase in emissivity over the galactic background. However, more recent high-resolution observations may require local regions of high emissivity, implying high compression ratios.

However, young remnants in which cooling shocks seem totally absent, such as Tycho, Kepler, and SN 1006, are too bright to be explained by this mechanism. Tycho's mean emissivity is larger than the ambient galactic value by a factor of about 10^5, vastly too large to be accounted for by a factor of 4 compression. From Chapter 1, the synchrotron emissivity j_ν is proportional to $kB^{(p+1)/2}$, so for a typical value of p of 2, $j_\nu \propto kB^{3/2}$. Compression of the electrons raises their energy density as the $\frac{4}{3}$ power of volume, or $(\rho_2/\rho_1)^{4/3}$. If the field is entirely parallel to the shock, it will increase as (ρ_2/ρ_1), since it is "frozen into" the fluid; so $j_\nu \propto (\rho_2/\rho_1)^{17/6}$. So a compression ratio of 4, if applied to both field and cosmic-ray electrons, could at most boost the emissivity by a factor of about 50. Thus in Tycho and the other young remnants, the relativistic particles and/or magnetic field must have a different origin to explain the observed strength of the radio emission.

Two classes of process have been invoked to generate relativistic particles in young shell SNR: turbulent acceleration at the unstable contact discontinuity and acceleration at the outer shock itself. Both classes rely on variants of the Fermi mechanism, originally designed to accelerate cosmic rays in the ISM. In it, particles collide with randomly moving interstellar clouds and are reflected by the (presumed) higher magnetic field in the clouds. To first order in $v(\text{cloud})/c$, the rate of energy gain from approaching collisions balances the rate of loss from receding ones, but the second-order contribution is not zero; approaching collisions are slightly more frequent, resulting in a net acceleration. Thus, this mechanism of reflection from random scattering centers is referred to as "second-order Fermi acceleration." The acceleration inferred to be necessary in Cas A may be of this nature; the irregular radio shell, with no smooth edge, suggests turbulent processes behind the shock. The efficiency can be improved if the turbulence is imagined as due to the turbulent wakes of the fast-moving knots, behind which regions of largely converging cloud velocities might be found. Such a picture can account for the flattening of Cas A's radio spectrum with time (Scott and Chevalier 1975), requiring that a percent or so of turbulent energy appear as relativistic electrons and magnetic field, an efficiency consistent with the calculations of Gull (1973).

However, in a shock front, material enters rapidly and leaves slowly, resulting in a systematically converging velocity field. If the fluid contains embedded scattering centers such as magnetic turbulence, particles can be reflected back and forth across the shock front between always-covering scattering centers, until they are convected away behind the shock, resulting in much more efficient, more rapid acceleration ("first-order Fermi acceleration"). This mechanism has the highly attractive feature that, independently of the details of the diffusive scattering, the accelerated particles have a distribution that is a power law in energy, with the power law index depending only on the shock compression ratio r: $p = 3r/(r - 1) - 2$. Therefore, a strong shock ($r = 4$) produces $N(E) \propto E^{-2}$, similar to spectra inferred in SNR, the galactic synchrotron background, quasars and radio galaxies, and cosmic-ray electrons observed at Earth. In fact, the mean values of p inferred for these sources are somewhat greater·than 2, but a multitude of processes are known which, when included in the simple calculation described above, have the effect of steepening the spectrum slightly. For a particularly lucid derivation of this result in terms of escape probabilities, see Bell (1978); Blandford (1979) also gives a valuable discussion. Estimates of the efficiency of the process differ but range as high as 50% or greater. This process is particularly attractive applied to the other young SNR, Tycho, SN 1006, and Kepler. All have highly limb-brightened radio emission, with smooth, steep outer contours, unlike Cas A, which are well explained by the creation of new fast particles immediately at the shock. In fact, there is no reason to suppose shock acceleration does not also contribute in cooling shock remnants; Blandford and Cowie (1982) describe a model for Cygnus Loop-type remnants in which a combination of compression and shock acceleration in an inhomogeneous interstellar medium provides the relativistic electrons.

As with relativistic electrons, two classes of process have been proposed by which remnants might manufacture magnetic field. In addition, the possibility exists that the newly shock-accelerated electrons simply radiate in the compressed ambient field. For Cas A, a picture involving turbulent amplification of magnetic field inside the main shock, at the contact discontinuity, seems plausible. Again, its irregular shell morphology supports such a picture. Observational limits on the gamma-ray flux from Cas A, applied to electron bremsstrahlung on the high-Z nuclei of the ejecta, set an upper limit on the electron energies, and hence a lower limit on the magnetic field of 8×10^{-5} gauss, a value too high to be a factor of 4 compression on the ambient field, and implying some sort of amplification (Cowsik and Sarkar 1980).

For the other historical shell remnants, the sharp emissivity increase at the shock suggests that the field may be amplified there. Such amplification would require a process not at all understood, though there is evidence that shocks in the solar wind may accomplish it. Observational evidence that such a process operates in young remnants comes from several sources. In the case of Kepler, comparison of X-ray and radio data allows the field to be calculated; it is near the equipartition value of 7×10^{-5} gauss (Matsui et al. 1984), too high to result from compression. While the required amplification could come from turbulent processes in the interior, Kepler's sharply bounded northern rim suggests that magnetic field and fast electrons both appear abruptly, presumably at the shock. However, detailed modeling of profiles is necessary to determine whether field amplification occurs there or at the contact discontinuity. Other arguments against the magnetic field in young SNR being swept-up ambient field include radio polarization observations that show that young remnants have predominantly radial magnetic fields. If the field were swept up from the ISM, from most aspects a SNR should appear to have a tangential field in its shell, as in fact some older remnants do. If one were looking directly down a galactic field line, the projected field would appear radial, but this is unlikely to be true for all young SNR. Thus, some process reorders the field significantly, and may well increase its magnitude.

The time evolution of radio flux from a remnant can be readily calculated for several of these models. Shklovskii (1960b) predicted the decline of Cas A assuming that the mean particle energy density u_e dropped adiabatically, thus as R^{-4}, and the magnetic field was frozen in, so $B \propto R^{-2}$, where R is the remnant radius. An individual particle energy obeys $E \propto R^{-1}$, so $u_e \propto kE^{2-p}$ implies $k \propto R^{-2-p}$. Then the synchrotron flux $S_\nu \propto R^3 kB^{(p+1)/2} \propto R^{-2p}$. Since Cas A seems not to have been appreciably decelerated, $R \propto t$; taking $p = 2.5$, appropriate for Cas A, then $S_\nu \propto t^{-5}$. This implies a decline now at the rate of 1.5% yr^{-1}, more than double the observed rate. Tycho's remnant is probably in the adiabatic phase, so $R \propto t^{2/5}$; then for $p = 2.2$ and an age of 414 yr, the prediction is a decline at 0.43% yr^{-1}, falling between two determinations (see Section 10.3.4). If particles were not all injected at $t = 0$, but continue to be accelerated at the shock, one needs to know the time dependence of the acceleration process. If the particle and field energy densities are both taken to be proportional to the postshock pressure, $P_2 \propto \rho_1 v_s^2 \propto R^{-3}$; so the flux from a remnant should drop as $t^{-3(p+1)/10}$. For Tycho, with $p = 2.2$, this predicts $\dot{S}_\nu/S_\nu = 0.25\%$ yr^{-1}. A model in which the particles are accelerated, but the field is compressed ambient field, then has constant flux with time. These evolutionary models predict paths in the $\Sigma-D$ plane as well, of course; for the Shklovskii model, with $p = 2.2$, $\Sigma \propto D^{-6.4}$, while for the acceleration model $\Sigma \propto D^{-4.4}$ (Reynolds and Chevalier 1981). Both of these predictions are steeper than the mean observed $\Sigma-D$ relation (Section 10.3.5). The Shklovskii prediction is sufficiently at variance with the observations that one can infer that some later input of particle or magnetic field energy must occur in young SNR. However, the exact nature of that input is far from clear at present. Cowsik and Sarkar (1984) develop the Cas A-type turbulent acceleration model, finding that the radio emission rises and peaks at ages of a few hundred years, thereafter declining rapidly as the remnant enters the Sedov phase. For older remnants to be visible as radio sources, then, they assert that shock acceleration must become effective at later times.

In Crab-like SNR, the problem of the origin of the magnetic field and relativistic particles responsible for the nonthermal emission is generally left in the laps of the pulsar theorists. In the Crab, the equipartition field is about $10^{-3.5}$ gauss; evolutionary models give a somewhat larger value of 10^{-3} gauss while a detailed magnetohydrodynamic (MHD) model of the shock gives essentially the equipartition result when averaged over the nebula (Kennel and Coroniti 1984). In this case, there is little question about the origin of the magnetic flux; the magnitude is too large and the geometry wrong for it to be compressed ambient field. As well, the medium just outside the observed nebula may not be undisturbed ISM but high-velocity ejecta. There is little direct evidence in other Crab-like SNR; for 0540-693 in the Large Magellanic Cloud, evolutionary models give a result similar to the Crab, again implying that the source of the flux is the pulsar. As well, the electron distribution responsible for the radio emission is flatter than that found anywhere else in astrophysics; all other nonthermal-particle spectra (radio galaxies, shell SNR, cosmic rays) have energy indices p between 2 and 3, whereas for Crab-like remnants p is between 1 and 2. This suggests a totally different origin from the shock or turbulent acceleration picture that so nicely accounts for the universality of the $E^{-2.5}$ spectrum. However, detailed calculations of the relativistic-particle spectrum accelerated by a pulsar do not exist.

Once a relatively flat distribution of particles is assumed to be injected into the bubble surrounding the pulsar, its evolution in time, along with the evolution of the field strength, can be straightforwardly calculated. The balance between pulsar energy input and radiative and adiabatic expansion losses leads to an evolution in which the field strength rises to a maximum at extremely early times after the supernova ($\sim$ days), declining thereafter. The particle distribution is steepened by one power of energy by radiative losses above some energy dependent on the magnetic field strength; this break energy rises with time as the field rapidly weakens. The bubble's radiated spectral flux at frequencies above the frequency corresponding to the break energy first rises with time (though it is likely to be obscured by free-free absorption in the surrounding ejecta at these early times); when the break frequency passes the observing frequency, the flux begins to decline, eventually dropping as the fourth or fifth power of time. Once the pulsar's energy input begins to decrease, the picture becomes slightly more complex, but the result remains that the maximum steepening of the particle spectrum is one power of energy (corresponding to one half power in observed frequency). Since all Crabs with a known X-ray spectrum have a larger difference than one-half between X-ray and radio spectral indices, one infers an injection spectrum that either intrinsically steepens at some energy or involves two completely separate components with different spectral indices. It is possible that shock acceleration is responsible for the steeper (X-ray-emitting) electron distribution; the wind shock, where the relativistic pulsar wind is decelerated, may be responsible. This picture predicts a rate of decay of the Crab Nebula's radio flux consistent with the observations mentioned above.

10.4.5 Radio Supernovae

In attempting to explain the very early appearance of radio emission from the supernovae mentioned above, one is faced with the same problem of the origins of field and particles. Initial conjectures focused on miniature versions of the two SNR types: a model involving a new pulsar, invoking its field and particles to produce the radio emission (Pacini and Salvati 1981), and a model making use of turbulent acceleration and/or field amplification at the shock where the supernova ejecta interact with an ambient circumstellar medium produced by presupernova mass loss (Chevalier 1982b). The former model possesses the major difficulty that the surrounding ejecta, if anywhere close to uniform, have an enormous free-free opacity at radio wavelengths at these early times; an instability must be invoked to begin operating immediately after the supernova to cause the ejecta to clump into filaments. The circumstellar model argues from analogy with Cas A that the turbulent interface between the shocked circumstellar medium and ejecta might deposit about one percent of the supernova energy into field and particles; since the emission occurs outside the ejecta, the opacity is much less. In fact, the appearance of the radio supernovae at higher frequencies first is ascribed to free-free absorption, with the frequency at which the optical depth becomes unity dropping in time. See Weiler et al. (1985) for a detailed discussion of the observations and the various models. Since a circumstellar wind presumably results in a medium with a density law of $\rho \propto r^{-2}$, the radio emission should fade rapidly with time as the interaction weakens. This has begun to occur with SN 1979c, but has not yet with 1980k; if the latter's radio flux does not begin to drop within a few years, the model may need major adjustment.

Thus nonthermal emission is expected, and has been observed, immediately after the supernova, and some hundreds of years later. Models for the early emission predict its rapid decay with time; models for full-fledged SNR require the initiation of deceleration of the ejecta due to interaction with an ambient medium to produce detectable emission, presumably resulting in an increase of emission with time. So far, there are only two candidate observations for objects in the middle of this gap: two radio sources in M83 identified with SN 1950 and 1957 (Cowan and Branch 1985). Extensive radio searches at the positions of other supernovae of the last few decades have yielded nothing; this intermediate-age transition from supernova to SNR could be a productive line of research in the near future.

10.5 Collective Interactions of Supernova Remnants with the Galaxy

10.5.1 Structure of the Interstellar Medium
The discussion above of the three-phase ISM has demonstrated that the global effects of supernovae can affect dramatically the structure of the ISM. If the typical interstellar pressure is low enough, SNR can come to occupy most of the volume of the ISM. If the probability of being inside an SNR with radius less than R is $Q(R)$, and R_* is the radius at which SNR reach equilibrium, then the volume filling factor of SNR, f, is given by

$$f = 1 - e^{-Q(R_*)} ; \tag{10.9}$$

$Q(R_*)$ is given by $(4\pi/3)R_*^3 S\tau$ where S is the supernova rate per unit volume and τ the lifetime of an SNR. If the latter is given by the sound crossing time of the SNR at radius R_*, then

$$Q(R_*) = \frac{0.5(E_0/10^{51} \text{ erg})^{1.3}}{n_1^{0.14}(\bar{P}_{1,4})^{1.3}} \left(\frac{S}{10^{-13} \text{ pc}^{-3} \text{ yr}^{-1}} \right) \tag{10.10}$$

where $\bar{P}_{1,4} = (P_1 k^{-1}/10^4 \text{ K cm}^{-3})$ and this value of S corresponds to one supernova per thirty years in a disk 15 kpc in radius and 200 pc thick (McKee 1982). Note that $Q(R_*)$ can be greater than 1 for sufficiently low ambient density or pressure. An argument like this stands behind McKee and Ostriker's (1977) assertion that most of the volume of the ISM is in the hot phase (i.e., SNR interiors). However, it is possible that f is much lower, perhaps 0.2; then SNR form a network of hot tunnels with some isolated bubbles.

10.5.2 Energy and Mass Input into the Interstellar Medium
By the time a supernova remnant has entered Phase IV and dissipated in the ambient medium, it has deposited its 10^{51} ergs of kinetic energy and perhaps 10^{49}

ergs of ionizing radiation into the ISM. At a supernova rate of one per twenty to thirty years, this is enough to supply all energy needs of the ISM: ionization, turbulent motions of clouds, losses to radiation. As McKee (1982) points out, if SNR dominate the energetics of the ISM, they must also dominate the pressure, requiring a large filling factor. Ultimately, the energy they inject into the ISM must be radiated away, convected away in a wind, or perhaps put into cosmic rays. It may be radiated by SNR interiors in situ, or by hot, buoyant gas that rises into a galactic corona, cools, and falls back down ("galactic fountain"; Shapiro and Field 1976). If the energy injection rate is sufficiently large, the energy might unbind gas altogether, driving a galactic wind. However, this is unlikely to be the case in our galaxy; the mass loss rate required for a steady wind is far too high to be supported by the mass injected by SNR and stellar winds (unless T is extremely high, violating X-ray observations; Chevalier and Oegerle 1979). The characteristic turbulent motions of "standard" HI clouds of 10 to 30 km s^{-1} can easily be driven by SNR; this energy either is radiated away in cloud-cloud collisions or helps accelerate cosmic rays.

Supernova remnants affect the mass balance of a multicomponent ISM by converting ambient warm material or clouds into hot-phase material, neglecting any effects of a weak galactic wind which would remove mass from the Galaxy altogether. They can do this by the initial shock heating or by evaporation. A steady state requires that the rate of creating hot-phase material equal that at which old SNR cool and join the cooler phases (presumably clouds). McKee (1982) argues that this requirement demands either that the filling factor of SNR is small and the effects of evaporation negligible, in which case the energy input from SNR is radiated away by the warm and cold phases, or that SNR dominate the volume of the Galaxy and evaporation is an important process; in this case, the SNR energy input is radiated away by the hot-phase gas.

10.5.3 Cosmic Rays

In the last two decades, the inferred supernova rate has climbed to the point where even a modest efficiency for cosmic-ray production by supernovae or SNR (of order one percent) can easily account for the observed cosmic rays. However, consistency of the energetics is a long way from a detailed mechanism. This problem is, of course, related to the issue of the origin of the nonthermal particles responsible for the radio emission of SNR, though the observations of the latter tell us about electrons while protons and nuclei carry the dominant energy of the cosmic rays. The adiabatic-loss problem in having the supernovae create the radio-emitting electrons also holds for nuclei. The model Fermi originally proposed, a second-order model involving scattering of particles from interstellar clouds, is far too inefficient, given known levels of interstellar turbulence. While a second-order model involving turbulent acceleration in young SNR (in the style of the Cas A nonthermal particles) can be more efficient, current conventional wisdom favors first-order acceleration in SNR shocks. Blandford and Ostriker (1978) show that in a dominantly hot ISM, SNR shocks traverse an enormous volume of space, encompassing some 10^{51} ergs of cosmic-ray energy, and are capable of roughly doubling the energy of the cosmic rays in that volume (thus spending a substantial fraction of the entire supernova energy budget on fast particles). Cosmic-ray observations, in particular the ratio of secondary (spallation-produced) nuclei to primary ones as a function of energy, indicate that most cosmic rays are not accelerated in a large number of shocks; however, one or two (per typical cosmic ray) will do to replenish the galactic store against radiative losses, catastrophic losses, and leakage.

One might expect to be able to infer the efficiency of fast electron production of SNR directly, from their radio emission. However, this requires independent knowledge of the magnetic field. SNR shocks can produce particles up to a maximum energy of order 10^{14} eV (at which energy the particles' Larmor radius becomes comparable to the size of the remnant); but if this is to be the case, some constraints are put on the magnetic field. Weak, ambient fields ($\sim 10^{-5}$ gauss) and energetic electrons would overproduce X-rays by orders of magnitude in young remnants, as well as requiring so much energy in radio-emitting electrons that nuclei with 100 times that energy density, as observed at Earth, could not be produced. Equipartition fields, requiring shock amplification, are more efficient but still overproduce X-rays. High fields can solve the X-ray problem at the expense of causing too little energy to be put in electrons (Reynolds and Chevalier 1981). Unless the relation between electron and nuclear components is much different than inferred at Earth (for instance, if the electron component cuts off above 3000 GeV to avoid the X-ray difficulty), young remnants seem a problematic source for cosmic rays. However, cooling shocks may also contribute to the acceleration of cosmic rays; they certainly do so by compression, and shock acceleration, unhampered by the above arguments concerning adiabatic-phase remnants, might also contribute.

10.6 Future Prospects

As should be evident from the preceding discussion, a great many questions remain unanswered concerning supernova remnants, their evolution, and the source of their nonthermal emission. Future theoretical and observational work will certainly address many of these issues. How can we best rationalize our simple picture of the hydrodynamic evolution of SNR with the $N(<D)$ data? What stage of evolution are most SNR in? The answer to these questions may involve deciding on the volume filling factor of the hot phase of the ISM, and thus has wide application. A related question concerns the apparent absence of indisputably radiative shells.

The statistics of supernovae and SNR are not in the most satisfactory condition. While there is crude agreement between the supernova rate inferred from other galaxies and the SNR formation rate inferred from galactic studies, there seems to be a paucity of very young SNR. Many supernovae in the Galaxy could have gone unnoticed in the last few centuries, but radio remnants resembling Cas A or the Crab could not possibly be missed. Even Tycho or 3C58, made somewhat older and placed across the galaxy, would probably have been found by galactic-plane surveys. The pulsar formation rate is in crude agreement with both the Type II supernova rate and SNR formation rate; but Crab-like remnants are too scarce, suggesting perhaps that many pulsars are born rotating considerably more slowly than the Crab pulsar. If the Crab and 3C58 resulted from normal Type II supernovae, where are their shells? Could a significant fraction of supernovae leave no detectable radio remnant, or could the turn-on time for shell emission be much longer sometimes than it seems to be for Cas A or Tycho?

A satisfactory consensus seems to be evolving around shock or turbulent acceleration as a source of energetic particles in young (and perhaps middle-aged) SNR. But the details are far from clear. What are the relative contributions of shock and turbulent processes? The unresolved edges of radio emission in Tycho and SN 1006 (see Figure 10.3) require that the emissivity be increased by a large factor immediately at the shock front. The spectrum of accelerated particles at high energies is not known. Does it roll off at intermediate energies corresponding to synchrotron radiation of IR-optical frequencies? Could a synchrotron-loss break occur in a detectable portion of the spectrum? This would enable the magnetic field to be determined. Is there a nonthermal component to SNR X-ray emission? If not, where do 10^{12}- to 10^{14}-eV cosmic rays come from? The issue of magnetic field amplification is of potentially great interest. If it is a common feature of astrophysical collisionless shocks, it clearly has very wide application. Probably theoretical investigation is the most fruitful approach at this time, though additional observational evidence from young remnants may be forthcoming.

Observational advances in radio astronomy promise to assist in the clarification if not the final disposition of many of these questions. Most of the 150 or so catalogued galactic SNR have not yet been observed at high resolution. The typical SNR is a far more complicated and confusing object than the best-known objects such as Tycho or Cas A. High-resolution spectral-index maps may clarify issues of particle acceleration; higher-frequency observations are important here. Studies of time variations in young remnants should be very helpful in understanding early SNR evolution. Second-epoch observations of Tycho and Cas A have already yielded useful information. In general, observations of supernova remnants will continue to be some of the most difficult and rewarding radio astronomical investigations.

Recommended Reading

Woltjer, L. 1972. "Supernova Remnants," Annu. Rev. Astron. Astrophys. **10**:129.

Chevalier, R.A. 1977. "The Interaction of Supernovae with the Interstellar Medium," Annu. Rev. Astron. Astrophys. **15**:175.

Trimble, V. 1983. "Supernovae. Part II: the Aftermath," Rev. Mod. Phys. **55**:511.

Raymond, J.C. 1984. "Observations of Supernova Remnants," Annu. Rev. Astron. Astrophys. **22**:75.

Rees, M.J., and R.J. Stoneham (eds.), 1982. Supernovae: A Survey of Current Research. Dordrecht: Reidel.

Danziger, J., and P. Gorenstein (eds.), 1983. Supernova Remnants and Their X-Ray Emission. Dordrecht: Reidel.

Kafatos, M., and R.B.C. Henry (eds.), 1985. The Crab Nebula and Related Supernova Remnants. Cambridge: Cambridge University Press.

Catalogues of Galactic Supernova Remnants

Clark, D.H., and J.L. Caswell. 1976. Mon. Not. R. Astron. Soc. **174**:267.

Green, D.A. 1984. Mon. Not. R. Astron. Soc. **209**:449.

Weiler, K.W. 1983. In J. Danziger and P. Gorenstein (eds.), Supernova Remnants and Their X-Ray Emission. Dordrecht: Reidel, p. 299.

References

Aller, H.D., and S.P. Reynolds. 1985a. Astrophys. J. (Lett.) **293**:L73.

Aller, H.D., and S.P. Reynolds. 1985b. In M. Kafatos and R.B.C. Henry (eds.), The Crab Nebula and Related Supernova Remnants. Cambridge: Cambridge University Press, p. 75.

Ashworth, W.B. 1980, J. Hist. Astron. **11**:1.

Baade, W. 1942. Astrophys. J. **96**:188.

Baade, W., and F. Zwicky. 1934a. Proc. Natl. Acad. Sci. USA **20**:254; **20**:259.

Baade, W., and F. Zwicky. 1934b. Phys. Rev. **46**:76.

Baars, J.W.M., and A.P. Hartsuijker. 1972. Astron. Astrophys. **17**:172.

Barbon, R., F. Ciatti, and L. Rosino. 1973. Astron. Astrophys. **25**:241.

Barbon, R., F. Ciatti, and L. Rosino. 1979. Astron. Astrophys. **72**:287.

Bell, A.R. 1978. Mon. Not. R. Astron. Soc. **182**:147.

Blair, W.P., and R.P. Kirshner. 1985. Astrophys. J. **289**:582.

Blair, W.P., R.P. Kirshner, and R.A. Chevalier. 1981. Astrophys. J. **247**:879.

Blandford, R.D. 1979. *In* J. Arons, C. McKee, and C. Max (eds.), Particle Acceleration Mechanisms in Astrophysics. New York: American Institute of Physics, p. 333.

Blandford, R.D., and L.L. Cowie. 1982. Astrophys. J. **260**:625.

Blandford, R.D., and J.P. Ostriker. 1978. Astrophys. J. (Lett.) **221**:L29.

Bolton, J.G., G.J. Stanley, and O.B. Slee. 1949. Nature **164**:101.

Branch, D. 1986. Astrophys. J. (Lett.) **300**:L51.

Braun, R. 1985. Ph.D. thesis, University of Leiden.

Caswell, J.L., and I. Lerche. 1979. Mon. Not. R. Astron. Soc. **187**:201.

Caswell, J.L., D.H. Clark, D.F. Crawford, and A.J. Green. 1975. Aust. J. Phys. (Astrophys. Suppl.), No. 37.

Chanan, G.A., D.J. Helfand, and S.P. Reynolds. 1984. Astrophys. J. (Lett.) **287**:L23.

Chevalier, R.A. 1976. Astrophys. J. **208**:826.

Chevalier, R.A. 1977a. Annu. Rev. Astron. Astrophys. **15**:175.

Chevalier, R.A. 1977b. *In* D.N. Schramm (ed.), Supernovae. Dordrecht: Reidel, p. 53.

Chevalier, R.A. 1982a. Astrophys. J. **258**:790.

Chevalier, R.A. 1982b. Astrophys. J. **259**:302.

Chevalier, R.A. 1982c. Astrophys. J. (Lett.) **259**:L85.

Chevalier, R.A., and W.R. Oegerle. 1979. Astrophys. J. **227**:398.

Chevalier, R.A., R.P. Kirshner, and J.C. Raymond. 1980. Astrophys. J. **235**:186.

Clark, D.H., and J.L. Caswell. 1976. Mon. Not. R. Astron. Soc. **174**:267.

Clark, D.H., and F.R. Stephenson. 1977. The Historical Supernovae. Oxford: Pergamon.

Cowan, J.J., and D. Branch. 1985. Astrophys. J. **293**:400.

Cowsik, R., and S. Sarkar. 1980. Mon. Not. R. Astron. Soc. **191**:855.

Cowsik, R., and S. Sarkar. 1984. Mon. Not. R. Astron. Soc. **207**:745.

Cox, D.P., and B.W. Smith. 1974. Astrophys. J. (Lett.) **189**:L105.

Danziger, J., and P. Gorenstein (eds.). 1983. Supernova Remnants and Their X-Ray Emission. Dordrecht: Reidel.

Dent, W.A., H.D. Aller, and E.T. Olsen. 1974. Astrophys. J. (Lett.) **188**:L11.

Dickel, J.R., and D.K. Milne. 1976. Aust. J. Phys. **29**:435.

D'Odorico, S., P. Benvenuti, and F. Sabbadin. 1976. Astron. Astrophys. **52**:93.

D'Odorico, S., M.A. Dopita, and P. Benvenuti. 1980. Astron. Astrophys. Suppl. Ser. **40**:67.

Fesen, R. 1983. Astrophys. J. (Lett.) **270**:L53.

Field, G.B., D.W. Goldsmith, and H.J. Habing. 1969. Astrophys. J. (Lett.) **155**:L49.

Fusco-Femiano, R., and A. Preite-Martinez. 1984. Astrophys. J. **281**:593.

Green, D.A. 1984. Mon. Not. R. Astron. Soc. **209**:449.

Gull, S.F. 1973. Mon. Not. R. Astron. Soc. **161**:47; **162**:135.

Gull, S.F. 1975. Mon. Not. R. Astron. Soc. **171**:263.

Hamilton, A.J.S., C.L. Sarazin, and A.E. Szymkowiak. 1986. Astrophys. J. **300**:698.

Helfand, D.J., and R.H. Becker. 1985. Nature **307**:215.

Hesser, J.E., and S. van den Bergh. 1981. Astrophys. J. **251**:549.

Hillebrandt, W., P. Höflich, J.W. Truran, and A. Weiss. 1987. Nature **327**:597.

Högbom, J.A., and J.R. Shakeshaft. 1961. Nature **189**:561.

Holt, S.S. 1983. *In* J. Danziger and P. Gorenstein (eds.), Supernova Remnants and Their X-Ray Emission. Dordrecht: Reidel, p. 17.

Huang, Y.L., and P. Thaddeus. 1985. Astrophys. J. (Lett.) **295**:L13.

Hubble, E. 1937. Mt. Wilson Ann. Rep., 1936–1937.

Hughes, J.P., D.J. Helfand, and S.M. Kahn. 1984. Astrophys. J. (Lett.) **281**:L25.

Jones, E.M., B.W. Smith, and W.C. Straka. 1981. Astrophys. J. **249**:185.

Kafatos, M., and R.B.C. Henry (eds.). 1985. The Crab Nebula and Related Supernova Remnants. Cambridge: Cambridge University Press.

Kamper, K.W., and S. van den Bergh. 1976. Astrophys. J. Suppl. Ser. **32**:351.

Kamper, K.W., and S. van den Bergh. 1978. Astrophys. J. **224**:851.

Kamper, K.W., and S. van den Bergh. 1983. Astrophys. J. **268**:129.

Kennel, C.F., and F.V. Coroniti. 1984. Astrophys. J. **283**:710.

Lundmark, K. 1921. Publ. Astron. Soc. Pacific **33**:234.

Lyne, A.G., R.N. Manchester, and J.H. Taylor. 1985. Mon. Not. R. Astron. Soc. **213**:613.

Mathewson, D.S., V.L. Ford, M.A. Dopita, I.R. Tuohy, K.S. Long, and D.J. Helfand. 1983. Astrophys. J. Suppl. Ser. **51**:345.

Mathewson, D.S., V.L. Ford, M.A. Dopita, I.R. Tuohy, B.Y. Mills, and A.J. Turtle. 1984. Astrophys. J. Suppl. Ser. **55**:189.

Matsui, Y., K.S. Long, J.R. Dickel, and E.W. Greisen. 1984. Astrophys. J. **287**:295.

McKee, C.F. 1982. *In* M.J. Rees and R.J. Stoneham (eds.), Supernovae: A Survey of Current Research. Dordrecht: Reidel, p. 433.

McKee, C.F., and J.P. Ostriker. 1977. Astrophys. J. **218**:148.

Mills, B.Y. 1983. *In* J. Danziger and P. Gorenstein (eds.), Supernova Remnants and Their X-Ray Emission. Dordrecht: Reidel, p. 551.

Milne, D.K. 1970. Aust. J. Phys. **23**:425.

Nomoto, K. 1987. *In* D.J. Helfand and J.H. Huang (eds.), Birth and Evolution of Neutron Stars. Dordrecht: Reidel, p. 281.

Oort, J. 1946. Mon. Not. R. Astron. Soc. **106**:159.

Pacini, F., and M. Salvati. 1981. Astrophys. J. (Lett.) **245**:L107.

Panagia, N., R.A. Sramek, and K.W. Weiler. 1986. Astrophys. J. **300**:L55.

Poveda, A. and L. Woltjer. 1968. Astron. J. **73**:65.

Reynolds, S.P., and R.A. Chevalier. 1981. Astrophys. J. **245**:912.

Reynolds, S.P., and R.A. Chevalier. 1984a. Astrophys. J. **278**:630.

Reynolds, S.P., and R.A. Chevalier. 1984b. Astrophys. J. (Lett.) **281**:L33.

Scargle, J.D. 1969. Astrophys. J. **156**:401.

Scott, J.S., and R.A. Chevalier. 1975. Astrophys. J. (Lett.) **197**:L5.

Scott, P.F., J.R. Shakeshaft, and M.A. Smith. 1969. Nature **223**:1139.

Sedov, L.I. 1959. Similarity and Dimensional Methods in Mechanics. New York: Academic Press.

Seward, F.D. 1985. Comm. Astrophys. **11**:15.

Seward, F., Gorenstein, P., and Tucker, W. 1983. Astrophys. J. **266**:287.

Shapiro, P.R., and G.B. Field. 1976. Astrophys. J. **205**:762.

Shklovskii, I.S. 1953. Dokl. Akad. Nauk. SSSR **91**:475.

Shklovskii, I.S. 1960a. Cosmic Radio Waves. Cambridge: Harvard University Press, p. 282.

Shklovskii, I.S. 1960b. Sov. Astron.-AJ **4**:243.

Shklovskii, I.S. 1965. Sov. Astron.-AJ **9**:224.

Shklovskii, I.S. 1968. Supernovae. London: Wiley Interscience.

Shull, J.M. 1983. *In* J. Danziger and P. Gorenstein (eds.), Supernova Remnants and Their X-Ray Emission. Dordrecht: Reidel, p. 99.

Spitzer, L. 1978. Physical Processes in the Interstellar Medium. New York: John Wiley and Sons.

Straka, W.C., J.R. Dickel, W.P. Blair, and R.A. Fesen. 1986. Astrophys. J. **306**:266.

Strom, R.G., W.M. Goss, and P.A. Shaver. 1982. Mon. Not. R. Astron. Soc. **200**:473.

Tammann, G. 1982. *In* M.J. Rees and R.J. Stoneham (eds.), Supernovae: A Survey of Current Research. Dordrecht: Reidel, p. 371.

Tan, S.M., and S.F. Gull. 1985. Mon. Not. R. Astron. Soc. **216**:949.

Trimble, V. 1982. Rev. Mod. Phys. **54**:1183.

Trimble, V. 1983. Rev. Mod. Phys. **55**:511.

Turtle, A.J., D. Campbell-Wilson, J.D. Bunton, D.L. Jauncey, M.J. Kesteven, R.N. Manchester, R.P. Norris, M.C. Storey, and J.E. Reynolds, 1987. Nature **327**:38.

van den Bergh, S., and K.W. Kamper. 1977. Astrophys. J. **218**:617.

van der Laan, H. 1962. Mon. Not. R. Astron. Soc. **124**:125.

Vinyajkin, E.N., and V.A. Razin. 1979. Aust. J. Phys. **32**:93.

Weiler, K.W. 1985. *In* M. Kafatos and R.B.C. Henry (eds.), The Crab Nebula and Related Supernova Remnants. Cambridge: Cambridge University Press, p. 227.

Weiler, K.W., R.A. Sramek, N. Panagia, J.M. van der Hulst, and M. Salvati. 1986. Astrophys. J. **301**:790.

Weisskopf, M.C., E.H. Silver, H.L. Kestenbaum, K.S. Long, and R. Novick. 1978. Astrophys. J. (Lett.) **220**:L117.

Woltjer, L. 1972. Annu. Rev. Astron. Astrophys. **10**:129.

Zwicky, F. 1965. *In* L.H. Aller and D.B. McLaughlin (eds.), Stellar Structure. Chicago: University of Chicago Press, p. 367.

11. Pulsars

Don C. Backer

11.1 Introduction

In August 1967 Jocelyn Bell and Antony Hewish detected pulses of radio emission from the constellation of Vulpecula during a comprehensive investigation of extragalactic radio sources. The pulses repeated every 1.337 s. Their announcement of this pulsating radio source, or pulsar, in early 1968 opened a new window for studying the end products of stars (Hewish et al. 1968). This pulsar was first named CP1919: CP for Cambridge Pulsar and 1919 for the hours and minutes of its right ascension. Pulsars are now uniformly named: a prefix of PSR followed by their 1950 coordinates in right ascension and declination. In dense parts of the sky, the full IAU source designation, which includes tenths of degrees of declination, is used. In this chapter the PSR prefix is dropped for brevity.

The low galactic latitudes of CP1919 (1919+21) and others pulsars that were soon discovered indicated that pulsars were galactic objects. Millisecond intensity fluctuations of the signals implied that their emitting regions were no larger than 300 km. The stability of their rotation periods required the inertia of a stellar mass. Thomas Gold (1968) concluded that pulsars must be highly magnetized, rapidly rotating neutron stars. Radio beams anchored in the rotating stars sweep past the Earth like signals from celestial lighthouses. Gold predicted both that even faster rotation periods would be found and that the periods would lengthen as the stars were decelerated by their energy emissions. The neutron-star hypothesis was soon validated by the discovery of a pulsar in the Crab Nebula supernova remnant (Chapter 10) with a spin rate of 30 Hz that is slowly decaying. The deceleration time scale matches the age of the remnant that was formed in 1054 A.D.

Goldreich and Julian (1969) analyzed the electrodynamics of rotating, magnetized neutron stars. A pulsar acts like a giant electric generator as a result of the rapid rotation of its dipolar magnetic field in a conducting medium. Their study has formed the basis for most later thinking about the astrophysics of pulsar magnetospheres. Gunn and Ostriker (1970) presented a model of the origin and evolution of pulsars that has guided more recent studies based on larger statistical samples.

11.1.1 Brief History of Neutron Stars
The discovery of pulsars followed an era of speculation concerning the possibility of neutron stars. Physicist Lev Landau and astronomers Fritz Zwicky and Walter Baade proposed the neutron star idea shortly after the discovery of the neutron by Sir James Chadwick in 1932. Zwicky equated the brightest optical outbursts in recorded history, supernovae, with the sudden deaths of stars. He surmised that an

implosive by-product, the neutron star, might also be created. By 1939 atomic theory had developed to the point that Robert Oppenheimer and George Volkoff could calculate the structure of a neutron star.

In the following decades, astrophysicists were pessimistic about detection of neutron stars since their small size of 10 km would not yield a significant optical luminosity. Nevertheless, in 1942 Baade and Rudolph Minkowski speculated that a peculiar star in the midst of the Crab Nebula might be a neutron star created in the 1054 A.D. supernova event. Their speculation was validated in 1968. In the year before the pulsar discovery, Franco Pacini (1967) realized that a magnetized, rotating neutron star could provide the continuous source of particle acceleration required to keep the Crab Nebula shining for 1000 years.

Detection of neutron stars at X-ray wavelengths became possible with the advent of space observatories. The Crab Nebula was detected with a rocket-borne instrument in 1964, and five years later X-ray pulses from the Crab pulsar were discovered. Satellite-based X-ray observatories discovered many galactic stars in the 1970s; some of these were later detected at radio wavelengths (Chapter 9). Pulsating X-ray stars with periods similar to the radio pulsar periods were shown to be in high-mass binary systems by the Doppler variations of their periods. In a binary X-ray pulsar, matter that is transferred to the neutron star from the evolved companion star radiates when it falls into the deep gravitational potential well of the neutron star. If neutron stars are always created in supernova explosions as proposed by Zwicky and Baade, then the existence of X-ray binaries shows that such explosions do not necessarily disrupt the binary. These and other X-ray binaries involving neutron stars, white dwarfs and, possibly, black holes provide important signposts for investigations of the death of stars.

11.1.2 New Types of Pulsars

In 1975 the first binary radio pulsar, 1913+16, was discovered by Russell Hulse and Joseph Taylor (1975). The narrow width of its pulses and the high stability of its 59-ms period have allowed Taylor and his colleagues to determine precisely the masses of both stars and the decay of the orbital period as a consequence of gravitational wave emission. Both stars are thought to be neutron stars which have remained gravitationally bound despite the energetic explosions accompanying their formation.

The most recent step in the investigation of neutron stars is the discovery of a pulsar, 1937+21, spinning at 642 Hz (Backer et al. 1982). This millisecond-period pulsar rotates nearly as fast as a neutron star can spin before its structure becomes unstable. The rotation speed at the equator of 1937+21 is one-tenth that of the speed of light. This object may have been produced by a neutron-star binary that swallowed its companion when the orbit collapsed as a consequence of gravitational wave emission. Measurement of its pulse arrival times over several years with microsecond precision shows that its stability is comparable to that of the best Earth atomic standards. This stability provides the best limit on the amplitude of a stochastic background of gravitational radiation from chaotic processes in the early universe.

Extensive surveys for pulsars have been conducted over the past ten years.

These have begun to define the galactic population of pulsars. The density of pulsars and their distribution in the Galaxy is required to connect these end products of stellar evolution to their parent population. The binary and millisecond pulsars demonstrate the multiple paths along which stars evolve to form nature's most exquisite object: the pulsating radio source.

11.2 Basic Properties

11.2.1 Intensity Spectra

Pulsars are faint radio sources with steep power law spectra above 400 MHz. Below 400 MHz their spectra either continue to rise or turn over and cut off sharply. Typical time-averaged flux densities at 400 MHz are around 10 mJy. Their faintness kept them hidden until the serendipitous discovery of Bell and Hewish. Only the pulsar in the Crab Nebula was known as a bright, pointlike source before 1968. No one suspected that its emission consisted of sharp, periodic pulses.

The total luminosity emitted in radio waves by a typical pulsar is 10^{30} erg s^{-1}, many orders of magnitude less then the radio emission from the Sun, and three orders of magnitude less than the total solar luminosity. However, pulsar radiation is similar to solar radio emission in other ways. Pulse intensities exhibit a bewildering array of random and systematic fluctuations (Section 11.4), and pulsar radiation is often highly elliptically polarized. The only other non-solar radio sources with similar emission characteristics are the planets and M-dwarf flare stars (Chapter 9).

11.2.2 Dispersion and Distance Scale

Pulsar pulses arrive at low frequencies later than at high frequencies. This dispersion of the signal is the result of the interaction of the radio waves with ionized thermal gas along the line of sight (Section 11.5). Pulsar observations require restricted bandwidths so that the dispersion within the band is less than the pulse width. This restriction makes viewing the already faint signals even more difficult. Searches for new pulsars overcome this bandwidth restriction by recording data from many narrow frequency bands. The resultant two-dimensional time series is first Fourier analyzed and then inspected for candidate pulsars of unknown period and unknown dispersion.

The distribution of 440 pulsars in galactic coordinates is shown in Figure 11.1. The concentration of pulsars toward the galactic plane indicates that pulsars are associated with Population I stars. Nonuniformities in the distribution result from variations in the extent and sensitivity of pulsar searches. The concentration of objects at low latitudes inside longitude 60° results from the deep searches conducted at the Arecibo and Jodrell Bank observatories. The concentration of objects in the first and fourth quadrants is a real effect and demonstrates that the density of pulsars is highest for galactocentric radii less than that of the Sun. The galactic distribution of pulsars can be studied by using path integrals of neutral and ionized hydrogen along with models for the distribution of these atoms. The nearest pulsar is about 100 pc from the Sun. The Galaxy contains about 10^5 pulsars if one

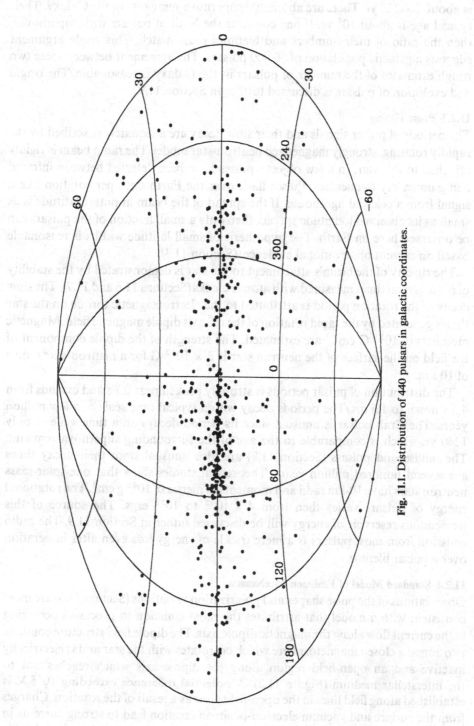

Fig. 11.1. Distribution of 440 pulsars in galactic coordinates.

estimates their density from our nearest neighbor. Their age based on period decay is about 3×10^6 yr. There are about 200 supernova remnants in the Galaxy. Their typical age is about 10^4 yr. If one connects the birth of pulsars with supernovae, then the ratio of their numbers and lifetimes must match. This crude argument suggests a galactic population of 50,000 pulsars. The agreement between these two rough estimates of the number of pulsars in the Galaxy is reasonable. The origin and evolution of pulsars is discussed further in Section 11.9.

11.2.3 Pulse Timing

The periods of pulsar signals and their slow decay are adequately described by the rapidly rotating, strongly magnetized neutron-star model. The radio beam is rigidly attached to this star. In a few objects, pulses have been detected between infrared and gamma-ray frequencies. Pulses flash over the Earth once per rotation like a signal from a celestial lighthouse. If the spread of the beam in pulsar latitude is as small as its observed longitude spread, then only a small fraction of the pulsars can be observed here on Earth. The hypothesis of small latitude widths is reasonable based on current observational evidence. (Section 11.3).

The rigidity of the beam's attachment to the star is demonstrated by the stability of pulse arrival times measured with atomic clocks (Sections 11.6 and 11.7). The slow decay of the rotation period is attributed to the electromagnetic torque on the star that is generated by the rapid rotation of the off-axis dipole magnetic field. Magnetic moments of 10^{31} G cm^{-3} are estimated. The strength of the dipole component of the field on the surface of the neutron star is 2×10^{12} G for a neutron-star radius of 10 km.

The distribution of pulsar periods is strongly peaked near 0.7 s and extends from 4.3 s down to 1.6 ms. The periods decay with a typical time scale of a few million years. The Crab pulsar is unusual since its period decays on a time scale of only 1240 yr, which is comparable to the age of the surrounding supernova remnant. The millisecond pulsars (Section 11.7) are also unusual since their decay times are several hundred million years. Theoretical studies show that one-solar-mass neutron stars have 10-km radii and moments of inertia of 10^{45} g cm^3. The rotational energy of pulsars varies then from 4×10^{45} to 10^{52} ergs. The source of this tremendous reservoir of energy will be discussed futher in Section 11.9. The radio emission from most pulsars is a mere trickle of energy loss even after integration over a pulsar lifetime.

11.2.4 Standard Model of Emission Mechanism

Observations of the pulse shapes and polarization variations (Section 11.3) are most consistent with a model that attributes the radio emission to processes occurring in the current flow along the magnetic dipole axis. The dipole field structure contains two zones: a closed magnetosphere which co-rotates with the star and is electrically inactive and an open-field region along the dipole axis which reaches out to the interstellar medium (Figure 11.2). A potential difference exceeding 10^{12} V is established along field lines in the open-field zone as a result of the rotation. Charges from the surface and vacuum electron-positron creation lead to strong currents in the open-field zone. The radio emission is commonly attributed to clumps of charges in the current flow which undergo slow acceleration as they exit the star. The low

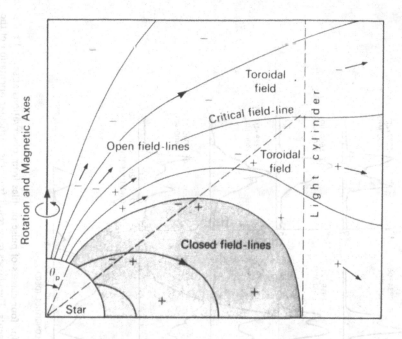

Fig. 11.2. Schematic cross section of a magnetized, rotating neutron-star model that displays regions of special interest for pulsar astrophysics. Magnetic field lines that leave the star are termed open, while those that loop back to the star are called closed. The maximum radius for the closed field lines is set by their rotation at the speed of light. Pulsar radiation is beamed along the open field lines. (After Goldreich and Julian 1969).

energy of the radio photons stands in dramatic contrast to the extreme properties of strong currents along intense magnetic field. While this standard model for radio emission is incomplete, and may be erroneous in parts, it serves a useful function for describing the wealth of pulsar properties in the succeeding sections. Pulsar magnetospheres and the emission mechanism are discussed further in Section 11.8.

11.3 Pulse Morphology and Polarization

11.3.1 Pulse Components

Average total-intensity profiles of most pulsars usually contain one to three distinct peaks. These peaks identify components of the pulse. Components result from a combination of higher than average intensities in a few pulses and lower than average intensities in many pulses. The profiles from a sample of pulsars (Figure 11.3) show that pulse widths are roughly independent of the rotation period when measured in degrees of rotational phase. In the standard model, the pulse width is related to the opening angle of dipole field lines that conduct ultra-relativistic current to the interstellar medium. The properties of closely observed pulsars that are summarized below suggest a qualitative model for the two-dimensional beam pattern. Knowledge of this pattern is essential both to determine the number of the

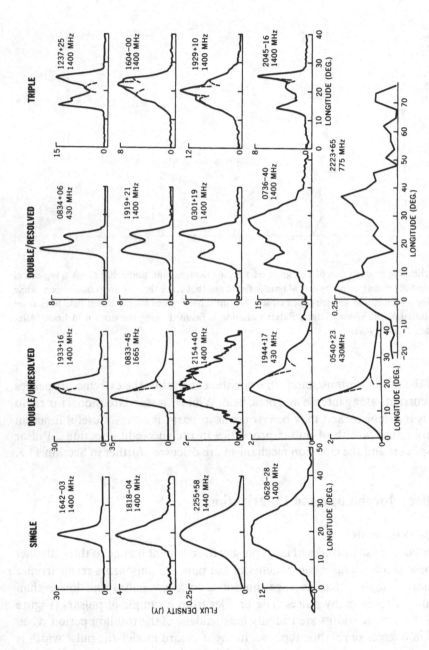

Fig. 11.3. Pulse morphology. The flux density as a function of rotational longitude is shown for four classes of pulse profiles from observations at several radio frequencies (From Backer 1976). The classes are distinguished by the number of pulse components. The classes may result from the particular orientation of the observer with respect to both the path of a standard beam and the spin axis.

pulsars in the Galaxy and to provide a basis for physical models of the radio-emission mechanism.

Most pulse profiles change slowly with observing frequency. Component widths and spacings decrease with increasing radio frequency up to about 1 GHz where variations either cease or decrease more slowly. The spectra of components in most pulsars are neither identical to each other nor simple monotonic functions of frequency. Often component spacings or relative amplitudes become too small to allow component resolution. There are a number of cases where a profile section appears to be missing over a few octaves of radio frequency, as if the signal was absorbed. Observation of a pulsar over several decades of radio frequency is essential to categorize its radiation pattern.

11.3.2 Polarization

Radhakrishnan and Cooke (1969) detected a rapid, monotonic swing of the linear polarization angle across the pulse of 0833−45, the pulsar related to the Vela supernova remnant. The swing rate was 6.2 degrees of position angle per degree of rotation. They proposed that pulsar radiation is produced by a distribution of particles streaming along curved magnetic field lines in the vicinity of the dipole axis. They speculated that curvature acceleration of charged particles would produce linearly polarized radio emission with the electric vector in the plane of curvature. The rotation of the star leads to the observed swing if the dipole axis passes close to the observer line of sight. This rotating-vector model (with the vector defined as that from the dipole axis at the emission altitude toward the observer) has received much confirmation from later polarization observations. The rapid swing has also been attributed to the relativistic transformation of a localized source of emission. A slow swing of position angle in the frame of the emitter is transformed to a rapid swing across a narrow beam concentrated in the direction of motion of the emitter. Relativistic-transformation models lack the specificity of the rotating-vector models since the kinematics of the emitting particles has not been discussed.

The rotating-vector model predicts a range of possible swings of polarization position angle (PA) for emission at a fixed altitude in a pure dipole field given by the following expression

$$\tan[\mathrm{PA}(t)] = \frac{\sin \alpha \sin \phi(t)}{\sin \zeta \cos \alpha - \cos \zeta \sin \alpha \cos \phi(t)} \tag{11.1}$$

where α and ζ are the stellar colatitudes of the magnetic axis and the line of sight, respectively, and ϕ is the longitude of the magnetic axis relative to the plane defined by rotation axis and the line of sight. The sweep of PA is limited to 180°. The range of observed position-angle sweep rates at the center of profiles indicates that most pulse emission occurs within $|\alpha - \zeta| \sim 10°$ of the dipole axis.

11.3.3 Hollow-Cone Beam Model

Consideration of the pulse component data (Figure 11.3) along with the linear-polarization swings suggests that emission is strongest in a hollow-cone beam centered on the dipole axis. For example, 0525+21 has two isolated components that are symmetrically located with respect to the S-shaped swing of linear polariza-

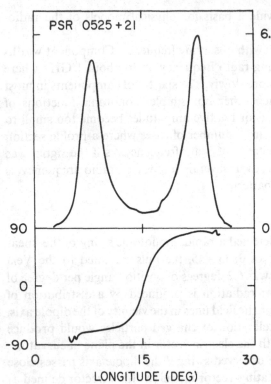

Fig. 11.4. Linear polarization variation of 0525+21. The rapid S-shaped sweep of the linear polarization angle with pulse longitude is accurately modeled by the formula discussed in the text. The minimum separation between the observer and the dipole axis is 1°. The pulse components suggest a hollow-cone pulse morphology.

tion position angle (Figure 11.4). The S-shape agrees precisely with the model discussed above if the magnetic pole axis passes within one degree of the line of sight.

It the emission beam is roughly circular with a hole in the middle, then we will observe a single component when the outer rim of the hollow cone just grazes the Earth-pulsar line of sight. The width of these single-component objects would be narrower than that of a two-component object viewed at a similar magnetic-pole inclination. Pulse-profile studies support this relation. Furthermore, if all pulsars emit roughly circular hollow-cone beams with similar dimensions, then the fraction of objects with single components should be comparable to the fraction of double-component objects. The observed ratio is near one-half.

The hollow-cone description of pulse profiles suggests that the latitude extent of the standard pulse beam is equal to that of the double-component pulse width, or 10°. Polarization-angle data can be used to infer the separation between the spin and dipole axes. However, this inference relies on the aforementioned dipole formula and can be misleading if the polarization angle is established in the outer magnetosphere where the dipole field can be heavily distorted. Narayan (1984) uses the polarization-angle data as evidence for elliptical-shaped beams whose ellipticity increases with decreasing pulse period.

A number of pulsars emit a centrally located third component in addition to what appears to be a typical hollow-cone double component (Backer 1976; e.g., 2045−16 in Figure 11.3). The properties of core components are clearly distinguished from those of the cone components discussed above (Rankin 1983). Cone intensity spectra

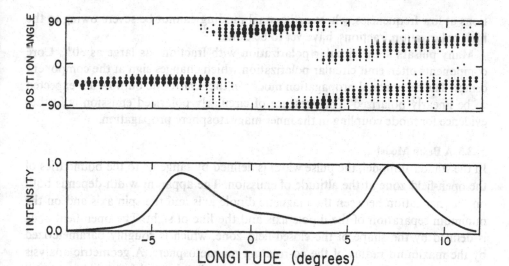

Fig. 11.5. Linear polarization in individual pulses of 2020 + 28 is shown in the top histogram of frequency of occurrence of polarization angle with longitude. Two angles separated by 90° are found at many longitudes. Each polarization mode follows monotonic variations that are consistent with the rotating-vector model.

are flatter than core spectra; cone components are often detectable only at high radio frequencies. Polarization and intensity fluctuation data also distinguish the two types of radiation. The presence of core emission along directions where the cone emission mechanism is absent is a puzzle. There may be a very different current structure along the dipole axis leading to this additional radiation beam. The relative locations of the two processes along the dipole axis are not known. The asymmetric locations of cores with respect to cones could result from retardation, aberration, and field-line distortion between emission at two different altitudes.

11.3.4 More Polarization Effects

The average polarization profiles of a few pulsars are wildly inconsistent with the rotating-vector model, and often show abrupt changes of position angle by 90°. Summaries of individual pulse data for these objects have revealed an underlying consistency with the rotating-vector model (Figure 11.5). The peculiarities in such objects result from the frequent occurrence at many longitudes of polarized signals at two orthogonal angles at different times. The average is thus dependent on the dominant mode, and can switch from one to the other at different points in the pulse. The mechanism responsible for these modes is not understood. Some authors attribute the presence of orthogonal angles to propagation effects in the dense relativistic outflow; if the emission frequency is below the local cyclotron frequency, then normal modes that propagate in different directions and at different speeds can produce odd polarization effects.

The fractional linear polarization often decreases monotonically with radio frequency. Near 100 MHz the fraction saturates at values near unity. Less is known about the frequency dependence of the underlying position-angle swings since detailed studies have been limited to a narrow range of radio frequencies. Observations at low frequencies, where orthogonal moding cannot be severe owing to the high polarization fractions, have yet to be done.

Many pulsars show circular polarization with fractions as large as 30%. Core components often emit circular polarization which changes sign at the component center. The orthogonal propagation modes in the relativistic outflow are expected to be linearly polarized. The presence of circularly polarized emission is further evidence for mode coupling in the inner magnetosphere propagation.

11.3.5 A Beam Model
In the standard model, the pulse width is defined by tangents to the boundaries of the open-field zone at the altitude of emission. The apparent width depends both on the inclination between the magnetic dipole axis and the spin axis and on the minimum separation of the dipole axis and the line of sight. The open-field zone is defined by the shape of the closed-field zone, which is roughly parameterized by the maximum radius of the co-rotating magnetosphere. A geometric analysis that relates the above quantities to the measured pulse width predicts a period dependence of the width with certain simplifying assumptions. Assume that all magnetospheres co-rotate out to the light cylinder and that emission occurs at a fixed altitude above the stellar surface. Assume that the dipole-axis inclination is random and does not depend systematically on age. Then the minimum pulse width Φ_{min} results when our line of sight and the dipole axis are in the neutron-star equatorial plane, and will scale with $P^{-0.5}$ for small angles.

$$\sin^2\left(\frac{\Phi_{min}}{2}\right) = 0.5\left(\frac{2\pi}{Pc}\right)^{0.5}, \tag{11.2}$$

where r is the emission radius. Widths at a given P for an ensemble of objects will spread above the minimum owing to the range of inclinations. This scaling is observed when objects of similar morphology are compared, a remarkable fact given the number of assumptions.

11.3.6 Interpulses and Steady Emission
Several pulsars have interpulse components situated nearly 180° away from the strongest component. The ratios of the peak interpulse intensity to that of the main pulse range from 0.85 (1055 − 51) to 0.005 (0823 + 26). Upper limits of 10^{-3} have been established for a dozen other objects (Hankins and Fowler 1986). Interpulses occur most frequently in pulsars with periods below 0.5 s, and have the largest ratios for the shortest-period objects. A notable exception to this is 0826 − 34 with period of 1.840 s and a pulse profile that has been interpreted as a broad main pulse (120°) and a strong interpulse spaced by 150° from the center of the main pulse. The mainpulse-interpulse separation is less dependent on frequency than the component separation dependence discussed above.

Interferometer observations at Jodrell Bank have established the true zero-emission level for 25 pulsars (Perry and Lyne 1985). Four objects show steady emission throughout the pulse period of 0.03 to 0.16 times the total. The ratio of peak intensity to the steady level is 50 to 1000. Most objects have ratios in excess of 1000. Shorter periods are favored for steady emission. These observations have also revealed faint new components not centered on 180°. Pulsar searches are biased against detection of profiles with steady, or nearly steady, emission.

These "offpulse" contributions to the pulsar emission are useful to constrain models of the emission mechanism. In the standard model, main pulses and inter-pulses are conical beams of full width Φ emitted from opposite poles. Observation of both poles requires the observer and the dipole axes to be oriented nearly perpendicular to the rotation axis. If axes are randomly aligned, then the expected fraction of two-pole, or interpulse, pulsars is 0.055 ($\Phi = 0.17$ rad). This is close to the observed fraction if the observed favoring of short periods is taken into account. The preference for short periods might result from a tendency for the dipole axis to align with age. The separation of the interpulse from 180° could indicate distortion of the magnetic field from a pure dipole. Interpulses may also result from large ellipticities of the beam; if the dipole-axis inclination is small enough, the observer could cross the beam twice.

11.4 Intensity Fluctuations

The radio emission of pulsars is modulated on a wide range of time scales ranging from 10 microseconds to 100 days. The variations with time scales less than 1000 s are attributed to effects intrinsic to the pulsar. These intrinsic variations are discussed below in order of increasing time scale.

11.4.1 Micropulses

Observations of the emission from individual pulses for a few pulsars with high time resolution have revealed intensity variations with 100 to 1000 μs scale called micropulses (Figure 11.6). High time resolution is achieved either by signal proces-sing to remove the smearing effects of interstellar dispersion within a band of frequencies (Section 11.5) or by choosing a high center frequency where the dispersion is reduced. Micropulses as narrow as 10 μs with intensities of 10^4 Jy have been observed; the corresponding brightness temperature, which is computed

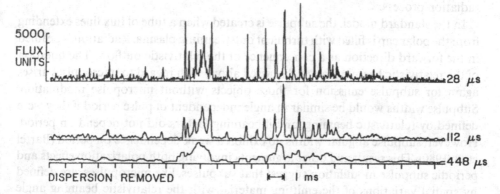

Fig. 11.6. Micropulses from 0950+08 displayed with varying degrees of smoothing. A number of the micropulses are periodically spaced.

using the variation time scale for a size estimate, is 10^{32} K. The physical process responsible for the observed chaotic jumble of micropulses has not been identified. Within the context of the standard model, there are two clear alternatives that are distinguished by the size of the relativistic beaming angle with respect to the observed micropulse scale expressed in degrees of rotational longitude. If the relativistic beaming is much smaller than the micropulse scale, then the micropulse must consist of many radiating elements distributed in the outflow current. On the other hand, individual radiating elements with beaming factors equal to the micropulse scale could explain the micropulse intensity fluctuations. Detailed studies of the statistics of micropulse emission show that the signal can be represented as noise pulses with Gaussian amplitude statistics and a time scale comparable to the inverse of the radio frequency whose average amplitude is slowly modulated by the envelope that delineates the micropulse. The amplitude-modulated noise model favors the former micropulse hypothesis. In this model, the relativistic beaming factor, γ, certainly exceeds 10^4, and may be as large as 10^5. Individual radiating elements either are generated by density enhancements in the turbulent outflow or result from more complex nonlinear radiation processes.

In a few of the pulsars with micropulse emission, pulses contain periodically spaced micropulses (Figure 11.6). A number of authors have noted that the characteristic vibration time scale in neutron-star crusts is similar to the micropulse periodicity scale. However, there are no models that explain how crustal vibrations might be excited or how such vibrations could modulate the radio emission.

11.4.2 Subpulses

The emission in many individual pulses is confined to a few subpulses with typical widths of a quarter of the pulse width. Subpulses may contain bursts of micropulses, or they may vary more smoothly. The average intensity at any longitude results from both the number of subpulses and the distribution of subpulse amplitudes. The statistical properties of subpulses change significantly across the pulse. Intensity modulation is often highest on the periphery of pulse components. This means that low amplitudes result in part from infrequent strong subpulses rather than from frequent weak emission. The absence of distinct subpulses in some objects is either from overlapping subpulses or from incomplete subpulse modulation of the radiation process.

In the standard model, the subpulse is created when a tube of flux lines extending from the polar cap is filled with turbulent outstreaming plasma. Radiation is beamed in the forward direction as a consequence of the relativistic outflow. The question of the relativistic beaming factor discussed above with regard to micropulses arises again for subpulse emission for those objects without micropulse modulation. Subpulse widths would be similar in angle independent of pulse period if they were defined by relativistic beaming, and if beaming factors did not depend on period. However, subpulse angular widths do exhibit a weak dependence on period (Bartel et al. 1980). This result and other arguments involving both polarization effects and periodic subpulse modulation suggest that subpulses, like micropulses, are defined by spatial variations of the emitting material with the relativistic beaming angle smaller than the subpulse scale. This leads to lower limits on γ of about 10^3.

11.4.3 Periodic Subpulse Phenomena

The occurrence of subpulses in a succession of pulse periods often follows a pattern. In the simplest cases, subpulses are regularly spaced in individual pulses at longitude intervals P_2 that range from 3 to 30°. From one pulse to the next, these subpulses appear to drift across the window defined by the average pulse profile. Subpulses disappear on one side of the pulse window and new subpulses appear on the other side. The pattern of drifting subpulses recurs with a period P_3 which ranges from 2 to 10 rotation periods. Drifting occurs equally in forward and reverse directions, although only drifts from the rear to the front of the profile were seen in the first few cases studied. Figure 11.7 displays the subpulse pattern for 1944 + 17. Values for P_2 and P_3 are 12 ms and $15P_1$, respectively. In other pulsars, the drift may be irregular in a myriad of ways. In 0031 − 07 the drift rate, P_2/P_3, jumps between values related by factors of 2. In 2016 + 28 the drift rate varies from moment to moment in an irregular manner. In 0826 − 34, which has a particularly broad pulse, the drift even reverses itself. The subpulse spacing P_2 is the most constant quantity throughout these variations.

In many pulsars there is a periodic pattern of subpulse emission, but not clear evidence for drifting of subpulses. In all cases the pattern period is evidence for a long-term memory process acting on the pulsar radio emission. In polar-cap models memory in the pulse-to-pulse fluctuations is difficult to explain since the radiating particles leave the system in a fraction of a pulse period. The stability, or Q, of the subpulse pattern periodicity ranges from a few to as high as 1000. However, even the lowest-Q periodicities require memory in the subpulse modulation mechanism since there is some correlation of the subpulse emission from one pulse to the next. Various locations and causes for the memory process have been proposed. These include hot spots in the stellar crust, active flux tubes in the open-field zone that carry excess current, and orbits in the closed magnetosphere with excess current. The first two are closely connected. Theoretical calculations of factors relating to the stability of structure on the polar cap and in the open-field zone are difficult, yet show some promise. Little work has been done on periodic structures within the closed magnetosphere and how they might couple to the observed emission patterns.

11.4.4 Pulse Nulling

On a time scale of 100 periods many objects switch off, or null, for intervals ranging from a few to several hundred periods. Figure 11.7 displays the pulse nulling in 1944 + 17. In 0826 − 34 and 1944 + 17 there are fewer periods with pulse emission than without emission. There may be objects which emit very small fractions of the time which have escaped detection because observers have looked at the wrong time. Nulling is common in pulsars that either display drifting subpulse activity or have more complex pulse profiles. In objects with triple-component profiles, all emission ceases in the null pulses despite the suggested different origins for cone and core components (Section 11.3). The correlation of age with nulling has suggested a connection between nulling and the cessation of pulsed emission.

In 0809 + 74 and a few other pulsars, there is an intimate relation between pulse nulling and subpulse drifting. After a null the subpulse pattern returns at nearly the

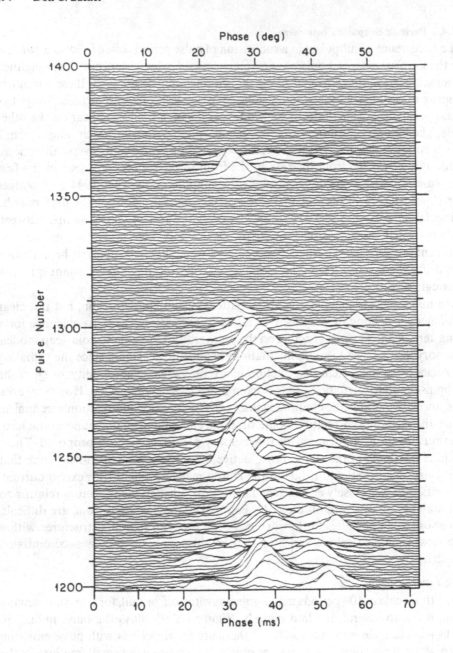

Fig. 11.7. Subpulses from 1944 + 17 in a 200-pulse sequence. Nulls where no radiation is detected are very frequent in this pulsar. The subpulses drift toward earlier times as pulse number increases in the first hundred pulses. (From Deich et al. 1986.)

same location as it had before the null. It appears that drifting ceases during the null and then begins again after the null with the location of the subpulse emitting regions remembered across the null. In $1944+17$ there is no memory of the subpulse pattern phase across nulls. The details of the subpulse/nulling phenomenon in $0809+74$ are remarkable. Nulls occur with variable duration. The location of the subpulse pattern after the nulls is offset slightly relative to the location before the null, and the offset depends on the null length. Furthermore, the drift rate after the nulls is initially slower than before and then gradually accelerates to its nominal value. Lyne and Ashworth (1983) show that the drift rate suddenly decreases at the onset of a null by an amount that depends on the null length. For long nulls, the new drift rate may even be of the opposite sign. Their analysis indicates that the drift rate recovers exponentially and that the emission returns after a fixed proportion of the perturbation has occurred. In the standard model, subpulse emission is defined by current filaments within the open-field zone and subpulse drifting arises from electrodynamic forces on the current structures. The continued drifting during a null necessarily requires that the current structure is preserved through the null, but is not generating intense radio emission beamed toward the observer for unexplained reasons.

11.4.5 Other Modulations

A small number of pulsars exhibit a curious sudden change of the pulse profile termed mode switching. The mode switch lasts for 100 to 1000 periods and then reverts back to normal. Several modes occur in a few objects. Mode switching is most prevalent in the multicomponent pulsars and appears to be a redistribution of the emitting directions. The multiplicity of subpulse pattern periods discussed above is related to mode switching since it occurs on similar time scales with similar abruptness and has accompanying changes in the average pulse profile. The memory time scale associated with mode changes is similar to that of pulse nulling, and is comparable to that of the highest-Q subpulse periodicities.

A fourth periodicity is revealed in the intensity fluctuation spectra of a number of pulsars. The periods, P_4, are typically a hundred times P_1 and the Q's are small. There have been few detailed investigations of these long-period effects. One suggestion is that the fourth periodicity is connected with the drifting subpulse process. If the subpulse modulation mechanism drifts at a constant rate for long periods of time, then a particular pattern will recur in a time $P_4 = P_1 P_3 / P_2$. This relation results in P_4 values considerably shorter than observed.

The average pulse profiles discussed in Section 11.3 are well defined despite the wealth of fluctuations discussed in this section. The rate of approach to stabilization has been studied for relatively few objects. Helfand et al. (1975) showed that the decorrelation between the average pulse profile and subaverage profiles of duration n periods was proportional to n^ζ with ζ ranging from -0.60 to -1.40. These slopes are all steeper than -0.5 which is expected for completely random fluctuations. Steeper slopes result from the many systematic variations of subpulses and entire pulse profile discussed above. In a few objects, no secular variations in the pulse profiles were found for periods extending over several years. Limits on the secular variation of pulse profiles provide important limits on both the possibility of free or geodetic precession in pulsars as well as on the evolution of the emission mechanism. Long-term studies of very young pulsars such as the Crab pulsar might display evolutionary effects.

11.5 Interstellar Propagation

Pulsar signals traverse interstellar paths with hydrogen column densities of roughly 3×10^{21} cm^{-2} kpc^{-1} and thermal electron column densities of 10^{20} cm^{-2} kpc^{-1}. The paths are threaded by magnetic fields whose strengths are a few μgauss. The ionization that produces the electrons comes from the ultraviolet flux of hot stars, cosmic rays, and shocks generated by star formation events and supernova explosions. The magnetic fields are generated by large-scale currents of interstellar plasma. The plasma is weakly turbulent on scale lengths of the Fresnel zone for pulsar signals, 10^{11} cm. The shocks mentioned above are the most likely driving source for this microscale interstellar turbulence.

11.5.1 Neutral Hydrogen Absorption
Detection of the absorption of pulsar radiation by atomic hydrogen has established the distance scale for pulsars. The observations yield a profile of signal absorption depth as a function of velocity of the 1420-MHz frequency. This is compared to the intensity of emission as a function of velocity. Models of the galactic distribution of hydrogen discussed in Chapter 7 are required to estimate the pulsar distance from the presence and absence of absorption at different hydrogen velocities. Hydrogen-absorption distance estimates calibrate the relation between electron column density and distance. Distances for all pulsars can then be estimated since electron column densities are easily measured. The uncertainties are large, but there are no systematic effects that will alter the overall distribution of pulsars. Distances for particular objects are also obtained by trigonometric parallax and by estimates of the distance to their binary companion or associated supernova remnant.

11.5.2 Thermal Plasma Dispersion
Microwave signal propagation through the interstellar electrons leads to three effects. The signals are dispersed as a result of the frequency-dependent index of refraction; pulses arrive later at long wavelengths than at short wavelengths. Birefringence is produced by the dependence of the index of refraction on magnetic field; the position angle of the pulsar signal's plane of linear polarization rotates with frequency (Faraday rotation). Diffraction and refraction of pulsar signals is produced by the turbulent structure of the interstellar plasma on the Fresnel-zone scale; the intensity scintillates and images are blurred. The compact size of pulsars, the short duration of their signals, and the stability of their rotation periods allow one to use the pulsar signals as tools to probe a variety of hitherto unknown properties of the interstellar medium.

The index of refraction for a dilute plasma threaded by a magnetic field is given by

$$n_{o,x}(v)^2 = 1 - \frac{v_p^2}{v^2 \left(1 \pm \frac{v_B}{v}\right)} \tag{11.3}$$

where $v_p^2 = N_e e^2/\pi m_e$ and $v_B = eB_\parallel/m_e c$. The plasma frequency v_p is 8980 Hz$\sqrt{N_e(\text{cm}^{-3})}$, and the cyclotron frequency v_B is 2.8 Hz $B_\parallel(\mu\text{gauss})$. The indices O and X and signs $+$ and $-$ correspond to the ordinary and extraordinary circularly polarized (right and left) modes of propagation in the dilute plasma, respectively. The subscript on the magnetic field indicates that the signal only interacts with the field component parallel to the propagation. Magnetic-field effects are ignored in the following discussion of dispersion. They are significant only for birefringence, which is discussed later.

The dispersive nature of the interstellar medium is seen by calculating the pulse arrival time as a function of frequency, $t_{N_e}(v)$. The arrival time is the integral of the inverse of the group velocity over the path. The group velocity nc is obtained from the derivative of the dispersion equation $\omega(k) = kc/n(\omega)$, $\omega = 2\pi v$:

$$t(v) = \frac{1}{c} \int_0^L dz \left[1 + \left(\frac{v_p}{v}\right)^2 \right] = \frac{L}{c} + t_{N_e}(v) . \tag{11.4}$$

$$t_{N_e}(v) = 4.148 \times 10^{15} \text{ s DM(pc cm}^{-3})v(\text{Hz})^{-2} . \tag{11.5}$$

The quantity DM is the dispersion measure, which is the column density of electrons along the path $\int N_e \, dl$ in units of pc cm^{-3}. A column of 10^{20} electrons results in DM of 30 pc cm^{-3} and a delay of 12 s for a signal at 100 MHz relative to infinite frequency. The pulsar data provide evidence for three components to the dispersion: (a) nearby, intervening HII regions; (b) a thin disk that is associated with ionization from hot, young stars; and (c) a halo whose scale height exceeds that of the pulsar population. These components are not well defined owing to the uncertainty in pulsar distances.

The spread of dispersion delays over a narrow band of signals b limits the resolution of pulse structure to a time τ_b that varies directly with the DM v^{-3};

$$\tau_b = 8.30 \text{ } \mu\text{s DM(pc cm}^{-3})b(\text{MHz})v^{-3}(\text{GHz}) . \tag{11.6}$$

This relation can be inverted to obtain the bandwidth b_τ required for a given resolution τ. The highest resolution for direct detection of power from a filter is obtained when the dispersion delay τ_b is equal to the impulse response of the filter $1/2b$. Higher resolution can be achieved only by sampling the complex voltage and removing the dispersive phase by convolution or Fourier transform techniques. This digital signal analysis is essential both to study pulse microstructure at low radio frequencies (Section 11.4) and to obtain high-resolution profiles of millisecond pulsars (Section 11.7).

11.5.3 Birefringence
The group delay for the two circularly polarized modes differs if the integral of $B_\parallel$ is nonzero. This difference is typically seven orders of magnitude less than t_{N_e} and is ignored in dispersion calculations. The phase-delay difference does produce observable effects for linearly polarized signals. A linearly polarized signal can be decomposed into opposing circularly polarized signals whose relative phase carries the position-angle information. The plane of polarization will rotate by an angle χ equal to the phase delay between modes $\omega(t_O - t_X)$. The Faraday rotation angle is derived from the path integrals of the inverse of the phase velocities, $c/n_{O,X}$:

$$\chi = \omega(t_O - t_X) = \frac{\pi}{cv^2} \int_0^L dz \, v_p^2 v_B \ . \tag{11.7}$$

$$\chi = \mathrm{RM}(\mathrm{rad}\ \mathrm{m}^{-2})\lambda(\mathrm{m})^2 \ . \tag{11.8}$$

$$\mathrm{RM} = 0.81\ \mathrm{rad}\ \mathrm{m}^{-2} \int_0^L dz(\mathrm{pc})N_e(\mathrm{cm}^{-3})B_{\parallel}(\mu\mathrm{gauss}) \ . \tag{11.9}$$

The parameter RM is the rotation measure. Rotation measures range from 0.1 to 1000 rad m^{-2}.

The ratio of rotation and dispersion measures gives a measure of the galactic magnetic field strength parallel to the signal path and averaged along the path with weighting by the electron density:

$$\langle B_{\parallel} \rangle = 1.232\ \mu\mathrm{gauss} \frac{\mathrm{RM}(\mathrm{rad}\ \mathrm{m}^{-2})}{\mathrm{DM}(\mathrm{pc}\ \mathrm{cm}^{-3})} \ . \tag{11.10}$$

The pulsar data provide unique measures of the strength and orientation of the local galactic magnetic field since they are distributed throughout the volume around the Sun. Figure 11.8 summarizes the global structure of magnetic fields derived from pulsar data. The systematic variation of the sign of the RMs with galactic longitude indicates the presence of a large-scale magnetic field in the solar neighborhood with strength of 2 μgauss directed toward longitude 90°. The random component has a comparable amplitude.

11.5.4 Secular Variations of Path Integrals
The path integral DM may vary with time owing to the large proper motion of pulsars and other secular variations. This effect can be estimated by considering the case of an intervening HII region with a radius of 10 pc and an electron density of 1 cm^{-3}. If the pulsar has a transverse motion of 100 km s^{-1}, then the DM will change at the rate of 10^{-4} pc cm^{-3} per year. This rate corresponds to an arrival-time change of 40 μs per year at 100 MHz. In most pulsars this change is too small to measure. However, the microsecond arrival-time accuracy of the 1.6-ms pulsar (Section 11.7) is sufficient to measure the effects of a variable DM, and such observations are now being recorded (Figure 11.13). While the HII-region example deals with a discrete object, variations in the DM could also arise from AU-scale turbulence in the interstellar medium. One pulsar with known DM variations is the Crab pulsar. The Crab DM fluctuations are probably the result of AU-scale turbulence within the nebula itself, rather than along path in the interstellar medium. The Crab DM also exhibits annual variations resulting from the solar-wind plasma. In other pulsars, correlation of microstructure (Section 11.4) over large bandwidths has been used to determine accurate dispersion measures; no variations have been reported.

Rotation measure variations have been detected in the Vela pulsar. No observations have been conducted to look for small-scale structure in neutral hydrogen as the line of sight moves rapidly through the interstellar medium.

11.5.5 Diffraction Caused by Interstellar Turbulence
The early pulsar observations showed that the signals were deeply modulated both in time with scales of minutes and in frequency with scales of 100 kHz. Figure 11.9

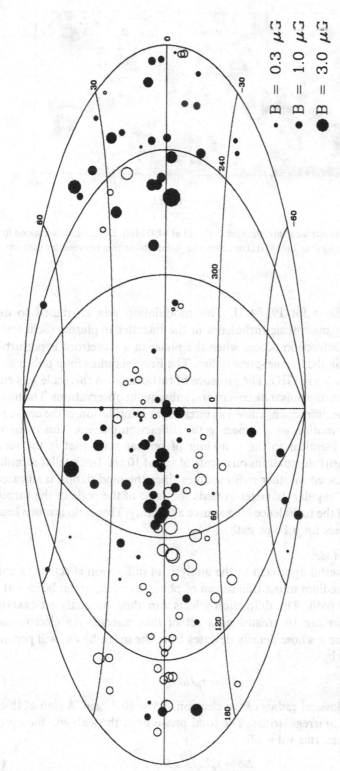

Fig. 11.8. Magnetic field components parallel to the line of sight of pulsars determined from rotation and dispersion measures. Only pulsars with dispersion measures less than 100 pc cm^{-3} are plotted; these objects have distance less than about 3 kpc. Positive (negative) fields are shown by filled (open) circles. The symbol size has a monotonic relationship to the field strength.

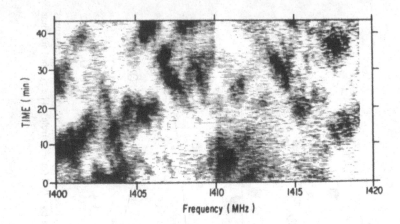

Fig. 11.9. Pattern of interstellar scintillation for $1937+21$ at 1400 MHz. Dark areas correspond to strong emission. A discontinuity near 1410 MHz comes from imperfect connection between two data sets. (From Bacher et al. 1982).

displays these effects for $1937+21$. This modulation was attributed to diffraction produced by microscale turbulence in the interstellar plasma (Scheuer 1968; Salpeter 1969). Diffraction occurs when the phase in a wavefront is perturbed by large values within the Fresnel-zone radius. The Fresnel radius for a pulsar at 1 kpc is $10^{11}\ v^{0.5}$ cm with v in GHz. The presence of turbulence on this scale was entirely unsuspected before the pulsar discovery and subsequent observations. The temporal modulation of the diffraction, known as interstellar scintillation, is the consequence of the observer's motion with respect to the diffraction pattern. This twinkling of radio signals is identical to the twinkling of optical stars that is the result of tropospheric density fluctuations on a scale of about 10 cm. Interstellar scintillation hat not been detected in other radio sources since the modulation is smeared out when the source angular diameter exceeds the ratio of the scale of the turbulence to the distance of the turbulence for a source at infinity. This ratio is much less than $2 \times 10^{-5}\ v^{+1}$ arcsec for a 1-kpc path.

(a) Thin-Screen Model
A naive, yet powerful approach to the analysis of diffraction effects is to collapse the perturbing medium into a thin screen of phase-changing irregularities at some point along the path. The diffraction effects can then be analyzed exactly. An extended medium can be treated as a set of thin screens. An electron-density irregularity of size a whose density deviates from the mean by δn_e will perturb the wavefront phase by

$$\delta\phi = r_e \lambda a \delta n_e \tag{11.11}$$

where r_e is the classical radius of the electron, 1.5×10^{-13} cm. A slab of thickness L will contain L/a irregularities. The total phase from the slab will then grow by random walk to an rms value of

$$\Delta\phi = r_e \lambda \sqrt{La}\,\delta n_e \ . \tag{11.12}$$

In the thin-screen analysis, an important role is played by the transverse length scale over which the rms random phase is one radian, l_ϕ. If we assume that $\Delta\phi$ is much greater than one radian,

$$l_\phi = \frac{a}{\Delta\phi} . \tag{11.13}$$

In fact, we can define l_ϕ without appealing to a scale size a simply by computing the transverse scale for which two paths will differ by a phase difference of one radian. The dependence of the phase difference on transverse scale is known as the phase structure function. A radian phase perturbation over a length l_ϕ will scatter the plane wave into a cone with width

$$\Theta_s = \frac{\lambda}{l_\phi} = r_e \lambda^2 \sqrt{\frac{L}{a}} \delta n_e . \tag{11.14}$$

This result, which is only slightly more sophisticated than dimensional analysis, tells us that a point source will appear enlarged to a size that depends on the wavelength squared and on the square root of the path length traversed in the turbulent interstellar plasma. Observations demonstrate that the wavelength dependence is closely followed. This is a reflection of the origin of the wavelength dependence in the plasma dispersion law. A typical scattering angle is 1 milliarcsecond at 1 GHz. The scattering angle for a scattering screen at a distance xD and source at distance D is $\theta_s = (1 - x)\Theta_s$; the $(1 - x)$ factor corrects for the spherical wavefront emanating from the source.

(b) Pulse Broadening
The scattering of pulsed radiation broadens a pulse when the group delay along the scattered path relative to the direct path exceeds the pulse width. Observations of the Crab pulsar radio source prior to the discovery of pulsars could not have detected the pulsed signal because the broadening below 100 MHz exceeds one pulse period. In the thin-screen approximation, the distribution of scattered radiation is assumed to be Gaussian in angle, and the resultant pulse is broadened by the convolution with a one-sided exponential with time scale τ_B:

$$\tau_B = x(1 - x)\Theta_s^2 \frac{D}{c} = x(1 - x)r_e^2 \lambda^4 L D \delta n_e^2 / ac . \tag{11.15}$$

Observations of pulsar broadening are summarized in Figure 11.10. The observed distance dependence is much stronger than expected from the above analysis. Evidently the medium is not uniformly filled with scattering material. The precise form of the convolving function is dependent on the distribution of scattering along the path. The finite thickness of any scattering medium will produce a convolving function which is not as discontinuous as the one-sided exponential.

(c) Scintillation
The arrival of two or more versions of a radio signal with typical time separation τ_B results in interference over a bandwidth B_d given by the Fourier transform relation $B_d = (2\pi\tau_B)^{-1}$. The interference in frequency is shown in Figure 11.9. The

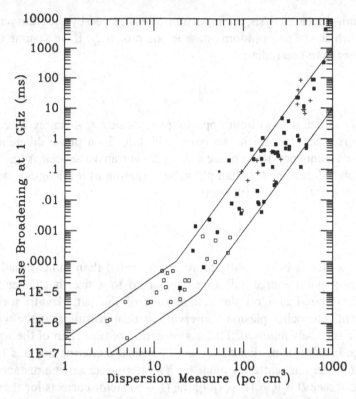

Fig. 11.10. Variation of the broadening of pulses by interstellar scattering with dispersion measure. Filled symbols show data obtained directly from pulse-broadening observations. Open squares are from measurements of the Fourier conjugate parameter, the correlation bandwidth. Plus symbols are upper limits from pulse-width observations.

distribution of intensities of the interference peaks is roughly exponential. The exponential distribution means that the most likely amplitude is zero. This severe modulation imposed on pulsar signals has been a nuisance for many observations, although it does provide a unique probe of the intervening medium.

The interference, or diffraction pattern, varies according to the motions of the source, the observer, and the medium. If nothing moved, then the signal would be modulated in frequency but would not change with time. In fact, everything moves. The pulsar has a proper motion transverse to the line of sight, and, for binaries, an orbital motion. The Earth has corresponding proper and orbital motions. Finally, the scattering irregularities are probably in shock fronts or other large-scale structures that have their own peculiar motions. The characteristic time scale for scintillation τ_d is the ratio of the diffraction-pattern scale size to the net transverse velocity of the signal path with respect to the scattering medium. The diffraction pattern of the thin screen has a spatial scale l_ϕ transverse to the path.

$$\tau_d = \frac{l_\phi}{V} = \frac{a^{0.5}}{r_e\lambda\sqrt{L\delta n_e V}}. \tag{11.16}$$

The modulation in Figure 11.9 can be analyzed to measure the quantities B_d and τ_d. Pulse broadening requires measurements over a wide range of frequencies to allow deconvolution of the intrinsic pulse width from the broadening function characterized by τ_B. Simultaneous measurement of the scintillation at two or more stations spaced across the globe can resolve the instantaneous diffraction pattern, and thereby measure the scale l_ϕ. High-resolution interferometry of pulsars with large scattering effects can provide measures of the scattering angle θ_d. Combined measurements of the scattering parameters for a single object are essential to constrain theoretical descriptions of the scattering medium.

(d) Scattering Measure
In all these expressions, a critical role is played by the product of the thickness of the scattering medium L and the ratio of the square of the density fluctuations to the scale size $\delta n_e^2/a$. This product is a scattering measure similar to the emission measure used in radio spectroscopy (Chapters 2 and 3). When the scattering measure is normalized by the thickness L, for which we can substitute D in the absence of other information, a quantity related to the power spectral density of electron-density fluctuations is obtained. Analysis of interstellar scattering can proceed assuming a power law spectrum of irregularities with results similar to those presented above. The power law index is constrained by observations to be near 4 (Cordes et al. 1985).

Comprehensive observations of the scattering parameters of any pulsar can provide an estimate for the magnitude of the transverse velocity of the pulsar. These easily obtainable velocity estimates can be used to study the correlation between velocity and other parameters. The errors in velocity estimates by the scattering technique are less sensitive to the distance of the pulsar than those from direct radio interferometry measurements (Section 11.6). In the case of a binary pulsar (Section 11.7), scattering measurements throughout the binary orbit can determine both the motion of the center of mass of the binary and the inclination of the orbit. This technique can be inverted to use the orbital velocity of the Earth to determine the vector components of the motion of a pulsar.

11.5.6 Refraction by Interstellar Turbulence
Plasma turbulence on a scale exceeding the Fresnel-zone size will produce refractive effects in signals that are correlated with the diffractive effects discussed above (Rickett 1986, Cordes et al. 1986, Blandford and Narayan 1985). Refraction is strongly dependent on the shape of the turbulence spectrum. The refractive effects result from large lenses of random shape which drift into the line of sight. The source intensity and apparent position will vary. The weak focusing and defocusing alters the diffractive parameters. Refraction can focus multiple ray paths from scales much larger than the Fresnel radius; this may provide sufficient resolution to distinguish between emitting regions of individual pulse components. Many investigations are in progress to unambiguously identify refractive effects. These include observations of extragalactic sources at long wavelengths since the refractive effects are not limited to minute diameter sources.

11.6 Timing and Astrometry

Precise measurements of both pulse arrival times and celestial coordinates of pulsars have many applications. Deviations of arrival times from a simple rotation model can both elucidate the nature of the torque acting on neutron stars and probe the internal structure of neutron stars. Dispersion of the arrival time with radio frequency measures the thermal-electron column density toward each star that provides a distance estimate. Both timing and interferometric measurements yield proper motions and trigonometric parallaxes that are essential to understanding the origin of pulsars. Measurements of arrival times and celestial coordinates of the pulsar ensemble provide precise data for uses other than pulsar astrophysics.

11.6.1 Time of Arrival Measurements

Arrival times are measured by first sampling the pulsar intensity received in a narrow band of radio frequencies and then averaging the samples synchronously with the apparent period. The typical sample spacing is 0.001 P, and the bandwidth is restricted so that dispersion smearing (Section 11.5) is less than 0.001 P. Data from many contiguous bands are combined for enhanced sensitivity. Orthogonal polarizations are summed to provide the total-intensity signal. The sample spacing is synchronized with the apparent period for each observation by computer predictions of the period of data and its Doppler offset at the observatory. The phase of the samples may also be set by a similar prediction. An atomic clock (e.g., rubidium) is required both as a frequency standard to maintain the synchronous sampling and as a phase standard to determine the epoch of the samples on the scale of Coordinated Universal Time (UTC). The fundamental data are then sequences of pulse profiles averaged over about one minute with a known UTC epoch of a sample during one pulse period at the midpoint of the average.

11.6.2 Time of Arrival Analysis

Comparison of one average with another requires determination of the epoch of some fiducial point of the pulse such as the pulse peak. This is accomplished by cross-correlation of each average with a standard template. A high-sensitivity average of the observed data is often used for a template. A least-squares fit to the peak of the cross-correlation products provides an offset relative to the UTC epoch of the average. The epoch plus offset reduces each observation to a single datum, the observatory UTC arrival time. A typical accuracy of this datum is about 0.001 period, 1 μs to 1 ms.

The periodic properties of the observatory arrival times can be studied only when variations of the observatory atomic clock and the effects of the Earth's motion are removed. The epoch of the observatory UTC scale requires correction first to a standard UTC via the GPS, or LORAN C, radio time-dissemination signals, and then to the uniform scale, International Atomic Time (TAI). The GPS, Global Positioning Satellite, service broadcasts near 1.5 GHz; the LORAN, Long-Range Navigation, signals are at 100 kHz and propagate as a ground wave over distances of 100 km or more. The UTC scale contains leap seconds to maintain rough consistency with solar time. A uniformly running clock on Earth such as TAI will not appear uniform to an external observer such as the pulsar. The scale of proper time on the Earth, TAI, is related to the scale of a clock in uniform circular motion about the Sun at 1 AU, Barycentric Dynamical Time (TDB) by the following expression.

$$\frac{d(\text{TAI})}{d(\text{TDB})} = 1 - \sum \frac{GM_i}{r_i c^2} - \frac{1}{2}\frac{v^2}{c^2} - \left\langle \sum \frac{GM_i}{r_i c^2} - \frac{1}{2}\frac{v^2}{c^2} \right\rangle \tag{11.17}$$

where the sum is taken over the masses M_i of all solar system bodies at distances r_i from the clock, v is the clock velocity with respect to the solar system barycenter, and the average $\langle\ \rangle$ is over a long time interval that is not officially defined. The first term is the net gravitational redshift and the second is a time dilation. The correction requires a numerical integration of this expression using the best solar system ephemeris. The elliptical orbit of the Earth leads to a 1.6-ms annual effect, while the Moon, which pulls the Earth in and out of the solar potential every 29.5 days, leads to a 0.5-μs effect. The variable relativistic delay of the pulsar signal as it traverses the gravitational potential of the Sun is also removed.

Next, the TDB observatory arrival time is corrected to solar system barycenter by taking the dot product between the current position of the pulsar and the position of the observatory interpolated from an ephemeris of the Earth's position. Groups at the Jet Propulsion Laboratory and at the Center for Astrophysics, Harvard University, continue to improve the precise ephemerides of the location and motion of solar system bodies. The accuracy of these Earth motion predictions is about 1 μs over a several-year interval. A final correction for the dispersion is made by computing the delay to infinite frequency at the true observing frequency (i.e., with the Earth's Doppler shift removed).

Barycentric arrival times are compared to a model for the pulsar's spin. The model typically contains polynomial terms for the rotational phase, rate, and acceleration. In some cases a fit is done including the third derivative of the phase. When estimates of the rotation parameters are sufficiently accurate, the pulse number of each arrival time is unambiguous and the data can be reduced to fractional period residuals. These residuals form the basis for improved estimation of the pulsar parameters. The eight basic parameters are: pulse epoch ϕ_0, rotation rate Ω_0, two derivatives of the rotation rate, $\dot{\Omega}_0$ and $\ddot{\Omega}_0$, right ascension α, declination δ, and proper motion in right ascension μ_α and declination μ_δ. The model for the residual arrival time $R(t)$ is:

$$\Omega R(t) = \Delta\phi_0 + \Delta\Omega_0(t - t_0) + \frac{1}{2}\Delta\dot{\Omega}_0(t - t_0)^2 + \frac{1}{6}\Delta\ddot{\Omega}_0(t - t_0)^3$$

$$+ \left[\Delta\alpha + \Delta\mu_\alpha(t - t_0)\right]\left[\frac{r\Omega}{c}\cos\delta_\oplus \cos\delta \sin(\alpha - \alpha_\oplus)\right]$$

$$+ \left[\Delta\delta + \Delta\mu_\delta(t - t_0)\right]\left[\frac{r\Omega}{c}(\cos\delta_\oplus \sin\delta \cos(\alpha - \alpha_\oplus) - \sin\delta_\oplus \cos\delta)\right] \tag{11.18}$$

where t_0 is an arbitrary reference time, and r, $\alpha_\oplus$, and $\delta_\oplus$ are the position and barycentric coordinates of the Earth. A fit for the dispersion measure requires

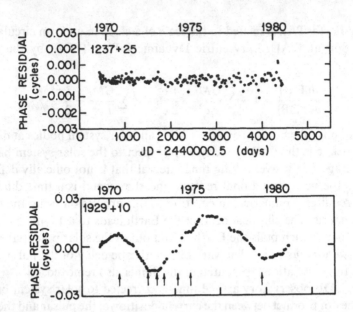

Fig. 11.11. Residual phase from timing observations of two pulsars. One cycle represents 1.382s for 1237+25 and 0.227s for 1929+10. The former object spins quietly, while the latter exhibits rotational noise.

multiple-frequency data. Binary pulsars require more parameters and will be discussed in the next section.

11.6.3 Interpretation of Pulsar Spin Properties

A number of theories for the secular deceleration of pulsar spins predict that the torque is proportional to a power of the rotation frequency:

$$L = I\dot{\Omega} = A\Omega^n \ . \tag{11.19}$$

The quantity n is called the braking index. Deceleration by gravitational radiation predicts $n = 5$ while loss by vacuum dipole radiation predicts $n = 3$. Measurement of $\dot{\Omega}$ along with lesser derivatives provides a solution for n:

$$n = \frac{\Omega\ddot{\Omega}}{\dot{\Omega}^2} \ . \tag{11.20}$$

The braking indices for pulsars 0531+21 and 0540−69 are in the range of 2.5 to 3.0. These values may result from dipole radiation by a centrifugally distorted magnetosphere coupled with possible action of internal torques between the crust and core.

The residuals from pulsars are characterized by random jitter, discontinuities (glitches), slow variations, and noise. Figure 11.11 displays residuals from quiet and noisy pulsars. The jitter arises from fluctuations in the individual pulse shapes that are attributed to an unstable coherent emission process (Section 11.4) and the limited sensitivity of the receiving system.

Glitches in pulse arrival times have led to the development of detailed models of the structure and dynamics of neutron stars. These events, now observed in seven pulsars, are characterized by fractional increases in the rotation rate of 10^{-6} to 10^{-9}. After a glitch, the spin deceleration increases such that the rate increment decays in roughly one year. This behavior suggests a two-component model for the neutron star. The two components are the neutron-rich superfluid core, which is strongly coupled both to nuclei in the crust and to the magnetosphere, and the superfluid component of the crust. The crustal superfluid is organized in angular momentum quanta called vortices. The vortices are forced out of their equilibrium rotation state by their interaction with the crystalline lattice of crustal nuclei. They are pinned to the nuclei and only drift slowly as a result of quantum tunneling. The interaction leads to a coupling between the two components. The crustal superfluid provides an acceleration torque on the star that is proportional to the product of its moment of inertia and the spindown rate. The glitch event is associated with an uncoupling of the two components when a disequilibrium threshold is reached. The sudden reduction in the moment of inertia of the star leads to a spinup under conservation of angular momentum. The loss of internal acceleration increases the rate of spin-down. On a time scale of weeks to months the coupling is reestablished. Vortex coupling theory predicts the surface temperature of a neutron star (Pines and Alpar 1985). Future pulsar timing observations and satellite-based X-ray detections of isolated neutron stars are essential to advance such detailed physical models of the internal structure of neutron stars.

Other slow systematic variations of the rotational phase of pulsars are called timing noise. In some objects the noise is characterized by a random walk in either phase, frequency, or acceleration (Cordes and Downs 1985). The strength of a random walk can be quantified by the product of the rate of events and the square of the event amplitude. The noise activity in pulsars appears to be correlated with the deceleration $\dot{\Omega}$ as expected in the vortex-line coupling theory. Minor events in coupling are expected to be the source of timing noise.

11.6.4 Astrometry

A comparison of pulsar positions derived from pulse timing and from radio interferometry yields a cross-check on the fundamental reference frames used in astronomy. The pulse timing reference frame is tied to the solar system ephemeris and therefore is an ecliptic system. Radio interferometry uses an equatorial system since it is based on the rotation of the Earth. The source of 0.5-arcsec discrepancies between the absolute coordinates of pulsars using the two techniques is an area of active investigation. Some of the difference results from the noise in pulsar timing observations and some is attributed to inaccuracies in the "old" set of astrometric parameters associated with the Besselian 1950 system relative to the "new" Julian 2000 system.

Both position determination techniques can be used to measure proper motions and trigonometric parallaxes. Proper motions for twenty-five pulsars are in agreement with the two techniques and indicate space velocities of 170 km s^{-1}. Pulse timing noise limits the accuracy of the timing results. The parallaxes of two pulsars have been measured using VLBI techniques: $0823+26$ (0.0028") and $0950+08$ (0.0079"). The parallax-derived distances of these two pulsars lead to electron density estimates consistent with those derived by other methods (Section 11.5).

These and future parallax measurements are essential to establish the distance scale of pulsars, and thereby improve our understanding of both the total population of pulsars and their rate of formation.

11.7 Binary and Millisecond Pulsars

11.7.1 The New Pulsars

In July 1974 a binary pulsar, 1913 + 16, was discovered at the Arecibo Observatory during a sensitive survey for new pulsars (Hulse and Taylor 1975). The discoverers found that the pulse period of this object was not stable in subsequent timing observations. The pattern of period changes soon revealed that the pulsar was in a highly elliptical orbit around a companion star with an orbital period of 7^h45^m and an ellipticity of 0.6. These parameters imply that the orbit of the pulsar ranges between 1 and 5 solar radii with respect to its invisible companion. The companion is another neutron star. The pulsar's rotation period of 0.059 s was the second shortest known at that time. This first binary pulsar led to many inquiries concerning the origin and evolution of binary pulsars. The effects of general relativity in this system were shown to be orders of magnitude larger than those in the solar system. The properties of 1913 + 16—short period, large spindown age of 2×10^8 yr, and absence of a supernova remnant—led to suggestions that this object was indeed old and that the short period was obtained by angular momentum transfer from the companion star long after the formation of the neutron star presently observed as the pulsar. In 1978 two new binary pulsars were detected with orbital periods of 1232 d and 1.03 d, and rotational periods of 0.864 s and 0.195 s, respectively.

In late 1982 radio pulsations with the remarkable period of 1.557 ms were observed in the midst of an enigmatic source complex called 4C21.53 (Backer et al. 1982). The lack of supernova remains and limits to its spindown rate indicated that this object, now known as 1937+21 or the 1.6-millisecond pulsar, was old despite its short period. The immediate suggestion was that it had achieved its fast spin by momentum transfer in a binary system as described above, but in this case the binary was disrupted after the transfer. Spinup in binaries became a more attractive hypothesis in 1983 when a second fast pulsar with the period of 6.1 ms was discovered in a very low eccentricity 117-d binary system (Boriakoff et al. 1983). Three other binaries were discovered in 1985 and 1986 during searches for new millisecond pulsars. In 1987 a 3.0-ms pulsar, 1821 − 24, was discovered in the globular cluster M28, and a 11.1-ms binary pulsar, 1620–26 was found in the globular cluster M4. The properties of the binary and millisecond pulsars are summarized in Table 11.1 and are explained below.

11.7.2 Timing a Binary Pulsar

The Doppler variations of a binary pulsar's rotation period provide data equivalent to that for a "single-line spectroscopic binary" in classical astronomy. Five elements of the system can be determined. The shape and orientation of the orbit are given by the projected semi-major axis $a_1 \sin i$, the ellipticity e and the longitude of periastron measured from the line of nodes ω. The location of the pulsar in the orbit

Table 11.1. Properties of binary and millisecond pulsars.

PSR	P (s)	$\dot{P} \times 10^{18}$ (s s^{-1})	P_b (d)	e	$a_1 \sin i$ (s)	$f(m_p)$ ($M_\odot$)	DM (pc cm^{-3})
0655+64	0.195671	0.63	1.029	<0.00005	4.12	0.0712	8.7
0820+02	0.864873	100.00	1232.340	0.0119	162.13	0.0030	23.7
1620−26	0.011076	—	205	—	—	—	62.
1821−24	0.003054	1.6	—	—	—	—	120.
1831−00	0.520947	<100.	1.811	<0.0050	0.72	0.0001	95.0
1855+09	0.005362	0.16	12.327	0.00002	9.23	0.0055	13.3
1913+16	0.059030	8.63	0.323	0.6171	2.34	0.1318	171.6
1937+21	0.001558	0.11	—	—	—	—	71.1
1953+29	0.006133	0.03	117.349	0.0003	31.44	0.0027	104.5
2303+46	1.066371	400.	12.339	0.6584	32.69	0.2463	61.

is given by the epoch of periastron passage T and the orbital period P_b. The mass function $f(m_1)$ given in Table 11.1 is defined as $(m_1 \sin i)^3/(m_1 + m_2)^2$, and is derived from $a_1 \sin i$ and P_b using Kepler's law. The mass function leads to a range of possible values for the pulsar companion mass m_2 given an assumed pulsar mass m_1 and orbital inclination i. The smaller mass functions in Table 11.1 suggest companion masses of 0.2 to 0.4 $M_\odot$. These companions are most likely white dwarfs. On the other hand, the larger mass functions are most consistent with one-solar-mass neutron stars.

The small size of the 1913+16 orbit leads to a variety of relativistic effects with amplitudes far exceeding those of any other astronomical system. The narrow pulse width, 7 ms, and the high stability of the pulsar rotation allow these effects to be measured with remarkable precision. Analysis of the idealized system of two orbiting point masses in the context of general relativity adds four binary elements to the system. The periastron precesses at an average rate of 4.2° yr^{-1}. This can be compared to an identical effect in the solar system, the contribution to the perihelion advance of Mercury resulting from general relativity, which is merely 0.43″ yr^{-1}. The transformation from pulsar proper time to TDB varies throughout the orbit through the combined effects of gravitational redshift and time dilation. This is the same effect discussed in Section 11.6 for the transformation from TAI to TDB, and has similar amplitude, 4 ms. The apparent pulse period decreases when closest to its companion. The third effect is the variable retardation of the pulsar signal as it escapes the gravitational potential of the companion star. This effect also occurs in the solar system when a pulsar signal traverses the solar gravitational potential field. The inclination of the binary orbital plane i can be derived from measurement of the retardation effect.

The fourth relativistic effect in a binary system is the decay of the orbital period from the reaction to the emission of gravitational waves by the system. Gravitational waves were predicted by Einstein for a system of masses with a time-varying quadrupole moment, but have never been observed by any Earth-bound detector. The binary pulsar acts as a indirect detector by displaying orbital period decay if there are no competing effects. In a few hundred million years, 1913+16 will spiral

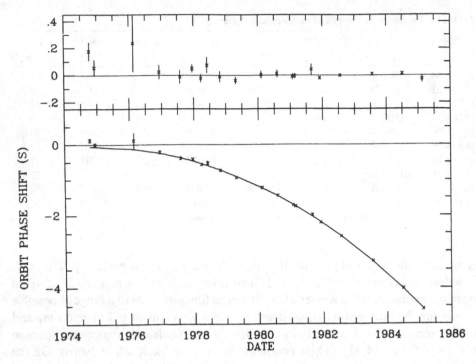

Fig. 11.12. Shift in orbital phase for 1913+16 as a result of emission of gravitational waves. As this binary pulsar loses energy the orbital period decreases and a quadratic phase error builds up. The lower plot displays a fit of the observations to the Einstein gravitational-wave emission model. The upper plot displays the residuals between the observations and the model.

in to a final cataclysm which may produce a massive neutron star or perhaps a black hole.

The first three relativistic effects are sufficient to determine the individual masses and the inclination which are indeterminate for the classical case. For 1913 + 16 the masses of the pulsar and its companion are remarkably close, 1.45 and 1.38 $M_\odot$. The proximity of these masses to the Chandrasekhar limiting mass for white dwarfs suggests critical role for the Chandrasekhar limit in the formation of neutron stars. The decay of the orbital period can be predicted from the derived parameters of the system and the concepts of general relativity:

$$\dot{P}_b = -\frac{96}{5} P_b \frac{1 + \frac{73}{24}e^2 + \frac{37}{96}e^4}{(1 - e^2)^{7/2}} \frac{G^3 m_1 m_2^5}{(m_1 + m_2)^3 a_1^4 c^5} . \tag{11.21}$$

The precise agreement between the observations of 1913 + 16 and this function (Figure 11.12; Weisberg and Taylor 1984) is the first and only experimental evidence for the existence of the gravitational waves that were predicted by Einstein.

The timing analysis discussed above shows that the orbit of 1913 + 16 is inclined by 45° to the line of sight. Consequently, one concludes that the magnetic axis is inclined by 45° relative to the rotation axis and that secular alignment of the magnetic axis is a small effect over the age of this pulsar.

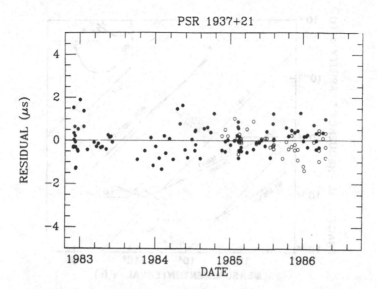

Fig. 11.13. Residual phase from timing observations 1937+21. The pulse period for this object is 1.556 ms. The filled (open) symbols are for 1400-MHz (2380-MHz) observations. Over a four-year interval the rotation of this star is extremely stable.

11.7.3 Timing a Millisecond Pulsar

The rapid rotation of 1937+21 leads to a centrifugal bulge that is approximately ten percent for most neutron-star models. Classical studies of rotating, uniform density, self-gravitating fluids show that as the body is spun faster the equilibrium figure progresses first through the Maclaurin series of ellipsoids with two axes of symmetry and then through the Jacobi series of ellipsoids with three axes of symmetry. This equilibrium series of ellipsoids provides a good approximation to the equilibrium figures of neutron stars since degenerate neutron matter has a nearly uniform density. Centrifugal forces would rip matter away from the equatorial region if the star is spinning faster than 1000 Hz. However, the time-varying quadrupole moment of inertia of the Jacobi ellipsoids leads to the radiation of gravitational waves that decelerate the rotation before the centrifugal limit is reached. The gravitational-wave damping leads to a maximum rotation rate for a neutron star that depends on the mass and the equation of state. Recent calculations predict upper limits in the range from 750 to 1000 Hz, just above the 642 Hz of 1937+21. If the fast rotation of 1937+21 is the result of spinup in a binary system, then that spinup may well have been limited by the radiation of gravitational radiation. Other pulsars with similar periods are likely. Magnetic dipole radiation acting over the age of the galaxy would lead to typical rotation rates several times smaller than the upper limit.

Timing observations of 1937+21 display a stability that is comparable to the best atomic time standards for intervals of a year or more. Figure 11.13 shows the microsecond-level residuals obtained over three years at the Arecibo Observatory (Davis et al. 1985). A comparison of these measurements of the rotation of 1937+21 with atomic time standards and other natural and constructed oscillators is shown

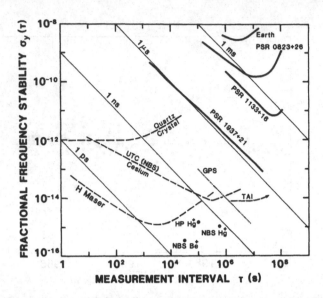

Fig. 11.14. Fractional frequency stability of earth and celestial timing systems. The ordinate is the root-mean-square difference between successive frequency measurements that have each been determined over a measurement interval T. Quartz and several atomic clocks are shown with dashed lines. The GPS time-transfer stability is given by the dotted line. Results from several experimental, trapped-ion clocks are shown with dots. The rotation stabilities of the earth and three pulsars are given by solid lines. (From Backer and Hellings 1986.)

in Figure 11.14. The statistic used is the fractional frequency stability, or Allan variance. The Allan variance measures fractional frequency changes determined over varying time-averaging intervals. The present data suggest that rotations of the pulsar may be as precise as TAI for intervals of one year. However, in order to compare these two clocks, corrections for the motion of the Earth (Section 11.6) need to be applied, and these may have microsecond errors over intervals of several years. A long-term comparison of pulsar data with atomic clocks will require a simultaneous solution for pulsar parameters and for solar-system dynamical constants such as the mass of Neptune.

The arrival times of pulsars will be perturbed by the presence of a background of gravitational waves. These waves may have been generated by chaotic processes such as the formation of galaxies and phase transitions in primordial matter during the early stages of the universe. The effect of a small-amplitude gravitational wave on electromagnetic signal propagation reduces to a time-varying index of refraction in Euclidean space. Consequently, precise pulsar timing can detect the presence of waves along the line of sight as an instability on a time scale comparable to the wavelength of the wave. Microsecond fluctuations of arrival time on time scales of one year would result from a background of waves with an energy density of 10^{-4} times the closure density, 10^{-29} gm cm^{-3}. Theoretical estimates are well below this limit. However, on longer time scales the expected effects are larger and improved pulsar observations may one day provide evidence to support or refute theories of the fundamental equations of physics that govern the evolution of our universe.

11.8 Radio Emission Mechanism

11.8.1 Basic Requirements

All models of the radio emission mechanism operating in pulsars are based on assumed properties of particles and fields in the vicinity of a rotating neutron star. There is, however, no self-consistent theory to describe pulsar electrodynamics. Nevertheless, the models do provide a mechanism for guiding both further theoretical development and further experimentation.

The salient features of the emission which any model must satisfy are:

1. extremely high brightness temperature;
2. stable mean pulse profile with one, two, or three components;
3. monotonic rotation of linear polarization vector with a signature that depends on the component structure;
4. slow decrease of the pulse width with increasing radio frequency, and with increasing period;
5. interpulse components located nearly halfway between main pulses and continuous emission at low level;
6. linear polarization angle modes spaced by 90° and circularly polarized emission;
7. memory in the pulse-to-pulse fluctuations which in some objects takes the form of periodically spaced subpulses drifting slowly with time;
8. erratic pulse-to-pulse amplitude fluctuations; and
9. micropulses within single pulses which in some objects are periodically spaced.

Unfortunately the detail in this list does not place strong constraints on the models since the radio emission is such a small part of the total energy loss.

In all models the high brightness temperature is attributed to the collective acceleration of charges with extremely high densities. The electromagnetic fields generated by such charges are coherent, or in phase. The emitted power is then proportional to the square of the number of charges; when motions of charges are independent, the power is proportional to the number. The high-brightness-temperature radio photons have an extreme ratio of $kT/h\nu$, 10^{28}. The channel within which this emission is generated must lack any degree of freedom other than the creation of radio photons. Otherwise, the energy would be emitted at much higher frequencies with $kT/h\nu$ near unity. Gold once summarized this condition with the comment, "There can be no dirt in the waveguide." The absence of coherent radio emission in X-ray-emitting binaries containing a neutron star is then readily explained by the many degrees of freedom offered by infalling matter from the secondary companion.

We can estimate the number of charges involved in collective motion, N, by the ratio of the observed brightness temperature, T_B, to the kinetic temperature of the (electron) charges in collective motion, $N\gamma mc^2/k$. At meter wavelengths this number is as large as 10^{19} for γ equal to 10^3 as suggested by micropulse beamwidths (Section 11.4). Densities in this plasma then reach 10^{13} cm^{-3} assuming a single clump. Masering processes are also possible wherein the same number of charges are periodically spaced with the period of the radiated wavelength.

11.8.2 Emission Site

Where are these charges? Gold (1968) suggested a location in the outer regions of the magnetosphere where the co-rotation speed approaches c. Radiation would occur primarily as a consequence of the centripetal acceleration arising from co-rotation, not from the charge motion in the frame of the star. The particle energy density derived above far exceeds the magnetic energy density near the light cylinder in dipole models for slowly rotating pulsars. This discrepancy is difficult to understand since pulsar emission is extremely stable both in average profiles and in periodic fluctuations. For this reason, there has been little development of light-cylinder models over the past decade.

The vacuum-dynamo model for a rotating, dipole magnetic field developed by Goldreich and Julian (1969) contains a large electric potential between the star and the interstellar medium, 10^{15} V, along field lines that cross the light cylinder. Radhakrishnan and Cooke (1969) suggested that the monotonic swing of linear polarization angle within a pulse results from charge bunches accelerated by this potential with emission vectors in the plane of curvature of the field lines. The origin on the star of the magnetic field lines with this accelerating potential is near the pole, so this model is termed polar cap. Sturrock (1971) proposed that the source of the charges was an electron-positron avalanche triggered by vacuum pair creation. Figure 11.15 illustrates this process. The remainder of this section will discuss developments of polar-cap models.

11.8.3 Polar-Cap Models

The electron-positron plasma generated near the polar cap travels at relativistic velocities and is constrained to follow magnetic field lines. Ruderman and Sutherland (1975) developed many possible relationships between the Sturrock model and the observations. Arons (1979) presents an alternative view of the relationship between his model of the pulsar magnetosphere and the prominent features of the observations. The net charge of the outflowing plasma is the source of the difficulty in developing a consistent model for the field and current structures.

The narrow and stable pulse profile results from the confinement of this flow to open-field lines and to relativistic beaming of the radiation in the forward direction. The coherent motion of charge bunches may result from a streaming instability in the outflow. The processes involved in the conversion of the free energy in this outflowing plasma-wave spectrum into vacuum electromagnetic waves are an area of continuing investigation. Components in the pulse profile are attributed to characteristic patterns in the outflow and the conversion efficiency around the magnetic dipole axis. A hollow cone of emission surrounding a centrally located pencil beam satisfies most observations (Section 11.4). The linear polarization signatures suggest strongly that observed electric fields are either parallel or perpendicular to the magnetic field in the vicinity of the dipole axis.

The formation of subpulse patterns has been a live issue in pulsar astrophysics. In the polarcap model there is no clear reason how one "spot" can remain active through many rotations of the star. A hot spot on the surface is a reasonable suggestion, but calculations suggest that the thermal conductivity is too high to isolate a small region. A further idea is that a backflow of current hits the surface

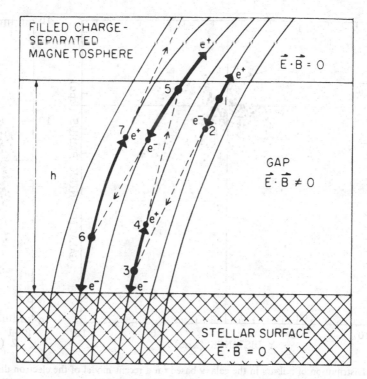

Fig. 11.15. Electron-position avalanche in the polar-cap model for pulsar emission. An electron and a positron are produced from a stray gamma ray. Acceleration of these charges leads to more gamma rays, and then to more pairs. (From Ruderman and Sutherland 1975.)

and thereby stabilizes an active region. The subpulse patterns may result from the slow evolution of the active region. The properties of subpulse patterns around pulse nulls (Lyne and Ashworth 1983) place strong constraints on the models.

11.9 Origin and Evolution

11.9.1 Introduction

The detection of a large number of pulsars in our galaxy has provided us with concrete data on one final product of stellar evolution, the neutron stars. The first goal of this section is to specify this population of galactic objects: where are the pulsars located, what are their kinematic properties, and how do they evolve with time? The basic understanding of the pulsar population then allows a discussion of the relation between these stellar remnants and their progenitors, both post-main-sequence objects, such as supernova remnants and X-ray-emitting objects, and then main-sequence stars themselves. General features of the galactic population of pulsars have reached some stability in current literature (e.g., Lyne et al. 1985). The connection between pulsars and their main-sequence progenitors is poorly understood. The difficulty in making this connection comes both from the limited ability to follow stellar evolution models to their final state as a collapsed, or

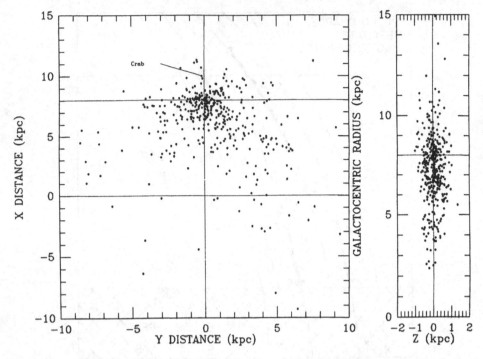

Fig. 11.16. Distribution of pulsars in the galaxy based on a recent model of the electron distribution. The Sun is assumed to be at a galactocentric distance of 8 kpc (X).

exploded, object as well as the multiplicity of paths that stars may follow to reach the neutron-star state. The latter complication arises from the possibility of mass transfer in binary systems.

11.9.2 Galactic Distribution
The distribution of known pulsars in galactic coordinates (Figure 11.1) shows that pulsars are associated with the galactic plane. Dispersion measures yield distance estimates for the full body of pulsars based on a model of the electron-density distribution. The set of pulsars whose distances are known from other means constrain the electron-density model. These other distance estimates are from direct trigonometric parallax observations, from pulsars associated with supernovae, globular clusters or white-dwarf campanion stars, and, most importantly, from neutral-hydrogen absorption observations. Recent studies conclude that the average electron density is 0.04 cm^{-3} in the plane of the galaxy near the Sun. The density decreases to 0.025 cm^{-3} above the plane with a scale height of 70 pc, and increases toward the interior of the galaxy. The increase in the inner galaxy is related to the increase of matter and ionizing radiation at galactocentric radii near 4 kpc. Figure 11.16 displays 440 pulsars in two projections of galactic coordinates. One immediately can see the planar distribution of the known pulsars with a Z-scale height of a few hundred parsecs and the concentration of the population within the solar circle. These are real effects and not the consequence of observational selection effects.

The model for the dispersing electrons is further constrained by proper-motion observations. Lyne et al. (1982) find that all distant high-latitude pulsars are leaving the plane. This key observation suggests strongly that pulsars are born near the plane, receive momentum impulses corresponding to velocities of 170 km s^{-1} at birth, and die in a few million years. The lifetime estimate is the observed scale height divided by the vertical velocity component. The absence of downward motions places an upper limit to pulsar lifetimes of roughly 10^7 years since the lifetime must be a fraction of the oscillation period about the galactic plane, 10^8 years.

Further observations of proper motions, trigonometric parallaxes, and neutral-hydrogen absorption for pulsars can be combined with developments in galactic dynamics to produce consistent models of the pulsar and electron-density distributions. New surveys for distant pulsars will play a pivotal role in this study.

Careful analysis of the galactic distribution of pulsars leads to an estimated population of pulsars which numbers 70,000 (Lyne et al. 1985). This population is restricted to objects whose radio luminosity at 400 MHz is above 0.3 mJy for a distance of 1 kpc. In addition to the objects enumerated above, there is an unknown population of pulsars with their beams directed away from the Earth. This beaming factor can be inferred from the properties of the beam cross sections which we do observe (Section 11.3). Uncertainties in such inferences allow the factor to range from 0.5 to 0.1. If most pulsars live for three million years, then the estimates discussed above lead to a pulsar birth rate of one per four to twenty years. Detailed studies that consider selection effects revise the birth rate downward to about one per fifty years.

11.9.3 Period Evolution

The evolution of pulsars can be summarized in a log-log diagram relating the pulse period with its first derivative (Figure 11.17). The ratio of these parameters is the age estimate based on spindown. Objects in the upper left are therefore the youngest, and this is confirmed by the location of the Crab and Vela pulsars in this region. Both pulsars have supernova remains surrounding them with ages less than 10,000 years. The absence of supernova remains around other pulsars likewise confirms that their ages are greater than the 10^5 years required to dissipate completely the supernova remains. If all pulsars were born with periods between 1 and 100 ms and if they evolved by magnetic dipole braking with constant magnetic moment, then we would expect a pileup of pulsars at long periods. The limiting period would indicate the lifetime of the pulsating radio source phase of the neutron star. The indication from the discussion above is that this phase lasts a few million years, and therefore the pileup should occur at periods of a few seconds. After that time one assumes that the radio luminosity either has faded away or has stopped abruptly. The distribution in Figure 11.17 certainly does show a maximum period that observers are convinced is not the result of selection. However, the distribution does not conform to expectations based on formation of Crab- or Vela-like pulsars and period evolution by magnetic braking with constant dipole moment, even if the luminosity is allowed to vary with an arbitrary function of period and period derivative. There are too many objects with small derivatives. A more complex evolutionary model is required.

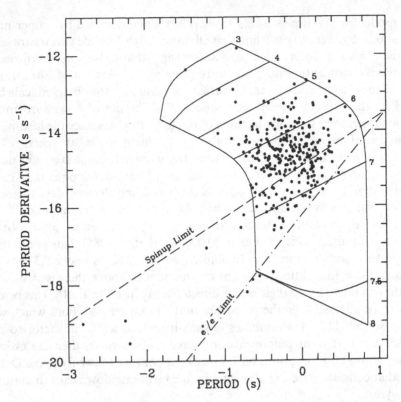

Fig. 11.17. Distribution of pulsars in period and period derivative. The evolution of pulsars described in the text is shown by two extrema: one for low original magnetic torque and short asymptotic period, and one for high torque and long period. Isochrones across the pulsar distribution are labeled with the log of time in years. A residual torque decay leads to all pulsars evolving toward a point near the label 8. Limits for electron-positron production in some models and for spinup by mass accretion from a companion star are given by dashed lines.

The solid lines in Figure 11.17 present an evolutionary model that explains most of the observed pulsar population. In this model, the perpendicular dipole magnetic moment decays with an exponential time scale of six million years. A residual moment is reached after 10^8 years. The decay presumably results from resistive decay of field in the neutron-star crust, rather than mere alignment of the dipole and spin axes. An evolutionary corridor in Figure 11.17 is defined by lines that follow weak-magnetic-moment objects and strong-moment objects whose asymptotic periods are 0.25 s and 7.0 s, respectively. An initial period of 10 ms is assumed. Isochrones are drawn across the corridor at logarithmic intervals. Evolution from 10^3 to 10^6 years is with constant field, and both the period and period derivative decrease. Later evolution is with constant period as the exponentially decaying, magnetic dipole moment reduces the derivative. If there is a residual moment, then all pulsars evolve to the lower right-hand corner of the diagram. Evolution to longer periods is extremely slow. The requirement for luminosity evolution is clear in the context of the model presented in Figure 11.17.

Without luminosity evolution each age decade in the corridor would have ten times more objects than the previous decade. This scaling clearly stops between one and ten million years.

Many authors have suggested that pulsars will become radio-quiet neutron stars when the rotation-induced potential along open-field lines falls below the 10^{12} volts required to make electron-positron pairs. In the simplest dipole models, this potential is proportional to the ratio of the surface field to the square of the period (Sturrock 1970) which leads to a line of slope 3 in Figure 11.17. Various models of the electron-positron production yield slopes with a range from 2.3 to 3.0 (Barnard and Arons 1982). The locus of this pulsar "death" line in Figure 11.17 has been adjusted within theoretical constraints to match the cutoff in the observations. While the data do suggest a pulsar cutoff, the main factor that reduces the number of detectable pulsars along the outlined corridor in Figure 11.17 is the diminishing luminosity.

11.9.4 Binary Pulsar Evolution

The short-period objects outside the evolutionary path discussed above are probably the result of spinup in the late stages of binary evolution (van den Heuvel 1984). Binary systems allow for multiplicity of evolutionary paths as a result of a range of masses, mass ratios, and orbital parameters. Extremes of low- and high-mass systems are discussed below.

In a low-mass binary, first the primary evolves to a white dwarf and then the secondary evolves to a mass-shedding red giant. The upper mass limit for evolution of a main-sequence star into a white dwarf is around 8 $M_\odot$, and the lower limit is around 1 $M_\odot$ since stars with less mass do not evolve off the main sequence in a Hubble time. The white dwarf can form a neutron star if it captures sufficient mass to exceed the Chandrasekhar limit, 1.4 $M_\odot$. A Type I supernova accompanies this transformation, but little mass is lost from the system and the binary system can remain bound. If the mass-shedding phase of the secondary continues for 10^8 years, the neutron star will evolve first toward long periods while its surface field is strong, and then to shorter periods as its field decays. The spinup proceeds along the spinup line in Figure 11.17 as matter is accreted from a disk surrounding the neutron star. The radius at which the neutron star is coupled to the accretion disk, the Alfven radius, is defined by the balance between magnetic and matter pressures. The star is spunup to the Keplerian speed at this Alfven radius. The Alfven radius can be related to the surface field B and stellar radius R if a dipole geometry is assumed. The equilibrium period is

$$P_{eq} = 2.4 \text{ ms } B_9^{6/7} R_6^{18/7} M^{-5/7} \dot{M}^{-3/7} , \tag{11.22}$$

where B is in units of 10^9 gauss, R is in units of 10^6 cm, M is the neutron-star mass in solar units, and $\dot{M}$ is the mass accretion rate in units of the maximum allowable rate, the Eddington limit. When accretion stops, the neutron star can turn on as a pulsar. The secondary star cools to become a white dwarf. Measurement of the temperature of pulsar white-dwarf companions provides an independent age estimate since the neutron star formed since the cooling curve of white dwarfs is known independently (Kulkarni 1986). The orbit is circular owing to the long development of the system in the mass-accreting phase. The proper motion expected for this system is comparable to that of the old disk population, 20 km s^{-1}.

The evolution of a massive binary can follow two initial paths when the primary star makes its transition to a neutron star in a Type II supernova event. The binary may be disrupted, or it may remain bound. The members of a disrupted binary continue their evolution independently. Both the pulsar and the main-sequence secondary star are expected to have large proper motions indicative of the large binding energy of the parent binary. Massive stars that appear to be running away from nearby OB associations may be the result of this evolution. If the system remains bound, then subsequent evolution has many possible paths depending on the mass ratio and period. Models show that mass transfer to the neutron star can tighten the orbit until the neutron star and the core of the evolving secondary are rotating inside the common envelope of the secondary. Secondaries with initial mass exceeding 8 $M_\odot$ will produce cores that will collapse to produce a second neutron star in the system and a second supernova event. If the second supernova occurs after the field has decayed on the primary neutron star, the primary may be spunup to short periods. If the resulting NS-NS system remains bound, a binary pulsar will be produced. The eccentricity of this system will be large as a result of the final supernova explosion. Pulsars $1913 + 16$ and $2303 + 46$ were probably formed in this manner, the former with significant spinup and the latter without.

Discovery of new pulsar binary systems will help specify the paths that stars can evolve. Observations of interacting binaries at all wavelengths need to be integrated to obtain a comprehensive view of the "death of stars." The X-ray-emitting, interacting binaries catch systems in a state that is intermediate between the main sequence and the collapsed state of pulsars. Further theoretical studies of the evolution of binary systems are also necessary to understand how stars pass from contained main-sequence stars to degenerate embers and expelled remnants.

Recommended Reading

Backer, D.C., and R.W. Hellings. 1986. Annu. Rev. Astron. Astrophys. **24**:537.

Helfand, D.J., and J.-H. Huang. 1987. The Origin and Evolution of Neutron Stars, Int. Astron. Union Symp. 125. Dordrecht: Reidel.

Kennel, C.F. 1979. NASA/JPL Workshop on the Physics of Planetary and Astrophysical Magnetospheres, Astrophys. Space Sci. **24**:1.

Manchester, R.N., and J.H. Taylor. 1977. Pulsars. San Francisco: W.H. Freeman.

Michel, F.C. 1982. Rev. Mod. Phys. **54**:1.

Radhakrishnan, V. 1982. Contemp. Phys. **23**:207.

Reynolds, S.P., and D.R. Stinebring. 1984. Birth and Evolution of Neutron Stars: Issues Raised by Millisecond Pulsars. Green Bank, West Virginia: National Radio Astronomy Observatory.

Rickett, B.J. 1977. Annu. Rev. Astron. Astrophys. **15**:479.

Shapiro, S.L., and S.A. Teukolsky. Black Holes, White Dwarfs, and Neutron Stars. New York: John Wiley and Sons.

Sieber, W., and R. Wielebinski. 1981. Pulsars, Int. Astron. Union Symp. 95. Dordrecht: Reidel.

Taylor, J.H., and D.R. Stinebring. 1986. Annu. Rev. Astron. Astrophys. **24**:285.

References

Arons, J. 1979. Space Sci. Rev. **24**:437.

Backer, D.C. 1976. Astrophys. J. **209**:895.

Backer, D.C., S.R. Kulkarni, C. Heiles, M.M. Davis, and M.M. Goss. 1982. Nature **300**:615.
Barnard, J.J., and J. Arons. 1982. Astrophys. J. **254**:713.
Bartel, N., W. Sieber, and D.A. Graham. 1980. Astron. Astrophys. **87**:282.
Blandford, R., and R. Narayan. 1985. Mon. Not. R. Astron. Soc. **213**:591.
Boriakoff, V., R. Buccheri, and F. Fauci. 1983. Nature **304**:417.
Cordes, J.M., and G.A. Downs. 1985. Astrophys. J. Suppl. **59**:343.
Cordes, J.M., J.M. Weisberg, and V. Boriakoff. 1985. Astrophys. J. **288**:221.
Cordes, J.M., A. Pidwerbetsky, and R.V.E. Lovelace. 1986. Astrophys. J. **310**:737.
Davis, M.M., J.H. Taylor, J.M. Weisberg, and D.C. Backer. 1985. Nature **315**:547.
Deich, W.T.S., J.M. Cordes, T.H. Hankins, and J.M. Rankin. 1986. Astrophys. J. **300**:540.
Gold, T. 1968. Nature **218**:731.
Goldreich, P., and W.H. Julian. 1969. Astrophys. J. **157**:869.
Gunn, J.E., and J.P. Ostriker. 1970. Astrophys. J. **160**:979.
Hankins, T.H.H., and L.A. Fowler. 1986. Astrophys. J. **304**:256.
Helfand, D.T., R.N. Manchester, and J.H. Taylor. 1975. Astrophys. J. **198**:661.
Hewish, A., S.J. Bell, J.D.H. Pilkington, P.F. Scott, and R.A. Collins. 1968. Nature **217**:709.
Hulse, R.A., and J.H. Taylor. 1975. Astrophys. J. **195**:L51.
Kulkarni, S.R. 1986. Astrophys. J. (Lett.) **306**:L85.
Lyne, A.G., and M. Ashworth. 1983. Mon. Not. R. Astron. Soc. **204**:519.
Lyne, A.G., B. Anderson, and M.J. Salter. 1982. Mon. Not. R. Astron. Soc. **201**:503.
Lyne, A.G., Manchester, R.N., and Taylor, J.H. 1985. Mon. Not. R. Astron. Soc. **213**:613.
Narayan, R. 1984. In S.P. Reynolds and D.R. Stinebring (eds.), Birth and Evolution of Neutron Stars: Issues Raised by Millisecond Pulsars. Green Bank, West Virginia: National Radio Astronomy Observatory, p. 279.
Pacini, F. 1967. Nature **216**:567.
Perry, T.E., and A.G. Lyne. 1985. Mon. Not. R. Astron. Soc. **212**:489.
Pines, D., and M.A. Alpar. 1985. Nature **316**:27.
Radhakrishnan, V., and D.J. Cooke. 1969. Astrophys. Lett. **3**:225.
Rankin, J.M. 1983. Astrophys. J. **274**:333.
Rickett, B.J. 1986. Astrophys. J. **307**:564.
Ruderman, M.A., and P.G. Sutherland. 1975. Astrophys. J. **196**:51.
Salpeter, E.E. 1969. Nature **221**:31.
Scheuer, P.A.G. 1968. Nature **218**:920.
Sturrock, P.A. 1971. Astrophys. J. **164**:529.
van den Heuvel, E.P.J. 1984. J. Astrophys. Astron. **5**:209.
Weisberg, J.M., and J.H. Taylor. 1984. Phys. Rev. Lett. **52**:1348.

12. Extragalactic Neutral Hydrogen

RICCARDO GIOVANELLI and MARTHA P. HAYNES

12.1 Introduction

The 21-cm hyperfine transition of neutral hydrogen was first detected in the Milky Way in 1951 by Ewen and Purcell. Two years later (Kerr and Hindman 1953), the HI emission of the Magellanic Clouds was observed from Australia. Largely as a result of the pioneering efforts of M.S. Roberts, by the time of his comprehensive review (Roberts 1975; the reader is referred to this source for the early development of the field), HI in about 140 extragalactic objects had been detected. Since then, mostly with the Green Bank 91-m, the Nançay, the Effelsberg, and more recently the Arecibo telescopes, the line has been observed in emission in thousands of galaxies, to distances 10^4 times greater than that of the Magellanic Clouds, and in absorption out to redshifts greater than 2. For the first twenty years, single-dish telescopes provided the vast majority of the HI data. By the mid-seventies, the techniques of spectral line aperture synthesis, developed especially at Cambridge, Green Bank, and Owens Valley, bloomed in full maturity at Westerbork and more recently at the VLA. The contributions of single-dish instruments have, however, remained important, especially for survey-oriented projects that require high-sensitivity observations of many noncontiguous regions of sky, and for pure detection experiments.

12.1.1 The Role of Neutral Hydrogen in Galaxies

Neutral-hydrogen concentrations, or clouds, provide the starting point for the collapse of matter into stars. Abundant HI in a galaxy indicates potentially active star formation processes, while lack of HI is a guarantee of a barren galaxy and an ineluctably aging stellar population. Thus, from an evolutionary viewpoint, the role of HI is of primary importance. Dynamically, that is, in terms of the fractional contribution to the mass, the perspective is different and, as discussed in detail in Section 12.5, quite dependent on morphological type. The Milky Way provides a reference point. Our Galaxy is an intermediate- to late-type spiral, perhaps an Sbc. The optical surface brightness of the disk decreases exponentially with distance from the center, with one-half of the spatially integrated luminosity falling within an effective radius of 4 or 5 kpc; the luminosity of the entire disk is about $1.1 \times 10^{10} L_\odot$, twice that of the nuclear bulge. Although ionized gas is found in the halo, the bulk of the interstellar gas is in the disk; similarly to the light, but with the exception of the very central regions, the total hydrogen gas $(HI + HII + H_2)$ component mimics an exponential disk, with an effective radius close to that of the optical emission. While molecular hydrogen is the dominant constituent of the interstellar medium within the solar circle, atomic hydrogen predominates towards the periphery of the disk. In fact, HI extends farther out than any other galactic disk tracer. By mass, the atomic hydrogen resides mostly in diffuse clouds and in the envelopes surrounding molecular clouds. The thickness of the HI disk is nearly constant within the inner 10 to 12 kpc, the averaged density falling to half its peak value at the plane at about 150 to 200 pc from it. A density cross section perpendicular to the plane does, however, show weak, broad wings, associated with a warmer, less clumped HI component of the interstellar medium, more than twice as thick as the main disk component. The total amount of interstellar hydrogen in the inner 16 kpc of the Galaxy is about $6 \times 10^9 \, M_\odot$, with roughly half in molecular and half in atomic form. While the atomic gas extends well outside that radius, the contribution of the gas (as well as that of all luminous matter) to the total mass decreases sharply with distance, and probably the total gaseous mass of the Galaxy does not exceed $10^{10} \, M_\odot$. As the luminous mass fades rapidly in the outer regions of the Galaxy, the dynamical mass traced by the sytem of globular clusters and by the dwarf spheroidal satellites of the Milky Way continues to grow almost linearly with distance from the center. Within 16 kpc from the center, this mass is $1.5 \times 10^{11} \, M_\odot$, rising to about $10^{12} \, M_\odot$ inside 100 kpc. Atomic hydrogen thus represents less than 10% of the total mass within the easily discernible stellar disk, and about 1% of the total mass out to 100 kpc. Its importance relative to other constituents grows for galaxies of later morphological type than our own; within the region mapped by the HI in some extremely optically-faint blue galaxies, HI can be the principal constituent of the visible mass. By contrast, very early spiral and lenticular galaxies frequently have less than 0.01% of their mass in the form of neutral hydrogen gas, ellipticals sometimes even less.

12.1.2 The Information in the 21-cm Line

Although in most cases HI is just a tracer component of the makeup of a galaxy, it constitutes an important tool in the study of the structure and dynamics of galaxy evolution. Notable reasons for the popularity of the line as such a tool are the ease of detection and the relative reliability of the column densities and other physical parameters inferred from the line profiles.

The 21-cm line results from a magnetic dipole transition between the two levels that characterize the ground state of the hydrogen atom. Depending on whether the electronic and nuclear spin vectors are parallel or antiparallel, a slightly higher (triplet) or lower (singlet) hyperfine level is obtained. The statistical equilibrium between the populations of the singlet and triplet levels can be expressed by means of the Boltzmann equation, in which the effective temperature that regulates the relative populations is referred to as the "spin temperature," T_s. Several processes contribute to T_s; the most important, as originally described by Field (1959), are: (i) absorption of and emission stimulated by photons of an existing radiation field; (ii) collisions of H atoms with other particles, especially with other H atoms and with electrons; (iii) "pumping" by Lyman α photons. As an example of the last process, a H atom in the singlet state of $n = 1$ can be excited to the $n = 2$ state by a Lyman α photon; subsequent deexcitation to the $n = 1$ level will leave the H atom in the triplet level in a fraction of the cases. Spontaneous transitions from the triplet to the singlet level occur with a transition probability $A_{10} = 2.868 \times 10^{-15}$ s^{-1}; in most cases of interest, however, a H atom will not be left alone for as long as eleven million years, and so T_s is largely determined by the above-mentioned extrinsic influences.

A cloud of optical depth τ and spin temperature T_s, bathed in a radiation field of temperature T_r, will yield a brightness temperature

$$T_b(v) = T_r e^{-\tau(v)} + T_s[1 - e^{-\tau(v)}] \qquad (12.1)$$

where the optical depth $\tau(v)$ varies with velocity v across the line profile. The optical depth $\tau(v)$ can be expressed as the ratio of the HI column density to the spin temperature (see Spitzer 1978). If one assumes that the gas is optically thin (i.e., $\tau \ll 1$), the column density of HI can then be obtained as

$$N_H = 1.82 \times 10^{13} \int \left(\frac{T_s}{T_s - T_r}\right) T_{bl}(v)\, dv \qquad \text{atoms cm}^{-2} \qquad (12.2)$$

where $T_{bl}(v)$ is the brightness temperature of the line profile at velocity v, in cm s^{-1}, above the continuum or baseline level, and the integral is over the line profile. Over most of the HI mass of a galactic disk, collisions determine the population of the two $n = 1$ hyperfine levels, and T_s closely approaches the kinetic temperature of the gas. Since this is usually much larger than T_r, the term in square brackets in Equation (12.2) is about unity. In the outer regions of the disks, and in envelopes of tidal debris after galaxy-galaxy interactions, where gas densities are low and collisions rare, T_s will be regulated by the radiation field at 21 cm and at 1216 Å (Lyman α). The former will be dominated by the cosmic background radiation at 2.7 K; in the absence of a sufficient ionizing flux, the term $[T_s/(T_s - T_r)]$ could be significantly larger than 1, and the determination of N_H would require a knowledge of T_s, which is difficult to obtain. Watson and Deguchi (1984) have shown, however, that the Lyman α flux in intergalactic space may be large enough that Lyman α pumping regulates the populations of the hyperfine levels and tends to push T_s to values well above 2.7 K, thus suggesting that even in very-low-density clouds outside of disks, $[T_s/(T_s - T_r)]$ remains close to 1. A simple integral across the line then will give a close estimate of the atomic hydrogen column density. It should be stressed again that this conclusion rests on the assumption of a sufficient value for a relatively uncertain quantity: a pervasive intergalactic ultraviolet flux. If current estimates of that quantity were to be found too high, column densities of HI in the outskirts of galaxies and in diffuse clouds outside of disks would have to be revised.

We address next the validity of the assumption of optical transparency. That assumption is invalid in a number of cases: dense regions of the interstellar medium, whence originate absorption lines, are opaque. The integrated line profile of a galaxy results from the collective emission—and absorption—by all its HI. In order to correctly convert the line integral into a total HI mass, we need to know the fraction of the HI gas in the disk which is found in optically thick regions. That fraction in turn depends on the intrinsic properties of the gas distribution in the galaxy and on the geometrical perspective with which it appears to us. The practical approach to this problem consists in computing column densities in the optically thin approximation and then applying a correction that accounts for self-absorption.

The self-absorption correction can be estimated (a) from statistical studies on large samples of galaxies, covering a broad range of disk inclinations and morphological and possibly luminosity class, (b) from HI absorption line studies in our Galaxy, or (c) by modeling the distribution of cold gas in the interstellar medium. A detailed description of these methods is given in Appendix B of Haynes and Giovanelli (1984). The results indicate that when a disk like that of our Galaxy is seen face-on, the assumption of optical transparency leads us to underestimate the column density by a small amount, probably less than five percent. When disks are seen close to edge-on, however, substantial corrections to the column densities and masses inferred using Equation (12.2) become necessary. The corrections appear to be larger for intermediate-type spirals. It is estimated that for the majority of galaxies, such column density corrections amount to factors less than 1.5.

If the distance is known, the total HI mass of a galaxy can be obtained by integrating Equation (12.2) over the effective disk area (i.e., the approximate ellipse obtained by the inclined view of the disk with respect to the line of sight), after correcting for self-absorption. In the case of single-dish observations where the galaxy is unresolved by the beam, Equation (12.2) yields simply

$$M_H = 2.356 \times 10^5 \, d^2 \int S_c(v) \, dv \; M_\odot \tag{12.3}$$

where d is the distance in Mpc and $\int S_c(v) \, dv$ is the integral, over the line, of the flux density corrected for self-absorption and other instrumental biases, with $S_c(v)$ expressed in Jy and v in km s^{-1} [see Roberts (1975) for details of the derivation of Equation (12.3)].

Standard radio spectroscopy techniques allow for spectral resolutions better than 0.1 km s^{-1} at 21 cm. Extragalactic observations are however preferentially done with broad bandwidths, covering typically 2000, 4000, or 8000 km s^{-1}; spectral resolutions are then more typically on the order of 5 to 20 km s^{-1}, although they can be much better than that when the signal strength so permits. The fine spectral resolution achievable in the HI line ordinarily allows very precise measurements of radial and rotational velocities of large samples of galaxies. In face-on systems, accurate measurements of the velocity dispersion of the interstellar gas perpendicular to the plane of the disk are possible. In inclined systems, the determination of the disk velocity field allows a detailed study of the large-scale galaxian dynamical characteristics. The maximum rotational velocity, as indicated, for example, by the half-width of a galaxy's integrated line profile, has been found to be a good luminosity indicator, thus providing a powerful method for the determination of extragalactic distances and hence the expansion rate of the universe.

The HI content of a galaxy has been shown to be vulnerable to environmentally driven gas removal mechanisms, and may thus be used to probe the effects of the latter on galaxy evolution. The fragile outer layers of the disk are easily disrupted by close encounters of galaxies, especially of those located in small groups, and HI observations have provided spectacular data on this sort of perturbation. As for the physics of disks, HI observations provide valuable insights on the dynamic and thermal equilibrium of the interstellar medium.

In addition, the combination of large apertures and modern receivers has made the execution of large-scale redshift surveys in the 21-cm line a practical endeavor.

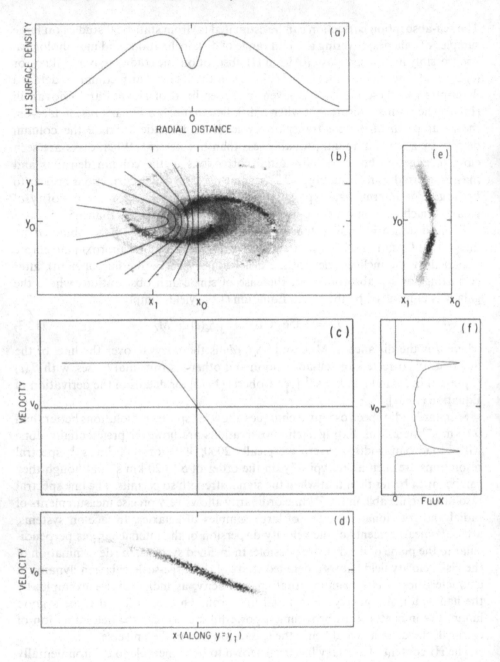

Fig. 12.1. Pictorial representation of HI data. Panel (b) simulates the distribution of HI in a galaxian disk inclined 60° with respect to the plane normal to the line of sight; the shade intensity is visualized as being proportional to the HI column density across the disk. Superimposed on the left part of the image is a family of curves which identifies isovelocity contours. Because this is a symmetric idealized representation, isovelocity contours on the right part of the galaxies are mirror images of those on the left. Panel (a) illustrates the azimuthally averaged HI surface density, as a function of distance from the center of the galaxy. Panel (c) describes the rotational velocity of the galaxy, as measured along the major axis $y = y_0$; correction of this function by a factor $1/\sin i$, where i is the inclination (60° here) of the disk,

Spiral galaxies within 100 to 200 Mpc are easily detectable in the line with modest investments of telescope time, yielding recessional velocities with errors of only a few km s^{-1}. While optical redshift surveys preferentially favor objects of high central optical surface brightness, galaxies of low optical surface brightness tend to have abundant HI and, at equal apparent magnitude, are more easily detected by radio surveys. For example, while redshifts of galaxies in clusters—where lenticular and elliptical morphologies are most abundant—are usually obtained by optical means, the vast majority of redshifts available for galaxies in the low-density regions of superclusters have been obtained using radio methods.

12.1.3 Pictorial Presentation of HI Data

In the best cases, the body of HI data on a galaxy can be thought of as a three-dimensional array of intensities, expressed in terms of two angular coordinates in the plane of the sky and radial velocity, which may be referred to as a "data cube." The array values are set by the six-dimensional position-velocity distribution of the galaxian HI. In an idealized form, one can assume that the HI is confined to a thin disk and that axial symmetry holds for both its distribution and velocity field. In panels (a) and (c) of Figure 12.1, both the radial dependence of the HI surface density and the rotation curve are idealized averages of those observed in intermediate-type spirals. Integration of the data cube along the radial velocity axis will yield a map of the column density distribution in the sky; in panel (b) of Figure 12.1, the disk is simulated at a viewing inclination of 60°. The degree of shading represents the column density map. Each member of the superimposed family of lines identifies the locus of points characterized by a constant radial velocity. If instead of integrating over radial velocity, we slice the data cube at a constant value of one of the sky coordinates, for example, slightly off and parallel to the major axis of the tilted disk, we obtain a position-velocity map (d) which mimics the rotation curve (b), blurred by the limitations of spatial and spectral resolution. A slice of the cube at constant velocity will yield a "channel map," as in panel (e): the angular distribution of all galaxian HI whose velocity falls within a narrow velocity range, for example, the span of a single receiver channel, now mimicking one of the loci of panel (b). Finally, observation of the galaxy with a single dish that does not resolve the HI disk will yield an integrated profile of the type shown in panel (f). Of course, real data will be cursed (or blessed...) with asymmetries, spiral features, small-scale irregularities, flared or disrupted disks, etc. Sub-arminute-resolution HI maps are now available for several dozen galaxies, and

◁ ——

yields the galaxy's rotation curve. The systemic velocity of the galaxy is V_0; the straight, vertical isovelocity contour in panel (b) identifies the locus of points at $V = V_0$. A slice of the galaxy along $y = y_1$ gives a position-velocity map as in panel (d), which in shape (but not in slope) mimics the curve in panel (c); again, intensity of shade is proportional to column density; the smoothing effects of a nonzero angular and spectral resolution have been introduced and are responsible for the breadth of the region with significant emission in panel (d). A "moment map" (or a "channel map"), i.e., the surface density distribution of HI within a narrow-velocity interval (e.g., a single spectral channel) is illustrated in panel (e); the chosen velocity interval in panel (e) corresponds roughly to that between the second and third isovelocity contours plotted in panel (b). Finally, panel (f) shows a spectrum of the whole galaxy, as seen by a single dish that cannot resolve its HI disk.

single-dish profiles for approximately 6000. A catalog of extragalactic HI observations (Richter et al. 1983) is maintained, updated, and available to the astronomical community.

12.2 The Distribution of HI in Galaxies

12.2.1 Morphology

As mentioned in Section 12.1, the optical morphology of a galaxy certainly bears some relevance to the HI content and distribution, but it is important to remember that the optical light and the HI emission do not necessarily arise from the same locations. Most of the detected HI in our own galaxy represents a galaxian population clearly associated with the disk but not coextensive with other conspicuous large-scale components of the interstellar medium, such as molecular clouds or HII regions. The neutral, atomic hydrogen is typically found farther out in the disk than any of its other directly measurable components.

The HI distribution in galaxies with a well-established disk in the stellar component is also disk-shaped. The outlines of spiral arms are often conspicuous in the HI gas, as beautifully illustrated by the two-armed spiral pattern of the map of M81 of Rots and Shane (1975). In nearby galaxies for which the linear resolution of observations is particularly high, details that bear on the structure and dynamics of the interstellar medium are discernible. For example, in the case of M31, bubblelike features are observed, with sizes ranging from the map resolution limit (80 pc) up to 1 kpc (Brinks and Bajaja 1986).

A notable aspect of galaxian HI disks is the occurrence of strong central surface density depressions. Such depressions tend to appear more conspicuous in high-luminosity early-type spirals which possess large bulges. In lenticular galaxies the central depressions are frequently larger than the optical disks themselves, in which case the term "HI disk" tends to be modified to "HI ring." Although these outer HI rings are usually concentric with the optical disks, they can be strongly inclined to the latter. The azimuthal distribution of HI within the rings can be rather irregular, to the point that segments of the ring may not even be detected. In some S0's, diffuse polar rings are seen, orthogonal to the plane of the normal stellar disk, and are occasionally rich in HI. The trend of central depressions growing with earlier optical morphology is not entirely systematic, as summarized by van Woerden et al. (1983). The presence of the central depression in M31, an intermediate spiral, can be appreciated in Figure 12.2; notice how in M31 the region dominated by the stellar bulge is nearly devoid of HI. In our galaxy, the HI surface density starts to decrease at approximately the radius at which that of molecular hydrogen reaches a peak. In the case of early-type disks, not enough molecular gas is found to compensate for the locally anemic HI distributions. At the other extreme of the Hubble sequence, irregulars and gas-rich dwarfs exhibit an HI distribution similar in character to that of their light: clumpy and disorganized.

A frequently used azimuthally symmetric model for the HI surface density $\sigma_H(r)$ which gives a useful rough representation for most spirals is the sum of two

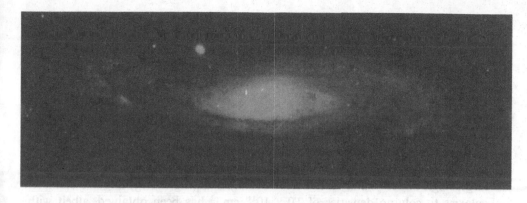

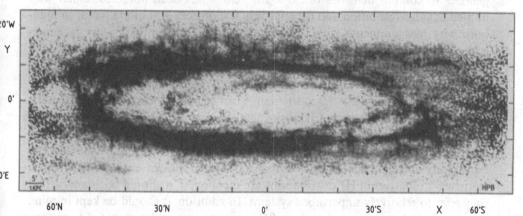

Fig. 12.2. HI distribution in the disk of M31, obtained by Brinks and Bajaja (1986) with the Westerbork Synthesis Radio Telescope; the linear resolution on the plane of the sky is about 80 pc.

Gaussians, the inner one with amplitude and width respectively -0.6 and 0.5 of the corresponding values for the outer one. This simplified model glosses over individual details and asymmetries and some large-scale features that may be common to most spiral disks. Sancisi (1983) has shown that a common feature seen in the outer regions of spirals is an HI "shoulder": a steep drop in $\sigma_H(r)$ occurs usually near the optical radius (normally the "Holmberg radius" at 26.5 mag arcsec^{-2}), followed by a gentle decrease at larger radii. Such shoulders are especially noticeable among galaxies that have very extended HI distributions. The onset of warps (see Section 12.2.3) tends to coincide with the "shoulder" radius. Gas in the shoulder may differ dynamically from the inner HI, being characterized by offset eccentric orbits instead of circular ones, and may form an envelope to the whole galaxy. Sancisi thinks that very large HI rings, as seen in lenticulars, may represent extreme examples of "shoulders." Interpretations of those features in evolutionary terms are not widely accepted. In special cases, episodes of accretion of gas-rich systems by a dominating galaxy are invoked, as in the case of polar rings. However, the relatively frequent occurrence of some features—such as the "shoulders" described by Sancisi, which have in several instances been observed to begin at a truncation edge of the stellar disk—suggests that internal causes should perhaps be sought within the framework of a slowly infalling disk model or of one where at some critical radius the onset of dynamical instabilities precipitates noncircular motions.

12.2.2 Sizes

The "size" of a galaxy as seen in HI is usually equated with the diameter of the furthest detectable HI isophote; this definition leads to a measurement which increases as observing hardware improves. Currently, routine observations of galaxies with synthesis instruments reach column densities of 1.0×10^{20} cm^{-2}. With large, single dishes (mainly the Arecibo 305 m), mapping of large-angular-diameter galaxies to column densities of 2.0×10^{18} cm^{-2} has been obtained, albeit with limited angular resolution and for relatively few objects. A limiting factor in high-sensitivity measurements with single dishes is the contamination by radiation collected by sidelobes of the antenna beam; while well understood, this contamination can be difficult to remove at very low HI surface density levels if bright HI emission is found just outside the field of interest. In the absence of such sources, however, measurements to 5.0×10^{17} cm^{-2} can be reliably conducted. Systematic analysis shows that the size of HI disks, like the integral HI content, is dependent not only on the internal characteristics of the galaxy, but also on its environment, as discussed in Section 12.6. The outer regions of a disk are very fragile and can be severely affected, thermally or dynamically, by a violent environment. When statistical properties of HI distributions are mentioned, it should be assumed that they refer to relatively unpertubed systems. In addition, it should be kept in mind that relatively low column densities of HI may become undetectable. To see that, note first that typically, for a given line width, the HI column density is proportional to the line brightness temperature. However, as seen in Section 12.1.2, if the opacity of the gas becomes very low, the fractional deviation of the spin temperature from the 2.7 K of the microwave background, $(T_s - 2.7)/T_s$, is depressed. For a given column density, then, the line brightness temperature diminishes in the same measure, as described by Equation (12.2) and discussed in more detail by Watson and Deguchi (1984).

More readily obtained estimates of the size of the HI distribution are those which express it as an isophotal or as an effective radius, the latter identifying the radius within which a fraction of the total HI mass is contained. When expressed in terms of the isophote at $N_H = 1.82 \times 10^{20}$ cm^{-2}, the average ratio of the HI to the Holmberg radius is about 1.5; alternatively, 70% of the galaxy's HI mass is contained, on average, within 1.2 Holmberg radii. Dependences of HI radius on morphological type are very weak and difficult to discern; the distribution of sizes around the mean is strongly skewed, relative to the Gaussian. For reasons that are very poorly understand, some galaxies exhibit enormously extended HI disks. Notable are the cases of Mk 348, illustrated in Figure 12.3, and of NGC 628, in which HI is detected out to several Holmberg radii. In the nearby system M33, HI is detected out to three Holmberg diameters. The exceptional HI objects are not systematically unusual in any other respect. Still, the incidence of very large HI disks has important implications. Dynamically, it allows inspection of the gravitational

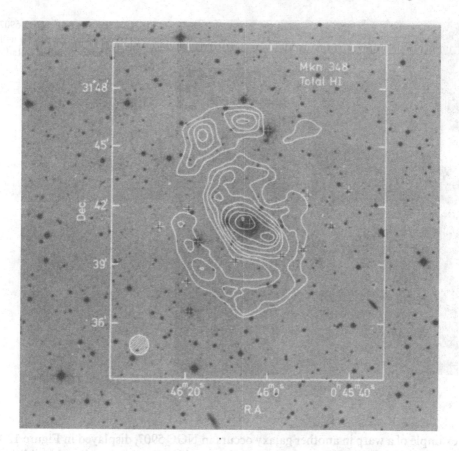

Fig. 12.3. HI distribution in Mk 348, obtained by Heckman et al. (1982) with the Westerbork Synthesis Radio Telescope. Reprinted with permission from the Royal Astronomical Society.

potential of a galaxy at very large distances from its center; from a cosmological viewpoint, the "intervening galaxy" explanation of high-redshift absorption lines seen along the line of sight to QSOs demands large cross sections of the intervening galaxian disks in order to account for the number of such lines observed (see Section 12.7.4). Huchtmeier et al. (1981) have noted that a significant fraction of irregular galaxies contain extended HI components not correlated with their optical size; the role of the extended gas component in the evolution of a galaxy is however still unclear.

12.2.3 Warps
The outer regions of the HI distribution of galaxies tend to exhibit strong departures from the regularity usually seen in the inner regions. It has been known for decades that the plane of the Milky Way displays an elegant hat-brim effect; the deviation of the centroid of the outer disk from the plane defined by the inner disk reaches 1.5 kpc at $r = 18$ kpc from the center. This effect was long attributed to a close tidal encounter with the Magellanic Clouds. However, perhaps the most spectacular

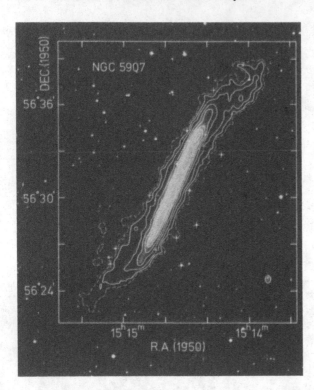

Fig. 12.4. HI warp of NGC 5907, as observed by Sancisi (1976) with the Westerbork Synthesis Radio Telescope.

example of a warp in another galaxy occurs in NGC 5907, displayed in Figure 12.4, which has no visible nearby neighbor that could serve as perturber. In addition, the galactocentric azimuth of the maximum height of the warp in our galaxy does not appear to depend on galactocentric distance, i.e., the warp does not precess under galactic rotation. This condition suggests the existence of self-maintaining mechanisms. A large fraction of the edge-on galaxies mapped in HI show evidence of warps. The onset of a warp usually coincides roughly with the Holmberg radius of the disk; the optical light also often falls steeply past this point. The warp of NGC 5907 shows only in the HI disk; in the case of NGC 4565, however, a slight warp is also observed in the stellar disk. While warps are easy to detect in maps of the HI distribution in edge-on galaxies, they can also be inferred from the characteristics of the velocity field in more face-on disks. Deviations from circular rotation, associated with the gas in warps, produce characteristic signatures in isovelocity contour maps. Warps thus identified are referred to as "kinematic" warps. Examples of kinematic warps are illustrated in some of the panels in Figure 12.7 and discussed in Section 12.3.2. A review of the theoretical problems associated with the interpretation of warps is given by Toomre (1983).

12.2.4 Appendages
In nearby loose groups of galaxies, HI appendages are often seen to extend outward from the disks of member galaxies: peninsular "plumes" and "tails," connecting "bridges" and other such features. A number (but not all) of these HI appendages

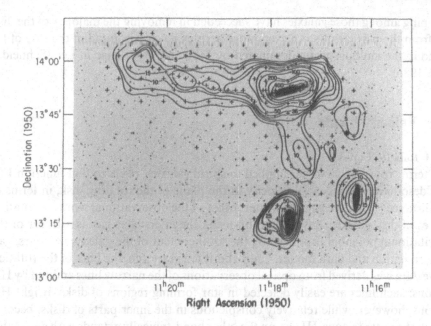

Fig. 12.5. HI distribution in the Leo triplet of galaxies—NGC 3623, NGC 3627, and NGC 3628—obtained with the Arecibo 305-m reflector by Haynes et al. (1979). Angular resolution of the map is 3.3'. The contours are of neutral hydrogen column density in units of 3.1×10^{18} cm^{-2}.

have been explained in terms of tidal interactions among neighboring galaxies. In the Local Group, the linear feature known as the Magellanic Stream is probably the result of tidal forces in the Milky Way-LMC-SMC system. Examples of HI tidal streams which have been successfully modeled include those seen in the systems NGC 4631/56, NGC 4038/9, and M81/M82/NGC 3077. Figure 12.5 shows the HI distribution in the Leo triplet NGC 3623/7/8 obtained at Arecibo. In this example, the overall characteristics of the observed HI distribution can be well reproduced by the interaction caused by postulating a hyperbolic passage of NGC 3627 past its neighbor NGC 3628 with a perigalactic distance of about 20 kpc. The disruption caused to NGC 3627 is much reduced because the sense of its orbital motion is opposite to that of its rotation; in NGC 3628, on the other hand, the two vectors are aligned and the damage is more severe, resulting in the conspicuous tail drawn mainly from the periphery of that galaxy. NGC 3623 has not recently had a close interaction with the other two.

Several systems which contain substantial amounts of HI outside of galactic disks are less easily modeled as transient phenomena. Stefan's quintet (Shostak et al. 1984) is a group of five objects located close together on the sky. One of the galaxies is likely to be in the foreground; the other four lie at the same redshift and are morphologically peculiar. In fact, the HI at the velocity of the suspected foreground galaxy is regular while at least three distinct velocity systems are seen at the higher redshift. The bulk of the HI at the higher velocities lies outside the optical boundaries of any of the remaining quartet, and it appears that tidal and collisional stripping among those galaxies have succeeded in removing the majority of the disk gas from the participants. An even more perplexing case is found in the ring of HI found in the multiple group dominated by M96 and M105, also in Leo (Schneider et al. 1986).

12.3 Velocity Fields

12.3.1 Rotation Curves

The term "rotation curve," as applied to spiral galaxies, refers to that function $V(r)$ that describes the tangential velocity in the plane of the rotating disk, in terms of the distance r from the galactic center. Most of the observational work on rotation curves of disk galaxies has focused on using the interstellar gas as a tracer of the gravitational potential produced by the total content of the galaxy, i.e., stars, gas, dust, radiation and dark matter. Until about ten years ago, the bulk of the rotation curve data was derived from optical observations of the narrow lines emitted by HII regions; such lines are easily detected in star-forming regions of disks. Bright HII regions, however, while relatively conspicuous in the inner parts of disks, become rare in the outer regions. HI gas, on the other hand, typically extends with detectable column densities beyond the outer boundary of the HII region distribution, so that velocity measurements from HI maps provide the rotation curve and, by inference, the mass in the outer parts of spiral galaxies. Luminous galaxian material fades exponentially with radial distance, and until the early 1970s, it appeared reasonable to expect that the luminous matter was a good tracer of the total mass. Therefore, rotation velocities were expected to start falling in the periphery of luminous disks, as a result of the corresponding decline in mass density. If the mass of a galaxy were strongly concentrated, one would in fact expect the rotation curve to approach asymptotically a Keplerian $r^{-1/2}$ decrease. Seminal work on the stability of disks, which invoked the presence of massive, dark halos, and early single-dish 21-cm measurement of large nearby galaxies indicated the fallacy of such an expectation.

Many rotation curves have now been measured well outside the optical disk. Perhaps the most interesting characteristic of the HI rotation curves is that the rotational velocity does not fall even at large distances from the center, as can be appreciated in Figure 12.6, showing several examples obtained by aperture synthesis observations. In practice, both optical and HI line measures are used in studying rotation curves. Optical spectra outline the rotation in the bright inner disk while the velocity field in the outer portion is derived from 21-cm mapping.

Given the rotation curve $V(r)$, simple balance of gravitational and centrifugal force in spherical symmetry gives for the total mass $M_T(r)$ contained within r

$$M_T(r) = 2.33 \times 10^5 r V^2(r) \, M_\odot \qquad (12.4)$$

with $V(r)$ in km s^{-1} and r in kpc. [If the assumption of spherical symmetry for $M_T(r)$ is relaxed, the constant in Equation (12.4) diminishes slightly.] A rotation curve which becomes flat in the outer regions implies a linearly rising $M_T(r)$ and a mean density that decreases as r^{-2}; since the luminous matter decreases exponentially, the ratio of dynamical to luminous mass grows with r. In general, rotation curves

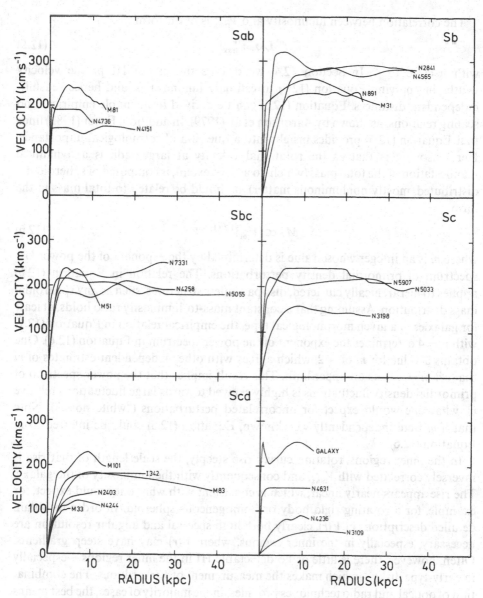

Fig. 12.6. Selection of rotation curves of spiral galaxies, assembled by Bosma (1981). (Reprinted with permission from The Astronomical Journal).

of spiral disks remain flat, or perhaps rise very mildly as $r^{+0.1}$ at large radii. The maximum rotational velocity reached within a disk, V_{max}, is a function of both luminosity and morphological type. For a given type, higher peak rotation velocities —as derived from long-integration optical measurements—are attained by brighter systems. At a fixed luminosity, earlier spirals are faster rotators than late ones. (The largest rotational velocity observed thus far in a disk is about 500 km s^{-1}, in the S0a galaxy UGC 12591.)

The correlation between luminosity and V_{max} is of the form

$$L \propto V_{max}^n \qquad (12.5)$$

with n close to 4. In Section 12.4, we discuss the use of HI profile velocity widths in applying Equation (12.5) to estimate luminosities, and hence redshift-independent distances. Equation (12.5) can be derived from simple empirical and scaling relations, as shown by Aaronson et al. (1979). In addition, Faber (1982) finds that Equation (12.5) provides insight into a question of cosmological importance. Her reasoning is that as the rotational velocity at large radii is a dynamical manifestation of the total mass (which, to a large extent, is composed of spheroidally distributed, mostly nonluminous matter), it should be related to total mass in the form

$$M_T \propto (V_{rot})^{12/(1-m)} \qquad (12.6)$$

where m is an integer whose value is determined by the exponent of the power law spectrum of primordial density perturbations. The relation in Equation (12.6) applies to hierarchically clustered, dissipationless structures, such as any spheroidal mass distribution. Assuming that a constant mass-to-luminosity ratio holds, at least for galaxies of a given morphological type, the empirical relation in Equation (12.1) with $n = 4$ determines the exponent of the power spectrum in Equation (12.6). One obtains a value for m of -2, which agrees with other independent estimates of m that set $m = -1$ as an upper limit. This result implies that the power spectrum of primordial density fluctuations is highly skewed towards large fluctuations relative to what one would expect for uncorrelated perturbations ("white noise"). Note that if m were independently well known, Equation (12.5) could be inferred from Equation (12.6).

In the inner regions, rotation curves rise steeply; the scale length of that rise is inversely correlated with V_{max}, and consequently with the luminosity of the galaxy. The rise appears nearly linear, which is consistent with what one would expect, for example, for a rotating rigid body or homogeneous spheroid. In order to obtain detailed descriptions of $V(r)$, clearly both high spectral and angular resolution are necessary, especially in the inner regions, where $V(r)$ may have steep gradients. Often, however, there is little or no detectable HI in the inner regions—especially in early-type disks—which makes the measurement of $V(r)$ arduous. The combination of optical and radio techniques provides, in the majority of cases, the best means of carefully determining $V(r)$ throughout the disk. Good quality rotation curves are known for several dozen systems. Single-dish profiles of the type shown in Figure 12.1(f) are much easier to obtain and thus more common; from them it is possible to extract a value of V_{max}, which can then provide an indication of the total mass contained within the region observed in the HI line. Unless the galaxy is mapped, the exact extent of this region is unknown and the estimate relies on statistical considerations.

12.3.2 Distortions in the Velocity Field

In many galaxies the velocity field, as described in Figure 12.1(b), is observed to deviate from axial symmetry in such a way that the HI cannot be rotating in circular, coplanar orbits. Figure 12.7 illustrates the variety of observed velocity fields. M83 and NGC 5383 illustrate two kinds of deviation that maintain a sort of twofold symmetry in the velocity field. One of them, the kinematic warp, was briefly discussed in Section 12.2.3 and is exemplified by M83: a change in the position angle and inclination of the warp produces a characteristic S-shaped distortion. The same effect can also be discerned, although less dramatically, in NGC 5055 (see Figure 12.7). M31 is perhaps the system where the most detailed rendition of a warped disk is kinematically observable but it is by no means typical. As the inclination of the disk is high (about 77°), the line of sight can cross the disk in three different locations, including two in the upward and downward brim of the warp, yielding triple-peaked velocity profiles. A warp is a perturbation that involves the outer parts of the disk; the second kind of distortion, an "oval distortion," arises in the inner parts and results from structures that depart from the symmetry of the optical image of the outer regions of the disk, such as bars or ovals (i.e., "fat" bars). The resulting perturbation in the velocity field causes the major and minor symmetry axes in the inner parts of the galaxy not to be perpendicular, as can be seen in NGC 5383 in Figure 12.7.

12.3.3 The Mass Distribution in Spiral Galaxies

By combining optical photometry data with the mass derived from 21-cm observations via Equation (12.4), we can determine total mass-to-luminosity ratios—although one should note that values thus obtained depend inversely on the assumed distance scale. Arbitrarily choosing some limiting radius, e.g., r_{25}, measured at the blue isophote of 25.0 mag arcsec^{-2}, the integral of Equation (12.4) and the blue luminosity within r_{25} yield a measure of the global mass-to-light ratio. Using this system, Rubin et al. (1985) find that ratio to be *6.2, 4.5, and 2.6 for Sa, Sb, and Sc galaxies*, respectively, with an error on those mean values of about 10%. Since the mass is proportional to $r_{25} V_{max}^2$, and since it is found that, in the blue, the relationship between luminosity and radius is independent of morphological type, the variation in (M_T/L_B) with type is essentially one of V_{max} with type. The mean values of V_{max} as a function of type bear out that conclusion. The values of (M_T/L_B) are virtually constant within a given Hubble type, over a range of several magnitudes. Now, what fraction of the total dynamical mass does the luminous mass constitute? To answer that question, we must first obtain an estimate of the fraction of the total luminous mass represented by L_B. Work based on model stellar populations indicates that reasonable values for the ratio between the total mass in all stars and L_B are 3.1, 2.0, and 1.0 for Sa, Sb, and Sc types, respectively. Although these numbers are relatively uncertain, they indicate that the ratio of total to luminous mass within the 25-mag arcsec^{-2} isophote is likely to be on the order of 2 for spirals of all types. As the mass grows more or less linearly and the light fades exponentially, the mass-to-light ratio grows rapidly outwards from r_{25}.

The analysis of rotation curves yields information on the mass distribution as a function of distance from the galactic center, but none whatsoever on the distribution as a function of distance from the plane of the disk. One would like to know which fraction of the total mass resides in the disk and which in the halo, and also how the total mass-to-light ratio of the disk alone varies with radius. In order to

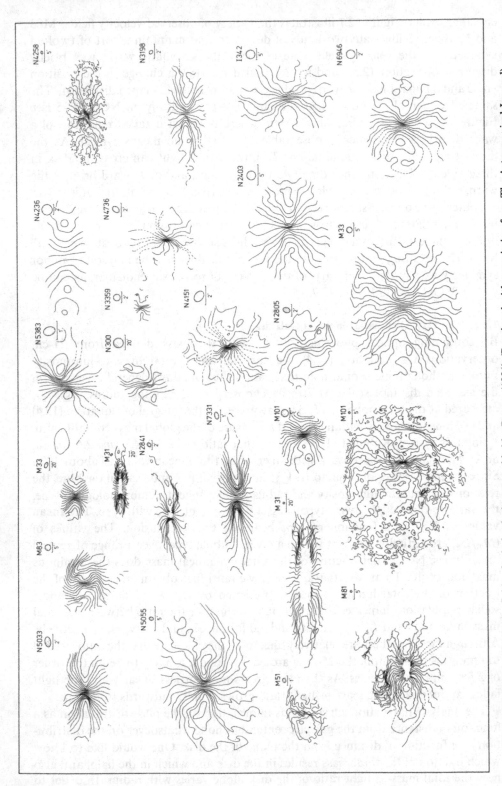

Fig. 12.7. Selection of velocity fields of spiral galaxies, assembled by Bosma (1981). (Reprinted with permission from The Astronomical Journal).

answer these questions, it is necessary to measure the characteristics of some tracer of the gravitational potential in the z-direction (i.e., perpendicular to the disk); handily, such is the HI. It is possible to obtain the distribution and the velocity dispersion, in the z-direction, of the galaxian HI. The former can be measured from high-resolution maps of edge-on galaxies, the latter from spectral profiles of isolated regions in face-on disks.

Following van der Kruit and Shostak (1983) and references therein, let us assume that a spiral disk can be approximated by a self-gravitating sheet which is locally isothermal (i.e., the velocity dispersion of any of its components is independent of z). Then the total mass density can be expressed as

$$\rho(r,z) = \rho(r,0) \operatorname{sech}^2 \left(\frac{z}{z_0} \right) \tag{12.7}$$

where z_0 may be a function of the distance from the center (in the plane), r. The HI disk can be assumed to be effectively massless, in comparison with other dynamically important components; it can then be shown that the HI density decreases to half of its midplane value at a height z_H

$$z_H = 0.85 \langle v_z^2 \rangle_H^{1/2} [2\pi G \rho(r,0)]^{-1/2} \tag{12.8}$$

where $\langle v_z^2 \rangle_H^{1/2}$ is the z-velocity dispersion of the gas and G is the gravitational constant. We can rewrite Equation (12.8) as

$$z_H(r) \propto [\langle v_z^2 \rangle_H(r)]^{1/2} \left[\frac{M_d(r)}{L(r)} \right]^{-1/2} [L(r)]^{-1/2} \tag{12.9}$$

where $M_d(r)$ is the disk mass within radius r and the luminosity profile $L(r)$ is obtained from major-axis photometry. Measurements of $\langle v_z^2 \rangle_H^{1/2}$ have been made directly in a few face-on galaxies; all yield values in the range of 7 to 10 km s^{-1} and appear to vary relatively little with r. One can thus apply Equation (12.9) to a well-mapped edge-on object, such as NGC 891 or NGC 7814, for which $z_H(r)$ is then known. By assuming a value of $\langle v_z^2 \rangle_H^{1/2}$ as measured in face-on spirals, and by photometrically determining $L(r)$, we can obtain the details of the (M_d/L) ratio within the disk. Such an operation yields the following results:

- The mass-to-luminosity ratio of the disk is independent of r, and hence the luminosity profile of the spiral disk is also a profile of the disk's mass.
- Only one-third of the total mass of NGC 891 within the distance to the edge of the optical disk actually resides in the disk itself, the rest being distributed in a much thicker component, the halo.
- At $r = 10$ kpc, the mass density of the disk at $z = 0$ exceeds that of the halo (assumed spherical) by a factor of 4; at $r = 21$ kpc, that ratio decreases to 2.
- In the edge-on Sab galaxy NGC 7814, only a small fraction of the total mass, interior to the optical radius of 22 kpc, is in the disk; its light distribution, furthermore, is dominated by the spheroidal bulge. The rotation curve then can be used to sample the bulge's mass-to-light ratio; it is found to increase by a factor of 10 between the inner regions ($r < 10$) and 22 kpc.

In conclusion, the dynamical masses of spiral galaxies are found to be still growing linearly beyond the edges of their optical disks, at least out to the greatest radii at which gravitational potential tracers like HI are detectable. Within the optical radius, no more than one-third to one-half of the mass resides in the disk. The ratio between luminous and dynamical mass within the disk is about constant with radius and independent of morphological type, while that of the spheroidal component grows rapidly with distance from the galactic center.

12.3.4 Dark Matter in Dwarf Galaxies

Although most of the galaxies that we see are of relatively high luminosity, it is not yet clear that they are in fact good tracers of the mass distribution. It is already known that low-luminosity galaxies are found to cluster less strongly than higher-luminosity ones (see Section 12.7.2). Since the number density of galaxies still rises at low luminosities, it is critical to understand whether the presence of dark matter—the so-called "missing mass"—is required to the same extent in the low-luminosity but numerous dwarf galaxies. In particular, we need to have some idea of whether these objects contribute more, less, or the same amount to the matter distribution as do galaxies of higher luminosity.

Dwarf counterparts of elliptical, lenticular, and irregular galaxies exist and are in fact the dominant population, by number, of the Virgo cluster. The presence of dwarf spiral galaxies is controversial, and an absence of low-mass spirals is predicted by some theories of the origin of spiral structure. While the dE's may be the most common galaxy in Virgo, dI's may represent the most widely distributed objects in the universe (see Section 12.7.2). Although dI's are typically of low optical luminosity and surface brightness, they also contain significant fractions of their total mass in the form of HI (Fisher and Tully 1975). Their global 21-cm profiles are often Gaussian rather than two-horned, indicating that random motions are significant and possibly dominant.

Recent synthesis observations have been made to investigate the HI distribution and velocity field of a number of very-low-luminosity dI's (Sargent and Lo 1986). The HI distributions vary from annular to totally without organized structure, but are generally larger than the stellar distributions. In all of the mapped dI's, the amplitude of the rotation curve is comparable to or smaller than the random velocity dispersion, and several of the objects show no sign of systematic velocity fields. While the kinematics of the brighter dI's ($M_{pg} < -14$) tends to be dominated by rotation, the chaotic motions are more important at lower luminosities. While there is a trend that the HI fraction of the mass of the dI's rises with decreasing optical luminosity, the gas only rarely dominates the total mass. Still, in no case is the rotation curve observed to turn over at large r; in most cases, the rotation curve is linear and still rising (solid-body rotation) even outside the optical image. It is difficult, if not impossible, to estimate the inclination of a dI, unless its velocity field is regular enough that the inclination can be derived from the best-fit kinematic model. Despite the resulting uncertainties in mass estimates, it is clear that in all cases, the total mass-to-light ratio M_T/L is high, in the range of 10 to 30: dI's have dark halos also.

Although true dwarf spirals do indeed seem rare, at least one low-luminosity Scd galaxy, UGC 2259, shows a symmetrical HI distribution and a regular velocity field indicating a flat rotation curve similar to that seen in more luminous spirals (Carignan et al. 1987). The interpretation of the mass distribution for this galaxy is model dependent, but even the assumption of a minimum-mass dark halo implies that at the optical (Holmberg) radius, the luminous disk and the dark halo contribute equal amounts of mass. Thus it appears that the ratio of nonluminous to luminous matter is independent of the Hubble type.

12.4 The Velocity Width as a Distance Indicator

It is well know that there is a tight correlation between galaxy morphology and angular momentum, dominated in spirals by rotation. As a distance-independent quantity, the velocity width can be used in conjunction with some other observable to estimate the distance to a galaxy via a method that does not involve the redshift. Comparison of this "luminosity distance" with that predicted from the redshift serves as a means of deriving the Hubble constant and in identifying deviations from a smooth Hubble expansion. Calibration of the method relies on the assumption of known distances to a few nearby galaxies, obtained via primary indicators such as variable stars.

12.4.1 The Velocity Width–Magnitude Relation

As discussed in Section 12.3.1, the maximum rotational velocity reached within a galaxian disk depends on the luminosity of the galaxy, as described by Equation (12.5). In 1977, B. Tully and R. Fisher were the first to realize that such a circumstance lent itself to the measurement of distance moduli. They expressed the relationship in the form

$$M_{pg} = -a \log W_c - b \qquad (12.10)$$

where M_{pg} is the absolute photographic magnitude, corrected for both the extinction by dust in our galaxy and by dust in the target galaxy itself. The latter is generally parameterized by an inclination and a morphological-type dependence. W_c is the velocity width of the 21-cm line profile, integrated over the whole solid angle subtended by the HI in the galaxy after correction by a factor (csc i), where i is the inclination of the galaxian disk to the line of sight. The coefficients a and b were initially found to be about 6.25 and 3.5, respectively. Numerous authors successively applied the method, but doubts were cast on the legitimacy of the application of Equation (12.10). Brosche (1971) noted that the maximum velocity width observed is different for galaxies of different morphological type; later, Roberts (1978) pointed out that the velocity width–magnitude relation depends strongly on the morphological type, a result later confirmed by the detailed studies of systematic properties of rotation curves by Rubin and coworkers (1985). The slope of the relation, in fact, when considered for individual morphological types, is closer to 10; it was the combination of galaxies of several morphologies, each with a different offset b, which initially yielded a flatter relation. The offset between Sa's and Sc's is, according to Rubin et al., about 2 magnitudes, in the blue. Important contributions towards

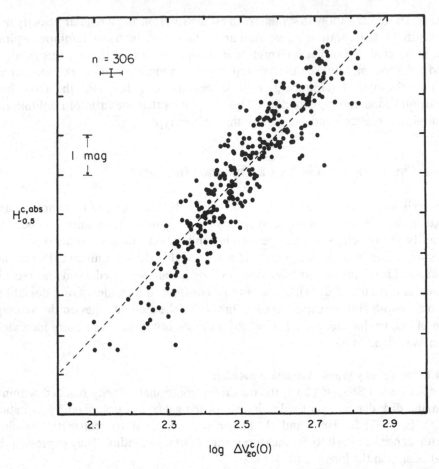

Fig. 12.8. Infrared magnitude–velocity width relation for 306 nearby galaxies, after Aaronson et al. (1982). Absolute magnitudes were calculated assuming a uniform Hubble flow. Dashed line has a slope of 10. Error bar represents the typical uncertainty for a single point, as estimated by the authors.

understanding the biases in the use of the velocity width—magnitude relation have been made by Bottinelli *et al.* (1986 and references therein), particularly in connection with the determination of the Hubble constant.

The sources of uncertainty in establishing values of *a* and *b* can be serious. The circumstance that optimizes the determination of W_c, i.e., high inclination, conspires to render more uncertain both the correction for internal extinction to M_{pg} and the determination of the morphological type, necessary to decide which values of *a* and *b* in Equation (12.10) should apply. Aaronson and coworkers first proposed in 1979 that most of the uncertainties associated with the inclination corrections are minimized if infrared rather than blue or photographic magnitudes are used. The interstellar extinction at 1.6 μm is practically negligible, and hence the uncertainty in the magnitude correction is practically eliminated. In addition, Aaronson and coworkers found no dependence in the coefficients of Equation (12.10) with morphological type, when the absolute magnitudes at 1.6 μm replace the photographic ones, and a reduced scatter about the best fit to the data. The quality of this correlation is illustrated in Figure 12.8, an infrared magnitude–velocity width plot for 306 nearby galaxies. Aaronson and coworkers have interpreted this apparently fortuitous and most convenient of circumstances as a natural consequence of using infrared magnitudes, given the properties of the stellar population they trace, but their interpretation has been disputed by Burstein (1982). Rubin et al. (1985) find that Equation (12.10) is not independent of morphological type even at infrared wavelengths. As with blue magnitudes, the slope of about $a = 10$ is virtually independent of type, but an offset of about 1.3 magnitudes separates Sa's from Sc's. They also find that the scatter in the relation is not significantly reduced (rms of 0.68 with blue magnitudes, 0.67 with infrared magnitudes) by adopting 1.6-μ magnitudes. The differences between the results of Rubin et al. and those of Aaronson et al. probably reside in different biases inherent in the choice of samples.

The question of whether the infrared version of Equation (12.10) is truly a single relation, valid for all spirals, remains open; the possibility that a second parameter, in addition to W_c (e.g., type), may be necessary to infer luminosities appears likely. There is general consensus that the slope of the relation is close to 10, both in the blue and in the infrared, when samples of a given type are chosen. The velocity width is also seen to correlate well with optical size and with the HI mass. Large samples of good-quality HI profiles are becoming available to distances in excess of 200 Mpc. The method relies on the absolute determination of distance for a small number of local calibrators, obtained with traditional means. The quality and number of these calibrators should dramatically improve after launch of the Hubble Space Telescope. As the wrinkles of the method are slowly smoothed away, the velocity width–luminosity relation may provide one of the best tools with which to measure the distance scale of the universe.

The physical basis for the velocity width–magnitude relation is poorly understood. (A possible interpretation was given in Section 12.3.1.) A similar relationship between the luminosity and velocity dispersion is seen among ellipticals, $L \propto \sigma^4$, with a third parameter, the integrated surface brightness, providing the needed scale length to minimize the scatter. Indeed, equilibrium disk models with constant surface brightness and constant mass-to-light ratio predict $L \propto V^4$, but the application to all real galaxies is not clear.

12.4.2 Deviations from Hubble Flow

Despite the current uncertainties in its application, the velocity width–magnitude relation has been shown to be a promising tool in the study of deviations from the smooth Hubble expansion. Once the relation is established for a certain sample of galaxies, a predicted redshift, V_p, can be estimated for each galaxy. V_p can then be compared with the observed redshift, V_0. Systematic variations over the celestial sphere of the residuals $V_0 - V_p$ imply a large-scale bulk flow. Hart and Davies (1982) have used HI observations of nearby Sbc galaxies to determine the motion of the Sun and the Local Group relative to the more distant galaxies. Their method uses the HI mass as the standard candle for galaxies in this single Hubble type. Further extension of this technique to larger distances will investigate the scale over which peculiar motions can be measured beyond the boundaries of the Local Supercluster.

The velocity width–infrared magnitude method has also been used to derive relative distances to nearby clusters of galaxies. Since the galaxies in an individual cluster can be assumed to be at the same distance, a determination of the relation for galaxies in separate clusters yields a measure of the Hubble ratio (velocity/distance) for each cluster. Aaronson et al. (1986) have derived Hubble ratios for ten nearby clusters; the scatter in their results again implies significant deviation of the Local Group motion from what would be expected if the local Hubble flow were simply a smooth expansion.

It is clear that future refinement of this technique will provide a critical link in our understanding of the magnitude and direction of the peculiar motion and the scale size of inhomogeneity in the local universe.

12.5 HI Content and Other Global Properties

The term "HI content" of a galaxy has often lent itself to some confusion. The simplest and most obvious meaning is the mass of HI within the entire galaxy. As with absolute magnitude, however, this quantity is usually uncertain by a factor that goes with the square of the assumed distance scale. The ratio of the hydrogen mass to optical luminosity, M_H/L, being equal to the ratio of the related fluxes, is conveniently distance independent, and the label "HI content" has frequently been applied also to this quantity, contributing a measure of ambiguity. Here we shall maintain the term for the total HI mass, M_H, or, more properly, for the scaled quantity $h^2 M_H$, where h is related to the distance scale parameter in the form $H_0 = 100h$ km s^{-1} Mpc^{-1}. Remember that the total mass, derived from the maximum rotational velocity and a disk radius, scales linearly with h, and hence we will utilize the quantity hM_T. In this section, we will address the following question: how do global properties of a galaxy which are inferred from 21-cm observations, i.e., $h^2 M_H$, hM_T, and the scale length of the HI distribution, correlate with global properties, such as luminosity, linear size, and morphological type, which are measured at other wavelengths? A caveat is necessary at this stage: it has been found that the relationships between those parameters which are intrinsic to a galaxy's structure are not unambiguously defined, unless account is taken of the peculiarities of the environment surrounding that galaxy. Galaxies that live in high-population-density regions exhibit notable deviations in their behavior and possibly in their evolution. In Section 12.6, we discuss the evidence that such perturbations occur. Here we will discuss the properties of unperturbed galaxies—those that have most likely spent their entire existence in relative isolation—and will assume, somewhat arbitrarily, that they define the standards of normalcy.

12.5.1 Relations Between Global Properties for Spirals (Sa and Later)

Derived from the sample of isolated galaxies of Haynes and Giovanelli (1984), Table 12.1 lists the average values of log $(h^2 M_H)$ separately for various morphological-type groupings. The units of mass and luminosity are solar; the units of diameter are kpc and the errors are the standard deviations appropriate for individual objects in each sample (not for the mean). The number of objects used in each determination is also

Table 12.1. Average global properties.

Type(a)	N	$\log(h^2 M_H)$	$\log(M_H/L_B)$	$\log(M_H/D^2)$
Sa, Sab	38	9.37 ± 0.47	-0.55 ± 0.41	6.69 ± 0.32
Sb	74	9.67 ± 0.42	-0.44 ± 0.37	6.83 ± 0.26
Sbc	38	9.54 ± 0.43	-0.32 ± 0.32	6.85 ± 0.19
Sc	72	9.40 ± 0.40	-0.28 ± 0.31	6.79 ± 0.19
>Sc	40	8.93 ± 0.73	-0.04 ± 0.33	6.87 ± 0.17

given. The HI mass, like other global properties of galaxies within a morphological-type group, exhibits wide variations that correlate reasonably well with blue luminosity, and less so with near-infrared ones. The scatter about the mean values of the ratio between HI mass and blue luminosity is reduced, with respect to that about HI mass means alone. As shown in Table 12.1, these two parameters exhibit variation not only within a single morphological class but also between different classes. For example, the M_H/L_B ratio is about twice as large for Sc's as for Sa's. However, the difference between these mean values for even the two extremes of spiral morphology (Sa, later than Sc) is only of the same order as the rms scatter within the various type groups. This large scatter is intrinsic to each statistical grouping, not the result of measurement errors. It arises in part because integrated luminosities contain the contributions of both a disk stellar component and a bulge, which are quite different from both a dynamical and an evolutionary viewpoint; the HI mass, on the other hand, is strictly a disk, extreme Population I component. Galaxies with widely different disk-to-bulge luminosity ratios coexist even within groups of objects sharing the same morphological label. The size of the major axis of the optical image of a galaxy, on the other hand, most likely represents a disk property, and one may expect a tight correlation with the HI mass. The logarithm of the ratio of the HI mass to the square of the linear blue major diameter, also a distance-independent ratio, exhibits a much reduced scatter about the mean for each type grouping, as listed in Table 12.1. In addition, the strong morphological-type dependence seen in M_H/L_B averages is greatly reduced. Involving as it does an HI mass and an optically defined surface area, the M_H/D^2 ratio is a hybrid form of mean HI surface density. As discussed in Section 12.1, there is little evidence for a morphological-type dependence of the ratio between HI and blue radius; thus, the relative type independence of M_H/D^2 suggests that the mean surface density of HI is nearly invariant (within a factor of 2) among spiral galaxies.

HI content can also be measured as a fraction of the total dynamical mass, the latter as obtained from the inclination-corrected HI velocity width and an assumed width-mass relation (the rotation curve). The mean values of this ratio for the various morphological classes are not well determined. Inclinations are sometimes difficult to estimate, samples are reduced by the exclusion of nearly face-on galaxies, and the adoption of standard rotation curves to estimate total masses may be inappropriate for very late galaxies (such as dwarfs and irregulars), the gravity of which may be balanced by disordered, random motions rather than by rotation. Despite these caveats, the HI-to-total mass ratio varies between about 1.3% for Sa galaxies and 2.2% for Sc's, more or less monotonically through the Hubble sequence.

The dispersions around these mean values are rather large, on the order of a factor of 2, comparable with the difference between mean values at the extremes of the spiral sequence. Because the current data do not conclusively indicate that the HI-to-total mass ratio is either type invariant or a good discriminator among the spiral classes, this ratio is not presently as useful an observational parameter reflecting HI content as M_H/D^2.

12.5.2 HI Content of Early-Type Galaxies

Elliptical galaxies are usually not detected in the HI line, even at sensitivity levels of a few mJy. While this may not be totally unexpected, as their stellar content is old and their angular momentum low, some gas should be present as a result of mass loss from aging stars and perhaps even more so in the fraction of ellipticals that appear to have dust lanes, central HII regions, or active nuclear sources that might result from gas inflow. In fact, some detections of HI in ellipticals have been obtained. Figure 12.11 illustrates one such detection, that of NGC 1052. Sanders (1980) proposed that the HI content of E's is bimodally distributed: most E's contain little or no gas, while a few are gas rich. Weighing carefully the information content even of non-detections, Knapp et al. (1985a) have analyzed a much larger sample—approximately 150 objects, of which 23 were detected—and have concluded that the ratio $k = (M_H/L_B)$ is distributed for ellipticals roughly as

$$\log[N(k)] = -1.9(\pm 0.2) - 1.48(\pm 0.13)\log k \qquad (12.11)$$

where $N(k)\,dk$ is the number of objects with (M_H/L_B) between k and $k + dk$. Note this distribution is *not* bimodal. This behavior is very different from that of spirals, which are generally distributed rather in Gaussian fashion than according to a power law. The concept of a "typical" value of M_H/L_B for ellipticals may thus be meaningless. The distribution of HI content for spirals and its correlation with disk properties indicates a tight relationship between the gas and the stellar population; on the other hand, the light of the elliptical bulges and the HI content of E's have little regard for each other. Knapp and coworkers suggest that the HI in E's has an external origin, obtained by accretion of a surrounding envelope or of a gas-rich companion. This proposal is also supported by the tendency for the HI, when found, to be located well outside the optical image of the galaxy and to be characterized by peculiar kinematics.

While it is generally accepted that in the more luminous ellipticals, the HI gas has an external origin and is not produced by evolutionary stellar mass loss, the origin of HI disks in low-luminosity galaxies has been more controversial. Further evidence that the cool gas seen in all ellipticals comes from a source other than mass loss from the stars seen in the optical image is offered by VLA observations of the HI distribution and kinematics in four low-luminosity ellipticals (Lake et al. 1987). The velocity field of these galaxies is roughly regular, implying the presence of a rotating disk. In the brighter two galaxies, the HI distribution is annular; in the fainter two, it is centrally concentrated. In all four cases, however, the HI emission extends to twice the optical (Holmberg) radius. It is hard to see how such large HI disks could accumulate from stellar mass loss over a volume much large than the stellar component.

In the very nearby galaxies, even small amounts of atomic gas can be detected.

VLA observations of the HI in the dwarf elliptical companions to M31, NGC 185 and NGC 205, show the presence of a few times 10^5 solar masses of atomic gas in contrasting configurations (Johnson and Gottesman 1983). Both galaxies show evidence of dust patches near their nuclei and small populations of blue, presumably young stars. In NGC 185, the centroid of the HI distribution is not coincident with either the nucleus or the dust clouds. On the other hand, the HI distribution in NGC 205, while elongated, reveals a rotating disk roughly coincident with the dust and blue star distribution. It seems that ellipticals in fact do contain small but measurable quantities of not only the atomic gas but also the expected associated quantities of dust derived from far-infrared observations (Jura et al. 1987).

The HI distribution in lenticular (S0) galaxies, recently reviewed by Wardle and Knapp (1985), appears to be intermediate between the cases of ellipticals and spirals. While S0a's appear to resemble spirals, in the sense that some properties of the stellar population are correlated with the gas content and that it appears sensible to estimate mean values of M_H/L_B and M_H/D^2, S0's are more like ellipticals. As for ellipticals, the values of k for S0's cover a very wide range and appear to be unrelated to the stellar population.

Knapp and Wardle propose that the HI in S0's may also have an external origin. Perhaps the strongest arguments in favor of such a hypothesis come from the special examples of the polar ring galaxies. These objects are otherwise-normal S0's, viewed edge-on and surrounded by tilted rings of luminous material. The current coexistence of two nearly orthogonal orbital planes suggests the formation of the ring by an unusual event, probably the capture of a gas-rich companion. Observations of the rotation curves of the ring material allow one to probe the *shape* of the dark halo surrounding these S0 galaxies (Schweizer et al. 1983). In combination with clues derived from optical spectroscopy and photometry, the derivation of the HI distribution and velocity field from synthesis maps determines an estimate of relatively young ages, about 1×10^9 to 3×10^9 years, for the observed polar rings. VLA maps are now available for several polar ring systems (van Gorkom et al. 1987). These objects appear to be relatively rich in HI, which is generally aligned with the ring rather than the S0 disk. Furthermore, the HI distribution is sometimes asymmetric and extends well beyond the optical image, suggesting that there has not been enough time for the gas to settle into the ring or for the induced star formation to be completed. The S0's with neutral hydrogen rings but no optical counterpart may be more extreme examples of the same phenomenon (Knapp et al. 1985b). Although ellipticals and S0's are indeed found typically in regions of higher galaxy density than spirals, where the accretion or merger rate is expected to be similarly elevated, it is not yet clear that accretion of gas-rich companions can explain all seemingly anomalous cases of HI in early-type systems.

12.6 Environmental Effects

While some of the differences in the HI content and distribution observed in galaxies of separate morphological classes undoubtedly result from the circumstances of the era of galaxy formation, the present capacity of a galaxy for star formation may be influenced to some extent by its current environment. The content and distribution of the neutral component of a galaxy's interstellar gas serves then as a useful probe of possible external influences on galaxy evolution, as the neutral gas represents a long-lived but vital component for star formation. A recent review of this subject is given by Haynes et al. (1984).

12.6.1 Tidal Interactions

As illustrated in the tidal disruption model of the Leo triplet presented in Figure 12.5, close encounters between neighboring galaxies in loose groups can result in the removal of a significant portion—observed to be as high as 50%—of a galaxy's initial HI mass. Tidal models such as those presented by Toomre and Toomre (1972) illustrate two significant consequences of tidal encounters, consequences that may dramatically alter galaxy evolution. First, if the relative orbital angular momentum vector is at least roughly aligned with the rotational angular momentum of the target disk, then particle capture is enhanced by the prolonged tidal acceleration so that mass transfer from one galaxy to another occurs. Additionally, after the collision, gas either captured from the companion or raised to large z-heights above the plane may be gravitationally pulled back toward the bottom of the target's potential well. The removal of angular momentum from the gaseous component may even cause such gas to fall into the center of the galaxy. This gas influx may fuel nuclear activity; it is well known that many (although by no means all) active galaxies have close companions. Excess infrared and radio emission, indicative of a recent burst of star formation, is observed in obviously interacting galaxies. Such activity may be relatively short-lived. HI observations of active galaxies are discussed in Section 12.7.3.

Because the outer portions of a galaxy are those most likely to be disrupted in an encounter, HI makes an excellent tracer of tidal events. While ongoing tidal interaction can be inferred from observations of disturbed optical morphology, strong optical emission lines, and large infrared, radio, or X-ray flux, the details of an encounter can be more readily deduced from the HI spatial distribution and its velocity field. Cottrell (1978) has claimed that the coexistence of early-type stellar spectra implying recent star formation with colors characteristic of an underlying older population in Irr II galaxies is likely produced by binary collisions in which mass transfer occurs if at least one of the progenitor galaxies contains a significant amount of interstellar gas. A number of nearby Irr II systems such as NGC 3077 and NGC 4747 show HI tidal tails. As mentioned in Section 12.5.2, mass transfer may play an important role in providing the HI presently seen in some elliptical and lenticular galaxies.

Although the density of galaxies is much higher in a rich cluster, the potential for dramatic tidal effects may be greater in loose groups. Since the tidal force varies as $1/r_p^3$ and the effective duration of the encounter roughly as r_p/v_r, the "disruption damage" is just $1/r_p^2 v_r$ in the impulse approximation, where r_p is the perigalactic distance and v_p is the relative velocity of the two galaxies. Because the typical velocity dispersion in a small group is much lower than that found in a rich cluster, a collision in a dense cluster will produce only about one-tenth the damage done by one with the same impact parameter and mass ratio in a small group.

12.6.2 Gas Deficiency in Cluster Spirals

Although tidal encounters may not appear as dramatic as in low-velocity dispersion groups, the cores of rich clusters are nonetheless very harsh environments for the fragile galaxian disks which pass through them. Not only is the density of galaxies much higher than in loose groups, but also cluster cores are often pervaded by a hot ($T \simeq 10^8$ K) intracluster medium responsible for the observed extended X-ray emission. A variety of mechanisms have been proposed which could sweep the interstellar gas from the disk of a spiral moving through the cluster: these include galaxy-galaxy collisions, tidal interactions, ram pressure sweeping by the intra-cluster medium, and evaporation.

Evidence that the circumstance of cluster residence affects galaxy evolution is offered by a variety of observations. The intracluster gas is believed to exert sufficient pressure to bend radio source jets into a variety of head-tail morphologies. Galaxies in clusters have been shown to be much less likely than are field galaxies to exhibit the optical emission lines associated with active star formation. The segregation of early-type galaxies into cluster cores, in contrast with the more widespread distribution of spirals throughout low-density regions, has led to speculation about the removal of gas from galaxies in clusters. Yet the bulk of the evidence, particularly concerning lenticulars, has led to the conclusion that much of the observed morphological segregation was introduced early in the galaxy formation era.

As a tracer of potential star formation, the neutral hydrogen content of galaxies serves as a probe of the efficiency of proposed gas removal mechanisms. In 1973, Davies and Lewis concluded that galaxies in the Virgo cluster possessed a lower HI surface density than did their counterparts in the field. Although that initial study was limited because of selection effects, particularly the Malmquist (luminosity) bias, its conclusions have been borne out by more recent observations. At present, data suitable for analyses of the comparative HI content of spirals now exist for more than ten clusters and for the field. The clusters studied generally contain a sizable population of spirals and are at redshifts less than $z = 0.04$. Some of the clusters have a high galaxian density, conducive to galaxy-galaxy interaction mechanisms, while others are characterized by high cluster X-ray luminosity, implying the presence of a healthy intracluster medium which is producing thermal bremsstrahlung emission. Thus, not only can we test the reality of the HI deficiency in cluster spirals, but also we can investigate the nature of the gas removal process.

As the nearest rich cluster and center of the Local Supercluster, the Virgo cluster provides a laboratory for the investigation of environmental influences on the HI distribution and content of its member spirals. Virgo is dynamically still evolving. Its ellipticals and spirals form separate populations, and it appears that while the ellipticals form a relaxed, collapsed cluster core, the spirals are still infalling. The spirals in the Virgo core are large and near enough that the HI distribution can be mapped in some and the HI content can be measured in many. Numerous authors have carried out such observations of Virgo spirals using Arecibo, Nançay, Effelsberg, Westerbork, and the VLA and consistently find that spiral galaxies in the Virgo core are HI poor with respect to field galaxies. A substantial fraction of galaxies covering a wide range in luminosity and morphological type exhibit HI deficiency by factors which exceed ten. Furthermore, the Virgo core spirals have shrunken HI disks; in many, the HI disk is actually smaller than the optical extent. The latter observation, implying that the outer portions of the HI disk have been swept, is a forceful argument in support of external gas removal mechanisms. A number of Virgo spirals have now been mapped in the CO 2.6-mm line. CO emission traces the distribution of the dense molecular material in spiral disks. Kenney and Young (1986) find that the molecular disks, which are more centrally confined than the HI, show no trace of perturbation, reinforcing the finding based on HI data that sweeping affects primarily the outer gas layers.

The two Abell clusters A2151 and A2147 in the Hercules supercluster provide the opportunity to compare predictions of galaxy-galaxy interaction sweeping models with those of galaxy–intracluster gas ones. Lying at the same distance, the two clusters are quite different in their morphology. A2151 is a denser, and yet more loosely organized, cluster and is associated with a relatively weak X-ray source, while A2147 appears azimuthally symmetric and exhibits a higher X-ray luminosity. The degree of HI deficiency is more pronounced in A2147. Further comparison of the HI deficiencies observed in nine clusters by Giovanelli and Haynes (1985a) shows that the degree of HI deficiency seen in a cluster correlates with the presence of a hot intracluster medium, as implied by cluster X-ray emission. The exact nature of the gas removal process, either ram pressure sweeping by the intracluster gas as the galaxy moves through the cluster, evaporation, or a more complicated combination of the two, is yet elusive. However, it seems that spiral disks which pass through the hostile cluster core lose as much as 90% of their initial HI mass but retain most of their denser molecular clouds. The HI-poor galaxies furthermore show evidence for reduced star formation rates; that is, they are redder than field galaxies of similar morphology. At comparable galaxian densities, the fraction of galaxies which are classified as lenticular is significantly higher, while that of spirals is correspondingly lower, in clusters with high-X-ray luminosity; this suggests that a causal relationship might exist between morphological type and current cluster conditions. While not all cluster S0's need be swept spirals, the observed HI deficiency in cluster spirals does argue that some reinforcement of the initial morphological segregation occurs in environments hostile to diffuse interstellar gas.

12.7 Cosmological Studies

Observations of 21-cm line radiation allow the exploration of cosmological questions because the processes leading to either emission or absorption are generally well understood. The redshifts measured with commonly available HI line spectrometers are among the most accurate. While galaxies of high optical surface brightness dominate volumes of high galaxy density, the low-surface-brightness objects found in the regimes of groups and supercluster peripheries are usually HI rich and thus are easily studied by 21-cm techniques. Furthermore, since the HI signature of an HI disk is distinctive, the observation of the characteristic two-horned 21-cm profile can testify to the presence of a gas-rich disk structure even in objects not visible optically. The 21-cm line thus serves not only as the complement of observations made at other wavelengths but can, in some cases, provide information not obtain-able any other way.

In this section, we discuss further the use of 21-cm line studies to address questions of cosmological relevance. The use of the velocity width–magnitude relation to measure deviations from the Hubble flow has already been discussed in Section 12.4.2. In conjunction with measures made at other wavelengths, 21-cm line ob-servations are critical to our understanding of the structure and evolution of the universe.

12.7.1 21-cm Redshift Surveys

In contrast with studies at optical wavelengths, HI redshift surveys are especially applicable to objects that cluster least and are therefore useful for studying the large-scale structure of the galaxy distribution. The spectacular surge of interest in the study of the large-scale structure of the universe that has occurred in the last decade has stimulated the undertaking of massive surveys to determine the radial velocities of galaxies at progressively fainter magnitude levels. While the emphasis at the beginning of the 1980s was mainly on the rough mapping and the determina-tion of the scale of the inhomogeneities in the distribution of luminous matter, attention has more recently shifted to the topology of the galaxian distribution and its segregation properties. Entering an area traditionally the monopoly of optical instruments, large-aperture radio telescopes have become important as redshift machines. In this section, we will review the technical aspects of redshift surveys, and in the next one, some of their results will be presented.

Following Table 12.1, we can assume that the mean expectation value for the HI mass of a spiral galaxy is on the order of $3 \times 10^9 \, M_\odot$ (for $h = 1$, as will be assumed throughout this section). Via Equation (12.3), that mass translates into a flux integral of $1.3 \times 10^4 \, d^{-2}$ Jy km s^{-1}, where d is the distance to the galaxy in Mpc. For a most probable inclination of the disk, an observed velocity width of about 400 km s^{-1} should be expected; if we express the distance in terms of a radial velocity v_3 (in 1000 km s^{-1} units), the average flux density of the galaxy over the line profile will be $320v_3^{-2}$ mJy. For comparison, a typical HI observation of 5 minutes on source plus 5 minutes off source (for subtraction of sky and instrumental effects) yields an rms noise of about 1 mJy, at 20-km s^{-1} resolution, with the Arecibo 305-m telescope (using a GaAs FET receiver and the current line feed system, which give a system temperature of 35 to 65 K, depending on the zenith distance of the source). Because the signal will be spread over many channels, an average signal-to-noise ratio of 2 will be more than sufficient to ensure detection. In other words, a "typical" spiral can be detected in one "on-off" pair, at Arecibo, out to a distance corresponding to $+15,000$ km s^{-1}. Such a radial velocity is larger than the characteristic redshift "depth" of most galaxy catalogues, such as the *Uppsala General Catalogue* (Nilson 1973) or the *Catalogue of Galaxies and Clusters of Galaxies* (Zwicky et al. 1961–68), which, combined, provide listings for about 30,000 galaxies in the northern hemi-sphere. Improvements in receiver technology and other instrumental upgrades promise to expand the number of accessible sources by one order of magnitude in the next few years.

Unlike continuum surveys with single-dish radio telescopes, 21-cm redshift surveys are not confusion limited. Adding the third dimension provided by the radial velocity guarantees that the vast majority of sources will be separable, even in high-density environments such as clusters. Confusion—or the inability to separate line blends—only occurs in very tight pairs or groups with small velocity dispersions and affects less than 1% of all surveyed sources.

Currently, other instrumental limitations, besides sensitivity, limit the scope of 21-cm redshift surveys. Standard autocorrelation spectrometers allow instantaneous coverage of velocity windows on the order of $+8000$ km s^{-1}. Thus, in spite of the fact that angular size is, to the first order, a good distance indicator, the search for the 21-cm line emission of a given galaxy may take several observations in contiguous velocity ranges. Currently, effective redshift searches of 21-cm emission are limited to velocities below $+25,000$ km s^{-1}. Man-made interference constitutes a severe handicap at some frequencies of interest for the redshifted 21-cm line, and special precautions, such as lateral shielding of the receiving feed sytem or software excision of interfering signals, are becoming necessary.

The first extensive 21-cm survey of cosmological significance was conducted by Fisher and Tully (1981), on a sample of approximately 2000 galaxies in the Local Supercluster. Currently, a sample about twice as large is nearing completion in the Pisces-Perseus supercluster region (Giovanelli and Haynes 1985b), and somewhat smaller efforts are under way in the Cancer and Hercules regions (Bicay and Giovanelli 1986). Figure 12.9 displays partial results of the Pisces-Perseus survey, including the mapping of a long filamentary structure which probably extends well beyond the boundaries of the surveyed region. The Local Supercluster is also being studied to fainter levels, with current attention concentrating on the characteristics of the dwarf galaxy population (Hoffman et al. 1987).

12.7.2 Voids and the Segregation of Galaxian Properties

Progress in the collection of radial velocity data, both by optical and radio means, has made clear that the distribution of optically luminous matter may be outlined by regions of enhanced density mixed with "voids," regions largely evacuated of luminous matter. Density contrasts on the order of 10 to 100, between the large-scale high- and low-density regions, are common. Currently popular cosmological models require the universe to have a mean density equal to the critical value (i.e., one that will asymptotically reduce the expansion rate of the universe to zero). The majority of matter necessary to achieve critical density is not visible and probably not in baryonic form. Virial analysis of groups and clusters of galaxies indicates that if the universe has indeed the critical density, the dark matter is not as clumped as are the luminous galaxies. It then appears difficult to understand how the baryonic component—in this scheme, a dynamically minor component of the universe—would have separately evolved to the high degree of clustering that is observed. One possible solution to the problem is that the baryonic mass distribution, in spite of the highly contrasted picture offered by the bright galaxies, is actually only slightly nonhomogeneous. In that case, one must postulate that galaxy formation is a threshold process. If bright galaxies only formed in 2- to 3σ peaks of the density fluctuations of pregalactic material, bright galaxy formation thrived in regions dominated by large-scale perturbations slightly above the average density.

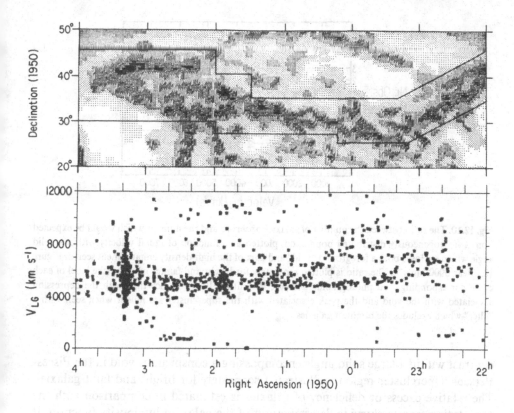

Fig. 12.9. The upper panel displays the density of galaxies brighter than $m = 15.7$ per unit solid angle, as contours of different shade intensity. The lower panel illustrates the radial velocity distribution, as a function of Right Ascension, of the galaxies which are contained within the jagged contour outlined in the upper panel, emphasizing the existence of a three-dimensional structure which coincides with the enhancement in the surface density visible in the upper panel (From Giovanelli et al. 1986).

That process was inhibited, however, in large-scale regions with mean density slightly below the universal average. This scheme would result in a large apparent amplification of the density contrast; such "biasing" of the galaxy formation process is well described by Rees (1985).

We may then address the observational questions of whether "voids" are actually devoid of baryonic matter or are populated by less conspicuous conglomerates such as very-low-luminosity galaxies. The answers require the study of samples of intrinsically faint galaxies over large volumes. Because their surface brightness is also low, such objects become increasingly difficult targets of optical surveys. A large number of them are HI rich, however, because a large fraction of their mass is in the form of HI (See section 12.5).

Recent studies (Bothun et al. 1986, Oemler 1986) indicate that dwarf galaxies are far from filling the voids. Studies of larger samples (such as that in the Pisces-Perseus supercluster, mentioned in Section 12.7.1) suggest that the density contrast between low- and high-density regions diminishes with the intrinsic luminosity of the galaxian population. This effect is sketched in Figure 12.10, where the density

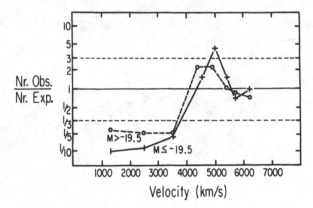

Fig. 12.10. The ratio between the number of galaxies observed and the number which would be expected from a homogeneously distributed population, plotted as a function of radial velocity within a solid angle which encloses both a foreground "void" and part of the high-density enhancement seen in Figure 12.9 near 5000 km s^{-1}. The ratio is plotted separately for bright and fainter galaxies (the label of each curve corresponding to a photographic absolute magnitude limit). Notice that both the depression associated with the void and the peak associated with the supercluster are milder when seen with a "filter" which excludes the brightest galaxies.

contrast within a large solid angle encompassing a conspicuous void in the Pisces-Perseus supercluster region is illustrated separately for bright and faint galaxies. The relative excess or deficiency of galaxies is estimated in comparison with the expectation from a skywide determination of the galaxian luminosity function. It shows that the density contrast between high- and low-density regions diminishes for fainter galaxies, a trend in the direction indicated by biased galaxy formation schemes.

The effect illustrated in Figure 12.10 indicates that there is luminosity segregation between different local density regimes. Other forms of segregation according to morphological type, gas content, and total mass are present. Such variations are intimately related to the physical processes that led to galaxy formation. Some of these effects, such as the local density dependence of the gas content, can be exacerbated in particularly active environments, such as the cores of clusters, as was described in Section 12.6. Those secularly occurring effects, however, may be small-scale alterations to the inbred characteristics of the galaxian population.

As an alternative component of baryonic matter capable of making a contribution to the mass density distribution, the existence of not fully collapsed intergalactic clouds of HI has been postulated. Searches, both directed and serendipitous, for such objects (Krumm and Brosch 1984, Altschuler et al. 1987) have been unsuccessful. Whether diffuse HI is abundant in intergalactic space is, however, a relatively open question because the dominating excitation process of the hyperfine levels and the spectral signature of possible HI features are still largely uncertain.

An alternative mode of investigating the dynamic mass contrast, i.e., whether the overall mass distribution is traced by that of luminous galaxies, consists in studying the magnitude of large-scale peculiar motions of galaxies. If voids are indeed devoid of mass, then substantial deviations from a smooth Hubble flow, as discussed in Section 12.4.2, should be visible in the population of galaxies near their edges.

12.7.3 HI in Active Galaxies

Some of the most luminous galaxies in the universe are those that are undergoing an enhanced phase of activity. The term "active" is applied to galaxies that are undergoing a variety of processes that increase the total luminosity of the host galaxy. Most often, the activity is confined to the nucleus or the inner few kiloparsecs of the galaxy. Seyfert galaxies contain bright, starlike nuclei and exhibit broad emission lines. The survey performed by the Infrared Astronomical Satellite (IRAS) in the wavelength range from 10 to 100 μm has revealed a new category of objects that are ultraluminous in the far-infrared bands. More often than not, these objects show signs of interaction with companions, and their enhanced infrared luminosity results from the heating of dust by ultraviolet photons emitted in a burst of star formation [see the review by Soifer et al. (1987)]. Even the Milky Way harbors energetic phenomena in its own nucleus, and it is likely that many galaxies will be active for some portion of their lifetime.

The Seyfert phenomenon occurs most often in early-type spirals and is proposed to result from the powering of nuclear activity by the accretion of gas onto a massive object in the center of the galaxy (Weedman 1986). HI surveys of Seyfert galaxies as have been conducted by Heckman et al. (1978) and Mirabel and Wilson (1984) help to establish the nature of the galaxy containing the Seyfert nucleus. While for the majority of Seyfert galaxies, the ratio of hydrogen mass to optical luminosity, M_H/L, is similar to that expected for galaxies of similar optical morphology, some Seyferts appear to be usually gas rich. The redshifts obtained from the 21-cm line observations are systematically larger than those derived from the nuclear optical emission lines, suggesting a net outflow of gas in the narrow-emission line regions. A significant percentage, 40% or so, of detected Seyfert galaxies show highly asymmetric HI profiles, not the characteristic two-horned ones expected from quiescent spiral disks. It is not yet known whether these deviations arise from interactions with companions, ionization of significant HI in selected regions of the disk, or blending of more than one emitting galaxy within the single-dish beam.

Because active galactic nuclei are often the hosts of compact radio continuum sources, 21-cm absorption as well as emission may often be seen both in Seyfert galaxies and in others that are radio sources. As illustrated in Equation (12.1), an HI cloud of total optical depth τ which is bathed in a radiation field with a significant flux at 21-cm wavelength will produce a spectrum that is the combination of the emission by (and self-absorption within) the cloud and the absorption of the continuum radiation originating behind the absorbing cloud. Because of the magnetic dipole nature of the 21-cm line transition, and of the fact that a significant correction of stimulated emission must be applied to τ, the 21-cm line optical depth is normally very low. Because of this small opacity, the detection of HI in absorption is possible only where the line of sight to a source of radio continuum emission passes through an HI region with a large column density, N_H, of neutral hydrogen. The resulting opacity τ is determined by the ratio N_H/T_S, where T_S is the spin temperature, as described in Section 12.1.2. If N_H can be derived from measurements at other spectral regimes, τ can then provide an estimate of the excitation characteristics of the gas. Furthermore, the ability of a source to appear in absorption, given some value of optical depth for the intervening HI region, depends only on its continuum flux density, not on its distance. Thus, HI absorption can be detected in objects at very high redshift; such objects need not be particularly massive in HI, but only optically thick along the line of sight and favorably positioned in front of a relatively strong continuum source.

Obviously, absorption lines are produced against sources which emit strongly in the radio continuum, and so information is gleaned only about the line of sight to such sources. HI absorption has been detected in some two dozen galaxies, with typical optical depths of a few hundredths found. Most typically, absorption is seen against a continuum source which is located in the galaxy's center. Absorption arises in clouds contained in the same galaxy, perhaps in its disk. Optical depths are derived on the assumption that the absorbing cloud (or clouds) covers the continuum source completely. If the cloud actually covers only a portion of the illuminating source, the optical depths will be underestimated.

In fact, only a small fraction of radio galaxies, including spirals and ellipticals with prominent dust lanes, exhibit HI absorption. Figure 12.11 shows the HI emission and absorption observed in the lenticular galaxy NGC 1052; the absorption profile is obtained in the direction of its central continuum source. The geometry of the source relative to potential absorbers plays a critical role. While narrow HI absorption features are seen in many active galaxies, those observed in others are broad and resemble features seen in lines of the OH radical. The HI absorption is often found offset from the galaxian systemic velocity, as would be expected if the absorbing material is infalling into the nuclear region. However, exceptions do occur; several cases of HI absorption show features that are blue-shifted by 100 km s^{-1} or more with respect to the galaxy's line emission centroid (Dickey 1982). Detailed maps of the absorption and (when detectable) emission are required in order to establish the location and kinematics of the gas.

The far-infrared emission detected by IRAS arises from both normal and peculiar galaxies. The dominant population of galaxies that are bright in the far infrared are spirals. The infrared emission is likely to arise from two dust components in the disk: a component closely associated with star formation activity in HII regions and molecular cloud complexes, and a second that is heated by the diffuse interstellar radiation field. The latter is identified with the so-called "infrared cirrus" seen in the Milky Way. The exact correlation of these two components with HI in our own galaxy and external galaxies is still under investigation.

Some of the galaxies that are overluminous in the far infrared and particularly identified to have very high star formation rates earn the designation of "starburst" galaxies. The high occurrence of multiplicity and peculiar morphology in these systems suggests that the current burst of star formation is somehow triggered by the interaction process between close companions. Observations of atomic hydrogen in such systems identify them as spirals because their hydrogen masses are generally large, typical of spirals but not earlier-type systems. In a significant fraction, the shape of the HI profile indicates the presence of global disturbances. In the most peculiar of these systems, such as the merger candidate Arp 220, there is a striking absence of atomic gas in locations where molecular hydrogen is found, indicating an enhancement of the atomic-to-molecular conversion process that must ultimately then lead to a higher star formation rate.

Low-optical-luminosity blue compact dwarf (BCD) galaxies, sometimes referred

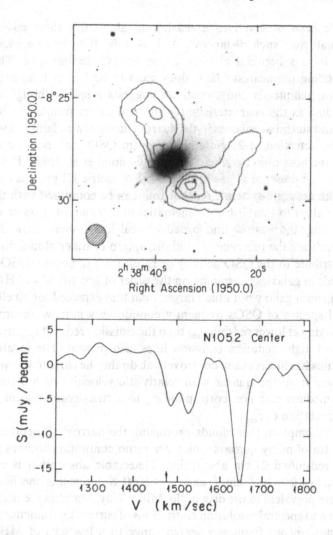

Fig. 12.11. *Top*: HI distribution in the early-type galaxy NGC 1052, obtained with the Very Large Array by van Gorkom et al. (1986). Contours correspond to HI column densities of 0.15, 0.6, 1.07 and 1.83×10^{20} cm^{-2}. *Bottom*: absorption profile obtained against the galaxy's nuclear continuum source.

to as "extragalactic HII regions," also seem to be undergoing active star formation. Like the blue star light, the HI emission is patchy and the HI clumps actually avoid the brightest optical features. Like the normal dI's, the HI usually extends well beyond the optical image. It is not at all clear what triggers star formation in some dwarfs and not in others, even though the HI column densities in both types may remain in excess of $N_H = 10^{20}$ cm^{-2} even beyond the optical radius.

12.7.4 HI in Quasars and in Their Spectra

The host galaxies of quasars (QSOs) are difficult to observe at most wavelengths because the line and continuum radiation from the underlying galaxy is usually overwhelmed by the central QSO itself. And, although the optical galactic envelopes of QSOs have been observed to a redshift of about 0.5, these envelopes are sufficiently small over such distances that it is difficult to choose whether they are better fit by a spheroidal $r^{1/4}$ law or by an exponential disk. The optical "fuzz" surrounding the nearest QSOs does seem to be fainter than expected for typical luminous ellipticals and sometimes suggests a spiral, though deformed, morphology. Just as the characteristic two-horned 21-cm profile can be used to place Seyfert and starburst galaxies in the heart of an otherwise faint or even unseen spiral disk, the detection of 21-cm emission from QSOs can be used as further evidence that the host object is likely a spiral (Condon et al. 1985). The detection of HI indicates the presence of gas in the host. Symmetric HI profiles can be used to determine an accurate systemic redshift that can be compared with the optical emission line values to establish the kinematics and radiation transfer within the regions producing the narrow and broad optical lines. Asymmetric 21-cm line profiles may indicate the presence of tidal disruption or other global disturbances that may contribute to the QSO activity or result from it. Several QSOs are now known to reside in galaxies containing on the order of $7 \times 10^9 \, M_\odot$ of HI, a typical HI mass for a spiral galaxy but much larger than that expected for an elliptical.

The optical spectra of QSOs frequently contain very narrow absorption lines nearly always lying at lower redshift z_{abs} than the emission redshift z_{em}, itself derived from the broad high-excitation emission lines also present. The location of the absorbing clouds has always been controversial: do they lie close to the quasar itself and move away from the quasar with relativistic velocities or are they located at some intermediate distance, corresponding to a strict cosmological recession-velocity interpretation of z_{abs}?

Under the assumption that clouds producing the narrow absorption lines are cold, the spectra of many quasars which are radio continuum sources have been searched for redshifted 21-cm absorption. Detectable absorption is expected if a cloud of $N_H > 1.0 \times 10^{20}$ cm^{-2} and $T_s < 1000$ K intercepts the line of sight. Such clouds are prevalent in the disk of the Milky Way. Searches are made difficult by bandwidth and spectral resolution restrictions of current instrumentation, which limits the instantaneous frequency search range to a few tens of MHz, and by the presence of man-made interference—an increasingly severe handicap—at frequencies below 1 GHz. Nevertheless, absorption of redshifted HI has been detected in about ten QSOs.

The first detection of the 21-cm line in a quasar spectrum was made by Brown and Roberts (1973), who detected it in absorption at 839 MHz in 3C286, corresponding to a redshift of $z_{abs} = 0.69$. The half-width of the line in the rest frame was observed to be 8.2 km s^{-1}. VLBI observations reveal the existence of two narrow-velocity features of dispersion 1.6 and 3.0 km s^{-1}, each located in front of a compact component in the continuum source. The inferred column densities are high, in excess of $8.5 \times 10^{19} \, (T_s/100$ K) cm^{-2}, reminiscent of those seen in the disk of the Milky Way. After the discovery of the absorption at 21 cm, weak absorption features identified with transitions in MgI, MgII, and FeII were detected at optical wavelengths. In Milky Way clouds, the element Mg appears mainly in its singly ionized form; therefore, it appears sensible to search for MgII and HI arising from the same absorber. In such a search, Briggs and Wolfe (1983) found only two

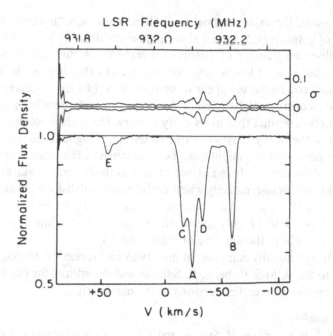

Fig. 12.12. HI absorption line against the BL Lac object AO 0235 + 164, obtained with the Arecibo telescope by Wolfe et al. (1982).

coincidences among eighteen redshifted clouds searched; most optical redshift systems do not contain highly opaque HI gas. Briggs and Wolfe concluded that there is no correlation between 21-cm optical depth and such optical properties as MgII equivalent width, MgII doublet ratio, or MgI equivalent width. This lack of correlation between the radio and optical properties can be explained if the opacity is only occasionally high enough to produce HI absorption, and even in those cases, the HI clouds do not contribute significantly to the optical equivalent widths. Two gas phases are inferred: one showing only optical absorption, similar to Milky Way halo clouds, and one showing also 21-cm absorption characteristic of clouds in a galaxian disk similar to our own.

Perhaps the most intriguing 21-cm absorption line system is the one exhibited by the BL Lac object AO 0235 + 164, discovered by Roberts et al. (1976). This system is the first in which both radio and optical high-redshift absorption lines were measured at coincident redshifts. Figure 12.12 shows the $z_{abs} = 0.52$ line, which occurs at 932 MHz. This spectrum, observed with a resolution of 1.6 km s^{-1}, reveals five separate HI clouds in front of the source. Two gas phases are required to explain both the distinct narrow features and the overall absorption. The narrow features represent high-column-density clouds, each characterized by a velocity dispersion of about 3 km s^{-1} spread over a range in mean velocity of 105 km s^{-1}. In order to fit adequately the optical curve of growth, however, an additional component, optically thin at 21 cm and having a velocity dispersion of about 40 km s^{-1}, must also be present.

If intervening galaxian disks are to be invoked to explain absorption lines in the spectra of quasars, the extrapolation to high redshift of a "normal" luminosity function and of commonly observed sizes of gaseous disks at low redshift cannot explain the observed number of absorption features. A different population of objects from those seen at low redshift or spiral disks that are much larger than their present-day counterparts appear to be necessary. Models of galaxy evolution that have been constructed to explain both the present metallicity and stellar luminosity function predict that in its early history, the typical column density of HI in the disk of the Milky Way was three to ten times higher than it is today; the velocity dispersion in the gas might have been as much as ten times larger than the current value of 10 km s^{-1}. It is not yet clear whether galaxy disks existed at a redshift of 2 or more, or alternatively whether the growth of disks was a slow process taking more than 10^{10} years after collapse to reach the Holmberg radius. As this field of research is still in a highly speculative stage, the 21-cm line appears to be a most promising probe of the universe at high redshifts.

This chapter was initially compiled in mid-1985 and revised in March 1987. The authors wish to thank M.S. Roberts, R. Sancisi and the editors for careful reading and useful suggestions on earlier versions of the manuscript.

Recommended Reading

Roberts, M.S. 1975. *In* A. Sandage, M. Sandage and J. Kristian (eds.), Galaxies and the Universe. Chicago: University of Chicago Press, p. 309.

Sancisi, R. 1981. *In* S.M. Fall and D. Lynden-Bell (eds.), The Structure and Evolution of Normal Galaxies. Cambridge: University Cambridge Press, p. 149.

Haynes, M.P., R. Giovanelli, and G.L. Chincarini. 1984. Annual Rev. Astron. Astrophys. **22**:445.

References

Aaronson, M., J. Huchra, and J. Mould. 1979. Astrophys. J. **229**:1.

Aaronson, M., J. Huchra, J.R. Mould, R.B. Tully, J.R. Fisher, H. van Woerden, W.M. Goss, P. Chamaraux, U. Mebold, B. Siegman, G. Berriman, and S.E. Persson. 1982. Astrophys. J. Suppl. Ser. **50**:241.

Aaronson, M., G. Bothun, J. Mould, R.A. Schommer, and M.E. Cornell. 1986. Astrophys. J. **302**:536.

Altschuler, D.A., M.M. Davis, and C. Giovanardi. 1987. Astron. Astrophys. **178**:16.

Bicay, M.D., and R. Giovanelli. 1986. Astron. J. **91**:732.

Bosma, A. 1981a. Astron. J. **86**:1791.

Bosma, A. 1981b. Astron. J. **86**:1825.

Bothun, G.D., T.C. Beers, J.R. Mould, and J.P. Huchra. 1986. Astrophys. J. **308**:510.

Bottinelli, L., L. Gouguenheim, G. Paturel, and P. Teerikorpi. 1986, Astron. Astrophys. **156**:157.

Brinks, E., and E. Bajaja. 1986. Astron. Astrophys. **169**:14.

Brinks, E., and W.W. Shane. 1984. Astron. Astrophys. Suppl. Ser. **55**:179.

Briggs, F.H., and A.M. Wolfe. 1983. Astrophys. J. **268**:76.

Brosche, P. 1971. Astron. Astrophys. **31**:205.

Brown, R.L., and M.S. Roberts. 1973. Astrophys. J. (Lett.) **184**:L7.

Burstein, D. 1982. Astrophys. J. **253**:539.

Carignan, C., R. Sancisi, and T.S. van Albada. 1987. Astron. J. (to be published).

Condon, J.J., J.B. Hutchings, and A.C. Gower. 1985. Astron. J. **90**:1642.

Cottrell, G.A. 1978. Mon. Not. R. Astron. Soc. **184**:259.

Davies, R.D., and B.M. Lewis. 1973. Mon. Not. R. Astron. Soc. **165**:231.

Dickey, J.D. 1982. Astrophys. J. **293**:87.

Ewen, H.I., and E.M. Purcell. 1951. Nature **168**:356.

Faber, S. 1982. *In* H.A. Bruck, G.V. Coyne, and M.S. Longair (eds.), Astrophysical Cosmology. Citta' del Vaticano: Pont. Acad. Scient., p. 191.

Field, G.B. 1959. Astrophys. J. **129**:536.
Fisher, J.R., and R.B. Tully. 1975. Astron. Astrophys. **44**:151.
Fisher, J.R., and R.B. Tully. 1981. Astrophys. J. Suppl. Ser. **47**:139.
Giovanelli, R., and M.P. Haynes. 1985a. Astrophys. J. **292**:404.
Giovanelli, R., and M.P. Haynes. 1985b. Astron. J. **90**:2445.
Giovanelli, R., M.P. Haynes, and G.L. Chincarini. 1986. Astrophys. J. **300**:77.
Hart, L., and R.D. Davies. 1982. Nature **297**:191.
Haynes, M.P., and R. Giovanelli. 1984. Astron. J. **89**:758.
Haynes, M.P., R. Giovanelli, and M.S. Roberts. 1979. Astrophys. J. **229**:83.
Haynes, M.P., R. Giovanelli, and G.L. Chincarini. 1984. Annu. Rev. Astron. Astrophys. **22**:445.
Heckman, T.M., B. Balick, and W.T. Sullivan III. 1978. Astrophys. J. **224**:745.
Hoffman, L.G., G. Helou, E.E. Salpeter, J. Glosson, and A. Sandage. 1987. Astrophys. J. Suppl. Ser. **63**:247.
Huchtmeier, W.K., J. Seiradakis, and J. Materne. 1981. Astron. Astrophys. **102**:134.
Johnson, D., and S. Gottesman. 1983. Astrophys. J. **275**:549.
Jura, M., D.W. Kim, G.R. Knapp, and P. Guhathakurta. 1987. Astrophys. J. (Lett.) **312**:L11.
Kenney, J., and J. Young. 1986. Astrophys. J. **301**:L13.
Kerr, F.J., and J.V. Hindman. 1953. Astron. J. **56**:218.
Knapp, G.R., E.L. Turner, and P.E. Cunniffe. 1985a, Astron. J. **90**:454.
Knapp, G.L., W. van Driel, and H. van Woerden. 1985b, Astron. Astrophys. **142**:1.
Krumm, N., and N. Brosch. 1984. Astron. J. **89**:1461.
Lake, G., R.A. Schommer, and J. van Gorkom. 1987. Astrophys. J. **314**:57.
Mirabel, I.F., and A.S. Wilson. 1984. Astrophys. J. **277**:92.
Nilson, P. 1973. Uppsala General Catalogue of Galaxies. Uppsala Astron. Obs. Ann. **6**.
Oemler, A. 1986. In S.M. Faber (ed.), Nearly Normal Galaxies: From the Planck Time to the Present Time. New York: Springer-Verlag, p. 213.
Rees, M.J. 1985. Mon. Not. R. Astron. Soc. **213**:75P.
Richter, O.-G., W.K. Huchtmeier, H.-D. Bohnenstengel and M. Hausschildt. 1983. ESO preprint nr. 250.
Roberts, M.S. 1975. In A. Sandage, M. Sandage, and J. Kristian (eds.), Galaxies and the Universe. Chicago: University of Chicago Press, p. 309.
Roberts, M.S. 1978. Astron. J. **83**:1026.
Roberts, M.S., R.L. Brown, W.D. Brundage, A.H. Rots, M.P. Haynes, and A.M. Wolfe. 1976. Astron. J. **81**:293.
Rots, A.H., and W.W. Shane. 1975. Astron. Astrophys. **45**:25.
Rubin, V.C., D. Burstein, W.K. Ford, Jr., and N. Thonnard. 1985. Astrophys. J. **289**:81.
Sancisi, R. 1976. Astron. Astrophys. **53**:159.
Sancisi, R. 1981. In S.M. Fall and D. Lynden-Bell (eds.), The Structure and Evolution of Normal Galaxies. Cambridge: Cambridge University Press, p. 149.
Sancisi, R. 1983. In E. Athanassoula (ed.), Internal Kinematics and Dynamics of Galaxies, Int. Astron. Union Symp. 100. Dordrecht: Reidel, p. 55.
Sanders, R.H. 1980. Astrophys. J. **242**:931.
Sargent, W.L.W., and K.-Y. Lo. 1986. In D. Kunth, T.X. Thuan, and J. Tran Thanh Van (eds.), Star-Forming Dwarf Galaxies and Related Objects. Gif-sur-Yvette: Editions Frontières, p. 253.
Schneider, S.E., G. Helou, E.E. Salpeter, and Y. Terzian. 1986. Astron. J. **91**:13.
Schweizer, F., B.C. Whitmore and V.C. Rubin. 1983. Astron. J. **88**:909.
Shostak, G.S., W.T. Sullivan III, and R.J. Allen. 1984. Astron. Astrophys. **139**:L5.
Soifer, B.T., J.R. Houck, and G. Neugebauer. 1987. Annu. Rev. Astron. Astrophys. (to be published).
Spitzer, L. 1978. Physical Processes in the Interstellar Medium. New York: John Wiley and Sons.
Toomre, A. 1983. In E. Athanassoula (ed.), Internal Kinematics and Dynamics of Galaxies, Int. Astron. Union Symp. 100. Dordrecht: Reidel, p. 177.
Toomre, J., and J. Toomre. 1972. Astrophys. J. **178**:623.
Tully, R.B., and J.R. Fisher. 1977. Astron. Astrophys. **54**:661.
van der Kruit, P.C., and G.S. Shostak. 1983. In E. Athanassoula (ed.), Internal Kinematics and Dynamics of Galaxies, Int. Astron. Union Symp. 100. Dordrecht: Reidel, p. 69.
van Gorkom, J., G.R. Knapp, E. Raimond, S.M. Faber, and J.S. Gallagher. 1986. Astron. J. **91**:791.

van Gorkom, J., P. Schechter, and J. Kristian. 1987. preprint.

van Woerden, H., W. van Driel, and U.J. Schwartz. 1983. *In* E. Athanassoula (ed.), Internal Kinematics and Dynamics of Galaxies, Int. Astron. Union Symp. 100. Dordrecht: Reidel, p. 99.

Wardle, M., and G.R. Knapp. 1986. Astron. J. **91**:23.

Watson, W.D., and S. Deguchi. 1984. Astrophys. J. (Lett.) **281**:L5.

Weedman, D. 1986. Quasar Astronomy. Cambridge: Cambridge University Press.

Wolfe, A.M., M.M. Davis, and F.H. Briggs. 1982. Astrophys. J. **259**:495.

Zwicky, F., E. Herzog, M. Karpowicz, C.T. Kowal, and P. Wild. 1961-68. Catalogue of Galaxies and Clusters of Galaxies, 6 vols. Pasadena: California Institute of Technology Press.

13. Radio Galaxies and Quasars

KENNETH I. KELLERMANN and FRAZER N. OWEN

13.1 Introduction

All galaxies and quasars appear to be sources of radio emission at some level. Normal spiral galaxies such as our own galactic system are near the low end of the radio luminosity function and have radio luminosities near 10^{37} erg s^{-1}. Some Seyfert galaxies, starburst galaxies, and the nuclei of active elliptical galaxies are 100 to 1000 times more luminous. Radio galaxies and some quasars are powerful radio sources at the high end of the luminosity function with luminosities up to 10^{45} erg s^{-1}.

For the more powerful sources, the radio emission often comes from regions well removed from the associated optical object, often hundreds of kiloparsecs or even megaparsecs away. In other cases, however, particularly in active galactic nuclei (AGN) or quasars, much of the radio emission comes from an extremely small region with measured dimensions of only a few parsecs. The form of the radio-frequency spectra implies that the radio emission is nonthermal in origin; it is presumed to be synchrotron radiation from ultra-relativistic electrons with energies of typically about 1 GeV moving in weak magnetic fields of about 10^{-4} gauss (see Section 1.1).

The extended radio sources constitute the largest known physical structures in the universe. Their energy content is very large, up to 10^{60} erg or more. The origin of this energy and the manner in which it is converted into relativistic particles and magnetic fields has remained one of the most challenging problems of modern astrophysics. High-resolution radio images generally show a very compact component which is coincident with an active galactic nucleus or quasar, and which is thought to reflect the "central engine." Long thin "jets" extend away from the compact central core toward the outer radio lobes. Often the jets end up at *hot spots* (Figure 13.1).

13.1.1 Optical Counterparts

The optical identification of the discrete radio sources is important for two reasons. First, it is not possible from the radio measurements alone to determine the distance to a radio source. Only if there is an optical identification can the redshift and, from the Hubble law, the distance be determined. In this way it is then possible to calculate the absolute radio luminosity, linear size, and energy content from measurements of radio flux density and angular structure. Second, optical as well as X-ray and infra-red studies of the radio source counterparts may give some insight into the problem of the origin of the intense radio emission.

The coordinates of radio sources may be routinely measured with an accuracy

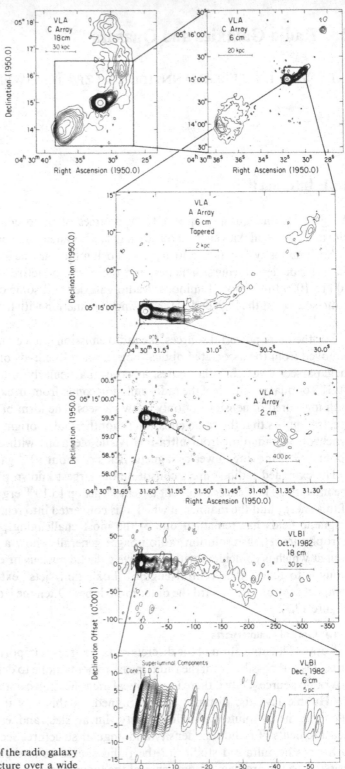

Fig. 13.1. Contour maps of the radio galaxy 3C120, showing the structure over a wide range of angular scales (Walker et al. 1987).

better than one second of arc, generally permitting the unambiguous association with optical counterparts as faint as about 24th magnitude, which is reached with large reflectors and modern instrumentation. Nevertheless, due to the large amount of telescope time required for systematic studies of color and spectra, only for the strongest few hundred radio sources are the optical identifications reasonably complete (Spinrad et al. 1985). But even for very weak radio sources, optical identifications are usually possible.

(a) Historical Background

The discrete sources of radio emission were first distinguished from the general background radiation as a result of their rapid amplitude variations at low frequencies, which were thought to be due to fluctuations in intrinsic intensity (Hey et al. 1946), but which we now recognize are due to scintillations in the earth's ionosphere. Considering that the dimensions of a variable source cannot greatly exceed the distance traveled by light during a characteristic variability time which is typically about one minute, it was generally thought that the discrete "variable" sources must be galactic stars; thus they were originally referred to as "radio stars."

When two of the strongest radio sources were identified by Bolton and Stanley (1949) with the nearby galaxies M87 and NGC 5128, it became clear that at least some of the discrete sources were extragalactic. Later the position of the powerful radio source Cygnus A was measured by Smith (1951) with sufficient precision to permit the identification by Baade and Minkowski (1954) with a relatively faint 15th-magnitude galaxy having redshift of 0.06. After this the extragalactic nature of the discrete sources was widely recognized, although the use of the term "radio stars" for extragalactic radio sources persisted for many years.

Other radio sources were identified with galaxies during the 1950s, but progress was slow because of the poor accuracy of the radio source positions. By the early 1960s, however, the increased use of interferometer systems led to position accuracies of the order of a few arcseconds, and many sources were identified with various galaxy types from inspection of the Palomar Sky Survey, which reaches a limiting magnitude about 20.

(b) Radio Galaxies

Galaxies which are identified with strong radio sources in the range of 10^{41} to 10^{46} ergs s^{-1} are generally referred to as "radio galaxies." For the most part, radio galaxies are giant ellipticals with absolute visual magnitude about -21.*

Many intermediate-luminosity radio galaxies are found in rich clusters of galaxies. X-ray observations show that these clusters often contain a hot (10^8 K), relatively dense (10^{-3} particles per cm^3) intracluster medium. Many of the radio sources in rich clusters show bends or distortions apparently associated with their interaction with this medium.

Many of the more powerful radio galaxies show bright optical emission lines in their nuclei whose strength appears to be correlated with the strength of the compact radio core (Heckman et al. 1983a). Radio galaxies with narrow emission lines in

---

*$H = 100$ km s^{-1} Mpc^{-1}.

their spectrum are referred to as *narrow-line radio galaxies* (NLRG). Typical line widths are of the order of 1000 km s^{-1} and include both forbidden lines of O, N, S, and Fe as well as the Balmer lines of He. *Broad-line radio galaxies* (BLRG) have very broad H and He features, with velocities up to 25,000 km s^{-1}, and may also include narrow emission features as well.

Because of their bright emission lines, it is relatively easy to determine the redshift of many radio galaxies. Indeed, the most distant galaxy redshifts measured are those of radio galaxies. However, for redshifts much greater than unity, the optical K correction becomes very large. Due to the rapid decrease in the brightness of elliptical galaxies toward the blue part of the spectrum, highly redshifted radio galaxies appear very faint at visual wavelengths, but can often be observed in the near infrared.

(c) Quasars

In 1960, the relatively strong, small-diameter radio source 3C295 was identified with a 20th-magnitude galaxy having a redshift of 0.46, and was the most distant galaxy known at the time (Minkowski 1960). Continued efforts to identify distant galaxies concentrated on radio sources of small diameter and high surface brightness since these positions could be measured with high accuracy. In 1961, the small-diameter radio source 3C48 was identified with what appeared to be a 16th-magnitude *stellar* object. The subsequent discovery of night-to-night variations in the light intensity led to the reasonable conclusion that 3C48, unlike the other identified discrete radio sources, was indeed a true *radio star* in our galaxy. Soon, the optical counterparts of two other relatively strong small-diameter radio sources, 3C196 and 3C286, were also found to appear stellar, and it appeared that as many as twenty percent of all high-latitude radio sources were of this type.

Early efforts to interpret the emission line spectrum of the three known *radio stars* were unsuccessful although by 1962 some apparent progress was being made in associating many of the lines in the 3C48 spectrum with highly excited states of rare elements. However, lunar occultation measurements gave an accurate position of the strong compact radio source 3C273 (Hazard et al. 1963). Shortly afterward, Maarten Schmidt (1963) identified 3C273 with a 13th-magnitude stellar object, and he noted that the relatively simple spectrum could be interpreted as a redshifted ($z = 0.16$) Balmer series plus MgII.

A reinspection of the 3C48 spectrum indicated that if the bright line at 3832 Å was identified with the MgII line at 3727 Å in the rest frame, its redshift would be 0.37. Other lines in the 3C48 spectrum could then be identified with OII, NeIII, and NeIV. Additional spectra of other similar objects led to the identification of CIII, CIV, and finally Lyα, permitting much larger redshifts to be easily measured. The word *quasar* is now most often used to describe the entire class of highly redshifted *quasi-stellar objects* or QSOs.

Assuming that the measured redshifts are cosmological and the distance is given by the Hubble law with $H = 100$ km s^{-1} Mpc^{-1}, then the absolute visual magnitudes of quasars range from about -24 to -31. Thus, at optical wavelengths, quasars are up to a few hundred times brighter than the most luminous galaxies. Some relatively low redshift (nearby) quasars, which are strong radio sources, appear to

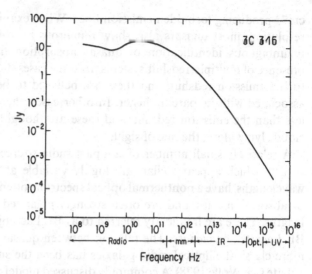

Fig. 13.2. Spectrum of the quasar 3C345 from the radio through the ultraviolet part of the spectrum. Adopted from Bregman et al. (1986) with additional data from the radio literature.

be surrounded by a faint fuzz whose dimensions, color, and brightness are typical of giant elliptical galaxies, thus supporting the idea that quasars are the extremely active nuclei of galaxies. Further down the optical (as well as radio) luminosity function are the Type II *Seyfert galaxies*, which have relatively bright nuclei with broad emission lines and absolute magnitudes of -20 to -23, the so-called *Markarian galaxies*, which were originally isolated on the basis of their large UV excess on photographic plates, and the blazars and BL Lacs. The literature is not always consistent on nomenclature. We will use the term *AGN* to describe galaxies with prominent (active) nuclei and *quasar* for objects where the starlike component dominates although there may be a faint underlying galaxy.

The optical spectrum of quasars is clearly nonthermal with a typical spectral index $\alpha \sim -1$ [(flux density) $\propto$ (frequency)$^\alpha$] which may continue to the near infrared (2.2 μm) as well as to X-ray wavelengths (see Figure 13.2). Becase quasars have an ultra-violet excess compared with the spectra of galaxies, moderate redshifts will cause quasars to appear blue when measured by multi-color photometry or when the color is estimated from the "red" and "blue" plates of the Palomar Sky Survey. This property has proved useful in making optical identifications of radio sources with quasars, without taking individual spectra, but it is not infallible. For very large redshifts, the color may appear neutral or even "red" when compared with stars and galaxies, so the identification of high-redshift quasars depends on position coincidence alone and requires both radio and optical position accuracies of the order of an arcsecond.

Generally, quasar spectra show intense broad emission lines characteristic of a highly ionized gas with $T \sim 10^4$ K and $n_e \sim 10^8$ cm^{-3}, line widths corresponding to velocities of 10,000 km s^{-1}, or more, and dimensions of the order of a parsec. The most commonly observed lines are those of Lyα (1216 Å), CIII (1909 Å), CIV (1549 Å), MgII (2798 Å), OIII (4363 Å, 4959 Å, 5007 Å), and the hydrogen Balmer series.

In addition, there is a larger narrow-line emission region with densities $n_e \lesssim 10^7$

cm^{-3} producing forbidden emission lines. When examined with sufficient spectral resolution, most quasars also show numerous narrow absorption lines, but the unambiguous identification of quasar absorption lines is complicated by the presence of multiple-redshift systems. In some cases the absorption redshift is close to the emission redshift, and these are believed to be intrinsic to the quasar or associated with its parent cluster. In other cases, the absorption redshift is much less than the emission redshift, and these are thought to originate in intervening clouds lying along the line of sight.

A relatively small number of compact radio sources are identified with optical objects which appear stellar, are highly variable at optical as well as at radio wavelengths, have a nonthermal optical spectrum, often very steep, with no emission or absorption lines, and are often strongly polarized at optical and radio wavelengths. These are frequently referred to as BL Lac objects (the prototype object is BL Lacerte), or *blazars*. The relation between quasars and blazars as well as the more classical elliptical radio galaxies has been the subject of much research and debate (see Wolfe 1978). A commonly discussed model considers the BL Lac objects as quasars with enhanced continuum emission which overrides the emission line spectrum. This picture is supported by the detection of faint emission lines in a few BL Lac objects at the time when the strength of the continuum emission is near a minimum.

Not all quasars are strong radio sources. Optical surveys using objective prisms to spot the characteristic bright emission line spectra of quasars or surveys of UV excess objects show that *radio-quiet* quasars are about ten times more numerous than the *radio-loud* quasars. Quasars with very broad emission lines, in particular, do not seem to be strong radio sources. High-sensitivity radio observations indicate that most optically selected quasars however are weak radio sources. The underlying or "host" galaxies of the radio-quiet quasars appear to be spirals, whereas the "hosts" of radio-loud quasars are probably elliptical galaxies. Quasars which are strong radio sources are usually strong X-ray sources.

13.1.2 Radio Source Properties

The radio-frequency spectra and polarization properties of radio galaxies and quasars are characteristic of synchrotron radiation from relativistic electrons having a power law distribution of electron energies with a Lorentz factor, $\gamma \sim 1000$ and a magnetic field strength $B \sim 10^{-5}$ gauss. The radio emission can be conveniently divided into two categories: the *extended* structure, which is *transparent*, and the *compact* structure, where the density of relativistic electrons is so great that the source becomes *opaque* to its own radiation. There is no simple relation between the structure or dimensions of the radio-emitting region and the dimensions of the associated optical galaxy or quasar, although there are clear statistical differences. Most compact sources are identified with QSOs or with active galactic nuclei. However, less powerful compact sources are also found in normal-looking elliptical galaxies as well (see Ekers 1978, 1981 for a more complete discussion). The extended sources are typically associated with galaxies, but many are quasars with no visible optical extent. Most extended sources, particularly quasars, when examined with sufficient sensitivity and resolution, are found to contain a compact central radio component. The central components are particularly prominent in quasars (e.g., Owen and Puschell 1984). On the other hand, most compact sources, when examined with high sensitivity and dynamic range, exhibit weak extended radio structure. Because the compact sources are affected by self-absorption, their spectra are flat (Section 13.1.3). They are therefore most easily detected by radio surveys made at short wavelengths, whereas the steep-spectrum extended sources with their transparent spectra are characteristic of long-wavelength surveys. The terms *extended* (or lobe-dominated) source and *compact* (or core-dominated) source are often used to describe sources where the extended or compact structure, respectively, is most pronounced.

In the less powerful radio galaxies, the radio emission is often confined to the region of optical emission, or about 10 kpc, but in the more powerful radio galaxies, the radio emission comes from two well-separated regions hundreds of kiloparsecs across. In the giant radio galaxies, radio source dimensions larger than 3 Mpc have been observed. The compact features have dimensions typically ranging from 1 to 100 pc, although in a few nearby galaxies radio nuclei as small as 0.01 pc have been observed.

13.1.3 Radio Spectra

With the exception of the 21-cm line of neutral hydrogen, H_2O and OH found only in relatively nearby galaxies, there are no sharp features in the radio spectra of galaxies and quasars, and the observations are confined to measurements of the continuous spectra. Since individual radio telescopes generally operate only over a limited range of wavelengths, the determination of broadband spectra requires the combination of data obtained by many observers using many different types of telescopes. Because radio telescopes may differ widely in their characteristics, each antenna and radiometer system must be separately calibrated at every wavelength where observations are made. Generally, this is done by observing one or more sources whose intensity is known on an "absolute" scale. Standard sources calibrated with an absolute accuracy of a few percent are available over a wide range of wavelengths from short millimeter wavelengths to wavelengths of a few meters. The determination of relative intensities is much easier, at least at the shorter wavelengths where confusion from the galactic background is less of a problem.

Thousands of extragalactic sources have now been observed at decimeter and centimeter wavelengths, and for several hundred sources the spectra are complete over a range of wavelengths extending from a few millimeters to a few tens of meters (10 MHz to 100 GHz). In a few cases, the spectra extend to 1 mm (300 GHz), but measurements at short millimeter wavelengths are difficult due to the variable opacity of the atmosphere.

Radio spectra are usually displayed in the form of a logarithmic plot of flux density versus frequency (see Figure 13.3). Sources with power law spectra are then represented by a straight line, with slope equal to the spectral index, α. Although the radio spectra of only a few sources follow such a simple power law accurately, a spectral index may be defined at any frequency as the derivative $d(\log S)/d(\log v)$ or by the measurement of flux density at two arbitrarily selected frequencies. The observed spectra of extended sources generally show negative curvature in the

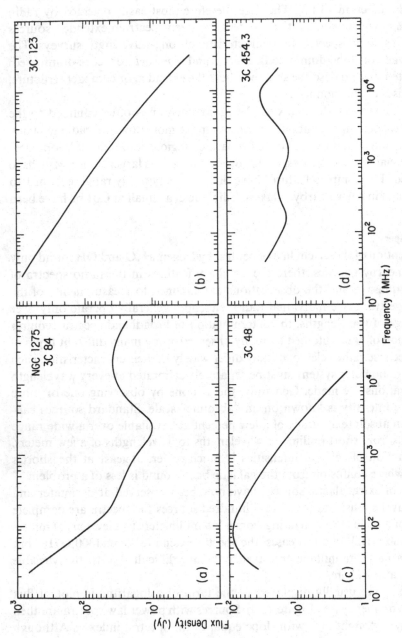

Fig. 13.3. Typical radio-frequency spectra: (a) the radio galaxy 3C84, showing low-frequency power law component which comes from the large-scale structure, an intermediate-size feature which becomes opaque at a frequency of a few GHz, and the small nuclear source which is opaque below 20 GHz; (b) the radio galaxy 3C123 which is transparent throughout the observed range of frequency but has a spectrum which steepens at high frequencies; (c) the quasar 3C48 which has a self-absorption cutoff near 100 MHz and is transparent at higher frequencies; and (d) the quasar 3C454.3 which has multiple peaks due to the superposition of several features which become opaque at widely different frequencies.

$\log S$–$\log v$ plane, that is, the spectrum becomes more steep at high frequency. Typically the region of curvature extends over a decade or so of frequency, At frequencies significantly removed from the maximum curvature, the spectrum can be represented by two well-defined power laws. Values of the spectral index of the extended radio features are in the range $-1.3 < \alpha < -0.5$, and over a wide range of frequencies show a strong concentration near -0.8 with a dispersion of only 0.15. The steepest spectral index which is observed is about -2.0 and the flattest about -0.5.

Radio sources or components of sources with spectra flatter than -0.5 are nearly always very compact, and are coincident with a quasar or AGN. In these sources, the flat spectrum is thought to be the result of self-absorption, rather than a flat electron energy distribution. In some sources, particularly quasars and BL Lac objects, the spectra remain opaque least up to a few hundred GHz but steepen at the infrared and optical wavelengths to a spectral index of about -1 (e.g., Ennis et al. 1982, Landau et al. 1983, Bregman et al. 1986, Roellig et al. 1986). Some sources show a flattening toward the near UV, which is often referred to as the "3000 Å bump." In general, only about ten percent of the radiation from quasars and AGNs is emitted at radio wavelengths, with most of the power being radiated at submillimeter wavelengths.

In the extended regions, where the relativistic plasma is transparent (optically thin) to its own radiation, and the observed spectral flux density is merely the sum of the radiation from the individual electrons and reflects the distribution, $N(E)$, of relativistic particle energy (see Chapter 1). In the case of a power law distribution of particle energies, $N(E) = KE^{-p}$, the radiation spectrum is a power law with $S \propto v^{\alpha}$, where the spectral index $\alpha \sim (1 - p)/2$ (Equation 1.14). The characteristic spectral index $\alpha \sim -0.8$ frequently found in the extragalactic sources then corresponds to a value of $p \sim 2.6$, which is close to the index of primary cosmic-ray particles in the Galaxy.

Even if relativistic electrons are initially produced with a power law distribution, differential energy losses will alter the energy spectrum, so that it is steeper at higher energy. Relativistic electrons lose energy by synchrotron radiation and by the inverse Compton effect, which are both proportional to the square of the energy; by ordinary bremsstrahlung and adiabatic expansion, which are directly proportional to the energy; and by ionization, which is approximately proportional to the logarithm of the energy. Approximating the logarithmic term by a constant, the rate of energy loss may be written

$$\frac{dE}{dt} = aE^2 + bE = c \ . \tag{13.1}$$

If the electrons are being supplied to the source at a rate $N(E, t)$, then the equation of continuity describing the time dependence of the energy distribution $N(E, t)$ is

$$\frac{\partial N(E,t)}{\partial t} = \frac{\partial}{\partial E}\left(\frac{dE}{dt}\right) N(E,t) + N(E,t) \ . \tag{13.2}$$

It is of interest to consider the case where synchrotron losses dominate ($b = c = 0$). Then from Equation (1.10), $a = -120B^2$. If the initial particle distribution is a power law of the form $N(E) = KE^{-p}$ between E_1 and E_2, and zero elsewhere, and if there is no continual injection or acceleration, then the energy distribution will remain a power law with the same slope, but with an amplitude which decreases with time according to

$$N(E, t) = \frac{KE^{-p}}{(1 - 120B^2Et)^2} \qquad \text{for } E'_1 < E < E'_2 \qquad (13.3)$$

where $E' = E/(1 + 120B^2Et)$. Thus, even with an initial energy distribution extending to unlimited energy, after a time t years, there will be an upper energy cutoff at

$$E_c = \frac{1}{120B^2t} \text{ GeV}.$$

From Equation (1.8) there is a corresponding cutoff in the synchrotron radiation spectrum at a frequency $v_b \sim B^{-3}t \text{ (yr)}^{-2}$ GHz.

If the distribution of electron pitch angles is random, the cutoff frequency for each pitch angle differs. At low frequencies where energy losses are not important, the spectral index, α, remains equal to its initial value $\alpha_0 = (1 - p)/2$. If the pitch angle distribution is conserved, then for $v \gg v_b$, $\alpha = \frac{4}{3}(\alpha_0 - 1)$ (Kardashev 1962). If, on the other hand, the pitch angle distribution is continuously made random, for example, by irregularities in the magnetic field, then all the electrons see the same effective magnetic field and the spectrum will show the same sharp cutoff which is observed with a single pitch angle. No such cutoff is observed, even for those sources whose spectra are determined out to 100 GHz or more.

If relativistic electrons are continuously injected with $Q(E) = KE^{-p_0}$, then for $v \ll v_b$ the spectral index again remains constant with $\alpha = \alpha_0$ [$\alpha_0 = (1 - p_0)/2$]. But for $v \gg v_b$ where the rate of energy loss is balanced by the injection of new particles, the equilibrium solution of Equation (13.2) with $(\partial N/\partial t) = 0$ gives $\alpha = (\alpha_0 - \frac{1}{2})$.

Observations over the frequency range 10 MHz to 100 GHz show curvature of the form expected from synchrotron radiation losses, with $v_b \sim 1$ GHz. Typically, $\Delta\alpha \sim \frac{1}{2}$, as expected if relativistic electrons are continually supplied. If a few sources $\Delta\alpha \sim 1$, suggesting that in these sources particle acceleration may have ceased.

Quantitative analysis is difficult, since the spectra may vary across the source, particularly if the magnetic field is not constant. Generally, the hot spots and jets appear to have flatter spectra than the more extended diffuse components, apparently reflecting their younger ages and correspondingly smaller synchrotron radiation losses. The very diffuse components associated with clusters have the steepest observed radio spectra with indices generally steeper than -1.

13.1.4 Energy Considerations

The problem of the origin and evolution of extragalactic radio sources is a formidable one; in particular, the source of energy needed to account for the large power output and the manner in which this energy is converted to relativistic particles and magnetic flux is a subject of considerable debate. Assuming only that synchrotron radiation from ultra-relativistic electrons is responsible for the observed radiation, the necessary energy requirements were shown by Burbidge (1958) to be as much as 10^{60} ergs or more.

Following Burbidge, if the relativistic particles have a power law distribution with an index p between E_1 and E_2, then for $p \neq 2$, the energy contained in relativistic electrons is

$$E_e = \int_{E_1}^{E_2} EN(E)\, dE = \frac{K}{(2-p)} [E_2^{(2-p)} - E_1^{(2-p)}] . \tag{13.4}$$

The constant K can be evaluated if the distance to the source is known. The total luminosity, L, is given by integrating Equation (1.10), or

$$L = \int_{E_1}^{E_2} N(E) \frac{dE}{dt}\, dE = \int_{E_1}^{E_2} 120B^2 E^{(2-p)}\, dE$$

$$= 120 \frac{KB^2}{(3-p)} [E_2^{(3-p)} - E_1^{(3-p)}] . \tag{13.5}$$

For $p = 2.5$, and $q_0 = +1$, $L \sim 10^{44} z^2 S$, where S is the flux density at 1 GHz.*
Eliminating K between Equations (13.4) and (13.5) we have

$$E_e = \frac{L}{120} \left(\frac{3-p}{2-p} \right) \left(\frac{E_2^{(2-p)} - E_1^{(2-p)}}{E_2^{(3-p)} - E_1^{(3-p)}} \right) \text{ ergs} . \tag{13.6}$$

Using Equation (1.8) to relate E_2 and E_1 to the cutoff frequency and grouping all the constant terms together,

$$E_e = C_e(p)LB^{-3/2} . \tag{13.7}$$

The total luminosity L of the source may be estimated by integrating the observed spectrum between 10 MHz and 100 GHz. The magnetic energy is obtained from

$$E_B = \int \frac{B^2}{8\pi}\, dV = C_B B^2 V . \tag{13.8}$$

The total energy in fields and particles ($E_c = E_e + E_B$) is minimized when $dE/dB = 0$ or when

$$B_{min} = \left(\frac{3}{4} \frac{C_e}{C_B} \frac{L}{V} \right)^{2/7} \sim 1.5 \times 10^{-4} \theta^{9/7} z^{-2/7} S^{2/7} \text{ gauss} . \tag{13.9}$$

The value of B estimated in this way must be treated with caution. It depends almost entirely on the angular size, θ, and is relatively insensitive to the flux density, distance, or spectral index.

From Equations (13.7), (13.8), and (13.9), if θ is expressed in arcseconds, then

$$(E_e)_{min} \sim \tfrac{4}{3}(E_B)_{min} \sim 10^{59} \theta^{9/7} z^{17/7} S^{4/7} \text{ ergs} \tag{13.10}$$

and depends only weakly on p. Thus, the minimum energy is given when the energy is nearly equally distributed between relativistic particles and the magnetic field, and this is usually referred to as the *minimum energy* or *equipartition* case.

---

*Numerical expressions given in this chapter are evaluated for $H = 100$ km s^{-1} Mpc^{-1}, $q_0 = 1$, and $p = 2.5$ ($\alpha = 0.75$). More general formulations for other values of γ and more complex geometries are given in Moffet (1975), Jones et al. (1974a,b), and Marscher (1983).

Typically, the total energy contained in the extended sources is estimated to be in the range of 10^{57} to 10^{61} ergs and the magnetic field between 10^{-6} and 10^{-4} gauss. Under these near equilibrium conditions ($E_e \sim E_B$), the total energy depends to a large extent on the size of the source ($E \propto r^{9/7}$). Thus the larger sources with low surface brightness and low luminosity, such as Centaurus A, are calculated to contain almost as much energy as the smaller but much more powerful high-surface-brightness objects such as Cygnus A or 3C295. However, high-resolution observations indicate that in many sources the observed radio emission comes from only a small fraction, Φ, of the projected volume. The minimum total energy calculated from Equations (13.9) and (13.10) is then multiplied by a factor of $\Phi^{3/7}$, and the corresponding magnetic field is increased by the factor $\Phi^{-2/7}$. It is by no means clear that minimum energy, or equivalently equipartition, conditions hold in extragalactic radio sources. Moreover, the value of the fill-in-factor Φ is very uncertain, and calculation of energy content or magnetic field strength based on minimum energy or equipartition arguments must be treated with caution.

For some years it was widely thought that the relativistic electrons were secondary particles produced as the result of collisions between high-energy protons. If the ratio of the number of protons to electrons is k, then the minimum total energy is increased by a factor of $(1 + k)^{4/7}$ and the magnetic field by $(1 + k)^{2/7}$. Some estimates of the value of k were as high as 100, with a corresponding increase in energy requirements by about an order of magnitude. Elimination of the factor k and inclusion of the fill-in factor Φ can easily reduce the minimum-energy estimates by two or more orders of magnitude.

A more direct, although not necessarily more accurate, method of determining the magnetic field in extended extragalactic radio sources is based on the scattering of the microwave background radiation by the relativistic plasma. The ratio between the synchrotron radio flux and Compton scattered X-ray flux depends on the magnetic field strength and is given approximately by (Miley 1980)

$$B \sim \{6.6 \times 10^{-40}(4800)^{\alpha}(1 + z)^{3-\alpha}[S(v)/S(x)]v^{-\alpha}E(x)^{-\alpha}\}^{1/(1-\alpha)} \text{ gauss} \quad (13.11)$$

where $S(v)$ is the radio flux density at frequency v and $S(x)$ is the X-ray flux density at energy $E(x)$. Values of B estimated from Equations (13.9) and (13.11) are in the range of 10^{-6} to 10^{-4} gauss, so that the energy content appears to be close to the minimum requirement. But estimates of B from Equation (13.11) are subject to error if the thermal X-ray flux is not negligible compared with the Compton scattered X-ray flux.

In a 10^{-4}-gauss field, electrons radiating at $v > 1$ GHz are expected to decay in about 10^6 years (Equation 1.8). Thus, the absence of any observed spectral cutoff even at $v \gg 10$ GHz suggests continued acceleration of relativistic particles. Even more restrictive limits are imposed by the observation of both radio and optical ($v \sim 10^{14}$ Hz) synchrotron emission from the M87 jet at least 10 kpc away from the nucleus (Biretta et al. 1983a).

Because an adiabatically expanding cloud of relativistic particles loses energy as R^{-2p} (e.g. Kellermann and Pauliny-Toth 1968), it has been argued that if the observed relativistic electron clouds had expanded adiabatically from a much smaller region, the initial energy requirements would be prohibitively large. Although a variety of models have been discussed which allow some of the kinetic energy of expansion to be converted into relativistic particle energy, the adiabatic loss problem has been a prime motivation for models which consider the primary energy source to be in a compact region at the quasar or AGN with the energy being transported from this "central engine" to the outer lobes by a relativistic *beam*. (Section 13.4.1). However, it has also been pointed out by Bicknell (1986) and others that if the source has expanded to some final size and another burst of particles occurs, without significantly expanding the source, then no further adiabatic losses occur.

13.2 Low-Luminosity Sources

The powerful radio galaxies and quasars have been recognized largely as a result of the optical identification of radio sources that have been catalogued in various radio surveys. By contrast, sources at the low end of the radio luminosity function are observed primarily by measuring the radio emission from known optical objects. The most extensive surveys are those of Sadler (1984), Kotanyi (1980), Hummel (1980), and Dressel and Condon (1978). These so-called "normal galaxies" typically have a radio luminosity between 10^{36} and 10^{39} erg s^{-1}. They are primarily ellipticals, but some spirals are also detected as weak radio sources (e.g., Ekers 1978, 1981, Sadler 1984, Preuss 1987).

The observed radio emission may come from an extended component comparable in size to the optical image or from a bright compact nucleus. Although the luminosities of the weak radio nuclei are some million times less than those found in radio galaxies and quasars, the surface brightness, magnetic field strength, structure, and time scale of observed intensity variations are all comparable to those of their more luminous counterparts discussed in Section 13.3.

13.2.1 Spiral, Seyfert, and Irregular Galaxies

For normal spiral galaxies, much of the radio emission comes primarily from the disk. In a few nearby spirals where the distribution of radio emission can be mapped, the spiral structure is clearly evident. This diffuse component of the nonthermal emission appears to be unrelated to that of the much more luminous radio galaxies and quasars but rather to the optical luminosity and morphology of the galaxy. High-resolution observations indicate that the characteristic size of radio nuclei of spirals is about 100 pc, but significant emission sometimes comes from a region less than a parsec across (Dressel and Condon 1978, Hummel 1980, van der Hulst 1981, Heckman et al. 1983a).

These latter sources are probably related to the powerful compact sources found in AGNs and quasars. The discovery of flat-spectrum, compact, variable sources in the nuclei of M81 and M104 (de Bruyn et al. 1976) gave the first evidence of activity in the nuclei of normal spirals of the type characteristic of quasars and AGNs. High-resolution VLBI observations of M81 show that the radio nucleus is only about 0.01 pc in extent. Based largely on the presence of narrow emission lines in the optical spectrum, M81 is sometimes classified as a Seyfert galaxy. But most

Seyfert galaxies are not strong radio sources, although a few, such as NGC 1068, have prominent radio cores and jets which have been extensively studied (e.g., Wilson 1982). Often the cores of Seyfert galaxies show a low-frequency spectral turnover due to free-free absorption by ionized gas in the emission-line region, with emission measures in the range of 10^4 to 10^6 cm^2. Meurs and Wilson (1984) discuss the radio emission from Seyfert galaxies and their relation to other radio sources.

A small fraction of spiral and irregular galaxies show enhanced radio emission which is closely correlated with the 10-μm infrared flux density and apparently corresponds to regions where there are bursts of star formation (e.g., Condon 1982). At one time the dramatic visual image of M82, particularly in Hα, was interpreted as evidence of explosive activity characteristic of the powerful radio galaxies. More recent work, however, has shown that the radio emission from M82 comes from a number of small (less than a few parsecs) discrete components probably related to supernovae events or to regions of intense star formation with intrinsic luminosities as much as 100 times that of Cas A. The strongest of these is known to be variable and is only a few hundred AU in size (Kronberg et al. 1981).

Observations by Stocke (1978) indicate enhanced radio emission from interacting pairs of galaxies, although, curiously, the excess emission comes primarily from a compact core, and not from the disk (e.g., Hummel 1980).

13.2.2 Elliptical Galaxies

Weak compact radio sources are frequently found in the nuclei of elliptical galaxies, particularly those with bright emission-line nuclei and those in which 21-cm observations show the presence of significant amounts of HI. These are high-surface-brightness compact sources which are frequently variable on time scales of months to years. When examined with high angular resolution, the weak radio nuclei typically show the same characteristic asymmetric core-jet structure which is observed in the nuclei of the more powerful radio galaxies and quasars (e.g., Jones et al. 1984, Wrobel et al. 1985). Often there is also large-scale structure on a scale of a kiloparsec to a few hundred kiloparsecs comparable to that found in the powerful radio sources (e.g., Wrobel and Heeschen 1984).

13.3 Compact Sources

13.3.1 Self-Absorption

When the apparent source brightness temperature approaches the equivalent kinetic temperature of the relativistic electrons, synchrotron self-absorption becomes important, and part of the radiation is absorbed by the relativistic electrons along the propagation path. Below the self-absorption cutoff frequency, the spectrum is just that of a blackbody with an equivalent temperature $T_k = E/k$.

From Equation (1.8), $E \propto \nu^{1/2}$, so in an opaque synchrotron source the flux density $S \propto \nu^{2.5}$, rather than the ν^2 law found in thermal sources. In other words, an opaque synchrotron source may be thought of as a body whose equivalent temperature depends on the square root of the frequency. Self-absorption occurs below a frequency ν_c where the kinetic temperature

$$T_k = \frac{E}{k} \sim 8 \times 10^3 B^{1/2} v^{1/2} k^{-1} \text{ K} ,$$

is equal to the brightness temperature

$$T_B = S\theta^{-2} c^2 (2k)^{-1} v^{-2} \text{ K} .$$

Assuming uniform source parameters, the magnetic field can then be estimated from observation of v_c and surface brightness from

$$v_c \sim f(p) B^{1/5} \theta^{2/5} S_m^{-4/5} (1 + z)^{1/5} \delta^{-1/5} \text{ GHz} \tag{13.12}$$

where S_m is the maximum flux density in janskys, v_c the cutoff frequency in GHz, and Θ the angular size in milliarcseconds. The quantity δ is a correction for the relativistic Doppler shift if the source is moving with high velocity (see Section 13.3.7). If $v \ll c$, $\delta \sim 1$. The function $f(p)$ only weakly depends on geometry and the value of p, and is about 8 for $p \sim 2.5$. Variations in opacity throughout the source give an overall spectrum that can be considered as the superposition of many simple regions described by Equation (13.2) and can give rise to the so-called *flat* or undulating spectra typically observed in compact radio sources (e.g., Condon and Dressel 1978).

The magnetic field in a compact radio source can be determined directly from the observables θ, S_m, and v_c by Equation (13.12). The magnetic energy, E_B, can be estimated from Equation (13.8) to be

$$E_B \sim 2.5 \times 10^{48} S_m^{-4} \theta^{11} v_c^{10} (1 + z)^{-2} \delta^2 z^3 \text{ ergs} . \tag{13.13}$$

Similarly, from Equation (13.6), the energy in relativistic electrons is given approximately by

$$E_e \sim 3.1 \times 10^{62} \left(\frac{S_{\max}}{\theta^2} \right)^{5.5} v_c^{-10.5} z^2 \text{ ergs} . \tag{13.14}$$

Synchrotron radiation losses lead to a characteristic half-life at a frequency v_m of

$$t \sim B^{-3/2} v_m^{-1/2} \sim 10^7 S^3 \theta^6 v_c^{-8} \delta^{3/2} (1 + z)^{3/2} \text{ years} . \tag{13.15}$$

In practice, the use of Equations (13.13) and (13.14) to derive the magnetic and particle energy is difficult due to the strong dependence on angular size and cutoff frequency. However, in almost every case where measurements exist, the particle energy appears to greatly exceed the magnetic energy. But note the dependence on θ and v_c to about the tenth power; so small changes in the geometry may lead to other conclusions. The ratio of E_c/E_B may also be reduced considerably with even modest values of δ. See Section 13.3.6.

13.3.2 Inverse Compton Radiation

In very compact sources, in which the radiation energy density is comparable to the magnetic energy density, inverse Compton scattering will cause additional electron energy losses. For a homogeneous isotropic source

$$\frac{L_c}{L_s} = \frac{U_{\text{rad}}}{U_B} = \frac{24L}{d^2 B^2 c} \tag{13.16}$$

where L_c is the power radiated by inverse Compton scattering, $L_s \sim 4\pi R^2 S_m \nu_c$ is the radio power radiated by synchrotron emission, $U_{rad} = 3L/4d^2c$ is the energy density of the radiation field, $U_B = B^2/8\pi$ is the energy density of the magnetic field, R is the distance to the source, θ the angular size, and $d = R\theta$ is the source diameter. Then, from Equation (13.16), recognizing that $S_m/\theta^2\nu^2$ is proportional to the peak brightness temperature, T_m, and including the effect of second-order scattering, we have

$$L_c/L_s \sim \frac{1}{2}\left(\frac{T_m}{10^{12}}\right)^5 \nu_m \left[1 + \frac{1}{2}\left(\frac{T_m}{10^2}\right)^5\right] \qquad (13.17)$$

where ν_m is the spectral upper cutoff frequency. Taking $\nu_m \sim 100$ GHz, when $T < 10^{11}$, $L_c/L_s \ll 1$, inverse Compton scattering is not important. But when $T > 10^{12}$ K, the second-order term becomes important, $L_c/L_s \sim (T_m/10^{12})^{10}$, and inverse Compton losses become catastrophic. The exact value of the peak brightness temperature which corresponds to the case where $L_c \sim L_s$ is somewhat dependent on the specific geometry, the value of p, and the spectral cutoff frequency, ν_m, but the strong dependence of L_c/L_s on T_m means that the maximum brightness temperature, T_m, cannot significantly exceed 10^{12} K, independent of wavelength. This places a lower limit to the angular size of

$$\theta \gtrsim 10^{-3} S_m^{1/2} \nu_c^{-1} \text{ milliarcseconds} . \qquad (13.18)$$

Observations show that the peak brightness temperature of compact radio sources measured by VLBI is almost always in the range of 10^{11} to 10^{12} K. Thus, the angular size of an opaque source can be estimated from the peak flux density, S_m, and the self-absorption cutoff frequency, ν_c, to give

$$\theta \sim S_m^{1/2} \nu_c^{-1} (\text{GHz}) \text{ milliarcseconds} . \qquad (13.19)$$

The observed angular size is generally in good agreement with that expected from Equation (13.19) and the measured peak flux density and cutoff frequency, and there is no evidence that the peak brightness temperature ever exceeds 10^{12} K. This is strong evidence that the compact radio sources indeed radiate by the synchrotron process, and that the radio emission is limited by inverse Compton cooling.

The inverse Compton scattered flux density, S_c, at an energy E is given by Marscher (1983) as

$$S_c \sim \ln(\nu/\nu_m)\theta^{-2(p+2)}\nu_m^{-1/2(3p+7)}S_m^{(p+3)}E^{-1/2(p-1)}(1+z)^{(p+3)}\delta^{-(p+3)} \mu\text{Jy} . \qquad (13.20)$$

where ν_m is the upper cutoff frequency of the synchrotron radiation spectrum.

Near the $E = 1$ keV band of the Einstein Observatory, this becomes (Biermann and Zensus 1984)

$$S_c \sim T_B^{p+2} S\nu^{(p+1)/2}(1+z)^{p+3}\delta^{-(p+3)} \mu\text{Jy} \qquad (13.21)$$

where the effective brightness temperature, T_B, is approximately $\nu^2\theta^2 S/1.22$ when T_B is expressed in units of 10^{12} K.

Observations at millimeter wavelengths, where the effect of self-absorption is small, do indeed show a correlation between measured radio and X-ray flux density in the sense expected if the X-ray emission is due to inverse Compton scattering from the radio photons (Owen et al. 1981)

13.3.3 Polarization

As described in Section 13.1, in a uniform magnetic field with a power law distribution of electron energies having an index p, the linear polarization, $P_t(p)$, of a transparent source is perpendicular to the direction of the magnetic field and is given by

$$P_t(p) = \frac{3p + 3}{3p + 7} \qquad (13.22)$$

which is of the order of 70% for typical values of p.

In the opaque portion of the spectrum, the polarization is parallel to the magnetic field and is given by

$$P_0(p) = \frac{3}{6p + 13} \qquad (13.23)$$

so that P_0 is typically only about 10% or less. Observations of integrated polarization are typically much less than that predicted for a uniform field by at least an order of magnitude, presumably due to the random orientation of the magnetic fields.

The observed polarization position angle is modified by Faraday rotation (Section 1.2.2). Although most of the rotation occurs primarily within our galaxy (Section 1.5.4), some is internal to the sources. In a few cases, the internal rotation measure is surprisingly large with measured values up to several thousand radians per meter^2.

There appears to be no evidence for internal Faraday rotation in the compact radio sources. This places strict limits on the number of thermal electrons that can coexist with the relativistic electron cloud (Wardle 1977, Jones and O'Dell 1977). Indeed, the absence of Faraday rotation requires that the number of relativistic electrons must greatly exceed the number of cold (thermal) electrons, or that there are both electrons and positrons. Moreover, the number of low-energy relativistic electrons must be much less than the number expected from a power law extrapolation of the high-energy population. This has important implications for possible acceleration mechanisms, and would appear to exclude most stochastic processes.

The synchrotron radiation from a single electron is elliptically polarized and has a large circular component which is mostly canceled if the pitch angle distribution is isotropic. However, there is still a small net circular polarization, since there are more electrons in the solid angle defined by $\theta + d\theta$ than in the one defined by $\theta - d\theta$. This effect is particularly important if the cone of radiation of a single electron $(\theta \sim mc^2/E)$ is large, which will occur at very low frequencies or in regions of high magnetic field strength (small values of electron energy).

In a uniform magnetic field of B gauss, and isotropic distribution of electron pitch angles, the integrated circular polarization is $P_c \sim (3B/v)^{1/2}$ at a frequency v (Sciama and Rees 1967). In a few sources the degree of circular polarization has been

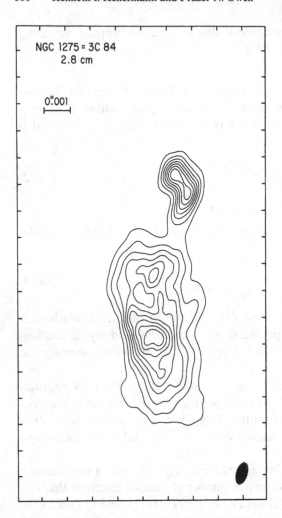

NGC 1275 = 3C 84
2.8 cm

0".001

Fig. 13.4. Milliarcsecond structure of the radio galaxy NGC 1275 (3C84) measured by VLBI (Romney et al. 1984).

measured to be ~0.01% to 0.1% near 1 GHz. This corresponds to magnetic fields of ~$3 \times 10^{-5 \pm 1}$ gauss—in good apparent agreement with the values derived from the synchrotron self-absorption cutoff frequency and the angular size.

13.3.4 Structure

Recent improvements in image formation techniques using interferometer baselines of thousands of kilometers (VLBI) now permit images of compact radio sources to be made with resolutions better than one milliarcsecond. Detailed radio pictures of quasars and galactic nuclei are now possible on a scale which is typically of the order of a few parsecs even for the most distant objects. For nearby galaxies, it is considerably less.

When mapped in detail (see Figure 13.4), the compact sources show a variety of structural forms. The great majority have asymmetric structure containing a bright region plus an elongated feature which resembles the jets seen on larger scales. These jetlike features often break up into a number of distinct components with different surface brightness and self-absorption cutoff frequency.

Equation (13.19) indicates that for a wide range of magnetic field strength, there is a characteristic size of compact radio sources which varies with the wavelength of observation. Although there may be a big spread of the opacity in each source, individual components are most readily observed at the wavelength where the flux density is near a maximum (opacity of the order of unity). Thus, when observed over a range of frequency, individual sources show structural features ranging from a few tenths of a milliarcsecond or less at short centimeter wavelengths to a few hundredths of an arcsecond or more at longer wavelengths. The integrated spectrum, which is the sum of many peaked self-absorbed components, often is remarkably flat and shows an average spectral index near zero. Comparison of component sizes and self-absorption cutoff frequency, v_c, indicate magnetic field strengths in the range of 10^{-4} to 10^{-2} gauss (Equation 13.12), but the observational uncertainties are very large as the derived value of B depends on v_c^5 and θ^2.

When there is an extended jet (see Section 13.4.1), it always lies on the same side of the core as the compact jet. Characteristically, the compact features are curved through an angle of a few tens of degrees, although in some cases the curvature extends through more than ninety degrees. The curvature is most pronounced near the inner region of the jet as it emerges from the core. The outer parts of the compact jets are usually aligned with the larger-scale jets, which are up to hundreds of kiloparsecs away. The alignment of these features over size scales ranging up to a factor of 100,000 means that the large jets are focused and collimated within a region less than a parsec across. This remarkable feature of extragalactic radio sources implies a unique axis which extends from a parsec to a few hundred kiloparsecs, and a current activity with a "memory" extending back at least 10^5 to 10^6 years.

Some sources have a well-defined self-absorption cutoff frequency, usually at a relatively low frequency of a few hundred MHz. Above this frequency, the spectra are characteristic of transparent sources. VLBI observations often show that these sources have complex angular structure with overall angular sizes about 0.1 arcsecond, corresponding to the relatively low self-absorption cutoff frequency. Other sources of this type which are generally referred to as *Steep-Spectrum Compact Sources* have two similar well-separated components with no evidence for any jet structure.

There appears to be no obvious difference in structure between the compact components in sources with weak extended structure and the weak compact sources which are located near the center of strong double-component extended sources.

13.3.5 Variability

Nearly all compact radio sources, when observed over a sufficient period of time and with sufficient precision, show variability on time scales ranging from a few days to a few years and with fractional flux density changes ranging from a few percent to about 100 percent. The most rapid variations occur in BL Lac objects. In general, the observed variations may be described as outbursts which are strongest at the highest frequencies and propagate toward lower frequencies with reduced amplitude (Figure 13.5) (e.g., Dent et al. 1974, Dent and Kapitsky 1976, Altschuler and Wardle 1977, Andrew et al. 1978, Fanti et al. 1981, 1983, Epstein et al. 1982, Aller et al. 1985). However, the variations which are observed at frequencies less

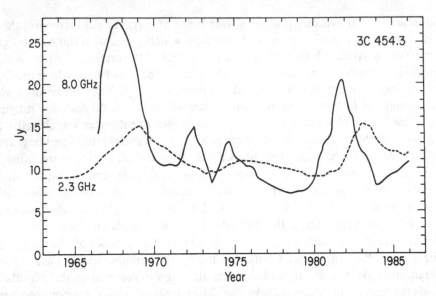

Fig. 13.5. Variations in the flux density of the quasar 3C454.3 observed over a wide range of wavelengths. [Data taken from Aller et al. (1985); Altschuler and Wardle (1977), with permission from the Royal Astronomical Society; and Pauliny-Toth et al. (1987). Reprinted by permission from Nature, Vol. 328, No. 6133, pp 778. Copyright(c) 1987, Macmillan Magazines Ltd.]

than 1 GHz or so for the most part appear unrelated to those observed at centimeter wavelengths and, as discussed below, are probably due to a different phenomenon.

The observation of variations in polarization is difficult since the degree of linear polarization is typically only a few percent, and the time scale for significant changes appears to be more rapid than for variations in total flux density (e.g., Aller et al. 1985). There appears to be no clear pattern to the variations in observed polarization. In some sources, there is a preferred orientation and the position angle remains constant throughout several flux density outbursts; in others, the direction may change even in the absence of obvious changes in total intensity.

Except for the most rapid variations, it is convenient to discuss the observed variations in terms of an expanding cloud of relativistic particles which is initially opaque out to short wavelengths but which becomes optically thin at successively longer wavelengths. In its simplest form, the model assumes that the relativistic particles are homogeneously distributed, that they initially have a power law spectrum, that they are produced in a very short time, in a small space, that the subsequent expansion occurs in three dimensions at a constant velocity, and that during the expansion the magnetic flux is conserved. Thus $\theta_2/\theta_1 = t_2/t_1$, and $B_2/B_1 = (\theta_1/\theta_2)^2 = (t_1/t_2)^2$, where θ is the angular size, t the elapsed time since the outburst, B the magnetic field, and the subscripts 1 and 2 refer to measurements made at epochs t_1 and t_2. The discussion below follows that of Kellermann and Pauliny-Toth (1968) and van der Laan (1966), and is based on ideas first described by Shklovsky (1965).

The observed flux density as a function of frequency, ν, and time, t, is shown in

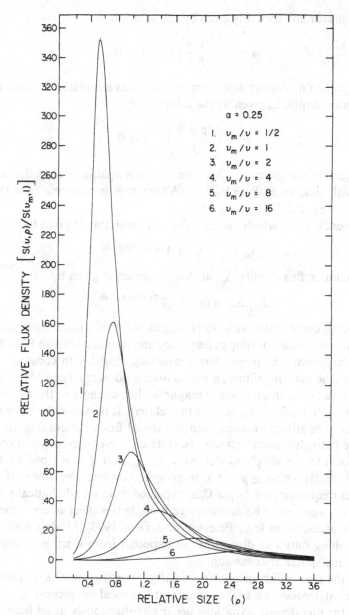

Fig. 13.6. Variation in flux density and frequency for adiabatically expanding homogeneous source (van der Laan, 1966. Reprinted by permission from Nature, Vol. 211, No. 5054, pp. 1131. Copyright(c) 1966. Macmillan Magazines Ltd.)

Figure 13.6 and is described by

$$\frac{S(v,t)}{S_{m_1}} = \left(\frac{v}{v_{m_1}}\right)^{5/2}\left(\frac{t}{t_1}\right)^3 \times \left\{\frac{\exp[-\tau(v/v_{m_1})^{-(p+4)/2}(t/t_1)^{-(2p+3)}]}{1-\exp(-\tau)}\right\} \quad (13.24)$$

where S_{m_1} is the maximum flux reached at frequency v_{m_1} at time t_1.

If the optical depth, τ, is taken as the value at the frequency, v_m, at which the flux density is a maximum, then it is given by the solution of

$$e^{\tau_v} - \left(\frac{p+4}{5}\right)\tau_v - 1 = 0 \ . \tag{13.25}$$

The maximum flux density at a given frequency as a function of time occurs at a different optical depth, τ_t, given by the solution of

$$e^{\tau_t} - \left(\frac{2p+3}{3}\right) - 1 = 0 \ . \tag{13.26}$$

In the region of the spectrum where the source is opaque ($\tau \gg 1$), the flux density increases with time as $S_2/S_1 = (t_2/t_1)^3$. Where it is transparent ($\tau \ll 1$), the flux density decreases as $S_2/S_1 = (t_2/t_1)^{-2\gamma}$.

The frequency, v_m, at which the intensity is a maximum is given by

$$v_{m_2}/v_{m_1} = (t_2/t_1)^{-(4p+6)/(p+4)} \tag{13.27}$$

and the maximum flux density, S_m, at that frequency is given by

$$S_{m_2}/S_{m_1} = (v_{m_2}/v_{m_1})^{(7p+3)/(4p+6)} \ . \tag{13.28}$$

Quantitative comparison with observations is difficult since most sources have multiple outbursts which overlap in frequency and time (e.g., Figure 13.6). Moreover, while a homogeneous, isotropic, flux-conserving model with constant expansion velocity is mathematically simple, more realistic models must consider nonconstant expansion rates, nonconservation of magnetic flux, changes in the electron energy index, p, the finite acceleration time for the relativistic particles, the inhomogeneous distribution of relativistic plasma, and the initial finite dimensions. In those few cases where individual outbursts may be isolated, the observed variations qualitatively conform to the simple model, with S_m, t_m, and v_m described by Equations (13.27) and (13.28) with $1 \lesssim p \lesssim 1.5$. In general, however, the observed variations at the lower frequencies are larger than expected from an adiabatically expanding source, and it appears to be necessary to include the effect of continued particle injection or acceleration (e.g., Peterson and Dent 1973). Models which consider expansion along only one dimension, as expected from a jet, may also be more realistic than a spherical expansion.

Because the source dimensions are initially finite, the initial spectrum is always transparent at frequencies higher than some critical frequency, v_0. Above this frequency, the flux density variations occur simultaneously at all frequencies and reflect only the rate of relativistic particle production or decay due to synchrotron and inverse Compton radiation losses. Characteristically, v_0 is in the range of 10 to 30 GHz. From Equation (13.18) this gives an initial size of $\sim 10^{-3}$ arcseconds for $B \sim 1$ gauss, corresponding to linear sizes of a few parsecs at $z \sim 1$. In those sources where good data exist in the spectral region $v > v_0$, the observed variations occur with roughly equal amplitude at all frequencies, indicating an initial spectral index $\alpha \sim 0$, or $p \sim 1$, in reasonable agreement with the value of p derived from Equation (13.27) or (13.28).

The greatest theoretical difficulty in interpreting the observations of variable compact radio sources in terms of conventional synchrotron models comes from the excessively high brightness temperatures implied from the observations of rapid variations. The problem arises because causality arguments require that if variability is observed on a time scale τ, then the dimensions of the radiating region must be less than $c\tau$, since otherwise differential signal travel time over the source would blur any variations. Using the distance obtained from the redshift, z, an upper limit to the angular size, θ, may be calculated. This value of θ often leads to brightness temperatures well in excess of 10^{12} K, in apparent conflict with the maximum value allowed for an incoherent synchrotron source, particularly for variability observed at frequencies $\nu < 1$ GHz or on time scales $t \ll 1$ year (e.g., Jones and Burbidge 1973).

For some years the variability observed at very low frequencies aroused considerable speculation about the reality of the observations, or about the validity of accepting quasar redshifts as a measure of distance. It now appears, however, that the low-frequency variations are most easily interpreted as the result of scintillations in the ionized interstellar medium (Shapirovskaya 1978, Rickett et al. 1984). The very rapid variations observed at centimeter wavelengths, with time scales of the order of one day (Heeschen 1984), are probably also unrelated to the "classical" variability and may also be due to the same scintillation phenomenon (Blandford et al. 1986).

However, causality arguments applied to the variations which occur on time scales of the order of one year at centimeter wavelengths also predict apparent brightness temperatures which often exceed the inverse Compton limit by a factor of 10 to 100, as well as an X-ray flux which is many orders of magnitude above the values actually observed. Shortly after the implication of inverse Compton scattering was first appreciated (e.g., Hoyle et al. 1966), it was realized that the problem could be avoided if the source of radiation was moving toward the observer with a velocity close to the speed of light (Rees 1966). In this case, the apparent time scale seen by an observer at rest is shortened and the apparent brightness temperature is enhanced by "Doppler boosting." Support for the so-called "relativistic beaming" model comes from the very long baseline (VLBI) observations of "superluminal" radio sources discussed in Section 13.3.7.

13.3.6 Source Dynamics and Superluminal Motion

Not unexpectedly, the compact variable radio sources show changes in their angular structure on time scales corresponding to the intensity variations. The observed motions can usually be described as an increase in separation between the core and one or more components which make up the jetlike feature. Of those sources which have been studied in any detail over a period of time, well over half show an apparent velocity of separation which appears to be five to ten times the speed of light (e.g., Cohen and Unwin 1984). This phenomenon is usually referred to as *superluminal motion*. In fact, there are very few sources where it has been clearly established that the transverse velocity is less than the speed of light. For one of these, the nucleus of NGC 1275 (3C84), there is a well-established subluminal motion with an apparent transverse velocity of about half the speed of light (Romney et al. 1984). NGC 1275 shows the same core-jet morphology seen in the superluminal sources. One of the most intensively observed superluminal sources is 3C273, shown in Figure 13.7.

For some years there was considerable debate about whether to take the observa-

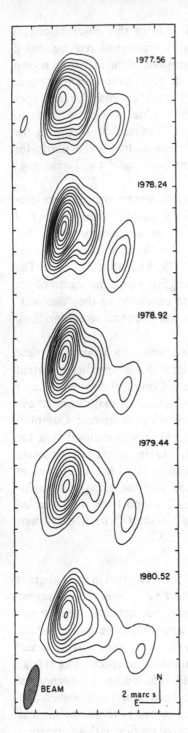

Fig. 13.7. Changes in the structure of the quasar 3C273 observed between 1977 and 1981 (Pearson et al. 1981. Reprinted by permission from Nature, Vol. 290, No. 5805, pp. 365. Copyright(c) 1981, Macmillan Magazines Ltd.

tions of superluminal motion seriously. Today the situation is very much improved. Multielement interferometer systems are used together with sophisticated image-restoring algorithms, and there is now little doubt about the reality of superluminal motion. The observed properties of the superluminal sources may be summarized as follows (e.g., Cohen and Unwin 1984, Kellermann 1985):

1. Superluminal motion is observed primarily in asymmetric sources with extended jet features and a strong core. Most are identified with bright quasars, but a few with relatively nearby AGNs.
2. Only increases in overall separations are observed, never decreases. However, in several sources there are both stationary and moving components, and in these sources, the separation of some component pairs may decrease.
3. Typical transverse velocities are of the order of 4c to 10c (e.g., Figure 13.7).
4. Component separations are typically in the range of 10 to 50 pc.
5. The cores superluminal sources show "flat" or inverted radio spectra and variable flux densities. Since the core features are more opaque than the moving components, they have flatter or more inverted spectra and are more prominent at shorter wavelengths.
6. Close to the core, the structure is often curved through an angle of several tens of degrees or more.
7. In many cases, large-scale jets also extend hundreds of kiloparsecs from the core and are continous with the much smaller superluminal features. These large-scale jets always lie on the same side of the core as the superluminal features.
8. Different components in the same source may show different velocities. The individual components fade with time and their spectra steepen as they move away from the core.
9. In general, each component moves with a constant velocity, although there is evidence in the quasar 3C345 that one component has accelerated.
10. The motion is generally radial, except possibly in 3C345 where either the origin is displaced from the core or the direction of motion has changed.
11. Extrapolation back to the time of zero component separation often coincides with the beginning of a flux density outburst.

The observations of course give only the angular separation and its rate of change. The linear velocity is calculated from $v = R(d\theta/dt)(1 + z)$ where R is the "angular size" distance and the factor $(1 + z)$ corrects for the relativistic time dilation due to the cosmological redshift. It has been argued that if the quasars are much closer than indicated by their redshift, then of course the linear velocities may be less than c. Moreover, the apparent inverse Compton catastrophe implied by the rapid flux density variations is then no longer a problem. However, this argument does not effect the discrepancy between the observed inverse Compton flux and the distance-independent value predicted from Equation (13.20) and the measured brightness temperature (e.g., Marscher 1983).

At least one superluminal source, the AGN 3C120, is found in a relatively low redshift galaxy ($z = 0.03$). Several others are identified with quasars associated with nebulosities which have measured redshifts; thus, interpretations based on noncosmological redshifts appear unsatisfactory unless noncosmological redshifts are accepted for galaxies as well as quasars.

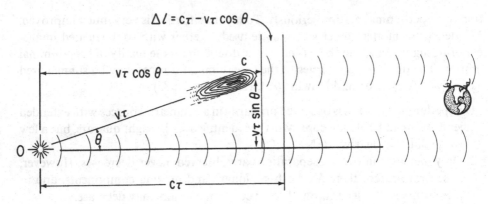

Fig. 13.8. Apparent superluminal motion results when the radiating source is moving so fast that it nearly catches up with its own radiation. Assume that a radiating plasma cloud is ejected from the origin, O, with a velocity v in a direction θ with respect to the line of sight. After a time t, the cloud has moved a distance vt. The motion, projected along the line of sight is $vt \cos\theta$, and projected perpendicular to the line of sight, $vt \sin\theta$. A distant observer sees the emission delayed by a time $t(c - v\cos\theta) = ct(1 - \beta\cos\theta)$ compared to the "signal" radiated when the cloud was at O. The apparent transverse velocity seen by the observer is then $(vt \sin\theta)/[ct(1 - \beta\cos\theta)] = \beta\sin\theta/(1 - \beta\cos\theta)$.

Many models have been considered to explain superluminal motion including:

1. Appropriately phased intensity variations in fixed components—the so-called "Christmas Tree" or "Movie Marquee" model.
2. Noncosmological redshifts.
3. Gravitational lenses or screens.
4. Variations in synchrotron opacity.
5. Synchrotron curvature radiation in a dipole magnetic field.
6. Light echoes.
7. Real tachyonic motion.
8. Geometric effects of relativistically moving sources.
9. Bulk relativistic motion along the line of sight.

For one reason or another, all but the last of these have been shown to be unsatisfactory (e.g., Marscher and Scott 1980) and most of the discussion in the literature centers around models based on bulk relativistic motion.

13.3.7 Relativistic Beaming

If the source of radio emission is moving near the speed of light along a direction which lies close to the line of sight, then the source nearly catches up with its own radiation. This can give the illusion of apparent transverse motion which is greater than the speed of light. As shown in Figure 13.8, if the true velocity is v and is at an angle, θ, with respect to the line of sight, then the apparent transverse velocity, v_a, is given by

$$v_a = \frac{v \sin\theta}{1 - \beta\cos\theta} \tag{13.29}$$

where $\beta = v/c$.

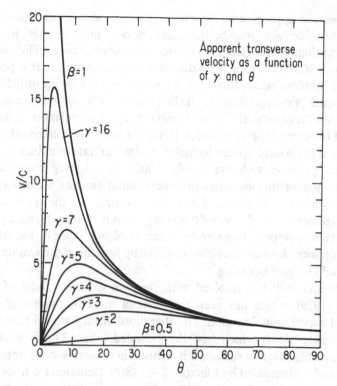

Fig. 13.9. Apparent transverse velocity as a function of the Lorentz factor, γ, and the inclination to the line of sight, θ.

The apparent transverse velocity has a maximum value $v_m \sim \gamma c$, which occurs at an angle $\theta = \sin^{-1}(1/\gamma)$, where $\gamma = (1 - \beta^2)^{-1/2}$.

The Doppler shift due to the motion of the source is given by

$$\delta = \gamma^{-1}(1 - \beta \cos \theta)^{-1} \ . \tag{13.30}$$

If account is taken of the cosmological redshift, z, the total Doppler shift is $\delta/(1 + z)$. The observed radiation from a relativistically moving body is enhanced by an amount $[\gamma^{\alpha - 3}(1 - \beta \cos \theta)]^{\alpha - 3}$, which is often referred to as "Doppler boosting." For $\gamma \gg 1$, the radiation is concentrated within a small cone of half-width $\sim 1/\gamma$. When $\theta \sim \sin^{-1}(1/\gamma)$ (i.e., $v_a \sim v_m$), then for $\alpha \sim 0$, $\delta \sim \gamma$ and the observed emission is enhanced by a factor of $\sim \gamma^3$. The relation of the observed quantities v_a and δ and θ is conveniently represented by the diagram shown in Figure 13.9.

For an approaching component viewed "head on," the boosting factor is about $8\gamma^3$, while the receding component ($\cos \theta \sim -1$) is suppressed by a factor about $1/8\gamma^3$ and is essentially invisible. The probability that a randomly oriented source is beamed toward the observer within an angle $\theta \sim 1/\gamma$ is $\sim 1/2\gamma^2$ for $\gamma \gg 1$.

For values of $\gamma \sim 7$, only about one percent of a randomly oriented sample is expected to show superluminal motion, yet the observed fraction is well over one-half. The large fraction of core-dominated sources with superluminal motion can be rationalized as the result of differential Doppler boosting which preferentially selects sources with appropriate geometry in flux-limited samples. Lobe-dominated sources, on the other hand, may be assumed to be randomly oriented. But, although the statistics are limited, superluminal motion in the central cores of lobe-dominated sources such as 3C179 (Porcas 1981) does not seem uncommon. It is possible that the relativistic outflow occurs throughout a wide cone, but we see only that portion of the cone which is moving close to the line of sight. This would give an increased probability over the canonical $1/2\gamma^2$ of observing superluminal motion, but the good alignment of the compact and extended jets and the highly collimated appearance of the extended jets would appear to make this interpretation unlikely.

An obvious problem with the simple relativistic beaming model is that the observed component flux densities of superluminal sources are always roughly comparable, whereas the expected flux density ratio of the approaching and receding components is $\sim\gamma^6$. Even if one component is stationary, the approaching component should appear brighter by a factor of $\sim\gamma^3$ unless, fortuitously, the intrinsic component luminosities always differ by just the right amount to cancel the differential Doppler boosting.

This apparent conflict is resolved with the *twin exhaust* model of Blandford and Konigl (1979), which has been the basis of most discussion of relativistic beaming and superluminal motion. The Blandford-Konigl model postulates symmetric relativistic beams which feed the extended lobes. The receding beam is essentially invisible since its radiation is focused in a narrow cone opposite to the line of sight and is attenuated by a factor of $\sim 1/(8\gamma^3)$. Emission from the stationary core is seen at the point where the approaching relativistic flow becomes opaque, and so it appears to be Doppler boosted by the same amount as the approaching components. Superluminal motion is observed between this stationary point in the nozzle and moving shock fronts or other inhomogeneities in the relativistic outflow. In the one source where the appropriate measurements exist, 3C345, the core component is indeed found to be stationary with respect to a nearby quasar (Bartel 1986).

Relativistic beaming has received considerable attention because with a minimum of assumptions it provides a simple interpretation of

(a) superluminal motion
(b) rapid flux density variations
(c) lack of inverse Compton scattered X-rays.

In view of the apparent absence of thermal plasma in compact sources (Section 13.3.3), containing highly relativistic electrons ($\gamma \sim 1000$), the possibility of bulk relativistic motion with $\gamma \sim 10$ does not seem unreasonable.

Various "unified schemes" have been discussed which attempt to explain the difference between "core-dominated" (e.g., compact) and "lobe-dominated" (e.g., extended) sources (Orr and Browne 1982) or between "radio-loud" and radio-quiet quasars (Scheuer and Readhead 1979) as the effect of Doppler boosting of a randomly oriented parent population which causes a wide range in apparent core strength depending primarily on the orientation of the motion. However, as discussed in Section 13.4.1, these unified models lead to problems with understanding the extended radio structure and large-scale one-sided radio jets, as well as the optical and X-ray emission.

The correlation between the compact radio emission and X-ray (Owen et al. 1981), infrared, and optical continuum emission suggests that if the radio emission is Doppler boosted, the continuum emission throughout the spectrum may be similarly enhanced (Konigl 1981). In many ways this would be attractive since it provides a convenient interpretation of the large dispersion in the luminosity of quasars which appear to be up to about 5 magnitudes brighter than first-ranked elliptical galaxies. But if the strength of the optical continuum depends primarily on geometry, it is difficult to understand the small spread in the ratio of emission line to continuum brightness, since the line-emitting regions do not show large blue shifts (e.g., Heckman et al. 1983b).

The trivial ballistic model described above is surely too simple. If the actual motion is in the form of a continuous flow rather than the motion of discrete components, then the Doppler boosting factor $\delta = [\gamma^{\alpha-2}(1 - \beta^2)]^{\alpha-2}$. More generally, Lind and Blandford (1985) have emphasized that the actual flow velocity may differ from the shock front velocity, which may be moving obliquely to the main flow. Since it is the relativistic flow velocity which causes the Doppler boosting and the shock front velocity which is seen as superluminal motion, the apparent constraints discussed above may be relaxed. However, in the one object where there is a direct Doppler measure of the flow velocity, SS433 (see Chapter 9), it is equal to the measured radio component velocity.

Realistic models will also be affected by variations in the opacity and dispersion in the actual velocity γ and in the intrinsic radio luminosity. Attempts to explain the wide range of properties of compact AGNs and quasars as simply geometric effects are probably unrealistic, but there is good evidence that the effect of relativistic beaming is relevant to quasars and AGNs, at least at radio wavelengths.

The importance of relativistic beaming and Doppler boosting of the radio, optical, infrared, and X-ray continuum is one of the central problems of current extragalactic research and may have profound implications for our understanding of quasars and AGNs.

13.4 Extended Sources

During the 1950s and 1960s, the imaging of extragalactic radio sources steadily improved, and most resolved sources appeared to be simple double sources surrounding either an optical galaxy or quasar. This led to theories involving explosions in the parent object which ejected clouds of magnetized plasma and relativistic particles (e.g., De Young and Axford 1967). It soon became clear, however, that in the simple versions of these models, radiation and adiabatic expansion losses were too large (e.g., Scheuer 1974). Thus, some sort of continuous supply of energy and, in some cases, in situ particle acceleration were required, and an external medium surrounding the radio-emitting region appeared to be necessary to keep the source from too dispersing rapidly.

Models involving ejection of multiple blobs of plasma (or plasmons) were suggested by Christiansen (1973) and others to overcome the rapid losses. A second class of model, supported by Saslaw et al. (1974) involved the ejection of supermassive objects, via three-body interactions in the nucleus, which then supplied relativistic particles continuously to maintain the source. A third class of models, first proposed by Rees (1971) and colleagues, involved resupply of the radio lobes by a continuous beam of particles or waves.

13.4.1 Jets, Lobes, and Hot Spots

The apparently simple properties of powerful, (i.e., $L > 10^{40}$ erg s^{-1}) extragalactic radio sources have been complicated by the high-resolution maps which have become available over the past fifteen years with ever increasing detail. However, at the same time, the nature of the physical processes necessary to explain these sources have become clearer.

For some time it has been clear that a general description of an "extragalactic source" includes a central component and some sort of extended double structure. During the 1980s, it has become evident that virtually all such sources also have narrow elongated tubes of radio emission connecting the central source to the outlying extended structure, suggesting that energy, magnetic field, relativistic particles, and probably thermal gas are being transferred away from the nucleus to form the observed extended structures.

The detailed morphology illustrated in Figure 13.10 has resulted in an updated form of the Rees (1971) beam model being generally accepted as the working picture of extragalactic radio sources. The models of Rees envisioned an invisible, relativistic flow which terminated in a shock where the flow energy was randomized and emerged, at least partly, as the flux of relativistic particles necessary to maintain the extended emission. Unlike the original model, however, the radio jets now commonly observed, emit radiation. Thus, the transport process must not be entirely efficient. Also many jets are observed to bend and wiggle, probably indicating an interaction with an external medium. In nearby sources, this medium is often observed through its X-ray emission and appears to be in pressure equilibrium with the plasma in the jet. Properties of radio jets are discussed more fully by Bridle and Perley (1984).

Extended extragalactic sources also show a basic change in their morphology at an absolute luminosity of $P(14\,\text{GHz}) \sim 10^{25}\,\text{W Hz}^{-1}$. Sources weaker than this level appear limb darkened, that is, they slowly fade away in brightness as one looks further away from the nucleus, while brighter sources have limb-brightened outer structures. The two classes are referred to as Fanaroff and Riley (1974) classes I and II, respectively (or FR I and FR II). FR II sources also often show small "hot spots" either at the farthest edges of the source or sometimes apparently embedded in more diffuse structure. It is believed that the entire limb-brightened structure is due to supersonic jets terminating at a boundary with an external medium surrounding the source. FR I sources, on the other hand, may be subsonic, at least in their outermost regions. They are thought to be in thermal pressure balance with the external medium and possibly to have entrained a great deal of external gas. FR I sources are often found in nearby rich clusters of galaxies and are often distorted by processes in the clusters.

Because of their higher luminosity and relatively low space density, FR II sources tend to be identified with distant radio galaxies and quasars, often near the edge of the observable universe. However, some examples do exist relatively nearby, such as Cygnus A which has a redshift of 0.057. Its parent galaxy has long been known to have very strong emission lines. The galaxy is also very bright at optical wavelengths but unfortunately lies within ten degrees of the galactic plane, which makes it difficult to study optically. It lies in the center of a very dense ($n_e \sim 10^{-2}$ cm^{-3}) cloud of very hot (10^8 K) gas. Lower-luminosity examples of the class, which are more common nearby, do not generally show such extreme X-ray properties and, except for some nuclear emission lines, resemble normal giant elliptical galaxies.

The dominant morphological characteristic of FR II radio galaxy emission is the brighter outer lobes. Often embedded in the lobes are more compact hot spots, perhaps as small as one kiloparsec. This structure is usually accompanied by a compact core in the center of the galaxy, although the core is sometimes too weak to be seen with present maps. Finally, in nearby examples which have been studied very extensively with the VLA, a faint jet can be seen connecting the nucleus with the outer hot spots and lobes, at least on one side of the double. Such jets have been seen in only a few cases, and it is unclear whether the jets are strongly one-sided as in the quasars discussed below.

The general properties of this morphology have recently been convincingly reproduced in numerical simulations of low-density, supersonic jets traveling through a denser external medium. This type of work is just beginning but suggests that we are on the right track and may ultimately be able to understand a great deal about these sources.

Quasars also produce FR II sources. However, the relative importance of the distinct features of the general morphology often is different. The bright outer lobes and hot spots are still visible but the central component and the jets are much brighter, both relative to the lobes and in absolute terms. The luminosity of the central component and jet of quasars is usually one to two orders of magnitude higher than for the galaxies. Quasar jets almost always appear to be one-sided. Furthermore, these jets often show many bends and wiggles, sometimes by as much as ninety degrees. They often seem to be made up of many small knots rather than a continuous brightness distribution.

The origin of the one-sidedness is, at present, still unclear. On the one hand, many of the properties seem consistent with relativistic beaming in the line of sight, as is used to explain the observed compact jets (Section 13.3.7). The ratio of brightness from one side to the other in any given case is consistent with fairly small values of γ's. Also the pronounced wiggles and bends can be more easily understood if they are intrinsically small wiggles, which when inclined to the line of sight, appear to be large bends in projection on the sky. Moreover, where VLBI and VLA observations exist for the same source, the one-sided jets on both scales lie on the same side of the source, suggesting a common origin for the observed one-sidedness.

However, it is hard to understand how this can be the case since in radio quasars we see only one-sided jets and we almost always can detect a jet. Also, quasars with one-sided jets and bright radio cores exist which appear as large as all but a few of the largest radio galaxies, so they are unlikely to be appreciably foreshortened (e.g., Schilizzi and de Bruyn 1983). Alternatively, the high-luminosity jets may be intrinsically one-sided. Since the diffuse lobe emission is usually found on both sides

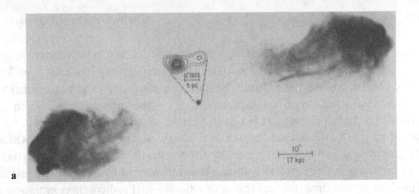

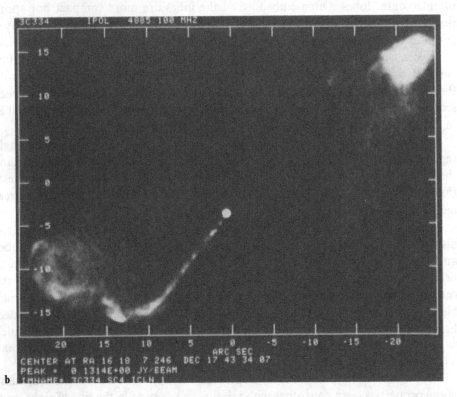

Fig. 13.10. (a) Cygnus A is the strongest extragalactic radio source in the sky and is the prototypical example of FR II radio structure. The image of the extended structure was made using the VLA (From Perley et al. 1984) and shows a faint radio jet which apparently feeds the outer hot spot on one side and the filamentary radio lobes. The structure shown for the compact core is based on VLBI data (From Downes et al. 1981). (b) 3C334 has a curved knotty jet emanatung from the bright unresoved source that coincides with the quasar. Both lobes contain some filamentary substructure. The thin filament extending back toward the quasar from the northwestern lobe may be part of a weak counterjet, which is rarely seen in powerful sources except on images of very high dynamic range. Counterjets are very common, however, in low-power sources. (Observers: Owen and Hines. Courtesy NRAO/AUI.) (c) M87 is a relatively weak radio galaxy in the Virgo cluster. The jet, shown here, is about 2 kpc in extent and its emission extends from the radio into the optical and X-ray regions of the spectrum. (d) The unresolved bright spot near the center of this 5GHz image is coincident with a quasar at a red shift of 0.77. The long, narrow, one sided radio jet is typical of powerful double lobed sources. There are prominent hot spots in both lobes suggesting that they have both been recently supplied with relativistic particles despite the appearance of only a single jet. (c and d: Observers: A. Bridle, I. Browne, J. Burns, J. Dreher, D. Hough, R. Laing, C. Lonsdale, P. Scheuer, J. Wardle).

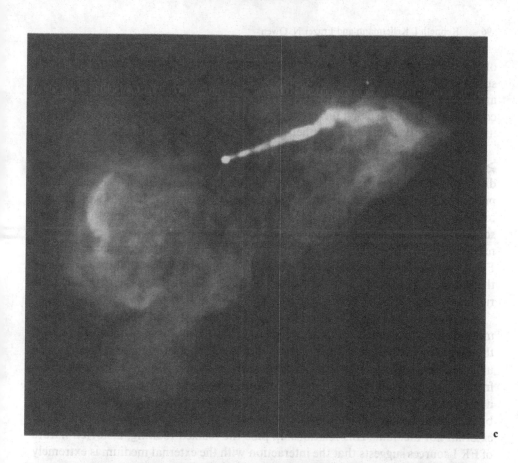

c

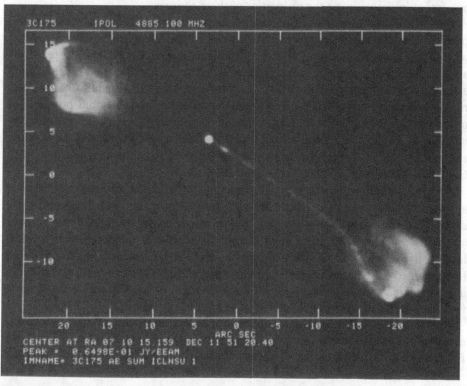

d

of the nucleus, this seems to imply that either the missing jet is not radiating as strongly for some reason or that the jet "flip-flops" between the two sides (Rudnick and Edgar 1984). Also, since there is strong evidence for relativistic motion in the core (Section 13.3.7), such a picture implies that the jet slows down a great deal on its journey to the outer lobe. Neither picture is entirely satisfactory at this time.

Both FR I and FR II sources can exhibit large degrees of linear polarization, $\gtrsim 50\%$ locally; however, the jets in the two types of sources usually show very different field geometry. Most straight FR I sources show either magnetic fields predominantly perpendicular to the jet axis or perpendicular fields which change to predominantly parallel fields at some point down the jet. Exceptions to this trend sometimes occur in very bent sources, where stretching and shearing near the bend may cause an apparent parallel-to-perpendicular flip in the magnetic field. In FR II sources, the magnetic field is usually parallel to the jet axis all the way along the radio jet. The lobes of FR II sources, however, usually have the magnetic field running along the outer edge of the lobe.

The lower-luminosity FR I sources which have been studied up to now are much more nearby on average, and thus we know more about the environment in which they exist. Virtually all are found in some sort of galaxy clustering from poor groups up to the richest clusters. In many of these cases, especially in rich clusters, we know from X-ray observations that they are surrounded by a hot (10^7 to 10^8 K), relatively dense (10^{-2} to 10^{-4} cm^{-3}) medium. The pressures inside the radio sources implied by minimum-energy calculations are often equal to or less than the pressure of the external hot medium. This relationship plus the relaxed-looking, distorted nature of FR I sources suggests that the interaction with the external medium is extremely important in determining the properties of these sources.

FR I sources take on a variety of morphological shapes. However, some general patterns can be recognized. The most luminous FR I sources are usually associated with bright D or cD galaxies located in the center of their associated cluster. They are usually one or two magnitudes brighter than the giant elliptical galaxies usually associated with (nearby) FR II sources. These galaxies also usually have a less rapidly dropping light distribution, suggesting a flatter gravitational potential of much larger extent than is found for the FR II sources. Their radio morphology can usually be described as a twin jet with a gradually widening and often bending channel. Many of these sources, especially those in rich clusters, are bent into C-shapes. These sources are called *wide-angle tails*.

At lower luminosities in rich clusters, one often finds sources which have apparently been even more distorted with jets which have been bent by ninety degrees on each side of the galaxy and merge into long diffuse tails. These sources are called *narrow-angle tails*. Their parent galaxies are intermediate in optical luminosity between *wide-angle tails* galaxies and FR II sources. They are believed to be formed by a normal radio galaxy moving through the hot, tenuous medium in a cluster (e.g., Owen et al. 1979), although this may be hard to reconcile with the very longest-tailed sources (Burns 1981).

Almost all FR I sources have a prominent central component, although this is mainly due to the lower surface brightness of the extended emission compared with FR II sources. Most show two-sided jets. These jets are often close to being equal in brightness, and in a few cases the orientation of dust lanes actually suggests that the slightly brighter jet is pointed away from us. Thus, relativistic motion is not indicated for these sources, at least far away from the nucleus.

13.4.2 Jet Physics

Since we have no direct way of estimating the density and velocity of a radio-emitting jet, a maze of indirect arguments and physical assumptions is made to deduce the nature of the phenomena. However, within the framework of a set of physical assumptions, many deductions can be made, and if we do not make a completely incorrect assumption, (for example, that the radio brightness is due to the incoherent synchrotron process), we can sometimes limit the physical conditions to a fairly narrow range of possibilities. Basically, we have the radio brightness and its linear polarization at one or more frequencies to work with. We also may have some knowledge of the external conditions from X-ray or optical observations.

(a) Straight Jets

Let us assume that a jet is a collimated flow consisting of thermal and relativistic gas initially moving with some velocity v and some radius r. The brightness is then affected by (1) radiation losses, (2) adiabatic gains or losses, and (3) other energy gains or losses by the relativistic electrons. As a jet expands, the particles in the jet will gain or lose energy, consistent with their equation of state. In particular, the relativistic particle energy density, ε, in a volume V will change as $\varepsilon \propto V^{-4/3}$, or the the total energy of a single particle will vary as $E \propto V^{-1/3}$.

Thus, in the cylindrical geometry of a jet, as the radius of the jet, r_j, increases, each relativistic electron should lose energy to the expansion as $E \propto r_j^{-2/9}$.

If magnetic flux is conserved then $B_{\parallel} \propto r_j^{-1}$, and $B_{\perp} \propto r_j^{-2}$.

If the velocity of the jet, v_j, remains constant and no energy is added to the particles or magnetic field from other sources, the luminosity of all observed jets would decrease much faster than is observed (e.g., Bridle and Perley 1984). Thus one of these assumptions must be incorrect. If the velocity decreases, then the density of particles and the perpendicular magnetic field strength will increase, thus counteracting the effects of any expansion. Combining both effects for a power law energy spectrum, the intensity, I_v, varies as

$$I_v \propto r_j^{-(5p+4)/3} v_j^{-(p+2)/3} \qquad (B_{\parallel})$$

or

$$I_v \propto r_j^{-(7p+5)/6} v_j^{-(5p+7)/6} \qquad (B_{\perp})$$

Thus, the jet can actually brighten with certain combinations of parameters. However, if v_j decreases sufficiently, then radiation losses can become important. Also, particles lose energy through inverse Compton scattering to the 3 K background, so the net rate at which particles lose energy reaches a minimum at a magnetic field strength of a few microgauss. Thus, it is not possible to explain the brightness of jets by simply letting v decrease idefinitely. If adiabatic effects alone cannot explain the brightness distribution, then some nonadiabatic effect must be contributing to the energy in the particles and/or fields. The most obvious source is probably the energy in the bulk flow of any thermal plasma in the jet. This could be transferred to the particles through interactions with shocks or through plasma waves in a turbulent plasma. These processes, however, seem to work best when adding energy to already relativistic particles. Theoretical calculations and in situ space observations show they are very inefficient in accelerating thermal particles, especially electrons, to relativistic energies (Lee 1983). Since these and other processes are uncertain in their details, usually it is simply assumed that a fraction, η, of the kinetic energy in the jet is converted to relativistic electron energy. Thus,

$$L_{\text{rad}} = \frac{1}{2\eta r_j \rho_j v_j^3} \tag{13.31}$$

where L_{rad} is the total emitted radiation and ρ_j is the density of thermal particles in the jet.

Equation (13.31) is called the kinetic luminosity equation. Rough estimates for jet or total-source requirements are often made by simply using the total luminosity of (half) of the source as L_{rad} and estimating ρ from the observed Faraday depolarization or the density of the background gas. Clearly, a better approach would be to combine at least adiabatic effects with particle acceleration but this has rarely been done.

(b) Bent Jets

If as in the wide-angle tails or narrow-angle tails, the jet is bent, an additional constraint exists, since the time-independent Euler's equation should apply or

$$(v \cdot \nabla)v = \frac{1}{\rho}\nabla P \ . \tag{13.32}$$

If R is the scale length over the jet bends, then $(v \cdot \nabla)v \simeq v_b^2/R$. Then

$$\frac{\rho_j v_j^2}{R} \simeq \nabla P \ . \tag{13.33}$$

A galaxy moving with velocity v_g through an intracluster medium with density ρ_{icm} experiences a ram pressure $\rho_{\text{icm}}v_g^2$. This pressure is exerted over a scale length h. If the jet is directly exposed to the intracluster medium, then $h = r_j$. On the other hand, the jet may be inside the interstellar medium of a galaxy. Then h is the pressure scale height in the galaxy. In any case, one can write

$$\frac{\rho_j v_j^2}{R} \simeq \frac{\rho_{\text{icm}}v_g^2}{h} \ . \tag{13.34}$$

Combining the kinetic luminosity equation (13.31) with the Euler's equation in the form (13.32), we can eliminate one of the common variables. For example, eliminating v_b we can get

$$v_g = \left[\frac{2\rho_j^{1/2} L_{\text{rad}}(r_j/R)^{3/2}}{\pi r_j^2 \eta \rho_{\text{icm}}^{3/2}} \right]^{1/3} \ . \tag{13.35}$$

For cases involving narrow-angle tails moving at 10^3 km s^{-1} with respect to the external medium, one can find acceptable applications of this equation. However, for the wide-angle tails, one has a higher luminosity to explain and strong evidence in some cases that the parent galaxy is moving very slowly or not at all with respect to the intracluster medium. Thus, a simple picture of motion causing the bending of wide-angle tails appears to fail. More complete models including adiabatic effects or other energy sources for the particles such as turbulence in the gas entrained from the intracluster medium appear to be necessary.

13.5 Summary

There is convincing quantitative evidence that all of the extragalactic radio sources radiate by the commonly accepted incoherent synchrotron process. This evidence includes:

1. The shapes of the spectra of the extended (transparent) sources are power law or dual power law and their detailed shapes are in agreement with synchrotron models where the relativistic particles both gain and lose energy.
2. In the compact sources, the spectral peak occurs at shorter wavelengths in the smaller sources, as predicted by the synchrotron model, and the measured angular sizes are in good agreement with those estimated from the observed self-absorption cutoff wavelength.
3. The maximum observed brightness temperature is $\sim 10^{12}$, as is expected from an incoherent synchrotron source which is "cooled" by inverse Compton scattering.
4. The variations in intensity and polarization and their dependence on wavelength and time are in qualitative agreement with those expected from an expanding cloud of relativistic particles.

The source of energy, the so-called "central engine," is thought to be the associated quasar or AGN. The observed correlation between 21-cm observations of HI and the strength of the nuclear radio source supports the concept that the radio source is fueled by the accretion of gas onto a massive collapsed object, possibly a black hole.

Energy from the central engine appears to be transported to the outer lobes via a highly collimated beam of relativistic particles, but there has been little progress in understanding how the potential energy of the condensed object is converted into an apparently stable particle beam. Near the core, VLBI measurements, as well as observations of rapid flux density variations and the absence of inverse Compton scattered X-rays, are interpreted as the result of Doppler beaming by a highly relativistic outflow from the nucleus with Lorentz factors typically in the range of 5 to 10. But the evidence for bulk relativistic motion in the extended jets is ambiguous.

Recommended Reading

Begelman, M.C., R.D. Blandford, and M.J. Rees. 1984. Rev. Mod. Phys. **56**:225.
Bridle, A.H., and R.A. Perley. 1984. Annu. Rev. Astron. Astrophys. **22**:319.
De Young, D.S. 1984. Physics Reports **111**:373.
Dyson, J.E. (ed.). 1985. Active Galactic Nuclei. Manchester: University of Manchester Press.
Fanti, R., K.I. Kellermann, and G. Setti. (eds.). 1984. IAU Symposium No. 110, VLBI and Compact Radio Sources. Dordrecht: Reidel.

Heeschen, D.S., and C.M. Wade. (eds.). 1982. IAU Symposium No. 97, Extragalactic Radio Sources. Dordrecht: Reidel.
Kellermann, K.I., and I.I.K. Pauliny-Toth. 1981. Annu. Rev. Astron. Astrophys. **19**:410.
Moffet, A. 1975. *In* A. Sandage, M. Sandage, J. Kristian. (eds.), Stars and Stellar Systems, Vol. 9. Chicago: University of Chicago Press, p. 211.
Swarup, G., and V.K. Kapahi. 1986. IAU Symposium No. 119, Quasars. Dordrecht: Reidel.
Zensus, J.A. and S. Unwin. (eds.). 1987. Superluminal Radio Sources. Cambridge: Cambridge University Press.

References

Aller, H.D., M.F. Aller, G.E. Latimer, and P.E. Hodge. 1985. Astrophys. J. Suppl. **59**:513.
Altschuler, D.R., and J.F.C. Wardle. 1977. Mon. Not. R. Astron. Soc. **179**:153.
Andrew, B.H., J.M. MacLeod, G.A. Harvey, and W.J. Medcl. 1978. Astron. J. **83**:863.
Baade, W. and R. Minkowski. 1954. Astrophys. J. **119**:206.
Bartel, N. 1986. Nature **319**:733.
Begelman, M.C., R.D. Blandford, and M.J. Rees. 1984. Rev. Mod. Phys. **56**:255.
Bicknell, G.V. 1986. Astrophys. J. **300**:591.
Biermann, P., and J.A. Zensus. 1984. *In* W. Brinkmann and J. Trumper (eds.), Proceedings of Conference on X-ray and UV Emission from Active Galactic Nuclei. Garching: Max Planck Institute für Extraterrestrische Physik, p. 275.
Biretta, J.A., F.N. Owen, and P.E. Hardee. 1983a. Astrophys. J. (Lett.) **274**:L27.
Biretta, J.A., M.H. Cohen, S.C. Unwin, and I.I.K. Pauliny-Toth. 1983b. Nature **306**:42.
Blandford, R.D., R. Narayan, and R.W. Romani. 1986. Astrophys. J. (Lett.) **301**:L53.
Blandford, R., and A. Konigl. 1979. Astrophys. J. **232**:34.
Bolton, J., and G.J. Stanley. 1949. Aust. J. Sci. Res. **2A**:139.
Bregman, J.N., A.E. Glassgold, P.J. Huggins, G. Neugebauer, B.T. Soiffer, K. Matthews, J. Elias, J. Webb, J.T. Pollock, A.J. Pica, R.J. Leucock, A.G. Smith, H.O. Aller, M.F. Aller, P.E., Hodge, W.A. Dent, T.J. Bulonek, R.E. Barvanis, T.P.L. Roellig, W.Z. Wisniewski, G.H. Bieke, M.J. Lebofsky, B.J. Wills, D. Wills, W.H.-M. Ku, J.D. Bregman, F.C. Witteborn, D.F. Lester, C.D. Impey, and J.A. Hackwell. 1958. Astrophys. J. **129**:841.
Burns, J.O. 1981. Mon. Not. R. Astron. Soc. **196**:523.
Christiansen, W.A. 1973. Mon. Not. R. Astron. Soc. **164**:211.
Cohen, M., and S.C. Unwin. 1984. *In* R. Fanti et al. (eds.), Proceedings of IAU Symposium 110, VLBI and Compact Radio Sources. Dordrecht: Reidel, p. 95.
Condon, J.J. 1982. Astrophys. J. **252**:102.
Condon, J.J., and A. Dressel. 1973. Astrophys. Lett. **15**:203.
de Bruyn, G. 1976. Astron. Astrophys. **52**:439.
Dent, W.A. et al. 1974. Astron. J. **79**:1232.
Dent, W.A., and J.E. Kapitsky. 1976. Astron J. **81**:1053.
De Young, D.S. 1984. Physics Reports **111**:373.
De Young, D.S., and W.I. Axford. 1967. Nature **216**:129.
Downes, A., et al. 1981. Astron. Astrophys. **97**:L1.
Dreher, J.W., C.C. Carilli, and R.A. Perley. 1987. Astrophys. J. **316**:611.
Dressel, A., and J.J. Condon. 1978. Astrophys. J. Suppl. **36**:53.
Ekers, R.D. 1978. Phys. Scripta **17**:171.
Ekers, R.D. 1981. *In* S.M. Fall and D. Lynden-Bell (eds.), Structure and Evolution of Normal Galaxies Cambridge, England: Cambridge University Press, p. 169.
Ennis, D.J., G. Neugebauer, and M. Werner. 1982. Astrophys. J. **262**:460.
Epstein, E.E., W.G. Fogarty, J. Mottmann, E. Schneider. 1982. Astron. J. **87**:449.
Fanaroff, B.L., and J.M. Riley. 1974. Mon. Not. R. Astron. Sci. **167**:31.
Fanti, C., R. Fanti, F. Ficarra, F. Mantovani, L. Padvielli, and K. Weiler. 1981. Astron. Astrophys. Suppl. **45**:61.
Fanti, C., R. Fanti, A. Ficarra, L. Gregorini, F. Mantovani, and L. Padvielli. 1983. Astron. Astrophys. Suppl. **118**:171.
Hazard, C., M.B. Mackey, and A.J. Shimmins. 1963. Nature **197**:1037.

Heckman, T.M., W. Van Breugel, G.K. Mcley, and H.R. Butcher. 1983a. Astron. J. **88**:1077.

Heckman, T.M. 1983b. Astrophys. J. **271**:L5.

Heeschen, D.S. 1984. Astron. J. **89**:1111.

Hey, J.S., S.J. Parsons, and J.W. Phillips. 1946. Nature **158**:234.

Hoyle, F., G. Burbidge, and W. Sargent. 1966. Nature **209**:751.

Hummel, K. 1980. Astron. Astrophys. Suppl. **41**:151.

Jones, T.W., and G.R. Burbidge. 1973. Astrophys. J. **186**:791.

Jones, T.W., and S.L. O'Dell. 1977. Astron. Astrophys. **61**:291.

Jones, T.W., S.L. O'Dell, and W.A. Stein. 1974a. Astrophys. J. **188**:353.

Jones, T.W., S.L. O'Dell, and W.A. Stein. 1974b. Astrophys. J. **192**:261.

Jones, D.L., J.M. Wrobel, and D.B. Shoffer. 1984. Astrophys. J. **276**:480.

Kardashev, N.S. 1962. Sov. Astron.-AJ **6**:317.

Kellermann, K.I. 1985. Comments Astrophys. **11**:69.

Kellermann, K.I., and I.I.K. Pauliny-Toth. 1968. Annu. Rev. Astron. Astrophys. **6**:417.

Konigl, A. 1981. Astrophys. J. **243**:700.

Kotanyi, C.G. 1980. Astron. Astrophys. Suppl. **41**:421.

Kronberg, P.P., P. Biermann, and F.R. Schwab. 1981. Astrophys. J. **246**:751.

Landau, R., T.W. Jones, E.E. Epstein, G. Neugebauer, B.T., Soifer, M.W. Werner, J.J. Puschell, and T.J. Balonek. 1983. Astrophys. J. **268**:68.

Lee, M.A. 1983. Rev. Geophys. Space Phys. **21**:324.

Lind, K.R., and R.D. Blandford. 1985. Nature **314**:424.

Marcaide, J.M., N. Bartel, M.V. Gorenstein, I.I. Shopiro, B.E. Corey, A.E.E. Rogers, J.C. Webber, T.A. Clark, J.D. Romney, and R.A. Preston.1985. Nature **314**:424.

Marscher, A.P. 1983. Astrophys. J. **264**:296.

Marscher, A.P., and J.S. Scott. 1980. Publ. Astron. Soc. Pacific **92**:127.

Meurs, E.J.A., and A. Wilson. 1984. Astron. Astrophys. **136**:206.

Miley, G. 1980. Annu. Rev. Astron. Astrophys. **18**:165.

Minkowski, R. 1960. Astrophys. J. **132**:908.

Moffet, A. 1975. *In* A. Sandage, M. Sandage, and J. Kristian (eds.), Stars and Stellar Systems, Vol. 9. Chicago: University of Chicago Press, p. 211.

Orr, M.J.L., and I.W.A. Browne. 1982. Mon. Not. R. Astron. Soc. **200**:1067.

Owen, F., J.O. Burns, L. Rudnick, and E.W. Greisen. 1979. Astrophys. J. (Lett.) **229**:L59.

Owen, F., D. Helfand, and S.R. Spangler. 1981. Astrophys. J. (Lett.) **250**:L55.

Owen, F.N., and J.J. Puschell. 1984. Astron. J. **89**:932.

Pearson, T.J., S.C. Unwin, M.H. Cohen, R. Linfield, A.C.S. Readhead, G.A. Seielstad, R.S. Simon, and R.C. Walker. 1981. Nature **290**:365.

Perley, R.A., J.W. Dreher, and J. Cowan. 1984. Astrophys. J. **285**:L35.

Peterson, F.W., and W.A. Dent, 1973. Astrophys. J. **186**:421.

Porcas, R.W. 1981. Nature **294**:47.

Preuss, E., W. Alef, and A. Pedlor. 1987. *In* K. Fricke (ed.), Observational Evidence of Activity in Galaxies Proceedings of IAU Symposium No. 121. Dordrecht: Reidel, p. 269.

Rees, M.J. 1966. Nature **211**:468.

Rees, M.J. 1971. Nature **229**:312.

Rickett, B.J., W.A. Coles, and G. Bourgois. 1984. Astron. Astrophys. **134**:390.

Roellig, T.L., et al. 1986. Astrophys. J. **304**:646.

Romney, J., et al. 1984. *In* R. Fanti et al. (eds.), IAU Symposium 110, VLBI and Compact Radio Sources. Dordrecht: Reidel, p. 137.

Rudnick, L., and B.K. Edgar. 1984. Astrophys. J. **279**:74.

Sadler, E. 1984. Astron. J. **89**:53.

Saslaw, W.C., M.J. Valtoner, and S.J. Anrseth. 1974. Astrophys. J. **190**:253.

Scheuer, P.A.G. 1974. Mon. Not. R. Astron. Soc. **166**:513.

Scheuer, P.A.G., and A.C.S. Readhead. 1979. Nature **277**:182.

Schilizzi, R.T., and A.G. de Bruyn. 1983. Nature **303**:26.

Schmidt, M. 1963, Nature **197**:1040.

Sciama, D.W., and M.J. Rees. 1967. Nature **216**:147.

Shapirovskaya, N. Ya. 1978. Sov. Astron.-AJ **22**:544.

Shklovsky, I.S. 1965. Sov. Astron.-AJ **9**:22.

Smith, F.G. 1951. Nature **168**:555.

Spinrad, H., et al. 1985. Publ. Astron. Soc. Pacific **97**:932.

Stocke, J. 1978. Astron. J. **83**:348.

van der Hulst, J.M., P.C. Crane, and W.C. Keel. 1981. Astron. J. **86**:1175.

van der Laan, H. 1966. Nature **211**:1131.

Walker, R.C., J.M. Benson, and S.C. Unwin. 1987. Astrophys. J. **316**:546.

Wardle, J.F.C. 1977. Nature **269**:563.

Wilson, A. 1982. Highlights of Astronomy **6**:467.

Wolfe, A. (ed.). 1978. Pittsburgh Conference on BL Lac Objects. Pittsburgh, Pennsylvania: University of Pittsburgh.

Wrobel, J.M., and D.S. Heeschen. 1984. Astrophys. J. **287**:41.

Wrobel, J.M., D.L. Jones, and D.B. Shaffer. 1985. Astrophys. J. **289**:598.

14. The Microwave Background Radiation

JUAN M. USON and DAVID T. WILKINSON

14.1 Introduction

14.1.1 The Discovery

The microwave background radiation was first detected in 1965 by A. Penzias and R. Wilson, who received the 1978 Nobel Prize for Physics for this measurement [see Wilson (1983b) for his delightful account of this discovery].

A background of blackbody radiation had been predicted by Gamow (1948) as a necessary component of the Big Bang theory and a temperature of 5 K was estimated from the observed abundances of helium and deuterium, a lower temperature corresponding to a higher helium content and vice versa (Alpher and Herman 1948). The idea was not pursued even though the instruments to make such a measurement existed. Indeed, Dicke and his collaborators (1946) had already set a limit of 20 K on the temperature of such a background radiation. Doroshkevich and Novikov (1964) did attempt to check Gamow's theory against observations but they apparently misread a report by Ohm (1961) on tests done with the Bell Labs horn antenna (the same one which later serendipitously discovered the radiation) and concluded that the amount of background radiation in the Universe was negligible.

Also in 1964, Dicke (who at that time had forgotten his earlier measurement) reinvented the notion of a microwave background as a necessary ingredient for a cyclic model of the Universe. Here, high-energy photons ($T > 10^{10}$ K) were needed to dissociate the heavy nuclei in each successive contraction. This would allow each subsequent cycle to start out with only elementary particles and radiation; no nuclei would accumulate over many cycles. He also predicted a residual background of thermal radiation and urged Roll and Wilkinson to set up an experiment to search for such a background (Wilkinson and Peebles 1983). Before they were ready to observe, word came that Penzias and Wilson had found an unexpected source of noise apparently coming from the sky. They had been carefully calibrating the Bell Labs horn antenna in order to make an absolute-temperature map of the galactic radiation at a wavelength $\lambda \sim 7$ cm. As soon as both groups discussed their work, joint papers were published announcing and interpreting the discovery of a new component of cosmic microwave radiation (Penzias and Wilson 1965, Dicke *et al.* 1965).

14.1.2 Cosmological Setting

The consensus of observational results supports a homogeneous and isotropic model of the Universe, which is expanding as described by Hubble's law (Weinberg 1972). This is the standard Big Bang model which assumes as well that the early

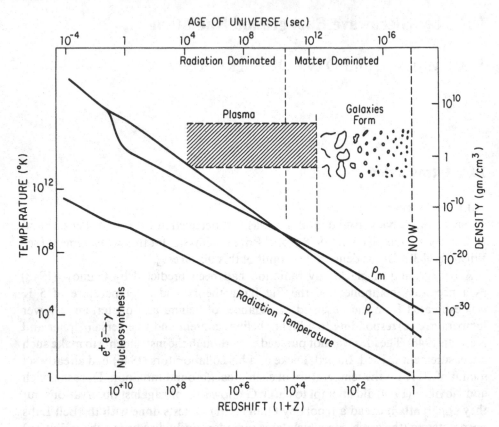

Fig. 14.1. Sketch of the evolution of key cosmological parameters in the standard hot Big Bang model, showing processes associated with the microwave background radiation. Electron-position pairs decay to photons just before nucleosynthesis of light elements begins. Radiation density, ρ_r, decreases more rapidly with the expansion than the matter density, ρ_m; they cross over at a redshift $z \sim 10^4$. At a redshift $z \sim 10^3$, the Universe has cooled to $T \sim 4000$ K and the matter becomes neutral and decouples from the radiation. The correspondence between opposite axes in this figure is only approximate and depends on the value of key cosmological parameters: the expansion rate (H_0), the deceleration parameter (q_0), and a possible nonvanishing cosmological constant (Λ_0).

Universe was hot and in thermal equilibrium, so that matter was immersed in blackbody radiation. Figure 14.1 shows the evolution of key cosmological parameters in this model. A starting time of 10^{-4} seconds is chosen so that the figure emphasizes physics associated with the microwave background radiation. (The intriguing physics of yet earlier times is currently under intense study.) Several processes are important in the evolution of the microwave background radiation, and the exploration of these processes is the main objective of most of the measurements discussed in this chapter.

The early Universe (at redshifts $z > 10^4$) was radiation–dominated; i.e., it was mainly the radiation density that decelerated the expansion. As the Universe grew older, the expansion reduced the radiation density [proportional to $(1 + z)^4$] faster than the matter density [proportional to $(1 + z)^3$], and at about $z \sim 10^4$ the matter density became larger than the radiation density and the Universe became matter-dominated.

At the epoch around $z \sim 10^3$, the Universe cooled below $T \sim 4000$ K and photons were no longer energetic enough to keep matter ionized: hydrogen and helium formed and the cross–section for Thomson scattering became negligible so that the Universe became essentially transparent to the radiation. At this stage, the radiation had a blackbody spectrum due to its prior state of thermal equilibrium with the ionized matter, and the subsequent expansion of the Universe preserved the thermal nature of the spectrum with the sole effect of lowering the radiation temperature in proportion to the redshift,

$$T_{CBR}(z) \propto (1 + z) . \tag{14.1}$$

The credibility of the hot Big Bang model rests heavily on two predictions: the abundances of light elements produced at the nucleosynthesis epoch and the thermal spectrum of the background radiation. As discussed in Section 14.2, accurate measurements of the brightness of the microwave background over three decades of wavelength are consistent with a thermal spectrum with a thermo-dynamic temperature,

$$T_{CBR}(z = 0) = 2.76 \pm 0.03 \text{ K} \tag{14.2}$$

with no measured deviations. The abundances of deuterium (^{2}H), helium (^{3}He and ^{4}He), and lithium (^{7}Li) predicted by the conventional nucleosynthesis theory for the model in Figure 14.1 agree with the observed cosmic abundances for the accepted range of cosmological parameters, i.e., a Hubble constant $H_0 \sim 50$ to 100 km s^{-1} Mpc^{-1}, a current background temperature $T \sim 2.7$ K, and a Universe in which the density contributed by the baryonic component is less than 0.2 times the closure density—that density needed to asymptotically stop the current expansion. Indeed, the ^{4}He abundance (24%) determines the temperature of the background radiation to within 10% and conversely. For a detailed review, see Boesgaard and Steigman (1985).

The hot Big Bang model has been used to predict a number of weak effects leading to perturbations of the spectrum of the background radiation. To date only one of these, the Sunyaev-Zel'dovich effect, has been observed. This effect is seen by switching on and off the line of sight to a perturbing cloud of hot plasma (Section 14.5). Direct measurements of the absolute brightness of the microwave background have not yet detected deviations from a blackbody spectrum, except for an excess flux at $\lambda \sim 0.5$ mm (Matsumoto et al., 1987) which, if confirmed (it might be due to a number of systematic effects), could be due to emission by high–redshift ($z \sim 100$) interstellar dust, or to inverse Compton scattering of the background photons by hot electrons at very high ($z > 1000$) redshift (see Section 14.2.3 below).

14.2 The Spectrum of the Microwave Background

14.2.1 Summary

The search for distortions in the thermal spectrum of the microwave background is motivated by the theoretically predicted effects of various physical phenomena which might have been important in the early Universe. [For reviews, see Peebles

(1971), Danese and De Zotti (1977), Jones (1980), Sunyaev and Zel'dovich (1980).] The expanding Universe, whose evolution is sketched in Figure 14.1, departs from thermal equilibrium in many ways; but the effects on the radiation spectrum are small because in the early Universe, the heat capacity is mostly in the radiation. The epochs probed by measurements of the spectrum lie between redshifts $z \sim 10^9$ and $z \sim 10^3$; at earlier times, thermal equilibrium is established quickly by electron-positron pairs and later the radiation has decoupled from neutral matter.

Distortions due to different cooling rates for radiation and matter, and from the decoupling process, are very small, $\Delta T/T < 10^{-5}$. However, larger effects are predicted if the electrons are heated by an auxiliary process such as dissipation of turbulence or annihilation of antimatter. The hot electrons then have two effects: (1) inverse Compton scattering shifts photons to higher frequency, making the observed temperature higher at short wavelengths and lower at long wavelengths; (2) low-frequency photons are emitted through bremsstrahlung and the radiative Compton effect, which make the radiation hotter at very long wavelengths. The magnitude of these perturbations depends on the amount of energy supplied by the heating process, which is very uncertain. Currently, the absence of spectral distortions in the microwave background can be used to limit the extent of possible heating mechanisms in the early Universe, although these limits are not yet very strong.

Spectral measurements of the microwave background radiation have been reviewed by Peebles (1971) and Weiss (1980). Each concluded that the measurements were in agreement with a blackbody spectrum and that the temperature was about 2.73 ± 0.08 K. By 1980, the important measurements over the spectral peak had been done, and the most sensitive of these indicated a possible deviation from the expected thermal spectrum (Woody and Richards 1979, 1981); this experiment is discussed in Section 14.2.3. Since 1980, a number of new measurements have been reported over a wide range of wavelengths and with improved sensitivity. These show a blackbody spectrum with a thermodynamic temperature of 2.76 ± 0.03 K, with no evidence for spectral distortions, except for a tentative detection of an excess flux at $\lambda \sim 0.5$ mm (Matsumoto, et al. 1987). Given that these experiments are subject to relatively large systematic errors, the results of measurements of the temperature of the microwave background radiation show a remarkable constancy with time.

Figure 14.2 shows the results of spectral measurements of the microwave background. The dashed curves represent galactic and atmospheric foreground radiation that have important influence on experimental strategies. Long-wavelength measurements must concentrate on separately determining and subtracting the contribution of the Galaxy, while experiments near the peak must measure, or avoid, atmospheric emission. Most measurements at wavelengths longward of 3 mm have been made with ground-based heterodyne (Dicke) radiometers. Direct measurements shortward of 3 mm are usually made using wideband bolometric detectors flown in balloons or rockets. Finally, interstellar CN molecules have rotational states which can be excited by photons with wavelengths near that of the peak of the microwave background spectrum ($\lambda = 2.64$ mm and 1.32 mm). The populations of these states are measured by optical absorption of background

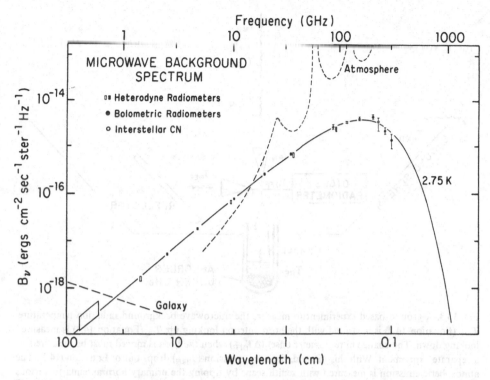

Fig. 14.2. Measurements of the spectrum of the microwave background radiation. Results published since 1980 are plotted, but for comparison, a few earlier results are shown as open boxes. The important measurements at long wavelength, where spectral distortions are most likely, have not been repeated.

starlight, and the excitation temperature can be determined with high precision. All three methods have now been used and refined by several groups; the more recent work is described in the rest of this section.

14.2.2 Heterodyne Radiometer Methods ($\lambda \geqslant 3$ mm)

Any reasonable radio telescope has sufficient sensitivity to detect the microwave background, but two special features are needed to make a spectral measurement. Since the radiation is diffuse, its signal cannot be isolated from that of other diffuse radiation by the usual "on–off" technique. Competing radiation must be shielded from the antenna (as is ground radiation), or it must be separately measured and subtracted from the total (as are atmospheric and galactic emission). Besides, the receiver output must be directly related to an absolutely known flux at the input. The need for such a "zero-point" determination is unusual in radio astronomy. Indeed, the low sidelobe antenna used by Penzias and Wilson, and the Bell Laboratories' general interest in low-noise receivers were primarily motivated by research in the area of satellite communications, where it is important to understand all sources of noise.

The general approach to such a measurement is shown in Figure 14.3. The radiometer uses a technique invented by Dicke (1946), who realized that a rapid comparison of the wanted signal with a stable reference source ("Dicke switching")

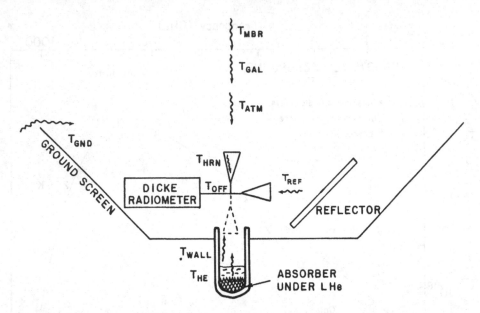

Fig. 14.3. A ground-based experiment to measure the microwave background radiation temperature. T_{SKY} (Equation 14.3) is measured with the horn antenna looking up; T_{CL} (Equation 14.4) is measured looking down. The change in radiometer offset (ΔT_{OFF}) when the horn is moved must be measured by a separate experiment. With this method, the horn emission (T_{HRN}) drops out of Equation (14.5). The atmospheric emission is measured with zenith scans by tipping the primary horn antenna to various elevations on both sides of the zenith. The radiometer rotates around the symmetry axis of the reference horn which accepts circularly polarized radiation reflected off the mirror. T_{REF} should therefore remain constant as the radiometer rotates (Friedman et al. 1984). An alternate method which avoids moving the instrument and therefore avoids ΔT_{OFF} is described by Wilkinson (1967).

filters out the low-frequency components of the noise; moderate sensitivity (system noise) is sufficient, but constancy of instrumental parameters (gain and offset) is important. The radiometer compares the radiation seen through the main horn with that seen through the reference horn. With the main antenna pointed to the zenith, the antenna temperature* is T_{SKY} and consists of several diffuse components: ground radiation (T_{GND}) diffracting over the screened enclosure, atmospheric and galactic emission (T_{ATM} and T_{GAL}), and the microwave background radiation (T_{MBR}). In this position the radiometer measures

$$T_{SKY} - T_{REF} = T_{MBR} + T_{ATM} + T_{GAL} + T_{GND} + (+T_{HRN}) + T_{OFF} \qquad (14.3)$$

where T_{OFF} is an instrumental offset which must be removed. If the horn antenna is used to view the cold load (as shown), then the emission from the horn walls (T_{HRN}) need not be separately evaluated.

The cold load in Figure 14.3 is used to find a zero-point reading (T_{CL}) for the radiometer output with a known temperature at the input. This is not a calibration

---

* Since these experiments deal with low-temperature blackbody sources, it is important to distinguish between antenna temperature (T) and thermodynamic temperature ($\mathcal{T}$). They are related by: $T = x(e^x - 1)^{-1}\mathcal{T}$, where $x = h\nu/k\mathcal{T}$.

of the gain (Kelvin/Volt) of the radiometer; that is accomplished in the usual way by changing the input temperature by a known amount, usually with a room-temperature absorber over the horn. If T_{CL} is close to T_{SKY}, the calibration and the linearity of the radiometer need not be known to great accuracy, so the zero-point absorber is usually immersed in liquid helium. A complication arises because the warm metallic walls at the top of the cold load emit a significant amount of radiation (T_{WALL}) which adds to the known antenna temperature of the cold absorber, T_{HE} (calculated from $\mathcal{T}_{He}$). T_{WALL} must be calculated from measurements of the temperature and emissivity of the inner surface of the metal pipe. This was a major source of error until cold loads were built using oversized waveguides to give $T_{WALL} \leqslant 1$ K. Looking into the cold load, the radiometer measures

$$T_{CL} - T_{REF} = T_{He} + T_{WALL} + T'_{OFF} + (+T_{HRN}) . \tag{14.4}$$

The last term is not present if the horn is replaced by the cold load.

The experiment measures the difference between the sky temperature and the cold-load temperature. Care has to be taken that the reference signal T_{REF} remains constant throughout the measurement. The device in Figure 14.3 uses a reference horn that views the sky reflected on a mirror. The instrument is rotated about the symmetry axis of this horn which accepts circular polarization in order to minimize the change in T_{REF} due to the rotation; any residual variation in T_{REF} can be lumped with the difference between the offsets T_{OFF} and T'_{OFF}.

Combining both measurements, we have

$$T_{SKY} - T_{CL} = T_{MBR} + T_{ATM} + T_{GAL} + T_{GND} + (+T_{HRN})$$
$$- T_{He} - T_{WALL} + \Delta T_{OFF} . \tag{14.5}$$

An important design consideration in these experiments is to minimize the change in the instrumental offset,

$$\Delta T_{OFF} = T_{OFF} - T'_{OFF} . \tag{14.6}$$

An older method (Wilkinson 1967) avoided moving the instrument by using a main horn antenna looking down (at 45°) to view the sky radiation reflected off a large aluminum plate. The cold load could then be moved into view without touching the instrument, and the instrumental offset due to strains (due to gravity) and magnetic effects on the Dicke switch (due to the Earth's magnetic field) is the same for the measurements of T_{SKY} and T_{CL}. A disadvantage of this method is that the contribution due to emission from the reflector ($T_{REFL} \sim 0.1$ K) must be separately measured. Alternatively (Smoot et al. 1985), one can avoid this problem by moving the main horn as shown in Figure 14.3 (the contribution of the reflector in front of the reference horn is added to T_{REF} and need not be computed since it is the same in both measurements). The disadvantage of this method is that critical parts of the radiometer are moved and T_{OFF} may change. Auxiliary measurements are needed to determine ΔT_{OFF}, called the flip offset.

However, the main limitation in the accuracy of ground-based radio measurements of T_{MBR} is the evaluation of the other terms on the right-hand side of Equation (14.5). At wavelengths shortward of $\lambda = 10$ cm, T_{ATM} is the main problem. The zenith angle dependence

$$T_{\text{ATM}}(z) = T_{\text{ATM}} \cdot \sec Z \qquad (14.7)$$

is used to evaluate T_{ATM}, the zenith value of $T_{\text{ATM}}(Z)$. The design in Figure 14.3 scans the atmosphere by moving the main horn (again rotating the instrument about the symmetry axis of the reference horn, with the possibility of a change in T_{OFF}). Alternatively, the horn can look into a reflector which is moved to accomplish the atmospheric scans; here one again has to compute the contribution of the reflector but changes in T_{OFF} are avoided. At longer wavelengths, the main problem for these experiments is the T_{WALL} component to T_{CL}. Finally, at very long wavelengths ($\lambda > 30$ cm), T_{GAL} is large and difficult to evaluate, and T_{GND} is more of a problem because antennas and ground screens must be very large. Representative values and error estimates for the quantities in Equation (14.5) are listed in Table 14.1 for some sensitive experiments that have used radio techniques. The trends in the sizes of the errors with wavelength are easy to see.

Table 14.1 includes the results of a systematic experiment at five wavelengths performed at White Mountain (California, an altitude of 3.8 km) by a U.S.-Italian collaboration (Smoot et al. 1985). The method of Figure 14.3 was used with a large cold load of aperture 0.7 m. Four radiometers performed atmospheric scans while the fifth one used the cold load. Furthermore, a sixth radiometer was dedicated to monitor and measure the atmosphere at a wavelength of $\lambda = 3.2$ cm (Partridge et al. 1984). Even so, fluctuations in T_{ATM} account for most of the final uncertainty in T_{MBR} for the four measurements at $\lambda < 10$ cm. Variations in T_{ATM} might be smaller at other sites (perhaps at Antarctica?) or by waiting for special conditions (winter on a high, dry site?), but until more is known, ground-based measurements are limited in accuracy to about ± 100 mK by atmospheric radiation ($\lambda < 10$ cm) or by emission from the walls of the cold load ($\lambda > 10$ cm).

The early and important measurements of Howell and Shakeshaft (1966, 1967) at different wavelengths longward of 20 cm have not been repeated. Because diffraction effects increase with wavelength, the horn antennas and ground screens must be large, but the main problem is rapidly increasing galactic radiation temperature ($T_{\text{GAL}} \propto \lambda^{2.8}$), as shown in Figure 14.2. Values and error estimates for measurements at 20.7 cm, 49.2 cm, and 73.5 cm are given in Table 14.1. The horn, too large to cover with a cold load, was removed in order to measure T_{CL}, so the term T_{HRN} in Equation (14.5) had to be evaluated. In these experiments it was calculated. In order to remove T_{GAL} from the observations at $\lambda = 49.2$ cm and 73.5 cm, it was assumed that the isotropic background had a blackbody spectrum. Thus, the resulting value of T_{MBR} applies to both wavelengths.

A new method designed to reduce the larger errors in Table 14.1 is sketched in Figure 14.4 (Johnson and Wilkinson 1987). The experiment was performed from a balloon at an altitude of 26 km to virtually eliminate T_{ATM}. A correlation radiometer and cryogenic isolators are used so that the microwave front end is passive and cold, thus minimizing the effects of reflections from the horn and loads. T_{CL} is measured as shown, and T_{SKY} is measured by sliding out the zero-point load and lowering the horn to couple with the radiometer input. As seen in Table 14.1, this basic measurement gives $T_{\text{SKY}} = 2.90 \pm 0.02$ K at $\lambda = 1.2$ cm, and the correction

Table 14.1. Representative values and errors for quantities in Equation (14.5).

λ (cm)	T_{CL} (K)	$(T_{SKY} - T_{CL})^a$ (K)	T_{ATM} (K)	T_{GAL} (K)	T_{GND} (K)	T_{HRN} (K)	ΔT_{OFF} (K)	T_{MBR}^b (K)	$\mathcal{T}_{MBR}^c$ (K)
73.5[d]	5.6 ± 0.2	21.0 ± 0.7	1.3 ± 0.1	20.6 ± 0.9[g]	0.6 ± 0.4	0.4 ± 0.2			3.2 ± 1.2
49.2[d]	5.7 ± 0.2	7.9 ± 0.5	1.95 ± 0.1	6.7 ± 0.7[g]	0.8 ± 0.4	0.4 ± 0.2		3.2 ± 1.2[k]	3.2 ± 1.2
20.7[d]	5.9 ± 0.2	0.9 ± 0.1	2.2 ± 0.2	0.5 ± 0.2	0.05 ± 0.05	1.3 ± 0.2		2.8 ± 0.6	2.8 ± 0.6
12.0[e]	5.52 ± 0.13	−1.56 ± 0.06	0.95 ± 0.05	0.20 ± 0.03	0.01 ± 0.01		[j]	2.81 ± 0.15	2.87 ± 0.16
6.3[e]	3.68 ± 0.01	−0.05 ± 0.01	1.00 ± 0.07	0.04 ± 0.03	0.02 ± 0.02		0.00 ± 0.01	2.59 ± 0.08	2.70 ± 0.08
3.0[e]	3.56 ± 0.01	0.04 ± 0.01	1.20 ± 0.13	0.004 ± 0.002	<0.003		0.00 ± 0.03	2.41 ± 0.14	2.64 ± 0.14
1.2[f]	2.60 ± 0.01	−0.30 ± 0.01	<0.004	0.002 ± 0.001	<0.004[h]	0.80 ± 0.02[i]		2.22 ± 0.03	2.78 ± 0.03
0.91[e]	3.07 ± 0.01	3.54 ± 0.03	4.53 ± 0.09	0.001 ± 0.001	<0.002		−0.03 ± 0.03	2.09 ± 0.13	2.81 ± 0.14
0.33[e]	2.08 ± 0.04	8.78 ± 0.06	9.87 ± 0.09	0.000 ± 0.001	<0.002		0.01 ± 0.01	0.99 ± 0.11	2.57 ± 0.14

[a] Statistical measurement error.
[b] Antenna temperature.
[c] Thermodynamic temperature.
[d] Howell and Shakeshaft (1966, 1967).
[e] Smoot et al. (1985). Values from 1983; similar results were obtained in 1982.
[f] Johnson and Wilkinson (1987).

[g] Includes discrete sources.
[h] Includes balloon emission.
[i] Includes window emission.
[j] Included in T_{CL}.
[k] Spectrum assumed blackbody to remove T_{GAL}.

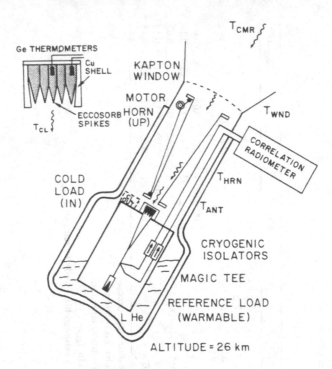

Fig. 14.4. Schematic drawing of the experiment of Johnson and Wilkinson (1987) at a wavelength of 1.2 cm. As shown, the radiometer is measuring T_{CL} from a "zero-point" load whose temperature is accurately measured. On remote command, motors remove the load and lower the horn antenna to measure $T_{SKY} + T_{HRN} + T_{WND}$ (window emission). A ground screen and a tilting device are not shown. Data were taken at $Z = 10°, 23°, 32°,$ and $45°$ to check for atmospheric, ground, and balloon emission. See Table 14.1 for values of various contributions and the result.

terms are relatively small. The horn emission T_{HRN} is kept to 49 ± 12 mK by cooling the lower portion, and the radiometer was calibrated by heating the reference load by a known amount, typically 1.30 K. The experiment was performed at several zenith angles to check for T_{ATM}, T_{GND} and radiation from the balloon. The result, $\mathcal{T}_{MBR} = 2.783 \pm 0.025$ K, $(\pm 0.079$ K if all errors are added linearly), is the most sensitive determination to date.

14.2.3 Bolometric Measurements ($\lambda \leqslant 3$ mm)

At wavelengths shorter than 2 mm, the brightness of the background radiation drops and the noise temperature of heterodyne radiometers increases. Wide-band bolometric techniques, developed by infrared astronomers, become more attractive. Observations must be made from balloons or rockets, and the entire instrument is usually operated at cryogenic temperatures. Spectral resolution is achieved by using a Michelson interferometer or bandpass filters. Because of the large response of bolometers to infrared radiation, spectral blocking is important. Weiss (1980) reviewed the early work on these difficult experiments, including the results of Woody and Richards (1979, 1981) and Gush (1981), both of which indicated deviations from a blackbody spectrum. However, the deviations have not been

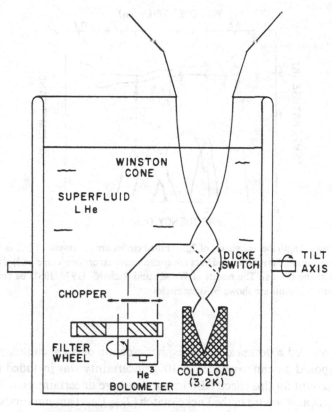

Fig. 14.5. Radiometer used by Peterson et al. (1985) to measure T_{MBR} on the short-wavelength side of the blackbody peak. Identical apertures were used on either side of the Dicke switch. Experimental errors are mainly due to uncertainty in the measurement of T_{ATM} (by tilting the instrument) and in possible changes in the asymmetry of the Dicke switch (T_{OFF}). Results are given in Table 14.2 and Figure 14.6.

confirmed by subsequent experiments, and Weiss was able to present reasonable explanations for the observed effect. For example, allowing a 27% calibration error in the experiment of Woody and Richards gives a good fit of their data to an undistorted 2.79 K blackbody spectrum.

A subsequent experiment by the Berkeley group (Peterson *et al.* 1985) used the balloon-borne instrument sketched in Figure 14.5. The (liquid-helium) cooled antenna made from two Winston concentrators was similar to the one used by Woody and Richards; otherwise the instrument was different. Filters were used for spectral discrimination, and a mechanical Dicke switch allowed the radiometer to view the sky or a zero-point load of known temperature. A 10-Hz chopper was used to avoid excess low-frequency noise in the detector—a composite bolometer cooled to 0.3 K by a ^{3}He refrigerator. Inflight measurements were made of atmospheric emission (by tipping), antenna wall emission (by heating), radiometer linearity, and the antenna beam pattern (by scanning the moon). An unexpected drift ($\sim 2\%$ per hour) in the measured sky temperature could not be explained; the

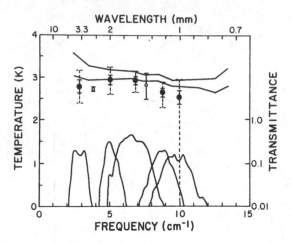

Fig. 14.6. Short-wavelength measurements of T_{MBR}. Filled circles are the results of Peterson et al. (1985). Solid bars are conventional errors; dashed bars are conservative errors (see Table 14.2). The bandpasses used are shown at the bottom. The results of Woody and Richards (1979, 1981) lie between the line segments. CN measurements are shown by open circles.

authors conjectured a decrease in atmospheric ozone or an accumulation of solid air on the exposed antenna surfaces. A 10% uncertainty was included in the error analysis to account for this effect. Also important were uncertainties in atmospheric emission (T_{ATM}), asymmetry in the Dicke switch (T_{OFF}), and antenna emission (T_{HRN}). Ground emission and galactic dust emission were shown to be negligible.

The results of the first flight of this experiment are shown in Figure 14.6 and listed in Table 14.2. The conventional error bars (solid) should be compared with the results of other experiments; however, the authors give their results as $\mathcal{T}_{MBR} = (2.78 \pm 0.11)$ K, pointing out that this range is included in all of their conservative error limits. They concluded that the results were consistent with a blackbody spectrum. The results of Woody and Richards are also shown in Figure 14.6, lying systematically above the newer results; the temperature (2.96 K) derived by Woody and Richards is not consistent with 2.78 ± 0.11 K. However, Peterson *et al.* made the interesting remark that the two data sets show the same trend with wavelength.

At the time of this writing, preliminary results from a collaboration between a group at the University of Nagoya (Japan) and the Berkeley group indicate excessive flux at wavelengths $\lambda = 0.7$ mm ($T = 2.963$ K ± 0.017 K) and at $\lambda = 0.5$ mm ($T = 3.150 \pm 0.026$ K) while they measure a flux corresponding to $T = 2.795$ K ± 0.018 K at $\lambda = 1.16$ mm (Matsumoto *et al.* 1987). Three data points measured at still lower wavelengths were clearly affected by galactic dust emission. This excess flux might be due to some systematic effect such as a possible contamination by the rocket exhaust cloud, or emission by frozen contaminants in the optical path. (Indeed, a large heat pulse was noticed when the cover of the instrument was removed). In spite of careful analysis by the experimenters, these questions have not yet been answered. Further analysis and laboratory measurements should resolve these issues. Of course, independent confirmation should settle these points.

Table 14.2. Measurements of the spectrum of the background radiation at short wavelengths.

Wavelength (mm)	Thermodynamic temperature (K)	Error Limits (K) Conventional[a]	Error Limits (K) Conservative[b]
4.35–2.94[c]	2.80	±0.16	+0.37, −0.39
2.80–1.89[c]	2.95	+0.11, −0.12	+0.29, −0.33
1.75–1.28[c]	2.92	±0.10	+0.23, −0.25
1.22–1.06[c]	2.65	+0.09, −0.10	+0.24, −0.34
1.12–0.91[c]	2.55	+0.14, −0.18	+0.40, −2.55
2.64[d]	2.66[f]	±0.06	±0.10
2.64[d]	2.72[g]	±0.05	±0.10
2.64[d]	2.70[h]	±0.08	±0.15
2.64[e]	2.74[f]	±0.05	±0.07
1.32[d]	2.76[f,g,h]	±0.20	[i]
1.32[e]	2.75[f]	+0.24, −0.29	[i]
0.559[j]	<5.23		
0.359[j]	<7.35		

[a] Independent systematic errors added in quadrature.
[b] Independent systematic errors added linearly.
[c] Peterson et al. (1985).
[d] Meyer and Jura (1985).
[e] Crane et al. (1986).
[f] Source: ζ-Ophiuchi.
[g] Source: ζ-Persei.
[h] Source: o-Persei.
[i] Result dominated by statistical errors.
[j] Thaddeus (1972).

14.2.4 Measurements Using Interstellar Molecules

An indirect method of measuring the temperature of the microwave background radiation uses CN, CH, or CH^+ as a thermometer. The equivalent widths of optical absorption lines are used to measure the relative populations of several low-energy rotational states in these molecules and the populations are then used to calculate the excitation temperatures. The method deserves a brief discussion since it has allowed high-sensitivity measurements at wavelengths near that of the peak of a 3 K blackbody spectrum. In particular, observations of CN clouds yield the temperature of the microwave background at wavelengths of $\lambda = 2.64$ mm and 1.32 mm. Recent measurements by Meyer and Jura (1985) and Crane et al. (1986) have produced the results shown by open circles in Figure 14.6 and listed in Table 14.2.

Two systematic errors must be taken into account in this method. The lines are narrow, requiring very high dispersion to resolve them. However, the linewidths are important in order to correct the measured equivalent widths for saturation effects. The early work of Hegyi et al. (1972) had determined a linewidth of 28 mÅ for an interstellar CN cloud on the line of sight towards ζ Ophiuchi, which is the cloud used by Meyer and Jura. They used this linewidth to correct the equivalent width of the $R(0)$ absorption line at $\lambda_0 = 3874.608$ Å which showed some saturation.

Recently, Crane *et al.* (1986) have used higher dispersion and measured a width of 19 mÅ for the CN lines of this cloud on the line of sight to ζ Oph. Reanalyzing the data of Meyer and Jura with this width lowered the derived excitation temperature by 0.05 K, nearly the length of one conventional error bar (Crane *et al.* 1986). Clearly, direct observations of the linewidths and shapes are important. The saturation analysis assumes that the lines are single-component gaussians; this should be justified with higher dispersion measurements.

A second correction to the data is needed to account for possible local excitation of the CN molecules by collisions with electrons. A density of $n_e = 0.1$ cm^{-3} contributes an equivalent temperature of 0.04 K to the measured excitation temperature. However, n_e is not known at the location of the CN molecules. Models of the CN cloud used in the measurement, based on the observed ionization fractions for several elements, give values for n_e ranging from 0.06 cm^{-3} in the outer envelope, to 0.25 cm^{-3} in the core. Both groups adopt $n_e \sim 0.1$ cm^{-3}, which gives $\Delta T = -0.06 \pm 0.04$ K for the correction to account for collisional excitation. Again, the systematic error in the correction is a large fraction of the total error. Crane *et al.* suggest that a useful limit on collisional excitation might be set by a very sensitive search for CN emission from the same molecular cloud; excess excitation would make it appear warmer than the microwave background radiation (Penzias *et al.* 1972).

The "molecular-thermometer" method gives direct evidence that the microwave background is universal. The same temperature is found in several sources (on the line of sight towards different stars: see Table 14.2), and no source has been found showing a strong absorption from the ground state and no excitation of the first rotational level. Such a finding would be devastating for the cosmological interpretation of the background radiation in the standard Big Bang model. To improve the sensitivity of the method beyond current levels will require more sensitive, high-resolution spectroscopy and an independent measurement of the contribution of collisional excitation.

14.2.5 Summary and Future Prospects

Measurements of the brightness of the background radiation over nearly a factor of 1000 in wavelength agree with the Big Bang model's prediction of a black-body spectrum. The results from heterodyne radiometer measurements (Table 14.1) range from $\lambda = 75$ cm to $\lambda = 0.33$ cm and give a weighted mean temperature of 2.76 ± 0.03 K (with a Chi-square of 4.3 with 7 degrees of freedom). Bolometric and molecular (CN) methods (Table 14.2) give a weighted mean temperature of 2.73 ± 0.03 K (Chi-square of 10.4 with 10 degrees of freedom). Since most errors are systematic, these errors in the mean values are too small; the values for Chi-square indicate that observers have succeeded in estimating their errors at about the one-standard-deviation level. New results around the spectral peak do not support the deviations indicated by Woody and Richards (1979, 1981); but more work is needed and under way. A new method for radiometric measurements has reached a sensitivity of 1% (Johnson and Wilkinson 1987) and shows the importance of avoiding atmospheric emission and using cryogenic instruments. This method should achieve comparable sensitivity over a range of wavelengths, perhaps from $\lambda \sim 3$ cm to $\lambda \sim 0.3$ cm.

A possible distortion has been suggested (Matsumoto *et al.* 1987). If further tests and independent measurements confirm their results, deviations from a pure blackbody spectrum will have been detected. The authors discuss the implications of their measurement when compared to the ones described above. Compton scattering by hot electrons (Zel'dovich and Sunyaev 1969) gives an effect at shorter wavelengths like that shown by the rocket data. The best fit gives a Rayleigh-Jeans value of $T = 2.705 \pm 0.015$ K and a comptonization parameter $y = 0.018$ but the fit is poor (a Chi-square of 33 with 19 degrees of freedom). Ignoring the measurement by Johnson and Wilkinson (1987) discussed in section 14.2.2 improves the fit significantly but then this latter measurement falls above the fitted curve by 120 mK, 4.8 times the estimated error.

An alternative explanation is that the rising flux at short wavelengths could be due to dust emission in galaxies at large redshift; the dust would have been heated by the hypothetical population-III stars. As the required energy density is proportional to $(1 + z)^4$, it is hard to place the dust at very high redshifts. Conversely, the authors point out that if the dust is placed at modest redshifts, the models have trouble fitting the much lower flux observed at a wavelength $\lambda = 0.26$ mm.

Ultimately, the Far-Infrared Spectrometer on the Cosmic Background Explorer (COBE) satellite should give the nearly ideal measurement of the spectrum of the microwave background from $\lambda = 8$ mm to $\lambda = 1$ mm (Mather 1982). Statistical errors are expected to be about ± 1 mK on each $7°$ patch of sky, and systematic errors should be less than ± 10 mK.

14.3 Polarization of the Microwave Background

This has not been an active field, probably because theoretical predictions are either discouragingly small or based on speculative models. Although in the standard Big Bang model the background radiation is essentially unpolarized at decoupling time, a number of cosmological processes could leave some degree of polarization imprinted on the radiation. Rees (1968) showed that an anisotropy in the expansion of the Universe would induce a net polarization in the background radiation through Thomson scattering. The magnitude of the polarization on large angular scales depends on the degree of anisotropy of the expansion and the ionization history of matter since the decoupling time—both unknown. Some ionization scenarios lead to a degree of polarization of the same order of magnitude as the residual anisotropy in the temperature of the radiation (Negroponte and Silk 1980, Basko and Polnarev 1980). Circular polarization can be generated by primordial magnetic fields or by an intrinsic rotation of the Universe. Depolarization by random Faraday rotation could occur if the Universe went through a reionization phase (see Section 14.4.1 below) or if there is a global magnetic field; the effect is likely to be small.

One expects comparable polarization and temperature fluctuations at the angular scale corresponding to the thickness of the last scattering surface (a few degrees). A few scatterings effectively produce polarized radiation from an initial temperature anisotropy.

At small angular scales, the density perturbations (at decoupling, $z \sim 10^3$) which lead to clusters of galaxies also produce temperature and polarization anisotropies. Kaiser (1983) has shown that $P \sim 0.2$ $(\Delta T/T)$ for the adiabatic scenario (see Section 14.4), and the expected angular scale is $3\Omega^{1/2}$ arcminutes.

The original discovery paper by Penzias and Wilson (1965) included a limit of 10% to the possible linear polarization of the radiation. Nanos (1979) carried out the first search for linear polarization at a wavelength $\lambda = 3.2$ cm. The apparatus used a switched Faraday rotator to compare two orthogonal, linear polarizations along a path at the local zenith (Dec $\sim 40°$). By rotating the instrument, both components of linear polarization were measured, and rotation was also used to subtract out a small instrumental offset. The antenna had a 15° beamwidth, so only effects at large angular scale could be seen. The axisymmetric model of Rees (1968) predicts that 12- and 24-hour periodic components would be seen along a circle of constant declination. Nanos saw no such effects at a level of 0.8 mK (90% confidence).

Limits on linear polarization in the 0.33-cm to 0.05-cm band have been set with a balloon-borne bolometer coupled to a telescope with fields of view of 0.5° or 15° through a rotating linear polarizer. The observations were limited by atmospheric contamination, and limits on linear polarization at the 1% to 0.1% were set on angular scales from 1.5° to 40° (Caderni et al. 1978). The most sensitive limits on linear polarization have been reached by a group at Berkeley (Lubin et al. 1983) who are also the only ones to have searched for circular polarization. Their experiment used a radiometer with a room-temperature mixer operating at a wavelength $\lambda = 9$ mm with a Faraday rotator switch; the apparatus was rotated to measure both components of linear polarization. The instrument was made sensitive to circular polarization by inserting the microwave equivalent of a quarter-wave plate in front of the linear polarization switch. Lubin et al. place 95% confidence level upper limits of 0.1 mK on any linear polarization component in an axisymmetric anisotropically expanding Universe. Circular polarization was not detected either, and upper limits of 12 mK on a variable component and of 20 mK on a constant (with sky position) component were set. The sensitivity for each circular patch of sky 7° in diameter was as good as 0.2 mK for linear polarization.

Future measurements with better receivers could improve on these measurements, although galactic contamination will be a problem at longer wavelengths ($\lambda > 1$ cm at current sensitivity levels). Systematic (instrumental) problems are more important in measurements of circular polarization because an offset cannot be removed by rotating the instrument. Measurements at small (minutes of arc) angular scale have not been done. These could be important as a probe of density fluctuations at decoupling and as a diagnostic for a possible reionization epoch.

14.4 Anisotropy Searches

14.4.1 Small-Scale Measurements

(a) A Scenario
Soon after the discovery of the background radiation, Peebles (1965) pointed out that it provides a picture of the Universe at the decoupling era ($z \sim 1000$) which

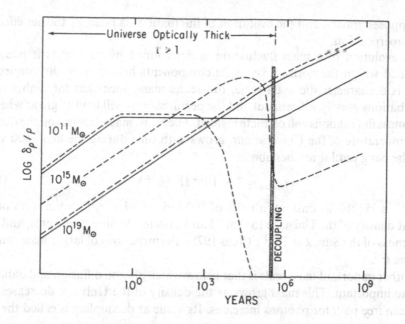

Fig. 14.7. Sketch of the evolution of density perturbations in a baryon-dominated Universe. The masses of the perturbations are indicated. Gravity increases $\delta\rho/\rho$ as $t^{2/3}$ until pressure prevents further collapse when the Jeans mass exceeds that in the perturbation. Adiabatic fluctuations (dashed lines) can be damped through photon diffusion (the Silk effect) whereas isothermal fluctuations (solid lines) are not affected by this mechanism. After the Decoupling era, matter ceases to interact with the radiation and the Jeans mass drops to about 10^5 to 10^6 $M_\odot$; fluctuations can subsequently grow until a time corresponding to the redshift $1 + z \sim \Omega^{-1}$.

might show the initial conditions: power spectra and characteristic lengths and masses, from which the matter in the Universe evolved into galaxies and clusters of galaxies. Subsequent scattering could modify or blur this picture if the Universe went through a phase in which an appreciable fraction of the mass was reionized (Dautcourt 1969). Such a phase has been suggested to occur at a variety of redshifts from $z = 300$ to fairly recent epochs.

The expected small-scale anisotropy in the background radiation can be inferred from the presently observed structure in the Universe, and the connection has been widely studied. However, predictions of the "minimal" anisotropy on various scales needed to produce today's Universe have been lowered whenever improved observations have failed to detect it.

A typical scenario is considered in Figure 14.7. This corresponds to a model Universe in which baryons provide the dominant component of the mass and no as-yet-unseen particles are included. During its early evolution, before the Universe cools below $T \sim 4000$ K, it is optically thick; matter is mostly ionized and in thermal equilibrium. Matter and radiation can be treated as a single fluid with pressure mainly due to the photons. (Earlier, the neutrino background did contribute to the pressure as well but this component decoupled when the Universe had cooled to a temperature of about 5×10^9 K.) The photon pressure provides a restoring force that opposes gravity, and the evolution of fluctuations is ruled by the net effect of their superposition.

The evolution of a mass fluctuation is determined by its size as it becomes contained within the horizon, i.e. as its components become causally connected. There is a characteristic mass scale, the Jeans mass, such that for higher mass perturbations gravity will win out and the perturbations will tend to grow; whereas lower mass fluctuations will oscillate instead. The Jeans mass depends on the density and temperature of the Universe and grows with time during the first 1000 years after the bang until it reaches about

$$m_{J,\,\mathrm{max}} = 3 \times 10^{14}\,\Omega^{-2} h^{-4} M_{\odot} \qquad (14.8)$$

where h is Hubble's constant in units of 100 km s^{-1} Mpc^{-1}, Ω is the ratio of the present density of the Universe to that of an Einstein–de Sitter* universe, and $M_{\odot}$ is the mass of the Sun, 2×10^{33} g (Rees 1978). Perturbations of larger mass tend to grow as $t^{2/3}$.

Another important mass scale is that below which photon diffusion and damping become important. This mass grows as the density of the Universe decreases and the mean free path for photons increases. Its value at decoupling is called the Silk mass,

$$M_s \sim 10^{12}\, h^{-5/2} \Omega^{-5/4} M_{\odot}\,. \qquad (14.9)$$

Adiabatic fluctuations of lower masses will be severely damped and essentially erased before decoupling occurs whereas isothermal perturbations will have evolved uninfluenced by this mechanism (Silk 1968).

At decoupling, the Silk mechanism ceases to operate and, besides, the Jeans mass drops to 10^5–$10^6\, M_{\odot}$, so that after this epoch, matter fluctuations are free to evolve under gravity and gas dynamics, independently of the background photons. Density fluctuations $(\delta\rho/\rho)$ will have grown at most by a factor $(1 + z_{\mathrm{dec}})^{-1}$ since the decoupling time (which corresponds to the redshift z_{dec}). Detailed calculations show that for the adiabatic scenario this requires fluctuations in the background radiation at a level of

$$\left(\frac{\Delta T}{T}\right)_{\mathrm{adiab}} \sim 3 \times 10^{-5} \qquad (14.10)$$

in order to account for present-day structures. In the isothermal case, matter fluctuations do not produce anisotropy in the radiation; nevertheless, fluctuations are induced by Doppler shifts at the last scattering of each photon at a level

$$\left(\frac{\Delta T}{T}\right)_{\mathrm{isoth}} \sim \frac{v}{c} \sim 10^{-5}\,. \qquad (14.11)$$

The connection between angular scales and physical scales at decoupling depends mostly on the density parameter Ω; a sphere whose diameter subtends an angle of

---

* This is a universe where the cosmological constant vanishes, and with an average density such that the expansion will stop asymptotically, i.e., when $t \to \infty$.

1 minute of arc contains a mass of about

$$M \cdot 7 \times 10^{11} h^{-1} \Omega^{-2} M_{\odot} ; \qquad (14.12)$$

a mass of $10^{15} M_{\odot}$ subtends an angle of the order of 5 minutes of arc for $h = 0.75$ and $\Omega \sim 0.3$.

As indicated above, a significant fraction of the matter could become ionized again at a much later epoch associated with galaxy formation. The initial burst of star formation that is assumed in order to produce the heavy elements that are seen in ordinary stars could have released enough energy to achieve a significant degree of ionization and sufficient optical thickness. This could have erased fluctuations on angular scales corresponding to areas causally connected during this phase, although the process would have enhanced fluctuations on scales corresponding to the horizons at this stage; besides, some degree of polarization would result due to protogalactic magnetic fields (Hogan *et al.* 1982). But Silk (1984) has argued that such a phase cannot have affected the small-scale structure of the background radiation in any relevant way due to an insufficient optical thickness and that we are indeed getting a clear picture from the decoupling epoch.

Detailed theoretical predictions of anisotropy at different angular scales require an assumption on what species are important in the problem: "ordinary" baryons, massive neutrinos, or other more exotic as-yet-undetected particles. Furthermore, one has to adopt a model for their initial distribution: for simplicity, most authors assume fluctuations with a spectrum of the form

$$\frac{\delta\rho}{\rho} \propto k^n \qquad (14.13)$$

for the dominant contribution to the mass at the decoupling epoch, and $n = 1$ is preferred as the most "natural" spectrum. One also has to include suitable cutoffs to prevent divergences and, finally, one has to choose the relevant cosmological parameters (h, Ω). The calculation then produces an evolved picture of the Universe at the present epoch, and matching the autocorrelation of this "evolved" picture of the matter distribution to the available data from studies of galaxy clustering fixes the amplitude of the initial spectrum, thus fixing the details of the assumed Universe at decoupling and producing a prediction of the structure that should be seen in the background radiation. Figure 14.8 shows a sample of such predictions for a number of cosmological models (Wilson and Silk 1981, Wilson 1983a, Bond and Efstathiou 1984, Vittorio and Silk 1984).

(b) Measurements
Measurements of small-scale anisotropy in the microwave background require the use of large telescopes in order to obtain the necessary narrow beams on the sky. This makes a measurement at a level of $\Delta T/T \sim 10^{-5}$ necessarily hard. The main experimental problems are again atmospheric emission at wavelengths $\lambda < 1$ cm, galactic dust emission at $\lambda < 1$ mm, confusion due to radio sources at $\lambda > 1$ cm, interference and variable instrumental effects, and ground radiation at all wavelengths. This last problem is very hard to eliminate as the telescopes are sensitive to radiation from a large solid angle through back and sidelobes. Spillover around

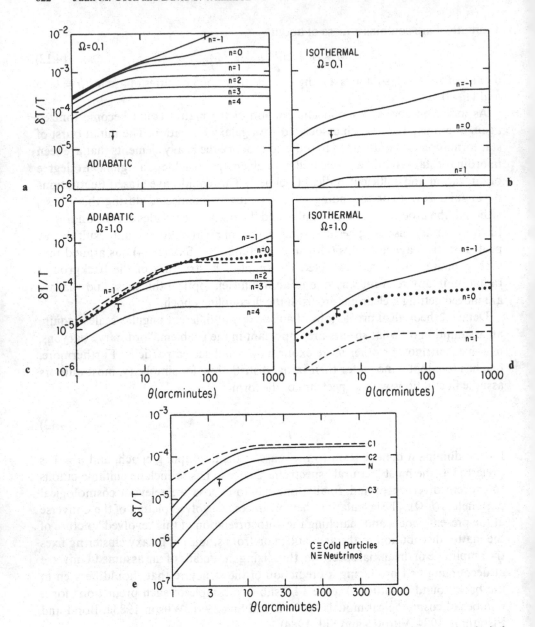

Fig. 14.8. Predicted small-scale anisotropy in the microwave background for various cosmological models. Those in (a) through (d) correspond to baryon-dominated universes, whereas in (e) the dominant contribution to the mass is in weakly interacting particles. Model (a) is a Universe in which $\Omega = 0.1$ and the initial perturbations are adiabatic; in (b) $\Omega = 0.1$ and the perturbations are isothermal; (c) corresponds to adiabatic perturbations with $\Omega = 1$; and (d) corresponds to isothermal perturbations with $\Omega = 1$. The curves are parameterized by the exponent n in Equation (14.13). The models in (e) correspond to $n = 1$, and different species provide the dominant contribution to the mass: cold as-yet-unseen particles (curves C_1 for which $\Omega = 0.2$, $h = 0.5$; C_2 for which $\Omega = 0.2$, $h = 0.75$; and C_3 with $\Omega = 1.0$, $h = 0.75$) or massive neutrinos (curve N, for which $\Omega = 1.0$ and $h = 0.75$). The indicated arrows correspond to the upper limit in Uson and Wilkinson (1984b). The predictions are statistical and the confrontation between them and the measurement relies on a number of assumptions, the most important being that the fluctuations are Gaussian distributed.

the primary reflector and diffraction off the secondary reflector or feed supports cause a large variable "ground effect." Ideally, one would like to have ground shields to reduce this problem, but the large area to shield makes this prohibitively expensive. Finally, one has to deal with receiver (Johnson) noise, which limits the sensitivity of any measurement of finite duration. For example, the 140-ft telescope at the NRAO site in Green Bank (West Virginia) has one of the most sensitive systems available; at a frequency of 19.5 GHz, it features a 36 K noise temperature and a 400 MHz bandwidth. Allowing for blanking time ($\sim 30\%$) and taking into account the main-beam efficiency of the telescope ($\sim 55\%$), this gives a sensitivity of

$$\Delta T_{\text{rms}} \sim 8 \text{ mK} \cdot s^{1/2} \ . \tag{14.14}$$

One could reach a sensitivity of $\Delta T/T \sim 10^{-5}$ on the comparison of two sky patches of the size of the telescope beam in about 23 hours of observation provided that all of the problems mentioned above have been reduced below this level and that atmospheric fluctuations are negligible; which requires hard work and luck and generally (almost) perfect weather conditions. Reaching this sensitivity on a dozen of spots requires maintaining such exceptional conditions for about two weeks! (One has to allow time for pointing, for calibrations, and for atmospheric monitoring with tipping scans.) This imposes rigorous requirements on the stability and repeatability of receivers and telescopes.

In high-sensitivity radiometry a crucial parameter is the magnitude and stability of the receiver power near the Dicke switch frequency. In general, electronic interference and gain instability in any of the various components of the radiometer might add noise at specific frequencies or add power to the radiometer's intrinsic $1/f$ spectrum. Noise and gain fluctuations which are synchronous with the Dicke-switching frequency must be scrupulously avoided because variations in such effects mimic a sky signal. In general, estimates of the sensitivity from system parameters alone (T_{sys}, bandwidth, integration time) are hard to achieve and only provide a worthwhile goal. In practice, the receiver power spectrum has to be monitored and kept as clean and stable as possible and the optimum Dicke-switching frequency should be determined by this power spectrum (Uson and Wilkinson 1984a).

One then has to decide on an observing strategy, i.e. how to actually conduct the experiment. Again, this decision has to be made after determining which are the problems associated with the particular telescope that one is using. The worst sources of systematic errors have to be found and the observing procedure tailored to minimize their effect.

Figure 14.9 shows the results of one such measurement (Uson and Wilkinson 1984b). Each data point results from the comparison of a ($\sim 2'$) field with the average of two reference fields lying 4.5 arcminutes away on opposite sides (close to the East-West direction). This pattern arises from a standard double subtraction procedure. The average of all the data points is $(26 \pm 39) \times 10^{-6}$ K, which is an indication that systematic offsets have been canceled out to the statistical noise level. Looking at the graph, it is clear that none of the data points deviates significantly from zero, so that it is clear that no anisotropy has been detected at a 2-σ level of about 0.4 mK ($\Delta T/T < 1.5 \times 10^{-4}$). A stronger limit can be set using all the data in a statistical argument, although this requires adopting a model for the expected

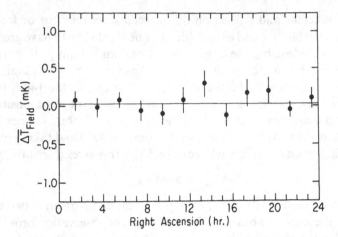

Fig. 14.9. Results of an unsuccessful search for small-scale anisotropy in the background radiation. ΔT_{FIELD} is the difference in thermodynamic temperature between a field the size of the telescope beam (~ 2 arcminutes in this case) and the average of two other such fields (which in this case lie 4.5 minutes of arc away from the first one, on opposite sides). The solid line indicates the overall weighted average, $\Delta T_{\text{AVE}} = 26 \pm 39 \ \mu\text{K}$, which indicates a good cancellation of systematic effects such as (variable) ground pickup. The low scatter of the points ($\chi^2 \sim 7.6$ with 11 DOF; a 28% chance if all the deflections are due to measurement noise alone) gives no indication of any anisotropy in the background radiation (see text).

anisotropy. The underlying sky signal is expected to be Gaussian distributed between the twelve data points and can be characterized by its standard deviation σ_{rms}. A sky signal, being uncorrelated with the measurement noise, would have inflated the scatter of the points in Figure 14.9 from that expected from the measurement noise alone. A Chi-square test can be used to decide the most likely value for σ_{rms} as well as upper and lower limits at any wanted confidence level. If the errors of the various data points are not the same (due to varying atmospheric conditions or different integration times), the Neyman-Pearson lemma can be used to find the optimal test; which will essentially be a modified Chi-square test with the data weighted somehow by their estimated uncertainty (Boynton and Partridge 1973). The details again depend on the assumed characteristics of the sky signal and, in general, one would have to perform a Monte-Carlo estimate of the statistic used.

Because one is comparing the scatter of the data with respect to the assumed noise, $\sigma_{\text{TOT}} = (\sigma_{\text{measurement}}^2 + \sigma_{\text{rms}}^2)^{1/2}$, this yields an estimate of the most likely value of σ_{TOT} together with upper and lower limits. In order to translate this to measure σ_{rms}, it is crucial to estimate the measurement noise accurately. Underestimating it could yield a spurious "detection" of a sky signal and overestimating it could mask such a signal and produce upper limits on σ_{rms} below the actual value!

Finally, translating σ_{rms} into an estimate of the σ_{sky} associated with one sky patch involves again an assumption about the correlation between the three fields that are involved in the double-subtraction scheme (Uson and Wilkinson 1984a). Most observers assume for simplicity that these are uncorrelated in which case $\sigma_{\text{sky}} = 0.82 \ \sigma_{\text{rms}}$.

A number of observers have tried to detect small-scale anisotropy in the back-

Table 14.3. Upper limits on small-scale anisotropy in the background radiation.

Angular scale	Wavelength (cm)	$\Delta T/T$ (upper limit)	$\Delta T/T$ corrected[a]	Reference
6″	6.0	3.2×10^{-3}		Knocke et al. (1984)
16.5″	6.0	1.6×10^{-4}		Kellermann et al. (1987)
18″	6.0	1.2×10^{-3}		Knocke et al. (1984)
18″	6.0	2.8×10^{-4}		Kellermann et al. (1986)
50″	6.0	5.0×10^{-5}		Kellermann et al. (1987)
60″	6.0	6.2×10^{-5}		Kellermann et al. (1986)
1.5′	1.5	2.4×10^{-5}		Uson and Wilkinson (1984b)
3.6′	0.9	2×10^{-4}	6×10^{-4}	Partridge (1980)
4′	6.3	6×10^{-4}		Ledden et al. (1980)
4′	6.2	4×10^{-4}		Lasenby and Davies (1983)
5′	3.9	8×10^{-5}	1.7×10^{-4}	Pariiskii et al. (1977)
7′	0.9	8×10^{-5}	1.9×10^{-4}	Partridge (1980)
8′	2.0	2.3×10^{-4}		Rudnick (1978)
10′	6.2	3.0×10^{-4}		Lasenby and Davies (1983)
10′	3.9	4.5×10^{-5}	9.5×10^{-5}	Pariiskii et al. (1977)
11′	2.8	3.0×10^{-4}		Seielstad et al. (1981)
16′	2.0	2.7×10^{-4}		Rudnick (1978)
20′	6.2	4.5×10^{-4}		Lasenby and Davies (1983)
20′	3.9	3.3×10^{-5}	7.5×10^{-5}	Pariiskii et al. (1977)
24′	2.0	2.8×10^{-4}		Rudnick (1978)
30′	6.2	4.7×10^{-4}		Lasenby and Davies (1983)
30′	0.13	1.2×10^{-4}		Caderni et al. (1977)
40′	6.2	5.4×10^{-4}		Lasenby and Davies (1983)
50′	3.9	2.4×10^{-5}	6×10^{-5}	Pariiskii et al. (1977)
75′	3.9	1.9×10^{-5}	5×10^{-5}	Pariiskii et al. (1977)

[a] Corrections due to Partridge and Lasenby; see Lasenby and Davies (1983).

ground radiation but, so far, nobody has succeeded. Table 14.3 shows the various limits placed by these measurements. A number of them have been examined by Partridge and Lasenby and the results have been revised by them (Partridge 1983, Lasenby and Davies 1983). In general, different measurements use different observing patterns and the data are analyzed with different statistical techniques. To establish limits on a particular model, each measurement should be examined separately. The results in Table 14.3 assume that the underlying anisotropy is gaussian noise on the scale of the various telescope beams used, and uncorrelated over the angular scales used in the observing scheme (beam separation on single or double subtraction procedures).

(c) Constraints on Cosmological Models

The upper limits in Table 14.3 can be compared to the predictions of various theoretical models. Some of the early results were instrumental in ruling out alternative models of the background radiation, such as the "discrete source" scenario in which the radiation was generated by a superposition of a number of different populations of radio sources (Boynton and Partridge 1973).

In the context of the standard Big Bang model, the measurements constrain the scenarios discussed in Section 14.4.1(a). The data in Figure 14.9 conflict with the predictions of those model universes in which the initial fluctuations are adiabatic and dominated by baryonic matter (except if $\Omega = 1$ and $n = 4$; see Figure 14.8). Baryon universes in which the initial perturbations are isothermal are presently out of fashion but only somewhat constrained by these measurements; for values of $n < 0$, they need to satisfy $0.1 < h \cdot \Omega < 0.3$, whereas for $n > 0$ no restriction is imposed. Models in which the mass is dominated by massive neutrinos require $\Omega \sim 1$, although they are more constrained by other reasons (White 1984). More exotic possibilities include those in which some as-yet-undiscovered particles such as axions, photinos, gravitinos, and the like provide the dominant part of the mass. These models predict lower degrees of anisotropy as these particles are only weakly coupled to the background radiation and can provide seeds for the rapid development of structure in the Universe from a very smooth distribution of baryons (and photons) at the decoupling epoch. Nevertheless, these models would still require that $\Omega \cdot h > 0.2$.

One should keep in mind that these conclusions are only formal as they depend on a number of restrictive assumptions which are likely not to be strictly true. The most important one is that the underlying fluctuations are gaussian distributed; which is untested because of the small sky coverage of the observations. This is hard to improve with a single-dish telescope as the required sensitivity makes the observation of a large sample of sky positions prohibitive; on the other hand, as the overall sensitivity of an experiment goes only as the fourth-root of the number of fields observed, there is a tendency to concentrate on a few fields instead of dividing the time between many fields. As the theoretical predictions are at least uncertain by a factor of 2, it is probably sensible to conclude that there is an interesting confrontation between measured limits and theoretical estimates; although the fate of the various models is far from settled: a baryon-dominated Universe is perhaps in trouble but not yet definitely ruled out.

(d) Near-Future Improvements

Most likely improvements should come from the use of new techniques. Some schemes under consideration by a number of groups involve using multiple receivers either on different antennas or on the same one, perhaps as focal plane arrays. Alternatively, several groups are developing broad-band receivers, especially at a wavelength of 3 mm.

Another alternative is to use a dedicated system and integrate for several months, perhaps at a cold and dry site such as the Antarctic; several groups have considered this possibility.

In any case, an increase in sensitivity by at least one order of magnitude would be desirable but is not certain to work out: one has to push systematic effects below the wanted level of sensitivity, which will be hard.

Furthermore, an anisotropy signal would be hard to interpret since it could be due to confusing radio sources; although this problem might not occur at

$\lambda = 3$ mm. Nevertheless, a convincing detection would require confirmation at a different wavelength

14.4.2 Measurements of Large-Scale Structure

(a) A Scenario

The microwave background is isotropic to better than one part in 10^4 on all angular scales that have been probed, except for the dipole component discussed below. Indeed, a long-standing problem in cosmology is how this isotropy came about because the radiation is coming from regions that have apparently never been in causal contact. The inflationary models of the very early Universe provide a possible explanation of this isotropy. Nevertheless, one expects departures from isotropy at large scales to occur at some level; although the most important one is not intrinsic to the radiation but a local effect instead. Soon after the discovery of the background radiation, Peebles pointed out that if it appeared isotropic to any given observer, any other observer moving with respect to the first one would see a dipole anisotropy in the radiation, which would appear hotter in the direction of motion. This is the "aether drift" effect, and it is straightforward to show that the second observer would see blackbody radiation from any direction at a temperature

$$T(\theta) = T_{\text{MBR}}\left(1 + \frac{v}{c}\cos\theta\right) \tag{14.15}$$

provided that $v/c \ll 1$. T_{MBR} is the temperature measured by the first observer and θ is the angle between any given direction and the direction toward which the second observer is moving. The spectrum of the radiation is therefore still that of a blackbody in all directions although the temperature is modulated by a dipole anisotropy.

Cosmological alternatives to the standard Big Bang model include models in which the Universe has an intrinsic rotation which generates large-scale aniso-tropies in the background radiation. These arise as well if the expansion of the Universe is not quite isotropic.

Even within the context of the standard model, one expects some intrinsic large-scale anisotropy due to anisotropies in the large-scale mass distribution. As explained by Peebles (1981), this is best described in terms of the Sachs-Wolfe (1967) effect as due to random fluctuations in the distribution of mass at decoupling which induce fluctuations in the gravitational potential. Photons are gravitationally redshifted on their way to an earthbound detector as they climb out of whatever potential well they are emitted from, so photons coming from different parts of the Universe might have been redshifted (blueshifted) by a different amount. Roughly speaking, the mass inside a sphere of radius r is proportional to its volume, so

$$M(r) \propto r^3 \tag{14.16}$$

and fluctuations on that scale are proportional to $r^{3/2}$, which induces fluctuations in the gravitational potential ϕ of order

$$\delta\phi(r) \sim \frac{\delta M(r)}{r} \sim r^{1/2} \tag{14.17}$$

which (formally) diverge at large distances, whereas fluctuations in the gravitational acceleration g are of order

$$\delta g(r) \sim \frac{\delta \phi(r)}{r} \sim r^{-1/2} \tag{14.18}$$

and vanish at large distance.

The fluctuations in $\phi(r)$ induce large-scale fluctuations in the background radiation

$$\frac{\delta T}{T} \sim \delta \phi \sim \theta^{1/2} \tag{14.19}$$

that should grow as the square-root of the angular separation θ between two random lines of sight. This effect should contribute as well to the dipole anisotropy discussed above, even though it does not have the same dependence with angular separation. Silk (1984) gives some estimates of this contribution as well as of the contribution of this effect to a quadrupole component. The estimated values of $\Delta T/T$ for an $\Omega = 1$ Universe are 1.5×10^{-4} (dipole) and 0.7×10^{-4} to 2.1×10^{-4} (quadrupole) for a baryon-dominated model; 1×10^{-3} to 5×10^{-3} and 3×10^{-5}, respectively, if the dominant mass is in the form of massive neutrinos; and 1×10^{-3} to 2×10^{-3} and 1×10^{-5}, respectively, if the mass is mainly due to cold, weakly interacting dark matter. As indicated below, the observed dipole is about $(\delta T/T)_{max} \sim 10^{-3}$, whereas current limits on the amplitude of a quadrupole anisotropy are below 8×10^{-5}.

Intermediate-scale anisotropy could have been produced by the reionization mechanism discussed above, which could induce a signal on scales of a few to about ten degrees. The best upper limits in this range are $\Delta T/T < 5.5 \times 10^{-5}$ at the 95% confidence level at $\theta \sim 6°$ (Melchiorri et al. 1981).

A possible detection of anisotropy at an angular scale of 8° has been announced by Davies and collaborators (Davies et al. 1987). Their measured signal, $\Delta T/T = 3.7 \times 10^{-5}$ at a wavelength $\lambda = 3$ cm, is about the expected signal due to contamination from galactic radiation (from an extrapolation of the fluctuations measured in long-wavelength maps). Further measurements by these authors at $\lambda = 6$ cm have failed to detect the signal that would be expected if their detection is due to galactic radiation. The authors are proceeding with further, more sensitive, measurements at $\lambda = 6$ cm; they should soon detect fluctuations if the 3 cm signal is intrinsic to the microwave background. A correlation analysis of their 3 cm and 6 cm data will be most interesting.

On the other hand, there are several possible spurious sources for the signal that they measure at 3 cm. For example, the radio sources Cygnus-A and the Galactic Center are strong and could be leaking into "switched" sidelobes due to the reflector edge; other possibilities are contamination by the moon in far sidelobes, synchronous heating of the reflector (a gradient of only 0.01 K across of it would have been sufficient).

Independent measurements, possibly at a shorter wavelength, on these intermediate angular scales should eventually settle this issue.

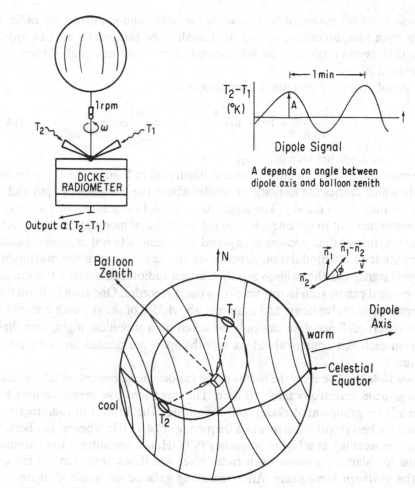

Fig. 14.10. Large-scale structure has to be measured from balloon altitudes to avoid variable atmospheric emission. The package is set into rotation (typically at 1 rpm) and Earth's rotation is used to scan the sky. The observed signal corresponds to a dipole anisotropy.

(b) Measurements

Experiments to measure large-scale anisotropy in the background radiation were initiated by Partridge and Wilkinson (1967). Atmospheric inhomogeneities are a serious source of non-statistical errors and have forced observers to give up on ground-based measurements; even though, in retrospect, it seems that Conklin (1969) detected the dipole anisotropy from a high-altitude site. His experiment measured the projection of the anisotropy on the (minor) circle defined by a fixed declination of $+32°$.

Henry (1971) obtained a $4\text{-}\sigma$ detection of the dipole anisotropy by taking the radiometer above most of the atmosphere on a balloon and pioneered a technique that has proven to be very successful. The first convincing detection of the dipole anisotropy was made from a U-2 aircraft by Smoot et al. (1977).

A number of different experimental schemes are used, but most are differential, using two horns viewing widely separated ($\theta \sim 60°$ to $90°$) points on the sky.

The radiometer is Dicke-switched between the horns and compares the radiation coming from two directions, $\mathbf{n_1}$ and $\mathbf{n_2}$. Usually, the radiometer rotates around the vertical (symmetry) axis to interchange horn positions and subtract out instrumental effects.

The signal produced by a dipole component is

$$\frac{T_1 - T_2}{T_{AVE}} = (\mathbf{n_1} - \mathbf{n_2}) \cdot \frac{\mathbf{v}}{c} = |\mathbf{n_1} - \mathbf{n_2}| \frac{|\mathbf{v}|}{c} \cos \phi \qquad (14.20)$$

where ϕ is the angle between $(\mathbf{n_1} - \mathbf{n_2})$ and $\mathbf{v}$.

The most successful technique to date is illustrated in Figure 14.10. The balloon gondola which carries the radiometer rotates about the vertical at 1 rpm and the earth's rotation scans the sky. The signal has a period determined by the rotation of the instrument; although one does not get a sinusoidal modulation since ϕ is not a linear function of time because $\mathbf{n_1}$, $\mathbf{n_2}$, and $\mathbf{v}$ are generally not coplanar. Besides, the amplitude of the modulation depends on the angle between the maximum of the dipole signal and the balloon zenith. Current radiometer sensitivities are such that the signal can be seen in real time on a chart recorder. One such balloon flight lasts from ten to twelve hours and examines about 35% of the sky with a somewhat uneven coverage. Since data can only be taken on a moonless night, four flights (two from each hemisphere) about six months apart are needed for complete sky coverage.

Tables 14.4a to 14.4c show the results of the various measurements of intermediate- and large-scale anisotropy made to date. The most sensitive measurements have been made by groups at Berkeley and Princeton. The Princeton radiometer has achieved the best statistical errors at a frequency of 24.8 GHz whereas the Berkeley radiometer, working at a higher frequency (90 GHz), is less subject to contamination due to galactic emission. Each radiometer was flown three times, only once from the southern hemisphere. After removing galactic contamination from the Princeton data, the dipole components found by the two groups are in reasonable agreement: $\Delta T_{max} = 3.18 \pm 0.17$ mK towards RA $= 11.18 \pm 0.05$ h, Dec $= -8.0° \pm 0.7°$ is measured by the Princeton group (Fixsen *et al.* 1983); and $\Delta T_{max} = 3.44 \pm 0.17$ mK towards RA $= 11.2 \pm 0.1$ h, Dec $= -6.0° \pm 1.5°$ by the Berkeley group (Lubin *et al.* 1985). This is an important check, as any intrinsic anisotropy as well as the motional effect should be frequency independent if the spectrum of the background radiation is that of an undistorted blackbody. The discrepancy in the amplitudes is small and most likely due to systematic calibration errors.

Assuming that the dipole distortion is entirely motional implies that the Sun is moving towards RA $= 11.2 \pm 0.1$ h, Dec $= -7.0° \pm 1.0°$ at a speed of 372 ± 25 km s^{-1}, which translates to a motion of the Local Group of galaxies at 600 ± 50 km s^{-1} towards RA $= 10.2 \pm 0.4$ h, Dec $= -27° \pm 5°$ (Wilkinson 1986). This direction is 49° from the direction towards the center of the Virgo supercluster, but only 17° away from the direction of motion of the Local Group with respect to a distant shell of 78 Sbc galaxies with recession velocities between 1000 and 5500 km s^{-1} (Hart and Davies 1982).

Two of the Princeton measurements were made at times when the Earth's motion was in the same and opposite direction to the motion of the Sun. This alignment

Table 14.4a. Upper limits on intermediate-scale anisotropy in the background radiation.

Angular scale	Wavelength (cm)	$\Delta T/T$ (upper limit)	Reference
2°–3°	3	5.6×10^{-4}	Mandolesi et al. (1986)
5°	3	6.8×10^{-4}	Mandolesi et al. (1986)
6°	0.05–0.3	5.5×10^{-5}	Melchiorri et al. (1981)
10°–90°	1.2	9×10^{-5}	Fixsen et al. (1983)

Table 14.4b. Measurements of dipole and quadrupole components.

Component[a] (mK) (thermodynamic temperatures)								Wavelength (cm)	Reference
D_x	D_y	D_z	Q_1	Q_2	Q_3	Q_4	Q_5		
-3.07 ± 0.17	0.67 ± 0.09	-0.45 ± 0.09	0.15 ± 0.08	0.15 ± 0.11	0.14 ± 0.07	0.06 ± 0.11	-0.01 ± 0.07	1.2	Fixsen et al. (1983)
-3.37 ± 0.17	0.69 ± 0.09	$-0.42 = 0.09$	0.10 ± 0.09	0.15 ± 0.10	0.12 ± 0.11	-0.09 ± 0.09	0.09 ± 0.08	0.33	Lubin et al. (1985)

[a] $\Delta T = D_x \cos \alpha \cos \delta + D_y \sin \alpha \cos \delta + D_z \sin \delta + Q_1(\frac{3}{2}\sin^2 \delta - \frac{1}{2}) + Q_2 \cos \alpha \sin 2\delta + Q_3 \sin \alpha \sin 2\delta + Q_4 \cos 2\alpha \cos^2 \delta + Q_5 \sin 2\alpha \cos^2 \delta$, where $\alpha \equiv$ Right Ascension and $\delta \equiv$ Declination.

Table 14.4c. Dipole anisotropy in the background radiation.

Dipole amplitude (mK)	Direction of maximum:		Wavelength (cm)	Reference
	Right Ascension (h)	Declination (°)		
3.18 ± 0.17	11.18 ± 0.05	-8.0 ± 0.7	1.2	Fixsen et al. (1983)
3.16 ± 0.12	11.3 ± 0.16	-7.5 ± 2.5	0.81	Strukov et al. (1987)
3.44 ± 0.17	11.2 ± 0.1	-6.0 ± 1.5	0.33	Lubin et al. (1985)
2.8 ± 0.8	9.6 ± 1.5	-9 ± 20	0.10–0.33	Weiss (1980)
$2.9(+1.3, -0.6)$	11.4 ± 0.7	3 ± 10	0.05–0.3	Fabbri et al. (1980)

allowed them to detect the Earth's orbital motion at a signal-to-noise level of 7 standard deviations.

Two tentative detections of a quadrupole anisotropy in the microwave background have not been confirmed, and it seems that systematic errors, probably due to incorrectly subtracted galactic emission, produced the early results (Wilkinson 1984). The upper limit on a quadrupole component is currently of order $(\delta T/T)_Q < 8 \times 10^{-5}$ at the 95% confidence level.

(c) Likely Near-Future Improvements
The Princeton group has retuned its radiometer to 19.5 GHz and modified the instrument which now features a single horn with a lens which gives a beam with half-power beamwidth of about 4°. The radiometer switches the sky signal against a liquid helium load. The main advantages of this design are improved stability, higher angular resolution, and less contamination from galactic radiation which was hard to avoid with the dual-horn design. The radiometer, which has already been flown twice, will be flown a total of four times to produce a whole-sky map with a sensitivity $\Delta T_{rms} \sim 0.1$ mK on each independent beam on the sky.

The COBE satellite will have on board a set of instruments, the Differential Microwave Radiometers (DMR), which will map the sky at three wavelengths: 3.3, 5.7, and 9.6 mm. Each radiometer will have two antennas with half-power beamwidths of 7° separated 60°, 30° away from the spin axis of the satellite. These should produce all-sky maps which, after one year, should have reached sensitivities of 0.3 mK at $\lambda = 9.6$ mm and of 0.15 mK at $\lambda = 3.3$ mm and 5.7 mm on each (7°) sky patch. The overall sensitivity to dipole and quadrupole components is expected to be $\Delta T \sim 10\mu$K (Mather 1982).

Figure 14.11 shows the measurements published at the time of this writing.

14.5 The Sunyaev-Zel'dovich Effect

14.5.1 Concept and Cosmological Consequences
Dense clusters of galaxies are powerful emitters of X-radiation. It has been shown that thermal bremsstrahlung due to an ionized, hot intergalactic gas is the main source of this radiation. The X-ray spectra show emission lines which are attributed to highly ionized iron and yield gas temperatures as high as 10^8 K for the denser and hotter clusters. It is perhaps surprising that the mass of this ionized gas is as large as that of the stellar component of the galaxies in the cluster and, furthermore, that it contains heavy elements with approximately solar abundances. This shows that this gas is heavily processed material which has likely been ejected by the galaxies through collisions with other galaxies or through ram pressure due to the hot intergalactic gas (Forman and Jones 1982, Sarazin 1986).

As the cross-section for interaction of the microwave background photons with ionized material is high, one expects them to interact; and, because the gas is so much hotter than the radiation, the interaction will be through inverse Compton scattering. This cools the gas but also distorts the spectrum of the radiation. The interaction conserves the number density of photons but the scattered photons emerge with higher energy than the incoming ones. The radiation becomes "cooler"

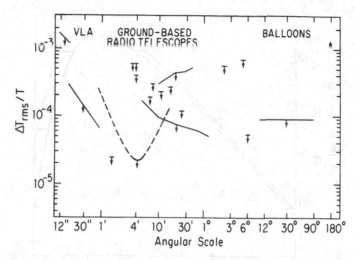

Fig. 14.11. Upper limits, at the 95% confidence level, on anisotropy of the microwave background at various angular scales. Some of the points between 1′ and 1° have been revised from the values given in the original references (see Table 14.3). The limits are derived assuming that patches of sky the size of the various beams are uncorrelated. The dashed line limits the amplitude of a sinusoidal distortion of half-wavelength given by the particular angular scale plotted (a random superposition of such signals was assumed) (Uson and Wilkinson 1984b. Reprinted by permission from Nature, Vol. 312, pp. 427. Copyright(c) 1984 Macmillan Magazines Limited).

on the Rayleigh-Jeans side and hotter on the Wien side with no effect at a frequency of about 218 GHz, slightly higher than that where the undistorted spectrum peaks (~ 162 GHz; see Figure 14.12). This effect was predicted by Sunyaev and Zel'dovich (1972) who calculated that at any given frequency, the change in spectral energy density, $F(v)$, of the background radiation is

$$\Delta F(v) = \chi v \frac{d}{dv}\left\{v^4 \frac{d}{dv}[v^{-3}F_{\text{blackbody}}(v)]\right\} \tag{14.21}$$

with $\chi = (kT_e\tau/m_ec^2)$; where T_e is the temperature of the electron gas, m_e is the mass of the electron, k is Boltzmann's constant, and c is the speed of light, whereas τ is the line integral

$$\tau = \int N_e\sigma_T\,dl \tag{14.22}$$

where N_e is the electron density, σ_T is the cross-section for Thomson scattering, which is on the order of 10^{-28} m², and the integral is taken along a line of sight across the cluster.

In the Rayleigh-Jeans part of the spectrum, where $F(v) \propto v^2$,

$$\Delta F(v) \simeq -2\chi F_{\text{blackbody}}(v) \tag{14.23}$$

and for the densest and hottest clusters one expects a distortion in temperature on the order of $\Delta T \sim -1$ mK towards the cluster centers in the Rayleigh-Jeans spectral region.

Let us consider for example the well-studied Coma cluster of galaxies. X-ray

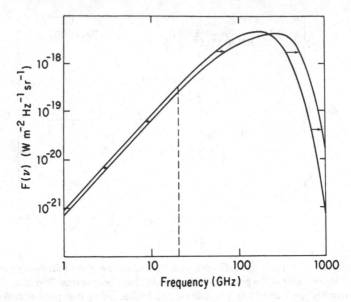

Fig. 14.12. Some background radiation photons are shifted to higher frequencies by inverse Compton scattering as they go through hot ionized gas in dense clusters of galaxies. Photon number is conserved but the blackbody spectrum is distorted. Even though this mechanism heats the radiation, in the Rayleigh-Jeans region the spectral energy density of the radiation is decreased by this "Sunyaev-Zel'dovich" effect (see text).

observations yield a gas temperature $T_{gas} = 7.9 \pm 0.3$ keV (Mushotzky and Smith 1980) and a surface brightness in the 0.5 to 4.0 keV band which follows

$$S_x \sim S_0 \left(1 + \frac{\theta^2}{\theta_x^2}\right)^{-1} \qquad (14.24)$$

with $S_0 = (1.2 \pm 0.1) \times 10^{-8}$ W m^{-2} and a core radius $\theta_x = 12.8 \pm 0.5$ arcminutes (Abramopoulos et al. 1981). The cluster is at a redshift $z = 0.023$ and the line-of-sight velocity dispersion of the member galaxies is $\sigma_v = 905 \pm 40$ km s^{-1}. However, the velocity dispersion within the angular distance given by the X-ray core is about 1200 km s^{-1} (from Kent and Gunn 1982). If the plasma scale height is of the same order of that of the galaxies, the gas temperature would be

$$T \sim \frac{m_p \sigma_v^2}{2k} \sim 7.7 \text{ keV} \qquad (14.25)$$

where m_p is the proton mass. The agreement with the X-ray temperature (7.9 keV) is likely fortuitous but encourages a simple isothermal model for the cluster. The central electron density is then $N_e \sim 3 \times 10^3$ m^{-3}, which gives an optical depth $\tau \sim 0.015$ and a (Rayleigh-Jeans) Sunyaev-Zel'dovich decrement

$$\Delta T_{RJ} = \Delta T_0 \left(1 + \frac{\theta^2}{\theta_x^2}\right)^{-1/4} \qquad (14.26)$$

with a central (maximum) decrement $\Delta T_0 \sim -1.2$ mK (Birkinshaw 1986, Uson

1986). Notice that the half-width of the Sunyaev-Zel'dovich effect is about four times the size of the X-ray core radius. The effect becomes larger as one approaches the blackbody peak and then vanishes and changes sign on the Wien side.

Sunyaev and Zel'dovich pointed out that if one chooses to observe at a single frequency and compares the radiation coming through a dense cluster with the undistorted flux coming from nearby blank sky, the effect appears as an anisotropy at the level of $\Delta T/T \sim 5 \times 10^{-4}$, which is one order of magnitude higher than the best upper limits discussed in Section 14.4.1 and therefore should be observable. One has to be careful with the slow angular dependence of the effect (Equation 14.26). If the reference measurement is made at $\theta \sim \theta_x$, which for the Coma cluster implies beam-switching by 13 minutes of arc—a hard experimental problem—one would only see 16% of the central decrement. More distant clusters are easier to detect but most observations will not see the full effect. This has to be taken into account when comparing observations made by different groups, which in general use different beamthrows.

A convincing measurement of the Sunyaev-Zel'dovich effect would prove that the background radiation comes from high redshifts but could also allow a determination of the two most uncertain key parameters in the standard Big Bang scenario: Hubble's constant, H_0, and the deceleration parameter, q_0 (Gunn 1978). This comes about because the X-ray emission from the cluster depends on the electron temperature, T_e (which can be obtained from the X-ray spectrum), and the density of electrons, N_e, through the combination $N_e^2 (cz/H_0)\theta^3$, where θ is the angle subtended by the cluster and z is its redshift. On the other hand, the Sunyaev-Zel'dovich effect depends linearly on the product $N_e T_e \theta cz/H_0$ in the Rayleigh-Jeans region. A measurement of the Sunyaev-Zel'dovich effect provides a second relation between N_e and Hubble's constant and would allow a determination of H_0 without the accumulation of uncertainties that is a consequence of the conventional stepwise determination of the extragalactic distance scale. Observations of the effect in clusters at widely separated redshifts would then yield a determination of the deceleration parameter q_0.

14.5.2 Measurements

A number of observers have attempted to detect the Sunyaev-Zel'dovich effect. Early efforts were restricted by the sensitivity of the receivers, which had to be pushed to their limits. This did not allow enough time to be spent in order to track systematic errors [see Section 14.4.1(b)] which plagued these early attempts. Therefore, it is not surprising that observers obtained discrepant results. Besides, unlike anisotropy measurements that can be done on whatever patch of blank sky is more convenient or easy to observe, the observations of clusters of galaxies have to be done wherever they are, which in general forces the motion of the telescopes over a broad range of elevation angles and tracking rates. Spillover effects can be large and variable and therefore are hard to cancel. Observing at low frequencies ($\nu < 15$ GHz), confusion due to radio sources becomes important. Besides, one can also pick up a contribution due to bremsstrahlung from a colder gas component in the cluster (Tarter 1978). Observations of a cluster on both sides of the blackbody peak would be important, as theory predicts a distinctive frequency signature for

Table 14.5. Measurements of the Sunyaev-Zel'dovich effect done since 1980.[a]

Cluster	Frequency (GHz)	HPBW(')	ΔT_{sz} (mK)	Reference
0016+16	10.7	3.3	-1.36 ± 0.28	Birkinshaw et al. (1981a)
0016+16	10.7	3.3	-0.26 ± 0.24	Birkinshaw and Gull (1984)
0016+16	20.3	1.8	-0.37 ± 0.16	Birkinshaw and Gull (1984)
0016+16	20.3	1.8	-0.65 ± 0.12	Birkinshaw et al. (1984)
0016+16	20.3	1.8	$-(0.315 \text{ to } 0.490) \pm 0.055$	Birkinshaw (1986)
0016+16	10.7	2.0	-0.9 ± 0.9	Andernach et al. (1983)
0016+16	19.5	1.5	-0.74 ± 0.22	Uson and Wilkinson (1984c)
0016+16	19.5	1.5	-0.48 ± 0.12	Uson (1986)
0016+16	89.6	1.2	-0.8 ± 0.9	Radford et al. (1986)
Abell 401	19.5	1.5	-0.88 ± 0.41	Uson and Wilkinson (1984c)
Abell 401	19.5	1.5	-0.64 ± 0.18	Uson (1986)
Abell 545	19.5	1.5	$+0.51 \pm 0.43$	Uson and Wilkinson (1984c)
Abell 665	10.6	4.5	-0.53 ± 0.22	Birkinshaw et al. (1981b)
Abell 665	10.7	3.3	$+0.03 \pm 0.25$	Birkinshaw and Gull (1984)
Abell 665	20.3	1.8	-0.55 ± 0.13	Birkinshaw and Gull (1984)
Abell 665	20.3	1.8	-0.24 ± 0.07	Birkinshaw et al. (1984)
Abell 665	20.3	1.8	$-(0.395 \text{ to } 0.550) \pm 0.045$	Birkinshaw (1986)
Abell 665	19.5	1.5	-0.50 ± 0.33	Uson and Wilkinson (1984c)
Abell 665	19.5	1.5	-0.37 ± 0.14	Uson (1986)
Abell 1763	19.5	1.5	-0.36 ± 0.25	Uson and Wilkinson (1984c)
Abell 1795	90–300	5.0	0.2 ± 0.9	Meyer et al. (1983)
Abell 2218	10.6	4.5	1.05 ± 0.21	Birkinshaw et al. (1981b)
Abell 2218	10.7	3.3	-0.38 ± 0.19	Birkinshaw and Gull (1984)
Abell 2218	20.3	1.8	-0.31 ± 0.13	Birkinshaw and Gull (1984)
Abell 2218	20.3	1.8	-0.24 ± 0.07	Birkinshaw et al. (1984)
Abell 2218	20.3	1.8	$-(0.255 \text{ to } 0.400) \pm 0.045$	Birkinshaw (1986)
Abell 2218	10.7	3.0	-1.84 ± 0.33	Schallwich (1982); withdrawn
Abell 2218	5.0	$8' \times 10'$	$+0.23 \pm 0.77$	Lasenby and Davies (1983)
Abell 2218	19.5	1.5	-0.29 ± 0.24	Uson and Wilkinson (1984c)
Abell 2218	89.6	1.2	0.16 ± 0.43	Radford et al. (1986)
Abell 2218	100–109	1.7	0.41 ± 0.32	Radford et al. (1986)

[a] Measurements are difficult to compare; indeed, some of the scatter with time for a given group is due to improving overall calibration.

the effect. Unfortunately, observations at frequencies above 30 GHz are once again hard due to fluctuations in atmospheric emission.

Table 14.5 shows the results of measurements done since 1980. Although there are still some problems and inconsistencies and a tendency to underestimate the measurement errors, the measurements are in reasonable agreement [except for the measurement of Schallwich (1982) on Abell 2218, which has since been withdrawn by the author]. The effect is seen in the clusters 0016 + 16, Abell 401, Abell 665, and Abell 2218 at a frequency of about 20 GHz.

Although the rough agreement is encouraging, one should not look for a complete agreement as a valid test of the accuracy of the different measurements. Different observers use instruments with various beamwidths, and given the angular dependence of the effect (Equation 14.26), the efficiencies of the various measurements are not the same. Besides, it is conceivable that radio sources at the positions of the clusters and associated reference fields could modify the results by a few tenths of a millikelvin. Here again, different measurements would suffer different contaminations. Interferometric observations at several, lower frequencies can identify and measure contaminating radio sources. These measurements are presently being done by several groups (Birkinshaw 1986, Uson 1986).

The only measurement attempted at the peak of the blackbody curve and on the Wien side is that of Meyer *et al.* (1983). They were severely limited by bad weather as well as by a low-efficiency coupling of their instrument to the telescope. They used four bands, two of which monitored atmospheric emission. Because of the problems encountered, their sensitivity was only $\sigma_{\Delta T} \sim 0.9$ mK, but one could hope for improvement here.

Measurements of the Sunyaev-Zel'dovich effect should reach sensitivities of $\sigma_{\Delta T} \sim 0.1$ mK in a few days of integration with currently available equipment. This should allow a much better understanding of the systematic problems, and the consistency of the results should rapidly improve. Besides, as this sensitivity becomes available at higher frequency (perhaps 90 GHz) where the possible contamination from radio sources would be negligible, a test of the variation of the effect with frequency will become available and this should provide a conclusive test on the Sunyaev-Zel'dovich effect.

14.5.3 Cosmological Applications

Let us consider the well-studied cluster $0016+16$. Combining the recent results of Birkinshaw (1986) and Uson (1986), which are in excellent agreement, yields $\Delta T = -0.45 \pm 0.08$ mK. Taking into account the beam patterns and beam separations (which are similar in both experiments) yields a central decrement

$$\Delta T_0 \simeq -0.90 \pm 0.17 \text{ mK} \tag{14.27}$$

for $0016+16$ at about 20 GHz.

As indicated above, one could try to combine these measurements with X-ray observations to try to derive a value for Hubble's constant. A rough estimate for this cluster yields

$$30 \text{ km s}^{-1} \text{ Mpc}^{-1} \leqslant H_0 \leqslant 200 \text{ km s}^{-1} \text{ Mpc}^{-1} \tag{14.28}$$

which is encouraging because the method is completely independent of the conventional way to estimate H_0. The main problem is that the temperature of the gas is uncertain, as the X-rays normally provide only a lower limit. Besides, the clumpiness of the gas is unknown but important (as the X-ray emission is proportional to N_e^2 and the Sunyaev-Zel'dovich effect is linear in N_e).

Among the solutions, determining the cluster dynamics by measuring a number ($\gtrsim 100$) of redshifts for the galaxies will certainly help: one can get an alternative estimate of T with the usual isothermal model for the clusters, and detailed dynamical studies could constrain the form of the gravitational potential and provide reasonable limits to the clumpiness of the gas. Besides, current sensitivities allow "mapping" the Sunyaev-Zel'dovich effect, i.e., measuring a few positions within a cluster. In the future, improved X-ray data will give T_e through good spectra and provide slightly better resolution. This should allow a determination of H_0 to within a factor of 2, but through a totally independent path from the usual one, which will provide an interesting consistency check.

Meanwhile, one can assume a value for H_0 and use the measured decrements to learn about the intracluster gas. Assuming a value for H_0 of 50 km s^{-1} Mpc^{-1} as well as an isothermal distribution yields $T_e = 11$ keV and central density $N_e(0) = 3 \times 10^3$ m^{-3} and pressure $p(0) = 6 \times 10^{-2}$ N m^{-2}. A similar calculation for Abell 665 and Abell 2218 also yields high central densities and pressures, and temperatures $T_e(\text{A665}) = 18$ KeV and $T_e(\text{A2218}) = 10$ keV (Birkinshaw 1986). Thus, the clusters for which the effect is seen are among the densest and hottest known.

Recommended Reading

Lynden-Bell, D. (ed.). 1982. The Big Bang and Element Creation. London: The Royal Society.

Peebles, P.J.E. 1971. Physical Cosmology. Princeton, New Jersey: Princeton University Press.

Weinberg, S. 1972. Gravitation and Cosmology. New York: John Wiley and Sons.

Weiss, R. 1980. Annu. Rev. Astron. Astrophys. **18**:489.

Wilkinson, D.T. 1986. "Anisotropy of the Cosmic Blackbody Radiation," Science **232**:1517.

References

Abramopoulos, F., G. Chanan, and W. Ku. 1981. Astrophys. J. **248**:429.

Alpher, R.A., and R.C. Herman. 1948. Nature **162**:774.

Andernach, H., D. Schallwich, G.B. Sholomitski, and R. Wielebinski. 1983. Astron. Astrophys. **124**:326.

Basko, M.M., and A.G. Polnarev. 1980. Mon. Not. R. Astron. Soc. **191**:207.

Birkinshaw, M. 1986. In C.P. O'Dea and J.M. Uson (eds.), Radio Continuum Processes in Clusters of Galaxies, NRAO-Green Bank workshop, p. 261.

Birkinshaw, M., and S.F. Gull. 1984. Mon. Not. R. Astron. Soc. **206**:359.

Birkinshaw, M., S.F. Gull, and A.T. Moffet. 1981a. Astrophys. J. (Lett.) **251**:L69.

Birkinshaw, M., S.F. Gull, and H. Hardebeck. 1984. Nature **309**:34.

Birkinshaw, M., S.F. Gull, and K.J. Northover. 1981b. Mon. Not. R. Astron. Soc. **197**:571.

Boesgaard, A.M., and G. Steigman. 1985. Annu. Rev. Astron. Astrophys. **23**:319.

Bond, J.R., and G. Efstathiou. 1984. Astrophys. J. (Lett.) **285**:L45.

Boynton, P.E., and R.B. Partridge. 1973. Astrophys. J. **181**:243.

Caderni, N., R. Fabbri, V. De Cosmo, B. Melchiorri, F. Melchiorri, and V. Natale. 1977. Phys. Rev. **D16**:2424.

Caderni, N., R. Fabbri, B. Melchiorri, F. Melchiorri, and V. Natale. 1978. Phys. Rev. **D17**:1908.

Conklin, E.K. 1969. Nature **222**:971.

Crane, P., D.J. Hegyi, N. Mandolesi, and A.C. Danks. 1986. Astrophys. J. **309**:822.

Danese, L., and G. De Zotti. 1977. Riv. Nuovo Cimento **7**:277.

Dautcourt, G. 1969. Mon. Not. R. Astron. Soc. **144**:255.

Davies, R.D., R. Watson, E.J. Daintree, J. Hopkins, A.N. Lasenby, J. Beckman, J. Sanchez-Almeida, and R. Rebollo. 1987. Nature **326**:462.

Dicke, R.H. 1946. Rev. Sci. Instr. **17**:268.

Dicke, R.H., R. Berynger, R.L. Kyhl, and A.V. Vane. 1946. Phys. Rev. **70**:340.

Dicke, R.H., P.J.E. Peebles, P.G. Roll, and D.T. Wilkinson. 1965. Astrophys. J. **142**:414.

Doroshkevich, A.G., and I.D. Novikov. 1964. Sov. Phys-Doklady **9**:111.

Fabbri, R., I. Guidi, F. Melchiorri, and V. Natale. 1980. Phys. Rev. Lett. **44**:1563; erratum, ibid. **45**:401.

Fixsen, D.J., E.S. Cheng, and D.T. Wilkinson. 1983. Phys. Rev. Lett. **50**:620.

Forman, W., and C. Jones. 1982. Annu. Rev. Astron. Astrophys. **20**:547.

Friedman, S.D., G.F. Smoot, G. De Amici, and C. Witebsky. 1984. Phys. Rev. **D29**:2677.

Gamow, G. 1948. Phys. Rev. **74**:505.

Gunn, J.E. 1978. In A. Maeder, L. Martinet, and G. Tammann (eds.), Observational Cosmology. Geneva Observatory, SAAS-FEE, p. 1.

Gush, H.P. 1981. Phys. Rev. Lett. **47**:745.

Hart, L., and R.D. Davies. 1982. Nature **297**:191.

Hegyi, D.J., W. Traub, and N. Carleton. 1972. Phys. Rev. Lett. **28**:1541.

Henry, P.S. 1971. Nature **231**:516.

Hogan, C.J., N. Kaiser, and M.J. Rees. 1982. Phil. Trans. R. Soc. London **A307**:97.

Howell, T.F., and J.R. Shakeshaft. 1966. Nature **210**:1318.

Howell, T.F., and J.R. Shakeshaft. 1967. Nature **216**:753.

Johnson, D.G., and D.T. Wilkinson. 1987. Astrophys. J. (Lett.) **313**:L1.

Jones, B.J.T. 1980. Phys. Scripta **21**:732.

Kaiser, N. 1983. Mon. Not. R. Astron. Soc. **202**:1169.

Kellermann, K.I., E.B. Fomalont, J.V. Wall, and D. Weistrop. 1986. Highlights Astron. **7**:367.

Kellermann, K.I., E.B. Fomalont, J.V. Wall and D. Weistrop. 1987. Proceedings of IAU Symposium # 130, Balatonfured (Hungary). Ed: J. Audouze (in press).

Kent, S.M., and J.E. Gunn. 1982. Astron. J. **87**:945.

Knocke, J.E., R.B. Partridge, M.I. Ratner, and I.I. Shapiro. 1984. Astrophys. J. **284**:479.

Lasenby, A.N., and R.D. Davies. 1983. Mon. Not. R. Astron. Soc. **203**:1137.

Ledden, J.E., J.J. Broderick, J.J. Condon, and R.L. Brown. 1980. Astron. J. **85**:780.

Lubin, P.M., P. Melese, and G.F. Smoot. 1983. Astrophys. J. (Lett.) **273**:L51.

Lubin, P.M., T. Villela, G. Epstein, and G.F. Smoot. 1985. Astrophys. J. (Lett.) **298**:L1.

Mandolesi, N., P. Calzolari, S. Cortiglioni, F. Delpino, G. Sironi, P. Inzani, G. De Amici, J.-E. Solheim, L. Berger, R.B. Partridge, P.L. Martenis, C.H. Sangree, and R.C. Harvey. 1986. Nature **319**:751.

Mather, J.C. 1982. Opt. Eng. **21**:769.

Matsumoto, T., S. Hayakawa, H. Matsuo, H. Murakami, S. Sato, A.E. Lange and P.L. Richards. 1988. Astrophys. J. (Lett.): in press.

Melchiorri, F., B.O. Melchiorri, C. Cecarelli, and L. Pietranera. 1981. Astrophys. J. (Lett.) **250**:L1.

Meyer, D.M., and M. Jura. 1985. Astrophys. J. **297**:119.

Meyer, S.S., A.D. Jeffries, and R. Weiss. 1983. Astrophys. J. (Lett.) **271**:L1.

Mushotzky, R.F., and B.W. Smith. 1980. Highlights Astron. **5**:735.

Nanos, G.P., Jr. 1979. Astrophys. J. **232**:341.

Negroponte, J., and J. Silk. 1980. Phys. Rev. Lett **44**:1433.

Ohm, E.A. 1961. Bell Syst. Tech. J. **40**:1065.

Pariiskii, Yu. N., Z.E. Petrov, and L.N. Cherkov. 1977. Sov. Astron. J. (Lett.) **3**:263.

Partridge, R.B. 1980. Astrophys. J. **235**:681.

Partridge, R.B. 1983. In B.J.T. Jones and J.E. Jones (eds.), The Origin and Evolution of Galaxies, VIIth course of the International School of Cosmology and Gravitation (Erice). Dordrecht: Reidel, p. 121.

Partridge, R.B., and D.T. Wilkinson. 1967. Phys. Rev. Lett. **18**:557.

Partridge, R.B., J. Cannon, R. Foster, C. Johnson, E. Rubinstein, A. Rudolph, L. Danese, and G. DeZotti. 1984. Phys. Rev. **D29**:2683.

Peebles, P.J.E. 1965. In J. Ehlers (ed.), Proceedings of Cornell Summer School: Lectures in Applied Mathematics, Vol. 8. Providence: American Mathematical Society, 1967, p. 274.

Peebles, P.J.E. 1971. Physical Cosmology. Princeton, New Jersey: Princeton University Press.

Peebles, P.J.E. 1981. Astrophys. J. (Lett.) **243**:L119.

Penzias, A.A., and R.W. Wilson. 1965. Astrophys. J. **142**:419.

Penzias, A.A., K. Jefferts, and R.W. Wilson. 1972. Phys. Rev. Lett. **28**:772.

Peterson, J.B. P.L. Richards, and T. Timusk. 1985. Phys. Rev. Lett. **55**:332.

Radford, S.J.E., P.E. Boynton, B.L. Ulich, R.B. Partridge, R.A. Schommer, A.A. Stark, R.W. Wilson, and S.S. Murray. 1986. Astrophys. J. **300**:159.

Rees, M.J. 1968. Astrophys. J. (Lett.) **153**:L1.

Rees, M.J. 1978. In A. Maeder, L. Martinet, and G. Tammann (eds.), Observational Cosmology. Geneva Observatory, SAAS-FEE, p. 618.

Rudnick, L. 1978. Astrophys. J. **223**:37.

Sachs, R.K., and A.M. Wolfe. 1967. Astrophys. J. **147**:73.

Sarazin, C.L. 1986. Rev. Mod. Phys. **58**:1.

Schallwich, D. 1982. Ph.D. Thesis, Univesity of Bochum, Federal Republic of Germany.

Seielstad, G.A., C.R. Masson, and G.L. Berge. 1981. Astrophys. J. **244**:717.

Silk, J.I. 1968. Astrophys. J. **151**:459.

Silk, J. 1984. In E.W. Kolb, M.S. Turner, D. Lindley, K. Olive, and D. Seckel (eds.), Inner Space/Outer Space, Fermilab workshop. Chicago: University of Chicago Press, p. 143.

Smoot, G.F., G. DeAmici, S.D. Friedman, C. Witebsky, G. Sironi, G. Bonelli, N. Mandolesi, S. Cortiglioni, G. Morigi, R.B. Partridge, L. Danese, and G. DeZotti. 1985. Astrophys. J. (Lett.) 291:L23.

Smoot, G.F., M.V. Gorenstein, and R. Muller. 1977. Phys. Rev. Lett. 39:898.

Strukov, I.A., D.P. Skulachev, and A.A. Klypin, 1987. Proceedings of IAU Symposium #130, Balatonfured (Hungary). Ed: J. Audouze (in press)

Sunyaev, R.A., and Ya. B. Zel'dovich. 1972. Comm. Astrophys. Space Phys. 4:173.

Sunyaev, R.A., and Ya. B. Zel'dovich. 1980. Annu. Rev. Astron. Astrophys. 18:537.

Tarter, J.C. 1978. Astrophys. J. 220:749.

Thaddeus, P., 1972. Annu. Rev. Astron. Astrophys. 10:305.

Uson, J.M. 1984. In J.L. Sanz and L.J. Goicoechea (eds.), Observational and Theoretical Aspects of Relativistic Astrophysics and Cosmology. Singapore: World Scientific, p. 269.

Uson, J.M. 1986. In C.P. O'Dea and J.M. Uson (eds.), Radio Continuum Processes in Clusters of Galaxies, NRAO-Green Bank workshop, p. 255.

Uson, J.M., and D.T. Wilkinson. 1984a. Astrophys. J. 283:471.

Uson, J.M., and D.T. Wilkinson. 1984b. Nature 312:427.

Uson, J.M., and D.T. Wilkinson. 1984c. Bull. Amr. Astron. Soc. 16:513. See also Uson (1984).

Vittorio, N., and J. Silk. 1984. Astrophys. J. (Lett.) 285:L39.

Weinberg, S. 1972. Gravitation and Cosmology. New York: John Wiley and Sons.

Weiss, R. 1980. Annu. Rev. Astron. Astrophys. 18:489.

White, S.D.M. 1984. In E.W. Kolb. M.S. Turner, D. Lindley, K. Olive, and D. Seckel (eds.), Inner Space/Outer Space, Fermilab workshop. Chicago: University of Chicago Press, p. 228.

Wilkinson, D.T. 1967. Phys. Rev. Lett. 19:1195.

Wilkinson, D.T. 1984. In G.O. Abell and G. Chincarini (eds.), Early Evolution of the Universe and Its Present Structure, Int. Astron. Union Symp. 104. Dordrecht: Reidel, p. 143.

Wilkinson, D.T. (1986). Phil. Trans. R. Soc. London A320:595.

Wilkinson, D.T., and P.J.E. Peebles. 1983. In K.I. Kellermann and B. Sheets (eds.), Serendipitous Discoveries in Radio Astronomy. Green Bank, West Virginia: National Radio Astronomy Observatory, p. 175.

Wilson, M.L. 1983a. Astrophys. J. 273:2.

Wilson, M.L., and J. Silk. 1981. Astrophys. J. 243:14.

Wilson, R.W. 1983b. In K.I. Kellermann and B. Sheets (eds.), Serendipitous Discoveries in Radio Astronomy. Green Bank, West Virginia: National Radio Astronomy Observatory, p. 185.

Woody, D.P., and P.L. Richards. 1979. Phys. Rev. Lett. 43:925.

Woody, D.P., and P.L. Richards. 1981. Astrophys. J. 248:18.

Zel'dovich, Ya. B. and R.A. Sunyaev. 1969. Astrophys. Space Sci. 4:302.

15. Radio Sources and Cosmology

JAMES J. CONDON

15.1 Introduction

Since the realization in the 1950s that the majority of catalogued radio sources are extragalactic and that some of the strongest are at cosmological distances, the application of radio observations to cosmological studies has been a central theme in radio astronomy.

The ratios of flux density to luminosity and angular size to projected linear size at different redshifts can, in principle, distinguish between world models by their geometrical properties and measure directly such basic parameters as the Hubble constant H_0 and the deceleration parameter q_0. The sensitivity and resolution of modern radio telescopes are more than adequate to make such measurements on a variety of sources, even at redshifts $z \gg 1$. Unfortunately, no "standard candle" or "standard rod" sources with luminosities or linear sizes known in a model-independent way have yet been recognized. Not only do the luminosities and sizes of extragalactic radio sources cover a wide range, but also their median values in radio-source populations evolve with cosmological epoch.

A less direct, but so far more successful, approach is to obtain complete samples of radio sources and model statistics describing radio-source populations. For example, the flux-density distribution of sources led to the discovery of evolution on cosmological time scales and the rejection of the steady-state world model. Such cosmological studies of radio-source populations might be used to constrain astrophysical theories of radio sources, their parent galaxies, their environments, and their evolution. Radio-selected samples are particularly well suited to these cosmological investigations for several reasons. Modern radio surveys are statistically complete and quite reliable, and radio flux densities are routinely measured with high accuracy. Radio samples are not badly confused by galactic stars, and they are not affected by dust obscuration in our galaxy, intervening material at high redshifts, or the host galaxies of the sources themselves (Ostriker and Heisler 1984). Radio sources have fairly smooth power law spectra, so there is no sharp redshift cutoff or discrimination against sources in certain redshift bands. Finally, many radio sources that have been found at redshifts $z \approx 1$ are so luminous that they could easily be detected in moderately sensitive surveys even at very high redshifts ($z \approx 10$). The biggest limitation of radio-selected samples is their dependence on optical identifications and spectroscopy for information about distance and host-galaxy morphology (elliptical galaxy, spiral galaxy, quasar, etc.). Only for the $N \approx 200$ strongest radio sources are the optical identifications and spectroscopy complete. The redshift distributions of most faint sources can only be estimated with the aid of evolutionary models constrained by statistical data—source counts, angular-size distributions, etc. This chapter describes some of the tests of world models and cosmological evolution that can be made using discrete radio sources.

15.2 Basic Relations

The flux density S as a function of frequency v and the angular size θ of a radio source are the observables directly relevant to most cosmological problems. They are related to the intrinsic source luminosity L and projected linear size d as described below.

Consider an isotropic source at redshift z with spectral luminosity L at frequency v (measured in the source frame). Its spectral flux density S measured at the same frequency v (in the observer's frame) will be

$$S = \frac{L}{A(1 + z)^{1+\alpha}},\tag{15.1}$$

where A is the area of the sphere centered on the source and containing the observer and $\alpha \equiv -\ln(S/S_o)/\ln(v/v_o)$ is the two-point spectral index between the frequencies v and $v_o = v/(1 + z)$ in the observer's frame. (Note that the *negative* sign convention for α is used throughout this chapter.) The $(1 + z)^{1+\alpha}$ term expresses the special relativistic Doppler correction; the geometry and expansion dynamics of the universe appear only in A. An "effective distance" D (Longair 1978) can be defined by $A \equiv 4\pi D^2$. Since the area of the sphere centered on the observer and containing a source at redshift z is always $A/(1 + z)^2$, the relation between (projected) linear size d and measured angular size θ is

$$\theta = \frac{d(1 + z)}{D}.\tag{15.2}$$

The "angular size" distance is defined by $D_\theta \equiv d/\theta = D/(1 + z)$. The "bolometric luminosity distance" D_{bol} defined by $S_{bol} \equiv L_{bol}/(4\pi D_{bol}^2)$ is given by $D_{bol} = D(1 + z)$.

In Friedmann models (cosmological constant $\Lambda = 0$) with zero pressure, density parameter $\Omega = 2q_0$, and current Hubble parameter H_0, the effective distance is traditionally given (Mattig 1958) as

$$D = \frac{2c\{\Omega z + (\Omega - 2)[(1 + \Omega z)^{1/2} - 1]\}}{H_0\Omega^2(1 + z)}.\tag{15.3}$$

However, this formula is numerically unstable for small Ωz, the transformation (based on Terrell 1977)

$$D = \left(\frac{cz}{H_0}\right)\left(\frac{1}{1 + z}\right)\left[1 + \frac{z(1 - \Omega/2)}{(1 + \Omega z)^{1/2} + 1 + \Omega z/2}\right]\tag{15.4}$$

is better for numerical calculations. For particular values of Ω, D reduces to the simpler forms:

$$D = \frac{cz}{H_0}\left(\frac{1 + z/2}{1 + z}\right) \qquad (\Omega = 0) \tag{15.5a}$$

$$D = \frac{2c}{H_0}[1 - (1 + z)^{-1/2}] \qquad (\Omega = 1) \tag{15.5b}$$

$$D = \frac{cz}{H_0(1 + z)} \qquad (\Omega = 2) \ . \tag{15.5c}$$

To describe the distributions of sources in space and time, we also need the comoving volume dV of the spherical shell extending from z to $z + dz$. It is $dV = 4\pi D^2\, dr$, where the comoving radial coordinate element is $dr = -(1 + z)c\, dt$. In a Friedmann universe the expansion rate is

$$\frac{dz}{dt} = -H_0(1 + z)^2(1 + \Omega z)^{1/2} \tag{15.6}$$

so

$$dV = \frac{4\pi D^2 c\, dz}{H_0(1 + z)(1 + \Omega z)^{1/2}} \ . \tag{15.7}$$

15.3 The "World Picture" and Source Evolution

The interval during which radio observations have been made is much shorter than the active lifetimes of individual sources and the time scales on which populations of radio sources evolve, so at best the data can give only a "world picture" covering the surface of our past light cone. The luminosity functions, size distributions, etc. of different source populations at different lookback times can be compared to reveal evolution, but we cannot directly observe any changes actually taking place. One consequence of this limitation is illustrated by Figure 15.1, showing the radio luminosity functions of elliptical galaxies at two different epochs. The luminosity functions do not overlap, so cosmological evolution must occur. The arrows in Figure 15.1(a) indicate one way in which the data might be interpreted— the comoving density ρ_m of sources was higher in the past, with the greatest changes being experienced by the most luminous sources. Such evolution is called "luminosity-dependent density evolution," a term that suggests an evolutionary mechanism capable of distinguishing between weak and strong sources. The arrows in Figure 15.1(b) show a very different interpretation of the *same* two luminosity functions—the luminosities of *all* sources were higher in the past, by an amount independent of luminosity. This "luminosity evolution" interpretation is consistent with evolutionary mechanisms that affect weak and strong sources alike. Since the active lifetimes of individual radio sources are generally shorter than the evolutionary time scales, which are, in turn, shorter than the ages of elliptical galaxies, descriptions of evolution based on associating points or features in the luminosity functions from different epochs probably oversimplify the actual changes

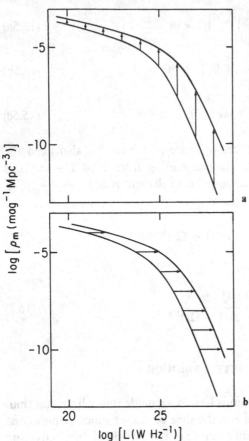

Fig. 15.1. The 1.4-GHz luminosity functions of elliptical radio galaxies at $z = 0$ and $z = 0.8$ with arrows illustrating (a) luminosity-dependent density evolution and (b) pure luminosity evolution. These particular luminosity functions are from the "shell model" described in Section 15.9. Abscissas: log luminosity (W Hz^{-1}). Ordinates: log comoving density (mag^{-1} Mpc^{-3}).

occurring on the individual source level. In any case, the data cannot distinguish between them.

Because luminosity functions $\rho_m(L|z)$ have dimensions of comoving source density, evolution has historically been described in terms of density changes. The "evolution function"

$$E(L, z) \equiv \frac{\rho_m(L|z)}{\rho_m(L|z = 0)} \tag{15.8}$$

is an example. Consequently, there is a widespread misconception that the data *imply* "luminosity-dependent density evolution," leading to unjustified conclusions like "In view of the lack of evolution of the low-luminosity sources, it seems implausible that their spectra should change with redshift." Even though the evolution function completely specifies the changes of mean source density with luminosity and epoch, it cannot completely describe the course of evolution.

Existing data do not even determine our world picture completely. The "generalized luminosity function" $\rho_m(L, \alpha, \ldots | z, v)$ of sources with luminosity L, spectral index α, and other relevant properties (e.g., type of galaxy) indicated by $\ldots$ at redshift

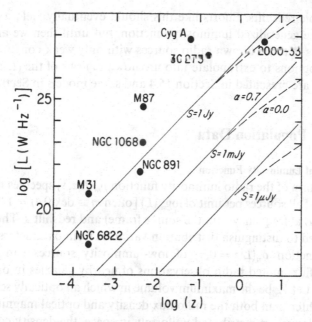

Fig. 15.2. Luminosities and redshifts of representative radio sources found at $v = 1.4$ GHz. Most nearby sources with $L < 10^{23}$ W Hz^{-1} are associated with spiral galaxies. The faintest are found in dwarf irregulars like NGC 6822; the luminosity of M31 is typical for full-sized spiral galaxies. An unusually high star formation rate probably accounts for the relatively high radio luminosity of NGC 891, and the Seyfert/starburst galaxy NGC 1068 is the strongest radio emitter of the nearby spiral galaxies. Elliptical galaxies such as M87 dominate the radio-source population at higher powers, reaching $L \approx 10^{28}$ W Hz^{-1} in the case of Cygnus A. Radio-selected quasars are concentrated near the upper end of this luminosity range and have known redshifts between $z = 0.158$ (3C273) and $z = 3.78$ (2000-33). Lines of constant flux density are shown for sources with spectral indices $\alpha = 0.7$ (solid lines) and $\alpha = 0.0$ (broken curves) in an Einstein–de Sitter universe ($\Omega = 1$) with $H_0 = 100$ km s^{-1} Mpc^{-1}. Nearly all radio galaxies selected at $v = 1.4$ GHz have spectral indices near $\alpha = 0.7$, while many quasars have $\alpha \approx 0$. The faintest discrete sources detected at $v = 1.4$ GHz have flux densities $S \approx 100$ μJy, and the sky density of sources as faint as $S \approx 10$ μJy has been estimated statistically. Abscissa: log redshift. Ordinate: log spectral luminosity (W Hz^{-1}).

z and frequency v is only partially determined. Most sources in the flux-limited sample found by any single radio survey complete to some level S are confined to a narrow diagonal band in the luminosity-redshift plane (Figure 15.2). Known radio sources span ten decades in luminosity and five in redshift, so surveys with a wide range of limiting flux densities S made with a number of different radio telescopes are needed to fill in the (L, z)-plane. More difficult than this is obtaining the optical identifications and redshifts needed to locate individual sources on the (L, z)-plane. Spectroscopic redshifts are available for most sources with $S \geq 2$ Jy at $v = 1.4$ GHz, but fainter sources with known S and unknown z could lie almost anywhere on the diagonal lines of constant S. The Leiden-Berkeley Deep Survey (Windhorst 1984, Windhorst et al. 1984b, Kron et al. 1985, Windhorst et al. 1985) is a major project to find sources as faint as $S \approx 1$ mJy at $v = 1.4$ GHz, identify nearly all of them on very deep photographic plates or CCD images, and obtain photometric or spectroscopic redshifts. Efforts like this should eventually yield a *direct* determination of the generalized luminosity function, but until then we are limited to making models that use known radio sources with only weak constraints on their redshift distributions to extrapolate into unknown regions of the (L, z)-plane. The available data are presented in Section 15.4 and some models in Section 15.5.

15.4 Source Population Data

15.4.1 The Local Luminosity Function

The simplest form of the radio luminosity function $\rho_m(L|z, v)$ specifies the comoving space density of all sources per unit of $\log_m(L)$ [often $m \equiv \text{dex}(0.4) = 1$ "magnitude"] in luminosity L at frequency v (in the source frame) and redshift z. The vertical bar notation is used to distinguish distribution variables from parameters. The "local" luminosity function $\rho_m(L|z = 0, v)$ of low-luminosity sources can be obtained directly from flux-limited radio observations of nearby galaxies in optically complete samples. Let V_m be the maximum volume in which an optically selected galaxy would be brighter than both the radio flux density and optical magnitude limits. If the sample galaxies are distributed uniformly in space, the density contributed by the N radio-detected galaxies with luminosities in the luminosity bin of width m centered on L is

$$\rho_m(L|z = 0, v) = \sum_{i=1}^{N} \left(\frac{1}{V_m}\right)_i \tag{15.9}$$

with an rms statistical uncertainty

$$\sigma = \left[\sum_{i=1}^{N} \left(\frac{1}{V_m}\right)_i^2 \right]^{1/2}. \tag{15.10}$$

The actual errors in ρ_m are larger at the lowest luminosities because clustering in the accessible volumes V_m associated with intrinsically faint galaxies is significant. Since radio sources of intermediate luminosities have space densities too low for them to be numerous in the small volumes covered by most optically complete samples, sources identified with bright low-redshift galaxies found in complete radio surveys must be used (e.g., Auriemma et al. 1977).

At the very highest luminosities the source density is so low that the nearest sources are already at cosmological distances and lookback times affected by evolution; their "local" luminosity function can only be estimated with the aid of evolutionary models. The density of evolving sources is uniform in volume elements dV' that have been weighted by the evolution function $E(L, z)$ (Equation 15.8). The weighted volume V'_m replacing V_m in Equation (15.9) is

$$V'_m = \int_0^{z_m} E(L, z) \, dV \tag{15.11}$$

for a source of luminosity L that could be seen out to a redshift z_m.

The distribution of

$$\frac{V}{V_m} \equiv \frac{\int\limits_0^z dV}{\int\limits_0^{z_m} dV} \tag{15.12}$$

should be uniform in the interval $(0, 1)$ if the sources in any sample have a constant (comoving) density (cf. Schmidt 1968). Incompleteness is usually revealed by a deficit of V/V_m values near unity, and monotonically increasing evolution by $\langle V/V_m \rangle > 0.5$ for the whole sample.

The local luminosity function is best determined at $v = 1.4$ GHz. The local luminosity function of spiral galaxies has recently been derived from sensitive VLA observations (Condon 1987) of all spirals north of $\delta = -45°$ and brighter than $B_T = +12$ mag, the completeness limit of the *Revised Shapley-Ames Catalog* (Sandage and Tammann 1981). The corresponding luminosity function for low-redshift E and S0 galaxies was obtained by Auriemma et al. (1977) from both optical and radio samples as described above. There are essentially no local radio-selected quasars. The contributions of Seyfert galaxies and optically selected quasars to the local luminosity function fall within the high-luminosity extension of the spiral galaxy component (Meurs and Wilson 1984), and the radio-loud quasars are presumed to be in elliptical galaxies. The resulting 1.4-GHz local luminosity function $\rho_m(L|z = 0, v = 1.4 \text{ GHz})$ is plotted in Figure 15.3(a) for a Hubble parameter $H_0 = 100 \text{ km s}^{-1} \text{ Mpc}^{-1}$ (the value used throughout this chapter).

Other forms of the luminosity function are sometimes useful. One is the co-moving density $\rho(L|z, v) dL$ of sources with luminosities L to $L + dL$. Since $\rho_m(L|z, v) d[\log_m(L)] \equiv \rho(L|z, v) dL$,

$$\rho_m(L|z, v) = \ln(m) L \rho(L|z, v) . \tag{15.13}$$

The weighted luminosity function, or "visibility function,"

$$\phi(L|z, v) \equiv L^{5/2} \rho(L|z, v) \tag{15.14}$$

introduced by von Hoerner (1973) emphasizes the contributions of sources in different luminosity ranges to the weighted source count $S^{5/2} n(S|v)$ (Section 15.4.2). In the low-redshift (static Euclidean) limit $L = 4\pi D^2 S$ and $dV = 4\pi D^2 dD$, the number $n(L, D|v) dL\, dD$ of sources per steradian with luminosities L to $L + dL$ and distances D to $D + dD$ is related to the local luminosity function by $n(L, D|v) dL\, dD = \rho(L|v) dL \times D^2 dD$. The corresponding number $n(L, S|v) dL\, dS$ in the flux-density range S to $S + dS$ is given by $n(L, S|v) = n(L, D|v)|dD/dS|$. The total number $n(S|v)$ of sources per steradian with flux densities S to $S + dS$ is

$$n(S|v) = \int\limits_0^\infty n(L, S|v) dL = \frac{1}{2(4\pi)^{3/2} S^{5/2}} \int\limits_0^\infty L^{3/2} \rho(L|v) dL . \tag{15.15}$$

This can be rearranged to yield

$$S^{5/2} n(S|v) = \frac{\ln 10}{2(4\pi)^{3/2}} \int\limits_{-\infty}^\infty \phi(L|v) d[\log(L)] . \tag{15.16}$$

The local weighted luminosity function $\phi(L|z = 0, v = 1.4 \text{ GHz})$ is plotted in

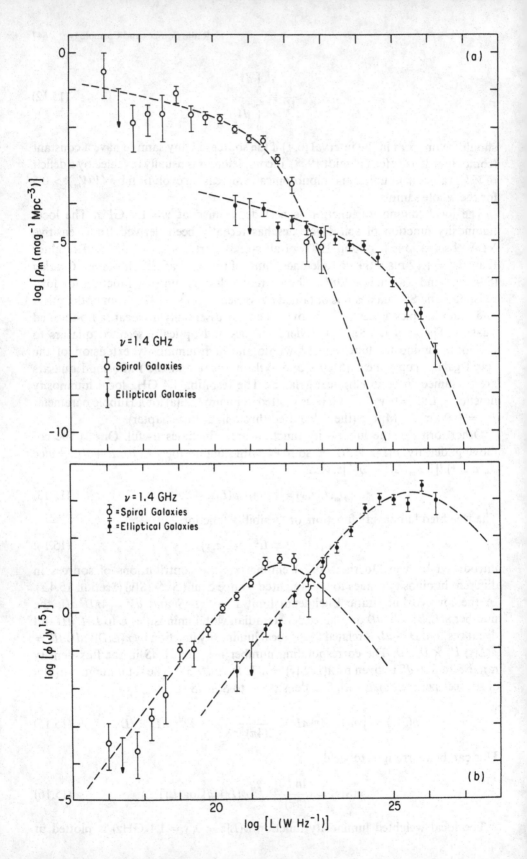

Figure 15.3(b). With Equation (15.16) it shows that, in the low-redshift limit, most radio sources selected at $v = 1.4$ GHz have luminosities $L \approx 10^{24}$ to 10^{27} W Hz^{-1} and are found in elliptical galaxies. Spiral galaxies contribute only about 1% of the sources; and radio-selected spiral galaxies have luminosities $L \approx 10^{22}$ W Hz^{-1}, about an order of magnitude higher than the typical radio luminosity of an optically selected spiral galaxy.

15.4.2 Source Counts

Because of the wide range of flux density and source density involved, no individual radio telescope can provide complete data, even at a single frequency (Figure 15.4). Pencil-beam instruments, large steerable dishes, and phased arrays are typically used to survey large regions of the sky to obtain statistically significant counts for the stronger sources with relatively low surface densities. Separate surveys made from the northern and southern hemispheres are necessary to cover the whole sky, and all-sky catalogues at $v = 0.408$, 2.7, and 5 GHz have been complied from large-scale radio surveys (Robertson 1973, Wall and Peacock 1985, and Kühr et al. 1981, respectively). A number of pencil-beam surveys go much deeper than the all-sky surveys over limited areas. Synthesis instruments provide the most sensitive surveys, but only in very small regions of the sky, typically 10^{-5} to 10^{-3} sr. Counts of very faint sources based on only a few such small fields may be subject to error if there is significant clustering. Nevertheless, in contrast to the early radio source surveys (cf. Jauncey 1975), modern data obtained by different observers using very different kinds of radio telescopes are in good agreement with respect to individual source positions and flux densities as well as surface densities.

The number $n(S|v)\,dS$ of sources per steradian with flux densities S to $S + dS$ found in a survey made at frequency v is called the *differential* source count; the total number per steradian stronger than S, $\int_S^\infty n(s|v)\,ds$, is called the *integral* source count. Integral counts are rarely used any more because they smear rapid changes of source density with flux density and the numbers are not statistically independent from one flux-density level to the next (Jauncey 1967, Crawford et al. 1970). The steep slopes of the differential source counts tend to obscure features in graphical presentations, so the counts are usually either weighted (simply multiplied by $S^{5/2}$) or normalized [divided by the count $n_0(S|v) = k_0 S^{-5/2}$ expected in a static Euclidean universe; the constant k_0 is usually set so that $n(S|v)/n_0(S|v) \approx 1$ at $S = 1$ Jy] before plotting. Historically, this normalization has been used to facilitate comparisons with the static Euclidean count—level portions of the actual normalized counts are said to have a "Euclidean slope," for example. Such comparisons can be misleading, however, because the static Euclidean approximation has surprisingly little relevance to the actual source counts except at the very highest flux densities (cf. Section 15.9). In particular, a Euclidean slope does not signify that the sources in that flux-density range have low redshifts or are not evolving.

◁ ——

Fig. 15.3. (a) Local luminosity function $\rho_m(L|z = 0, v = 1.4$ GHz). Open circles represent radio sources associated with spiral galaxies (Condon 1987), and the filled circles are based on the Auriemma et al. (1977) luminosity function for E and S0 galaxies. Abscissa: log luminosity (W Hz^{-1}). Ordinate: log comoving density (mag^{-1} Mpc^{-3}). (b) Local weighted luminosity function $\phi(L|z = 0, v = 1.4$ GHz). Abscissa: log luminosity (W Hz^{-1}). Ordinate: log weighted luminosity function (Jy$^{1.5}$).

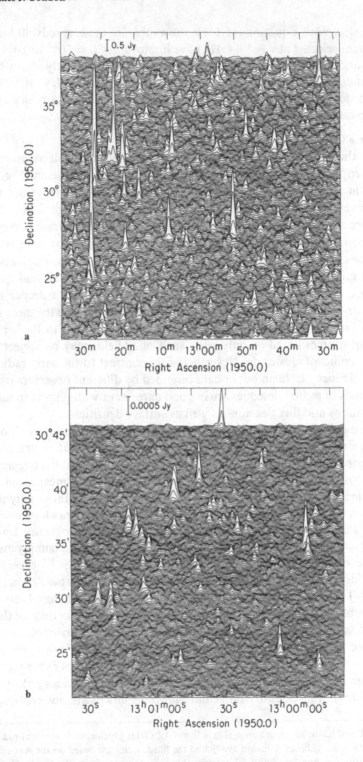

Fig. 15.4. Profile plots of the sky near the north galactic pole mapped with (a) the NRAO 91-m telescope (beamwidth $\approx$ 12′) (Condon and Broderick 1985) and (b) the VLA (beamwidth 17″.5) (Mitchell and Condon 1985) illustrate the range of source intensities and sky densities that go into the $\nu = 1.4$ GHz source count. (From Condon 1984.)

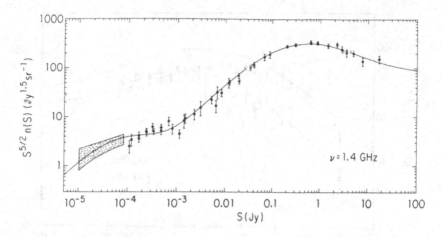

Fig. 15.5. Weighted source count at $v = 1.4$ GHz. Abscissa: flux density (Jy). Ordinate: weighted source count $S^{5/2}n(S)$ $(\text{Jy}^{1.5}\ \text{sr}^{-1})$.

Source counts covering a wide range of flux densities are currently available at $v = 0.408, 0.61, 1.4, 2.7,$ and 5 GHz (cf. Condon 1984b). The most extensive is at $v = 1.4$ GHz and is shown in Figure 15.5. The NRAO 91-m telescope was used to measure the flux densities of sources stronger than $S = 2$ Jy at $v = 1.4$ GHz (Fomalont et al. 1974) and also in the $0.175 \leq S < 2$ Jy range (Machalski 1978). The fainter levels are based on VLA "snapshot" surveys (Condon et al. 1982b, Mitchell 1983; Coleman et al. 1985), the WSRT deep survey of the Lynx area (Oort 1987), and the deepest VLA survey (Mitchell and Condon 1985). The sky densities of sources too faint to be detected and counted individually in the latter survey were estimated statistically from their contribution to the map fluctuation or "P(D)" distribution (Scheuer 1957, 1974, Condon 1974) and are indicated by the shaded region extending down to $S = 10\ \mu$Jy. The integrated emission from extragalactic sources can be used to constrain the source count at even fainter levels. After subtracting galactic emission, Bridle (1967) obtained $T \approx 30$ K at $v = 178$ MHz, corresponding to $T \approx 0.1$ K at $v = 1.4$ GHz. The contribution $T(S) = [c^2/(2kv^2)]\int_S^\infty sn(s)\,ds$ from sources stronger than $S = 10\ \mu$Jy is about 0.08 K, so the bulk of the extragalactic background can be accounted for by known source populations.

Most sources found in low-frequency surveys have power law spectra with spectral indices near $\alpha \approx +0.8$, but some have the more complex spectra and lower spectral indices ($\alpha \approx 0$) indicative of synchrotron self-absorption in compact ($\theta < 0\rlap{.}''01$) high-brightness ($T \approx 10^{11}$ K) components. These two source types are effectively distinguished by the simple criterion $\alpha \geq 0.5$ ("steep-spectrum" source) or $\alpha < 0.5$ ("flat-spectrum" source). The flat-spectrum sources can usually be identified with quasars, while most steep-spectrum sources are associated with galaxies (or empty fields if the galaxies are too distant). Many flat-spectrum sources vary in both intensity and structure on time scales of years, and their apparent luminosities may be affected by relativistic beaming (see Chapter 13). The evolutionary histories of these two source types may also differ. Being so compact,

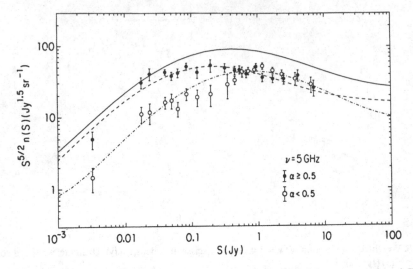

Fig. 15.6. Weighted counts of steep-spectrum ($\alpha \geqslant 0.5$) (filled symbols) and flat-spectrum ($\alpha < 0.5$) (open symbols) sources found at $\nu = 5$ GHz, along with model predictions (From Condon 1984b) (dashed and dot-dashed lines, respectively). Since the weighted counts of these two spectral populations are comparable but peak at slightly different flux densities, their sum (solid line) has a very broad peak.

flat-spectrum sources are probably less sensitive than extended steep-spectrum sources to changes in the average density of the intergalactic medium or in the energy density of the microwave background radiation with cosmological epoch (Rees and Setti 1968). Finally, flat-spectrum sources can be seen at greater redshifts because they are not so strongly attenuated by the $(1 + z)^{1+\alpha}$ Doppler term in Equation (15.1). For these reasons, it is worthwhile to separate the steep- and flat-spectrum sources and count them independently when possible. The numbers of flat-spectrum ($\alpha < 0.5$) and steep-spectrum ($\alpha \geq 0.5$) sources are comparable in high-frequency samples, and their counts at 5 GHz are plotted separately in Figure 15.6. The data were taken from Pauliny-Toth et al. (1978), Condon and Ledden (1981), Owen et al. (1983), and Fomalont et al. (1984).

15.4.3 Spectral-Index Distributions
Two-point spectral indices $\alpha(\nu_1, \nu_2)$ have been measured between $\nu_1 \approx 1.4$ GHz and $\nu_2 \approx 5$ GHz for a number of flux-limited source samples. The integral number $N(\alpha|S, \nu)\, d\alpha$ of sources per steradian with spectral indices α to $\alpha + d\alpha$ and flux densities $\geq S$ at frequency ν is shown for $S \geq 0.8$ Jy at $\nu = 5$ GHz in Figure 15.7(a). This (unnormalized) spectral-index distribution consists of a narrow steep-spectrum component with $\langle \alpha \rangle \approx 0.7$ and a broader flat-spectrum component centered on $\langle \alpha \rangle \approx 0.0$. As the sample selection frequency ν is lowered, the number of steep-spectrum sources increases rapidly and the median spectral indices $\langle \alpha \rangle$ of both components increase. The increase in $\langle \alpha \rangle$ of each spectral component is proportional to the square of its width (Kellermann 1964), so the median spectral index of the flat-spectrum component changes more rapidly with frequency. These effects can be seen by comparing Figure 15.7(a) with the spectral-index distribution of sources stronger than $S = 2$ Jy at $\nu = 1.4$ GHz [Figure 15.7(b)]. The fraction of

flat-spectrum sources may also change with the sample flux-density limit S at a given frequency v. Figure 15.7(c) gives the spectral-index distribution of fainter ($S \geq 0.035$ Jy) sources selected at $v = 4.8$ GHz.

15.4.4 Redshift/Spectral-Index Diagrams

A plot of the integral number $N(\alpha, z|S, v)\, d\alpha\, d[\log(z)]$ of sources per steradian stronger than S at frequency v with spectral indices α to $\alpha + d\alpha$ and redshifts $\log(z)$ to $\log(z) + d[\log(z)]$ shows the correlation of α on z (or L) for steep-spectrum sources (Laing and Peacock 1980, and references therein). Such a redshift/spectral-index diagram for sources with $S \geq 2$ Jy at $v = 1.4$ GHz is plotted in Figure 15.8.

15.4.5 Redshift and Luminosity Distributions of Strong Sources

Spectroscopic redshifts exist for nearly all 3CR sources stronger than 10 Jy at 178 MHz (Laing et al. 1983, Spinrad et al. 1985) and for most of the stronger flat-spectrum sources found in high-frequency radio surveys (Kühr et al. 1981, Véron-Cetty and Véron 1983, Wall and Peacock 1985). Photometric redshifts of the remaining strong sources identified with galaxies can be estimated from Hubble relations such as $\log(z) \approx (m_v - 22.5)/6$ valid for first-ranked cluster galaxies and 3CR and 4C radio galaxies (van der Laan and Windhorst 1982, Windhorst 1986); empty-field sources can be treated as galaxies just fainter than the plate limit. Redshift distributions of quasar candidates may be approximated by the redshift distributions of known quasars or by a broad Hubble relation (Wall and Peacock 1985). Let $N(z|S, v)d[\log(z)]$ be the integral number of sources per steradian with redshifts $\log(z)$ to $\log(z) + d[\log(z)]$ and stronger than S at frequency v. Such an (unnormalized) redshift distribution of 202 extragalactic sources stronger than $S = 2$ Jy at $v = 1.4$ GHz is shown in Figure 15.9.

Since most of the sources in a flux-limited sample are within a factor of two of the flux-density limit, the integral luminosity distribution $N(L|S, v)d[\log(L)]$, or number of sources per steradian with luminosities $\log(L)$ to $\log(L) + d[\log(L)]$ that are stronger than flux density S at frequency v, can be used almost interchangeably with the integral redshift distribution. However, both of these distributions bin the (S, z)-data and hence do not make the most efficient possible use of the strong-source data (Peacock 1985).

15.4.6 Optical Constraints

The magnitude distributions of galaxies identified with faint radio sources are often used to estimate redshift distributions. These estimates can be refined if broadband colors and optical morphologies are available, since the absolute magnitudes of the giant elliptical galaxies associated with the most luminous radio sources are somewhat brighter than those of elliptical galaxies identified with less luminous sources (Auriemma et al. 1977) and the "blue" population of galaxies investigated by Kron et al. (1985). Windhorst et al. (1984a) have shown that the fraction of radio sources indentified with galaxies brighter than $J \approx 23.7$ is about 50%, nearly independent of 1.4-GHz flux density in the range of 1 to 100 mJy. Thus, the median redshift in this flux-density range is roughly equal to the redshift of a giant elliptical galaxy at their plate limit, $z \approx 0.8$. Unless the amount of evolution changes discontinuously at redshifts just beyond $z \approx 0.8$, most of the remaining objects

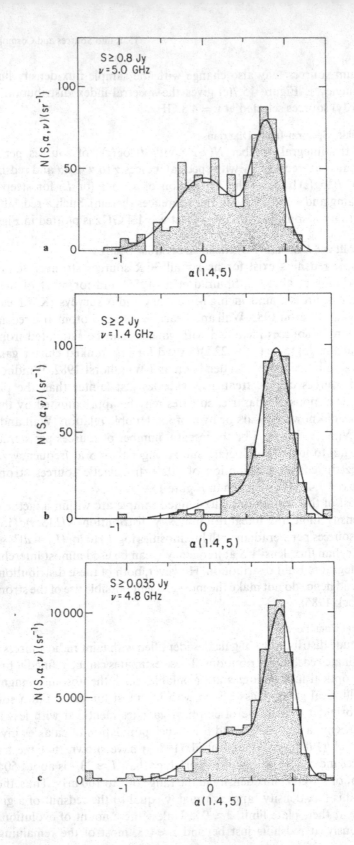

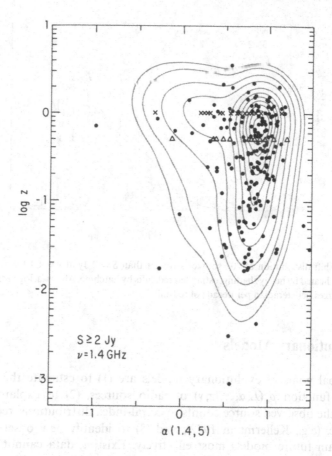

Fig. 15.8. Redshift/spectral-index diagram for a sample of 202 sources in 3.86 sr stronger than $S = 2$ Jy at $v = 1.4$ GHz. Quasar candidates are indicated by crosses at $z \approx 1$, and empty fields by triangles at $z \approx 0.5$. Abscissa: spectral index between 1.4 and 5 GHz. Ordinate: log redshift. Contours: 1, 2, 5, 10, 20, 30, 40, 50, 60, and 70 sr^{-1}.

fainter than $J \approx 23.7$ must have redshifts only slightly larger. In fact, the identification rate approaches 100% at $V \approx 26$ (Windhorst 1986), and few galaxies in this flux-density range should have redshifts greater than 1.9.

The identification content appears to change below $S \approx 1$ mJy from primarily red elliptical galaxies to bluer, morphologically distorted galaxies with fainter absolute magnitudes (Windhorst 1984, Kron et al. 1985). The nature of these galaxies and their redshift range is still uncertain (Wall et al. 1986, Windhorst 1986, Weistrop et al. 1987).

◁ ————————————————————————————————————

Fig. 15.7. (a) Spectral-index distribution of 320 sources stronger than $S = 0.8$ Jy at $v = 5$ GHz (Witzel et al. 1979). (b) Spectral-index distribution of 202 sources stronger than $S = 2$ Jy at $v = 1.4$ GHz as compiled by Condon (1984b). (c) Spectral-index distribution of 479 sources stronger than $S = 0.035$ Jy at $v = 4.8$ GHz (Owen et al. 1983). Abscissas: spectral index between 1.4 and 5 GHz. Ordinates: number of sources per steradian per unit α.

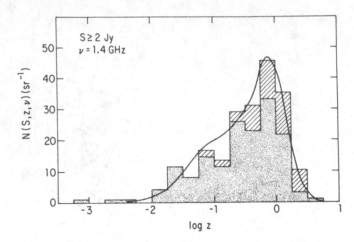

Fig. 15.9. Redshift distribution of 202 sources stronger than $S = 2$ Jy at $v = 1.4$ GHz. Spectroscopic redshifts are indicated by heavy shading, estimated redshifts by hatching. Abscissa: log redshift. Ordinate: number of sources per steradian per decade of redshift.

15.5 Evolutionary Models

The principal goals of evolutionary models are (1) to estimate the generalized luminosity function $\rho_m(L, \alpha, \dots | z, v)$ of radio sources, (2) to explain significant features in the observed source counts, spectral-index distributions, redshift distributions, etc. (e.g., Kellermann 1972), and (3) to identify new observations that will constrain future models most effectively. Existing data cannot fully determine the evolving luminosity function, so practical models incorporate additional *assumptions* about source evolution. The computations must be simplified by *approximations* as well—source spectra are taken to be simple power laws, spectral-index distributions may be replaced by Gaussians or even δ-functions, and evolution may be described by smooth functions with a limited number of free parameters.

A typical model-generating procedure is:

1. Choose a particular world model to fix the effective distance D and the volume element dV (Section 15.2). The Einstein–de Sitter ($\Omega = 1$) model is most commonly picked, but changing Ω has only a small, easily estimated effect (Peacock 1985).
2. Guess a form for the evolving luminosity function.
3. Compute the model local luminosity function, source counts, spectral index distributions, etc. and compare them with the data.
4. Revise the luminosity function.
5. Repeat steps (3) and (4) until satisfactory agreement is reached.

Differences between the models produced by different authors reflect their different data sets, assumptions about the type of evolution occurring, computational approximations, and methods for deciding that a model is satisfactory. Steps (2) and (4) are not straightforward, although simplified models (Section 15.9) can make the relations between the input luminosity function parameters and the output observables clearer.

15.5.1 Source Distribution Equations

The actual spectra of radio sources are normally approximated by power laws so that the spectral luminosity function at all frequencies v is determined by the spectral luminosity function at any one frequency v_0 from

$$\rho_m(L, \alpha | z, v) = \rho_m\left[\left(\frac{v}{v_0}\right)^\alpha L, \alpha | z, v_0\right] \tag{15.17}$$

for α measured between *any* two frequencies. Models based on this approximation generally work well at $v = 408$ MHz and higher frequencies, but they overestimate the 178-MHz source count significantly (Peacock and Gull 1981, Condon 1984b). Both 178-MHz flux-density scale errors and genuine spectral curvature caused by synchrotron self-absorption may contribute to this discrepancy.

The total number of sources with luminosities L to $L + dL$, spectral indices α to $\alpha + d\alpha$, at frequency v, and lying in the spherical shell with comoving volume dV at redshift z is

$$\rho(L, \alpha | z, v)\, dL\, d\alpha\, dV . \tag{15.18}$$

The number of sources in this shell equals the total number $\eta(S, \alpha, z | v)\, dS\, d\alpha\, dz$ of sources with flux densities S to $S + dS$, spectral indices α to $\alpha + d\alpha$, and redshifts z to $z + dz$ found in a survey of the whole sky (4π sr) at frequency v. Weighting by $S^{5/2}$ and eliminating both dV and dL/dS yields

$$S^{5/2}\eta(S, \alpha, z | v) = \frac{c\phi(L, \alpha | z, v)\, dz}{(4\pi)^{1/2} H_0 D(1 + \Omega z)^{1/2}(1 + z)^{5/2 + 3\alpha/2}} . \tag{15.19}$$

Integrating over redshift and dividing by 4π gives the weighted (spectral) source count $S^{5/2}n(S, \alpha | v)$, where $n(S, \alpha | v)\, dS\, d\alpha$ is the number of sources per steradian with flux densities S to $S + dS$ and spectral indices α to $\alpha + d\alpha$ found at frequency v:

$$S^{5/2}n(S, \alpha | v) = \frac{c}{(4\pi)^{3/2} H_0} \int_0^\infty \frac{\phi(L, \alpha | z, v)\, dz}{D(1 + \Omega z)^{1/2}(1 + z)^{5/2 + 3\alpha/2}} . \tag{15.20}$$

The distribution equations (15.19) and (15.20) can be integrated numerically to give the observables described in Section 15.4. Using the weighted luminosity function ϕ to calculate the weighted source count directly minimizes the interpolation errors that can be significant in numerical integrations of the more rapidly varying luminosity function ρ_m (cf. Danese et al. 1983, Peacock 1985) to obtain the unweighted source count.

The weighted differential source count $S^{5/2}n(S | v)$ at frequency v (Section 15.4.2) is

$$S^{5/2}n(S | v) = \int_{-\infty}^\infty S^{5/2}n(S, \alpha | v)\, d\alpha . \tag{15.21}$$

The (unnormalized) spectral-index distribution $N(\alpha | S, v)$ (Section 15.4.3) is obtained by integrating the differential spectral count:

$$N(\alpha|S, v) = \int\limits_{S}^{\infty} n(s, \alpha|v)\, ds \ . \tag{15.22}$$

The redshift/spectral-index diagram (Section 15.4.4) shows the values of $N(\alpha, z|S, v)$ given by

$$N(\alpha, z|S, v) = \frac{\ln 10}{4\pi} \int\limits_{S}^{\infty} z\eta(s, \alpha, z|v)\, ds \ . \tag{15.23}$$

The (unnormalized) integral redshift distribution $N(z|S, v)$ (Section 15.4.5) is found by integrating over α:

$$N(z|S, v) = \int\limits_{-\infty}^{\infty} N(\alpha, z|S, v)\, d\alpha \ . \tag{15.24}$$

Most radio sources in any flux-limited sample have flux densities only slightly higher than the flux-density limit, so redshift and luminosity are strongly correlated. Thus the integral luminosity distribution $N(L|S, v)$ can be used instead of the integral redshift distribution $N(z|S, v)$. Let $n(L, S, \alpha|v)d[\log(L)]\, dS\, d\alpha$ be the differential number of sources per steradian with luminosities $\log(L)$ to $\log(L) + d[\log(L)]$, flux densities S to $S + dS$, and spectral indices α to $\alpha + d\alpha$ at frequency v. Then, $4\pi n(L, S, \alpha|v)d[\log(L)] = \eta(S, \alpha, z|v)\, dz.$ and

$$n(L, S, \alpha|v) = \frac{\ln(10)\eta(S, \alpha, z|v)}{4\pi}\left[\frac{L}{(dL/dz)}\right] \tag{15.25}$$

where

$$\left[\frac{L}{(dL/dz)}\right] = \left(\frac{1 + \alpha}{1 + z} + \frac{2}{D}\frac{dD}{dz}\right)^{-1} \tag{15.26}$$

and

$$\frac{2}{D}\frac{dD}{dz} = \frac{2\Omega[1 + (\Omega/2 - 1)(\Omega z + 1)^{-1/2}]}{\Omega z + (\Omega - 2)[(1 + \Omega z)^{1/2} - 1]} - \frac{2}{1 + z} \ . \tag{15.27}$$

Integrating Equation (15.25) over flux density and spectral index yields the (unnormalized) integral luminosity distribution

$$N(L|S, v) = \int\limits_{S}^{\infty} \int\limits_{-\infty}^{\infty} n(L, s, \alpha|v)\, d\alpha\, ds \ . \tag{15.28}$$

15.5.2 Evolutionary Models of Radio Luminosity Functions

Ryle and Clarke (1961) showed that their 178-MHz source count above $S = 0.25$ Jy is incompatible with nonevolving Einstein–de Sitter and steady-state models. They also recognized that "the introduction of evolutionary effects which appear to be necessary will make the selection of a unique [world] model difficult." In a paper introducing many of the features found in subsequent model calculations, Longair (1966) modeled the 178-MHz source count, strong-source luminosity distribution, and integrated emission from discrete sources in an evolving Einstein–

de Sitter universe. "Power-law" evolution proportional to $(1 + z)^n$ yielded satisfactory fits only if restricted to the most luminous sources and truncated at high redshifts. "Exponential" evolution proportional to $e^{-t/\tau}$, where t is the cosmic time and τ the evolutionary time scale, was proposed by Rowan-Robinson (1970) because it does not diverge at high redshifts. Although this parametric form shows that the data do not require a real truncation of evolution at large z, Rowan-Robinson also considered *physical* factors that must ultimately truncate evolution, such as the time needed to form the parent galaxies of radio sources and electron energy losses by inverse Compton scattering off the microwave background radiation. A parametric model explicitly constrained by astrophysical assumptions was tried by Grueff and Vigotti (1977) to explain the 408-MHz source count and the luminosity distributions of sources stronger than $S = 10$ Jy and $S = 0.9$ Jy. They assumed that quasars form at $z = 2.5$ and evolve into galaxies whose radio-emitting lifetimes are inversely proportional to their radio luminosities. One difficulty with this model is that the evolution of low-luminosity radio sources can be minimized only if their radio emitting lifetimes are comparable with the Hubble time.

Extensions of the source counts to lower flux densities and the availability of more complete redshift data for strong sources eventually justified reexamination of the first parametric models. Wall et al. (1980) found that the 408-MHz source count extending to $S = 0.01$ Jy and the "all-sky" luminosity distribution of sources stronger than $S = 10$ Jy were sufficient to show that "power law" models are a poor representation of the cosmological evolution of powerful radio sources. They also investigated "exponential" evolution of the form $\exp[M(1 - t/t_0)]$, where M specifies the strength of the evolution and t_0 is the present age of the universe. Successful models were constructed in which M depends on luminosity [e.g., $M = 0$ for $L < L_1$, $M = M_{max}$ for $L > L_2$, and $M = M_{max}(\log L - \log L_1)/(\log L_2 - \log L_1)$ for $L_1 < L < L_2$] or redshift. Robertson's (1978, 1980) "free-form" analysis of essentially the same data did not assume a functional form for the redshift dependence of evolution, but used the data to *solve* for it. However, these data cannot fully determine both the redshift and luminosity dependence of the evolution, so Robertson did assume a parametric form for the luminosity dependence that is similar to the one specified above. An artifact of the rather sharp changes of evolution with luminosity implied by this parametric form is a markedly bimodal redshift distribution at low flux densities (cf. Figure 10 of Wall et al. 1980).

The preceding models approximate the spectral-index distributions of all sources by a single δ-function centered on $\alpha \approx 0.8$, and they work well for data selected at any one low ($\nu < 1$ GHz) frequency. Extensive sky surveys made at 2.7 and 5 GHz in the late 1960s revealed significant populations of sources with $\alpha \approx 0$ and led to models accounting for both the steep- and flat-spectrum sources simultaneously (Schmidt 1972, Fanaroff and Longair 1973, Petrosian and Dickey 1973). The decline in the fraction of flat-spectrum sources as the 5-GHz sample flux-density limit is decreased below $S \approx 1$ Jy (Figure 15.6) can be reproduced if (1) the local spectral luminosity function is separated or "factorized" into independent spectral-index and luminosity functions at some frequency ν_f lower than 5 GHz and (2) the rate of evolution is the same for both steep- and flat-spectrum sources. Then there is an inverse correlation induced between α and L at higher frequencies, and the weighted

5-GHz count of flat-spectrum sources peaks at a higher flux density than the weighted count of steep-spectrum sources.

There have been some indications that flat-spectrum quasars may evolve less than steep-spectrum or optically selected quasars. The weighted count of flat-spectrum sources (mostly quasars) peaks near $S \approx 1$ Jy (Figure 15.6), so the *average* count slope is nearly Euclidean for the flat-spectrum sources found in the first large-scale 5-GHz surveys that are complete down to $S \approx 0.6$ Jy. Because the source count slope and $\langle V/V_m \rangle$ (Section 15.4.1) are closely related (Longair and Scheuer 1970), quasar identifications of flat-spectrum sources from these surveys have nearly static-Euclidean values $\langle V/V_m \rangle \approx 0.5$ (Schmidt 1976), much lower than the $\langle V/V_m \rangle \approx 0.7$ of quasars identified with primarily steep-spectrum 3CR quasars stronger than $S = 9$ Jy at $v = 178$ MHz (Schmidt 1968). While high $\langle V/V_m \rangle$ values indicate evolution, $\langle V/V_m \rangle \approx 0.5$ does not exclude evolution because the *distribution* of V/V_m may still be nonuniform. Evolution increasing at low redshifts ($z < 2$, for example) and decreasing at higher redshifts still in the sample volumes V_m could yield a nonuniform V/V_m distribution with $\langle V/V_m \rangle \approx 0.5$. Just this situation is probably occurring. The 3CR quasars can be seen only out to limiting redshifts $z_m \approx 2$, and their large $\langle V/V_m \rangle$ value reflects monotonically increasing evolution up to $z \approx 2$; flat-spectrum quasars stronger than $S = 0.6$ Jy can be seen at higher redshifts ($z_m \approx 3$ or 4) where their evolution has started to decline. Kulkarni (1978) produced models that allow the steep- and flat-spectrum populations to evolve independently, approximate the spectral-index distributions of each population by Gaussians, and include the correlation of α with L among steep-spectrum sources. In both the Kulkarni (1978) and Machalski (1981) models, the flat- and steep-spectrum sources evolve differently, but later models by Peacock and Gull (1981) and Condon (1984b) show that these two spectral classes may indeed evolve at the same rates without violating the data constraints.

Condon (1984b) searched for a single model to fit *in detail* a wide range of available radio data (the local luminosity functions of spiral and elliptical galaxies at $v = 1.4$ GHz; source counts at $v = 0.408, 0.61, 1.4, 2.7,$ and 5 GHz; counts of steep- and flat-spectrum sources at $v = 2.7$ and 5 GHz; spectral-index distributions of sources in a number of samples complete to different flux-density limits at $v = 1.4,$ 2.7, and 5 GHz; redshift/spectral-index diagrams and redshift distributions of strong sources selected at 1.4, 2.7, and 5 GHz). The local 1.4-GHz visibility functions of spiral and elliptical galaxies were approximated by hyperbolas [Figure 15.3(b)]. The spectral luminosity function was "factorized" at $v_f = 1.4$ GHz. The spectral-index function was approximated by two Gaussians, and the median spectral index of the steep-spectrum Gaussian varied with $\log(z)$. The evolution was constrained by the assumption that the *form* of the $v = 1.4$ GHz luminosity function be independent of redshift:

$$\rho_m(L|z, v) = g(z)\rho_m[L/f(z)|z = 0, v] \tag{15.29}$$

where $f(z)$ and $g(z)$ are "free-form" functions that describe "luminosity evolution" and "density evolution," respectively. This "translation evolution" [so named because the evolution can be represented by translating the local luminosity function in the $(\log L, \log \rho_m)$-plane] could result from evolutionary mechanisms

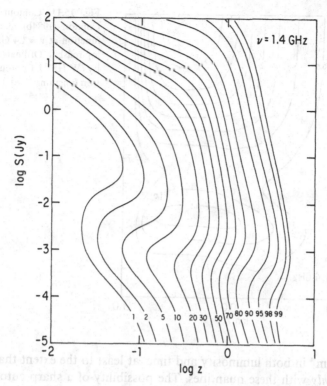

Fig. 15.10. Model redshift distributions at $v = 1.4$ GHz (From Condon 1984b). Abscissa: log redshift. Ordinate: log flux density (Jy). Parameter: percentage of sources at flux density S with redshifts less than z.

that do *not* discriminate on the basis of source luminosity. A simple model was found that fits the radio data (curves in Figures 15.3, 15.5–15.9) as well as predicting redshift distributions (Figure 15.10) consistent with the magnitude distributions of faint-source identifications. Large values of the evolution function $E(L, z)$ (Figure 15.11) are restricted to high luminosities, as they must be to avoid producing too many faint sources. The luminosity range in which $E(L, z)$ is large is not a free parameter in this model; it is determined by the location of the bend near $L \approx 10^{25}$ W Hz^{-1} in the local luminosity function [Figure 15.3(a)]. Since this model assumes that all sources evolve equally, it demonstrates by example that restricting large $E(L, z)$ values to high luminosities does not imply that "only powerful sources evolve." Such an overinterpretation of the evolution function has led, for example, to the incorrect belief that only the (relatively luminous) radio quasars evolve, but that the (less luminous) radio galaxies do not.

All of the models described above are based on strong assumptions about the form of evolution in the (L, z)-plane. They show that evolutionary forms consistent with the data exist, but these solutions are certainly not unique. In order to explore the *range* of luminosity functions consistent with the source counts at $v = 0.408$, 2.7, and 5 GHz, luminosity distributions, and optical identification data, Peacock and Gull (1981) generated a number of "free-form" models in which the evolution is described by power series in $\log(L)$ and either $\log(1 + z)$ or $\log(1 - t/t_0)$ so that

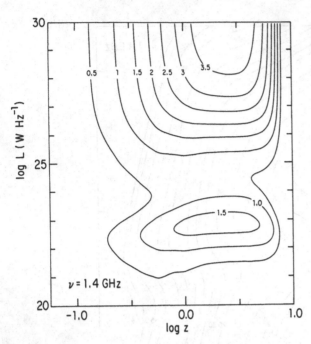

Fig. 15.11. Contour plot of the Condon (1984b) model evolution function at $v = 1.4$ GHz. Abscissa: log redshift. Ordinate: log luminosity (W Hz^{-1}). Contours: log evolution function.

it is "free-form" in both luminosity and time, at least to the extent that evolution varies *smoothly* with these quantities. The possibility of a sharp cutoff at high z was also considered. Steep- and flat-spectrum sources were allowed to evolve independently, and their spectral-index distributions were approximated by δ-functions. This approximation affects the accuracy with which the model can reproduce the radio data (Condon 1984b) but does not significantly increase the uncertainty of the derived evolving luminosity functions (Peacock 1985). Their successful models indicate that flat- and steep-spectrum sources evolve similarly, with only the most luminous sources exhibiting large changes in their comoving density with epoch. However, the density of sources in most areas of the (L, z)-plane is not well defined by the data. By locating the areas of greatest uncertainty, Peacock and Gull could specify the most important data still needed—source counts and redshift distributions of faint flat-spectrum sources in particular. The redshifts of forty-one flat-spectrum quasars with $S \geq 0.5$ Jy at $v = 2.7$ GHz were later added to the Peacock and Gull (1981) data base, and they allowed Peacock (1985) to suggest that the density of powerful flat-spectrum sources *declines* between $z \approx 2$ and $z \approx 4$ (unless the small number of quasars still lacking spectroscopic redshifts all have $z > 3$).

15.6 Source Size Evolution

15.6.1 The Angular Size–Redshift Relation

Even though radio sources are not good "standard rods," some statistic describing their size distribution might be stable enough to permit direct measurements of size evolution in populations of optically identified radio sources with known redshifts.

Fig. 15.12. Median angular sizes of samples of galaxies (open circles), galaxies in a narrow luminosity range (crosses), and quasars (filled circles) (Kapahi 1986). The relation $\langle\theta\rangle \propto 1/z$ suggests size evolution in Friedmann models. Abscissa: redshift. Ordinate: median largest angular size (arcsec).

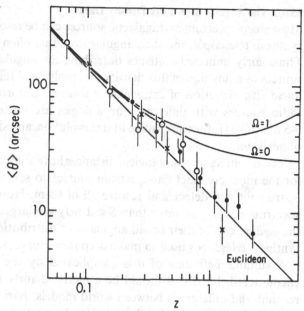

Since the apparent size of a double radio source is always reduced by projection onto the sky, the maximum θ_m of the distribution of angular sizes θ will minimize projection errors. Miley (1971) plotted the angular sizes of strong radio galaxies (nearly all with $z < 0.3$) and quasars (nearly all with $z > 0.3$) versus redshift and found that their upper bound obeys the "static Euclidean" relation $\theta_m \propto 1/z$. In any expanding Friedmann model, this result indicates either size evolution or an inverse correlation between luminosity and size for radio quasars.

These two effects can be distinguished by comparing radio quasars in a narrow luminosity range but widely separated in redshift; i.e., quasars found in surveys complete to different flux-density limits. Such a comparison of 3CR ($S \geq 10$ Jy at $v = 178$ MHz) and 4C ($S \geq 2.5$ Jy at $v = 178$ MHz) quasars (Hooley et al. 1978) was inconclusive because θ_m depends on small numbers of sources and hence is a fairly insensitive statistic. No size evolution was needed to fit the data, but size evolution of the form $d = d_0(1 + z)^{-N}$ with $N \leq 1.5$ could not be excluded.

A recent comparison of the median angular sizes $\langle\theta\rangle$ of radio galaxies with luminosities near $L \approx 10^{26}$ W Hz^{-1} found in three 1.4-GHz samples complete to $S = 2$, 0.55, and 0.01 Jy indicates size evolution in the range $1 < N < 2$ (Kapahi 1985). Although the formal statistical significance of this result is high, there remain two possible selection effects. (1) Only the faint-source sample was obtained from an aperture-synthesis survey that discriminates against sources larger than $\theta \approx 12''$. (2) Flux-limited surveys conducted at the same frequency v but covering different redshift ranges are effectively complete at different emitted frequencies $v(1 + z)$. A correlation between linear size and either spectral index or frequency in the source frame will mimic size evolution. This angular size–redshift relation for median angular sizes has recently been extended to include quasars (Kapahi 1986), as shown in Figure 15.12.

15.6.2 The Angular Size–Flux Density Data

Most steep-spectrum extragalactic sources can be resolved with sensitive aperture-synthesis telescopes, and their angular sizes θ are often by-products of deep surveys. Thus, fairly unbiased statistics describing the angular-size distributions of radio sources as a function of flux density can easily be obtained and used to estimate the linear size evolution of extragalactic sources. Unfortunately, the angular sizes of radio sources with differing morphologies are not easy to define precisely, the θ-S relation is rather insensitive to size evolution, and size evolution is either weak or nonexistent.

Radio sources at cosmological distances have angular sizes ranging from $< 0.''001$ for the most compact flat-spectrum sources to several minutes of arc, so no one instrument can detect and resolve all of them. However, the majority of steep-spectrum sources stronger than $S \approx 1$ mJy are larger than $\theta \approx 10$ arcsec, making the *median* $\langle \theta \rangle$ of their broad angular-size distribution accessible to the aperture-synthesis telescopes used to make deep radio surveys.

A suitable *definition* of θ is complicated by the variety of source structures encountered. Ideally, θ should be a metric diameter, which is more sensitive to redshift and differences between world models than the isophotal diameters used in optical astronomy (Sandage 1961). A good working definition of θ should be insensitive to details of the source brightness distribution and observational limitations, low dynamic range and limited resolution especially. The traditional definition of θ is the component separation of a double source because most strong sources selected in low-frequency surveys have this morphology and also because their component separations are easy to measure directly from contour maps. Unfortunately, the measured value of θ depends on the component intensity ratio if the double source is barely resolved or observed with low dynamic range, and this definition must be generalized to the "largest angular size" before it can be applied to sources with core-jet or more complex radio morphologies. However, the largest angular size is generally larger on maps of strong sources and can lead to an apparent change of θ with S. For example, the strong low-redshift quasar 3C273 is listed as having an angular size $\theta = 21$ arcsec (Kapahi et al. 1987) because its jet was mapped with high dynamic range. If 3C273 were moved to a high redshift and discovered as a faint source in an aperture-synthesis survey, its jet would not be distinguishable from the bright compact core, and its quoted angular size would be very small. Since the angular variances of the observing beam and the source brightness distribution $B(\phi)$ add under convolution, the angular diameter defined by $\theta^* \equiv 2[\int \phi^2 B(\phi)\,d\phi / \int B(\phi)\,d\phi]^{1/2}$ can be measured even for sources just large enough to broaden the beam (cf. Coleman 1985). It can be applied to any source morphology and it is independent of map dynamic range, except at very low signal-to-noise ratios where it may be overestimated by Gaussian fitting. Its main drawback is that it is more difficult to determine from contour plots.

Even with a good definition of θ, there are biases that affect the angular-size distribution of faint sources found in aperture-synthesis surveys. The survey maps are complete only above some *peak* flux density, so weak sources significantly larger than the synthesized beam (typically 10″ to 20″ FWHM) will be missed and must be corrected for. The sky density of sources fainter than a few mJy is so high that

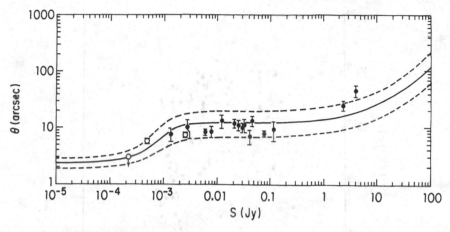

Fig. 15.13. Median angular size as a function of 1.4-GHz flux density. The filled circles are from the compilation by Windhorst et al. (1984); the open circles and the upper limit are from the Coleman and Condon (1985) high-resolution VLA survey. The solid line is the median angular size from the Coleman and Condon (1985) model ($\Omega = 1$, no size evolution), while the dashed lines mark the 60th- and 40th-percentile angular sizes. Abscissa: flux density (Jy). Ordinate: median angular size (arcsec).

distinguishing physically associated components of double sources from unrelated projected pairs of compact sources becomes a problem (Condon et al. 1982b).

15.6.3 Models of the Angular Size–Flux Density Relation

All evolutionary models of the observed $\langle\theta\rangle$-S relation (e.g., Figure 15.13) require as an input one of the models for the evolving radio luminosity function described in Section 15.5. Some estimate of the linear size distribution of sources covering a wide luminosity range (e.g., Figure 15.14) is also needed. The first detailed model (Kapahi 1975) was based on a luminosity function similar to that derived by Longair (1966). The approximation was made that the projected linear size d of a radio source is independent of its luminosity L, and a parametric "local" size distribution was obtained by a fit to the size distribution of low-redshift ($z < 0.3$) 3CR sources. Power law size evolution in which source sizes vary as $d = d_0(1 + z)^{-N}$ was tried, and the value $N \approx 1.5$ gave the best fit to the $\langle\theta\rangle$-S data for sources stronger than $S \approx 1$ Jy at 178 MHz.

The $\langle\theta\rangle$-S relation was extended to $S \approx 0.1$ Jy at 408 MHz by Downes et al. (1981). Their analysis was based on the improved Wall et al. (1980) radio luminosity functions. Instead of deriving a size distribution function, they assumed that individual 3CR sources are representative of the overall population of sources dominating the relevant epoch. Sources in their 3CR "parent population" were assigned weights by the weighted $1/V_m'$ method used to calculate the local luminosity function of an evolving population (Section 15.4.1). This method should automatically account for any possible correlation between linear size and luminosity in the parent sample, but spreading the parent population over a number of luminosity bins increases the statistical uncertainties in each. Also, the 3CR parent population does not correct for possible morphological differences between the 3CR sources and sources appearing elsewhere in the (L, z)-plane. If there is a substantial popula-

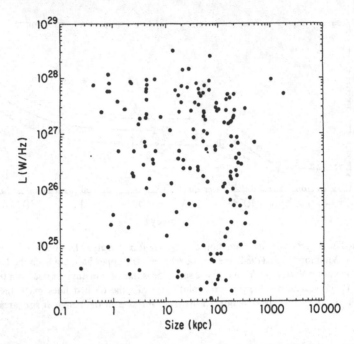

Fig. 15.14. Projected linear size distribution of sources stronger than $S = 2$ Jy at $\nu = 1.4$ GHz (From Coleman and Condon 1985) calculated for $\Omega = 1$. The correlation of size with luminosity is weak at best. Abscissa: projected linear size (kpc). Ordinate: 1.4-GHz luminosity (W Hz^{-1}).

tion of steep-spectrum compact sources (compact sources with steep spectra at high frequencies but relatively flat spectra at lower frequencies) among the faint ($S < 0.1$ Jy) high-redshift sources found at 408 MHz, it will be better represented in a (lower-redshift) parent population selected at some frequency *higher* than 408 MHz (e.g., Fielden et al. 1983, Allington-Smith 1984). Although Downes et al. (1981) found no value of the evolution exponent N reproduced the data with the 3CR parent sample, Kapahi and Subrahmanya (1982) used the same methods and data to find acceptable fits in the range $1 < N < 1.5$. Kapahi et al. (1987) attribute this discrepancy to a computational oversight by Downes et al. (1981). Using a parent sample selected at $\nu = 2.7$ GHz with the Peacock and Gull (1981) multi-frequency luminosity functions to predict the $\langle\theta\rangle$-S relation at other frequencies, Fielden et al. (1983) and Allington-Smith (1984) obtained good fits in the range 0.05 to 1 Jy at 408 MHz for $1 < N < 1.5$, but the stronger sources could not be accommodated simultaneously. Finally, Kapahi et al. (1987) modeled the $\langle\theta\rangle$-S relation above $S \approx 0.1$ Jy at 408 MHz using a variety of luminosity functions and parent populations selected at 178, 1400, and 2700 MHz. They concluded that size evolution is always required.

Most faint ($S < 1$ Jy at $\nu = 1.4$ GHz) sources probably have redshifts in the range $0.3 < z < 3$ for which the angular-size distance D_θ is nearly constant if $\Omega = 1$ (Figure 15.15). Without size evolution, changes in angular size with flux density reflect changes in linear size, not redshift. Flux density correlates more strongly with luminosity than with redshift for $S < 1$ Jy, so the flat portion of

Fig. 15.15. Most faint ($S < 1$ Jy at $v = 1.4$ GHz) radio sources fall in the shaded redshift band $0.3 < z < 3$, so they are at nearly the same angular-size distance D_θ if $\Omega = 1$. Abscissa: angular-size distance (Mpc). Ordinate: redshift.

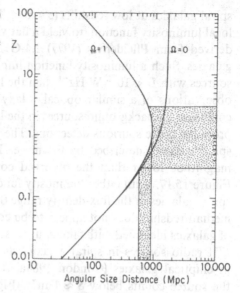

the $\langle\theta\rangle$-S curve (Figure 15.13) can easily be matched without evolution if there is no correlation of linear size with luminosity (Figure 15.14). Conversely, models requiring evolution to fit this flat region generally have parent populations in which low-luminosity sources have larger median linear sizes than high-luminosity sources. The rather sharp drop in $\langle\theta\rangle$ below $S \approx 1$ mJy at $v = 1.4$ GHz can be explained only by a correspondingly sharp drop in the median linear size of sub-mJy sources (Coleman and Condon 1985). This occurs naturally if the faintest sources are confined to the disks of spiral galaxies.

All of the models above have trouble matching the rather steep rise of $\langle\theta\rangle$ at high flux densities. This difficulty may be caused by the variation of θ with the dynamic range of the measurements. Using θ^* instead of θ for all sources reduces the rise above $S \approx 1$ Jy to the point that it can be fit without size evolution (Coleman 1985), although size evolution with $N \approx 1$ is still quite acceptable—the $\langle\theta\rangle$-S relation is just not very sensitive to size evolution.

15.7 The Faint-Source Population

Recent VLA and WSRT deep surveys (Mitchell and Condon 1985, Oort 1987) have extended the 1.4-GHz source count to $S \approx 100$ μJy directly and $S \approx 10$ μJy statistically (Figure 15.5). The weighted source count flattens below $S \approx 1$ mJy, suggesting the emergence of a significant low-luminosity source population— $L < 10^{24}$ W Hz^{-1} at $z \approx 1$, $L < 10^{22}$ W Hz^{-1} at $z \approx 0.1$. This luminosity range brackets the transition between elliptical and spiral galaxies in the local radio luminosity function (Figure 15.3), so the faint ($S < 1$ mJy) sources are likely to be quite different from the strong ones.

Subrahmanya and Kapahi (1983) associated the faint sources with *nonevolving* spiral galaxies at low redshifts ($z < 0.1$). Their conclusion depends on a very steep local luminosity function (to yield a flat visibility function ϕ) for $L < 10^{22}$ W Hz^{-1} derived from Pfleiderer's (1977) 1.4-GHz survey of bright spiral and irregular galaxies. Such a luminosity function implies much higher space densities of radio sources with $L < 10^{22}$ W Hz^{-1} than the luminosity function based on recent VLA observations of a similar optical galaxy sample (Figure 15.3), probably because confusion by background sources in the larger beam of the NRAO 91-m telescope produces some spurious detections. The 1.4-GHz source count from nonevolving spiral galaxies described by the newer local luminosity function is an order of magnitude lower than the observed count, as indicated by the dotted line in Figure 15.17. With either luminosity function, the median redshift of nonevolving spiral galaxies in the flux-density range 0.1 to 1 mJy is only $\langle z \rangle \approx 0.1$. Such a low median redshift does not appear to be consistent with the magnitude distribution of galaxies identified with faint sources selected at 1.4 GHz (Windhorst 1986).

If radio sources in spiral galaxies *do evolve* at about the same rate as those in elliptical galaxies (Condon 1984a, b), they can account for the flattening of the source counts below $S \approx 1$ mJy (Figure 15.5) as well as the higher redshifts (Figure 15.10) indicated by the optical-identification magnitude distribution. It should be emphasized that in this scenario, most of the faint radio sources are *not* the "normal" spiral galaxies found in optically selected catalogues. Typical radio-selected spiral galaxies are those with $L \approx 10^{22}$ W Hz^{-1} found at the peak of the visibility function, most of which are interacting "starburst" galaxies (e.g., Condon et al. 1982a), Markarian galaxies, and Seyfert galaxies (e.g., Meurs and Wilson 1984). Since there is a close correlation between the far-infrared ($\lambda = 60$ μm) and radio continuum flux densities of spiral galaxies, radio-selected spiral galaxies are more akin to galaxies found in the IRAS survey than to normal spiral galaxies.

A third explanation for the source-count flattening is that there is a "new population" of radio galaxies with $L \approx 10^{23}$ W Hz^{-1} at $\nu = 1.4$ GHz (Windhorst 1984) that was somehow missed by the radio luminosity functions derived from optically selected samples of spiral and elliptical galaxies (e.g., Figure 15.3) or Markarian and Seyfert galaxies (Meurs and Wilson 1984). Using spectroscopy and four-band photographic photometry of galaxies identified with radio sources stronger than $S = 0.6$ mJy (Kron et al. 1985), Windhorst (1984) estimated that the local space density of radio sources associated with his "blue galaxy" identifications exceeds that of radio sources in known spiral and elliptical galaxies by an order of magnitude in the luminosity range $L \approx 10^{22}$ to 10^{23} W Hz^{-1}. Only a moderate amount of evolution is needed for such a population to account for the source counts below $S \approx 5$ mJy (Windhorst et al. 1985).

The morphological composition of the blue radio-galaxy population is still unclear, and it may vary with both flux density and apparent magnitude. Those galaxies brighter than $F \approx 16$ mag are spiral galaxies; the fainter identifications "are often of peculiar compact morphology, sometimes interacting or merging" and have optical luminosities about equal to those of bright spiral galaxies (Windhorst 1984, Kron et al. 1985). Some of the faintest may be blue broad-line radio galaxies similar to those found in strong-source samples (Wall et al. 1986), photometrically misclassified elliptical galaxies (Kron et al. 1985), or misidentifications (Wall et al.

1986, Windhorst 1986, Weistrop et al. 1987). While the blue galaxies gradually displace red elliptical galaxies as the dominant population in faint radio-selected samples, it is not clear that they are a new population in the sense of surpassing the known local radio luminosity function near $L \approx 10^{23}$ W Hz^{-1}. If a weakly evolving population is significant in radio flux-limited samples at cosmological distances, it probably should be significant even in the small samples of optically bright galaxies used to construct the local luminosity function. Also, the radio identifications of the large ($N \approx 10^4$) UGC galaxy sample (Nilson 1973) south of $\delta = +82°$ with sources stronger than $S = 150$ mJy at $v = 1.4$ GHz (Broderick and Condon, manuscript in preparation) do not appear to be consistent with a factor-of-ten increase in the local luminosity function near $L = 10^{23}$ W Hz^{-1}. The UGC galaxies actually detected in this luminosity range are classified by Nilson as a mixture of active spirals, elliptical, S0, and "compact" galaxies. If the radio sources in this mixture evolve, they may account for most of the blue radio galaxies seen at cosmological distances in deep surveys. Because there is such a sharp decline in the luminosity function of spiral galaxies above $L \approx 10^{23}$ W Hz^{-1}, the blue galaxy population identified with sources in the 1 to 10 mJy range by Kron et al. (1985) probably contains a smaller fraction of spiral galaxies than sub-mJy samples do.

A sharp distinction between radio-emitting spiral and elliptical galaxies can be made by the relative strengths of their far-infrared emission—nearly all infrared-selected galaxies with detectable radio sources are spirals, not ellipticals. Furthermore, there is a remarkably tight correlation between the far-infrared and radio continuum flux densities of spiral galaxies. For *infrared-selected* spirals, the quantity $u \equiv \log(S_{60\,\mu m}/S_{1.4\,GHz})$ has a median value $\langle u \rangle = +2.15 \pm 0.15$ and a scatter $\sigma_u < 0.3$ (Condon and Broderick 1986). Biermann et al. (1985) used the infrared-radio correlation and a static Euclidean extrapolation of the $\lambda = 60\ \mu m$ source count to predict that spiral galaxies will contribute significantly to the 5-GHz source count below $S \approx 100\ \mu Jy$. Now that the $\lambda = 60\ \mu m$ source count has been extended to $S = 50$ mJy (Hacking et al. 1987), the contribution of *infrared-selected* spiral galaxies to the 1.4-GHz count of sources as faint as $S = 50$ mJy/dex$(u) = 0.35$ mJy can be estimated directly from the infrared data—the weighted count scales as $u^{-3/2}$, so $S^{5/2}n(S|v = 1.4$ GHz$) \approx 1$. At *least* 20% of the radio sources with $S \approx 0.35$ mJy at $v = 1.4$ GHz can be identified with spiral galaxies in this way; if $\sigma_u > 0$, the percentage rises. Finally, the $\lambda = 60\ \mu m$ source count is consistent with strong evolution of infrared-selected spiral galaxies (Hacking et al. 1987).

15.8 Isotropy and Homogeneity

Radio sky maps should be very sensitive to fluctuations in the sky density of sources for the reasons given in Section 15.1. Radio sources appear to cluster about galaxies at least as much as galaxies cluster with galaxies (Longair and Seldner 1979), so radio source clustering in space should reflect large-scale inhomogeneities in the distribution of galaxies in space. However, the faint radio sources in any area of sky are spread along a line of sight almost $cz/H_0 \approx 3000$ Mpc in length, reducing the angular sensitivity to clustering in space by averaging over many clusters in the line of sight. Consequently, existing radio surveys are sensitive only to clustering on very large scales ($d > 100$ Mpc) at cosmological redshifts ($z \approx 1$), in contrast to optical (Bahcall and Burgett 1986, de Lapparent et al. 1986) and far-infrared (Meiksin and Davis 1986, Rowan-Robinson et al. 1986) surveys that probe nearby ($z \approx 0$) clustering on scales up to $d \approx 100$ Mpc.

Several different techniques have been used to search for fluctuations in the sky densities of radio sources. One is plotting the distribution of angular distances to the nearest neighbors of all sources in a survey and comparing this distribution with the expected random distribution (Maslowski et al. 1973). The nearest-neighbor test is sensitive to clustering only in the small range of angular scales between the survey resolution and the typical separation between sources. Another simple procedure is to group the sources by position on the sky, flux-density range, etc. and compare their numbers with the expected Poisson distributions (Machalski 1977). A variation of this test for confusion-limited surveys is to divide the mapped region into small areas and compare the widths W of the $P(D)$ distributions in each (Hughes and Longair 1967). Then the effective number of sources sampled equals the number of beam areas in the whole map, potentially quite a large number. The distribution of widths W from the 480 squares, each covering $2° \times 2° \approx 91$ independent beam areas, from the Green Bank 1.4-GHz sky map overlapping the north galactic pole (Condon and Broderick 1985) is shown as a histogram in Figure 15.16. It is

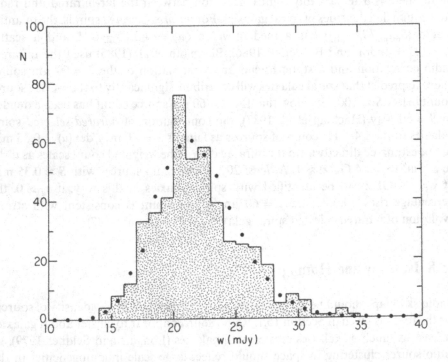

Fig. 15.16. The observed widths W of the $P(D)$ distributions in 480 $2° \times 2°$ squares (histogram) and the expected width distribution (filled circles) if sources are not clustered. Abscissa: width (mJy). Ordinate: number of maps.

indistinguishable from the distribution expected in the absence of clustering (filled circles). Such grouping tests are most sensitive to clustering on the grouping scale chosen, so they must be repeated on a variety of scales. But they can easily be applied to surveys with irregular boundaries. More powerful tests for clustering of discrete sources are power spectrum analysis (Webster 1976a, b) based on the Fourier transform of a map with the sources replaced by δ-functions, and its Fourier-transform relative, correlation function analysis (Masson 1979). A significant advantage of correlation function analysis is that confusion anticorrelation affects only the smallest correlation lags but essentially all Fourier components of the fluctuation power spectrum.

No convincing detection of anisotropy in the sky distribution of extragalactic radio sources has been made. The upper limits obtained are strong enough to rule out the "local hole" interpretation of the drop in the weighted source counts at high flux densities, but they do not yet strongly constrain clustering on scales $d < 100$ Mpc. This limitation is primarily statistical—the source density is too low in large-scale surveys. There are (very approximately) $N \approx [cz/(H_0 d)]^3 \omega/3$ clusters of comoving size d in the cone of solid angle ω. Only if the number n of sources in this solid angle is much greater than N can the statistical fluctuations be smaller than those caused by clustering. For $P(D)$ analysis of confusion-limited surveys, $n \approx \omega/\omega_b$, where ω_b is the beam area; and clustering on scales $d \gg (\omega_b/3)^{1/3}(cz/H_0)$ might be seen. The Green Bank 1.4-GHz sky map (Condon and Broderick 1985) has $\omega_b \approx 10^{-5}$ sr and so reaches $d \gg 50$ Mpc. Confusion-limited surveys with only moderately higher resolution may detect clustering on scales already known to exist in optical and infrared samples.

Finally, it should be noted that the speed v of the Earth relative to the extragalactic source frame produces a dipole anisotropy of amplitude $[2 + (\gamma - 1)(1 + \alpha)]/(v/c)$ in the differential count $n(S) \propto S^{-\gamma}$ of sources with spectral index α (Ellis and Baldwin 1984). This effect is just below current limits of detectability, requiring surveys covering $N \approx 2 \times 10^5$ beam areas with high gain accuracy.

15.9 Cosmology Made Simple: The Shell Model

The evolutionary models described in Section 15.5 are based on complex numerical calculations that tend to obscure connections between important features in the data and the calculated radio luminosity functions, redshift distributions, etc. In contrast, von Hoerner (1973) demonstrated with analytic approximations the importance of the broad visibility function ϕ (Equation 15.14) to the radio Hubble relation and to the form of the source count. In a uniformly filled, static Euclidean universe, the visibility function has no effect on the form of the source count (Equation 15.16) or the Hubble relation. The visibility function is important only if its width, $\Delta \log L$, is greater than twice the redshift range, $\Delta \log z$, containing most radio sources. Thus, the actual distribution of extragalactic radio sources in distance (lookback time) is so nonuniform that features in the source counts should not be interpreted as perturbations from a static Euclidean count. A much better starting model for the radio universe is actually a hollow shell centered on the observer!

This "shell model" reproduces many features of the data almost as well as the more elaborate models and clearly shows how they are related to the distribution of sources in space.

For sources with average spectral index $\langle \alpha \rangle$ the relation

$$S^{5/2} n(S|v) \approx \frac{1}{(4\pi)^{3/2}} \int_0^\infty \phi(L|z, v)(1 + z)^{-9/4 - 3\langle\alpha\rangle/2} \frac{dz}{z} \qquad (15.30)$$

is a good approximation to the exact Equations (15.20) and (15.21) for all $0 \leq \Omega \leq 2$, $z < 5$ (Condon 1984a). If most radio sources are confined to a thin shell of thickness Δz_s at redshift z_s,

$$S^{5/2} n(S|v) \approx \frac{\phi(L|z_s, v)}{(4\pi)^{3/2}} (1 + z_s)^{-9/4 - 3\langle\alpha\rangle/2} \left(\frac{\Delta z_s}{z_s} \right) . \qquad (15.31)$$

We assume translation evolution so $\rho_m(L|z, v) = g(z)\rho_m[L/f(z)|z = 0, v]$. Let $g_s \equiv g(z_s)$ be the amount of density evolution and $f_s \equiv f(z_s)$ be the amount of luminosity evolution at the shell redshift. Then, $\log[\phi(L|z_s, v)] = \log[\phi(L/f_s|z = 0, v)] + 3\log(f_s)/2 + \log(g_s)$ and

$$\log[S^{5/2} n(S|v)] \approx -\frac{3}{2}\log(4\pi) + \log\left[\phi\left(\frac{L}{f_s} \middle| z = 0, v \right) \right] + \frac{3}{2}\log(f_s) + \log(g_s)$$

$$+ \log\left(\frac{\Delta z_s}{z_s} \right) - \left(\frac{9}{4} + \frac{3}{2}\langle\alpha\rangle \right)\log(1 + z_s) . \qquad (15.32)$$

The redshift distribution of sources stronger than $S = 2$ Jy at $v = 1.4$ GHz (Figure 15.9) suggests $z_s \approx 0.8$ and $(\Delta z_s/z_s) \approx 1$; the spectral-index distribution [Figure 15.7(b)] gives $\langle \alpha \rangle \approx 0.7$. Substituting these quantities yields the following expression relating the weighted source counts, the local visibility function, and the evolution parameters at the shell redshift:

$$\log[S^{5/2} n(S|v)] \approx \log\left[\phi\left(\frac{L}{f_s} \middle| z = 0, v \right) \right] + \frac{3}{2}\log(f_s) + \log(g_s) - 2.49 . \qquad (15.33)$$

The values of f_s and g_s that satisfy Equation (15.33) can be found graphically by superimposing the observed source counts and local visibility functions, as shown in Figure 15.17. For sources with $z_s \approx 0.8$ and $\alpha \approx 0.7$, $\log[L(\text{W Hz}^{-1})] - \log[S(\text{Jy})] \approx 26.9$. Since the best fit of the local visibility function to the source counts occurs at $\log[L(\text{W Hz}^{-1})] - \log[S(\text{Jy})] \approx 25.7$ (Figure 15.17), we require luminosity evolution in the amount $\log(f_s) \approx 26.9 - 25.7 = 1.2$. This fit also implies $\log[S^{5/2} n(S|v = 1.4 \text{ GHz})] \approx \log[\phi(L|z = 0, v = 1.4 \text{ GHz})] - 0.65$, resulting in $\log(g_s) \approx 0.0$ (no density evolution). With these evolution parameters, the weighted source count predicted by the shell model corresponds exactly to the local visibility function plotted as the solid line in Figure 15.17. The model actually reproduces the entire observed source count from $S \approx 10 \ \mu\text{Jy}$ to $S \approx 10$ Ky.

Since the shell model ignores local sources, it must fail at the highest flux densities—the regime in which a nonevolving model is more appropriate. What is surprising is that the transition flux density is so high. An exact calculation based

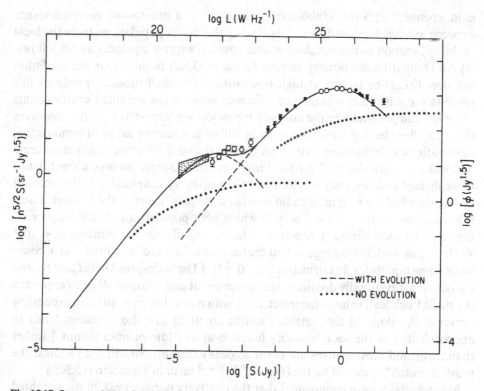

Fig. 15.17. Superposition of the weighted source count at 1.4 GHz (data points) and the hyperbolic fits to the 1.4-GHz local visibility functions for radio sources in spiral and elliptical galaxies (dashed lines). The combined local visibility function for all radio sources is indicated by the solid curve. This curve also plots the weighted source count predicted by the shell model. The source counts expected from *nonevolving* populations of spiral and elliptical galaxies described by this local luminosity function are shown as dotted curves. Lower abscissa: log flux density (Jy). Left ordinate: log weighted source counts $(sr^{-1} \, Jy^{1.5})$. Upper abscissa: log spectral luminosity $(W \, Hz^{-1})$. Right ordinate: log weighted luminosity function $(Jy^{1.5})$.

on the same local luminosity function without evolution yields $\log[S^{5/2} n(S|\nu = 1.4 \text{ GHz})] \approx 1.8$ at high flux densities (Figure 15.17), so the shell model and the nonevolving model predict the same source counts at $S \approx 20$ Jy. Thus, the static Euclidean approximation is reasonably good only for $S > 20$ Jy at $\nu = 1.4$ GHz; it applies only to the small number of sources in the very strongest flux-density bin plotted in Figure 15.17. It should not be used to describe features in the observed counts at lower flux densities. For example, the so-called "Euclidean" regions in which $S^{5/2} n(S|\nu = 1.4 \text{ GHz})$ is roughly constant near $\log[S(\text{Jy})] \approx 0$ and $\log[S(\text{Jy})] \approx -3$ do not indicate that the sources in these flux-density ranges are comparatively local—they only correspond to maxima in the visibility function of sources at $z \approx z_s$.

In the shell model, the median source redshift is $\langle z \rangle = z_s \approx 0.8$ for all $S \ll 20$ Jy, in good agreement with the observed redshift distribution of sources stronger than $S = 2$ Jy (Figure 15.9) and the magnitude distributions of galaxies identified with sources as faint as $S \approx 1$ mJy (Windhorst et al. 1984a, Kron et al. 1985). Since $\langle z \rangle$

is independent of S (no Hubble relation), there is a one-to-one correspondence between average luminosity and flux density that maps populations from the local visibility function to the weighted source count. Two consequences are as follows. (1) All standard evolutionary models (Section 15.5.2) require that the evolution function $E(L, z)$ be largest at high *luminosities*. The shell model reproduces this result (see Figure 15.1) because the difference between the weighted source counts observed and predicted by the nonevolving model are largest at high *flux densities*. (2) At any flux-density level, most sources will lie in a narrow range of luminosities; observations with that sensitivity look beyond the shell for more luminous sources and will not reach the shell for less luminous ones. Deeper surveys do not detect more distant sources, only feebler ones. Consequently, elliptical galaxies account for nearly all of the strongest radio sources and spiral galaxies the faintest. There is a transition region at $S \approx 1$ mJy in which both populations should be present. Because the local visibility function is falling rapidly for luminosities $L < 10^{21}$ W Hz^{-1}, this model also suggests that the (as yet unobserved) weighted source count will decline rapidly for flux densities $S < 10^{-5}$ Jy. [The widespread belief that nearby galaxies must eventually dominate the source count and cause its slope to approach the static Euclidean value is incorrect. Even with no evolution at all in an expanding universe, the slope of the weighted source count at low flux densities tends to approach that of the local visibility function at low luminosities (about $\frac{4}{3}$ rather than zero); and most sources are cosmologically distant, crowding up against the redshift "cutoff" imposed by the $(1 + z)^{-9/4 - 3\langle\alpha\rangle/2}$ term in Equation (15.30).]

Many authors have commented that the relatively narrow peak in the weighted source counts is difficult to model in terms of the relatively broad local luminosity function. It is inappropriate to compare these distributions because they do not have the same dimensions. The weighted source count $S^{5/2} n(S|\nu)$ should only be compared with the weighted local luminosity function $\phi(L|z = 0, \nu)$; the unweighted source count $n(S|\nu)$ is most appropriately compared with the unweighted local luminosity function $\rho(L|z = 0, \nu)$. Figure 15.17 shows that the weighted source counts and the local visibility function peaks actually have very similar widths at $\nu = 1.4$ GHz. The only conclusion that can be drawn from the fact that the weighted source count peak is not much broader than the local visibility function peak is that *some* form of evolution is restricting the lower end of the redshift range $\Delta \log z$ in which most radio sources are found. [The factor $(1 + z)^{-9/4 - 3\langle\alpha\rangle/2} = (1 + z)^{-3.3}$ for $\langle\alpha\rangle = 0.7$ in Equation (15.30) is quite effective at suppressing the contribution of high-redshift sources to the observed source counts, so the success of the shell model is not strong evidence that evolution stops or reverses at redshifts higher than z_s.]

The very similar forms of the *local* visibility function and the weighted source count (Figure 15.17) determined by the visibility function at $z \approx 0.8$ indicate that the form of the visibility function really does not evolve significantly; i.e., the "translation evolution" approximation is a good one. Pure luminosity evolution works in the shell model, and pure density evolution in a thin shell would also preserve the form of the local visibility function in the normalized source counts. The amounts of luminosity and density evolution actually required to fit the data are determined by the redshift of the shell, the difference between the luminosity of the local visibility function peak and the flux density of the weighted count peak, and the difference between the peak values of the local visibility function and the weighted source count, as described above. Pure luminosity evolution shifts the source-count curve along a line of slope $\frac{3}{2}$ in the $\{\log(S), \log[S^{5/2}n(S)]\}$-plane, and pure density evolution shifts it vertically. Thus, only one combination of luminosity and density evolution can match both z_s and the peak of weighted source count exactly.

The shell model emphasizes the insensitivity of the $\langle\theta\rangle$-S relation to source size evolution. Since there is no Hubble relation for $S \ll 20$ Jy, evolution of the projected linear size d with z affects sources of all flux densities equally. Furthermore, most sources with $S \ll 20$ Jy lie at redshifts within a factor of two of $z_s = 0.8$, so they are at very nearly the same angular-size distance if $\Omega = 1$ (Figure 15.15). Thus the $\langle\theta\rangle$-S plot really measures the variation of projected linear size d with luminosity. The flat region with $\langle\theta\rangle \approx 10$ arcsec extending from $S \approx 1$ mJy to $S \approx 1$ Jy indicates that $\langle d\rangle \approx 40$ kpc for all luminosities in the range $L \approx 10^{24}$ to 10^{27} W Hz^{-1} at $\nu = 1.4$ GHz. The sudden falloff to $\langle\theta\rangle < 3$ arcsec below $S \approx 1$ mJy cannot be caused by evolution; it reveals instead a dramatic decline in linear size to $\langle d\rangle < 10$ kpc among sources less luminous than $L \approx 10^{24}$ W Hz^{-1}. Such a decline is expected if most of the sources contributing to the flattening of $S^{5/2}n(S|\nu)$ below $S \approx 1$ mJy at $\nu = 1.4$ GHz are in the disks of spiral galaxies.

15.10 What Next?

(1) Despite our failure to find (or recognize) "standard candles" or "standard rods" relating L and S or d and θ, they would allow potentially definitive tests of world models and should not be forgotten. Every advance in instrumentation, every increase in astrophysical knowledge of radio sources should be considered an opportunity for determining world-model parameters. For example, recent high-sensitivity VLBI maps have found milliarcsecond jets with bulk relativistic velocities nearly equal to the speed of light, even in extended steep-spectrum sources whose orientations relative to the line of sight should be random. The angular separations between cores and jet knots in a complete sample of such sources might be combined with their nearly "standard velocities" $v \approx c$ to measure directly the angular-size distances corresponding to the source redshifts.

(2) The source counts $n(S|\nu)$ are generally better determined than most of the data needed for their interpretation. Little progress can be made at high flux densities because essentially the whole sky has been covered with accurate measurements. The sensitivity limits of the current generation of telescopes has nearly been reached. Source counts below $S \approx 10$ μJy at $\nu = 1.4$ GHz will be difficult to obtain directly, although some limits to the source counts at fainter levels might be obtained from background measurements. Increasing the sky area covered by deep surveys is still needed to reduce the statistical errors, especially at $\nu = 5$ GHz, where each map only covers about 10^{-5} steradians. Overlapping surveys at 1.4 and 5 GHz are badly needed to determine the spectral-index distributions of faint sources selected at these two frequencies.

(3) The weighted local luminosity function at $\nu = 1.4$ GHz has larger statistical

errors than the weighted source count, particularly near the peaks of the spiral-galaxy and elliptical-galaxy contributions, so it contributes significantly to the uncertainty in the evolution function $E(L, z)$. Improvements are especially needed near $L \approx 10^{23}$ W Hz^{-1} at $v = 1.4$ GHz to interpret the source count near $S \approx 1$ mJy and decide whether a "new population" of sources exists. This will require an optically complete sample with known redshifts filling a large volume of space, the UGC sample (Nilson 1973) for example. Radio observations of the same sample should be made at $v = 5$ GHz as well as at a lower frequency so that the local luminosity functions of steep- and flat-spectrum sources can be determined independently.

(4) Redshifts of complete samples of flat-spectrum sources stronger than $S \approx 0.1$ Jy should be measured to confirm suggestions that evolution peaks at $z \approx 2$. This may not be too difficult because most flat-spectrum sources stronger than $S \approx 0.1$ Jy can be identified with fairly bright quasars with strong emission lines. If very-high-redshift quasars are common, they will surely be found in such samples. If they are not, only radio-selected samples may be sufficiently immune to selection effects to prove it.

(5) Optical programs like the Leiden-Berkeley Deep Survey should be extended to sources as faint as $S \approx 0.1$ mJy at $v = 1.4$ GHz to determine the redshift distribution and host galaxy population of the faint-source population.

(6) High-resolution ($\approx 1''$) maps of sources fainter than $S \approx 1$ mJy should also be made to determine their angular-size distributions and improve their radio positions for reliable optical identifications.

(7) Isotropy tests at somewhat higher effective resolutions should be made to detect density fluctuations on scales less than 50 Mpc.

Recommended Reading

Annual Review Article

Jauncey D.L. (1975). Radio Surveys and Source Counts, Annu. Rev. Astron. Astrophys. **13**:23.

Books

Maeder A., Martinet L., and G. Tammann, eds. 1978. Observational Cosmology. Geneva: Geneva Observatory.
Sandage A., Sandage M., and J. Kristian, eds. 1975. Galaxies and the Universe. Chicago: University of Chicago Press.

Symposia Proceedings

Jauncey D.L., ed. 1977. Radio Astronomy and Cosmology. Proc. I.A.U. Symposium **74**.
Hewitt A., Burbidge G., and Li-Zhi Fang, eds. 1987. Observational Cosmology. Proc. I.A.U. Symposium **124**.

References

Allington-Smith, J.R. 1984. Mon. Not. R. Astron. Soc. **210**:611.
Auriemma, C., G.C. Perola, R. Ekers, R. Fanti, C. Lari, W.J. Jaffe, and M.H. Ulrich. 1977. Astron. Astrophys. **57**:41.
Bahcall, N.A., and W.S. Burgett. 1986. Astrophys. J. **300**:L35.
Biermann, P., A. Eckart, and A. Witzel. 1985. Astron. Astrophys. **142**:L23.
Bridle, A.H. 1967. Mon. Not. R. Astron. Soc. **136**:219.

Coleman, P.H. 1985. Ph.D. thesis, University of Pittsburgh.

Coleman, P.H., and J.J. Condon. 1985. Astron. J. **90**:1431.

Coleman, P.H., J.J. Condon, and C. Hazard. 1985. Astron. J. **90**:1437.

Condon, J.J. 1974. Astrophys. J. **188**:279.

Condon, J.J. 1984a. Astrophys. J. **284**:44.

Condon, J.J. 1984b. Astrophys. J. **287**:461.

Condon, J.J. 1987. Astrophys. J. Suppl. **65**, in press.

Condon, J.J., and J.J. Broderick. 1985. Astron. J. **90**:2540.

Condon, J.J., and J.J. Broderick. 1986. Astron. J. **92**:94.

Condon, J.J., M.A. Condon, G. Gisler, and J.J. Puschell. 1982a. Astrophys. J. **252**:102.

Condon, J.J., M.A. Condon, and C. Hazard. 1982b. Astron. J. **87**:739.

Condon, J.J., and J.E. Ledden. 1981. Astron. J. **86**:643.

Crawford, D.F., D.L. Jauncey, and H.S. Murdoch. 1970. Astrophys. J. **162**:405.

Danese, L., G. De Zotti, and N. Mandolesi. 1983. Astron. Astrophys. **121**:114.

de Lapparent, W., M.J. Geller, and J.P. Huchra. 1986. Astrophys. J. **302**:L1.

Downes, A.J.B., M.S. Longair, and M.A.C. Perryman. 1981. Mon. Not. R. Astron. Soc. **197**:593.

Ellis, G.F.R., and J.E. Baldwin. 1984. Mon. Not. R. Astron. Soc. **206**:377.

Fanaroff, B.L., and M.S. Longair. 1973. Mon. Not. R. Astron. Soc. **161**:393.

Fielden, J., A.J.B. Downes, J.R. Allington-Smith, C.R. Benn, M.S. Longair, and M.A.C. Perryman. 1983. Mon. Not. R. Astron. Soc. **204**:289.

Fomalont, E.B., A.H. Bridle, and M.M. Davis. 1974. Astron. Astrophys. **36**:273.

Fomalont, E.B., K.I. Kellermann, J.V. Wall, and D. Weistrop. 1984. Science **225**:23.

Grueff, G., and M. Vigotti. 1977. Astron. Astrophys. **54**:475.

Hacking, P., J.J. Condon, and J.R. Houck. 1987. Astrophys. J. **316**:L15.

Hooley, A., M.S. Longair, and J.M. Riley. 1978. Mon. Not. R. Astron. Soc. **182**:127.

Hughes, R.G., and M.S. Longair. 1967. Mon. Not. R. Astron. Soc. **135**:131.

Jauncey, D.L. 1967. Nature **216**:877.

Jauncey, D.L. 1975. Annu. Rev. Astron. Astrophys. **13**:23.

Kapahi, V.K. 1975. Mon. Not. R. Astron. Soc. **172**:513.

Kapahi, V.K. 1985. Mon. Not. R. Astron. Soc. **214**:19P.

Kapahi, V.K. 1986. In Proc. Int. Astron. Union Symp. 124. A. Hewitt, G. Burbidge, and L.Z. Fang (eds.), Dordrecht: Reidel, p. 251.

Kapahi, V.K., and C.R. Subrahmanya. 1982. In Proc. Int. Astron. Union. Symp. 97. D.S. Heeschen and C.M. Wade (eds.), Dordrecht: Reidel, p. 401.

Kapahi, V.K., V.K. Kulkarni, and C.R. Subrahmanya. 1987. J. Astrophys. Astron. **8**:33.

Kellermann, K.I. 1964. Astrophys. J. **140**:969.

Kellermann, K.I. 1972. Astron. J. **77**:531.

Kron, R.G., D.C. Koo, and R.A. Windhorst. 1985. Astron. Astrophys. **146**:38.

Kühr, H., A. Witzel, I.I.K. Pauliny-Toth, and U. Nauber. 1981. Astron. Astrophys. Suppl. **45**:367.

Kulkarni, V.K. 1978. Mon. Not. R. Astron. Soc. **185**:123.

Laing, R.A., and J.A. Peacock 1980. Mon. Not. R. Astron. Soc. **190**:903.

Laing, R.A., J.M. Riley, and M.S. Longair 1983. Mon. Not. R. Astron. Soc. **204**:151.

Longair, M.S. 1966. Mon. Not. R. Astron. Soc. **133**:421.

Longair, M.S. 1978. In Observational Cosmology. A. Maeder, L. Martinet, and G. Tammann (eds.), Geneva: Geneva Observatory, p. 125.

Longair, M.S., and P.A.G. Scheuer. 1970. Mon. Not. R. Astron. Soc. **151**:45.

Longair, M.S., and M. Seldner. 1979. Mon. Not. R. Astron. Soc. **189**:433.

Machalski, J. 1977. Astron. Astrophys. **56**:53.

Machalski, J. 1978. Astron. Astrophys. **65**:157.

Machalski, J. 1981. Astron. Astrophys. Suppl. **43**:91.

Maslowski, J., J. Machalski, and S. Zieba. 1973. Astron. Astrophys. **28**:289.

Masson, C. 1979. Mon. Not. R. Astron. Soc. **188**:261.

Mattig, W. 1958. Astron. Nachr. **284**:109.

Meiksin, A., and M. Davis. 1986. Astron. J. **91**:191.

Meurs, E.J.A., and A.S. Wilson. 1984. Astron. Astrophys. **136**:206.

678   James J. Condon

Miley, G.K. 1971. Mon. Not. R. Astron. Soc. **152**:477.
Mitchell, K.J. 1983. Ph.D. thesis, Pennsylvania State University.
Mitchell, K.J., and J.J. Condon 1985. Astron. J. **90**:1957.
Nilson, P. 1973. Uppsala General Catalogue of Galaxies. Uppsala: Uppsala Astronomical Observatory.
Oort, M.J.A. 1987. Astron. Astrophys. , submitted.
Oort, M.J.A., P. Katgert, F.W.M. Steeman, and R.A. Windhorst 1987. Astron. Astrophys. **179**:41.
Ostriker, J.P., and J. Heisler. 1984. Astrophys. J. **278**:1.
Owen, F.N., J.J. Condon, and J.E. Ledden. 1983. Astron. J. **88**:1.
Pauliny-Toth, I.I.K., A. Witzel, E. Preuss, H. Kühr, K.I. Kellermann, E.B. Fomalont, and M.M. Davis. 1978. Astron. J. **83**:451.
Peacock, J.A. 1985. Mon. Not. R. Astron. Soc. **217**:601.
Peacock, J.A., and S.F. Gull. 1981. Mon. Not. R. Astron. Soc. **196**:611.
Petrosian, V., and J. Dickey. 1973. Astrophys. J. **186**:403.
Pfleiderer, J. 1977. Astron. Astrophys. Suppl. **28**:313.
Rees, M.J., and G. Setti. 1968. Nature **219**:127.
Robertson, J.G. 1973. Aust. J. Phys. **26**:403.
Robertson, J.G. 1978. Mon. Not. R. Astron. Soc. **182**:617.
Robertson, J.G. 1980. Mon. Not. R. Astron. Soc. **190**:143.
Rowan-Robinson, M. 1970. Mon. Not. R. Astron. Soc. **149**:365.
Rowan-Robinson, M., D. Walker, T. Chester, T. Soifer, and J. Fairclough. 1986. Mon. Not. R. Astron. Soc. **219**:273.
Ryle, M., and R.W. Clarke. 1961. Mon. Not. R. Astron. Soc. **122**:349.
Sandage, A. 1961. Astrophys. J. **133**:355.
Sandage, A., and G.A. Tammann 1981. A Revised Shapley-Ames Catalog of Bright Galaxies. Washington, D.C.: Carnegie Institute of Washington.
Scheuer, P.A.G. 1957. Proc. Cambridge Phil. Soc. **53**:764.
Scheuer, P.A.G. 1974. Mon. Not. R. Astron. Soc. **166**:329.
Schmidt, M. 1968. Astrophys. J. **151**:393.
Schmidt, M. 1972. Astrophys. J. **176**:303.
Schmidt, M. 1976. Astrophys. J. **209**:L55.
Spinrad, H., S. Djorgovski, J. Marr, and L. Aguilar. 1985. Publ. Astron. Soc. Pacific **97**:932.
Subrahmanya, C.R., and V.K. Kapahi. 1983. G.O. Abell and G. Chincarini (eds.), Proc. Int. Astron. Union Symp. 104. Dordrecht: Reidel, p. 47.
Terrell, J. 1977. Am. J. Phys. **45**:869.
van der Laan, H., and R.A. Windhorst. 1982. H.A. Brück, G.V. Coyne, and M.S. Longair (eds.), Astrophysical Cosmology. Vatican City: Pontificia Academia Scientiarum, p. 349.
Véron-Cetty, M.P., and P. Véron. 1983. Astron. Astrophys. Suppl. **53**:219.
von Hoerner, S. 1973. Astrophys. J. **186**:741.
Wall, J.V., and J.A. Peacock. 1985. Mon. Not. R. Astron. Soc. **216**:173.
Walls, J.V., T.J. Pearson, and M.S. Longair. 1980. Mon. Not. R. Astron. Soc. **193**:683.
Wall, J.V., C.R. Benn, G. Grueff, and M. Vigotti. 1986. J.-P. Swings (ed.), Highlights of Astronomy. Dordrecht: Reidel, p. 345.
Webster, A. 1976a. Mon. Not. R. Astron. Soc. **175**:61.
Webster, A. 1976b. Mon. Not. R. Astron. Soc. **175**:71.
Weistrop, D., J.V. Wall, E.B. Fomalont, and K.I. Kellermann. 1987. Astron. J. **93**:805.
Windhorst, R. 1984. Ph.D. thesis, University of Leiden.
Windhorst, R. 1986. J.-P. Swings (ed.), Highlights of Astronomy. Dordrecht: Reidel, p. 355.
Windhorst, R., R.G. Kron, and D.C. Koo. 1984a. Astron. Astrophys. Suppl. **58**:39.
Windhorst, R., G.M. van Heerde, and P. Katgert. 1984b. Astron. Astrophys. Suppl. **58**:1.
Windhorst, R., G.K. Miley, F.N. Owen, R.G. Kron, and D.C. Koo. 1985. Astrophys. J. **289**:494.
Witzel, A., J. Schmidt, I.I.K. Pauliny-Toth, and U. Nauber. 1979. Astron. J. **84**:942.

Index